Rudenko International Conference "Methodological Problems in Reliability Study of Large Energy Systems" (RSES 2019)

E3S Web of Conferences Volume 139 (2019)

Tashkent, Uzbekistan
23 – 27 September 2019

Editors:

Nicolai Voropai
Sergey Senderov
Alexander Michalevich
Husein Guliev

ISBN: 978-1-7138-0031-6

Printed from e-media with permission by:

Curran Associates, Inc.
57 Morehouse Lane
Red Hook, NY 12571

Some format issues inherent in the e-media version may also appear in this print version.

For additional information, please contact EDP Sciences – Web of Conferences at the address below.

EDP Sciences – Web of Conferences
17, Avenue du Hoggar
Parc d'Activité de Courtabœuf
BP 112
F-91944 Les Ulis Cedex A
France

Phone: +33 (0) 1 69 18 75 75
Fax: +33 (0) 1 69 28 84 91

contact-edps@webofconferences.org

Additional copies of this publication are available from:

Curran Associates, Inc.
57 Morehouse Lane
Red Hook, NY 12571 USA
Phone: 845-758-0400
Fax: 845-758-2633
Email: curran@proceedings.com
Web: www.proceedings.com

TABLE OF CONTENTS

Statement of Peer review

In submitting conference proceedings to *Web of Conferences*, the editors of the proceedings certify to the Publisher that

1. They adhere to its **Policy on Publishing Integrity** in order to safeguard good scientific practice in publishing.
2. All articles have been subjected to peer review administered by the proceedings editors.
3. Reviews have been conducted by expert referees, who have been requested to provide unbiased and constructive comments aimed, whenever possible, at improving the work.
4. Proceedings editors have taken all reasonable steps to ensure the quality of the materials they publish and their decision to accept or reject a paper for publication has been based only on the merits of the work and the relevance to the journal.

Title, date and place of the conference

Methodological Problems in Reliability of Large Energy Systems (RSES 2019)
23-27 September 2019
Tashkent, Uzbekistan

Proceedings editor(s):

Prof. Nicolai Voropai, Melentiev Energy Systems Institute, Russia.
Prof. Sergey Senderov, Melentiev Energy Systems Institute, Russia.
Prof. Alexander Michalevich, Energy Institute of Belarus NAS, Belarus.
Prof. Husein Guliev, Azerbaijan Research and Design and Exploration Institute of Power Engineering, Azerbaijan.

Date and editor's signature

10.10.2019

E3S Web of Conferences **139**, 01001 (2019)
RSES 2019
https://doi.org/10.1051/e3sconf/201913901001

Key directions of development of measures to improve the reliability of electrical power systems

R.B. Jalilov[1*], *S.T.* Latipov[1], *G.N.* Aslanova[1], *A.A.* Akhmedov[1], *S.P.* Shayimova[1]

[1]Bukhara Engineering-Technological Institute

Abstract. The article covers the study of measures aimed to improving the reliability of electrical power systems at the present stage. It considers the integration of the global electricity industry. The article analyzes the latest technical work in the field of automation, such as Smart Grid and intelligent power systems.

Introduction

Patterns of changes in the conditions for the development and functioning of electric power systems (EPS) lead to significant transformations in the structure of systems and their modes of operation. These transformations are caused by a number of objective factors that determine the shape of EPS of the future. They continue to increase the scale of EPS, expanding the territories they serve, uniting various EPS for joint work with the formation of interregional, interstate and intercontinental energy associations [1,3,4,9]. The formation of interstate electrical connections (ISEC), associations (ISEA) and markets is a global integration process that has been going on in various regions of the world for almost a century. The chronology of the development of the ISEA since the first quarter of the twentieth century is detailed considered in [9-13 and 16-21].

Integration of the global electrical power industry

In studies [9-12], it is noted that the ISEA of the UES / EES is currently very poorly integrated. It should be expected that in frame of the formation of a common economic space in the integration projects of the Customs Union and the EAEU, the disintegration trend in the electric power industry of the post-Soviet space will finally change to integration trend, which, as was evident from the consideration of world experience, corresponds to the global trend.

Recently, the problem of the formation of the Global Super-Energy Association has been raised again, and promising electric power mega-projects that will make up its important elements are being considered [9]. In the zone of influence of the Unified Energy System of Russia and the ISEA of the OEC / UES, various interstate electric power projects, that actually determine the direction of Russia's integration into the electric power space of Eurasia, are being implemented (Fig. 1) [9]. Among them, the interesting one is the CASAREM project - Central Asia South Asia Regional Electricity Market (SHPP project between the countries of Central and South Asia).

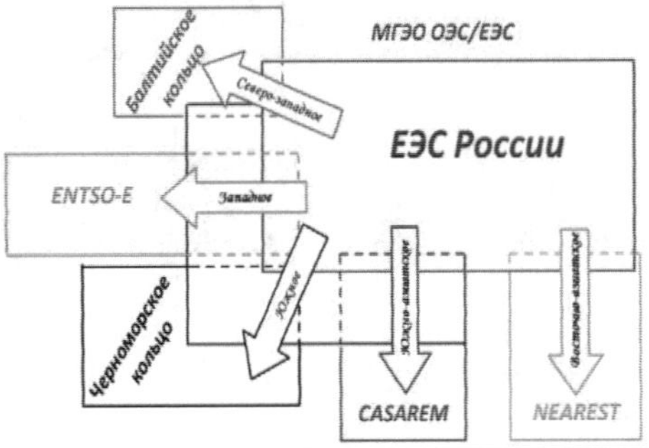

Source: ISEM SB RAS

Fig. 1. Directions for the integration of the UES of Russia into the Eurasian ISEA [9].

UES of Russia (as noted in [9]), as well as UES of Uzbekistan, participating in electric power integration projects with neighboring countries, can become the link between national and regional EPS in a significant part of Europe and Asia, contributing to the formation of the Eurasian transcontinental energy association. In the end, this helps to increase the reliability of power systems in the Central Asian region.

Given the new technologies in converting technology based on power electronics, reducing costs, increasing reliability and ensuring high controllability of direct current power transmission, they will receive significant development in the transmission electric network. At the same time, the widespread use of devices that form flexible AC power transmission (FACTS) based on the use of power electronics will radically increase the controllability of the AC transmission network [3]. New technologies, including the use of FACTS devices, will significantly increase the reliability and controllability of the distribution electric network.

The use of energy storage systems in modern energy systems

Another direction in the development of measures to increase the reliability of EPS is the active introduction of energy storage systems (ESS) [13]. Taking into account that the new economy of the future is neo-industrialization, based on a combination of large installations and distributed generation, power and information processes, physical and intellectual systems, in the near future energy will be combined systems designed for a wide range of energy services offered to consumers at the right time and necessary quality.

Especially small (in terms of unit capacity) generation is rapidly developing, which provides a more complete use of local (both renewable and other alternative) resources. At the same time, a consumer can simultaneously become a producer of electricity, which must either be accumulated on-site or delivered to the network. In recent years, the development of electric power transmission systems in the world is characterized by their significant complication, increase of their intellectual level, the active introduction of renewable energy sources (RES), distributed generators, end-users with controlled consumption, as well as communication systems and control systems. These changes are designed to provide benefits, such as the ability to meet the objectives of the state policy to diversify the electricity industry and provide consumers with greater choice while ensuring their own electricity supply [2]. However, these transformations encounter certain problems indicated in [13].

Implementation of energy storage systems

Active implementation of energy storage systems (ESS) in energy systems to some extent will overcome the above problems.

Each energy storage technology is characterized by capital expenditures for its implementation, as well as operating costs. The cost of various ESS and the benefits of their use depend to a large extent on their structure in terms of discharge power (MW) and energy capacity (MW·h). Independent system operators (ISO), energy companies, sellers and providers of energy storage technologies need actively formulating the rules of the emerging ESS market, as well as operational requirements, in order to achieve maximum economic efficiency from the use of ESS.

Ten key areas represent the predominant use of ESS and are most interesting for potential ESS owners and electricity market operators. Due to the high capital costs for most ESS, it is advisable to use separate devices to perform several tasks simultaneously, corresponding to those areas of application, indicated in the table 1.2.

Table 1. Definition of the main areas of application of ESS at the level of generation and power systems [13]

Application	Description
Services in the wholesale electricity market	ESS installed in the power system for bidding on the electricity and capacity market and on the system services market1)
RES integration in the energy system	ESS installed in the power system and providing a shift of the maximum load in time and system services to ensure parallel operation of renewable energy with the power system
Stationary ESS to support transmission and distribution of electrical energy	ESS to support the transmission and distribution of electricity, increasing the efficiency of the transmission and distribution network, as well as savings due to the rejection of the construction of additional network facilities
Mobile ESS to support power transmission and distribution	Mobile ESS to support power transmission and distribution while delaying the entry of transmission and distribution network facilities in many places as needed.
[1)] only large ESS operating simultaneously in the electricity market and in the system services market	

Table 2. Identification of the main ESS applications at the level of transmission and distribution network [13]

Application	Description
ESS in distribution networks	Centrally managed, modular systems providing increased reliability of power supply, support for transmission and distribution networks, and potentially system services
ESS aggregated by energy companies providing electricity in the retail electricity market.	ESS, installed and serviced living spaces, aggregated and centrally managed for distribution systems.
ESS to ensure the quality of electricity and reliability in the supply of electricity to commercial and industrial consumers	ESS to ensure the quality and reliability of electricity supply to commercial and industrial consumers
ESS for power management for commercial and industrial consumers.	ESS to reduce the time of use of the energy supplied from the power system during hours of maximum load of electricity by commercial and industrial consumers.

The European Association for the Development of Energy Storage Technologies, together with the European Alliance for Energy Research, has developed a roadmap for the development of ESS until 2030 [14]. The tables 5 and 6 [13] provide information from this roadmap for lithium-ion batteries and flow redox storage. These two types of electrochemical ESS are by far the most common in the world.

The development of smart energy systems taking into account the reliability factor

The idea of intellectualization of EPS is close to the principles of cybernetic control, and this is largely due to advances in engineering and technology [15-20, 22-24]. Initially, the concept of constructing intelligent EES (IES) in foreign countries - Smart Grid - was aimed at the development of power supply systems for end consumers with the installation of control systems and power consumption (Smart metering) and providing the ability to connect distributed power sources to the EES, first turn, autonomous wind turbines and photovoltaic panels. Today, the ideology of the development of Smart Grid includes almost all the main areas of activity in the electric power industry and the corresponding technological and information-communication bonds between them [15, 22, 24].

The term Smart Grid has not received a unified definition. To characterize the diversity of the meaning of this concept, a series of definitions of the leading organizations of the USA (EPRI, IEEE) and Western Europe in the field of developing the Smart Grid ideology are given in [15].

Smart Grid refers to the modernization of the power supply system aimed at the ability to monitor, protect, optimize the functioning of all elements of EPS - centralized and distributed generation, high-voltage transmission and distribution networks, industrial consumers and building management systems, energy storage devices, end-users, electrical transport, household appliances [17].

Smart grid is a combination of energy, communication and information technology for an improved power infrastructure that enables continuous evolution of end use devices [18].

Smart Grid is an electric network that can cost-effectively combine the modes and actions of all connected users - generators, consumers and their associations, to ensure a cost-effective and sustainable energy system with low losses, high quality and reliability of power supply and safety [19].

The Smart Grid functional model, presented by the USA National Institute of Standards and Technology (NIST) in 2009, identifies the main areas of activity in the electric power industry, represented by seven areas-domains, combined by technological and communication links (Fig. 2.) [19]:

- Bulk Generation;
- Transmission;
- Distribution;
- Operations;
- Customer;
- Markets;
- Service provider.

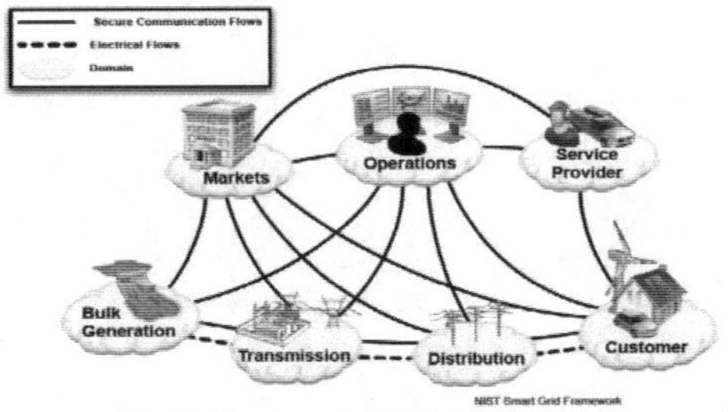

Fig. 2. Conceptual model of Smart Grid
—— secure communication currents
- - - electric currents

The issues of standardization in the formation of intellectual EPS are in the focus of attention of leading international organizations and institutions [15, 16, 19-21]:

- International Electrotechnical Commission (IEC);
- International Council for Large Electrical Systems (CIGRE);
- European Standards Committees (CEN and CENELEC);
- European Telecommunications Standardization Institute (ETSI);
- Institute of Electrical and Electronics Engineers (IEEE);
- US National Institute of Standards and Technology (NIST);
- Electric Power Research Institute (EPRI).

As noted in [15, 17], a number of international standards on which the design and operation of Smart Grid elements can be based already exist (Fig. 3.).

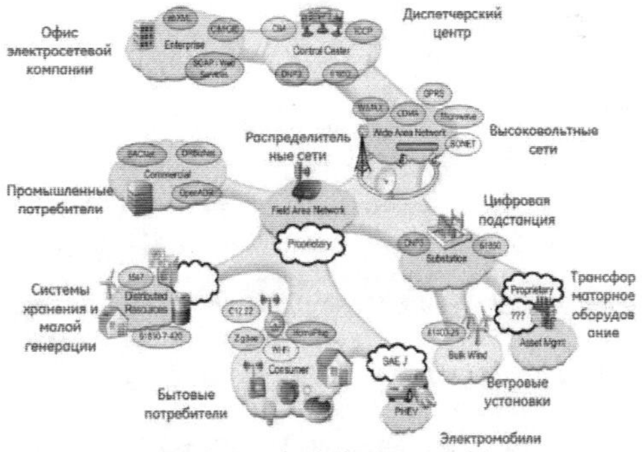

Fig. 3. Standards for Smart Grid.

For example, these are standards for the IEC 61850 series digital substation, IEC 61970/61968 standards for the CIM common information model, IEEE 802.15.4 standard for interacting with consumer devices using the ZigBee protocol, etc.

Technical Committee 57 of IEC develops international standards (Fig. 4) for EMS, SCADA, automation of distribution networks, information exchange, real-time control, etc., which will be used in the planning, use and operation of the IES [20].

Discussion

Another important new factor for future EPSs is the emergence of active consumers who independently manage their own energy consumption depending on price conditions in the retail electricity market by transferring electricity consumption by some power consumers from periods with a high price of electricity to periods with a low price.

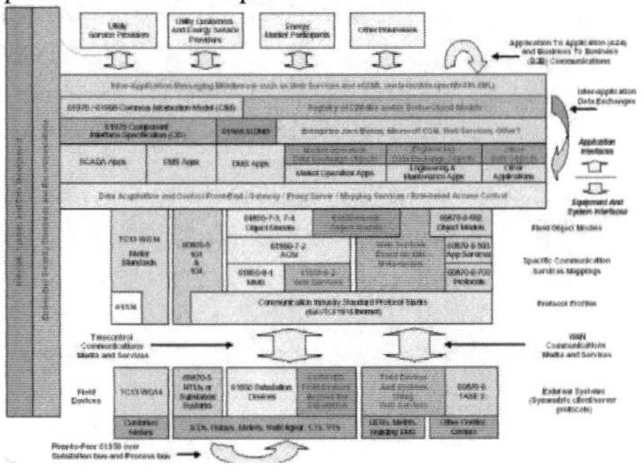

Fig. 4. IEC Standards for Control Systems.

Such a load control of active consumers, independent of the dispatch schedule, creates problems for managing the EPS modes due to the uncertainty in the power consumption of active consumers. Therefore, the interaction between EPS and consumers on joint control of system modes using the regulatory capabilities of consumers is promising [3].

Conclusion

1. The proposals given in this article open the prospect of further research on an extremely important problem in assessing the reliability and safety of energy facilities and systems.

2. With the development of future EPS on the technological base of the intelligent energy system, new problems will become aggravated in the future, related to the need to strengthen coordination of control of EPS modes at various levels, increase management efficiency, and ensure the reliability of the control system for EPS modes. The issues of information and cyber security during monitoring and control of EPS are especially acute [16].

3. All of the mentioned above requires serious in-depth research to study the properties of future EPS, to develop principles and methods for their formation, taking into account changing conditions, as well as dispatch and automatic control systems for future EPSs.

References

1. Allaev K.R. Power industry of Uzbekistan and the world. T.: Fan va texnologiya, 2009, p.463.
2. Fortov V.E. Energy of the future based on breakthrough technologies as the basis of a new energy civilization // Materials of the XXIV International scientific, technical and practical conference "Prospects for the development of the electric power industry and high-voltage electrical equipment. Switching devices, conversion equipment, microprocessor control systems and security". November 28 - 29, 2016. Innovative power industry – 21. [Electronic resource]. URL: http://www.energy-strategy.ru/projects/energy_21.htm (date of the application: 20.04.2019).
3. Voropay N.I., Osak A.B. New properties of future power systems. // The same address [Electronic resource]. URL:http:// www. energy strategy. ru/projects/-energy_21.htm (date of the application: 20.04.2019).
4. V.A. Ovseychuk, V.A. Nepomnyashi. Rationing the reliability and quality of power supply to consumers // Innovation and Expertise. -2016. Volume 1/16. C.178-185. [Electronic resource]. URL:http: //www.energystrategy. ru/projects/energy_21.htm (date of the application: 20.04.2019).
5. Voropay N.I., Domyshev A.V., Nepomnyashchy V.A. Models and methods for investigating the operational reliability of electric power systems // Reliability of energy systems: Problems, models and methods of their solvation. Novosibirsk: Science, 2014, p. 57-73.
6. Voropai N.I., Efimov D.N., Kurbatsky V.G., Panasetskiy D.A., Tomin N.V. ISEM SO AN // Algorithms of survivability and self-restoration of intelligent electric power systems [Electronic resource]. URL:http://www.energystrategy.ru/ projects/ energy_21. htm (date of the application: 20.04.2019).
7.Trends in microgrid control. IEEE-PES Task Force on Microgrid Control / D.E.Olivares, A.Mehrizi-Sani, A.H.Etemadi, e.a. // IEEE Transactions on Smart Grid, 2014, Vol.5, No.4, p. 1905-1919.
8. Kitushin. Reliability of energy systems. Part 1. Theoretical Foundations: Textbook.-Novosibirsk: Publishing House of NSTU. 2003, p.256.
9. Podkovalnikov S.V., Savelyev V.A., Chudinova L.Yu. Integration of global electricity industry is the path to creating a global energy associations // [Electronic resource]. URL:http://www. energystrategy. ru/projects/ energy_21.htm (date of the application: 20.04.2019).
10. Global Energy Network Institute – GENI – URL: http://www.geni. org/index.html .
11. Electricity of the CIS 2003-2013 // Executive Committee of the CIS Electric Power Council, 2015 – 172 c. – URL: http://energo-cis.ru.
12. 2013 Year's Annual Report of GSE // Georgian State Electro system, 2014. – 39 c. – URL: http://www.gse.com.ge/new/ wpcontent/uploads /2014/07/SSA_ ANNUAL_2013_FOR_WEB.pdf.
13. Kucherov Yu.N., Novikov N.P. The main trends in the development and application of energy storage systems in modern energy systems //[Electronic resource]. URL:http://www. Energy-strategy. ru/projects/ energy_21.htm (date of the application: 20.04.2019).
14. Joint EASE/EERA Recommendation for a European Energy Storage Technology Development Roadmap Towards 2030, *Final Report,* March 2013.

15. Clark W. Gellings. Let's Build a Global Power Grid // IEEE Spectrum.– 2015. – July – URL:http://spectrum.ieee.org/energy/the-smarter-grid/lets-build-a-global-power-g.
16. Kucherov Yu.N., Fedorov Yu.G. Features of the development of intelligent power systems taking into account the reliability factor // Electronic resource]. URL:http://www. Energy-strategy. ru/projects/ energy_ 21.htm (date of the application: 20.04.2019).
17. Estimating the Costs and Benefits of the Smart Grid / EPRI
Technical report No.1022519, 2011. – 162 c.
18. IEEE Std 2030-2011 Guide for Smart Grid Interoperability of
EnergyTechnology and Information Technology Operation with the Electric Power System (EPS), End-Use Applications, and Loads. – 126 c.
19. European Technology Platform Smart Grids – Strategic Deployment
Document for Europe's Electricity Networks of the Future, 2010. – 69 c.
20. Technical and Commercial Standardisation of DER – microGrid
Components / CIGRE TB 423, WG C6.10, 2010. – 129 c.
21. Reliability Standards Set / NERC, 2010. – 1074 c.
22. Jalilov R.B. Prerequisites for the creation of
automated industrial energy supply management systems // Chief Power Engineer, 2006, No.6, p.41-43.
23. Jalilov R.B. Digital relays: regulatory framework for use, brief description, market overview // Electrical equipment, 2008, No.2, p.7-11.
24. Jalilov R.B. A new approach to the creation of an intelligent automation substation // Industrial Energy, 2008, No.4, p.15-18.
Use A4 paper size (210 x 297 mm) and adjust the margins to those shown in Table 1. The final printed area will be 172 x 252 mm.

Prospects of diversification and ensuring energy safety of Uzbekistan

Kahraman Allaev[1], *Tokhir* Makhmudov[1*]

[1] Tashkent State Technical University named after Islam Karimov, Power plants, networks and systems Department, Universitetskaya 2, Tashkent, Uzbekistan

Abstract. The data on the current state of energy in Uzbekistan are given. The need to diversify the structure of the energy balance of the republic is shown, which ensures the energy security of the state in the medium and long term. It is argued that the construction of a nuclear power plant in Uzbekistan is not only expedient, but also necessary. In the future, renewable energy and nuclear power plants will become the basis of energy in Uzbekistan.

1 Introduction

The global energy transformation currently taking place in the world in the 21st century will have an impact on the technological, social, economic and other directions in the development of countries. The emerging new reality will be fundamentally different from the usual energy geopolitics, which has dominated for more than a hundred years [1-6].

Fossil fuels being the basis of the global energy system, economic growth and modern lifestyles, over the past two centuries of operation have increased the global use of energy fifty times, creating an environment in the modern world [1]. The geographical concentration of fossil fuels has had a significant impact on the welfare and security of nations. At present, one can observe the emerging energy of the future, which can be represented in the form of a square, the sides of which are the sun, wind, hydropower and atomic energy. Such a transformation of the energy sector can lead to as radical changes in its scale and impact as the industrial and agricultural revolution. The Republic of Uzbekistan is subject to such changes.

The Republic of Uzbekistan is a dynamically developing state located in the central part of Central Asia. It has common borders with Kazakhstan (2203 km), Turkmenistan (1621 km), Tajikistan (1161 km), Kyrgyzstan (1099 km), Afghanistan (137 km) [7].

The territory of Uzbekistan is 447000 km^2, of which 22000 km^2 is water. The Aral Sea (in 1998, the surface area was 28687 km^2) is located in the north-west of the country. The rivers of Uzbekistan are part of the Aral Sea basin, which is fed by two main rivers: the Amu Darya (length of 2580 km) and the Syr Darya (length of 2221 km).

The fuel and energy complex of the Republic of Uzbekistan is the largest association of enterprises in Central Asia for the extraction, processing, transportation of oil, natural gas, coal, as well as the production and distribution of electrical and thermal energy.

By the measures taken by the leadership of the country, Uzbekistan in 1995 achieved energy independence. At present, taking into account the self-sufficiency of the republic, measures are being taken for the medium and long-term satisfaction of its needs with the necessary high-quality energy resources based on the sustainable development of the industry.

Uzbekistan has concentrated about 50% of the generating capacity of the countries of Central Asia, which is more than 13 million kW of installed capacity, with the possibility of generating electricity of about 70 billion kWh per year. The Uzbek power system is the main link in the unbroken chain of production and transmission of electrical energy in the region. Geographical location, the presence of developed electrical networks allow you to successfully organize and be an active participant in the market of electrical energy and power [7].

The basis of the Uzbek energy system consists of large thermal power plants: Syrdarya thermal power station (TPP) (with a capacity of 3000 MW), Tashkent TPP (1860 MW), Novo-Angren TPP (2100 MW), Navoi TPP (1250 MW). 137 power units with unit capacity from 150 to 300 MW are installed on them. The 800 MW power unit was commissioned at the Talimarjan thermal power plant.

The electric grid economy includes more than 230 thousand km of high voltage power lines and transformer substations with a total capacity of about 45 million kVA.

Electricity generation in Uzbekistan in the years 2000-2018, on average annually amounted to 50-67 billion kWh. In 2018, more than 60 billion m^3 of natural gas, about 3 million tons of petroleum products were

* Corresponding author: tox-05@yandex.com

https://doi.org/10.1051/e3sconf/201913901002

produced, and coal production increased to 3.5 million tons.

2 Development perspective

The energy strategy of Uzbekistan sets the goal of ensuring energy independence and security, increasing energy efficiency, and reducing the impact of energy on the environment. To solve these problems requires the development and introduction of new technologies based on scientific advances that provide a safer, cleaner energy [8].

In the long term, Uzbekistan's energy sector has been tasked with increasing its electricity generation to 117 billion kWh by 2030.

The implementation of this ambitious task will be carried out in the following directions [6]:

1. Intensification of energy efficiency and energy saving (reserve up to 40% of energy savings). This will require accelerated modernization of the main means of production, diversification of energy sources, conducting an efficient energy-saving policy, assessing the real costs of energy supply in the country, etc.

2. The widespread introduction of combined-cycle technologies instead of steam at thermal power plants will increase the efficiency of the units from 32÷35% to 54÷60% and, accordingly, will make it possible by 2030 to use as much natural gas for energy purposes as and now.

3. The development of energy resources of wind and sun. In most countries of the world, great work is being done to diversify the energy structure, including by expanding the use of non-carbon renewable energy sources and atomic energy [1, 2]. According to estimates of domestic and foreign experts, the real technical potential of solar and wind energy in Uzbekistan is at least 10 million kW of power in each area.

4. Increasing the share of coal and shale in power generation using modern technologies for burning low-grade solid fuels. The reserves of brown coal in Uzbekistan are about 2 billion tons, and shale about 50 billion tons. At the same time it is important to comply with the highest environmental standards.

5. Introduction of hydropower and hydroaccumulation stations, flexible tariff system, as well as closer cooperation with the power systems of neighboring countries - to give the system more flexibility, leveling consumption schedules during peak hours (maximum).

6. Studying the possibility of building nuclear power plants by 2030 and beyond. Solving this problem in the medium and long term is the content of the reform, development and implementation of the strategy for the development of the energy sector of Uzbekistan, taking into account the preservation of its export potential.

Currently, when generating electricity, natural gas dominates in the energy balance of Uzbekistan - its use reaches 90%, i.e. there is a disproportion in the use of primary energy sources. In order to diversify the government of Uzbekistan expands the composition and base of energy resources. Consider the fuel base of the energy complex of Uzbekistan.

3 Oil and gas industry

The modern oil and gas industry of Uzbekistan is one of the largest sectors of the economy, the most important energy base of the country. The industry has created significant scientific and technical potential [14].

A large-scale action program has been developed for the intensive development of the oil and gas complex based on the use of high-quality equipment, materials, modern technologies, accelerating the pace of updating and replacing outdated equipment, implementing large-scale projects with foreign companies and solving a range of social problems. The development of this program will allow Uzbekistan to significantly strengthen its fuel and energy independence and expand its integration into the world economy.

Uzbekistan is ranked 8th in the world in gas production. In the period from 1991 to 2018, the oil and gas industry of Uzbekistan showed extremely successful development rates in all production, technical and economic indicators. By 2019, hydrocarbon production reached 60 million tonnes of oil equivalent (toe), which is more than 1.6 times more than that produced in 1991 [9].

The processing industry of hydrocarbon raw materials of Uzbekistan is successfully developing. The modern Bukhara and Fergana and other refineries are working effectively. Shurtan (2001), Ustyurt Gas Chemical Complexes (2016), Kandym Gas Processing Complex (2018) are industrial giants working on the basis of modern oil and gas processing technologies.

4 Coal industry

According to experts, in Uzbekistan the estimated resource of coal is more than 2 billion tons [7]. Industrial stocks amount to 1.9 billion tons, of which brown – 1.853 million tons, stone - 47 million tons.

It is planned to intensify coal mining, with the result that its annual volume should increase to 20 million tons by 2025, of which brown coal - to 9.200 million tons, hard coal - to 110 thousand tons (Fig. 1).

Currently, coal mining is carried out in three fields: Angren (Tashkent region), Baysun and Shargun (Surkhandarya region). There are still a number of coal fields in the republic, for example, the Terekli site in the Kashkadarya region with industrial reserves of 50 million tons.

 https://doi.org/10.1051/e3sconf/201913901002

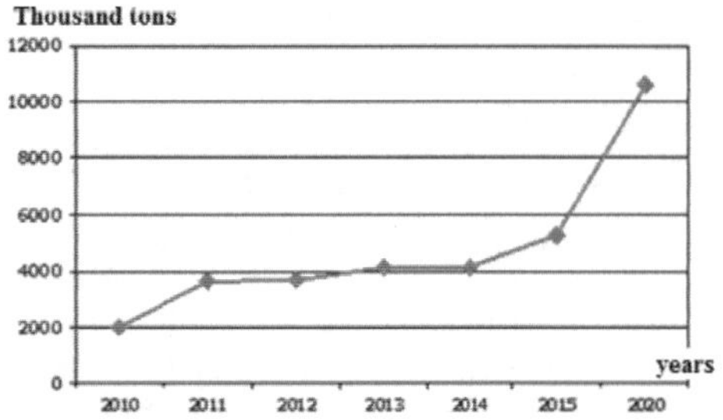

Fig. 1. Prospects for increasing coal production in Uzbekistan.

On the basis of the Angrenskoye deposit, coal is processed by the method of underground gasification of coal - by the Podzemgaz station (OJSC Erostigaz), commissioned in 1961 and still remaining one of the largest and unique facilities of its kind in the CIS. The design capacity of the station is 2.3 billion m^3 of gas per year or in terms of coal - 0.6 million tons.

Coal is used in the country for the development of motor fuel, calcium carbide, sorbents, activated coal and medicines. Coal from the Shargunskoye deposit is used as carbonaceous additives in ferrous and non-ferrous metallurgy, as well as coke briquettes at Uzmetkombinat, which reduces the volume of coke imports for foundry.

In the future, the issue of using oil shale is pending. The territory of Uzbekistan is rich in oil shale, with a specific heat of combustion of more than 1200 kcal / kg. Technically realizable forecast resources of the Kyzylkum and Amudarya combustible basins amount to 2 billion tons, and total geological reserves of 47 billion tons.

5 Electricity

The power industry of Uzbekistan is a basic sector of the economy of the Republic and, having a significant production and scientific and technical potential, has a significant impact on its development. Uzbekistan has become the largest energy power in the Central Asian region.

Over the past 30 years, electricity production in the country has increased by more than 3 times, with the possibility of generating electricity to 55-70 billion kWh. (Fig.2).

According to the CIS Electric Power Council, the total installed capacity of power plants in Uzbekistan at the end of 2017 was 14140 MW, including 12129 MW - the capacity of thermal power plants (85%) 1878.7 MW - hydroelectric power plants (13%), 133 MW - other (2%) [10].

The total length of electrical networks exceeds 235 thousand km, including the main transmission lines of 220 and 500 kV - 7.8 thousand km.

Currently in Uzbekistan, 83% of electricity is generated by gas, 13% by hydroelectric power plants, 4% by coal.

In Uzbekistan, about 16 billion m^3 of natural gas, 85 thousand tons of fuel oil and 2.3 million tons of coal are consumed annually for the needs of the power industry [11, 12].

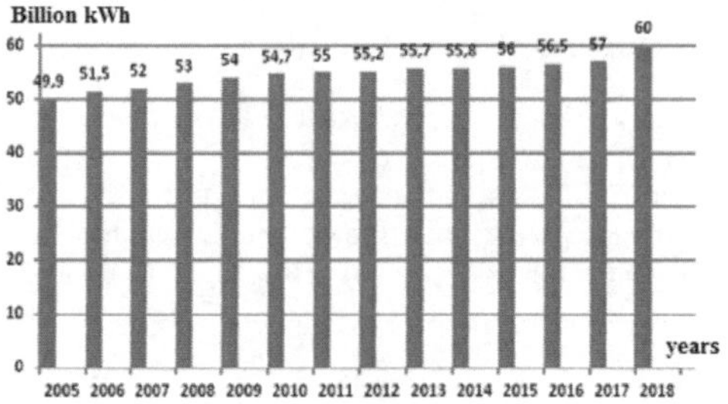

Fig. 2. Electricity production in Uzbekistan.

The problem of aging of fixed assets in power system, which is the basis of the Uzbek energy system, is most acute. 63 turbine generators with a unit capacity of 25 to 800 MW are installed at thermal power plants in Uzbekistan, of which 39 turbines with a capacity of 5.9 million kW (or 55%) have been in operation for more than 30 years.

Electricity generation at hydropower plants of Uzbekistan is more than 6.5 billion kWh per year. After the reconstruction and construction of new stations, the power generation will exceed 9 billion kWh. There is also the problem of aging equipment: 63 out of 69 hydraulic turbines were installed more than 30 years ago. As the world and domestic experience shows, the problem of increasing the energy efficiency of the production and consumption of electricity in the long term is solved in stages, through modernization, reconstruction, renovation of power and power grid equipment [1, 11].

The need to improve the energy efficiency of the power industry requires the introduction of the next level of automated information and measurement systems for the commercial accounting of electricity generation and consumption.

The power industry of Uzbekistan has begun reforming. By the decision of the President of the Republic of Uzbekistan dated March 27, 2019, the Ministry of Energy of Uzbekistan was established on the basis of state energy monopolies - Uzbekenergo JSC and Uzbekneftegaz JSC, Uzgidroenergo JSC, Ugol JSC, Uzatom Agency. Three joint-stock companies were organized: Thermal Power Plants, National Electric Networks of Uzbekistan, and Regional Electric Networks. The first will be responsible for the production of electric energy, the second - for transportation between regions and interaction with neighboring energy systems, and the third joint-stock company will sell electricity to end users.

The adopted document is aimed at developing not only the electric power industry, but the entire economy of Uzbekistan as a whole.

6 Potential of renewable energy sources (RES) of Uzbekistan

The reserves of renewable energy in the Republic of Uzbekistan are huge: the total volume (hydro, solar, wind, geothermal, biomass) is about 118 billion toe. Of these, only 179 million toe are economically viable, and currently only 0.31% of this potential has been developed.

The energy of the Sun, falling per square meter of the surface of the territory of the Republic, is in the range of 1500–1900 kWh / m^2, which is equivalent to 0.154 toe. If we consider that in the republic the number of sunny days is more than 250, then this is a large reserve.

Solar and wind energy can be the main driving force of the economy and the energy production sector in Uzbekistan. Under the optimistic scenario, the installed capacity of solar power plants can reach 10 GW by 2031, which will produce 17 TWh of energy, providing an annual income of about 1.8 billion dollars and create up to 36 thousand jobs.

Regions with average wind speeds of at least 4–8 m / s are located in the Aral Sea region, Navoi and Tashkent regions. According to experts, taking into account the rapidly developing equipment and technologies, the practically realizable potential of wind energy in Uzbekistan is at least 10 GW.

The development of hydropower in the republic will go at the expense of realizing the potential of small rivers, irrigation canals, reservoirs, watercourses.

Uzbekistan has significant geothermal and petrothermal resources. The explored reserves of hydrothermal resources with a temperature of 40–120 ^{0}C are 16 • 10^3 toe / year, and the forecasted reserves are much higher. Petrothermal resources (heat contained in dry heated rocks occurring at a depth of up to 3 km) are estimated to be about 6.7 • 10^{12} toe.

In addition, in Uzbekistan there are more than 260 large landfills with a total average annual accumulation of more than 7 million tons. Up to 250 m^3 of gas can be obtained from each ton of waste, which is approximately 1.5 billion m^3 of gas per year with its further use for the production of heat and electricity.

7 Nuclear power engineering of Uzbekistan

Nuclear power engineering is among the fastest growing areas in the world, as significant progress has been made by scientists in improving the safety of a new generation nuclear power plants (NPP) [2].

To date, 193 nuclear power plants with 454 power units are operated in 32 countries. Another 56 power units are under construction.

On December 29, 2017, Uzbekistan signed an intergovernmental agreement on cooperation in the field of peaceful uses of atomic energy with Russia [11-13].
In July 2018, the President of the Republic of Uzbekistan signed the Decree "On measures for the development of nuclear energy in the Republic of Uzbekistan".

It is obvious that the world reserves of economically and environmentally acceptable fossil hydrocarbons are limited, and in this case, nuclear energy, along with renewable energy sources, becomes an example of a carbon-free alternative to fossil energy resources [1, 5].

A modern NPP is a high-tech construction, high-tech-logical equipment and highly qualified personnel. That is, nuclear energy will give impetus to the development of science, education and the emergence of new professions.

In Uzbekistan, a nuclear power plant will be built as part of two power units with VVER reactors with a capacity of 1200 MW each. These are new generation units, so-called. "3+" generations, which differ from their predecessors by a high degree of reliability, simplicity in management due to the complete automation of the entire production cycle. At the same time, there is no negative impact on the environment, the design of the reactor has a high degree of seismological stability and takes into account all the IAEA safety requirements [11-13].

The planned construction of a nuclear power plant in Uzbekistan will ensure a stable supply of electricity in the country. This will be facilitated by a signed memorandum of cooperation between the Russian State Corporation Rosatom and the Academy of Sciences of Uzbekistan [11].

The introduction of atomic energy minimizes carbon dioxide emissions into the atmosphere, which is the main component of greenhouse gases that affect climate change.

Nuclear power plays a key role in ensuring the implementation of the Paris Climate Agreement, signed by Uzbekistan on April 19, 2017, under which it was decided to take measures to prevent temperature changes by more than 1.50 ° C. Uzbekistan has acceded to the Convention on the Safety of Spent Fuel Management and the Safety of Radioactive Waste Management and the Agreement on the Basic Principles of Cooperation in the Field of Peaceful Use of Atomic Energy [11-13].

The project for the construction of nuclear power plants in Uzbekistan is estimated at about $ 13 billion, of which $ 11.3 billion is provided by Russia as a state export credit [12].

The Republic of uranium reserves is on the 11th place in the world (131 thousand tons) and mined uranium is sold primarily in the United States, China and Korea [13].

8 Conclusion

The construction of a nuclear power plant will give a serious impetus to the development of energy, science, education and nuclear medicine in the country. Therefore, this is a serious leap in the development of the national economy and the development of national personnel. This is an opportunity to develop energy-intensive sectors of the economy in the country, while preserving the environment without having a harmful effect, because Atomic energy is a source of clean energy. Therefore, the construction of nuclear power plants in Uzbekistan is not only justified, but also extremely necessary!

The full-fledged operation of a nuclear power plant will make it possible to save 3.7 billion m^3 of natural gas annually. As a result of its processing, petrochemical products with high added value will be produced. If the saved gas, even without processing, is sent abroad for export, it will already provide the republic with an additional 550-600 million dollars a year [12].

In general, about nine thousand people will be employed in the construction of the nuclear power plant, and after commissioning, 3.5 thousand people, while engineering and technical workers of the highest qualification will be involved.

Consequently, nuclear and renewable energy combined with the development of energy storage systems is transformed into a force that creates a new, clean energy future in the world [1, 6]. In the second half of the 21st century, they will become the dominant factor in ensuring the energy balance in the world and in Uzbekistan.

According to the authors forecast, by 2030, as a result of the implementation of the measures planned, the following diversified structure of electricity generation should be expected: gas-47%, NPP-18%, hydro-electric power plants (HPP) - 10%, coal - 12%, renewable energy sources - 13% (Fig. 3 .) to ensure the energy security of the state and the growth of the investment attractiveness of the domestic power industry.

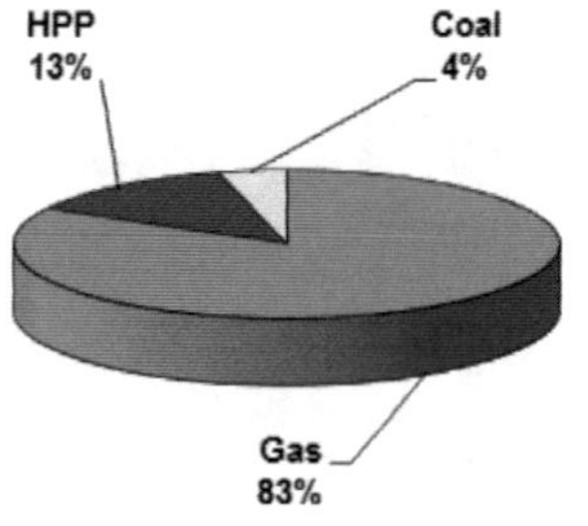

A

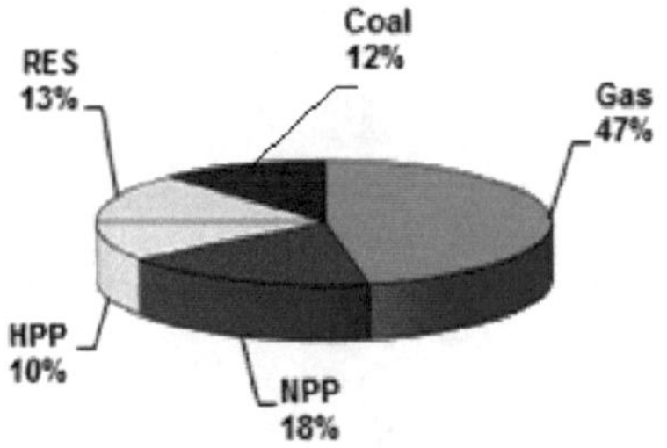

B

Fig. 3. Power generation in Uzbekistan:
A - 2017, production of 57 billion kWh.
B - 2035, production of 117 billion kWh (the forecast of the authors).

Thus, large-scale work carried out in the energy sector of Uzbekistan on the development of the fuel and energy complex will, in the future, make it possible to significantly diversify its structure, which will strengthen one of the most important indicators - the energy security of the republic.

References

1. A New World: The Geopolitics of the Energy Transformation, 88 p. (2019).
URL: https://www.irena.org/publications/2019/Jan/A-New-World-The-Geopolitics-of-the-Energy-Transformation

2. J. Buongiorno, J. Parsons, M. Corradini, D. Petti, The future of nuclear energy in a carbon-constrained world. An interdisciplinary MIT study, 248 p. (2018).
URL:https://www.ourenergypolicy.org/wp-content/uploads /2018/09/The-Future-of-Nuclear-Energy-in-a-Carbon-Constrained-World.pdf

3. M. Alverà, David C. Carroll, I. Marten, Global gas report 2018, 60 p. (2018).
URL:http://www.snam.it/export/sites/snam-rp/repository/file/gas_naturale/global-gas-report/global_gas_report_2018.pdf

4. Jan Ban, Jorge León Arellano, Haris Aliefendic and others. World Oil Outlook 2040, 412 p. (2017)
URL: https://www.opec.org/opec_web/flipbook/WOO2017/WOO2017/assets/common/downloads/WOO%202017.pdf

5. World Energy Council. World Energy. Issues Monitor, 211 p. (2019).
URL: https://www.worldenergy.org/ publications/2019/world-energy-issues-monitor-2019-managing-the-grand-energy-transition/

6. J. Pyper, Fossil Fuels Won't Go Away, Even if the Energy Transition Accelerates (2018).
URL: https://www.greentechmedia.com/articles/read/woodmac-fossil-fuels-wont-go-away-accelerated-energy-transition#gs.m6c8um

7. K.R. Allaev, Energy of the World and Uzbekistan. Tashkent: Molia, 388 p. (2007)

8. K.R. Allaev, Energy needs a strategy. *Economic Review*, **6**, 40–47 (2018)

9. The main indicators of Uzbekneftegaz (2019).
URL: http://www.ung.uz/investors-and-shareholders/information-disclosure/activities/

10. List of power plants in Uzbekistan.
URL: https://ru.wikipedia.org/wiki.

11. M. Djangirov, An expert from Uzbekistan on the safety of nuclear power plants for the region (2018).
URL:https://nuz.uz/intervyu/35771-ekspert-iz-uzbekistana-o-bezopasnost i-aes-dlya-regiona.html

12. A.M. Xodjaev, Does NPP need Uzbekistan? (2018).
URL: https://nuz.uz/kolumnisty/34567-nuzhna-li-aes-uzbekistanu.html

13. M. Djangirov, Atomic Renaissance. What will give Uzbekistan entry to the nuclear club? (2018). URL: https://nuz.uz/nauka-i-tehnika/33844-atomnyy-renessans-chto-dast-uzbekistanu-vstuplenie-v-yadernyy-klub.html

14. A.A. Abidov, P.K. Azimov, To the problem of increasing oil production in Uzbekistan. *Oil and gas*, **1**, 11–14. (2003)

https://doi.org/10.1051/e3sconf/201913901003

Strategic cooperation of electric power systems of Russia and Central Asia for the creation of common Eurasian electric power space

Lyudmila Chudinova [1], *Sergei* Podkovalnikov[1*]

[1]Melentiev Energy Systems Institute of Siberian Branch of the Russian Academy of Sciences, Irkutsk, Russia

Abstract. The paper considers electric power integration projects realized and to be implemented on the territory of Eurasia, that include Russia and countries of Central Asia, neighboring regions of Caucasus, South Asia, and others. Studies are focused on the effectiveness of electric power integration of electric power systems of Russia and Central Asia in the long-term perspective, with account of electric ties with neighboring countries.

1 Introduction

Processes of electric power integration are characteristic both of Asian region, of Eurasia as a whole, and of the entire world. Asian and Caspian power grids are under consideration, electric power integration within the projects of Eurasian Economic and Customs Unions are being developed on the post-soviet area, European Supergrid is being formed. All of the projects contribute to the creation of Common Eurasian Electric Power Space in the future [1-3 et al.].

Although the development of these integration processes requires the solution of certain problems (ensuring the energy security of the countries participating in the integration, harmonization of their energy and power legislation, the creation of a multilateral body to promote the interstate power grid (ISPG), the construction of expensive and long interstate electric ties (ISETs), etc.), as a result of electric power integration, significant economic, energy and other benefits arise for all its participants [4].

Russia has ISETs with Kazakhstan and further via Kazakhstan with other countries of Central Asia (CA) for effective power exchange and for raising the reliability of power supply of consumers. Development of integration projects requires revision of the main parameters and operating modes of Russia-Central Asia ISETs as they will become major components of sub-regional ISPG covering the European part of Russia, Ural and Siberia, Central Asia and Asia Minor (AM), a part of Southern Asia (SA), and Middle East (ME). For that matter, effective parameters of those ties, expedient operating modes and ways of their use need comprehensive study.

2 Electric ties between Russia and Central Asia

Integration of Russia and countries of the region started with construction of power transmission lines between Russia and Kazakhstan and via Kazakhstan to other countries of Central Asia as early as at development of Unified Power System (UPS) of the USSR. After the USSR disintegration, Kazakhstan, Kyrgyzstan, Tajikistan and Turkmenistan formed an interstate Central Asian Power System. Today this interstate power grid operates within the power interconnection formed in the post-Soviet space and named IPS/UPS. A diagram of IPS/UPS and its external electric ties with foreign countries is given in Fig. 1. Exchange of power flows along the ties is given for the year 2017.

As is seen from the Figure, as of 2017, Russia was a net importer of Central Asia (Kazakhstan, to be more exact) receiving (into Interconnected Power Systems – IPSs – of Ural and Siberia) about 3.5 TWh/y of electric power. Total transfer capacity of transmission lines (TL) between IPS of Siberia, Ural, Middle Volga and South, on the one hand, and EPSs of different regions of the North Kazakhstan, on the other hand, made about 6.7 GW. Owing to considerable transfer capacity of ISETs, ability to exchange power flows, the systems benefits attained by cooperating countries might have been even higher, as well as the effectiveness of such cooperation.

3 Problem statement and research scenarios

Study of cooperation between EPSs of Russia and countries of Central Asia implies feasibility analysis of perspective amounts and operating modes of power flows between the named above power systems. Commissioning of generating capacities and modes of

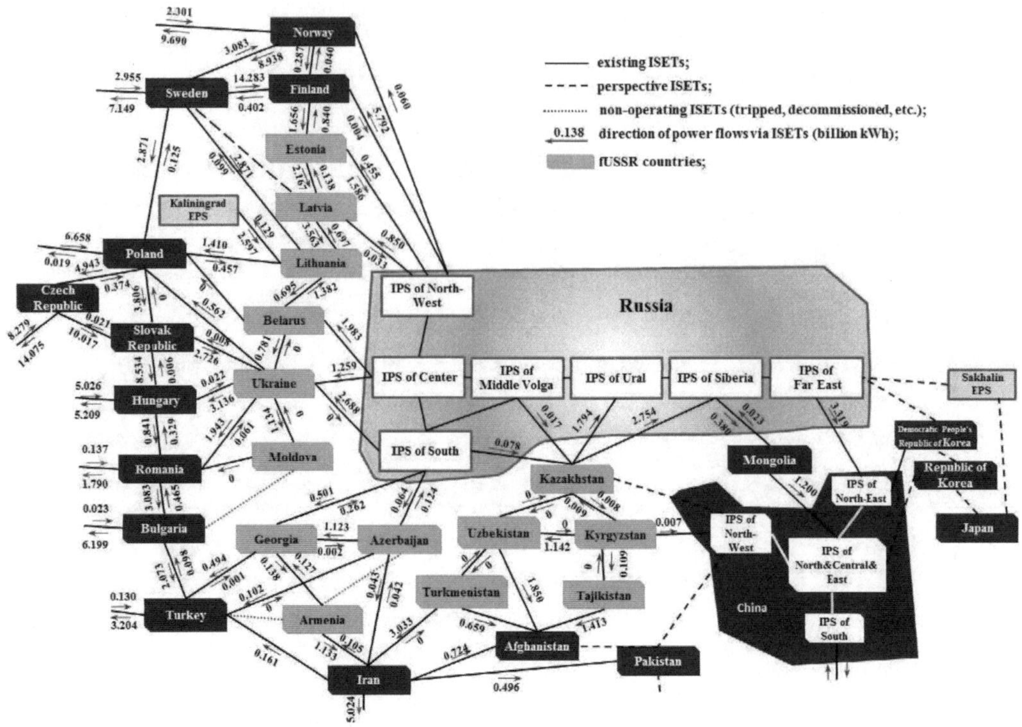

Fig. 1. A diagram of IPS/UPS and its external electric ties

their operation in the considered perspective are being optimized. The result is an optimum mix, amount and operating modes of generating capacities, optimum directions, amounts and modes of power flows between EPSs that ensure minimum costs (including capital, operating and fuel costs) for the grid as a whole.

Calculations were made using a mathematical model for EPS expansion and for its operating modes that was developed in Melentiev Energy Systems Institute (ISEM SO RAN) and widely used for studying the prospects and effectiveness of forming ISETs and ISPGs in the Asian region [4].

Consideration was given not only to EPSs of Russia and countries of Central Asia, but (taking into account rapidly developing electric power integration in Eurasia) to neighboring EPSs covering regions (and countries) of Caucasus (Armenia, Azerbaijan, Georgia), partially Southern Asia (Afghanistan, Pakistan), Asia Minor, and the Middle East (Turkey and Iran). A diagram that includes EPSs of the listed regions (countries) represented as nodes, and electric ties between them, is given in Fig. 2.

Subject of consideration was two scenarios of research that differ in the development level of interstate electric network infrastructure in the considered ISPG. In the first scenario the current values of transfer capacities of ISETs were assumed. The fact that electric tie between Turkey and Armenia (between nodes of Caucasus, Asia Minor and Middle East in Fig. 2) was tripped was taken into account. In the second scenario the transfer capacities of interstate electric ties that are being planned, constructed and will soon be commissioned were additionally taken into account. Those are TL projects Russia-Georgia (500 kV), Armenia-Iran (400 kV), Azerbaijan-Iran (400 kV), Armenia-Iran (400 kV), Turkmenistan-Iran (400 kV), Turkey-Greece (400 kV), CASA-1000 (Tajikistan-Kyrgyzstan-Pakistan-Afghanistan), and others. It was assumed that TL between Turkey and Armenia was in operation. Potentials for Siberia-Kazakhstan-Ural TL retrofitting (its changeover from 500 kV to initial rated 1150 kV voltage with appropriate increase in transfer capability) were also given due regard for.

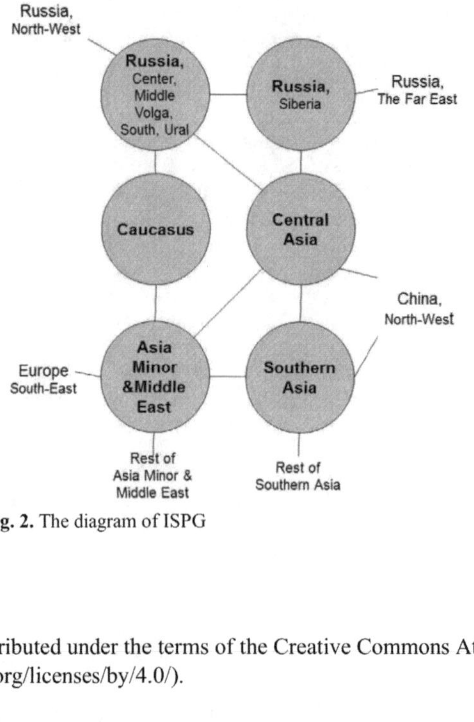

Fig. 2. The diagram of ISPG

Present-day state of power flow exchange given in Fig. 1 can actually be considered as 'zero' scenario. It can be a 'reference point' for scenarios 1 and 2.

Consideration period was taken to be up to 2035-2040 with account of uncertainty of power sector development rates in the countries under consideration.

4 Main input data

Table 1 presents forecast annual maxima of electric loads for the considered target year that are grouped by nodes on the diagram of ISPG (Fig. 2). They were obtained based on the analysis of materials on electric power sector expansion in the considered countries [5-10].

Table 1. Forecasted yearly peak load, 2040, GW.

Russia* (Center, Middle Volga, South, Ural)	Russia (Siberia)	Caucasus*	Central Asia*	Asia Minor and Middle East*	Southern Asia*
140.9	41.8	13.3	61.3	234.3	65.0

* The combined maximum

Table 2 presents present-day transfer capacities of transmission lines that form interstate electric network infrastructure of the considered ISPG..

Table 2. Present-day transfer capacities of interstate electric ties, MW.

Russia - CA	Russia-Caucasus	CA - SA	CA-AM &ME	SA-AM &ME	MA - Europe	Total
6700	1700	2300	460	80	1600	2400

Tables 3 and 4 present main economic indicators of power plants of different types and on different fuels [11-13].

Table 3. Specific capital investments, USD/kW

	Hydro	Pumped storage	Thermal, coal	Thermal, gas	Thermal, oil	Nuclear
Russia	3000	1100	1800-2000	1200		2800
Central Asia	2100	1600	2150	1250		4300
Southern Asia	2600		1700	1200	1400	5200
Caucasus	1500		2000	1000	1500	5500
Asia Minor and Middle East	2250	1000	1800	670	1400	4500

Table 4. Fuel costs, USD/kW

	Thermal, coal	Thermal, gas	Thermal, oil	Nuclear
Russia	0.014-0.25	0.027-0.030		0.004
Central Asia	0.014-0.017	0.034-0.038	0.099	0.004
Southern Asia	0.034	0.041	0.100	0.006
Caucasus	0.030	0.050	0.100	0.010
Asia Minor and Middle East	0.034	0.050	0.090	0.007

Costs of interstate ties were not accounted and, hence, transfer capacities of those ties were not optimized since projects of those ties, according to the plans available, are predetermined and will be commissioned prior to the target year. The above mentioned 500 kV TL Siberia-Kazakhstan-Ural previously operated at 1150 kV was an exception. In the optimization model, when minimizing the costs for ISPG as a whole, the optimum transfer capacity of the line was determined with account of costs for its retrofitting.

5 Results of studies and analysis

Comparison of results of optimization for Scenarios 1 and 2 shows that enlargement of transfer capacity of interstate grid infrastructure (Scenario 2) allows reduction of total ISPG costs by more than 2 billion USD a year. Capital investments into power plants in this case are 4.7 billion USD lower, and demand for commissioning new power plants for the considered time span falls by 2.6 GW. As is seen from Table 5, enlargement of transfer capacity of ISETs has different impact on capacities of various nodes, (difference of commissioning of generating capacities for Scenarios 2 and 1). Additional power plant capacities are commissioned in the Caucasus node (Scenario 2), which proves that in case of enlargement of transfer capacity of this node, higher power generation is profitable (providing it is transported beyond the region). Installed capacity of power plants in the Central Asia (see the Table) does not change.

Table 5. Generating capacity benefit, GW.

Russia	Central Asia	Southern Asia	Caucasus	Asia Minor and Middle East
1.3	—	1.1	-2.8	3.0

Table 6 presents change in the power generation by the nodes of the diagram in Scenario 2 vs Scenario 1. Russia, despite its capacity benefit, increases power generation in Scenario 2, similarly to CA that has no that benefit. Caucasus node knowingly increases power generation in the second scenario as its capacity benefit is negative (it increases its generating capacities in the second scenario). The mentioned additional power generation is transmitted to the nodes of SA and AM&ME.

Table 6. Change in the power generation, TWh/y.

Russia	Central Asia	Southern Asia	Caucasus	Asia Minor and Middle East
10.1	11.3	-4.9	10.9	-23.0

Tables 7 and 8 show the amounts of power exchanges between nodes for the considered scenarios. Power flows in the second scenario grow as compared to the first one, especially in the direction Russia - Caucasus - AM&ME. Total amount of export-import power flows in Scenario 1 made about 240 TWh per year, whereas in Scenario 2 it was as high as 320 TWh/y, i.e., increased by one third. It should be kept in mind that those values do not reflect the entire amount of interstate power flows. Nodes of the diagram (with the exception of nodes representing EPSs of Russia) include several countries of a certain region, and power flows between those countries cannot be accounted in the model.

Table 7. Power flows exchange, Scenario 1, TWh/y

Electric ties \ Nodes	Russia (Siberia)	Russia (Center, Middle Volga, South, Ural)	Central Asia	Southern Asia	Caucasus	Asia Minor and Middle East
Russia (Siberia) - Central Asia	-11.5	-	12.4	-	-	-
Russia (Center, Middle Volga, South, Ural) - Central Asia	-	51.6	-50.1	-	-	-
Russia (Center, Middle Volga, South, Ural) - Caucasus	-	17.9	-	-	-17.4	-
Central Asia - Southern Asia	-	-	18.6	-17.9	-	-
Central Asia - Asia Minor and Middle East	-	-	4.9	-	-	-4.7
Southern Asia - Asia Minor and Middle East	-	-	-	-0.4	-	0.4
Caucasus - Asia Minor and Middle East	-	-	-	-	12.8	-12.7
Asia Minor and Middle East - Europe	-	-	-	-	-	4.0

(-) power received by a node, (+) power supplied by a node

Table 8. Power flows exchange, Scenario 2, TWh/y

Electric ties \ Nodes	Russia (Siberia)	Russia (Center, Middle Volga, South, Ural)	Central Asia	Southern Asia	Caucasus	Asia Minor and Middle East
Russia (Siberia) - Central Asia	-13.1	-	14.1	-	-	-
Russia (Center, Middle Volga, South, Ural) - Central Asia	-	53.2	-50.0	-	-	-
Russia (Center, Middle Volga, South, Ural) - Caucasus	-	27.1	-	-	-26.3	-
Central Asia - Southern Asia	-	-	23.5	-22.5	-	-
Central Asia - Asia Minor and Middle East	-	-	9.5	-	-	-9.2
Southern Asia - Asia Minor and Middle East	-	-	-	-0.6	-	0.6
Caucasus - Asia Minor and Middle East	-	-	-	-	32.6	-32.3
Asia Minor and Middle East - Europe	-	-	-	-	-	4.0

(-) power received by a node, (+) power supplied by a node

The highest growth of power flows in both scenarios is observed in case of comparison to a 'zero' scenario, i.e., present-day power flows given in Fig. 1. Thus, power flows from Central Asia to Asia Minor and Middle East in Scenarios 1 and 2 grows almost threefold, that from Central Asia to Southern Asia grows actually tenfold.

As to cooperation between Russia and Central Asia countries, its power flows in Scenarios 1 and 2 were considerably enlarged as compared to the present-day ones. Thus, power flows from European Russia and Ural to Central Asia (Kazakhstan) grows to more than 50 TWh in the target year, and that from Central Asia (Kazakhstan) to Siberia makes 12-13 TWh/y. The obtained values of power flows are potentially feasible.

Table 9 gives transfer capacities of ISETs for Scenario 2 (for Scenario 1 they were given earlier in Table 2). As is seen, in Scenario 2 they are more than 5.5 GW higher and ensure corresponding growth of power flows exchange under this scenario (Table 8).

Table 9. Transfer capacities of interstate electric ties, Scenario 2, MW.

Russia - CA	Russia - Caucasus	CA - SA	CA - AM& ME	SA - AM & ME	Caucasus - AM&ME	AM - Europe	Total
7300	2580	3300	900	240	4230	2400	20950

6 Conclusion

Enlargement and effective use of interstate power network infrastructure between Russia and Central Asia, Caucasus, Asia Minor and Southern Asia, and partially Middle East contribute to intensification of power flows within this ISPG, and give systems benefits. ISETs in the study were limited by the existing, planned electric ties, and by ties under construction. Expansion of interstate network infrastructure is deemed to accelerate integration of the considered domestic EPSs and raise ISPG effectiveness in the area.

Further detailed and comprehensive feasibility studies are needed that may form the basis for developing the policy in the field of electric power integration of Russia, countries of Central Asia, and of neighboring regions, including Caucasus, Sothern Asia, etc.

Acknowledgments

The research was carried out under State Assignment, Project 17.6.2 (reg. no. AAAA-A17-117030310447-3) of the Fundamental Research of Siberian Branch of the Russian Academy of Sciences.

References

1. Belyaev L.S., Voropai N. I., Koshcheev L. A. Prospects for development of interstate grids in the Eurasia supercontinent. Proceedings of the Russian Academy of Sciences. Energy, no. 2, pp. 27-35. (2000) [in Russian]

2. Barinov V.A., Bushuev V.V., Samorodov G.I. Structural and technological foresight of UPS development in Russia and infrastructure of Eurasia. Energy Policy, no. 3, pp. 43-40 (2014) [in Russian]
3. Voropai N. I., Podkovalnikon S.V., Saveliev V.A., Chudinova L.Yu. Interstate grids of Eurasia and their interconnection with UPS of Russia. Electric power. Transmission and distribution, no. 3, p. 8-13. (2017) [in Russian]
4. Voropai N. I., Podkovalnikon S.V., Saveliev V.A., Chudinova L.Yu. Effectiveness of interstate electric ties. Novosibirsk, Science, 239 p. (2008) [in Russian]
5. Scheme and Program of Unified Energy System of Russia development for 2019-2025. RF Ministry of Energy, 273 p. [Electronic resource]. Access mode: https://minenergo.gov.ru/system/download-pdf/14828/97146. (2019) [in Russian]
6. General layout of power facilities in Russia for the period up to 2035. RF Ministry of Energy, 277 p. [Electronic resource]. Access mode:https://minenergo.gov.ru/system/download-pdf/14828/97146. (2017) [in Russian]
7. Ten Year Network Development Plan of Georgia 2019-2029. Georgian State Electric System, 2019. 438 p. [Electronic resource]. Access mode:http://www.gse.com.ge/sw/static/file/TYNDP_GE-2019-2029_ENG.pdf. (2019)
8. Electricity Demand Forecast: Based on Multiple Regression Analisis. Period 2014 to 2037. National Transmission & Despatch Company. 111 p. [Electronic resource]. Access mode:http://www.ntdc.com.pk/planning. (2014)
9. A concept of Kazakhstan Fuel and Energy Complex development till 2030 (Decree of Kazakhstan Republic dated 28/06/2014, no. 724). [Electronic resource]. Access mode: http://adilet.zan.kz/rus/docs/P1400000724. (2014) [in Russian]
10. Forecast balance of capacity and power of Kazakhstan EPS for the period to 2019-2025. Kazakhstan Electric Energy Management Company, 7 p. [Electronic resource]. Access mode: https://www.kegoc. kz. (2019) [in Russian]
11. Abtin Ataei, Khaled Fararian. Environmental and Economic Analysis of Scenarios for Transition to Renewable Energy in the Iranian Electricity Supply System. International Journal of Energy Science and Engineering, no 4 (1), pp. 153-162. (2015)
12. Central Asia Regional Economic Cooperation: Power Sector Regional Master Plan. Asian Development Bank, 65 p. [Electronic resource]. Access mode: https://www.adb.org. (2012)
13. Update of the Power Sector Master Plan of Azerbaijan 2013 - 2025. 48 p. [Electronic resource]. Access mode: http://www.inogate.org/documents. (2013)

E3S Web of Conferences **139**, 01004 (2019)
RSES 2019
https://doi.org/10.1051/e3sconf/201913901004

Allowance for capacity adequacy criterion in optimizing the prospective structure of electric power system

Nikolay Belyaev[1,*], *Andrey* Egorov[1], *Nikolay* Korovkin[2] and *Vladimir* Chudny[2]

[1] Russian Energy Agency Ministry of Energy of the Russian Federation, Russia
[2] Peter the Great St.Petersburg Polytechnic University, Russia

Abstract. This paper considers the optimization procedure of electric power system prospective structure with account made for two criteria: economic value (minimization of specific reduced energy costs for customers) and capacity adequacy (minimization of capacity shortage probability). The proposed procedure is based on genetic algorithm application. The results of procedure evaluation have been considered by the example of optimization of the structure of generating capacities within concentrated power system.

Introduction

The standard problem statement of planning electric power systems (EPS) reduces to cost-minimizing procedure in relation to total reduced costs З connected with electric power supply to consumers [1, 2]:

$$\text{З}(\mathbf{x}) = \sum_t (\text{З}_t^{\text{K}}(\mathbf{x}) + \text{З}_t^{\text{O}}(\mathbf{x}))\cdot(1+d)^{-t} \xrightarrow[\mathbf{x}]{} \min, \qquad \mathbf{x} \in R, \tag{1}$$

where З_t^{K}, З_t^{O} – capital and operational costs per year t respectively, d – discount rate. Variables $\mathbf{x}$ in problem (1) are major engineering solutions of planning electrical generation, transmission and distribution systems that determine their cost-based value and structure. The range of limitations R is specified by EPS reliability and security requirements influencing the choice of certain engineering solutions or their combinations. Such limitations include, among others, the EPS adequacy requirements which are given as regulated value of capacity adequacy index (CAI) or respective capacity margin [3, 4].

Such an approach of taking into account the capacity adequacy requirements has some disadvantages because it needs to be feasibly studied in terms of mentioned regulated CAI value. This kind of study is made difficult due to the absence of appropriate assessment of damages arising from capacity shortage and to complicated way of its acquisition defined by the variety of EPS customers' composition and their modes of operation. Moreover, known procedures of CAI regulated value justification [5] are based on the assumption of linear relation existing between cost-based margin capacity and its value that reveals to be unfair bearing in mind a real structure of generating capacities in electric power systems.

So, this paper proposes a two-criterion problem statement of EPS planning providing for low costing with CAI minimization - probability $J_Д$ of capacity shortage in EPS:

$$\text{З}(\mathbf{x}) \xrightarrow[\mathbf{x}]{} \min, \quad J_Д(\mathbf{x}) \xrightarrow[\mathbf{x}]{} \min, \quad \mathbf{x} \in R. \tag{2}$$

1 Source data, variables and constraints

When planning EPS, engineering solutions should be examined after having been integrated in technologies [2] under types and performance-based indicators. The potential structure of these technologies is given in Table 1. Technologies of generation development *(G)* cover power station units and power generation plants of various types of different unit capacity. Technologies covering the development of electrical networks are as follows: construction of new power transmission lines of various voltage classes as well as means of reactive power compensation making possible to expand their capacity. Each technology should be provided with performance indicators and reliability data so as to ensure the calculation of target functions (2). Parameters related to the demand of power supply in EPS must be given as source data: forecast power consumption, peak electrical demands and electricity load curves, nonconforming load indicators with details for every node of electric power system.

Variables in the problem are discrete values $x_{k,i} \in \mathbf{x}$ that correspond to the number of technology k measures to be implemented in node i which is understood as a part of power system (zone) where there is no limitation for power transmission [6].

Table 1. Structure of technologies at EPS planning.

Type	Name	Available (maximum) capacity, MW	Technological minimum, MW	Specific capital costs, th. Rub/kW (mln Rub/km)	Specific operational costs Rub/ kW·h
Generation					
TPP	ST-50	50	40	55	1,17
TPP	ST-100	100	80	50	1,11
CTPP	CTST-50	50	25	50	1,17
CTPP	CTST-100	100	50	45	1,11
CTPP	CTST-200	200	100	40	1,08
CTPP	CTST-300	300	150	37	1,05
CTPP	CTST-500	500	250	35	1,03
CTPP	CTST-600	600	360	50	0,70
CTPP	CTST-800	800	400	34	1,00
CTPP	CCP-150	150	60	60	0,77
CTPP	CCP-450	450	180	50	0,73
CTPP	CCP-800	800	240	40	0,69
CTPP	GT-50	50	0	65	1,24
CTPP	GT-100	100	0	60	1,17
HPP	Hydraulic unit-100	100	15	145	-
HPP	Hydraulic unit-300	300	30	120	-
NPP	LWGR-1200	1200	1200	110	-
WG	WG-50	*	*	150	-
Network					
TL	220 kV	249	-	12	-
TL	330 kV	748	-	13	-
TL	500 kV	1698	-	19	-
CSR	per 1 kVAr	-	-	3	-
CB	per 1 kVAr	-	-	0,4	-

* to be specified as probability distribution series [7]

Range R is described by different limitations:

maximum or minimum possible (with account made for existing EPS structure and solutions adopted for implementation) number ($\underline{x}_{k,i}$, $\overline{x}_{k,i}$) of measures for each technology:

$$\underline{x}_{k,i} \le x_{k,i} \le \overline{x}_{k,i}\,, \tag{3}$$

total volume of generating capacities required as a value of maximum electrical load $N_{\max}$:

$$\sum_i \sum_k P_k x_{k,i} \ge N_{\max}\,,\quad k \in G\,, \tag{4}$$

where P_k – available unit capacity of generating equipment under technology k,

technical constraints of total capacity ($\underline{P}_{k,i}$, $\overline{P}_{k,i}$) of certain types of power plants: as per technological minimum, fuel supplies, required volume of thermal energy supply (for TPP) etc.:

$$\underline{P}_{k,i} \le P_k x_{k,i} \le \overline{P}_{k,i}\,. \tag{5}$$

2 Solution

To solve the problem (2) with account made for constraints (3) – (5) while considering its discreteness and availability of two criteria a genetic algorithm is proposed for application whose principal provisions in terms of branch problems are given in [8]. The calculation of target functions on each algorithm step is carried out by applying Monte-Carlo technique in the following way:

1. For each formed version of solution multiple random states of EPS are examined with account made for EPS maintenance outages and emergency repairs, regular or irregular load changes, available capacity changes of seasonal and stochastic nature on the basis of indicators given as source data. The following should be specified for each described state: working capacity P_{Pi} of power plants in nodes, load N_i for customers in nodes, transmission capacity $Z_{i,j}$ of connections between nodes.

2. For every random state the problem of capacity shortage minimization D_i is to be solved:

$$\begin{aligned}&\sum_i D_i \xrightarrow[P_{Hi}]{} \min\\ &\sum_i (P_{Hi} - n_i) = 0\,,\\ &\forall i : D_i = N_i - n_i \ge 0\,,\\ &P_{Hi} \le P_{Pi}\,,\\ &\forall i,j : \sum_m k_m^{i,j}(P_{Hm} - n_m) \le Z_{i,j}\,,\end{aligned} \tag{6}$$

where P_{Hi} and n_i – respectively actual loads of power plants and customers in nodes, $k_m^{i,j}$ – linear dependency factor of power flow between nodes i,j in relation to node capacity m (mains factor). The procedure of mains factor calculation is given in [9].

3. To get values n_i by results of solution (6) the problem of operational costs minimization 3° is to be solved:

$$\begin{aligned}&3^O \xrightarrow[P_{Hi}]{} \min\\ &\sum_i (P_{Hi} - n_i) = 0\,,\\ &\forall i : P_{Hi} \le P_{Pi}\,,\\ &3^O = \sum_i \sum_k P_{Hk,i} 3_k^o\,,\\ &\forall i,j : \sum_m k_m^{i,j}(P_{Hm} - n_m) \le Z_{i,j}\,,\end{aligned} \tag{7}$$

where 3_k^o – specific operational costs of technology k ($k \in G$).

4. Following the results of solution of (6) and (7) for random sets the values of target functions (2) of a given solution option are defined. The probability of capacity shortage is defined as a relative number of states with capacity shortage obtained upon solution of (6), the value of total costs is according to (1) with account of average value 3°, got upon solution results of (7).

Based on the results of problem (2) solution the variety of Pareto optimal solutions is specified, its analysis provides for the definition of the influence of CAI regulated value change on the value of costs required for its obtaining and for selection of the most appropriate option of EPS planning.

It should be noted that to get a feasible accuracy of target function calculations (2) the number of random states for each alternate solution must make ~10^6. In such a case to ensure the algorithm convergence at specific number of variables of ~20 the number of single generation should be at least 10^3. Thus, when the number of generations is ~10^{2}, one must estimate ~10^{11} random EPS states by solving (6) and (7) in order to get the solution of studied problem.

3 Evaluation

Calculations have been performed for a single-node EPS. Maximum load of EPS makes 12,87 GW, the load curve irregularity factor being of 0,77 and the load factor – of 0,92. To carry out calculations there has been accepted the structure of technologies according to data of Table 1 while the equipment reliability indicators are in line with data [10]. The results are shown on Fig. 1.

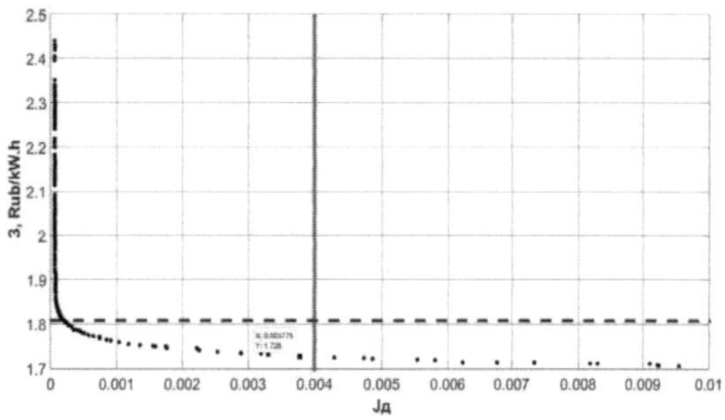

Fig. 1. Calculation results.

On Fig. 1 the values of target functions (2) are plotted on axes: on horizontal line – probability of capacity shortage, on vertical line – working cost of electric energy compatible with accumulated total costs. Points mark the results. The total number of obtained solutions is equal to 375.

During Soviet time the regulated value of CAI has been justified and until now it is applied at the level of 0,996 [1, 5], that is in line with $J_Д$ = 0,004. Thus, when using this rate one should choose the solution marked on Fig. 1, in this case the electricity cost will make 1,73 Rub./kWh. However this option does not seem to be optimal if other solutions are examined. For example, an insignificant price increase within 1 cop./kWh allows reducing $J_Д$ to the level of 0,00226. Moreover, it should be noted that from cause-and-effect view under circumstances when the operating conditions of EPS through the long run are uncertain, the parameter setting error for technologies applied in problem as source data, cost parameters, first of all, may reach up to 5 per cent and more that significantly affects the accuracy of cost assessment. In this case the price increase by 5 per cent as compared to the solution marked on Fig. 1 (up to 1,81 Rub./kWh, shown by horizontal dot lines) allows reducing $J_Д$ up to 0,00016, i.e. more than by an order of magnitude.

Further reduction of $J_Д$ may appear to be cost-ineffective as costs progress at a far quicker rate and an insignificant reduction of $J_Д$ results in price hikes. At a price of 1,97 Rub./kW·h the limit value of $J_Д \sim 6,3 \cdot 10^{-5}$, is reached, in practice it is insensitive to price increase.

The range of installed capacity variation upon got solutions was within 14,95 GW – 22,15 GW that is in line with the margin capacity value from 16,2% to 72,1% of EPS maximum load. For the solution marked on Fig. 1 which is in conformity with regulated CAI, the capacity margin was 19,7%. For the solution with the price of 1,81 Rub./kW and $J_Д$ = 0,00016 the capacity margin is equal to 29,8%.

Conclusion

In comparison with standard problem statement of EPS planning (1) the solution of proposed problem (2) allows not only taking into account regulated requirements as to EPS capacity adequacy but selecting an optimal level of capacity adequacy on the basis of evaluation of costs increase incurred to its provision with due account made for a specific EPS and potential engineering solutions. In this case based on performed calculations a better CAI improvement in relation to regulated level at insignificant cost increase within the range of source data setting accuracy.

It should be noted that there is the possibility to set the problem (2) by including therein additional criteria effective for EPS planning. Such criteria are minimization of environmental impacts (minimization of fuel plant emis-sions), ensurance of energy security (minimization of installed capacity shortage in certain EPS nodes) etc.

References

1. I.M.Volkenau, A.N.Zeiliger, L.D.Khabachev, *Economics of electric power system formation* (Energiya, Moscow, 1981)
2. F.V.Veselov, T.V.Novikova, A.V.Fedosova, *Fundamentals of electric power's economics: Guidelines* (Gubkin Russian State University of Oil and Gaz, Moscow, 2016)
3. Yu.Ya.Chukreev, M.Yu.Chukreev, *Models of evaluation of capacity adequacy indices when managing the development of electric power systems* (Komi scientific centre, Syktyvkar, 2014)
4. D.V.Iakubovskii, D.S.Krupenev, D.A.Boyarkin, E3S Web of Conferences, **114**, 03002 (2019)
5. D.S.Krupenev, L.M.Lebedeva, G.F.Kovalev, N.A.Belyaev, A.E.Egorov, R.E.Gromov, Energy policy, E **1**, 33 (2018)
6. N.Belyaev, A.Egorov, N.Korovkin, V.Chudny, E3S Web of Conferences, **58**, 01010 (2018)
7. D.Krupenev, E3S Web of Conferences, **58**, 01012 (2018)
8. N.V.Korovkin, A.A.Potienko, Electricity, E **1**, 2 (2002)
9. N.A.Belyaev, A.E.Egorov, N.V.Korovkin, V.S.Chudny, Electricity, E **5**, 11 (2018)
10. V.A.Nepomniaschiy, *Reliability of electrical power system equipment* (Electroenergiya. Transmission and distribution, Moscow, 2013)

Information support of intellectual energy systems and its impact on ensuring balance reliability

Yuri Chukreyev[1], *Mikhail* Chukreyev[2]

[1]Federal Research Center "Komi Scientific Center of the Ural Branch Russian Academy of Sciences", ISE and EPN, 167000, Russia
[2]Federal Research Center "Komi Scientific Center of the Ural Branch Russian Academy of Sciences", ISE and EPN, 167000, Russia

Abstract. The is UPS of the impact of accounting for the energy supply of hydroelectric power plants on the means of ensuring the balance reliability of the UPS of Russia for the conditions of conducting commercial power take-offs are considered. The analysis of predicted and reported data on the generation of electricity at hydroelectric power stations and their impact on the rationale for backup tools.

Introduction

The state of the electricity industry has undergone significant changes since the end of last century. In organizational, technological and informational aspects. The transition to market relations in country has led to a significant reduction in energy consumption and capacity. In modern conditions, the task of substantiating values normative reserve of power (NRP) in planning development UES of Russia should be considered in a completely different plane. In the pre-perestroika period, there was a problem of power shortages and the task was to justify the inputs of generating facilities. Currently, there are significant excess capacity. The task is to justify the equipment commissioning (in the market terminology of power supply contract) and its dismantling. Its need is identified as a result implementation of competitive bids during a closed auction for commercial selection of capacity (CSC).

Development planning at all times is associated with the formation of power and electricity balances UES of the country. In accordance with the decree of the Russian Federation Government[a] since 2010, the noted balances are formed in the work “Scheme and Program for the Development UES for a 7-Year Period”, annually carried out by JSC SO UPS and PJSC FGC UPS. Demand for power determines the expenditure side of the balance sheet and, in accordance with the Regulation on the procedure for determining this value[b] (hereinafter Regulation No. 431), significantly affects the cost indicators of power during CSC. The amount demand for capacity includes three components: the projected maximum load, export / import of capacity and the value of NRP. The most difficult for consumers and suppliers of power and electricity to understand is the third component.

The magnitude of the NRP depends on many factors, including randomly determined. During CSC procedure, the amount of NRP in accordance with the aforementioned Regulation No. 431 is determined by the reservation coefficients approved by the RF Ministry of Energy for the allocated price zones of the electricity market. The numerical values of these coefficients are based on the materials of guidelines for designing the development energy systems, the latest edition of which was completed in the mid-90s of the last century and approved by the Russian Ministry of Energy only in 2003 [1]. It is obvious that this document requires its updating under modern realities.

The rationale for the NRP is based on the solution of the problem of assessing indicators of balance reliability (IBR) as applied to design scheme UES of Russia with the allocation territorial reliability zones in it [2]. The UES was the last in the development of 2003 guidelines. It is possible to split UES into smaller territorial zones. However, due to the complexity of the content, such fragmentation is advisable only in cases of significant influence of limited bandwidth between the allocated territorial zones. The enlargement of territorial zones to the UES level can also be justified by the fact that the procedure for conducting CSC currently focuses only on two price zones - the European part with Ural and Siberia.

It is known that in the CSC procedure the selection of the most efficient existing units of power plants is

[a] Rules for the development and approval of schemes and programs for prospective development electric power industry, approved by Decree of the Russian Federation Government, October 17, 2009 No. 823.
[b] The Regulation on the procedure for determining the amount demand for long-term power take-off on a competitive basis in the wholesale market of electric energy (power) and the procedure for determining the planned coefficients of capacity reservation in the zones (groups of zones) of free flow of electric energy (power), approved by the Order of the Russian Ministry of Energy from 09/07/2010 No. 431.

carried out. It is not possible to take this aspect into account in models of ensuring balance reliability. It is clear that in the presence excess capacity of a hydroelectric power station in the economic aspect it is more attractive than a thermal one due to lower operational costs (lack of a fuel component). At the same time, their operating modes depend on weather conditions (dry years), and this aspect in modern conditions presence of excess capacity should be taken into account when substantiating backup means. Below are considered the issues of the possibility accounting for the energy supply of hydroelectric power plants during their joint work with thermal and nuclear power plants in the territorial zones UES of Russia.

1 Methodological principles of accounting for restrictions on production of electricity at hydroelectric power plants when assessing indicators of balance reliability EPS

When developing mathematical models for assessing the balance reliability indicators of multi-zone EPSs both in our country and abroad, approaches based on either analytical methods or statistical modeling methods are used. The latter approaches are most widespread due to the great potential to take into account various kinds of restrictions [2]. Regardless of the principles of managing the electric power industry (centralized, market), the methodology for solving the problem of assessing balance reliability indicators using statistical modeling methods is based on two interrelated stages:

– formation of load levels and random conditions of generating capacity caused by unplanned conclusions in emergency repairs of power plant equipment;
– assessments of the formed states with a view to ensuring the load of territorial zones.

The determination of balance reliability indicators is carried out sequentially for all time intervals during which the structure of generating capacities and the load remain unchanged [2-5]. The formation of random states for each territorial zone of the model of the calculation scheme UES of Russia is carried out by statistical modeling on the functions of the probabilities of reducing the generating capacities obtained by analytical methods due to unscheduled output of equipment for emergency repairs (regardless of the type of equipment) and load due to errors in predicting it, - caused mainly by temperature fluctuations. In order to evaluate the random state of the system formed in this way, due to many circumstances (information uncertainty, estimation of a large number of states, etc.), approaches based on the use of DC idealization are used [2, 5]. The summation indicators of balance reliability over all the allocated time intervals, taking into account the probabilities of their existence, allows us to determine their integral values.

To date, the models for evaluating balance reliability indicators of EPS have not specified the types of generating equipment involved in covering consumer demand. And this was justified, because when evaluating the balance reliability indicators of the entire set of randomly generated states of generating power and load, less than a percent of the states are scarce. Only in these states is the generating power fully utilized. In deficiency-free conditions, of which more than 99%, the generating capacity exceeds the load. These states do not affect the indicators of balance reliability and in them the generating power can be redistributed between different types of generating equipment (hydroelectric power stations, thermal power plants, nuclear power plants, renewable energy sources) from the standpoint of ensuring a power balance.

Taking into account the participation of hydroelectric power plants in covering consumer demand in deficit-free conditions can only be carried out by separately forming the functions for reducing capacities caused by equipment outages in unscheduled (emergency) repairs, for thermal, including nuclear power plants and for hydroelectric power plants for seasonal (annual) flow control. Moreover, studies show that renewable energy can be included in the first group. Methodological approaches for assessing each state generated by power generation at HPPs and TPPs should also be transformed. Considering that the redistribution of power between hydropower plants and thermal power plants can be carried out only in deficit-free conditions and in no way affects for indicators of balance reliability, the rule of maximum load of thermal and nuclear power plants can be applied. At the same time, the algorithm for evaluating indicators of balance reliability is constructed in such a way that the loading of the hydroelectric power station from one randomly generated deficient state to another changes. The process of change is dynamic, in which the end result is the achievement, if possible, of the planned indicators of electricity production for all time intervals for which information is available on the power supply of the hydropower plant (season, year).

The task of flux distribution in the models for evaluating indicators of balance reliability is called the problem of distribution of power shortages (DPS). As the objective function, the functional was used to maximize the load supply or minimize the power shortage [2, 3]:

$$\sum_{j=1}^{n} c_j^{\text{н}} (\overline{P}_j^{\text{н}} - P_j^{\text{н}}) \rightarrow min, \qquad (1)$$

Accounting for the energy supply of hydroelectric power plants slightly changes this target function, due to the introduction of two terms that characterize the loading of thermal (including nuclear) stations and hydroelectric power stations:

$$\sum_{j=1}^{n} c_j^{\text{н}} (\overline{P}_j^{\text{н}} - P_j^{\text{н}}) + c_j^{\text{эт}} (\overline{P}_j^{\text{эт}} - P_j^{\text{эт}}) + c_j^{\text{ээ}} (\overline{P}_j^{\text{ээ}} - P_j^{\text{ээ}}) \rightarrow min, \qquad (2)$$

where $P_j^{\text{н}}, \overline{P}_j^{\text{н}}, P_j^{\text{эт}}, \overline{P}_j^{\text{эт}}, P_j^{\text{ээ}}, \overline{P}_j^{\text{ээ}}$ – accordingly, covered demand and load demand capacity, used and available generating capacities of heat and hydroelectric power plants of the j-th territorial zone; $c_j^{\text{н}}$ – the load security factor is taken equal to unity; $c_j^{\text{эт}}, c_j^{\text{ээ}}$ – coefficients reflecting the

https://doi.org/10.1051/e3sconf/201913901005

need for loading in deficient states of thermal and nuclear plants ($P_j^{\text{г}m}$) and hydropower plants ($P_j^{\text{гг}}$).

The constraints in these expressions are the observance of power balances, as well as restrictions on the load capacities and generating capacities as applied to all j-th territorial zones [2]. The introduction of two additional terms (2) into the objective function (1) is caused by the need to redistribute the load power between thermal (including nuclear) power plants and hydroelectric power stations in non-deficient states. In order to exclude the influence of the introduced terms on the power deficit in the EPS, the magnitudes of the coefficients for these terms ($c_j^{\text{г}m}$ and $c_j^{\text{гг}}$) much less (100 times or more) coefficient $c_j^{н}$ are accepted for the main term of the functional.

When randomly simulating the states of generating power for each time interval with a given production of electricity at hydroelectric power stations, the coefficients $c_j^{\text{г}m}$ and $c_j^{\text{гг}}$ from one state to another are not constant. Their values depend on how it is necessary to redistribute the generating capacity between thermal power plants and hydroelectric power stations in each j-th territorial zone UES of Russia in non-deficient states, so that at the end of the simulation in the considered time interval we can get as close as possible but not exceed the production volume set on it electricity at hydroelectric power station.

2 Methodological principles for taking into account restrictions on the production of electricity at hydroelectric power stations when substantiating power reserves

It should be noted that the methodology for substantiating the NRP to compensate for the withdrawal of generating equipment to unscheduled (emergency) repairs (operational reserve as amended by the 2003 guidelines) remains unchanged. In modern conditions, unfortunately, there are no scientifically substantiated provisions on the application of decision-making criteria to ensure a particular level of reliability. Here you can focus on either normative indicators countries of Western Europe (LOLH = 3–8 hours / day), or North America (LOLE = 0.1 times a year), or the domestic standard for territorial zones [1, 3, 6] ($J_{д}$ = 0.004).

Important for all indicators of balance reliability listed is approximately the same methodological basis for their preparation and completely different information component, especially in terms of accounting for power consumption modes [2]. The European standard is focused on the accounting of hourly electricity consumption schedules for all 8760 hours of the year, the North American provides for accounting for the load only the maximum hour of the day in year (365 values). Domestic - focused on accounting for only one hourly average daily schedule for the month of December, assuming it is valid for all working days of the year. Comparison of these indicators, from the point of view of their influence on the justification of funds reservation, the event is quite complicated. In [2], for certain conditions, such studies were carried out that showed an acceptable coincidence of the results for substantiating the value of the operational component of the NRP.

The use separate modeling of random states at HPPs and TPPs with NPPs when assessing balance reliability indicators makes it possible to determine the generating capacities required to provide the load separately for these types of stations. This opens up the possibility of determining the necessary addition to the NRP value, caused by the insufficient energy supply of hydroelectric power stations in dry years. For this, it is necessary to carry out two calculations of determining the operational reserve of capacity, which is an integral part of NRP, with the forecasted amount of electricity production at hydroelectric power stations and for a dry year. In both calculations, the value of the operational power reserve remains unchanged, but due to a change in the energy supply of the hydroelectric power station, the redistribution of generating capacities involved in covering the load between hydroelectric power station and thermal power plant with nuclear power plants will occur. The difference in the redistribution will be an additional to operational, and therefore to the NRP. The most difficult when using this approach is uncertainty information on the power supply of hydropower plants.

3 Analysis of information on the energy supply hydroelectric power plants UES of Russia, presented in Design and Development Program UES

When planning the balances of electricity and capacity in the work of the S&D UES, the volume of electric power production at the hydroelectric power stations of the territorial zones is given for the most probable scenario of water supply. For Siberia and the Far East, where the share of electricity production at hydropower plants is significant (from 35% and higher), starting in 2012, the electricity balance is also given for a dry year. It should be emphasized that at the level of regional energy systems there is no such forecast. This once again indicates the complexity of the task content when using the model design scheme UES of Russia with fragmentation into territorial zones.

In the table 1 shows the actual and forecast values of electricity production at hydroelectric power stations in Siberia (the most probable for a dry year), performed with different time periods up to 7 years. Values are for 7 years (column 1). For six of the seven years of the forecast period (columns 4 through 9), the difference in forecasts between the most probable and dry years is from 10.6 to 14.4%. At the same time, when planning for the coming year (the 10th column), the forecast values coincide (except for 2014). There is no explanation for this phenomenon. It should be noted that the forecast for the coming year does not affect the conduct of CSC.

Table 1 Actual and forecasted values of electricity production at hydroelectric power stations of Siberia in million kWh (from the work of the S&D UES for the seven-year period 2010-2016 ... .2019-2025)

Year	Forecast Characteristic	Actual	The forecast lead time for the year given in the 1st column						
			7 years	6 years	5 years	4 years	3 years	2 years	1 year
1	2	3	4	5	6	7	8	9	10
2014	likely low water	94135,4	no forecast	no forecast	108800,0 no info	105420,0 no info	102720,0 91820,0	102310,0 91070,0	103050,0 89300,0
2015	likely low water	88274,0	no forecast	108800,0 нет инф.	106250,0 нет инф.	107380,0 95570,0	107610,0 95560,0	107350,0 91880,0	94696,0 94696,0
2016	likely low water	99845,1	108800,0 no info	106960,0 no info	107380,0 95570,0	107610,0 95560,0	108120,0 92540,0	108118,0 92537,0	92690,0 92690,0
2017	likely low water	93943,2	107190,0 no info	107380,0 95570,0	107610,0 95560,0	108120,0 92540,0	108118,0 92537,0	107377,0 95673,0	93556,0 93556,0
2018	likely low water	101864,3	107380,0 95570,0	107610,0 95560,0	108120,0 92540,0	108118,0 92537,0	107377,0 95673,0	107377,0 93556,0	89550,0 89550,0
2019	likely low water	–	107610,0 95560,0	108120,0 92540,0	108118,0 92537,0	107377,0 95673,0	107377,0 95673,0	107377,0 95673,0	91080,0 91080,0
2020	likely low water	–	108120,0 92540,0	108118,0 92537,0	107377,0 95673,0	107377,0 95673,0	107377,0 95673,0	107377,0 95673,0	no forecast

Noteworthy is the significant excess of the predicted values of the likely production of electricity at hydroelectric power plants (columns 4 through 9) of the actual values (column 3). For a dry year, they, as one would expect, are lower than the actual production of electricity (except for 2015), but insignificantly.

In order to take into account a possible reduction in the production of electricity at hydroelectric power plants in dry years in the task of substantiating NRP, in our opinion, it is necessary to use retrospective information on the actual production of electricity and relate forecasts for dry years to it. The average deviations of the predicted values of electricity production in dry years in accordance with the retrospective information of the table. 1 do not exceed 7.9% of actual values.

4 A critical analysis of regulatory acts CSC in terms of the rationale for reserve capacity

In accordance with Regulation No. 431 referred to in the above, power reserve coefficients in price zones are determined by the sum of the planned reservation coefficient, the predicted capacity underutilization coefficient, and the coefficient taking into account power export relative to the predicted combined maximum load increased by the coefficient of influence of the temperature factor. The value predicted underutilization of capacity, which depends on the average statistical value of unplanned power reductions, is taken into account when substantiating the planned redundancy ratio (17%) in 2003 guidelines. Thus, it can be stated that when determining the redundancy coefficient in Regulation No. 431, the same indicator is taken into account twice.

The methodology for determining the operational reserve, which is an integral part of the NRP, involves taking into account the temperature factor. CSC procedure also provides for the consideration of this factor. Thus, a double account of the same factor is also observed here. In CSC procedure, this leads to an increase in demand for capacity of more than 4%. The determination of the operational reserve of power without taking into account the temperature factor leads to a decrease in NRP by approximately the same amount (from 3 to 5%) [3], as well as an increase in demand for power. This leads to a decrease in the reserve ratio from 0.17 (17%) to 0.12–0.14.

5 Practical results of accounting for the limited power supply of hydroelectric power stations

Obtaining practical results on accounting for the power supply of hydropower plants was associated with the modernization of existing software ("Orion-M" [2]), including for additional information content. The content, in addition to the power supply of hydroelectric power plants, is also associated with a number of features of the assessment indicators of balance reliability used in domestic practice. This primarily relates to the description of the power consumption mode. It was noted above that in domestic practice, in determining indicators of balance reliability, the average daily hourly load graph of the month of December is used with the possibility of taking into account random deviations of the load caused by the temperature factor. It is clear that such a representation of the load can in no way be consistent with the indicators of electric energy consumption planned in the work of development scheme and program UPS. To eliminate this factor, some correction factors are introduced in the methodology. It should be emphasized that this is done not because of the impossibility of taking into account the existing or predicted modes of power consumption in the software for determining indicators of balance reliability and backup tools, but because of the lack of information available to researchers. This also applies to the energy supply of hydroelectric power stations in terms of time intervals at the level of hours, days, months.

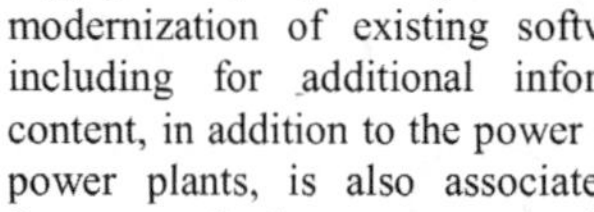

Fig. 1 - Model of the design scheme UES of Russia:
1 – Ural; 2 – Middle Volga; 3 – South; 4 – Northwest; 5 – Centre; 6 – Kazakhstan; 7 – Siberia.

The rationale for the operational reserve of capacity of the territorial zones of the UPS of Russia was carried out at the level of 2022. The

Table 2 Power and electricity balances UES of Russia for 2022 in context UES and some calculated indicators

Indicators	European part of UES of Russia						Siberia
	Ural	Middle Volga	South	Northwest	Centre	Total for UES	
From the power balance of work of development scheme and program UPS, MW							
DEMAND							
Combined maximum load	37390	17096	16831	15151	39266	125734	30126
Power export	290	10	350	1910	0	2560	300
Standard reserve of power, MW	6840 18,30%	2351 13,74%	2138 12,7%	3206 21,16	6840 17,42%	21375 17,00%	3615 12,00%
TOTAL power demand	44520	19457	19319	20267	46106	149669	34041
COATING							
Installed capacity, total	53610	27997	24740	25956	55616	187919	52029
including HPP	1872	7008	6342	3005	2639	20866	25291
Limitations and locked power	2494	2119	2337	2751	958	10568	12404
TOTAL coverage of demand	51116	25878	22403	23207	54658	177261	39625
Excess (+) / deficiency (-)	6596	6421	3084	2940	8552	27592	5584
From the balance of work of development scheme and program UPS, GWh							
Electricity production, total	265218	105638	99857	111041	242296	824050	209729
including HPP	4966	20310	21730	12629	4745	64380	107377
low water	-	-	-	-	-	-	95673
Estimated figures, MW							
Power used to cover the load, total, MW	42234	17574	18385	16805	43031	138029	33501
including HPP	1938	6983	6365	1800	2407	19493	20801
Operational reserve of power, MW / %	4844 12,96	478 2,8	1554 9,23	1654 10,92	3765 9,59	12295 9,78	3375 11,2

information content of task was carried out from the work of the S&D UES of Russia for 2016–2022. The design scheme (Fig. 1), the composition of the generating equipment, the power consumption regimes for December days, probabilistically determined data on random load deviations, equipment outages for unscheduled repairs were obtained during work and are partially presented[c] in the monograph [2].

In the table 2 shows the power balance for 2022, which indicates the presence of significant excess capacity: in the European part of the UES, they make up 21.94% (27592/125734), in Siberia – 18.53% (5584/30126). The purpose of CSC is to select the most efficient generating units of power plants from the existing excess composition of generating equipment to cover possible consumer demand. We emphasize once again that the demand for capacity, both during the development of 2003 guidelines [1], and in Regulation No. 431, depends on the size of the NRP. In the table 2, the regulatory reserve is given in accordance with [1].

The share of electricity production at hydroelectric power stations in the European part UES of Russia is only 7.8% (64380/824050), in Siberia 51.2% (107377/209729). This can explain the introduction of an additional component to the NRP during the CSC for the price zone of Siberia, caused by possible regime restrictions of hydroelectric power plants in dry years. Given in the table 3, the calculated information on the operational component of the NRP shows the contradictions of the NRP values used in compiling the power balances in the work of the S&D UES of Russia according to the recommendations [1] of modern conditions, especially for Siberia. The share of the strategic and repair component in the NRP value for the Siberian Energy System is only 0.6%. This circumstance can explain the adoption of the NRM during the CSC for the price zone of Siberia is not 12%, but 17%, as in the European part of UES.

Table 3. Additives to the normative power reserve caused by a possible decrease in electricity production at hydroelectric power plants of the Siberian Energy System

The percentage reduction in electricity production at HPP	Indicator		
	The percentage reduction in electricity production at HPP	HPP power in load coverage	Additive to the power reserve MW/ %
0	107377	20801	0/0
2.5	104692	20431	370/1.22
5.0	102008	20071	730/2.42
7.5	99324	19701	1100/3.65
10.9	95673	19196	1605/5.33
15.0	91270	18571	2230/7.4
17.27	88832	18225	2576/8.55

In the table 3 presents the results of the addition to NRP obtained as a result of applying the above procedure. It can be seen that an addition to NRP of 8.55%, adopted in accordance with paragraph 107 of the Wholesale Market Rules[d], is possible for the case where

[c] Report on scientific and research work Justification of the normative values of the components of the full reserve of power in the context of the UES and UES of Russia as a whole when planning their development. / Syktyvkar, 2016 - 66 p. (Contract ISEandEPN Komi SC UB RAS with JSC SO UES, No. 926 dated September 22, 2016)

[d] Decree of the Government of the Russian Federation No. 1172 dated December 27, 2010 (as amended on January 19, 2018) "On the Approval of the Rules for the Wholesale Market of Electric Energy and Power and on Amending Certain Acts of the Government of the Russian Federation on the Organization of the Operation of the Wholesale Market of Electric Energy and Power".

 https://doi.org/10.1051/e3sconf/201913901005

the difference in electricity production between the most probable and dry years is 17.27%. Such a difference was not observed in any edition of work of the S&D UES of Russia (table 1). For the year under review, she, in accordance with table 2, makes up only 10.9% (107377 and 95673 million kWh). The decline in electricity production at hydroelectric power stations in Siberia in dry years should not exceed 7.9% (section 4). This corresponds to the addition to the operational power reserve of a maximum of 4, but not 8.55 percent, taken during the development of CSC for 2022-2024[e]. A decrease in the value of the addition to the planned reserve ratio from 8.55% to 4.5% will lead to a decrease in demand for capacity by more than 1800 MW.

Thus, the contradictions identified in the previous section when determining the demand for power when using the information necessary for carrying out CSC for 2022–2024 and the results presented in this section on the effect of underutilization of Siberian hydroelectric power stations in dry years in relation to 2022 lead to:

– for the price zone European part UES, to a decrease in demand for capacity of more than 5 GW (3.5%), in monetary terms when extrapolating CSC 2020, to a decrease in the cost of capacity by more than 13 billion rubles;

– for the Siberia price zone, the decrease in demand for capacity is more than 1.5 GW (3.5%), and if we take into account the correction coefficient (8.55%), taking into account the limitations of hydroelectric power plants in dry years, then about 3.3 GW (7.8 %) or a decrease in the cost of capacity by more than 16 billion rubles.

References

1. Guidelines for designing development of energy systems. (Approved by Order of the Ministry of Energy of Russia dated June 30, 2003, No. 281). M .: Ministry of Energy of the Russian Federation, CO 153-34.20.118-2003.
2. Chukreev Yu.Ya., Chukreev M.Yu. Models assessing balance sheet reliability indicators for managing development of electric power systems. Syktyvkar: Komi Scientific Center, Ural Branch of the Russian Academy of Sciences, 2014, 207 p.
3. Chukreev Yu.Ya. Models ensuring reliability of electric power systems. Syktyvkar: Komi Scientific Center, Ural Branch of the Russian Academy of Sciences, 1995.176 p.
4. Kovalev G.F., Lebedeva L.M. Reliability of power systems. Novosibirsk: Nauka, 2015, 224 p.
5. Billinton R., Allan R.N. *Reliability Evaluation of Power Systems. Second Edition*. (New York and London, Plenum Press, 1996)
6. Volkenau I.M., Zeiliger A.N., Khabachev L.D. Economics of formation electric power systems. M .: Energoatomizdat, 1981, 320 pp.

[e] Website of SO UES JSC Competitive power take-off; monitor.so-ups.ru.

Approaches to establishing voltage control requirements for distributed generation units

Pavel Ilyushin[1*], *Petr* Berezovskiy [2], and *Sergey* Filippov[3]

[1]Petersburg power engineering institute of professional development, Aviation str., 23, Saint-Petersburg, Russia
[2]Study committee C6 of CIGRE Russian national committee, Aviation str., 23, Saint-Petersburg, Russia
[3]Energy research institute of the Russian academy of sciences, Nagornaya str., 31, bld. 2, Moscow, Russia

Abstract. Analysis of Russian national regulation and technical standards requirements for generating equipment participation in voltage and reactive power control in the power system carried out. Summary of some international experience in voltage control organization under conditions of distributed generation units' deep integration in public grid is presented in article. Advantages of voltage and reactive power control automation over decentralized and poorly coordinated control are substantiated. Problematic aspects of distributed generation units (including RES) integration, as well as ENTSO-E approaches to voltage and reactive power control in the synchronous zone of continental Europe are considered. Technical requirements of Deutsch and Danish regulations, regarding participation of distributed generation facilities in voltage control are given. Approaches to establishing voltage control requirements for distributed generation units' operation with national power system are presented.

1 Introduction

The Russian business community certainly seems interested in distributed generation (DG), as businesses seek cheaper energy, while DG is an effective tool to produce it. Another reason to use DG in industry is that some critical consumers need uninterrupted power supply, as the technology they use tolerates no blackout.

In Russia and worldwide, wind farms and solar farms are on the rise. As a result of contest-based selection of renewable-energy projects, Russia is expected to construct 5,278.3 MW of solar and wind generation until 2024 [1].

All the generator sets (GS) at DG facilities can generate reactive power, thus affecting the reactive power flow across grids as well as the voltage level at load nodes. Given how many DG facilities Russia has, it is crucial to refer to international experience of using DG for voltage and reactive power flow control to draft sufficient Russia-specific technical requirements (TR) to the GS designed for operating in parallel with the grid.

2 International voltage control experience

Connecting the generator sets at DG facilities, including renewable source-based ones, to the grid adds a significant number of new generators to the grid. This complicates voltage and reactive power flow control while also continually raising the dimensionality of the control problem. On the other hand, grid reliability and electricity quality requirements are ever higher on the consumer bus side.

In this context, decentralized barely coordinated grid control based on «manual» adjustment of automatic excitation controls (AEC) in electric machines, switching the static capacitor batteries (SCB) and power transformer under voltage regulation (UVR) by visual assessment of a limited set of local parameters does not seem efficient [2].

Coordinated control over all reactive power sources is necessary to maximize the positive effect, which can only be done by computing the steady states (SS) for each grid section in particular.

In some countries of Europe, secondary voltage and reactive power flow controls are based on splitting the grid into control zones on the basis of maximum reciprocal electrical remoteness. Each control zone has control plants connected to high voltage (HV) and ultra-high voltage (UHV) grids; such plants have greatest reactive power control range and maximum impact on the voltage in the local controlled grids. Zonal voltage controllers (ZVC) will automatically maintain a set voltage at the reference points (RP) of their control zones by reconfiguring the AEC thresholds in the controlling power units.

Each control plant has local plant-wide voltage control systems (PVCS), which maintain a set voltage on the plant buses and help ensure the non-ambiguous distribution of reactive power between the power units by affecting the AEC configurations.

As far as possible, ZVC provide independent control for their reference points and enable unambiguous

 https://doi.org/10.1051/e3sconf/201913901006

distribution of reactive power between control stations (plants) within each zone by adjusting the PVCS input setpoints. Thus, ZVC coupled with PVCS provide secondary voltage control for RP and control-plant buses. Besides, ZVC are also tasked to switch the SCP; to control the shunting reactors (SR), the transformer UVR, and the static reactive power compensators so as to release some range of control at zonal control power plants. In case a control zone only has one control plant, ZVC functionality is handled by the PVCS of this plant. This voltage and reactive power control structure for a single control zone is nearly identical to the Russian-designed automatic nodal voltage controller (ANVC) made to control concentrated generating nodes (CGN) [3].

Tertiary voltage control (TVC) is being studied as an alternative that will provide safer and more efficient grid-wide controls. TVC is expected to minimize the deviation in actual voltage readings from the optima for control points other than control-zone reference points. By controlling the adherence to the specified voltage curves, the TVC will trigger voltage adjustments; both tertiary and secondary controls operate in a closed circuit.

It is the targeted allocation of more or less independent control zones plus coordinating the performance of the above-mentioned control circuits that stabilizes the tri-circuit automatic voltage control system comprising the generator AEC, the ZVC, and the TVC. Thus, the generator AEC time constant is about 0.3 to 0.5 s; the time constant of secondary control circuits provided by PVCS must be about 3 to 5 s; the secondary control time constant for plant buses and control zone reference points must be about 30 to 50 s. TVC must affect ZVC at 5 to 20 minutes. The input lag between different levels of the control hierarchy should not exceed 2 s. The untied performance of different control circuits enables them to function separately and nearly independently while the transient response remain aperiodic [4].

Primary reactive power and voltage control is the task of AEC used in generators and synchronous compensators (SC), as well one of the automatic voltage controllers (AVC) controlled by substation-based reactive power compensators (RPC). Within the set reactive power control range, the AEC of synchronous machines will keep the output voltage readings within 2–4% of the design value. They operate continuously in any grid mode and are adjusted to the grid transfer functions and control algorithms.

3 ENTSO-E transition to requirements regulation in the synchronous zone of continental Europe

The pan-European system code, ENTSO-E, regulates the matter of connecting any generating node to the grid; as such, it has been approved as consistent with the Regulations of the European Commission [5]. The document does not specify the requirements to the synchronous zone of continental Europe in terms of allocating a reactive power range for generator sets with an installed capacity of up to 50 MW (type *B*); instead, national operators can specify such requirements for themselves. However, for type *B* synchronous GS, AEC is mandatory by law; reactive current must be injected in case of voltage dips caused by multiphase SC (the rule is mandatory for wind farms).

3.1 German experience

In Germany, DG facilities mostly use renewable source-based GS (both wind farms and solar farms), which are in most cases connected to the distribution grid. The historical vertically-oriented structure of the energy system is slowly but steadily switching to distributed generation as renewable energy evolves.

The basic regulatory documents that govern the DG GS involvement in voltage control are [6] for the GS in low-voltage (LV) grids and [7] for their medium-voltage counterparts. The TR [6] are mainly focused on getting and keeping the solar farms up and running. Such farms are mainly used in Southern Germany. The TR [7] are mainly intended for wind farms common in Northern Germany, as well as for sundry generation types.

For low-voltage energy systems and grids, the code mainly seeks to provide technical requirements to voltage maintenance and reactive power generation at solar farms; another area of focus is to reduce such facilities' active power generation shall the grid-wide frequency rise.

One of the most problematic parts of integrating DG facilities is to raise the voltage at DG nodes [8]. If more active power is generated in an LV grid than could consumed, whereby the resistance exceeds the reactance, the GS connection point has its voltage increased by >10% of the rated value (230 V). In case of excessively long feeders, the LV transformer substations must keep their low voltage above the rated values so that the consumer connection points always have a voltage of 207 V (minimum).

When using a conventional approach to voltage control in LV grids, solar farms connected to respective feeders trigger ever more actions involving LV/MV-transformer UVR, making them more expensive to maintain and repair [9]. Figures 1*a* and 1*b* clarify the above-mentioned problems of solar farms in a LV grid.

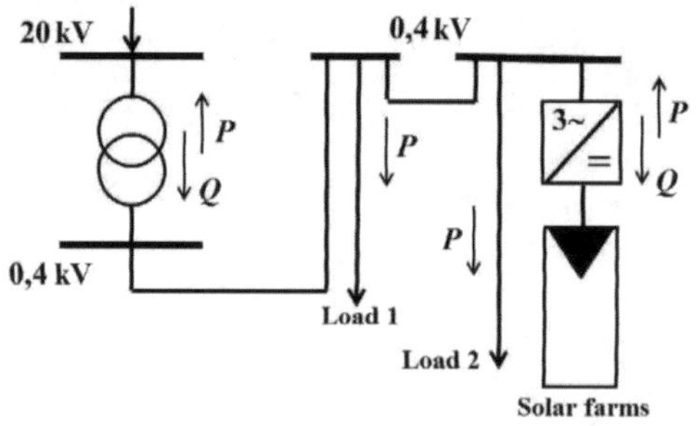

a)

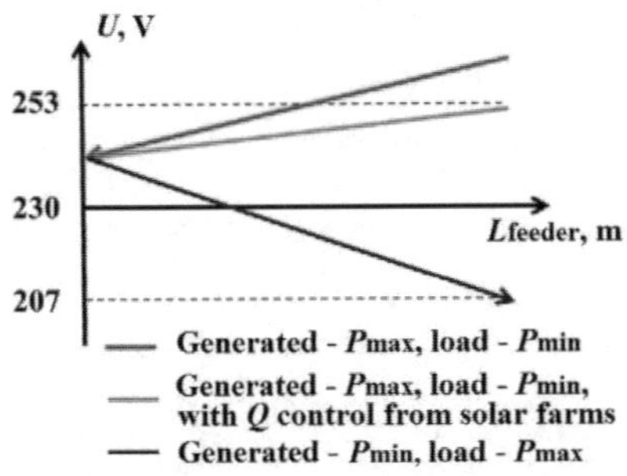

b)

Fig. 1. Solar-farms problems: (a) simplified single-line distribution grid circuit; (b) 0.4-kV grid curves in different operating situations.

The Code [5] presents requirements to keeping the GS up and running in a range from cosφ=0.95 (underexcitation) to cosφ=0.95 in case of overexcitation, whereby the active power generation must equal at least 20% of the rated value, see Figure 2*a*. In case the total installed capacity of a GS located at a common location point exceeds 13.8 kVA, the control range is from cosφ=0.9 for underexcitation to cosφ=0.9 for overexcitation, see Figure 2*b*.

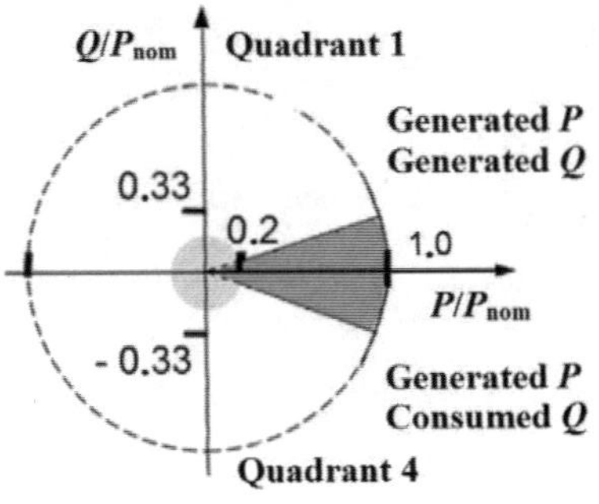

a)

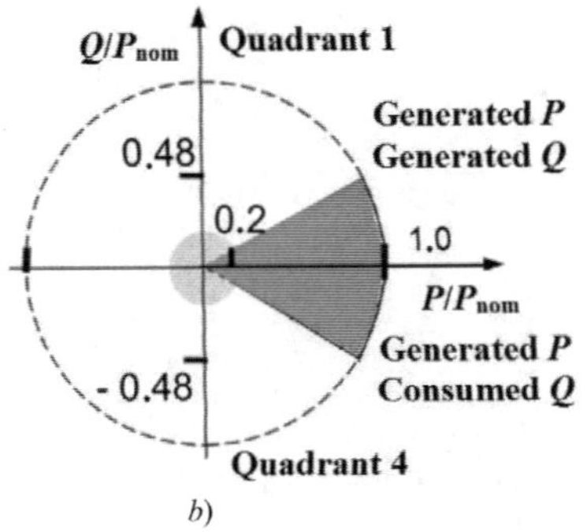

b)

Fig. 2. Requirements to GS in case of under- and overexcitation within the power factor range: (a) cosφ = 0.95; (b) cosφ = 0.9.

The GS reactive power must adjust automatically to the changes in active power in compliance with cosφ. The method for finding the required performance by reactive power (cosφ=const or active power-affected cosφ) shall be chosen by the National operator. The GS must reach the reactive power suitable for set performance within 10 seconds. This does not apply to direct-switch GS. Figure 3 presents typical reactive power control performance.

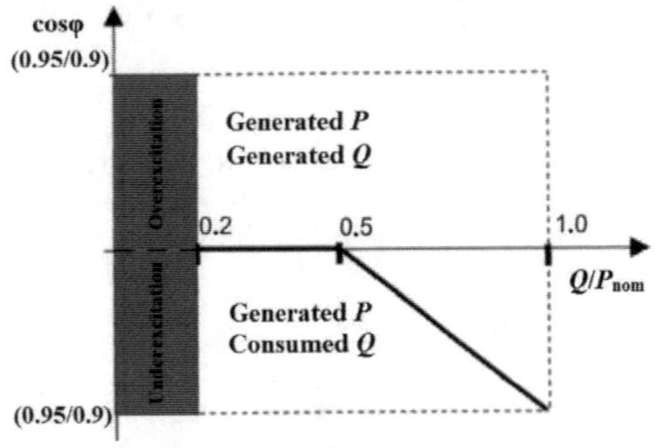

Fig. 3. Typical reactive power control performance.

Document [7] stipulates with respect to medium voltage (MV) grids and networks that the GS must be able to keep the facility running within a cosφ range from 0.95 (underexcitation) and 0.95 (overexcitation). For GS running at 0 to 10% of rated power, the reactive power output or consumption must not exceed 10% of the maximum coordinated active power, while reducing the active power to control the GS connection point is not subject to penalties unlike Iceland's configuration.

3.2 Danish experience

Unlike most foreign energy systems, the Danish grids have spent over two decades to build distributed infrastructure with large-scape DG implementation. Active integration of renewable source-based GS did not affect the energy development system; instead, it replacing one set of generation technologies with another set.

Electricity-related regulatory acts are developed by Energinet.dk, the sole regulator of Denmark's electricity transmission and gas transport infra-structure. Energinet.dk requirements to connecting to an energy grid are a set of five documents:

– Requirements to generation facilities up to 11 kW in power, type-agnostic [10],
– Requirements to solar farms > 11 kW [11],
– Requirements to combined heat and power plant (CHPP) > 11 kW [12],
– Requirements to wind farms > 11 kW [13],
– Requirements to energy storage systems [14].

Consider requirements to solar farms and wind farms. Any such farm is expected to be able to control at least one of its parameters: reactive power output at the grid connection point (regardless of the active power generation figures); cosφ (reactive power as a function of the active power output); or the control point voltage, which is affected after the FASs has received a specific command within 10 seconds.

Solar farms are also requirement to be able to automatically alter cosφ as soon as the control point reaches a certain threshold, or reactive power output does. Usually, the said function is triggered as soon as the control point voltage reaches 105% of the rated value; cosφ is configured as follows:

1 – (P/P_{nom} = 0.0; cosφ = 1.00);
2 – (P/P_{nom} = 0,5; cosφ = 1.00);
3 – (P/P_{nom} = 1.0; cosφ = 0.90).

At an equal installed power, the reactive power adjustment range is limited at cosφ = 0.90 for solar farms and cosφ = 0.995 for wind farms (50 kW to 1.5 MW) or cosφ = 0.975 (1.5 MW to 25 MW). Denmark has somewhat stricter requirements to reactive power control at solar farms.

Wind farms of the above-mentioned power class are not facing the control point voltage control requirements (only for wind farms > 25 MW), neither are they subject to automatic cosφ requirements. The specified requirement to solar farms only applies to low-power facilities (50 kW to 1.5 MW), which means this function is mainly intended to handle voltage control in low- and medium-voltage grids with numerous renewable source-based GS within non-dispatchable households.

4 Russian technical requirements to DG GS

Given the specifics of Russian 0.4-kV distribution grids, it is advisable to formulate the Russian technical requirements to DG GS with respect to their involvement in grid-wide voltage.

It is also advisable to set a lower limit of microgeneration facility power, to which voltage control TR will not apply at 1 kW. Technical requirements to DG facilities ranging from 1 kW to 0.5–1 MW connected to 0.4-kV grids should be structured as follows:

– Requirements to generation facilities 1 to 15 kW in power, type-agnostic,

– Requirements to solar farms, 15 kW to 0.5–1 MW,

– Requirements to wind farms, 15 kW to 0.5–1 MW,

– Requirements to CHPP, 15 kW to 0.5–1 MW.

It is advisable to set technical requirements to DG facilities connected to medium-voltage 6- to 35-kV grids, 0.5–1 MW to 5 MW in power, as well as to medium-voltage DG facilities, 5 to 25 MW in power, with a similar gradation.

Note that when connecting a > 5 MW DG facility, the applicant must design the power output circuitry and have it approved by the grid operator as well as by the energy dispatcher.

For setting technical requirements to DG facilities, it is necessary to consider both the DG GS and the operating features of distribution grids where such facilities are to be used. Here are some of them:

– GS excitation systems based on gas-turbine and gas-piston engines, equipped with automatic excitation controls that usually use four control algorithms. As such, the algorithm selection is subject to regulation,

– Lowering the grid voltage will drastically drop the maximum available GS reactive power, imposing a heavier load on the grid when operating in forced, repair, or post-emergency mode, whence lower voltage,

– GS involvement in secondary voltage control requires taking into account the maximum available GS reactive power and how it could be raised by partial active-power offload,

– Readjustment of AEC voltage setpoints at a single DG facility will trigger an opposite readjustment of reactive power at other DG facilities, which requires coordinated GS AEC readjustments,

– Using gas-turbine and gas-piston GS for secondary voltage control to reduce voltage spikes requires taking into account the constraints on minimum reactive power output, which is necessary for the prevention of static destabilization and overheating of stator ends,

– Using a load harmonization module as part of the GS AEC might trigger a voltage collapse at industrial-load nodes, which means its usability must be tested by calculating the electromechanical transients.

In the light of international TR standardization experience, the first regulatory document drafts might contain only the minimum necessary requirements to be further refined and amended with better and more extensive operational and project experience as more and more DG facilities are commissioned and can be studied in terms of their effects on the grid.

Thus, the TR to DG facilities must be developed in packages with due account for the Russian grid features, the distribution grid voltage and reactive power flow control principles, as well as DG GS features.

5 Conclusion

Integrating the DG generator units, including wind and solar farms, does increase the number of generators operating in the grid, which complicates the problem of voltage control and reactive power flow controls in the grid while also making this approach a high-dimensional one.

Barely coordinated grid control based on "manual" adjustment of AEC in electric machines, switching the static capacitor batteries (SCB) and power transformer UVR by visual assessment of a limited set of local parameters does not seem efficient.

A three-circuit automatic voltage control system can function reliably by targeted allocation of more or less independent control zones and coordinating the control circuit performance for their nearly independent functioning, in which case the transients in each circuit remain aperiodic.

The international experience of drafting the regulations on DG GS is mostly that of seeking to stabilize the operation of grids using numerous renewable sources for generation, as well as to maintain the high quality of electricity in low- and medium-voltage grids.

Europe's experience shows that the TR to allocating a reactive power range and to the involvement in voltage control are part of the mandatory requirements to DG GS for integration in grids; such TR are part of the European commission's regulations and can be elaborated at the national level.

Russian requirements to DG GS for their involvement in voltage control must be based on the international experience; however, they must also take into account the specifics of Russian grids and use breakdowns by voltage, GS type and power.

References

1. P. Ilyushin, A. Pazderin, R. Seit, *Proceedings 17th International Ural Conference on AC Electric Drives, ACED* (2018).
2. S. Makarovsky, Z. Khvoshchinskaya, Energetik, **6**, 18-21 (2002).
3. L. Rosman, *Means and control systems in the energy sector, Overview information, Issue 2* (1987).
4. P. Gorozhankin, A. Mayorov, S. Makarovsky, A. Rubtsov, Power Plant, **6**, 40-47 (2008).
5. Commission Regulation (EU) 2016/631 establishing a network code on requirements for grid connection of generators (2016).
6. VDE-AR-N 4105: 2018-11. *Power generation systems connected to the low-voltage distribution network* (2018).
7. BDEW, *Guideline for generating plants' connection to and parallel operation with the medium-voltage network* (2008).
8. Y. Wu, S. Tsai, M. Zou, Energy and Power Engineering, **5**, 209-214 (2013).
9. Y. Agalgaonkar, B. Pal, R. Jabr, IEEE Transactions on Power Systems, **29(1)**, 182-192 (2014).
10. Energinet.dk, *Technical regulation 3.2.1 for power plants up to and including 11 kW* (2016).
11. Energinet.dk, *Technical regulation 3.2.2 for PV power plants above 11 kW* (2016).
12. Energinet.dk, *Technical regulation 3.2.3 for thermal plants above 11 kW* (2017).
13. Energinet.dk, *Technical regulation 3.2.5 for wind power plants above 11 kW* (2016).
14. Energinet.dk, *Technical regulation 3.3.1 for battery plants* (2017).

On the system of key indicators for monitoring the compliance with requirements of the Russia's Energy Security Doctrine in terms of reliable fuel and energy supply to domestic consumers of energy resources

Sergey Senderov[1, 2,*], *Viktor* Rabchuk [1]

[1]Melentiev Energy Systems Institute of Siberian Branch of the Russian Academy of Sciences, 130 Lermontov str., Irkutsk, Russia
[2]Federal State Budget Educational Institution of Higher Education «Irkutsk National Research Technical University», 83 Lermontov str., Irkutsk, Russia

Abstract. The paper is devoted to the problem of forming a system of indicators for monitoring the fulfilment of the requirements of the Russian Energy Security Doctrine, approved by the President of the Russian Federation in 2019, in terms of ensuring reliable fuel and power supply to energy consumers within the country. The main threats to energy security are described from the standpoint of ensuring the reliability of fuel and power supply to the regions of Russia and the country as a whole. Relevant indicators of federal and regional review levels are presented.

1 Introduction

Presidential Decree No. 216 of May 13, 2019 approved the text of the new Doctrine of the Energy Security of the Russian Federation (hereinafter referred to as the Doctrine) [1]. On November 29, 2018, a meeting of the Security Council of the Russian Federation was held, where the draft of this Doctrine was considered. At this meeting, the President of Russia noted that "unique reserves of all key energy sources today give us the opportunity to completely close domestic needs for energy resources, and impressive exports - including to the countries of the European Union, the Asia-Pacific region, the CIS - brings Russia not only tangible incomes but also allows it to be one of the main guarantors of world energy security" [2]. These words are certainly true today. At the same time, without taking radical decisions to overcome some of the negative trends in the energy security (ES) of the country and its regions, there are certain concerns about this role of Russia already for the medium term.

The new Doctrine was approved instead of the 2012 Doctrine [3]. In the 2012 Doctrine and in the relevant supporting materials [4, 5, etc.], significant problems in ensuring the ES of the country and its regions have already been noted. It was shown that the share of hard-to-recover reserves of natural gas and oil is growing; The wear and tear of the basic production assets of the fuel and energy complex (FEC) is unacceptably high, the pace of their renewal is low; the technical resource of a significant proportion of energy utilities, oil refining, electrical and heat generating equipment ends; The level of scientific and technical development in the Russian FEC remains low. These problems are directly related to the increase in the frequency of emergency situations in the fuel and energy systems of the country and some regions. There are no major achievements in the field of energy conservation. Regional asymmetry is observed in the provision of the constituent entities of the Russian Federation with their own primary energy resources. The imperfection of pricing and tax policies persists, and the problems of efficiency and transparency of the activities of the fuel and energy complex organizations remain unresolved, all of which lead to a significant increase in energy prices for end users.

2 Energy Security Threats Analysis

In 2012 Doctrine, threats to energy security by the nature of occurrence and action were divided into internal (economic, socio-political, technological, natural) and external (foreign economic and foreign policy). The internal economic threats included: insufficient investment in the fuel and energy industries to upgrade production capacity, which in conditions of high equipment wear and its low technical level can lead to uncompensated retirement of production capacity of energy production facilities; the deterioration of the resource base of the fuel and energy sector and nuclear energy; the lack of the necessary diversification of the structure of the fuel and energy balance of the regions and the country as a whole; the shortage of energy resources and the breakdown of energy supply in some regions of Russia (the Far East and the Far North regions); high growth rates of prices (tariffs) for energy resources; high energy intensity of products; poor geological knowledge of the resources of the continental shelf of the Russian Federation; insufficient depth of hydrocarbon processing. Technogenic threats were directly related to economic ones: an increase in the share of obsolete and physically worn-out equipment in the fuel and energy complex due to insufficient investment; reduction of the technical level of energy facilities due to insufficient funding of research, development and technological works and poor implementation of their results; lack of highly qualified personnel, including in connection with the reduction in the training of specialists for the fuel and energy sector, while increasing the need for them; increasing the vulnerability of fuel and energy facilities associated with the complexity of systems and algorithms for managing these facilities.

Considerable attention was paid to the strategic threat of a high investment risk of developing gas fields in the Russian West Arctic Zone. In [6–8], it is shown that in the long term, with the rising cost of natural gas production in new hard-to-reach areas, the main volumes of Russian gas destined for export can be delivered to European markets with a cost price that is close to or even greater than the prices established by this time on gas. There may be a question about the effectiveness of such exports. But the question of refusing to export Russian gas in this situation is extremely complex: can a country without these foreign exchange earnings invest in the development of gas fields in the Western Arctic zone? But it is necessary to develop these new fields, because currently operating gas fields operate in a mode of falling production. At the same time, the share of gas in the balance of boiler-furnace fuel in the country exceeds 72%, and in a significant number of regions in the European part of Russia, this share is close to 90% and above.

3 The groups of the indicative analysis in the ES Doctrine 2012

In order to form a monitoring system for the main indicators of ensuring energy security requirements in terms of reliable fuel and power supply to consumers in the Doctrine 2012, among others, were formulated the relevant groups of indicative analysis. First of all, it is:

- the ratio of the increase in the balance reserves of primary fuel and energy resources to the volumes of their production;
- the share of natural gas in the structure of the balance of primary fuel and energy resources;
- the level of internal prices for gas, petroleum products, coal and electricity (capacity);
- the volume of non-payments in the retail and wholesale electricity markets;
- the pace of implementation of investment programs by fuel and energy companies;
- relative decrease in the specific energy intensity of the gross domestic product;
- stability of end-user energy supply (including regional level);
- the level of physical deterioration of basic production assets for the fuel and energy sector and related industries;
- number of emergencies at fuel and energy facilities.

To analyze the situation for each of these groups, in terms of ensuring the energy security of the country, in due time a corresponding list of ES indicators was formulated. These indicators were divided into relevant groups, a system for forming an integral indicative assessment of the condition of each group from the standpoint of ensuring the requirements of the ES was created and the same system for forming the integral assessment of the ES in the country as a whole. The results of this work and the methodology itself are quite clearly shown in [9].

The analysis showed that the identified negative trends that form an insufficient level of provision of the ES of Russia and its regions are fairly stable. The result of the implementation of these trends in recent years has been either the stagnation of the situation on most problems, or its aggravation. Accordingly, the main problem points indicated in the text of the previous Doctrine are reflected in the text of the Doctrine approved in 2019.

4 ES threats and the groups of indicative analysis in terms of ensuring the reliability of fuel and power supply in ES Doctrine 2019.

Considerable attention in the new Doctrine is paid to foreign economic and foreign policy threats to Russia's energy security. This is an important component of the problem of ES ensuring, but in this article we will dwell in greater detail on internal threats of a different nature, the implementation of which is directly related to the issues of reliable fuel and power supply to consumers within the country. From the positions indicated here among the most important internal threats presented in the new Doctrine, it makes sense to consider the following:

- mismatch of fuel and energy facilities with the needs of the socio-economic development of the Russian Federation (energy shortage or excess energy and energy infrastructure);
- reduction in the quality of the mineral resource base of the fuel and energy complex (depletion of existing deposits, reduction in size and decrease in the quality of the discovered deposits);
- insufficient supply of FEC organizations with labor resources, especially highly qualified personnel.

The new Doctrine is not explicitly stated, but among the most important ES threats is the impressive depreciation of the main production assets of the FEC.

In addition to these internal threats, the so-called cross-border threats are also considered in the new Doctrine. From the standpoint of reliability of fuel and power supply, in their list the most actual are:

- terrorist and sabotage activities that damage infrastructure and fuel and energy facilities;
- illegal use of information and communication technologies, including implementation of computer attacks on information infrastructure facilities and communication networks used to organize their interaction, which could lead to disruptions in the functioning of infrastructure and fuel and energy facilities;
- adverse and dangerous natural phenomena, environmental changes that lead to disruption of normal functioning and destruction of the FEC infrastructure and facilities.

The new Doctrine states that, in general, the realization of these threats may lead to the following negative consequences (including other):

- disruption of the normal functioning of organizations, including FEC organizations and branches of the economy of the Russian Federation;
- a rise in prices (tariffs) for FEC products and energy services;
- decrease in financial stability and termination of activity of FEC organizations.

In addition to the negative consequences indicated in the Doctrine, the slowdown of the specific energy intensity of the economy can also be expressed, as well as the lag of the vocational education system from the existing and future needs of the fuel and energy complex in qualified personnel with a corresponding increase in the threat of imperfect control of the functioning and development of the energy industries.

Based on the analysis of the main threats and their potential consequences for ensuring the energy security of the country and its regions, the Doctrine declares that the regulatory level of the ES should be ensured by the achievement of an acceptable situation in the following analysis groups:

1. Reproduction of the mineral resource base of the FEC.
2. Reliability and sustainability of the provision of Russian consumers with standard-quality energy resources and energy services.
3. The formation of a stock of products of FEC organizations in the state material reserve and maintaining it at the required level.
4. Ensuring the technical accessibility of the FEC infrastructure to various groups of consumers and the possibility of providing them with energy services.
5. Price (tariffs) regulation for the products of FEC organizations and energy services.
6. Investment activities in the field of energy, protection of investors' rights, control over foreign investments in Russian FEC organizations, which are of strategic importance for the defense and security of the country.
7. The implementation of antitrust regulation and the development of competition, including the development of organized (exchange) trade in products of FEC organizations.
8. Ensure energy saving and energy efficiency.
9. Ensuring antiterrorist security and safety of FEC infrastructure and facilities, including in emergency situations.
10. Ensuring the security of the critical information infrastructure of FEC facilities.
11. Exporting of products, technologies and services of FEC organizations.
12. Limiting the negative impact on the environment and ensuring the environmental safety of economic activities of FEC organizations.
13. Protection of the population and territories from emergencies at FEC facilities.
14. Applications of Russian technologies, equipment, materials, software in the implementation of investment projects in the FEC industries in the Russian Federation.

5 ES Indicators in terms of reliability of fuel and power supply to consumers for monitoring the requirements of ES Doctrine 2019

Considering that the monitoring of all these analysis groups is important for ensuring the ES of Russia and its regions, at least on the basis that they are highlighted in the text of the Doctrine, we will dwell in more detail on those that are directly related to ensuring the reliability of fuel and power supply to domestic consumers. In our opinion, groups 1, 2, 5, 6, 8 and 9 directly relate to such groups of analysis.

To monitor the situation with ensuring the requirements of the ES of Russia and its regions for the above groups, relevant indicators were formulated. At the same time, it makes sense to evaluate the values of some indicators only at the country level, and the values of others need to be assessed at the regional level. The values of the regional level indicators that signal a negative situation with the reliability of fuel and power supply to consumers in a particular region should be considered and taken into account when assessing the situation with the provision of ES in the country as a whole.

The table below shows the indicators of the federal and regional levels. The values of these indicators should be monitored when analyzing the situation with the state of a country's ES in terms of ensuring reliable fuel and power supply to consumers of final types of energy within the country. These indicators are grouped by the corresponding analysis groups (group numbers correspond to numbers from the above list of groups).

 https://doi.org/10.1051/e3sconf/201913901007

Table. ES indicators by groups of indicative analysis in terms of ensuring reliable fuel and power supply to consumers within the country (group numbers correspond to numbers from the above list of DL analysis groups)

<table>
<tr><td colspan="2">1. Reproduction of the mineral resource base of the FEC</td></tr>
<tr><td colspan="2">State level:
1.1_s. The ratio of the increase in the balance reserves of oil over a 5-year period to the total volume of oil production in the country for the same period, p.u.
1.2_s. The ratio of the increase in the balance reserves of natural gas over a 5-year period to the total volume of natural gas production in the country for the same period, p.u.
1.3_s. The ratio of the increase in the balance reserves of coal over a 5-year period to the total volume of coal production in the country for the same period, p.u.
1.4_s. The ratio of the increase in the balance reserves of uranium over a 5-year period to the total volume of uranium concentrate production in the country for the same period, p.u.</td></tr>
<tr><td colspan="2">2. Reliability and sustainability of the provision of domestic consumers with energy, energy and raw materials of the appropriate volume, range and quality, including the necessary reserve of capacity and reserves</td></tr>
<tr><td>State level:
2.1_s. The share of natural gas in the structure of the balance of primary fuel and energy resources, %.</td><td>Regional level:
2.1_r. The ratio of the sum of the available capacity of power plants and the capacity of the intersystem connections of the region with its neighbors to the maximum electrical load in its territory, p.u.
2.2_r. Relative ability to meet the total demand for boiler and furnace fuel from the internal sources of the region, %.
2.3_r. The share of the dominant resource in the total consumption of boiler and furnace fuel in the region, %.</td></tr>
<tr><td>2.2_s–2.6_s. Relative total undersupply of certain types of fuel and energy resources in the country as a whole for the analyzed year (gas, coal, fuel oil, light oil products as a whole, electricity), %.
2.7_s–2.13_s. Physical depreciation of basic production assets by FEC branches and related industries (FEC as a whole, the gas industry, the sphere of oil production and transportation, oil refining, the coal industry, electric power industry, power engineering), %.</td><td>2.4_r. The share of the largest power station in the installed electric power of the region (for the regions electrically isolated and weakly connected with the UES), %.
2.5_r–2.9_r. Relative total undersupply of certain types of fuel and energy resources in the region for the analyzed year (gas, coal, fuel oil, light oil products as a whole, electricity), %.
2.10_r. The level of potential endowment of the region with the furnace and boiler fuel during cold snap in the relevant federal district, %.
2.11_r. The level of potential endowment of the regional electrical load during cold snap in the relevant federal district, %.
2.12_r. The level of potential endowment of the region with thermal energy during cold snap in the relevant federal district, %.
2.13_r. The risk of shortage of the main types of fuel and energy resources of the region during the implementation of natural emergencies that are typical for this region, %.
2.14_r. The level of relative undersupply of the main types of fuel and energy resources in the region during emergency situations at critical energy facilities in the region or technologically related to it, %.
2.15_r. The degree of depreciation of the basic production assets of the energy sector of the region, %.
2.16_r. The ratio of the average annual input of installed capacity and the reconstruction of power plants in the region over the last 5-year period to the installed capacity of the region, p.u.</td></tr>
<tr><td colspan="2">5. The establishment of a predictable, diversified, depending on the quality of the goods or services and acceptable to producers and all groups of consumers of the level of tariffs and prices for energy resources, energy and energy services</td></tr>
<tr><td>State level:
5.1_s–5.6_s. The relative growth of domestic prices for the main types of fuel and energy resources (gas, coal, fuel oil, light oil products in general, electricity, heat energy), %.</td><td>Regional level:
5.1_r–5.6_r. The relative growth of prices in the region for the main types of fuel and energy resources (gas, coal, fuel oil, light oil products in general, electricity, heat energy), %.</td></tr>
<tr><td colspan="2">6. Investment activities, protection of the rights of investors in FEC, control over foreign investments in fuel and energy companies of strategic importance to ensure the defense and security of the state</td></tr>
<tr><td colspan="2">State level:
6.1_s. The ratio of the amount of non-payments to sales organizations (for the country as a whole) by retail consumers of electricity for electricity supplied to them (consumers) in the current year to the total sales value of all electricity supplied to retail consumers in the same year related to the same indicator for the previous year, p.u.</td></tr>
<tr><td colspan="2">6.2_s. The ratio of the amount of non-payments to electricity generating companies (for the country as a whole) by grid companies for electricity supplied to grid companies in the current year to the total wholesale cost of all electricity supplied by electricity generating companies to grid companies in the same year related to the same indicator for the previous year, p.u.
6.3_s. The ratio of the amount of non-payment to grid companies (for the country as a whole) by the sales organizations for the electricity supplied to these organizations in the current year to the total cost of all electricity supplied by the grid companies to organizations in the same year related to the same indicator for the previous year, p.u.
6.4_s. – 6.8_s. Implementation of investment programs by the FEC sectors (oil production and transportation, oil refining, gas industry, coal industry, electric power industry), p.u.</td></tr>
<tr><td colspan="2">8. Energy saving and energy efficiency of technologies, equipment and materials used in the FEC sectors and at energy facilities of consumers</td></tr>
<tr><td>State level:
8.1_s. The relative change in the specific energy of gross domestic product, %.</td><td>Regional level:
8.1_r. The relative change in the specific energy of gross regional product, %.</td></tr>
<tr><td colspan="2">9. Ensuring antiterrorist security and safety of FEC infrastructure and facilities, including in emergency situations</td></tr>
</table>

State level:
9.1s. The ratio of the number of acts of terrorism at the Russia's FEC facilities in the current year to the number of such acts in the previous year, p.u.
9.2s. The ratio of the number of emergencies in the Russia's FEC in the current year to the same value in the previous year, p.u.

It can be noted that in some groups of indicative analysis (in this table) the list of indicators is not complete. This applies mainly to group 6 in terms of ensuring the protection of investors' rights, controlling foreign investments in Russia's FEC organizations, which are of strategic importance for the defense and security of the state. There is a feeling that the indicators of this orientation are not directly related to the process of reliable, uninterrupted fuel and power supply to consumers, affecting this process indirectly. Such indicators in the general system of indicative analysis in accordance with the provisions of the new Doctrine, apparently, should be brought in by the specialized organizations dealing with these issues professionally.

The values of all indicators listed in this table except for indicators 2.10_r - 2.14_r should be calculated on the basis of official statistical information. The values of indicators 2.10r - 2.14_r are calculated using specialized mathematical models that take into account all the technological aspects of the operation of energy systems that are related to the reliability of the fuel and power supply to consumers in various conditions. In addition, the peculiarities of the interrelated work of these industries in a Russia's FEC should be taken into account. This approach will make it possible to take into account the internal potential (diversification of sources of fuel and energy resources, interchangeability of fuel and energy resources, etc.) of the FEC to minimize the negative effects of the realization of various ES threats. Such a model apparatus, for example, was created and is actively used to study issues of ES in ESI SB RAS [10–13].

The table shows that in group 2, concerning the issues of reliability and sustainability of the processes of fuel and power supply to consumers, most of the regional level indicators are concentrated. These indicators are as close as possible to the fuel and power supply systems of the regions. The values of these indicators together form the ability to assess the situation with the reliability of supplying all types of final energy to consumers in each region. Solving such a problem at the level of consideration of state indicators is impossible because of the too large level of aggregation.

6 Conclusion

In this paper, based on an analysis of the text of the Doctrine of Russia's Energy Security, approved in 2019, an attempt was made to highlight the fundamental indicators to characterize the situation in terms of the technological reliability of the fuel and power supply to energy consumers within the country. Monitoring the values of the indicators of the federal and regional levels formulated for each group of analysis will help to identify the main problem points in the organization of fuel and power supply to consumers in the country and its regions in relevant aspects. Such an approach will provide an idea of the directions and degree of transformation of existing trends in ensuring the ES of the country and its regions.

The work was carried out within the framework of a scientific project III.17.5.1 of program of fundamental research of the SB RAS, reg. number AAAA-A17-117030310451-0 and RFBR research project №18-58-06001, reg. number AAAA-A18-118050490009-5.

References

1. http://publication.pravo.gov.ru/Document/View/0001201905140010?index=0&rangeSize=1
2. https://minenergo.gov.ru/node/13311
3. Bushuev V.V., Voropai N.I., Senderov S.M., Saenko V.V., Economika regiona [Economy of Region], **2**, 40-50 (2012).
4. Voropai N.I., Pyatkova N.I., Rabchuk V.I., Senderov S.M. & oth., Energeticheskaya politika [Energy Policy], **1**, 45-53 (2006).
5. Rabchuk V.I., Senderov S.M., Energeticheskaya politika [Energy Policy], **3**, 25-34 (2010).
6. Rabchuk V.I., Senderov S.M., ECO vserosiyskiy economicheskiy jurnal [ECO All-Russian economic journal], **8**, 19-30 (2013).
7. Rabchuk V.I., Senderov S.M., Academia energetiki [Academy of Energy], **6**, 14-20 (2014).
8. Rabchuk V.I., Senderov S.M., Academia energetiki [Academy of Energy], **2**, 12-18 (2016).
9. Senderov S.M., Rabchuk V.I., Izv. RAN. Energetika [Proc. of the Russ. Acad. of Sciences. Pow. Engin.], **2**, 3-12 (2018).
10. Pyatkova N.I., Senderov S.M., Cheltsov M.B., Bondarenko A.N. Izv. RAN. Energetika [Proc. of the Russ. Acad. of Sciences. Pow. Engin.], **6**, 31-39 (2000).
11. S. Senderov, A. Edelev, Energy Systems Research, 1 (1), 35-43 (2018).
12. Senderov S.M., Rabchuk V.I., Izv. RAN. Energetika [Proc. of the Russ. Acad. of Sciences. Pow. Engin.], 1, 70-78 (2016).
13. Senderov S.M., Vorobev S.V., Pyatkova N.I., Izv. RAN. Energetika [Proc. of the Russ. Acad. of Sciences. Pow. Engin.], 6, 3-11 (2017).

Priority directions of development "Uzbekneftegas" jsc in the conditions of globalization of the world economy

G.Zh. Allaeva [1]

[1]Tashkent State Technical University named after Islam Karimov, 100095, Uzbekistan, Tashkent, University St. 2A

Abstract. The article considers the role of "Uzbekneftegas" JSC in the economic development of the fuel and energy complex of the Republic in the face of increasing global economic globalization. The structure of the company, the priority areas for the development of JSC activities are shown. The perspective directions in hydrocarbon production are considered. The data on the production, use and distribution of natural gas by sectors of the economy of Uzbekistan are presented, and the structure of the energy balance of the Republic of Uzbekistan is shown.

1 Introduction

The increasing demand for hydrocarbons in the world and rising prices for oil and gas raw materials make Uzbekistan one of the important participants in the global energy market. The energy policy of Uzbekistan in the context of the increasing globalization of the world economy is primarily aimed at ensuring energy independence and security, increasing energy efficiency and reducing the environmental impact of energy. In order to increase the level of independence, maximize profits and, accordingly, further economic development, our state seeks to diversify energy export flows. "Uzbekneftegas" JSC takes a leading place in the implementation of this policy.

For the effective economic development of the fuel and energy complex of the Republic of Uzbekistan, it is necessary to modernize the currently inefficient, morally and physically obsolete equipment in the industry with modern energy-efficient and energy-saving equipment, it is necessary to optimize the production process in large and medium-sized enterprises. In the energy sector, these are gas-turbine and combined-cycle plants, combined generation of heat and electric energy, an increase in the oil and gas extraction coefficient and an increase in the efficiency of their processing and transportation, enrichment of coal at the extraction site and the introduction of effective technologies for its combustion, reduction of losses for domestic needs, and an adjustable electric drive and etc.

2 Analysis of the main economic indicators of development of “Uzbekneftegaz” JSC

In Uzbekistan, “Uzbekneftegaz” Joint-Stock Company (JSC) deals with the production of hydrocarbons and providing with them industries and the population. “Uzbekneftegaz” JSC is a vertically integrated three-level company. The current company management system was approved in accordance with the resolution of the President of the Republic of Uzbekistan in 2006 [1]. The company includes 6 Joint Stock companies engaged in geological exploration, drilling of oil and gas wells, oil, gas and gas condensate, natural gas processing: "Uzburneftegas" JSC, "Uzneftegasdobycha" JSC, "Uztransgas" JSC, "Uznefteproduct" JSC, “Uzneftegazstroyinvest” JSC, "Uzneftegasmash" JSC.

The capacities of “Uzbekneftegaz” JSC allow for the production of natural gas in the amount of about 70 billion m^3 and liquid hydrocarbons in the amount of 8 million tons/year [1]. To further stabilize and increase the growth in oil production, “Uzbekneftegaz” JSC has launched the "Concept for the development of the oil and gas industry of Uzbekistan until 2020", which provides for an increase in hydrocarbon reserves by 2020 compared to 2010 by 1.25 times for gas and for oil in 1.65 times, condensate - 1.33 times. It is planned to increase the annual volumes of prospecting, exploration and parametric drilling 2.5 times and reach 300 thousand linear meters [1].

In the oil and gas industry of Uzbekistan, the main indicators characterizing the efficiency of enterprises are: the number of used scientific and technical measures, the costs of their implementation, the resulting economic effect, the additional volume of production through the use of scientific and technological developments, the mustering and production of new products and the conditional release of the number of employees. In 2005 – 2017 years were used 929 measures, the actual economic effect of which amounted to 550.5 billion UZS. The implementation of measures allowed to increase profit by 853.1 billion soums, including by reducing costs - by 115.4 billion UZS, of which in 2018 year, respectively, by 175.4 billion UZS and 19 billion UZS. Fuel and energy resources and material and technical resources were saved by 29.9 billion UZS, including 7.3 billion UZS in 2017 (Table 1.)[1].

Table 1. Technical and economic indicators of the use of scientific and technological achievements in "Uzbekneftegaz" JSC.

The realization of the intellectual potential of oil and gas industry enterprises was manifested in the development and commissioning of large strategic facilities: the Bukhara oil refinery and the Shurtan gas and chemical complex, the modernization of the Fergana oil refinery complex, the Ustyurt gas and chemical complex. The construction of a gas processing plant at the Kandym field was completed, together with the Russian oil company "LUKOIL". The design for the production of (LNG) liquefied natural gas and (GTL) synthetic liquid fuels is nearing completion (project feasibility study has been prepared on the basis of purified methane). The construction project of the Jizzak oil refinery is being developed. At various stages of implementation are investment projects that play an important role in the development of the fuel and energy complex of Uzbekistan.

Name of indicators	Unit meas.	2005 year	2011 year	2012 year	2013 year	2014 year	2015 year	2016 year	2017 year
1. The number of STP events used	Pcs	84	87	75	61	47	54	63	61
2. Actual economic effect	billion UZS	29	38,4	28,4	24,9	18,8	112	72,5	81,4
including self-supporting	billion UZS	28,9	38,3	28,4	24,9	18,8	103,6	50,2	64,3
3. Profit growth	billion UZS	29,1	28,8	27,5	24,9	18,8	210,5	164,7	175,4
including by reducing costs	billion UZS	28,8	5,0	5,3	0,1	0,9	2,8	6,8	2,9
4. Saving fuel and energy resources, and material and technical resources	billion UZS	2,4	3,0	0,7	0,02	2,8	2,5	4,7	7,3

As can be seen from the data, for all indicators except exports, there is a significant increase. The drop in export of products is due to the fact that in recent years one of the main priorities of the economy is to deepen the processing of natural raw materials and the production of alternative fuels to meet the needs of consumers of the Republic.

The effectiveness of using the intellectual capital of "Uzbekneftegaz" JSC in the system of innovative cooperation in education, science and production is evidenced by the dynamics of the final results of the industry for 2007-2016 years (Fig. 1.).

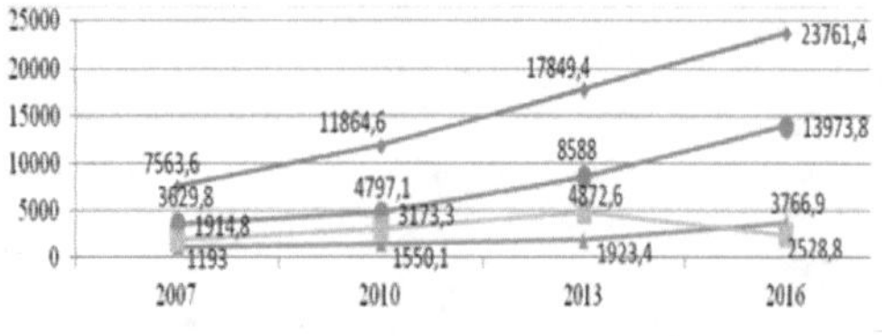

Fig. 1. Dynamics of changes in the volume of product realization, high-tech industrial products, export of products and investment in fixed assets (billion UZS).[2]

The current stage of development of the oil and gas industry of Uzbekistan is characterized by the introduction of new capacities for in-depth processing of hydrocarbon raw materials and bringing the quality of products to the level of world standards.

The Thomson-Reuters World Business Information Agency noted the uniqueness and innovation of the Ustyurt Gas Chemical Complex Construction project, and it was recognized as the best project of the year in the petrochemical industry. According to the rating published by the reputable consulting company "DILOGIC" (Great Britain), the project was included in the list of ten global investment projects in the world.

It should be noted that hydrocarbons mainly dominate in the structure of the energy balance of Uzbekistan - their share is more than 96%. Currently, a number of government decisions have been adopted [2-4], which will lead to diversification of the energy balance of Uzbekistan.

https://doi.org/10.1051/e3sconf/201913901008

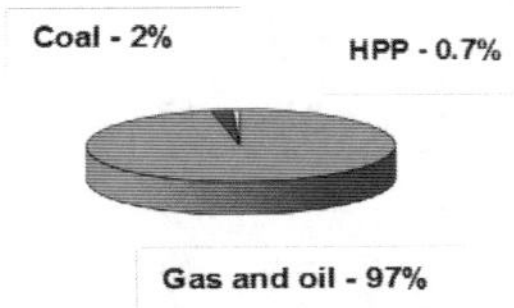

Fig. 2. The structure of the energy balance of the Republic of Uzbekistan [8]

Total energy production: 2018 - 82 million tons of oil equivalent

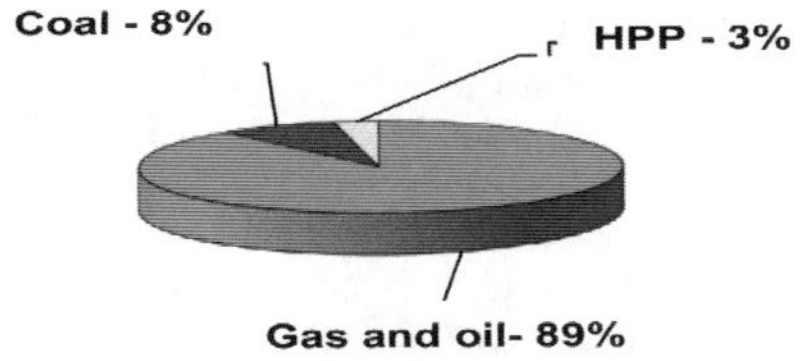

Fig. 2. Forecast of the structure of the energy balance of the Republic of Uzbekistan for 2020. (Forecast by the author)

"Uzbekneftegaz" JSC plans to direct USD 235 million by 2020 to increase oil production at long-developed fields in the Republic [4].

It is planned that by 2021 year the Holding will implement a deep drilling program for about 100 facilities with the participation of foreign companies that provide their experience in applying advanced technologies for enhancing oil recovery in fields with difficult to extract reserves. By 2021 year, the holding plans, due to geological and technical measures aimed at compensating for oil production losses, to annually extract an additional 300 thousand tons of oil [5].

It should be noted that investors are interested in joint processing of raw materials, with the release of new types for Uzbekistan, including gas-chemical products. The projects on production of modern oil and gas equipment, rendering of services are considered. Today, among the partners of "Uzbekneftegaz" JSC are the largest companies from China, Malaysia, Korea, Russia and other countries.

Together with a number of foreign partners of China and Russia, the possibilities of producing oil and gas equipment were considered with the prospect of creating several Joint Ventures on the basis of "Uzneftegasmash" JSC. enterprises and including them in the localization program [5].

"Uzbekneftegaz" JSC in 2018 – 2030 years will increase hydrocarbon reserves by 2.2 times - up to 5.6 billion tons of standard fuel. To create a reliable resource base and maintain the level of gas production for the period until 2030 year, it is necessary to annually increase the reserves of raw materials in volumes exceeding their production by 1.1-1.5 times. According to the Holding's calculations, within the framework of the concept of developing the oil and gas industry for the period until 2030 year, it is expected to increase 2.393 trillion m^3 of natural gas, oil reserves - 132.4 million tons, condensate - 102.6 million tons. To achieve these goals until 2030 year, it's planned to carry out prospective seismic surveys [6].

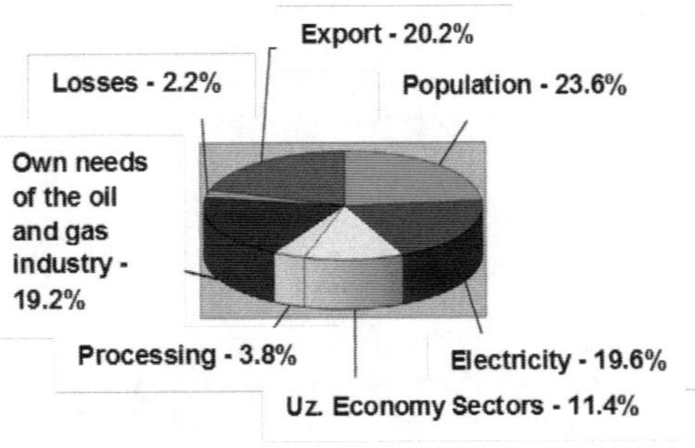

Fig. 3. Usage - the distribution of natural gas by sectors of the economy of Uzbekistan [8]. In fig. 3 shows the structure of natural gas consumption in the sectors of the economy of Uzbekistan.

3 Key findings and suggestions

The analysis shows that, with the exception of the Bukhara - Khiva region, in all others, the development in the main productive complexes doesn't exceed 35.5%, that is, today they are the main objects for placing prospecting volumes in the future. In particular, in the near future exploration work (geological exploration) will be intensified below the underlying strata in the Ustyurt, Bukhara-Khiva, Surkhandarya and Fergana oil and gas regions. To increase the production of petroleum products, a geological exploration program for the search and exploration of heavy oil deposits within the "Besharcha" area in the Surkhandarya region and a geological exploration program for the search and exploration of heavy oil and bitumen deposits within the Fergana region are under development [7].

To increase the production of petroleum products, the development stage includes the exploration program for prospecting and exploring heavy oil deposits within the "Besharcha" area of the Surkhandarya region and the exploration program for exploring and exploring heavy oil and bitumen deposits within the Fergana region [7].

According to the Joint-Stock Company: recoverable proven hydrocarbon reserves in Uzbekistan as on 1 January 2016 year amounted to 2.53 billion tons of oil equivalent tons, of which about 65% are gas reserves on 1 January 2016 year 242 hydrocarbon deposits were discovered in Uzbekistan, of which 192 were gas. Of the open fields, 104 are under development, 66 are prepared for development, exploration work continues at 72 fields [7].

It should be noted that one of the most promising areas in hydrocarbon production is the processing of oil shale into oil products. "Uzbekneftegaz" JSC plans to set up shale oil production on the basis of the Sangruntau field in the Navoi region of the Republic. The project will be implemented jointly with the Japanese company JGS Corporation [6]. Financing of the project with a preliminary cost of 600 million US dollars is planned to be carried out at the expense of the company's own funds, loans of the Fund for Reconstruction and Development of Uzbekistan and foreign investments in the amount of 300 million US dollars.

The company provides for the construction of a refinery with a capacity of up to 1 million tons of oil products per year. Estimated reserves of oil shale in Uzbekistan amount to 47 billion tons. In particular, reserves of the Baysun, Urtabulak, Sangruntau, Aktau, Uchkir, Kulbeshkak deposits were estimated in the Republic. In addition to hydrocarbons, in significant quantities there is a wide range of non-ferrous and rare metals.

The output of the diesel fraction from the resin of the shales of the Sangruntau deposit is about 30%, which is the highest indicator among the main deposits of this type in the Republic [7].

Thus, "Uzbekneftegas" JSC pays special attention to the introduction of modern technologies for advanced processing of raw materials and the development of competitive, high-quality products that are in demand on the domestic and foreign markets, contributes to the further diversification of economic structures, ensuring its sustainable and dynamic development, and also contributes to the expansion of export potential of our Republic.

References

1. http://www.gazovik-lpg.ru/about/news/2011/06/28/52/
2. "On measures for the further development of alternative energy sources." Decree of President I. Karimov of March 1, 2013.
3. Decree of the President of the Republic of Uzbekistan No. UP-4947 dated February 7, 2017 "On the action strategy for the further development of the Republic of Uzbekistan".
4. Allaeva G. J. Innovative potential of the fuel and energy complex of the Republic of Uzbekistan. //Monograph. - Tashkent, "Fan va texnologiyalar", 2017. - 160 p.
5. http://oilgasfield.ru/news/39490.html
6. http://ruspower.ru/news/202752233
7. Uvraimov I. A. Review of key trends in the global and Central Asian oil and gas market. The Uzbek Journal of Oil and Gas. T., Special Issue, May 2013, p. 28 - 35.
8. Allaev K. R. Energy of the world and Uzbekistan. T., Moliya, 2007, 388 p.
9. Alternative energy sources: opportunities for use in Uzbekistan. Analytical report. Center for Economic Research. UNDP. 2011, 74 p.

Problems of using pumped storage power plants to improve the reliability and controllability of electric power system of the republic of Uzbekistan

Murodilla Mukhammadiev[1,*], *Boborakhim* Urishev[2], *Shirin* Esemuratova[1], and *Nigina* Djumaniyozova[1]

[1]Tashkent state technical university, 2 Universitet street Tashkent, Uzbekistan
[2]Karshi engineering institute, 225 Mustaqillik street Karshi, Uzbekistan

Abstract. This article deals with the analysis and development perspectives of the use pumped storage power plants use to increase the reliability and regime controllability of electric power systems of the Republic of Uzbekistan.

1 Introduction

One of the main issues of increasing the reliability and operational controllability of electric power systems (EPS) is the issue of accumulation of generated energy.

Energy storage has been used in a wide variety of energy purposes, previously, it was used as a reserve power, which compensated the blackout disruption at the electricity transmission, and in the present time is used to solve a wide range of tasks, ranging from improving the overall reliability of the power system to the provision of system services on the market electricity. Recently, it has become clear that the effective development of the renewable energy use sources is impossible without energy storage, since the nature of its receipt from these sources is sharply variable [1-4].

Energy storage system can provide the advantages and flexibility for consumers, managers of energy companies and utilities, and can also participate in a number of market segments, especially in the electricity markets, acting as a direct supplier of energy for the EES, and to individual consumers [3, 4].

As you know, with the accurate unevenness of the load schedule, the presence of peaks and sharp decreases in the level of energy consumption create technical problems for energy producers related to the need to ensure compliance with the production and consumption of electricity, i.e. reduced electricity generation during periods of low consumption.

This leads to three groups of problems : firstly, the efficiency of using the installed capacity of power equipment is reduced, which is the most important factor affecting its payback; secondly, the frequent operation of power equipment in variable mode dramatically increases fuel consumption at thermal plants and leads to previously temporary wear of equipment; thirdly, the technical ability to reduce the power of generating equipment is often insufficient to compensate for the reduction in consumption, in this case, excess electric power is transferred to other regions [5 - 11].

To pass periods of nightly lows, power plants are forced to dump power. Furthermore, loss of increased fuel consumption and equipment due to uneven load, as well as loss of power transmission over long distances entail reducing economical profitability generating capacity.

In world practice operating large EPS solve uneven consumption of electricity is achieved either by creating special maneuvering power equipment (peak power plants, gas turbine power plant), or using the battery, consuming excess power during periods of general load reduction in EPS and outputs it to periods of increased consumption [1- 9].

Numerous foreign studies prove that the use of maneuverable power plants operating in a variable load mode leads to a significant consumption of fuel, a decrease in the efficiency of using installed capacity, a decrease in the service life of equipment, and other negative consequences [4 - 9].

The most promising way to solve this problem is the use of pumped storage power plants (PSPP), which are both a highly maneuverable peak power source and a consumer-regulator [5–9, 12–15].

PSPPs are the main type of storage systems; currently, they account for 96% of the accumulated energy in the world [16]. Composition PSPP consists of hydropower installations, as values for generating power and pumping water into the upper reservoir, which is the hydraulic energy storage. Often the volume of the upper and lower reservoirs in a fixed time does not match, because the time of the turbine operating mode does not match the temporary pumping of the second mode.

World experience in PSPPs exploitation in various countries convincingly proves their technical efficiency to ensure the profitability of energy systems, manifested in the increase and reliability of power supply and power quality [9, 12-15]. Currently, in the world there are more than 460 operating PSPP different layout with wide range installed capacity - from 100 M W 3,000 MW (PSPP Edison in the U.S.A and pressure of several tens of meters to 1,700 meters (PSPP Reyseek - Kraytsek in Austria), the total installed capacity of which exceeds 150 GW [9, 15]. According to the IRENA agency projections is expected to increase power PSPPP to 325 GW in 2030 [17].

2 Methods and models for the use of PSPPPs in the EPS of the Republic of Uzbekistan

Studying the operation experience of the electric power industry of Uzbekistan and mill shows that one of the main issues of reliability improve of power supply is necessary to ensure range and dynamic regulations of power generating, which would fully meet the requirements of energy consumption. As is known, this problem is solved by selecting the optimum structure generating capacity, which provides stability and profitability of the EPS.

Coefficient irregularity ratio curve of power consumption (daily load curve) EPS of the republic is K_{nm} =0,72-0,76, which causes certain difficulties to the peak load coating and to a greater extent to the pass of the night at least to have accurate load curves.

This problem is typical for the thermal power plants (TPP) of the republic, which is planned to reconstruct inefficient power units to more modern and efficient. The compelled decrease in capacity at night time hours minimum taking into account necessity of performance of the schedule of load curve and maintenance of uniform work of power units becomes a serious problem for them. In addition, in connection with sharp increasing of load problems of quality maintenance of the electric power (maintenance of normal of frequency and voltage values) in the evening become aggravated.

Russia and other foreign countries EPS experience with predominance TPP, both and in Uzbekistan shows that the proportion of highly mobile units fraction w in at least 25% of the total installed power of the energy of the system [9, 14]. Thus for optimum organization and reliable operation TPP proportion PSPPP (based on their ability to perform dual p e -regulation - both generation and load) should be 10-12% [9,10,14]. Unfortunately, this parity in the republic power supply system is not observed yet.

Insufficient proportion of highly maneuverable hydroelectric power plants (HPP) and PSPP in the structure of generating power of EES of the Republic forces power regulation with thermal power plants. This leads to deterioration of power supply, in particular the voltage and frequency of the power system, since those are thermal power plants cannot quickly change the power. Frequent starts and stops of turbo units results over-expenditure the fuel, reduce the durability of thermal power equipment and increase the costs of maintenance, repair, deterioration of the ecological situation in TPP located areas [9].

By the data of Unitary Enterprise "Talimardjan TPP" in Uzbekistan in 2017, the total fuel consumption for starts and stops power blocks was 1601.000 m^3, which corresponds to the loss of fuel at each start an average of 320.000 m^3 natural gas.

According to the JSC "Uzbekenergo" in 2018, the value of the average specific fuel consumption was 354.6 g.c.f. /kW·h, and in the EPS countries where the proper regulation of capacities, and with use of PSPPP, this figure does not exceeds 330.0-340.0 g.c.f./kW·hour [12, 13].

The main cause of accelerated wear of heating equipment at his frequent and short stops is uneven temperature state various hours by the elements of the turbine, boiler and steam, leading to their cooling at different speeds, and about 25% of emergency stops power occurs due to damage in the course of their starting modes [8, 9, 10, 18].

All the above-stated problems in the power system of the Republic indicate a need for the development of the construction of highly maneuverable power plants using renewable energy sources, primarily pumped storage power plants.

Currently, the Republic has adopted state programs for the construction of a number of small and medium-sized hydroelectric power stations [19]. However, it can solve only part of the problem, i.e., PSPP may cover only peak and booting providing vides the removal tension associated with the passage of the peak daily schedule by producing a cheap and clean energy. The other part of the problem, i.e. preventing failures generated loads can only be solved by construction PSPPP, which works on load curve failures in the pumping mode, providing a more favorable basis regime mode of TPP, and also promoting decrease in interstation overflows of energy.

In addition to the above advantages of PSPP may have a so-called fuel effect which is achieved due to the difference in fuel consumption during periods of energy storage and generation peak power. During periods of pumping mode, that correspond to minimal load curve, PSPPs consumes energy, which generated by basis power plants with the specific expense of conditional fuel G_n = 0.2 to 0.27 kg/(kW·h). Then generation mode of PSPP replaces load thermal power plant installations in peak mode with the specific expense of conditional fuel - G_g = 0,5 kg / (kW · hr). If the efficiency PSPP take η = 0,70- 0.75, the specific fuel savings equal to 0.16-0.187 kg / (kW · h) [6,10,12].

3 Discussion and results

As an example, we show how to efficiently use of PSPP in EES the Republic of Uzbekistan, using a daily schedule of electricity generation received from JSC "Uzbekenergo" for 23.07.2017 (Fig. 1).

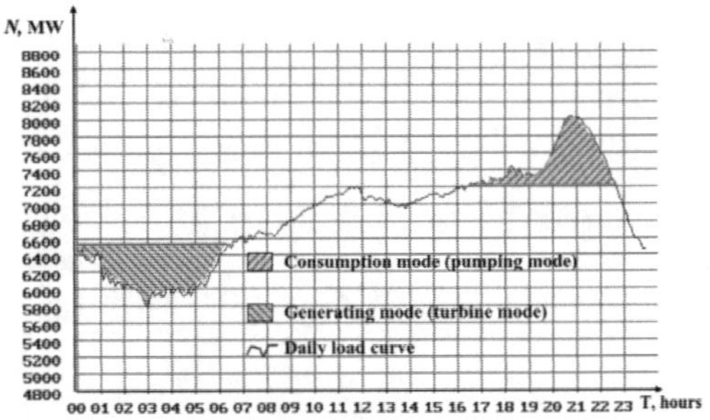

Fig. 1. Schedule of daily electricity generation EES of the Republic of Uzbekistan

It marked on the working zone of pumping stations of the PSPP power consumption regime in the minimum load and the working zone in the turbine mode of PSPP in the peak part of the curve. If such mode will adhere, the PSPPs in pumping mode will consume approximately 2,000 MW·hour electricity for 6 hours, and this power equivalent to volume of water will collect in the reservoirs. When taking efficiency PSPP 0.75, then the volume of water allows to generate 1,500 MW · hour electrical energy during 6 hours in a peak part of the curve. It means, that only at the expense of fuel effect it is possible will save 240-280 t.c.f. per day. In this case, the coefficient of non-uniformity of the daily load schedule varies from $K_{nm} = 0.72$ to $K_{nm} = 0.91$.

The daily load profile EES Republic most tense in winter n e IRS. The peak load in December - January month ah is about 22-25 % more than the average , but the maximum regulation capacity of HPP Republic for by covering of this load does not exceed 12 % of the installed capacity of all power plants.

All this points to the need to intensify the use of hydropower and other alternative energy sources, the parallel improvement management of these resources on energy-saving background.

In our opinion, the main prerequisites for using PSPP in Uzbekistan are:
- increasing demand in maneuver power for covering daily load peaks and compensation its short-term variations;
- the use of cheap night electrical power in the failures of the daily load schedule;
- need to optimize the basic power plants;
- fuel economy in the power system by comparing with other variation peak power;
- providing a quick operational reserve;
- participation in the regulation of operational parameters in order to ensure standard quality of electrical power.

One characteristic feature of PSPP in abroad is that they are usually located energy centers in industrial and urban areas of the country or near to non maneuverable powerful source of electrical power [20]. Under these conditions, control of the current produced and consumed energy will be significantly effective and the power losses will be less [19].

As an example, we can cite the complexes "TPP - PS - PSPPP", geographically located in proximity to each other.

Pump stations (PS) water systems are large energy consumers, and often their work has seasonal character and depends on the accepted order of irrigation works on under the command areas to them. At present, 17-18% of electric energy from the volume of the republican energy consumption is consumed by irrigation pumping stations [21].

Most of these pumping stations continuously work 8-10 months during the year in changing charge of water and less change head pressure. In the water management system of the republic, there are also PS of year-round operation, which, outside the irrigation time, serve to fill the reservoirs. Many of the PS, the servant of the melt during the growing season, are operated in full power 3-4 months a year (mostly in May - August), and in this connection their operating ratio of the installed power on time does not exceed 50-60 %.

In our opinion, in the current circumstances it would be appropriate to use the spare power of large PS to accumulate hydraulic energy by adapting their operation to the modes of operation of TPP, which are the source of power supply of pumping units.

As one of such examples of the possible use of PS, a diagram of the pumping station and PSPP is shown (Fig. 2.).

PS can work weekly, or even yearly regulation power load, but it needs more wide-ranging tasks and you are associated with an increase in additional volumes of reservoirs and equipments of PSPP. In this regard, the inclusion of the PS of water system into hydraulic energy storage is advantageously carried out to cover the daily load curve.

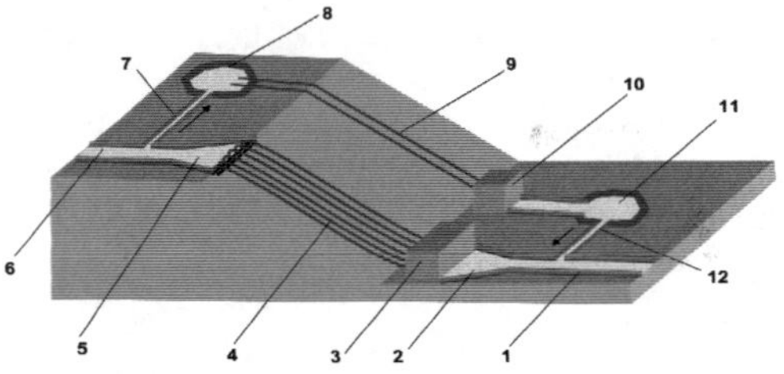

Fig. 2. PSPP position at the pumping station
1 - water supply channel; 2 – entrance bay; 3 - building of the pumping station; 4 - pressure pipelines; 5 - pressure head reservoir; 6 - outlet channel; 7 - feeding channel; 8 - upper reservoir; 9 - turbine pipelines; 10 - hydroelectric power station building; 11 - lower reservoir; 12 - water intake channel.

These PSPP have connection with high voltage power lines basic nodal distribution substations the consumer and large switchgear TPP. Related location PSPP allows:
- to carry out a normal operation, the energy system controller function characteristic for the PSPP in order to increase the reliability and quality of power supply;
- in an emergency, due to the commissioning of the PSPP in the power supply structure of large energy consumers, address and emergency backup of generating capacity;

 https://doi.org/10.1051/e3sconf/201913901009

- to compensate for the load that went out of service due to an accident in the turbine units of the TPP, while maintaining their rotating generating power .

As an example, we cite the option of using the PSPP-PS-TPP system, based on the Talimardjan TPP and six pumping stations of the Karshi main channel.

Pumping stations on Karshi main channel are large energy consumers with an annual rate of 2.0 - 2.2 billion kilowatt·hours that equal 3,5- 3,6% of the total production of energy in the republic. Placing PSPP near to current PS (in the vicinity of each of the six PS will settle the relevant PSPP), in our opinion, it is not only beneficial in terms of cost savings by eliminating the need to install pumps, and can improve operation Talimardjan TPP, which is a source of energy and is located near the PSPP. It can be illustrated the following example using the daily schedule power generation Talimardjon TPP shown in Fig. 3.

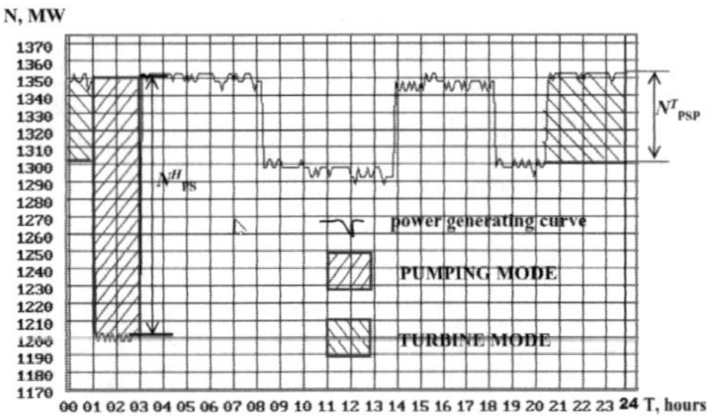

Fig. 3. Modes for PSPP daily schedule of power generation Talimardjan TPP

The decrease in energy production from 01^{00} to 03^{00} a.m. the power of one CCP-150 unit with a capacity of 150 MW is switched off , as a result of which the power of the station dropped to 1200 MW. This failure to use a daily schedule for charging the upper reservoir, increasing electricity generation to 1350 MW (i.e., without shutting down the unit at TPP) due to power consumption pumping mode 150 MW. The upper reservoir will be filled by existing PS, by supplying water through a connection channel (Figure 2).

In the period from 20^{30} to 01^{00} the PSPP operates in turbine mode, supplying water from the upper reservoir to the lower one, while the power of the generating regime is about 52.5 MW and the duration of operation is 4.5 hours. As a result of this, the operation mode of the state district power station is significantly improved, the unit shutdown is excluded, the unevenness coefficient of the daily schedule is changed, if this parameter had a value of $k = 0,889$, then in the new mode it can have a value of $k = 0,954$.

This PSPP option has the following indicators:

1. Power of PSPP on pumping mode $N^P = 150\ MW$ and turbine mode $N^T = 52,5\ MW$.
2. Working time: on pumping mode 2.0 hours and 4.5 hours on turbine mode.
3. The volume of water: V = 700000 m^3.

The calculation results showed the cost-effectiveness of the PSPP compared with TPPs of similar capacity with cost savings of $ 1,708 thousand per year, a payback period of 5.09 years and a cost of $ 0.07 / kWh.

Conclusion

1. As the most perspective direction of achievement of reliability and regime controllability of Republic Uzbekistan EES use PSPP, which simultaneously are highly maneuver a source of peak power and the consumer-regulator in the minimum load is offered.
2. The results of the calculations to determine the economic efficiency of pumped storage systems - PS - thermal power plant as an example Talimardjan TPP and six pumping stations Karshi main channel showed that economic effect of the substitution power of TPP, and also due to the fuel of the first effect and the provision of ancillary services does not output and so the scope of global indicators and pointing e t to achieve a significant economic effect.
3. The stated material and the received results of this paper points to the need for deeper studies on the possible uses of the PSPP in the power system of the Republic of Uzbekistan , the development in the principles and methods of work to changing conditions, control modes of operation of the EPS in the normal and emergency situations.

Figures and tables, as originals of good quality and well contrasted, are to be in their final form, ready for reproduction, pasted in the appropriate place in the text. Try to ensure that the size of the text in your figures is approximately the same size as the main text (10 point). Try to ensure that lines are no thinner than 0.25 point.

References

1. Electricity storage and renewables: Costs and markets to 2030 http://www.irena.org/publications/2017/Oct/Electricity-storage-and-renewables-costs-and-markets
2. Electric Power Research Institute: Electricity Energy Storage Technology Options. A White Paper Primer on Applications, Costs, and Benefits https://www.iea.org/publications/freepublications/publication/Technology RoadmapEnergystorage.pdf.
3. International Energy Agency: Technology Roadmap on Energy Storage, (2014) https://www.iea.org/publications/freepublications/publication/Technology RoadmapEnergystorage.pdf
4. European commission directorate-general for energy. DG ENER Working Paper. The future role and challenges of Energy Storage https://ec.europa.eu/energy/siTPP/ener/files/energy_storage.pdf
5. V.V. Elistratov. The use of renewable energy. 225 (2008)
6. Y.S.Vasiliev, G.A. Pretro. Pumped storage power plants. LPI. 76 (1984)
7. Electrical Energy Storage. IEC white paper; December 2011.

http://www.iec.ch/whitepaper/pdf/iecWP-energystorage-LR-en.pdf
8. V.V. Zubarev Accumulating power plant and uses them as of a power systems. issue **4**, 34-38 (1986)
9. V.Y. Sinyugin, V.I. Magruk, Rodionov V.G. Hydroaccumulative power plants in modern electric power industry. ENAS, 352 (2008)
10. L.B. Scheinman, Accumulating power plants. Energy, 184 (1978)
11. Energy storage — packing some power. The economist. http://www.economist.com/node/21548495?frsc = dg% 7Ca
12. C.J. Yang, Pumped hydroelectric storage: Technical report, (2011) https://wiki.opensourceecology.org/images/5/5d/Phs.pdf
13. Paul Denholm, Erik Ela, Brendan Kirby and Michael Milligan. The Role of Energy Storage with Renewable Electricity Generation. NREL / TP-6A2-47187 (January 2010)
14. Brandi A. Antal. Pumped Storage Hydropower: A Technical Review. 84 (2014)
15. Istvàn Tàczi. Pumped Storage Hydroelectric Power Plants: Issues and Applications, Energy Regulators Regional Association. Budapest, Hungary, 11 (2016)
16. Renewables 2017 Global Status Report. http://www.ren21.net/wp-content/uploads/2017/06/17-8399_GSR_2017_Full_Report_0621_Opt.pdf
17. Renewable nergy sources and storage of electrical energy (June 2015) https://www.researchgate.net/publication/314154863_Renewables_and_Electricity_Storage_A_technology_Roadmap
18. V.A. Zubakin is necessary to develop highly maneuverable powerful about STI // Energy market. 34-42, **9** (2004)
19. M.M. Muhammadiev, B.U. Urishev Hydroaccumulative power plants. Fan va texnologiya, 212 (2018)
20. C. Yang and R. B. Jackson. Renew. Sustain. Energy Rev., vol. **15**, no. 1, 839-844, (Jan. 2011)
21. M.M. Muhammadiev, B.U. Urishev, Energy efficient technologies in the operation of pumping stations. 115 (2012)

Patterning aspects of small solar power development in Uzbekistan

O.R. Sitdikov[2*], *A.M.* Mirzabaev[1], *T.A.* Mahkamov[3], *Sh.* Mirzabekov[3]

[1] "MIR SOLAR" LLC, Uzbekistan, Tashkent
[2] International Institute of Solar Energy (IISE), Uzbekistan
[3]Tashkent State Technical University, Uzbekistan

Abstract. The article describes the experience of "Mir Solar" LLC (Uzbekistan) in the use of patterning for the development and formation of small solar energy. The paper describes the patterning of the problem field of solar energy, which leads to effective practical solutions in the energy supply of rural and remote consumers of electric energy. The use of patterns shows that they contribute to the solution of many problems arising from the use of photovoltaic installations (PV), providing reliability and efficiency of power supply. Examples of effective implementation using patterning methods are given.

Introduction

Around the world, energy based on renewable sources is developing rapidly due to many reasons: the limitations of traditional fuel resources, environmental requirements, consumer demand, etc. [1]. The factors determining the development of renewable energy technologies are: long-term forecasting of electricity consumption; change in the structure of energy consumption, changing in developed countries towards decentralized consumption; low-carbon fuel energy, which is based on the rejection of coal, oil, gas and the transition to new alternative sources; automation and intellectualization of accounting, monitoring and management systems in the energy sector; environmental pollution and others.

General information about patterning

This term is derived from the English word "pattern", which translates as "pattern", "example", "form", "structure". Pattern (eng. pattern-pattern, pattern, system) is a borrowed word that is used as a term in many modern technologies [2].

The term pattern refers to a repeated pattern or a pattern that is predictable is reproduced. When we talk about patterns, we mean their presence or absence in the relevant field of activity. They often look the same, but do not always copy each other and can be diverse. [3]
Currently, there are many varieties of patterns, which in our case can be divided into two groups: the formation of continuation and reversal of the trend. For small energy, the continuation and reversal models indicate the further development of the trend and are signals to open new positions (patterns) in accordance with current trends in technical, technological, regulatory, organizational and other areas.

Methodological aspects of patterning in small power engineering

Trends in the development of small-scale energy and facts suggest that there will be a further transition to a combined system of centralized and decentralized energy supply, in which small, including solar energy, will occupy an important place in our Republic [4-6].

The analysis shows that the processes of its development and functioning require appropriate components of support: technical and technological, financial and innovative, regulatory, organizational, educational, scientific, training of relevant specialists, etc.at the same time, as experience shows, it is necessary to adhere to and use patterning as one of the effective mechanisms to ensure the efficiency of energy systems, including small.

Small-scale solar energy technology patterns need to be introduced in sectors and clusters such as the private sector; micro-grid formations; local generation; energy storage; energy supply to agriculture and remote consumers.

The “Mir Solar” LLC is active in the field of solar energy of the Republic of Uzbekistan more than 15 years, and use patterns in this area helped to design and implement effective technical and technological solutions allowing to combine the best properties of the most common types of solar technologies and installations, which is important to ensure remote from large electricity consumers, such as the water system of livestock and crop production, communications, meteorological system etc. We are talking about the use of those types of equipment that are produced on a large scale, have low market prices and are easy to operate.

The innovative role of the introduction of patterning techniques in the formation of small energy is that it allows you to select the best properties: reliability, efficiency, low prices, ease of installation and maintenance, durability, payback in an acceptable time.

The development of small-scale energy requires comprehensive state support, including the creation of an appropriate system of regulatory support, standardization, the introduction of various preferences, benefits to producers and consumers, "green" tariffs, etc.the use of solar photovoltaic plants (FES) together with wind, biogas, small hydropower plants, geothermal, hydrogen and other installations, as well as energy storage facilities is understood as a complex. It is necessary to use the experience of other countries where patterning is widely used. For example, Pattern Energy Group (PEGI) is an independent international renewable energy company, successfully engaged in solar and wind generation in a number of countries, in particular, in the United States, Canada, Chile and Japan. The company operates wind and solar power plants with a total capacity of 2942 MW [7-8].

On the basis of patterning specialists of "MIR SOLAR" LLC developed effective circuit solutions of FES for power supply of pumps of water lifting of agriculture, remote stations of cellular communication, rural medical posts, etc., providing reliability and continuity of their power supply.

Long-term operation more than 1500 objects assembled, "MIR SOLAR" LLC is confirmed by long-term use (5 to 15 years or more), and return – the costs of purchased equipment and operation of FES within 8-12 years [4-6,9].

In all applied technologies and the used types of the equipment and actions "MIR SOLAR" LLC has the "know-how". The quantitative ratios used simultaneously on PMT panels of monokristalline and polikristalline silicon, the corners of their attitudes to the countries of the world (with oatom of the wind rose), horizon properties used and developed special controllers, cooling systems, inverters, choice of AB system, warranty service, etc. All this allows to provide long-term reliability and efficiency of the installed equipment, to expand the market of FES [9].

As a necessary pattern, several groups of students of the energy faculty of Tashkent state technical University annually undergo practical training at the MIR SOLAR LLC enterprise. This contributes to the development of small solar energy in the Republic, and the company allows it to determine the necessary graduates and provide them with qualified work.

Example 1. Guaranteed energy supply of water supply systems

The most successful solar technologies based on patterning are implemented in agriculture, in particular for guaranteed water supply of remote areas [10].

The main way to solve the problem of providing the needs of the population and agriculture with fresh water in remote arid and arid areas is the use of artesian wells. Therefore, the issue of guaranteed power supply of electric drives of such pumps is very relevant for the arid regions of the Republic of Uzbekistan, as well as the countries of Central Asia.

In modern practice, two main schemes of using photovoltaic panels (FEP) for water lifting are successfully operated – accumulator-free and with a battery (AB). Such schemes and the corresponding equipment were selected using the patterning described earlier, which allowed to identify the most suitable variants of equipment and schemes, taking into account their further operation in specific local conditions.
In this case, conventional types of equipment are used, including typical inverters (table 1.).

Table 1. Main parameters of widely used pump inverters

Pump inverter type	rated voltage, V	kW rated power		kW recommended	
		P_{min}	P_{max}	U_{min}	U_{max}
Single phase	220	0,5	5	290	500
Three phase	380	3,0	75	500	900

1. Accumulator-free scheme of water lifting using photovoltaic panels-FEP.

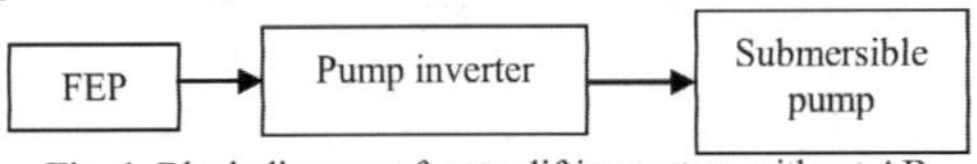

Fig. 1. Block diagram of water lifting system without AB

The main element of this system (Fig. 1.) is an inverter to drive pumps with asynchronous motors. The advantage of such a scheme is the relative simplicity, cheapness and durability of the installation, and the disadvantage is the possibility of using it only for water supply needs, and the continuity of water supply is provided by the accumulation of water.
2. The accumulator scheme of ensuring water lifting with use of FEP (Fig. 2.).

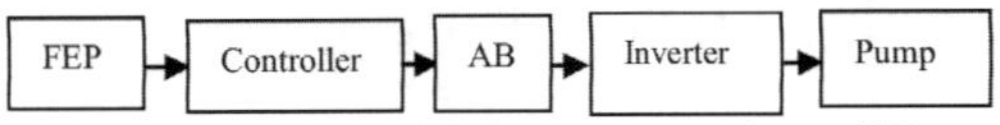

Fig. 2. Block diagram of the battery system of water lifting

This scheme is an example of using a classic GRID-OFF scheme (without connection to a common network), the main advantage of which is the ability to operate around the clock and feed other consumers, and the main drawback is the relative high cost due to the cost of installing AB and the need for periodic replacement.

The need for irrigation water is seasonal, so often FES remain unloaded. With the current shortage of electricity in remote rural areas, the use of these FES at full capacity becomes relevant.

Taking into account the advantages and disadvantages of the main power supply schemes of water lifting systems, scientists of Tashstu, MISE and specialists of "MIR SOLAR" based on the experience of developing photovoltaic systems, offer a more flexible connection scheme that guarantees the use of a minimum number of AB. The structure of such a scheme is shown in Fig. 3.

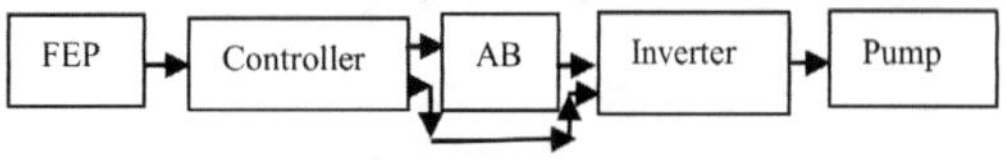

Fig. 3. Fes block diagram developed by MIR SOLAR LLC for remote power supply systems

The main advantage of the proposed scheme is the possibility of using a minimum number of expensive AB at high values of total power and, accordingly, the connected pump and / or other consumers.

The controller provides for disabling the charging current stabilization function, allowing it to be used as a normal controller. Thus, if necessary, AB is used as a stabilizing factor: in cloudy weather and to power household equipment in the dark.

It can be noted that such systems are able to fully provide water to the needs of livestock in remote and arid regions of our Republic, as well as other water users of agriculture.

To date, the Republic is very successfully operated several dozen solar systems of deep water lifting of different capacities under this scheme, mounted LLC "MIR SOLAR", and the need for them is expanding day by day (table 2).

Table 2. Dislocation of some water lifting stations equipped with equipment of "Mir Solar" LLC»

Location	kW FES Power	kW pump Power	Flow, not less than (m3 / h)	lifting Height, m
Kushrabat	8	5,5	25	50
Bektemir	8	5,5; 5,5	35	20
TIIMSH	9,2	7,5	40	40
TIIMSH	9,2	7,5	40	35
TIIMSH	9,2	7,5	40	35
Fergana	9,2	7,5	40	35
Samarkand	13,5	9,2; 7,5	60	40
Andijan	13,5	7,5	15	80
Samarkand	13,5	9,2; 7,5	40	40
Agricultural i-t	18,5	13	110	30
Agricultural i-t	18,5	13	110	30
Samarkand	20	15	50	50
Jizzakh	37	30; 26	80	80

Example 2. Photovoltaic plant for reliable power supply of base stations of cellular companies

Engineering and technical staff of one of the world's leading mobile operators working in Uzbekistan, together with scientists of Tashkent state technical University and specialists of "MIR SOLAR", studied the possibilities of application, advantages and disadvantages of different approaches to ensure uninterrupted power supply of base stations (BS) of cellular communication systems operating from FEU. Various options of backup power supply sources were considered:

- laying a backup power line from the transformer substation;
- increase the number of batteries BS;
- installation of wind turbines;
- use of diesel generators;
- installation of FES.

The first option disappeared immediately because of the high cost and impossibility of laying power lines over relatively long distances. And all nearby substations, as a rule, are disconnected at the same time.

The second option is rejected because of the high cost of the AB and powerful chargers and, most importantly, because of the inexpediency. In winter, AB simply do not have time to fully charge.

The third option was rejected due to the lack of stable wind at low altitudes and the need for maintenance of rotating parts. Recall that the optimal wind speed for low-power wind turbines is 12-15 m / s.

The fourth option was rejected because of the relative high cost of high-quality diesel generators, their limited service life compared to FES and, most importantly, the need for continuous maintenance and rising prices for fuel and lubricants.

On the basis of the calculations, their analysis and the experience of patterning, appropriate decisions were made, including the use of FES as both the main and backup power source of the BS, when using AB.

In addition, calculations and operating experience of such FES have shown that with a three-fold difference in solar radiation in summer and winter [2], it is most effective to mount photovoltaic panels (PV panels) at a certain angle, which allows to ensure the maximum power of the PV in the winter months and does not contribute to the formation of snow cover.

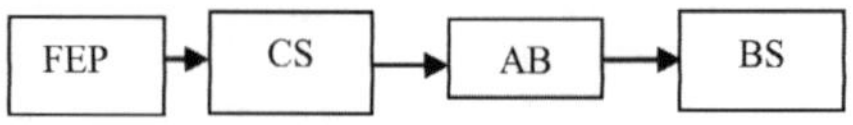

Fig.4. Block diagram of BS power supply

A distinctive feature of the proposed version of the FES 5 kW is the simultaneous use of both types of widespread solar panels, i.e. part of the power is generated on panels made of monocrystalline silicon, and the other part - on panels made of polycrystalline silicon; in the presence of a specially designed controller (CS).

The joint combination and use of a certain ratio of monocrystalline and polycrystalline solar panels gives the greatest reliability and efficiency when taking into account the peculiarities of operation of a particular station, including-economic, climatic, insolation, as well as load schedules BS and others. In table 3 the parameters of such a FES for the power supply of one of the cellular communication stations in Uzbekistan are given.

Table 3. Main parameters of FES produced by MIR SOLAR LLC»

Type of solar element (SE)	number of FEP SE	Maximum power W	Voltage with max is powerful,V	In Current at max	Voltage xx, V	In the short-circuit Current	And the Efficiency of the SE, %
Monocrystalline silicon	72	200	37,8	5,39	45,25	5,92	18,5
Polycrystalline silicon	60	265	30,5	8,17	36,9	9,15	16,5

In this case, monocrystalline PECs are installed at a certain angle in the south direction, and polycrystalline PECs in the south-east direction. This arrangement of the solar cells provides fast charging of discharged batteries during the night and the most complete charge in the evening.

The most important component of the FES is the KS - the charge-discharge controller AB, specially designed and manufactured by the specialists of the manufacturing enterprise LLC MIR SOLAR, taking into account the peculiarities of the operation of the BS (patent UZ.SMT.01.002.1806349).

The developed CS allows controlling the total charge current, i.e. the power of the photovoltaic panels is regulated depending on the current of the charger built into the BS. This factor significantly increases the battery life, because prevents the possibility of "boiling" of batteries. The SC allows, as necessary, to gradually turn on and off individual BS functional units, depending on the charge level of the battery, which maximizes the possibility of providing priority telecommunication services.

KS is designed taking into account local climatic conditions. The installed radiators of a special design allow you to abandon the vulnerable forced ventilation system and increase the life of the unit. KS allows you to adjust the value of the charge current and the shutdown threshold of the battery at the request of the customer.
The most vulnerable component of FES is AB. The specialists of MIR SOLAR LLC did a great job of choosing the type and manufacturer of the battery, their optimal number and connection schemes. Based on the criterion of "price-quality", helium batteries of the PW 100-12 type were selected (table 4) [11].

Table 4. Parameters of AB

Brand	Type	Weight kg	Capacity, A * hour	Voltage, V	in number of charge-discharge cycles		
					100 %	50 %	30 %
PW 100-12	GEL	30,5	100	12	600	1000	1400

The total capacity of AB 3200 A*h (32 pieces of 100 A*h) selected taking into account the BS power consumption allows to fulfill the manufacturer's requirement of inadmissibility of discharge below 50%.
It should be noted that the scientific and applied approach to solving the problem and the high qualification of specialists of "MIR SOLAR" LLC convinced the leadership of the largest mobile operator of Uzbekistan to choose the proposed FES as a reliable source of power supply to the BS. To date, 280 BS Uzbekistan successfully operated FES production of "MIR SOLAR" LLC.

Conclusion

Patterning has fully justified itself in the selection, installation and operation of solar installations with a capacity of 5-50 kW to ensure reliable and uninterrupted power supply to rural and remote facilities. This is confirmed by the experience of operation and monitoring of more than 1,500 facilities in Uzbekistan, equipped with FEU LLC "Mir Solar".

References

1. Allaev K.R., etc., The power sector of Uzbekistan and prospects of its development. - T.: Fan va texnologiyalar. 2016, -296 p.
2. https://xn----8sbfgf1bdjhf5a1j.xn--p1ai/1339-pattern-chto-znachit.html.
3. http://www.02s.ru/readarticleaf77.htm.
4. Mirzabaev A.M., Sytdykov O.R., Mirzabekov Sh.M., Dungbaev Sh. I., Nurimbetov S.P. Photovoltaic installation for guaranteed power supply of a remote object / in SB. "Modern problems of semiconductor physics and development of renewable energy sources", -Andijan. On April 20-21, 2018
5. Mirzabaev A.M., Makhkamov T.A. DEVELOPMENT DYNAMICS OF OOO "MIR SOLAR"./Applined Solar Energy, 2013, Vol.49 No4, Allerton Press, Inc (United States), pp 272-274.
6. Mirzabaev A.M., Kanonerov V.P., Makhkamov T.A., Sytdykov O.R., Mirzabekov Sh. M. Photovoltaic installation for power supply of base stations of cellular companies. //Solar technology. 2018. No. 1. Pp. 47-51.
7. http://www.patternenergy.com
8. https://ru.glosbe.com/en/ru/energy%20utilization%20pattern
9. Mirzabaev A.M., Sitdikov O.R., Mahkamov T.A. Mirzabekov, Sh.M. Integration of photovoltaic systems for water lifting in the electrical network. // Problems of energy and resource saving. 2018. No. 3-4. Pp. 313-317.
10. Alternative energy sources: opportunities for use in Uzbekistan. /Analytical report 2011/03. Economic research.
11. https://bigteh.ru/category/akb-gel-serija/.

https://doi.org/10.1051/e3sconf/201913901011

Digitalization of heat energy accounting as a means of improving the reliability of heat supply

I.G. Akhmetova [1]*, *K.V.* Lapin [2], *T. R.* Akhmetov [3], *E. Yu.* Balsamova [1]

[1]Kazan State Power Engineering University, Kazan, Russia
[2]JSC "Tatenergo", Kazan, Russia
[3]JSC "Kazenergo", Kazan, Russia

Annotation The paper considers the issue of improving the reliability of heat networks due to timely detection of deviations of actual operating conditions from the normative-sections of the heat network located in a humid environment, intensifying the processes of external corrosion of metal. The use of the software complex in real time analyzing and notifying about the operating conditions of thermal networks, deviations from the normative values is proposed. A method for automated calculation of normative values of coolant temperature is developed.

1 Introduction

Reliability is a complex property of an object, which, depending on the purpose of this object and the conditions of its operation, can include several single properties.

One of such single properties are survivability and long-eternity of heat networks. The survivability and durability properties of thermal networks are actually provided by three integral components: protection, detection and response. Here an important task is the timely detection of foci of corrosion in the pipelines.

In the conditions of urban development pipelines of heat networks are laid mainly in underground execution. Statistically, the most significant cause of failure of thermal networks is external corrosion of metal, which, in turn, most often leads to the presence of pipelines of thermal networks in a humid environment or simply in water when the channel is flooded or groundwater rises above the occurrence of pipelines laid by a canal-free method.

In turn, the humidity (flooding) of the cover layer of the thermal insulation of the pipelines of the heating network site is caused by factors, the main of which include: surface water leaks through leaky manhole covers, coolant leaks through the seals of valves and compensators, destroyed channel overlaps. At air convection on overlappings of thermal chambers, adjacent parts of the channel, and also on planes of the panel supports having temperature below dew point, there is a condensation of moisture with the subsequent formation of a drop therefore there is a humidification of the heat-insulating designs concentrated in separate places causing corrosion of metal of pipes.

Traditional methods of increasing the reliability of heat networks, such as hydraulic testing of networks for strength and density in the inter-heating period, major repairs and reconstruction of heat networks are not tools for the rapid search for places of humidification of thermal insulation, but only ensure the maintenance of regulatory reliability.

The solution of a problem of increase of reliability of work of thermal networks, preservation of their working capacity and durability should not be limited to replacement, reconstruction, modernization actually worn out, emergency, not passed hydraulic tests of sites of thermal networks.

It is equally important to maintain the regulatory conditions that determine the durability of pipelines during their operation, namely during the heating period.

An indicator indicating the presence of the pipeline in a humid environment can be a decrease in the temperature of the heat carrier at the end of the section below the normative values, as a result of increased losses of thermal energy. The influence of humidification of thermal radiation on the value of heat losses is given in [1,2] and is estimated by increasing the latter by 2-4 times.

For the purpose of operational search of places of increase of humidity of thermal insulation, the decision on development of a software product is offered.

For the development of software required are 2 conditions: 1) the presence of a heat source and consumers metering devices of thermal energy, equipped with means di-station transmission; 2) a system for the automated collection and storage of data.

In accordance with the requirements of [3] in recent years, the number of metering devices for consumers has increased significantly: in the service area of JSC Tatenergo (Kazan, Republic of Tatarstan) today, more than 98% of thermal energy consumption is determined by commercial metering devices, of which more than 60% are equipped with remote data transmission to the

 https://doi.org/10.1051/e3sconf/201913901011

company's automated data collection system.

The information delivered to the created software product consists of two temperature fields from the source of thermal energy to the consumers: on the supply (T1) and on the return (T2) pipeline.

For automated detection of abnormal operating conditions of pipelines in the online mode, it is necessary to set the temperature range of consumers corresponding to the norm, taking into account the time of passage of the temperature wave. The purpose of the calculation is to provide the possibility of automated determination of the normative temperature of the coolant at the consumer, taking into account the temperature of the coolant at the outlet of the heat source and the normative losses of thermal energy in thermal networks.

Other things being equal (constant coolant flow rate, constant design characteristics of the heat network), the following dependencies are observed:

$$Q_{\text{пот}} = f(T_{\text{тн}}, T_{\text{нв}}), \quad (1)$$

where $Q_{\text{пот}}$ – losses of thermal energy on the site of the heat network, Gcal/h; $T_{\text{тн}}$ – coolant temperature, °C; $T_{\text{нв}}$ – the actual outdoor air temperature, °C.

In turn:

$$T_{\text{тн}} = f(T_{\text{нв}}) \quad (2)$$

In accordance with [2] the temperature of the coolant in the supply and return pipeline of the heat network, as well as after the mixing unit of the consumer, depending on the outside air temperature are determined by the formulas:

$$t_1 = (1 + u_{\text{p}}) * t_3 - u_{\text{p}} * t_2 \quad (3)$$

$$t_3 = t_{\text{в}} + 0{,}5 * (t_{3\text{p}} - t_{2\text{p}}) * \frac{t_{\text{в}} - t_{\text{н}}}{t_{\text{в}} - t_{\text{нр}}} + 0{,}5 * (t_{3\text{p}} + t_{2\text{p}} + 2 * t_{\text{в}}) * \left(\frac{t_{\text{в}} - t_{\text{н}}}{t_{\text{в}} - t_{\text{нр}}}\right)^{\frac{1}{1+n}} \quad (4)$$

$$t_2 = t_3 - (t_{3\text{p}} - t_{2\text{p}}) * \frac{t_{\text{в}} - t_{\text{н}}}{t_{\text{в}} - t_{\text{нр}}} \quad (5)$$

where t_1, t_2, t_3 – the temperature of the coolant in the supply, return pipeline of the heat network, after the mixing unit at the input of the consumer at the temperature of the outside air corresponding to $t_{\text{н}}$, °C; u_{p} – the estimated coefficient of mixing; $t_{\text{в}}$ – estimated indoor air temperature, °C; $t_{\text{нр}}$ – estimated outdoor temperature for heating, °C; index "p" refers to the parameters for the calculation conditions.

In thermal networks of JSC Tatenergo (Kazan) the temperature chart of 135/65°C is applied, the table of temperatures is given on Fig. 1.

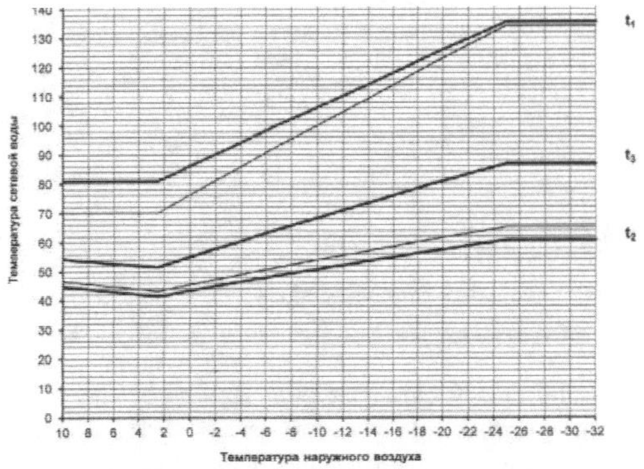

Fig.1-Temperature chart of thermal networks of JSC "Tatenergo", Kazan

The method of calculating the value of normative heat losses in the proper state of thermal insulation and normal operating conditions of thermal networks is given in [4].

The standards of technological losses include losses and costs of energy resources caused by the technical condition of heat pipelines and equipment and technical solutions for reliable provision of consumers with heat energy and the creation of safe operating conditions of heat networks, namely: losses and costs of the coolant (steam, condensate, water) within the established norms; losses of thermal energy by heat transfer through heat-insulating structures of heat pipelines and with losses and costs of the coolant.

Determination of standard technological losses of thermal energy, Gcal, caused by losses of the heat carrier, is made according to the formula:

$$Q_{\text{у.н}} = m_{\text{у.н}} * \rho_o * c * [b\tau_1 + (1 - b)\tau_2 - \tau_{\text{x}}] * n * 10^{-6} (6)$$

where $m_{\text{у.н}}$ - rate of coolant losses due to leakage, m³/h; ρ_o - the density of the coolant at the average temperature of the coolant in the supply and return pipelines of the heat network, kg/m³; b - share of mass flow rate of coolant lost by the supply pipeline of the heat network; τ_1 и τ_2 - average values of coolant temperature in the supply and return pipelines of the heat network according to the temperature schedule of heat load regulation, °C; τ_{x} - the average annual value of the temperature of the source water supplied to the heat supply source and used to feed the heat network, °C; c - specific heat of the coolant, kcal/kg °C; n - duration of functioning of heat networks, h.

The standard technological consumption of thermal energy for filling new sections of pipelines and after planned repairs, Gcal, is determined by the formula:

$$Q_{\text{зап}} = 1{,}5 * V_{\text{тр.з}} * \rho_o * c * [\tau_{\text{зап}} - \tau_{\text{x}}] * 10^{-6}, \quad (7)$$

where $V_{\text{тр.з}}$ - capacity of the filled pipelines of the heat networks operated by the heat network organization, m³; ρ_o - density of water used for filling, kg/m³; $\tau_{\text{зап}}$ - temperature of water used for filling, °C; τ_{x} - the temperature of the source water supplied to the heat energy source "x" during the filling period, °C.

Normative technological losses of thermal energy with drains from devices of automatic regulation and protection, Gcal, are defined by the formula:

$$Q_{а.н} = G_{а.н} * \rho_o * c * [\tau_{сл} - \tau_х] * 10^{-6}, \quad (8)$$

where $G_{а.н}$ - annual loss of coolant as a result of draining, m³; ρ_o - average annual density of the coolant depending on the place of installation of automatic devices, kg/m³; $\tau_{сл}$ и $\tau_х$- the temperature of the drained coolant and the source water supplied to the heat supply source during the draining period, °C.

The analysis of these formulas leads to the conclusion that under constant operating conditions, the value of thermal energy losses caused by losses and costs of the coolant is directly dependent on the temperature of the coolant.

Determination of normative technological losses of thermal energy by heat transfer through heat-insulating constructions of pipelines is made on the basis of values of hourly heat losses depending on year of design of heat pipelines, Gcal:

$$Q_{из.н.год} = \sum(q_{из.н} * L * Бета) * n * 10^{-6} \quad (9)$$

where $q_{из.н}$ – specific hourly heat losses by pipelines of each diameter determined by recalculation of tabular values of norms of hourly heat losses on average seasonal operating conditions, kcal/hm; L – the length of the pipe section thermal networks, m; Бета - the coefficient of local heat losses, which takes into account the thermal losses of shut-off and other valves, compensators and supports (is applied 1.2 at the diameter of pipelines up to 150 mm and 1.15 - at the diameter of 150 mm and more, as well as at all diameters of pipelines without channel laying, regardless of the year of design).

Thus, the change in the value of heat losses for a particular section of the heat network during the heating period is associated only with a change in the temperature of the coolant, due to a change in the temperature of the outside air.

The temperature of the coolant at the end of the heat network section, °C at the normative value of heat losses:

$$t_к = t_н - \frac{Q_{пот}}{G*n}, \quad (10)$$

where $t_н$ – coolant temperature at the beginning of the phase °C; $Q_{пот}$ – the value of the normative heat losses, converted to the actual temperature of the coolant at the beginning of the site, Gcal; G – design flow rate of the heat carrier on a site of a heat network, m³/h.

The value of the calculated coolant flow rate for each section is taken according to the results of hydraulic calculation.

In turn, the dependence of the value of heat losses on the temperature of the coolant is determined by the formulas (6)-(9) and the method of least squares:

$$Q_{пот} = f(t_н) = Q_{пот}^{max} * \frac{(-3,85+0,77*t_{тн}^{ф})+(37,57+0,46*t_{тн}^{ф})}{100*2} *$$
$$n \quad (11)$$

where $Q_{пот}^{max}$ – the calculated maximum value of heat losses on the site of the heat network at the calculated outdoor temperature for heating, taken by calculation in accordance with [4], Gcal/h; $t_{тн}^{ф}$- the actual temperature of the coolant at the beginning of the site at the current time, °C.

During the heating season, the temperature of the coolant varies depending on the temperature of the outside air. The temperature change of the coolant at the end of the section occurs with a delay in relation to the temperature change at the beginning of the section for the time of the temperature wave.

The formula takes the form:

$$t_к = t_{тн}^{ф} - Q_{пот}^{max} * \frac{(33,72+0,3542*t_{тн}^{ф})}{200*G} \pm \Delta t_н^{\tau} \quad (12)$$

where $\Delta t_н^{\tau}$ – the temperature difference of the coolant at the beginning of the section taking into account the time of passage of the temperature wave, G – coolant flow rate on the site, taken according to the hydraulic calculation, m³/h.

The time of passage of the temperature wave is determined by the statistical data of the actual temperature distribution of the heat carrier at different points of the studied heat network in Fig. 2, table. 1.

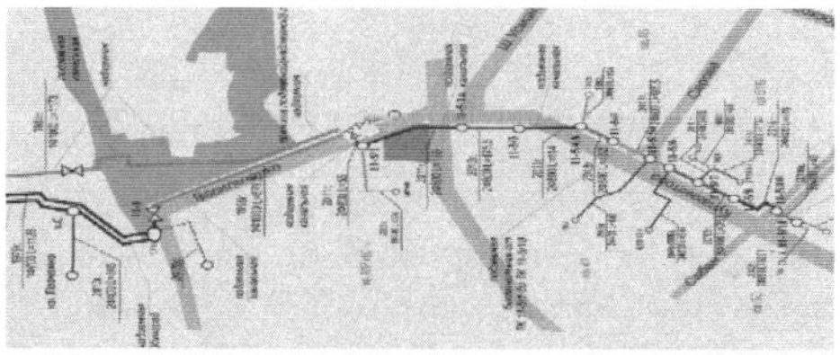

Fig.2-Scheme of the investigated section of the heating network

Table 1. Calculation of the coolant passage time (by fixing the passage of the temperature wave in time) from P-16

Measuring point	The transit time of the coolant, min.
Usmanov street, 33	58,3
Usmanov street, 33/1	35,1
Usmanov street, 33/2	59
Serov street, 6/1	45,66
Kulakhmetova street, 22	49,2
Baccalaureate street, 44a	54,09
Batyrshina street, 16	124,7
Batyrshina street, 22	225,1
Saban street, 7a	57,6

2 Description of digital product units (modules)

Schematically, a digital product can be divided into blocks (modules), each of which will perform a specific task.

Block 1. Input parameters of the heat carrier from a source of thermal energy and in points of release of thermal energy to consumers.

The source of thermal energy is understood as an

element of the heating network (CHP, boiler room, heat chamber, etc.), which is the reference point of the initial temperature of the coolant T1 in relation to which the program analysis of thermal energy losses (deviations of the coolant temperature from the normalized value) is performed.

In terms of collecting the parameters of the coolant: thermal insights CHP, thermal insights boiler room, pavilions of heat networks; heat chamber; an observation chamber; grips heating; objects consumers.

Input parameter: coolant temperature in the supply (T1) and return (T2) pipelines of the heat network.

Discreteness of automated parameters collection: depending on the type and scale of the heat supply system, but at least 1 RA-per day.

The array of input parameters is stored on the server of the system for remote collection of thermal energy parameter readings with connection to the consumer's heat supply facility (indicating the address of the facility).

Block 2. Interface of input parameters with objects of heat supply on the map of the district of heat networks.

The block provides an interactive map of heat supply facilities in the district of heat networks with reference to it: sources of heat energy, control devices for measuring the parameters of heat transfer in individual elements of the heat network, subscriber facilities.

Block 3. Analysis of the deviation of the current input parameters from the norm, assessment of the reliability of heat supply to consumers with the construction of a mathematical model of the parameters of the heating system (Fig.3).

The analysis of deviation of current input parameters from norm is made on each object of heat supply of consumers.

The analysis is based on the fact that during transportation of the coolant from thermal energy source to the object of a heat supply of consumers inevitably, technological losses of heating energy, resulting in reducing the temperature TEP of monocytes. The rate of decrease in the coolant temperature is calculated based on the level of normalized values of heat energy losses in the supply lines of the heating network through the insulation constructions of pipelines with a constant flow rate of the coolant in the network.

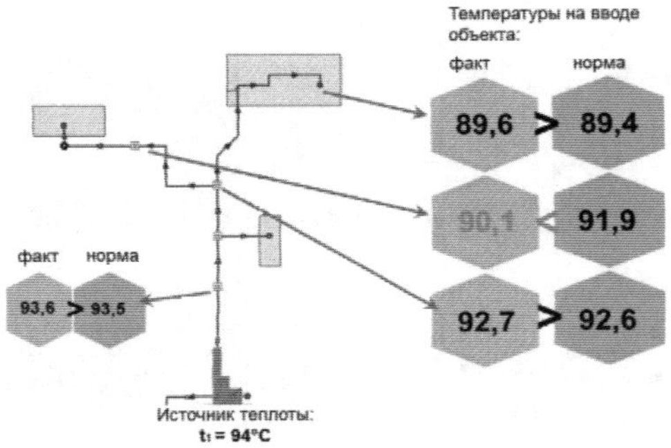

Fig.3-Scheme of analysis of deviation of the current input parameters from the norm

To group the set of obtained values of deviations of the actual temperature of the coolant from its normalized value, a dimensionless parameter φ is introduced, showing the materiality of the temperature deviation, determined by the formula:

$$\varphi = \left(t_{\text{к}}^{\Phi} - \text{t}_{\text{к}}\right)/\mu \qquad (13)$$

where $t_{\text{к}}^{\Phi}$ – actual coolant temperature at the end of the section, °C, μ - the coefficient of materiality of the deviation is numerically equal to the number of degrees Celsius, the deviation by the value of which is significant.

The following groups of object States are possible, presented in the table. 2.

Table 2. Verification of object States

Value φ	Condition heat supply
$\varphi \geq 0$	there is no violation of the thermal insulation layer
$-1 \leq \varphi < 0$	violation of the thermal insulation layer is available
$-2 \leq \varphi < -1$	humidification of the thermal insulation layer
$-\frac{\text{t}_{\text{к}}}{\mu} \leq \varphi < -2$	strong wetting of the thermal insulation layer

Block 4. Visualization of information on points of formation of excess losses of heat energy, indicators of reliability of heat supply of consumers and a problem site of a heating system

Each object in the interactive map should be displayed as a dynamically changing color element, for example,

- for the layer T1 according to the decrease of φ from black to red;
- for layer T2 in ascending order from green to yellow.

Visualization is provided on an Interactive map of the district heating networks with reference to the objects of heat supply.

For personnel shall be provided the opportunity to work in the "Interactive map" Kazan: by clicking courses-set on a necessary element of the thermal network to obtain necessary information in the current mode according to the actual parameters of the heat carrier on selected sections of heat networks collected with the con-monitoring of meters installed at the collectors of stations on the borders of the sections of teplovogo, the power districts, blocks, heat chambers, facilities, and regulatory parameters.

Block 5. Data archiving and reporting on sections of the heating system, where the parameters of the coolant

This block contains the statistics of the coolant parameters, data analysis parameters of the coolant and arose from the standing in heat networks and for a lot of consumers.

3 Conclusion

The task of timely identification of sections of pipelines of heat networks, which are in abnormal operating conditions, with the subsequent elimination of the threat of reliability is an urgent task for heat supply organizations.

It is proposed to introduce a software package, which, based on the readings of heat energy metering devices at consumers and at the source of heat energy, will allow real-time detection of abnormal operating conditions – humidification of pipelines of the heat network.

The method of determination of standard temperatures of the heat carrier at consumers is developed.

Application of the digital product allows to increase reliability of heat supply of consumers by timely identification of places of humidification of thermal insulation, and also can be used for preparation of the plan of pits of networks in the inter-heating period for research of degree of a thinning of pipelines of a heating system.

References

1. V. V. Ivanov, V. V. Vasilenko. Influence of insulation and soil moisture on heat losses of underground heating mains. Energosovet No. 7 (12), 2010, p. 27.
2. Nikolaenko R. A., Ermolenko M. V., Stepanova O. A. Influence of humidification of thermal insulation on the value of thermal losses of thermal networks. Young scientist. 2014. No. 6. Pp. 207-210. URL https://moluch.ru/archive/65/10655/ (accessed: 21.09.2019).
3. Federal law No. 261-FZ of 23.11.2009 "on energy saving and energy efficiency improvement and on amendments to certain legislative acts of the Russian Federation".
4. Order of the Ministry of energy of Russia of 30.12.2008 N 325 "About the statement of the order of definition of standards of technological losses at transfer of thermal energy, the heat carrier".
5. V. I. Manyuk, Ya. I. Kaplinsky, E. B. Khizh, A. I. Manyuk, V. K. Ilyin. Commissioning and operation of water heating networks 3rd edition revised and supplemented. Moscow. Stroizdat, 1988.

Analysis of reliability of power supply of consumers in non-full-phase modes using adaptive models of their calculation

T.N. Nasirov[1], *V.A.* Nepomnyashchiy[1], *O.V.* Radionova[1]

[1]Tashkent State Technical University named after Islam Karimov, 100095, Uzbekistan, Tashkent, University St. 2A

Abstract. Reviewed and analyzed issues related to the reliability of electricity supply in non-full-phase modes of electrical networks, to calculate which use adaptive methods, is built on one Central principle of modeling the modes of electric power systems (EPS). Special attention is paid to the calculation of damages from no full provided of electricity in non-full-phase modes.

Introduction

At damages (failures, incidents) of the main power supply equipment (power lines, groups of single-phase transformers, shunt reactors) in networks of 110 kV and above which are working with effectively grounded neutral, the main part of shutdowns is connected with steady single-phase short circuits which share reaches on the average 65%.

This creates an alternative:

1. For the period of repair of damaged equipment if you want to disconnect the damaged item at all, thus reducing network throughput with feasible energy-economic consequences in the form of limits to some consumers, the undersupply of electricity to them and thereby causing economic damage or
2. Using single phase control switches or highlighting with disconnectors' help of damaged phase, and translate the work of a damaged element in incomplete phase mode "two wires – ground", leading to the emergence of a network components of currents and voltages of inverse and zero sequences, with the corresponding consequences in the form of possible overloads in the remaining phases and overheating of transformers, generators and motors currents reverse sequence. To prevent these overloads, it will be necessary limits the loads of consumers with the appearance no full provided of electricity and economic damage from power supply disruptions.

The duration of incomplete phase modes can range from a few hours (for repair of the overhead line phase or replacement of the transformer's phase) to several days and months (for factory repair of the damaged transformer's phase in the absence of the reserve phase) [1]. Operation in such modes is one, and often the only way to improve the reliability of the operation of the EPS and power supply of consumers in post-accident and repair modes.

For estimation the admissibility of non-full-phase modes of operation of the electric network with a grounded neutral, it is necessary:

the first step - is to calculate the parameters of the normal mode and the parameters of the non-full-phase steady-state mode of the EPS, to determine the parameters of asymmetry in case of damage and subsequent shutdown of one of the phases;

the second step - is to identify the level of overcurrent of "healthy" phases, the degree of asymmetry of currents and the magnitude of the reverse sequence currents as the most dangerous for the occurrence of overheating of generators, transformers and motor load of consumers working in the network;

the third step - is the assessment of measures for unloading generating capacities and limiting the loads of consumers to remove re-loads from the network elements (generators, transformers, "healthy" phases of power lines).

To identify the economic efficiency of the use of non-full-phase modes of the EPS and to improve the reliability of the network, it is necessary:

the first - is to estimate the no full provided of electricity to consumers with a limited load and the resulting economic damage when operating in a non-full-phase mode;

the second - is to determine the probability and frequency arising of non-full phase regimes;

the third - is to determine the likely damage to consumers when the electric network operates in a non-full-phase mode: $\overline{З}_{уш}^{НПФР}$.

For comparison, you should do similar calculations for the alternative variant of the reliability of the network with disconnection of damaging element (transmission line, transformer) by, if it is necessary, measures to be introduced in allowable of mode settings area and entering the required limits of consumer loads.

Next, it is necessary to determine the under-supply of electricity and damage to consumers from violations of their power supply in the shutdown mode of the damaged network element, to assess the probability and

 https://doi.org/10.1051/e3sconf/201913901012

frequency of these situations and the likely damage to consumers: $\overline{З}_{\text{ущ}}^{\text{ав.откл}}$.

To select the option of doing network mode: unbalanced or disconnection of the damaged element, it is necessary to match the likely impact of the limitations of loads in unbalanced modes with the damages caused by outages of corrupted items:

$$\overline{З}_{\text{ущ}}^{\text{НПФР}} </> \overline{З}_{\text{ущ}}^{\text{ав.откл}} \quad (1)$$

and choose an option with minimal damage to consumers[2,3]:

$$\min\left\{\overline{З}_{\text{ущ}}^{\text{НПФР}}, \overline{З}_{\text{ущ}}^{\text{ав.откл}}\right\}. \quad (2)$$

Models of calculation of non-full-phase modes of EPS

Mathematical models and algorithms for solving the problems of calculation and analysis of the modes of EPS, in addition to the general requirements of reliability, speed, convenience of representation of the source and output information, are subject to increasingly high requirements for the adequacy of the display of real modes, adaptability to changes in the network structure, modes and types of disturbances.

On the basis of the conducted analysis of existent methods of calculation of unbalance EPS [4-10] the conclusion about expediency of application for calculation and analysis of substitute complex schemes (SCS) based on symmetrical components.

Linear mathematical models of calculation of non-full-phase modes of EPS are used for the purposes of relay protection and automation (RPA). Emergency modes of EPS caused both by longitudinal asymmetry, and modes of short circuits (SC) in the electric networks working in the incomplete phase mode are calculated.

The developed method of calculation of non-full-phase modes is based on the use of complex substitution schemes without direct electrical connections between the sequences.

For the formation of SCS, the substitution schemes of individual sequences of symmetric components in the places of asymmetries are connected to each other on the basis of boundary conditions. Modeling of the boundary conditions of asymmetry in the preparation of SCS is carried out with the help of multi poles communication (MPC) [6,7] or sets of ideal transformers (IT) [11,12] with transformation coefficients $K_{it} = 1, a, a^2$, where $a = exp(j2\pi/3)$, corresponding to the coefficients of the equations of boundary conditions. These two different methods of modeling boundary conditions are based on fundamentally different approaches to the representation of asymmetry locations. In the first case the matrix of self and mutual conductancies of the place asymmetry; in the second case, with set of IT compiled boundary conditions of unbalance.

The use IT for communication between schemes of direct (DS), reverse (RS) and zero (ZS) sequences allows to provide the uniform principle of modeling of elements of a network within SCS. In addition, if the numbering of the SCS nodes is carried out so that the IT nodes receive the last numbers, then the matrix of the node conductivities of the SCS will have a block-diagonal structure with a bordering part in which the conductivities of the IT connections with the nodes of asymmetry are located. The block-diagonal structure of the SCS matrix allows avoiding the procedure of end-to-end optimal renumbering of the complex circuit nodes in the analysis of non-full-phase modes and making maximum use of the results of calculations of the previous steady-state mode. At the same time, the advantages of calculating such matrices by methods of optimal elimination of variables are used to the maximum. The use of MPC for modeling of longitudinal type asymmetries violates the block-diagonal structure of the matrix of node conductivities of SCS due to the appearance of direct connections between the nodes of asymmetry of schemes DS, RS and ZS.

Based on the advantages of modeling the boundary conditions of IT over MPC, a universal algorithm for calculating incomplete-phase and complex-symmetric modes of EPS by complex substitution schemes is developed [12].

SCS, compiled for the calculation of non-full-phase and complex-symmetric modes of EPS, is described by a system of linear nodal equations:

$$\boldsymbol{Y}_{KC3}\dot{\boldsymbol{U}}_{KC3} = \dot{\boldsymbol{J}}_{KC3}, \quad (3)$$

where $\boldsymbol{Y}_{KC3}$ - is the matrix of nodal conductivities of SCS;

$\dot{U}_{KC3}$ - the desired vector of voltages of the nodes of SCS;

$\dot{J}_{KC3}$ - vector of nodal currents of SCS.

The structure of this system can be represented as follows:

$$\begin{bmatrix} Y^{(1)} & 0 & 0 & Y^{(1-D)} \\ 0 & Y^{(2)} & 0 & Y^{(2-D)} \\ 0 & 0 & Y^{(0)} & Y^{(0-D)} \\ Y^{(D-1)} & Y^{(D-2)} & Y^{(D-0)} & Y^{(D)} \end{bmatrix} * \begin{bmatrix} \dot{U}^{(1)} \\ \dot{U}^{(2)} \\ \dot{U}^{(0)} \\ \dot{U}^{(D)} \end{bmatrix} = \begin{bmatrix} \dot{J}^{(1)} \\ 0 \\ 0 \\ 0 \end{bmatrix}, \quad (4)$$

where $Y^{(1)}$, $Y^{(2)}$, $Y^{(0)}$ are the node conductance matrices of the DS, RS and ZS circuits, respectively;
$Y^{(i-D)}$, $Y^{(D-i)}$ – matrices of conductivity of connections MPC with nodes of asymmetry of PP, OP and NP (i = 1,2,0); $Y^{(D)}$ - matrix of nodal conductivities MPC;
$\dot{U}^{(1)}, \dot{U}^{(2)}, \dot{U}^{(0)}$ - vectors of node voltages of DS, RS and ZS circuits;
$\dot{J}^{(1)}$ - the vector of the driving currents in the nodes of the DS, determined by the values of the super-transient EMF generators obtained from the pre-calculated pre-emergency steady-state mode.

The system of linear equations of nodal stresses, describing SCS, with fixed values of super-transient EMF of generators and loads is solved by the method of ordered exclusion of unknowns (Gauss method). In addition, as a result of the solution, it is possible to directly determine the currents and voltages in the circuits of individual sequences.

The use of linear models of electrical systems used for the purposes of RPA for the calculation of non-full-phase steady-state regimes (NPSR) leads to large errors. NFRS are essentially load modes, so their calculation, as well as the calculation of the normal steady-state modes of power systems, should be carried out when setting the power values of load and generator units, that is, by nonlinear models.

Substitution schemes of individual sequences of DS, RS and ZS are formed on the basis of the same linear models, except for the account of loads and generators. As you know, in the coordinates of symmetric components synchronous generators are power sources only DS. Sources of currents and voltages of RS and ZS are places of asymmetries. Therefore, loads and generators in the calculation of NPSR are taken into account as follows: in the scheme of DS-the power of the normal (pre-emergency) mode, power losses in loads and generators caused by the flow of currents in them RS and ZS-the corresponding resistances.

The system of nonlinear nodal equations (4) is described. The equations of nodal stresses for DS nodes are nonlinear, since the driving currents in DS nodes are determined through the power of the nodes:

$$\dot{J}_i^{(1)} = \hat{S}_i / \hat{U}_i , \tag{5}$$

where $\hat{S}_i$ and $\hat{U}_i$ are the conjugate values of the power and voltage of the *i-th* node.

A method for solving a system of nodal equations describing the SCS is developed. Previously, the direct Gauss move excludes passive nodes of the RS and ZS circuits and additional it nodes. The result is a transformed system of nonlinear equations DS:

$$Y_*^{(1)} \dot{U}^{(1)} = (\hat{U}_{(diag)}^{(1)})^{-1} \cdot \hat{S} , \tag{6}$$

where $Y_*^{(1)}$ is the nodal conductance matrix of the extended (transformed) DS circuit;

$(\hat{U}_{(diag)}^{(1)})^{-1}$ - inverse diagonal matrix of conjugate stresses of DS nodes;

$\hat{S}$ - vector of conjugate powers of nodes.

Solution (6) is carried out by Newton-Raphson method for nodal equations in the form of current balance in a rectangular coordinate system. Further, the reverse Gauss stroke determines the voltage of the RS and ZS nodes.

The method of calculation of NPSR on complex schemes of substitution possesses sufficient simplicity of algorithm as is based on nodal equations, and universality for electric systems of any complexity. In addition, the obtained matrix $Y_*^{(1)}$ is a matrix of nodal conductivities of the power system taking into account non-full-phase inclusions and can be used for the study of static aperiodic stability of NPSR.

Probabilistic characteristics of non-full-phase modes and damage to consumers from violations of their power supply

In unbalanced mode, de-energizing phase "A" and conservation in the "healthy" phases "b" and "C" determination of the coefficient of over-current in the remaining phases "b" and "C" is carried out according to the following formula:

$$k_{nep}^{(1)} = \{abs(I^{(B)}) - I_{ВЛ}^{npeд}\}_{>0} / I_{ВЛ}^{npeд} , \tag{7}$$

where $I_{ВЛ}^{npeд}$ is the maximum permissible heating current in the "healthy" phase.

The maximum asymmetry of the currents will be equal to the ratio of the total current in phase "B"(or in phase "C") to the direct sequence current:

$$\mathrm{k}_{acuм}^{(1)} = I^{(B)} / I^{(1)} , \tag{8}$$

and the current asymmetry of the reverse sequence will be equal to:

$$\mathrm{k}_{acuм}^{(2)} = I^{(2)} / I^{(1)} . \tag{9}$$

These three factors will characterize the state of the network operating in the incomplete phase mode.

In [1] it is specified that at work of a network in the incomplete phase mode the asymmetry of currents in phases of generators shall not exceed the following values:

a) for turbogenerators - 12%;

b) for hydrogenerators with air cooling of stator windings with a capacity of 125 MVA and below - 20%, with a capacity of more than 125 MVA - 15%, and for hydrogenerators with water cooling of stator windings - 10%;

c) for synchronous compensators - 20%.

In this case, the reverse sequence current during long-term operation should not exceed 5% for turbogenerators and 10% for hydrogenerators.

If switching measures to reduce the asymmetry of currents and currents of the reverse sequence do not give the desired results, it is necessary to limit the loads of consumers and reduce the overall load of overloaded power lines and generating equipment.

Unloading of the network should be carried out purposefully with the choice to limit the loads of those nodes that lead to current limitation in the most overloaded transmission line operating in incomplete phase mode. For this purpose it is necessary to use well-known system of coefficients of distribution of currents $k_{(j)}^{(i)}$.

The current in the overloaded branch j is defined by the expression:

$$I_{(j)} = \sum_{i=1}^{n_y} k_{(j)}^{(i)} \times I_y^{(i)} , \tag{10}$$

where $I_y^{(i)}$ is the resulting current in the *i-th* node.

To limit loads , nodes with coefficients that affect the unloading of the most overloaded j-th branch before the condition is met are selected:

$$I_{(j)} = \mathrm{abs}(\sum_{i=1}^{n_y} k_{(j)}^{(i)} \times I_{y}^{(i)}) - I_{(j)}^{\text{пред}} = 0, \quad (11)$$

at the same time ensuring a minimum of damage to consumers according to the principles set out in [3, § 2.3].

The degree of load limitation of nodes to eliminate the overload of power lines operating in non-full-phase mode is defined as:

$$\varepsilon_i^{\text{огр}} = \mathrm{k}_{пер}^{(I)} = \{abs(I^{(B)}) - I_{ВЛ}^{пред}\}_{>0} / I_{ВЛ}^{пред}. \quad (12)$$

If the current asymmetry coefficients modulo (8) and the reverse sequence (9) are greater than their limit (normative) values, it will require additional limitation of consumer loads in the network nodes until the condition is reached:

$$\mathrm{k}_{асим}^{(I)} \le \mathrm{k}_{пред}^{(I)}; \quad \mathrm{k}_{асим}^{(2)} \le \mathrm{k}_{пред}^{(2)}. \quad (13)$$

With this in mind, the limited load nodes will be:

$$\Delta P_{нагр_i}^{\max} = \varepsilon_i^{огр} \mathrm{P}_{нагр_i}^{\max}, \quad (14)$$

and the system load limit will be equal to the sum of the node constraints:

$$\Delta P_{сист}^{огр} = \sum_i \Delta P_{нагр_i}^{\max}. \quad (15)$$

Slightly more complicated is the determination of the limits of the consumers in case of unbalance in open distribution equipment (ODE) nodes due to system failures or tire Assembly of switches, which leads to a transition in unbalanced mode all connected to this switchyard, power transmission lines and transformers. The difficulty lies in the fact that the restrictions of the loads of substations are fed from these power lines have an impact not just one but several working in unbalanced mode transmission lines. To do this, it is necessary to rank all substations connected with the damaged node according to the values of their influence on overload and current asymmetry in this node (ODE) and sequentially introduce load restrictions until the condition (13) is met.

Probabilistic characteristics of non-full-phase modes are:

1. The average annual frequency of single-phase steady-state power lines on the power lines entering the circuit and communication transformers (the latter, in principle, can be neglected, given their extremely rare damage relative to the power lines):

$$\lambda_{\text{уст}}^{\text{ав}(1)} = \mathrm{k}_{\text{к.з.}}^{(1)} \lambda_{\text{уст}}^{\text{ав}}, \text{ 1/year},$$

where $\mathrm{k}_{\text{к.з.}}^{(1)}$ - the share of single-phase SC in the total damage of the element (in the calculations, you can take the values of this coefficient for power lines-0.65 and for OPC-0.60).

2. Mean time to restore (repair) of single-phase failure of an element, $\tau_{\text{к.з.}}^{\text{ав}(1)} \approx$ 3-6 hours/off;

3. The probability of all emergency power lines, yelling at single-phase short-circuit:

$$\mathrm{q}_{к.з.}^{(1)} = \lambda_{уст}^{ав(1)} \tau_{к.з.}^{ав(1)} / 8760 \text{, Rel.units} \quad (16)$$

Undersupply of electricity to consumers will be determined by the expression:

$$\Delta Э_{к.з.}^{ав(1)} = 8760 \sum_{j=1}^{N} \varepsilon_i^{огр} \Delta P_{нагр_i}^{\max} q_{к.з.}^{(1)} \gamma_i(\varepsilon_i^{огр}),$$

million kWh, (17)

where $\gamma_i(\varepsilon_i^{огр})$ is the density of the bounded part of the load graph of the *i-th* node as a function of the degree of restriction of its load.

At the same time, the damage to consumers from violations of their power supply to ensure the functioning of the network with permissible asymmetry parameters will be:

$$З_{ущ.к.з.}^{ав(1)} = 8760 \sum_{j=1}^{N} \varepsilon_i^{огр} \Delta P_{нагр_i}^{\max} q_{к.з.}^{(1)} \gamma_i(\varepsilon_i^{огр})\, з_{ущ_i}^{ав}(\varepsilon_i^{огр}),$$

million rubles / year, (18)

Where $з_{ущ_i}^{ав}(\varepsilon_i^{огр})$ - specific damage from violations of power supply of consumers of the *i-th* node, rub/kWh, taken according to [3].

As shown by preliminary calculations for several circuits of electrical networks economically expedient to use in stable single-phase SC incomplete phase mode than to work with the disconnection of damaged elements.

Conclusion

1. The use of adaptive nonlinear models for calculation of non-full-phase steady-state modes of EPS allows to significantly increase the accuracy of calculations.

2. Proposed model of electric networks of 110 kV and above, taking into account the reliability of power supply allows to estimate the parameters of unbalance when single-phase short circuits in networks with grounded neutral and the resulting asymmetry of the currents in the healthy phases on the module and the currents reverse order and carry out the input modes allowable under asymmetry region by selectively targeted restrictions loads of customers in compliance with terms of at least economic damages.

3. Comparison of results of work of electric networks of 110 kV and above in incomplete phase modes and their work with shutdown of the damaged elements allows to justify economically the most effective mode of their functioning in the emergency situations arising at steady single-phase short circuits.

References

1. Methodical instructions on application of incomplete-phase modes of operation of the main electrical equipment of electrical installations of 330-1150 kV. M.: ORGRES, 1999.
2. Nepomnyashchiy V.A. Economic and mathematical model of reliability of power systems and their electric networks //Electricity, 2012, №2, pp.5-16.

3. Nepomnyashchiy V.A. Economic losses from power supply violations. -Moscow: Publishing house MEI, 2010, 190 p.
4. Losev S.B., Chernin A.B. Calculations of electrical quantities in asymmetric modes of electrical systems. -Moscow: EAI, 1983.
5. Dunaeva N.P., Kontorovich A.M., Shcherbachev O.V. Method of calculation of non-full-phase steady-state modes of complex electrical systems: Proceedings of LPI, 1980, No. 369, pp. 3-5.
6. Sokolov N.I. Construction and application of complex substitution schemes for complex asymmetric circuits //-Electricity, 1949, №8, pp. 21-28.
7. Popov V.A. Research and development of methods of calculation of electric quantities for relay protection at complex damages in electrical systems: diss. cand. tech. sciences. -Moscow: MEI, 1972, 32 p.
8. Berman A.P. Development and implementation of the method of analysis of asymmetric modes of complex electrical systems using phase coordinates: diss. cand. tech. sciences. -Moscow: VNIIE, 1984, 18 p.
9. Ragozin A.A., Misrikhanov M.Sh., Popov V.A., Yakimchuk N.N., Medov R.V. Modeling of electric power systems in solving problems of asymmetric modes // Improving the efficiency of power systems: Proceedings of IHEU, Vol. 4, -Moscow: Energoatomizdat, 2001, pp. 51-62.
10. Misrikhanov M.Sh., Ragozin A.A., Popov V.A., Kushkova E.I. Method of calculation of steady-state incomplete-phase modes of equipment 110 kV and above in electrical systems // Improving the efficiency of power systems: Proceedings of IHEU, Vol. 4, -Moscow: Energoatomizdat, 2001, pp. 62-69.
11. Fazylov H.F., Nasyrov T.H., Tyncherova E.L., Radionova O.V. Modeling of complex symmetric regimes in problems of emergency control and reliability of power systems // Mathematical modeling in power engineering: Abstracts of the all-Union scientific and technical conference, Part 4, Kiev, 1990, p. 142.
12. Nasirov T.H., Radionova O.V. Modeling of modes of electrical systems. -T.: Fan va texnologiya, 2016, 336 p.

E3S Web of Conferences **139**, 01013 (2019)
RSES 2019
https://doi.org/10.1051/e3sconf/201913901013

The element of digital transformation of regional network – an aggregator

Felix Byk[1,*] and *Lyudmila* Myshkina[1]

[1] Novosibirsk state technical university, Karla Marksa, 20, Novosibirsk, Russia

Abstract. PJSC "Rosseti" is forced to find new ways of getting additional income. It's caused by the insufficient financing of network. But a lot of known and supported by network organizations initiatives are leading to the increase of network component in electricity cost for consumers. By the way the new subject, called demand side response aggregator in the UPS, is appeared in power engineering. This problem can be partially solved by the large regional network companies in case of them being an aggregator. At the same time, the efficiency of the aggregator function increases in case of using local power systems as a subject of the retail market as a regulatory resource. It'll let to avoid the decrease of the power consumption and transmitted power volume, to increase the reliability and the quality of power supply due to reduce of power losses in network. It's obvious, that the appropriate automation is needed as an element of the digital transformation of the electric network.

1 Introduction

The implementation of functions related to the power supply to the terminal consumers was the main source of income for network organizations up to this day:

- Transmission and distribution of electricity;
- Technical connection of new consumers,

The known insufficient financing of network organizations is caused by the presence of state regulation and the lack of investment attractiveness. This aspect forces them to search for additional sources of income and new technical solutions to fulfill their own functions related to ensuring of reliable and high-quality power supply. It explains the following initiatives of PJSC "Rosseti":

- Initiation of the fee for reserved electricity;
- Inclusion of an investment component in the tariff for the consolidation of existing independent territorial network organizations;
- Initiation of differential transmission tariffs in the UPS.

These initiatives are supported by the Ministry of Power Engineering.

However, according to the Ministry of Economics, the introduction of fees for reserved electricity will lead to a slowdown of the economy and multibillion consumer losses. Federal Antimonopoly Service considers that the differentiation of the Federal Network Company tariff will make the industry leave the UPS. It will also lead to the decrease of consumption and the increase of cross subsidization and tariffs [1]. The increase of consumers' payments will not be more than 138 billion rubles. It will reduce the physical volume of industrial production on 0.5 percent and the production cost will increase on 5-9%. The total economic losses in 2020–2024 may reach the amount of 686 billion rubles (in prices of 2019) amid a decline in industrial investment by 0.18% till 2024.

Obviously, the containment of these network initiatives is an attempt to protect industry, communal services and public sector employees, and to force the network monopoly to increase its own efficiency, primarily due to the digital transformation of the industry. Otherwise, the UPS turns into a source of uninterrupted growth of the tariff load for the industry from a source of uninterrupted supply of electricity and stimulates the leaving to their own generation. Also it's obvious that the PJSC Rosseti got obligations to carry out the digital transformation of the network due to the "Digital Transformation" concept adoption in 2018. These obligations for sure require additional funds.

But the new subject, called demand side response aggregator in the UPS, is appeared in power engineering in 2019. Any subject of power engineering, including regional electric network companies, can perform its functions according to the [2]. The performance of aggregator functions by regional electric network companies may be accompanied by the extraction of significant technical and economic effects. The amount of these effects may be enough for the digital transformation.

2 Demand side response aggregator creation

Demand side response aggregators creation in Russia is considered as a solution of the insufficiency of managed resource problem to cover peak loads. This creation is carried out in accordance with the Decree of the Government of the Russian Federation adopted in March 2019 [2]. The project was initiated by the System Operator as part of the implementation of the EnergyNet roadmap and consists of two stages. Launch of pilot projects for the providing the system reliability by aggregator is supposed to be carried out during 2019-2020. At the same time, the volume of change in participants' demand in pilot projects is limited to 0.1% of the capacity demand [3, 4]. The direct participation of aggregators in the wholesale market is expected at the second stage. To embody it, during the pilot projects technical and technological solutions will be tested, demand side response aggregator for retail consumers will be formatted and active electricity consumers, able to adapt their consumption to the conditions in the energy system, will be created. It will ensure the participation of aggregators in the wholesale market of electricity and power from 2021 [5].

Substantially, demand side response aggregators are organizations that identify and coordinate the ability of a group of end users to manage their load and convert it into goods and services in the markets of electricity, power and system services. The main goals of managing demand for electricity are to reduce the peak load in the power system. It may be accompanied by decrease of prices on the electricity market and the prevention of excessive capital-intensive construction of power plants and networks. Demand management has become a full-fledged tool to ensure the balance of supply and demand in power systems of many countries over the past decade. According to Navigant Research, the global market volume for demand management goods and services in 2016 was 39 GW, and 28 of them are placed on the North America. In those 28 GW 21 GW are provided through demand management programs for commercial and industrial consumers and 7 are provided due to programs for household consumers and small businesses. The usage of this resource in the US power system avoids investments of 270 billion dollars (in prices of 2016) in the development of power engineering [6, 7].

The demand management resource in Russia may reach 13 GW. 6-10 GW are placed in the first price zone and 2-3 GW in the second one. The common economic effect of demand management in Russia is expected to be about 67-105 billion rubbles a year in the future.

System-wide effect will be shown by stabilization of electricity prices, alignment of the loading mode of generating and network equipment, which is manifested by increased durability, reduced requirements for controllability. Those factors are important for the digital transformation of control systems [8].

3 First results

Demand side response aggregators perform the commercial function of combining distributed demand management resources according to the developed rules [2]. Aggregators are acting as intermediaries between consumers providing their ability to change consumption and the buyer of demand management products. Aggregators acquire load reduction services from many consumers with different characteristics. Distributed aggregated load is one of the key benefits of aggregation.

Requirements for the aggregator [1]:

- The amount of controlled by the aggregator power must be at least 1 MW;
- Aggregator determines independently the number of consecutive hours per day in the amount of 2 or 4 hours. In these hours aggregator must be ready to provide the corresponding demand management services;
- The coefficient of accounting of the services volume will be 0.5 at work of 2 hours a day and 1 for 4 hours depending on the chosen number of hours.

Aggregators ensure their participation in several market segments by getting the necessary generalized resource for demand management. Those markets are electricity market, the day-ahead market, balancing market, the system services market of ensuring system reliability. The above is accompanied by the extraction of economic effects for the aggregator.

The composition of the aggregators is determined in the process of competitive selection. The criterion is to minimize the total cost of the total volume of the services provision for demand management of electricity [9]. According to the [2], the limited price of the provision of demand management services should not exceed the value of the price of power for the corresponding price zone.

The payment mechanism by the aggregator of the service for the resource provision is not fully formed and is developing as part of the pilot project.

Pilot aggregators' selection is held quarterly from the third quarter of 2019 (from July till September). The first selection showed interest in the new mechanism. The number of applications in the 1st price zone exceeded the common volume fixed by the Government of the Russian Federation for 2019. 56.18 MW for the first price zone and 8.5 MW for the second were submitted for selection, Table 1.

Table 1. Common data for the first competitive selection of demand side response aggregators.

Indicator	First price zone	Second price zone
Limit of managed demand, MW	39	11
Volume of applications, MW	56	8.5
Limit price, rubles/MW per month	795 659	622 841
Actual limit price of selected applications, rubles/MW per month	600 000	620 000
Average price of selected applications, rubles/MW per month	481 355	580 773

Demand management services contracts were concluded with 22 participants. There are representatives of various power energy subjects among the participants in the selection, including two network organizations – PJSC "Interregional Distribution Network Company of the South", PJSC "Interregional Distribution Network Company of the North Caucasus". 90% of the selected applications use the load of small and medium-sized enterprises as a regulatory resource [10].

The results of the load aggregators' work are quantitatively expressed as a decrease in the price of electricity on the day-ahead market: the first mechanism operation reduced the price by 4.7% [11]. At the same time, the value of actual obligations fulfillment amounted to only 72% on 15.07.2019 of the total value of the planned price-dependent decrease in the volume of electricity purchase. Readiness for unloading was declared for this obligations. It indicates the demand management effectiveness from the perspective of consumers and the system as a whole.

4 Key effects for network companies

It is important to note that demand management represents the unloading of consumer equipment at certain hours and not its redistribution in time. Therefore, this is accompanied by a decrease in power consumption and transmitted power through networks. It conflicts with the network companies interests if they do not assume the aggregator role. It will at least avoid a decrease in income.

According to the current regulatory legal acts, the aggregator is allowed to attract consumers with a generating capacity less than 25 MW. It is customary to refer them to local energy systems created based on the distributed generation which received the status of retail market entities due to technological connection to the grids of Distribution Network Company.

Attracting local energy systems' regulatory resources allows to redistribute the load schedule in time saving the volume of power consumption in the region and the size of electricity transmitted through distribution networks. Network companies will essentially allow crowding out the capacities of inefficient power plants and unloading Federal Network Company's networks by laying the peak load of small and medium-sized businesses and household consumers on the free generating capacities of local power systems. It will reduce the cost of network development.

According to the experts, reducing the construction of new network capacities need by reducing peak consumption can reach 120–180 MVA per year. It'll reduce the need of investment by 2-3 billion rubbles per year [8]. Obviously, this is in line with national economic development goals.

If the energy sales organizations perform the aggregator functions, it's possible to unload already lightly loaded substations. It contradicts with the interests of Distribution Network Company. Obviously, regional network companies are better informed than power supply companies on what 110/10 kV substations the peak load is running. The peak load causes an equipment overload. That's why Distribution Network Companies can determine the places where the availability of a resource is most appropriate for managing demand and reducing peak load better than other power engineering subjects.

Network companies have certain priorities for performing the function of demand side response aggregator as subject of the power engineering. In addition, they have an interest in obtaining additional "hidden" effects from attracting excess generating capacities of local power systems as an aggregated resource. Distribution Network Companies are able to indicate where the availability of the regulatory resource is effective, including a professional consumer in the retail market as the local power system. Therefore Distribution Network Companies can stimulate their appearance in the coverage area of "red" 110/10 kV substations as a demand side response aggregator. It will significantly increase the technical and economic efficiency of the networks functioning.

The "hidden" effects of Distribution Network Companies will not only be equalization of equipment loading. This is associated with the costs of their maintenance and services. Involving the generation of local power systems as an aggregated control resource will allow Distribution Network Companies to increase the reliability and quality of the services provided for the transmission of electricity ensuring their connection to the regional network [12]. It also provides several economic benefits. Connecting the local power systems provides next effects:

- The possibility of mutual reservation;
- The ability of using generator excitation regulators to maintain the voltage level in the nodes of the distribution network;
- The ability of performing multi-agent voltage regulation;
- Reducing the losses in network.

The growth of network technical efficiency allows to implement reliability and quality management and to take effect by increasing the tariff up to 2% [13], what is leading to income growth for Distribution Network Companies. It is important to note that using redundant local power systems capable of operating in island mode on principles of self-balance as a resource for aggregator implies the presence of appropriate automation, which ensures parallel operation of local and unified power systems. Their integration is beneficial not only for Distribution Network Companies, but also for distributed generation as the main energy source of local power systems. First of all, it's caused due to the increase of cogeneration source's Installed Capacity Utilization Factor. It reduces the cost of heat supply to consumers.

It should be noted that only the integration of the island local power system allows it to acquire the status of a retail market entity operating in the professional consumer mode, as a producer-consumer of electricity. This process depends largely on the Distribution Network Companies defining the technical conditions for connecting the local power system to the external network, which may include requirements for appropriate automation.

The specified requirements should allow to use regulatory resources of the local power system to perform the function of the aggregator, to control voltage in the distribution network nodes in the mode of parallel operation and to reduce restrictions in the power supply of responsible consumers in case of network equipment failures and power supply failure from the 110/10 kV substations [14, 15].

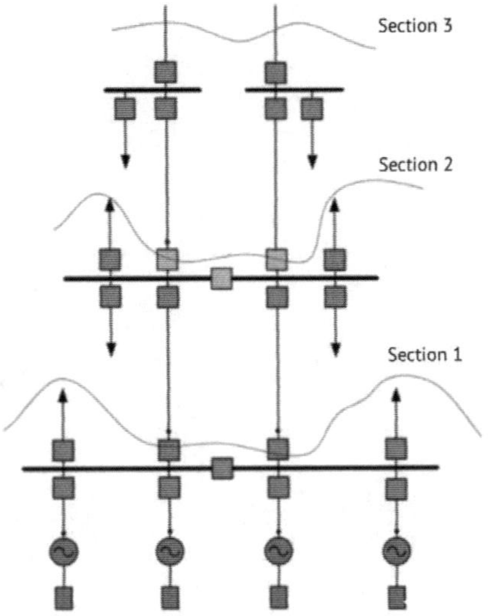

Fig. 1. Managed and controlled sections in integration

Three main sections for the automation system can be highlighted, each of them is used for management and accounting:

1. Operation mode and output electric power of the power source and its separation according to technological conditions from the local power system, afterwards power supply to consumers is carried out from the external network;
2. Supplied/consumed energy by the local power system, as a subject of the retail market operating in the professional consumer mode, and balanced separation of the local power system to the island mode according to the technological conditions of the external network operation;
3. Performing the requirements of the demand side response aggregator for the regulatory resources provision.

Nowadays, the manufacture and pilot operation of automatics ensuring the integration of the local power system to the external network is being completed [15]. Its development was carried out on the order of the network company JSC "Tyumenenergo" which is a 100% subsidiary of PJSC "Rosseti". Automation is introduced in JSC "Regional Electric Company" as a pilot object. JSC "Regional Electric Company" is Distribution Network Company in Novosibirsk region. It is obvious that this new technological solution corresponds to the technical policy of PJSC "Rosseti" aimed at digitalization of the network [16]. The indicated transition to intelligent energy will increase their economic efficiency due to additional income from the market of system services, competitive power selection and wholesale electricity and capacity market. Network companies will perform the functions of the aggregator in this transition. According to experts estimates expected economic effects will reach up to 67-105 billion rubbles per year and the network component won't increase in the final cost of electricity for consumers.

5 Conclusions

It will be possible to extract significant technical and economic effects that may be sufficient for the digital transformation when the demand side response aggregators will advent and network organizations will start acting as them.

Attraction of power systems as an adjustment resource will allow:

- Regional network companies to receive additional incomes from the market of system services, competitive power selection and wholesale electricity and capacity market;
- To increase income from improving the reliability and quality of their electricity transmission services;
- To reduce capital investment in development and operational costs.

Recoverable economic effects from cost reduction and additional income from the connection of local power systems will allow PJSC "Rosseti" to solve problems of digital transformation in the network without increasing the rate of tariff growth and the network component in the final price of electricity for consumers.

References

1. T. Dyatel Kommersant newspaper **143,** 1 (2019)
2. Government of the Russian Federation. (2019, Mar. 20), N. 287, On Amendments to Certain Acts of the Government of the Russian Federation on meeting the needs of the authorities in ensuring the unified energy system of Russia, as well as improving the price dependent mechanism for reducing energy consumption and providing energy supply services system reliability
3. EnergyNet national technology initiative roadmap [Online]. Available: http://www.nti2035.ru/markets/docs/DK_energynet.pdf (Accessed on: Jul. 20, 2019)
4. The system operator familiarized electricity consumers with the prospects of introducing demand management technologies in the retail market

[Online]. Available: http://www.odusv.so-cdu.ru/index.php?id=dr_news_view&tx_ttnews[tt_news]=12768&cHash=0133b4a449 (Accessed on: Jul.28, 2019)

5. N. Sidorova Energorynok **132**, 28-32 (2015)
6. Peak Load Management Alliance "Demand Response Acronyms & Glossary. Training Course Resource" [Online]. Available: https://www.peakload.org/assets/PLMADR%20_AcronymsGlossary_053117.pdf (Accessed on: Jul. 10, 2019)
7. Global Demand Response Capacity is Expected to Grow to 144 GW in 2025 [Online]. Available: https://www.navigantresearch.com/news-and-views/global-demand-response-capacity-is-expected-to-grow-to-144-gw-in-2025 (Accessed on: Jul. 05, 2019)
8. F. Opadchiy, D. Holkin *Demand management in the Russian Power Engineering: opportunities: expert and analytical report* [Online]. Available: https://onedrive.live.com/?authkey=%21AL1vFI6tP9alF9E&cid=7589EC31D3F642F1&id=7589EC31D3F642F1%218910&parId=7589EC31D3F642F1%217875&o=OneUp (Accessed on: Jul. 13, 2019)
9. Notification of competitive selection for electricity demand management services [Online]. Available: https://so-ups.ru/fileadmin/files/company/markets/dr/notice/notice_dr_110619-1.pdf (Accessed on: Jun. 26, 2019)
10. Decisions of the commission on selection of power engineering entities providing demand management services [Online]. Available: https://so-ups.ru/?id=markets_asm_com_dr_dec (Accessed on: Jul. 06, 2019)
11. First selection of market participants in demand management in Russia: results and conclusions [Online]. Available: https://medium.com/internet-of-energy/rus-demand-response-b2621247548a (Accessed on: Jul. 27, 2019)
12. F. Byk, L. Myshkina Rudenko International Conference "Methodological problems in reliability study of large energy systems", **69** book 2, 81-88(Irkutsk, 2018)
13. Government of the Russian Federation (2009, Dec. 31), N 1220, About definition of the long-term rates of reliability indicators and quality of the delivered goods and the rendered services applied at establishment
14. P. Ilyushin Electro, Electrotechnics, power engineering, electrotechnical industry **4**, 19-25 (2011)
15. E. N. Gezha, V. E. Glazyrin, G. V. Glazyrin, A. I. Marchenko, R. Iu. Semendiaev, A. G. Fishov [etc.] Releishchik **2**, 24-31 (2018)
16. Conception Digital Transformation 2030 [Online]. Available: http://www.rosseti.ru/investment/Kontseptsiya_Tsifrovaya_transformatsiya_2030.pdf (Accessed on: May 05, 2019)

https://doi.org/10.1051/e3sconf/201913901014

Assessment of reliability and safety of work large pumps of machine irrigation systems

Tolyagan Kamalov[1,*] , *Solixjon* Halikov[2]

[1]Institute of Energy of the Academy of Sciences of the Republic of Uzbekistan, 700100, Tashkent city, Republic of Uzbekistan
[2]Tashkent State Technical University, 700111, Tashkent city, Republic of Uzbekistan

Abstract. The article provides an assessment of the reliability and safety of large pumps of machine irrigation systems in case of violations occurring in abnormal conditions and mechanical malfunctions of pump elements.

In terms of energy consumption, pumping stations of machine irrigation systems (PSMIS) of the Republic of Uzbekistan are one of the energy-intensive facilities in the region. The installed capacity of electrical equipment at PSMIS is more than 2700 thousand kW with a consumption of 8 billion kW of electricity or about 17% of the total amount of electricity consumed in the republic.

Currently, there are more than 1300 pumping stations (PS) and over 9000 vertical drainage wells operating for irrigation, including such large PS of the Karshinsky main canal with PS No.1-No.6 with a capacity of 200 m^3/s each, Khamza-2 (150 m^3/s), Kizil-Tepa (150 m^3/s), Amu-Zang-1 (37.2 m^3/s), Amu-Zang-2 (37.2 m^3/s), etc. Along with large NS many medium and small pumping stations are in operation.

Long-term operation of pumps installed in irrigation pumping stations led to a deviation of their technical characteristics from the factory ones. Existing PSs and cascades of pumping stations (CPS) of the country's machine water-lifting systems are age-appropriate for the wear-out failure phase. This period of operation is characterized by an increase in the rate of failures and accidents, an increase in the volume of repair work and reconstruction and modernization of structures, structures and equipment and, accordingly, an increase in the costs and expenses of electric energy per unit volume of pumped water. And here is the question of determining the reliability and safety of large pumps.

The terms "reliability", "safety", "danger" and "risk" are often confused, and their meanings overlap. In [1], the terms "safety analysis" or "hazard analysis" are used as equivalent concepts. Along with the term "reliability analysis" they refer to the study of both operability, equipment failure, loss of operability, and so on the process of their occurrence. If, as a result of the analysis, it is necessary to determine the parameters characterizing safety, it is necessary, in addition to equipment failures and system malfunctions, to consider the possibility of damage to the equipment itself or other damage caused by them. If at this stage of the safety analysis the possibility of failures in the system is assumed, then a risk analysis is carried out in order to determine the consequences of failures in the sense of damage to equipment and the consequences for people near it. One of the goals of risk analysis is to assess the frequency (probability) of these or other possible subsequent other possible consequences due to failures in the system.

At present, methodological approaches and the corresponding regulatory documentation on the assessment, determination and standardization of safety and operational reliability of pumps have not been developed.

Risk assessment can be determined in monetary terms - arbitrary units or in the point system. In this case, we assess the risk assessment in monetary terms - conventional units, in terms of:

$$R_P = \sum_{i=1}^{n} \left\{ \left[Q_{Pi} + M \cdot \Delta Q_{Pi} \cdot T_i \right] \cdot P_{Pi} + Q_{Pi}^{EX} \right\} \frac{1}{T_{year}} \text{[(monetary unit)/year]} \quad (1)$$

where $P_{Pi} = f$ (EFR, P),

Q_{Pi} – damage caused to the pump by the *i*-th event [monetary unit (денежный единица)],

EFR – expected failure rate;

ΔQ_{Pi} – loss of water flow per hour of the pump from idle time at the *i*-th event [m^3/hour];

T_i – pump idle time at event *i*;

M – the cost of one m^3 of water [monetary unit / m^3];

P_{Pi} – probability of occurrence of the *i*-th event on the pump;

P – probability of an initial event;
Q_{Pi}^{EX} – external damage to crop yields due to water loss during the *i*-th event at the pump [monetary unit];
T_{year} - overhaul period of the pump unit for a year [year].

The resulting risk values are divided into 4 security categories in the amount of conventional monetary units, and the amount of economic damage is taken as:

Security category	Economic and social consequences
I	Economic damage can be very significant. R ≥ 100 million monetary unit/year
II	Economic damage can be significant. R ≥ 1 million monetary unit/year
III	The amount of possible damage is negligible. R ≥ 100 thousand monetary unit/year
IV	Minimal damage in the future is practically absent. R ≤ 100 thousand monetary unit/year

Thus, the derived expressions for calculating the risk of the pump allow us to assess the risk and monitor their safety indicators.

Studying the causes of disturbances in violations of normal operation (VNO) of large pump irrigation pumps indicates the need to develop an accident risk management methodology and safety assessment that will allow us to assess the balance between the extent of possible damage from potential accidents of this system and its technical-economic advantages.

In the existing method of safety analysis, in order to correctly introduce complex indicators of the risk type characterizing operational safety, a violation development model is used, which is represented by a right-handed dichotomous "event tree" [2].

In the considered risk calculation methods, in many cases, the conduct of a probabilistic safety analysis (PSA) (determining the function P_{Pi} = f (ОЧО, P)) in full can be difficult, especially in the absence of information about individual processes. Risk calculation is accompanied by a high degree of uncertainty that requires a significant investment of time.

From this point of view, it was proposed to use artificial intelligence systems (AI) based on artificial neural networks, genetic algorithms, expert systems, and systems of odd logic for conducting PSA pumps [2].

The basis of each artificial neural network is elements (cells) imitating the work of brain neurons (hereinafter referred to as a neuron an artificial neuron, an artificial neuron cell, an artificial neural network cell). Each neuron represents a pump element and events occurring as a result of VNO and is characterized by its current state by analogy with nerve cells in the brain that can be excited or inhibited. It has a group of synapses - unidirectional input connections connected to the outputs of other neurons, and also has an axon - an output connection of this neuron, from which (excitation or inhibition) goes to the synapses of the following neurons. Each synapse is characterized by the value of synoptic connection or weight W_{ij}, which is equivalent in physical meaning to electrical conductivity [3].

The current state of a neuron is defined as the weighted sum of its inputs:

$$S_{ij} = \sum_{\substack{i=1 \\ j=1}}^{n} X_{ij} \cdot W_{ij} \qquad (2)$$

The output of a neuron is a function of its state:

$$y = f\,(s) \qquad (3)$$

The nonlinear function *f* is called activation and can take the following forms: unit jump function; linear threshold (hysteresis); sigmoid.

For our case, the type of function is chosen - sigmoid - hyperbolic tangent, here the input of the function will be the running time t, the parameter of the function is the failure rate - λ, and the output is the probability of failure operation - P (t). The theory of fuzzy sets is used to evaluate criticality indices.

Violation of the normal operation of the pump elements occur due to abnormal operation of the pump and mechanical malfunctions of its elements [4].

Fig. 1 shows the existing 9 types of pump failures, and Fig. 2 shows 16 types of initial disturbance events during abnormal pump operation, which are indicated by neurons, respectively, S_{11}-S_{19} and S_{21}-S_{216}. Fig. 3 shows the existing 10 types of failures due to mechanical malfunctions of the pump elements, and Fig. 4, 5 - 26 types of initial events of pump disturbances during mechanical malfunctions of its elements, which are indicated by the neurons S_{31}-S_{310} and S_{41}-S_{426}, respectively.

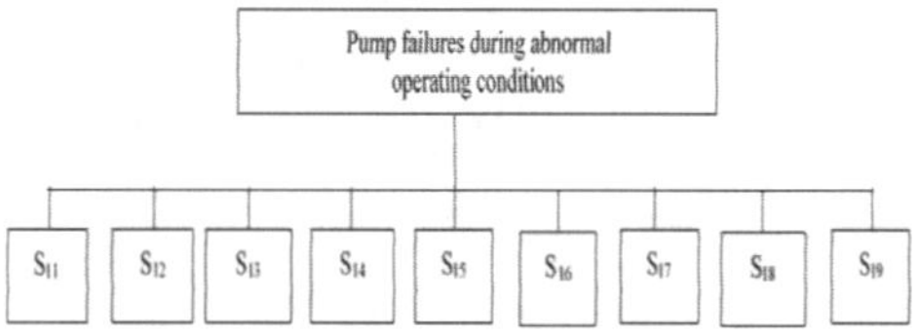

Fig. 1. Failures of the pump during abnormal modes of operation:

S_{11} - the appearance of vibration and an increase in the runout of the pump shaft, accompanied by shocks, knocks in the impeller; S_{12} - the pressure and pump flow pulsate and do not correspond to the operating mode; S_{13} - vibration with prevailing cavitation frequencies of 800-20000 Hz; S_{14} - the pressure pulsates and is higher than the permissible one, the supply is much less than the calculated one; S_{15} - the unit vibrates strongly with cavitation frequencies; S_{16} - the pump does not supply water when the motor is overloaded, permissible hydraulic resistance of the pipeline and backwater; S_{17} - enhanced vibration at frequencies that are multiples of the blade and rotation frequencies; S_{18} - the pump does not provide the required pressure, vibration at the blade frequencies; S_{19} - the pump does not provide the required flow. Vibration within acceptable limits.

 https://doi.org/10.1051/e3sconf/201913901014

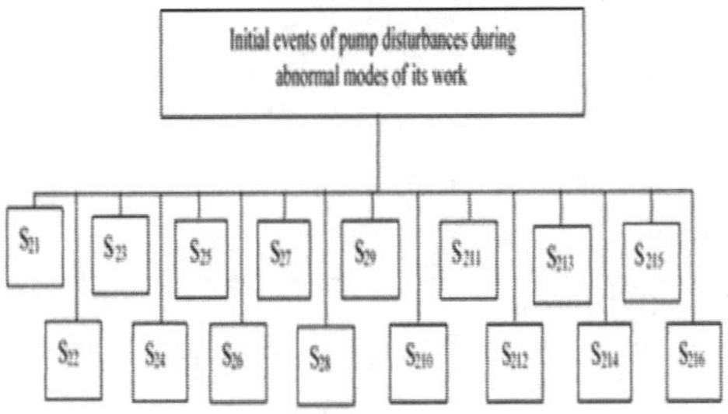

Fig. 2. Initial events of pump disturbances during abnormal modes of its operation:

S_{21} - clogging grids; S_{22} - air entering the suction pipe; S_{23} - the formation of air bags in the pipelines; S_{24} - siltation of the suction pipe; S_{25} - critical cavitation mode with characteristic knocks similar to stone blows against iron; S_{26} - a sharp curve of the decline in the level in the chamber with funnels at a low level of the lower buff; S_{27} - increased flow swirl in the advance chamber in the plan; S_{28} - surge phenomena; S_{29} - hydraulic resistance exceeds permissible; S_{210} - the pump operates in braking or reverse mode during reverse rotation; S_{211}- fragment of the blades and the alignment of the pump rotor; S_{212} - significant wear on the ends of the impeller blades and the chamber; S_{213} - wear of the sealing rings of the front wheel disc; S_{214} - the non-working angle of rotation of the axial pump vanes is set; S_{215} - the valve on the pipeline is covered or fully open; S_{216} - Vacuum stall valve is leaking.

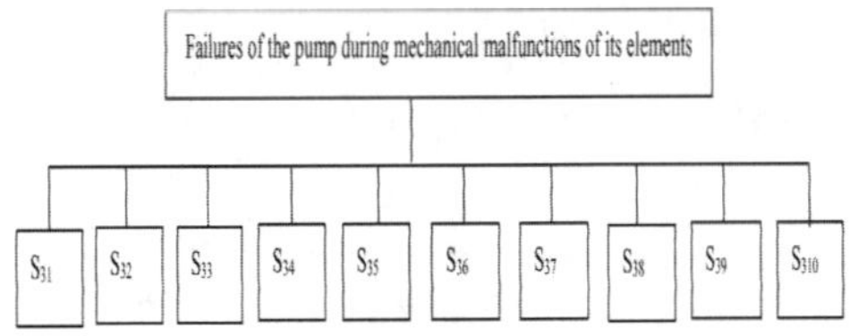

Fig. 3. Failures of the pump during mechanical malfunctions of its elements:

S_{31} - heating water in the reservoirs of water lubricated bearings; S_{32} - invalid shaft runout; S_{33} - increased power while providing working feed and pressure; S_{34} - oil seals allow water to pass above normal; S_{35} - the appearance of smoke or the smell of burning coming from ligno-foil and quick wear of the liners due to the increased content of abrasive suspended matter in water lubricant; S_{38} - Inadmissible heating of the thrust bearing; S_{39} - temperature increase in babbit bearings and heel with oil lubrication; S_{310} - the oil system does not provide the bearing with the necessary amount of oil (for pumps with forced lubrication).

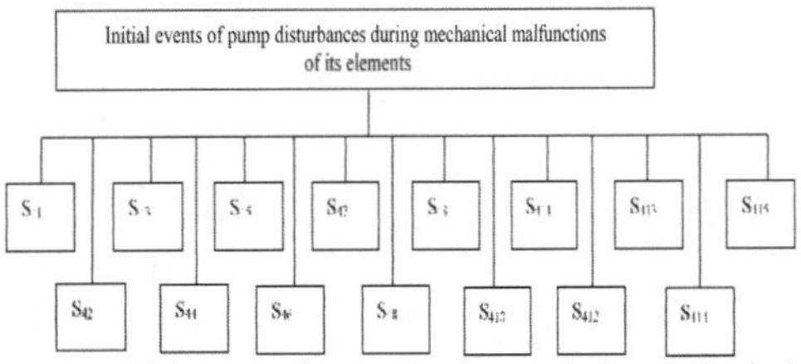

Fig. 4. Initial events of pump disturbances during mechanical malfunctions of its elements:

S_{41} - coagulation of the blades of the wheels of axial pumps at different angles; S_{42} - violation of the alignment of the axis of the shaft; S_{43} - inaccurate alignment of the shaft and bearings; S_{44} - getting foreign objects into the flow part; S_{45} - bearing displacement; S_{46} - the rotor of the pump (unit) is poorly balanced or the balance is broken; S_{47} - incorrect boring of couplings; S_{48} - wear and grazing of seals, uneven wear of the wheel blades; S_{49} - small clearances between the shaft and bearing shells; S_{410} - strong tightening of the seals or separation of the seal; S_{411}- a shirt or shaft surface has grooves due to the tightness of the gland; S_{412} - cessation of supply of technical clean water to bearings; S_{413} - grease ring jam, oil contamination or leakage; S_{414} - cessation of cooling water supply; S_{415} - small gaps in the bearing shells.

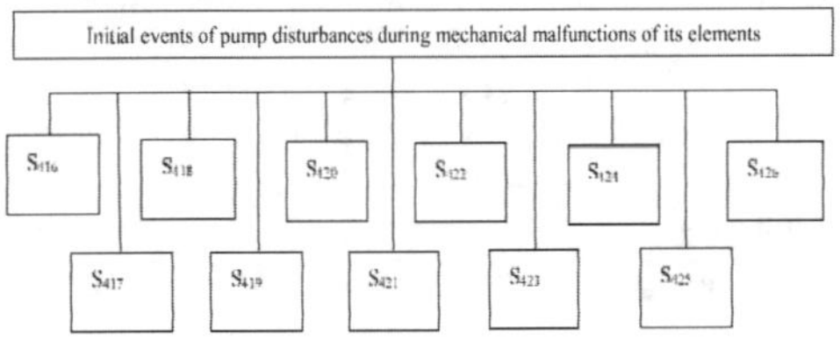

Fig. 5. Initial events of pump disturbances during mechanical malfunctions of its elements:

S_{416} - high abrasive content in water lubricant and pumped water, especially sulfate or chloride class; S_{417} - too tight pinching of the balls between the support rings; S_{418} - uneven fit of pads or ridges in thrust bearings; S_{419} - contamination or mismatch of the brand of oil, insufficient quantity; S_{420} - Incorrect clearance between guide bearings and shaft; S_{421} - misalignment between the bearings and the heel mirror; S_{422} - uneven loading of thrust bearing segments; S_{423} - the oil pump does not supply the required amount of oil, end clearances have been developed; S_{424} - the oil system is clogged (oil pipe, filter, oil cooler); S_{425} - insufficient supply of cooling water to the oil cooler; S_{426} - Water is detected in the oil, flow in the oil cooler.

Based on the above disturbance events, a block diagram of the PSA pump neural network is compiled for abnormal operating conditions, and for the pump neural network with violations of its operating mode, the expressions can be written:

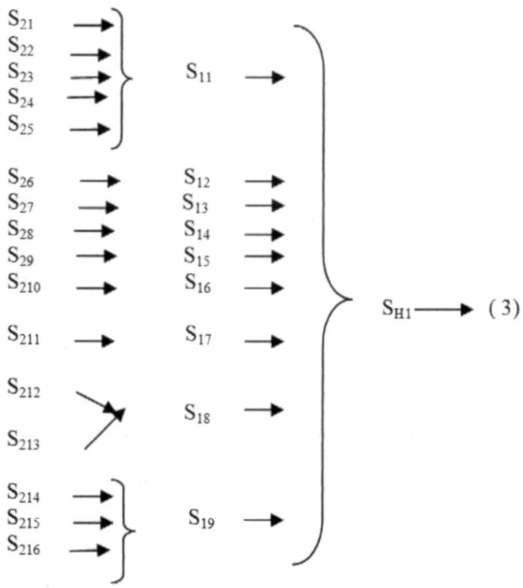

Here, each violation can be represented from series-connected neurons characterizing the path of the failure, for example, from the expression it can be imagined that one of the following events develops from the failure due to the appearance of vibration and increased runout of the pump shaft, accompanied by knocks, knocks in the impeller (S_{11}): driving in gratings (S_{21}); air entering the suction pipe (S_{22}); the formation of air bags in the pipelines (S_{23}); siltation of the suction pipe (S_{24}); critical cavitation mode with characteristic knocks similar to stone blows against iron (S_{25}). Here, the set of neurons S_{21} ... ... S_{25} characterizes the initial events, the set of signals X_{21} ... ... X_{25} is the output of these neurons, these output signals correspond to the signals arriving at the synapses of the biological neuron and each of them is multiplied by the corresponding weight W_{21} .... W_{25} and arrives at the input of a failure neuron. Each weight corresponds to the "weight" of one biological synaptic connection S_{11}. Neuron S_{11}, corresponding to the body of the biological element, adds the weighted inputs algebraically, creating an output, which we will call NET. In vector notation, this can be written as follows:

$$NET_{11} = X_{11}W_{11} \tag{4}$$

The signal NET_{11} then, as a rule, is converted into the activation function F_{12}

$$OUT_{11}=K(NET_{11}) \tag{5}$$

and gives an output neural signal.

The value of the function will be the probability of failure of the pump $P_{11}(t)$ during abnormal modes of its operation. From the expression (3) it follows that from the outputs of the neurons S_{12} - S_{19} we get the probability of failure-free operation P_{12} (t) - P_{19} (t). Here, many S_{12} - S_{19} neurons characterize failures, many X_{12} .....X_{19} signals are the output of these neurons. These output signals correspond to the signals arriving at the synapses of the biological neuron and each of them is multiplied by the corresponding weight W_{12} .... W_{19} and goes to the input of the failure neuron. Each weight corresponds to the "weight" of one biological synaptic connection S_P. The neuron S_{P1}, corresponding to the body of the biological element, adds the weighted inputs algebraically, creating an output, which we will call NET. In vector notation, this can be written as follows:

$$NET_{H1} = X_{H1}W_{H1} \tag{6}$$

From the output of the neuron S_{P1}, we obtain the values of the probability of failure-free operation of the pump $P_{P1}(t)$ under abnormal modes of its operation.

Knowing the probability of failure-free operation, we can diagnose the technical condition of the pumps, calculate the risk, determine the severity of each violation and evaluate the safety of each pump element, addressing each of them, display their status on the computer screen and effectively identify emergency factors, take the necessary emergency corrective measures aimed to increase the safety of the pump.

References

1. Henley E.J., Kumamoto H. Reliability and risk assessment: Per. from English V.S. Syromyatnikov, G.S.Demina. Edited by V.S. Syromyatnikov. - M.: Mechanical Engineering, 1984.- p.528
2. Statistical methods of safety analysis of complex technical systems: Textbook / L.N. Alexandrovskaya, I.Z. Aronov, A.I. Elizarov et al .; Ed. V.P.Sokolova.- Logos, 2001.
3. Kamalov T.S., Halikov S.S. A probabilistic safety analysis of large pumps of machine irrigation systems using a neural network // Uzbek Journal of Informatics and Energy, 2011, No. 1, p. 47-55.
4. Borisov E.S. The main models and methods of the theory of artificial neural networks- http: // mechanoid.narod.ru/nns/base/.
5. Kiselev I.I., German A.L., Lebedev L.M., Vasiliev V.V. Large axial and centrifugal pumps. Installation, operation and repair. Reference manual. M., "Engineering", 1977.

Procedure of selecting the priorities when constructing new most important facilities in the Russian energy sector in the context of materialization of strategic energy security threats

Viktor Rabchuk [1], *Sergey* Senderov[1, 2,*],

[1]Melentiev Energy Systems Institute of Siberian Branch of the Russian Academy of Sciences, 130 Lermontov str., Irkutsk, Russia
[2]Federal State Budget Educational Institution of Higher Education «Irkutsk National Research Technical University», 83 Lermontov str., Irkutsk, Russia

Abstract. The paper is devoted to finding ways to solve the problem of a scientifically based approach to the selection of priorities in the construction of new most important energy facilities in Russia for the future up to 15 years. The perspective scales of realization of strategic threats to the energy security of Russia are characterized. A list of subtasks that should be solved to form a methodology for solving the problem is shown.

1 Introduction

The paper concerns the ways to solve the problem of a scientifically-based selection of a list of most important facilities of the Russia's fuel and energy complex (FEC), which should be constructed as a priority in the medium term (up to 15 years). A most important facility from the standpoint of the development of the FEC in some perspective is the facility, the refusal to build which or its failure (in the case of construction) during operation can lead to an unacceptably significant reduction in domestic consumption of energy resources or in their export. The problem of choosing priorities, stated in the title of this paper, is becoming increasingly acute in the context of the active implementation of strategic threats to energy security (ES) in Russia. The current and expected scales of implementation of some of them are such that it is possible to talk about a possible process of containing economic growth from the energy sector in the medium term due to a significant deficit in the volume and duration of the supply of necessary types of energy [1, 2, etc.].

2 Strategic threats to Russia's energy security today and in the short term

In the list of ES strategic threats, it is worth highlighting the threat to a large extent determining the scale and dynamics of the development of other threats - the threat of a noticeable decrease in the investment opportunities of the national economy and energy sectors, in particular. Thus, the annual volume of investments during the period from 2011 to 2018 decreased: in the electric power industry - by 27% (from 893 to 648 billion rubles) [3, 4], in the gas industry - by 47% (from 647 to 340 billion rubles). Investments in the coal industry of Russia increased almost 1.5 times in 2018 compared with 2017 and amounted to 150 billion rubles [5], but earlier there was an active decline in their volumes, so in 2014 the decline in investment compared to 2012 was 37% and remained at this level until 2017 [6]. The situation is more stable in the sphere of oil production and transportation of oil and gas condensate, as well as in the field of oil refining (the total volume is about 1.4 trillion rubles [7]).

The uncertain situation with the investment opportunities of the country's energy industries is aggravated by the sharp deterioration in the capacity of large Russian companies (especially in the field of hydrocarbon production) to receive long-term low-cost borrowed funds. The increased negative consequences in the next 10-15 years will be caused by the introduction of the following strategic threats to the Russian ES:

- increase in the cost of production of Russian hydrocarbons and their transport to domestic consumers and for export (due to the depletion of the existing areas of their production and the need to enter new - more expensive - oil and gas production areas);
- the need for rapid growth in the coming years of the rate of replacement of physically and morally obsolete basic production assets (BPF) in the energy industries (in the last 10-15 years, the replacement rate of these BPF noticeably lagged behind the rate of aging, and their share today is clearly higher than the acceptable share);
- too high energy intensity of gross domestic product in Russia (it should be expected that in this perspective it will remain at the level of approximately one and a half times higher than in the Scandinavian countries and in Canada);

- deterioration of the situation for Russia, as an exporter of hydrocarbons, in world oil and gas markets (US shale hydrocarbons; expected large-scale participation in the world markets of Iran and Iraq with their hydrocarbons; continuous growth in the share of unconventional energy in total energy resources in major countries - importers of oil and gas (China, Western European countries) and the general desire of these countries to reduce hydrocarbon purchases).

In addition to taking into account the expected degree of implementation of the strategic threats to the Russia's ES listed above, the solution of the identified problem ("choosing priorities") requires a comprehensive analysis and consideration of the following factors and requirements:

- the interconnectedness and interdependence of the work of all energy systems;
- the possibility of various large-scale emergencies in the basic sectors of the Russian FEC (gas and electric power industry) in the future;
- the expected dynamics of changes in the country's domestic energy needs in the same perspective;
- the need to diversify directions and forms (pipeline gas, liquefied natural gas) of Russian gas export and maintain the volumes of this export at the highest possible level;
- the need for a significant increase in the added value of produced Russian hydrocarbons (including due to the depth of oil refining, improving the quality of oil products and the intensive development of oil and gas chemistry).

"Selecting priorities" is part of the overall task of managing the development of the country's FEC for the analyzed perspective. To a certain extent, this problem was solved in the past [8, 9] and should be solved constantly in the future. Since the middle of the current decade, the situation with investments has worsened significantly, in particular in the energy sectors. Other strategic threats to energy security began to intensify (reinforcing the situation with a lack of investment). Under these conditions, the problem of "choosing priorities" has actually become the main part of the overall task of managing the development of the Russian FEC.

3 Meaningful statement of the problem

A meaningful statement of the problem of "choosing priorities" can be formulated as follows: it is necessary to create a list of new most important facilities from the standpoint of the Russia's FEC strategic development. Such facilities should be constructed on a priority basis until 2035, taking into account the dynamics of changes in the degree of implementation of strategic threats to the country's ES. Along with this, factors mentioned above and which may also affect the composition of the selected facilities of their specified list should be taken into account too.

The task of "choosing priorities" is complex, because when solving it, along with taking into account the strategic threats of the ES of Russia and the influence of various factors and requirements. The fact of the interconnectedness of the work of all energy sectors of the country within the framework of the functioning of the Russian FEC is taken into account too. The semantic content of the requirements for the methodology for solving this problem, which needs to be developed, speaks about the scale and difficulties of solving the problem of "choosing priorities".

Firstly, this methodology should determine the methods for obtaining an adequate assessment of the degree of realization of the strategic threat "lack of investment in the energy sectors of Russia" in the future. The same applies to the assessment in relation to other strategic threats of electronic security, the implementation of which enhances the negative effect of the lack of investment.

The second thing that is required from this methodology is clear instructions on the selection mechanism of those factors and requirements that must be taken into account when solving the problem of "choosing priorities" and on methods for taking these factors and requirements into account. Among the factors and requirements taken into account, at least, there should be those that have already been indicated above (the interconnectedness of functioning of the energy industries, the expected dynamics of changes in the country's domestic energy needs, the need to diversify the directions and forms of export of Russian hydrocarbons).

The methodology should determine the principles of concretizing the concept of "unacceptably significant reduction in domestic energy resources consumption and their export", which was mentioned at the beginning of the paper when determining the meaning of the term "most important new FEC facility" for the certain perspective. Concretizing in this case will relate to quantitative indicators of the reduction in domestic consumption of fuel and energy resources and their exports in case of refuse to build this new FEC facility (or failure of an already constructed facility during its operation). The developed methodology should determine the general procedure for solving the entire considered full-scale task of "choosing priorities".

The task of "choosing priorities" is solved mainly through the use of formalized approaches with the use of expert analysis methods at certain stages. The use of formalized approaches requires a prepared mathematical formulation of both the entire problem to be solved and its individual stages, indicating the criteria and limitations used. As mentioned above, a reasonable choice of these criteria and limitations is the main requirement for the developed methodology for solving the problem.

4 The main subtasks of the general task of "choosing priorities"

Already today, in parallel with the development of the methodology mentioned above, we can begin to prepare for the fulfillment of the sub-tasks of the general task of "choosing priorities", which will have to be fulfilled. These subtasks include:

1. Formation (with appropriate justification) of the list of Russia's ES strategic threats, which will be implemented in the period up to 2035 and the scale of implementation of which can significantly affect the solution of the issues of creating new most important FEC facilities in the analyzed period.
2. The solution of the methodological issues of assessing the scope, nature of the transformation and the degree of implementation of Russia's ES strategic threats to the period up to 2035, including the development of new and adaptation of existing models to obtain this assessment. Substantiation of the procedure for taking this assessment into account in the "choosing priority" task when deciding on the creation of new most important FEC objects.
3. Assessment of the scales and nature of the transformation, until 2035, of the selected Russia's ES strategic threats using new developed and existing adapted models with an assessment of the degree to which each threat is realized in the analyzed prospect.
4. The solution of methodological issues of accounting for various factors and requirements (including the order of their choice) to solve the problem of "choosing priorities".
5. Adaptation of existing models to solve the problem of managing the Russian FEC development in order to obtain a formalized solution to the considered problem of "choosing priorities".
6. The information content of the adapted models in accordance with the essence of the substantive and mathematical formulations of the problem of "choosing priorities" and using the results of assessing the expected degree of implementation of Russia's ES strategic threats up to 2035. Here we should also select the necessary information on various factors and requirements that can impact on the "choice of priorities".
7. The use of adapted models after their information content to obtain a basic solution to the task of "choosing priorities".
8. Assessment of the possibilities of referring to the list of critical facilities (CF), facilities that in the formalized solution of the "choosing setting" task were included in the list of priority constructed new most important facilities of the FEC for the future until 2035 (done by modeling the functioning of these facilities in the context of large-scale emergency situations in the fuel and energy sectors). A number of issues related to the adaptation of the currently used approaches, methods and models with the analysis of the significance of a particular CF should also be preliminarily resolved.
9. Correction of the formalized solution of the task of "choosing priorities" using the results of expert evaluation. At the same time, the impact on the resulting solution of factors that are almost impossible to take into account by formalized methods (the uncertainty factor, a number of subjective factors, the feasibility of the solution, etc.) is evaluated. The final solution to the task of "choosing priorities" when considering the creation of new most important facilities of the Russia's FEC until 2035.

6 Conclusion

Until now, when solving the tasks of managing the development of the Russia's FEC, the effect of the strategic threats to the country's ES was not actually taken into account. Perhaps this was permissible in the first decade of the 21st century, when the degree of realization of these threats did not reach nowadays level. It is extremely important now to take into account the conditions for the implementation of factors such as: lack of investment, growth in unit costs for the production and transportation of hydrocarbons throughout Russia, a noticeable lag in the rate of replacement of fixed assets in the energy sector from the rate of their physical and moral aging, and others. Failure to take them into account when solving the problems of managing the development of the FEC leads, as a rule, to obtaining too optimistic solutions (especially regarding the expected production capabilities of the energy sectors). The result is a situation of unacceptably high risk of disruption of the process of reliable fuel and energy supply to consumers in emergency situations of various kinds and the formation of barriers for sustainable energy development in the future. The conventionally old approach to solving the problems of managing the development of the fuel and energy complex does not make sense not only for the future of 30-40 years (due to too much uncertainty in the future), but also for the medium term, mainly due to the lack of consideration of the effect of the ES strategic threats. The application of the approach proposed in this paper will increase the degree of confidence in the results of solving the problems of managing the Russia's FEC development and create a methodological basis for minimizing the negative consequences for consumers of fuel and energy resources within the country in the context of the implementation of ES strategic threats in the future.

The work was carried out within the framework of a scientific project III.17.5.1 of program of fundamental research of the SB RAS, reg. number AAAA-A17-117030310451-0 and RFBR research project №18-58-06001, reg. number AAAA-A18-118050490009-5.

References

1. Rabchuk V.I., Senderov S.M., Izv. RAN. Energetika [Proc. of the Russ. Acad. of Sciences. Pow. Engin.], **1**, 50-58 (2019).
2. V. Rabchuk, S. Senderov Proceedings International Conference Green Energy and Smart Grids (2018). DOI: 10.1051/e3sconf/20186902011.

3. S. Senderov, V. Rabchuk, N. Pyatkova, S. Vorobev Ensuring Russia's Energy Security: A Choice of Priorities (Nauka, Novosibirsk, 2017).
4. https://www.acra-ratings.ru/research/691.
5. https://tass.ru/ekonomika/6092746.
6. Volokitina V.M., Gedich T.G., Economicheskii analiz: teoria I practika [Economic analysis: theory and practice], 16 (7), 1260-1268, (2017).
7. https://rns.online/energy/Novak-nazval-ozhidaemii-obem-investitsii-v-TEK-Rossii-v-2019-godu--2019-04-05/.
8. Voropai N.I., Kler A.M. & oth. System studies in the energy sector: methodology and results (ERI RAS, Moscow, 2018).
9. Fillippov S.P., Dilman M.D. System studies in the energy sector: methodology and results (ERI RAS, Moscow, 2018).

Search of the most important combinations of gas industry objects from the positions of system operability

Sergey Vorobev[1*], *Elena* Smirnova[1]

[1]Melentiev Energy Systems Institute, 130 Lermontov str., Irkutsk, Russia

Abstract. The article is devoted to the search and determination of the most important combinations of gas industry objects from the standpoint of system operability. The study was carried out on the example of the Unified Gas Supply System of Russia. The study is a development of the topic of the search and determination of critical objects of the system. Cases of simultaneous failure of two disconnected and independent from each other system objects are possible. The article presents a list of the most important combinations of gas industry objects that are not included in the list of critical objects, failure of which can lead to a significant gas shortage among consumers. Their significance and impact on the performance of the Russian gas industry are shown. Conclusions have been drawn on the feasibility of searching for such paired combinations of gas industry objects from the standpoint of system operability.

1 Introduction

The increase in the number of major accidents in energy systems in recent years has been due to the significant depreciation of fixed assets, and the lack of significant financial investments in their reconstruction. Large-scale accidents in energy systems resulting from the failure of the most important system objects entail significant, sometimes irreparable, damage to consumers in the form of large short deliveries of final types of energy.

In addition, geographically dispersed energy systems are becoming more intelligent and complex every year, more closely integrated with other infrastructure systems. Therefore, the failure of the most important system objects will have a negative impact on other interconnected infrastructure systems.

Thus, today it is relevant to identify the most important objects and their combinations in energy systems with the subsequent development of measures aimed at reducing the importance of such objects.

2 The current state of the investigated problem

In world practice, the study of the most important objects and their combinations in energy systems, the following points can be noted.

In [1, 2], the authors analyzed the gas transmission network to determine its most important components. The methodological approaches used are based on topological network analysis with an emphasis on the study of reliability and manageability. This analysis allows you to quantify the reliability of the gas transmission network and determine the role of each component of the network at various time intervals. As an example, the authors consider a real combined gas transportation network of several EU countries. The article presents the results of an analysis of such a critical infrastructure, and shows the need to take into account physical characteristics, such as restrictions on the throughput of gas pipelines. To assess the impact of negative external influences on the normal supply of gas to consumers, a special flow model was developed. Vulnerability analysis was carried out from three points of view: global vulnerability analysis, demand reliability analysis and critical analysis of gas pipelines. A global vulnerability analysis was conducted taking into account possible disruptions in the operation of gas sources and gas pipelines.

In [3], a method for detecting and ranking critical components and sets of components in technical infrastructures is presented. The criticality of a component or set of components is defined as the vulnerability of the system to failures when a particular component or set of such components fails. This issue also addresses the issue of multiple simultaneous failures with synergistic consequences that complicate the problem. The proposed method solves this problem. As a case study, a method for analyzing a gas distribution system in a Swedish municipality was presented.

In [4], the authors propose a comprehensive model for assessing the impact of the interdependence between electric and gas systems on the reliability of power supply to consumers. The mode of operation of the gas network was modeled using a number of restrictions. Restrictions on gas supplies may affect changes in the operating mode of the electric power industry. Case studies reviewed by the authors have proven this.

In [5], an analysis of the possible effects in an integrated gas energy network is presented. Failures of the gas supply system, as has been shown, are considered more defining for an integrated energy supply system than failures in the energy supply subsystem itself. Accordingly, the authors drew attention to possible control actions aimed at minimizing the negative impact of failures in the gas supply system.

Studies [6, 7] are devoted to various issues of modeling energy systems as critical infrastructures. In [6], the authors propose an agent model of a typical regional power system, which includes the characteristics of specific types of plants and their cooling systems, which depend on adequate water supply at appropriate temperatures to support operation at full capacity. The study [7] presented a new approach to assessing the vulnerability of inter-city distribution gas pipeline networks caused by gas pipeline accidents. The effects of pipe-line failures are quantified based on the gas supply areas that they directly affect and the reduction in traffic efficiency caused by stops on the roads. This approach allows the identification of vulnerable links in a pipeline network that can not only have a significant impact on the pipeline network, but can also have a significant impact on the road network.

The study [8] examined and analyzed the problem of the vulnerability of critical energy infrastructures to terrorist acts on them. The study [9] offers a risk analysis methodology for systems of interdependent critical infrastructures in various extreme weather events.

The study [10] presents a probabilistic approach for identifying and ranking important components of the gas network in terms of supply security. The authors conduct a probabilistic risk analysis of the regional European gas transmission network for the selected attack scenarios. The results of 1 million Monte Carlo simulations in attack scenarios clearly indicate the various consequences for gas supply. Thus, the authors obtained a list of the most important infrastructure components.

In these articles, the authors come closest to determining the critical objects of the energy system, in this case the gas transmission network. At the same time, they assign different indexes to different objects of the system, which together determine the vulnerability of the system in case of disruption of this object.

A series of studies were conducted at ISEM SB RAS devoted to the identification of critical objects in the gas transmission network also were conducted studies focus on identifying critical facilities in the power industry [11, 12]. A list of crossings of main gas pipelines in the Unified Gas Supply System (UGSS) of Russia has been defined, the disruption of which will lead to a relative deficit of daily gas supplies through the system as a whole of 5% or more [13, 14]. Studies have been conducted to search for and determine combinations of individual sections of the main gas pipelines, the simultaneous disruption of which can lead to a significant shortage of daily gas supplies through the system (5% or more) [15, 16].

Studies have also been conducted that have identified critical combinations of gas industry facilities. Such objects were found, failure of which, coupled with one of the critical objects, will lead to a much greater gas shortage among consumers than if these objects fail separately [17]. After identifying possible critical combinations of objects in each case, the possibilities of bypassing bottlenecks by briefly increasing the throughput of individual sections of the main gas pipelines are analyzed. This technological measure was used to minimize gas shortages among consumers. As a result of such measures, the significance of a number of potential critical combinations of objects has been reduced. In this way, all possible critical combinations of gas industry objects are identified and their list is ranked by degree of importance.

At this stage of the research, it is proposed to find and analyze all the possible most important combinations of gas industry objects. The most important combination of objects in the framework of this study means a couple of unrelated, independent objects whose failure can lead to a significant gas shortage among consumers. Moreover, the objects under consideration should not be included in the list of critical objects, nor in the list of critical combinations of objects.

2 Definition of critical elements in energy systems

In this study, were used the flow model, which is the core of the "Russian Oil and Gas" software, to determine the critical objects itself and to search for critical combinations of objects [18, 19]. The use of this "Russian Oil and Gas" software allows user to determine the degree of satisfaction of gas needs within the country and ensure export supplies. In addition, the "Russian Oil and Gas" software allows user to identify bottlenecks — sections of network that in some cases limit the production capabilities of the system.

The flow distribution model in the Unified Gas Supply System of Russia in the "Russian Oil and Gas" software is designed to assess the production capabilities of the Unified Gas Supply System of Russia in conditions of various kinds of disturbances. The purpose of such studies is to minimize gas deficits at the consumption sites. The Unified Gas Supply System of Russia in the model is represented as a set of three subsystems: gas sources, main gas transport network and consumers.

When solving the problem of estimating the state of a system after a perturbation, the criterion of the optimality of the distribution of flows is the minimum gas deficit in the consumer with minimum costs for delivering gas to consumers. This problem can be solved by finding the maximum flow through the network, followed by minimizing the cost of gas delivery to consumers [20]. The mathematical formulation of this problem is described in [21].

In the flow distribution model in the Unified Gas Supply System of Russia, as already mentioned, the Basaker-Gowen algorithm is used to calculate the maximum flow of minimum cost, which as a result allows you to determine the possible level of gas consumer satisfaction. As a result of the implementation of various emergency situations, a gas shortage among consumers may occur due to a lack of flow capacity in certain sections of gas pipelines. Bypassing such narrow or limiting production possibilities of the system's places, in acceptable volumes, will allow reducing the gas shortage arising in the situation under consideration by consumers.

If there is a gas shortage among consumers caused by a lack of throughput capacities of the respective gas pipelines, other branches of the main gas pipelines not affected by the violation in question take on increased volumes of gas. In such a situation, the network congestion changes and a lack of throughput capacities in other sections of the main gas pipelines are possible. The subsequent breakdown of bottlenecks in the gas transportation system will minimize gas shortages among consumers and will make the assessment and determination of possible critical combinations of gas industry objects as adequate as possible.

The "Russian Oil and Gas" software has graphical capabilities, which allows the researcher to identify many potential bottlenecks. These are objects that do not have a reserve of production capabilities, when analyzing the calculation results. Most often, such objects are either, or may become in the future, the reason for the shortage of the necessary amount of gas to consumers.

To solve the problem of circumventing the bottlenecks found, the flow distribution model included the possibility of incrementing the gas flow along edges within 10% of their capacity. Such a short-term increase in the throughput capacity of the main gas pipelines section is possible with an increase in the working capacity of compressors at large main compressor stations [22]. As a result of increasing the working pressure in the gas pipeline, an increase in the throughput capacity of the main gas pipeline section is achieved up to 10%. As a result (by using the technical capabilities of the gas transmission network), the problem of minimizing gas shortages among consumers is being solved.

3 Identification of the most important combinations of gas industry objects

The scheme of the Unified Gas Supply System of Russia used for the calculations in this paper on the flow model takes into account all the main features of the operation of the Unified Gas Supply System of Russia and contains:
- 382 nodes, including:
• 28 gas sources;
• 64 gas consumers (subjects of the Russian Federation);
• 24 underground gas storage facilities;
• 266 nodal compressor stations (CS);
- 486 edges representing main gas pipelines and outlets to distribution gas networks.

Initial data, such as daily consumption, extraction, export and import of gas, throughput capacity of existing gas pipelines are adopted in accordance with official statistics [23-25] for 2017. In a specially conducted study [26], an appropriate analysis was carried out, as a result of which 61 objects of the gas industry were assigned to critical gas industry objects. Among these objects there are 25 edges between the node compressor stations and 36 nodes, including 30 nodal compressor stations, 5 head compressor stations at the outlets from large gas fields and a compressor station on one underground gas storage.

Taking into account the previously obtained 61 critical objects and 630 pairs of critical combinations, calculations were carried out for pairwise shutdown of all other objects of the calculation scheme, followed by a "breakdown" of bottlenecks - measures aimed at minimizing the gas shortage among consumers. These calculations were carried out using the software package [27] detailing the functioning of the gas transmission network of Russia and allowing simulating various operating conditions of its facilities, including complete shutdown. The calculations were carried out using the parallel computing methodology at [28].

As a result, 2865 pairs of objects were selected from the 207690 pair combinations of objects, failure of which leads to a total gas deficit of 5% or more in the system. After solving the problem of bypassing bottlenecks, 2555 pairs of objects remained.

In the table 1 presents the results of determining the most important combinations of Unified Gas Supply System of Russia objects and the impact on them of the implementation of measures to circumvent bottlenecks, structured by their influence on the system.

Table 1. The results of determining the most important combinations of Unified Gas Supply System of Russia objects

Number of combinations		Gas shortage, %
Before bypassing bottlenecks	After bypassing bottlenecks	
4	0	15
10	6	11
18	14	10
51	39	9
101	54	8
243	167	7
1074	406	6
1364	1869	5

Table 1 clearly demonstrates that bypassing bottlenecks allows, in some cases, to significantly reduce the gas shortage among consumers. Table 2 presents 20 combinations of objects, failure of which can lead to a gas shortage in the system as a whole of 10% or more.

Table 2. Combinations of Unified Gas Supply System of Russia objects, failure of which will lead to a maximum gas shortage in the system as part of the study

№ of pair	Object type №1	Object type №2	Gas shortage, %
1	CS	CS*	11
2	CS*	CS	11
3	Edge	CS*	11
4	Edge	CS*	11
5	CS	CS	11
6	CS	CS*	11
7	CS	CS*	10
8	Edge	CS	10
9	Edge	Edge	10
10	Edge	CS	10
11	CS	CS	10
12	CS	CS	10
13	Edge	CS	10
14	Edge	CS	10
15	CS	CS*	10
16	CS	CS*	10
17	Edge	CS*	10
18	Edge	CS*	10
19	Edge	Edge	10
20	Edge	CS	10

When analyzing the table 2, it is necessary to highlight an object - one nodal compressor station (CS*), which is not included in the list of critical objects. CS* is present in 10 combinations from the table 2. In addition, this CS* is present in 25% of all combinations leading to a total gas deficit of the system as a whole of 5% or more.

In general, the following should be noted from the results of the study. Violation of the functioning of the most important combination of objects can lead to a significant gas shortage among consumers (5-15%).

In this situation, measures to bypass bottlenecks lead to a slight decrease in the gas deficit in the system as a whole (by an average of 2-3%). This fact confirms the high importance of the identified combinations. It is worth noting that in the framework of this study, as a result of bypassing bottlenecks, the number of possible most important combinations of objects was reduced by 10%.

4 Conclusion

This study examined the search for the most important combinations of gas industry objects. The definition of such combinations of gas industry objects, failure of which can lead to a gas deficit in the system as a whole, is comparable to the deficit arising from the failure of individual critical object. To search for the most important combinations of gas industry objects the "Russian Oil and Gas" software is used. This model is designed to assess the production capabilities of the Unified Gas Supply System of Russia under various disturbances. The main key combinations of gas industry objects received are analyzed in detail.

Studies have shown that in the modern configuration of the Unified Gas Supply System of Russia, there may be situations when, when a pair of network objects that are not critically important fails, the total gas shortage among consumers can reach up to 15% of the total gas demand. The definition and accounting of such combinations of objects is necessary to increase the reliability of the Unified Gas Supply System of Russia during its development and reconstruction work. The main efforts should be aimed at reducing the conditional significance of the identified specific critical combinations of objects for the health of the entire system.

The study was carried out in the framework of the draft state assignment III.17.5.1 (reg. No. AAAA-A17-117030310451-0) of basic research of the SB RAS and with partial financial support from the Russian Federal Property Fund in the framework of scientific project No. 18-58-06001.

References

1. Han F., Zio E., Kopustinskas V., Praks P. Risk, Reliability and Safety: Innovating Theory and Practice, pp. 2565-2571. (2016). DOI: 10.1201/9781315374987-389.

2. Su H., Zio E., Zhang J. Li X. Reliability Engineering & System Safety, **Volume 175**, Pages 79-91, (July 2018). DOI: 10.1016/j.ress.2018.03.006.

3. Jonsson H., Johansson J., Johansson H. Proceedings of the Institution of Mechanical Engineers, Part O: Journal of Risk and Reliability. **Vol. 222**. No. 2. P. 235-243. (2008). DOI: https://doi.org/10.1243%2F1748006XJRR138.

4. Li T., Eremia M., Shahidehpour M., IEEE Transactions on Power Systems, **vol. 23**, p 1817-1824, (2008). DOI: 10.1109/TPWRS.2008.2004739.

5. Dokic S.B., Rajakovic N.Lj. Energy (2018), DOI: 10.1016/j.energy. 2018.04.165

6. Thompson, J. R., Frezza, D., Necioglu, B., Cohen, M. L., Hoffman, K., Rosfjord, K. International Journal of Critical Infrastructure Protection **Volume 24**, p. 144-165. (2019).

7. Kai L., Ming W., Weihua Z., Jinshan W., Xiaoyong Y. International Journal of Critical Infrastructure Protection **Volume 23**, p. 79-89. (2018).

8. Tichy L. International Journal of Critical Infrastructure Protection, (2019). DOI: 10.1016/j.ijcip.2019.01.003.

9. Tsavdaroglou M., Al-Jibouri S. H.S., Bles T., Halman J. I.M. International Journal of Critical Infrastructure Protection **Volume 21**, p. 57-71. (2018).

10. Praks P., and Kopustinskas V. "CRITIS 2018, LNCS 11260, pp. 3–16. (2019). DOI: 10.1007/978-3-030-05849-4_1

11. Iakubovskii D., Komendantova N., Rovenskaya E., Krupenev D., Boyarkin D. Geosciences (Switzerland). **Vol.9. №1**. ID: 54. (2019). DOI: 10.3390/geosciences 9010054

12. Krupenev D. E3S Web of Conferences. **Vol.58.** ID: 03009. (2018). DOI: 10.1051/e3sconf/20185803009

13. S.M. Senderov, V.I. Rabchuk, A.V. Edelev, Izv. Ros. Akad. Nauk. *Energetika*. № **1**. p. 70-78. (2016).

14. S. Senderov, A. Edelev, *Energy*. doi: 10.1016 / J.ENERGY.2017.11.063. (2017).

15. S. Vorobev, A. Edelev, *MLSD 2017*, IEEE, 2017. DOI 10.1109 / MLSD.2017.8109707. (2017).

16. S. Vorobev, A. Edelev, E. Smirnova, *RSES 2017*. E3S Web Conf. Vol. **25**, DOI 10.1051 / e3sconf / 20172501004. (2017).

17. Senderov S., Vorobev S., Edelev A. *RSES 2018*. E3S Web Conf. Vol. **58**, (2018). DOI: 10.1051/e3sconf/20185803002.

18. A.V. Edelev, S.M. Enikeeva, S.M. Senderov, *Computational technologies*. – Vol. 4, **№ 5**. – p. 30 – 35. (1999).

19. S.V. Vorobev, A.V. Edelev, *Software products and systems*. – **№3**. – p. 174 – 177. (2014).

20. L.R. Ford, D.R. Fulkerson. *Flows in networks*. Princeton University Press. 276p. (1966).

21. Vorobev S.V., Edelev A.V. Scientific Bulletin of NSTU **Vol. 62, No. 1**, p. 181-194. (2016).

22. Yu.P. Korotaeva, R.D. Margulova, *Extraction, preparation and transport of natural gas and condensate. Reference manual in 2 volumes. Volume II.* Nedra, 288 p., (1984).

23. *Export of the Russian Federation of the most important goods in 2011 - 2016 (according to the Federal Customs Service of Russia)* http://customs.ru/index.php?option=com_ newsfts&view =category&id=52&Itemid=1978&limitstart=60.

24. *InfoTEK Monthly oil and gas magazine*. **№1**, – P.154. (2017).

25. *Ministry of Energy of the Russian Federation. Statistics*. http://minenergo.gov.ru/activity/statistic.

26. S.M. Senderov, V.I. Rabchuk, S.V. Vorobev, *Proc. of the col. of rep. Methodological issues of reliability research of large energy systems. 90th meeting "Reliability of developing energy systems"*. July 1-7, Irkutsk. (2018).

27. Feoktistov A., Gorsky S., Sidorov I., Kostromin R., Edelev A., Massel L. Communications in Computer and Information Science. **Vol. 965.** P. 289-300. (2019).

28. Irkutsk Supercomputer Center SB RAS. URL: http://hpc.icc.ru

https://doi.org/10.1051/e3sconf/201913901017

The approach to determining vulnerable elements in critical energy infrastructures

Natalia Beresneva, *Natalia* Pyatkova[1]

Melentiev Energy Systems Institute, 130 Lermontov str., Irkutsk, Russia

Abstract. The paper presents the main provisions of the methodology for identifying vulnerable elements in the fuel and energy complex, which is studied as a set of industry critical energy infrastructures. This methodology is based on the use of industry models system in the fuel and energy sector, on the analysis of the interconnected critical energy infrastructure's work in normal functioning and during emergency situations. The paper proposes a schematic diagram for the formation of vulnerable elements list for fuel and energy complex based on critical elements of industries. The scheme of work with a two-level system of models is given. The results of testing the methodology and developed models are presented (illustrative example).

1 Introduction[a]

Critical infrastructures in accordance with the definitions [1, 2] with full confidence include industry energy systems and the fuel and energy complex (FEC), which is an interconnected infrastructure consisting of separate energy systems. This statement is confirmed by the importance of these infrastructures for the country's economy, supported by the need to ensure energy security, characterized as the state of protection of citizens, society, the state and the economy from threats of deficit in meeting their needs for energy resources of acceptable quality, from threats to disruption of uninterrupted energy supply. The main tasks to be solved in this case [3]:

– forecasting the conditions for the functioning and development of energy systems and the fuel and energy complex as a whole, taking into account possible critical and emergency situations;

– assessment of the condition under these conditions;

– identification of "bottlenecks" in the fuel and energy supply systems of consumers;

– the choice of alternatives, directions and specific measures to prevent emergency situations in systems or reduce their negative impact.

The consequences of the implementation of these emergency situations can lead to significant interruptions in reliable fuel and energy supply to consumers, and can cause a shortage of energy resources. This increases the relevance and need for models develop of critical energy infrastructures to identify potential bottlenecks (vulnerable elements of critical infrastructures), assess their criticality for guaranteed and timely supply of energy needs. To conduct research with their help, a methodology for determining the critical (vulnerable) elements for the fuel and energy complex as a whole was developed and tested, presented below.

[a] Corresponding author : nata@isem.irk.ru

2. Methodology for determining critical elements

Studies to identify the most vulnerable elements (objects) of interconnected critical energy infrastructures (energy systems) are carried out on the basis of optimization of territorial-production models of functioning of the country's fuel and energy complex. In this case, the hypothetical calculated (non deficient or indignant) states of the fuel and energy complex are analyzed. Perturbed states are formed on the basis of single shutdowns of fuel and energy facilities, or for shutdowns of their groups. These disconnections at the model level are realized by a sharp decrease in the production capabilities of the objects, namely, the zeroing of the upper restrictions of the model variables corresponding to these objects. A comprehensive assessment of these conditions is carried out on the basis of their model indicators for various territories and for the analyzed critical infrastructures. For this, the criterion for the effectiveness of each infrastructure is used - relative energy shortages that are developing among consumers. Additionally, it is possible to consider the using of the industries reserve capabilities - fuel reserves, reserves of generating capacities in electric and heat power engineering, and the possibility of diversifying fuel at double fuel supply facilities.

https://doi.org/10.1051/e3sconf/201913901017

To assess the criticality of the fuel and energy complex object in the framework of optimization calculations, the element significance criterion is used, which characterizes the relative changes in the analyzed model parameters when the functioning mode of this object changes over the entire set of states. Formally, this indicator is given by formulas (1):

$$ZO_i = \sum_{j=1}^{J} ZO_i^j * ZC^j \qquad (1)$$

ZO_i – significance of the i-th the fuel and energy complex object ;

ZO_i^j – assessment of states according to the j-th model indicator (relative energy shortage) when the i-th element is turned off ;

ZC^j – priority of the j-th indicator ;

J – number of analyzed indicators (energy shortage, depletion of reserves, capacity reserves, etc.) ;

I – plenty disconnected critical elements of the fuel and energy complex.

The research scheme (Fig. 1) is formally represented by three stages:

– the stage for the formation of the strategy for selecting critical objects;

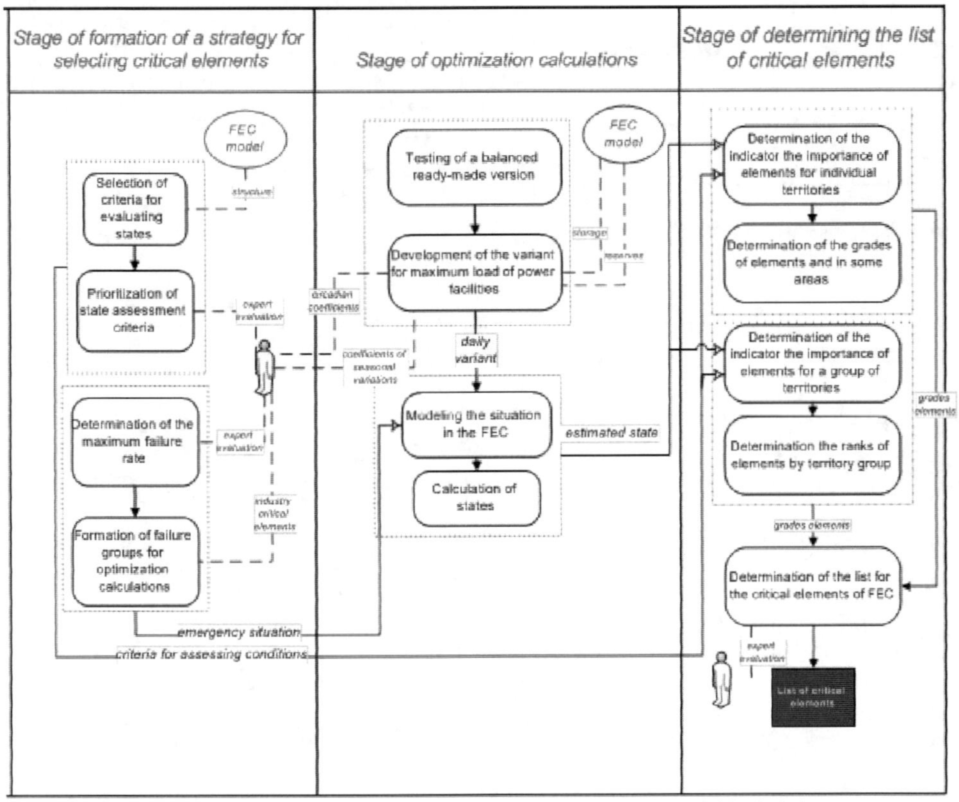

Fig. 1 A generalized search scheme for critical elements in the framework of model calculations

– the stage of carrying out optimization calculations;

– the stage of forming a list of critical objects based on their calculated significance indicators.

At the first stage

- objects of the simulated FEC territorial-production structure are determined;

- groups of disconnected elements are formed for their subsequent inclusion in simulated situations;

- a set of criteria for assessing states (relative model indicators) is formed, the significance of these criteria is determined

- categories of elements criticality and their threshold values are determined;

- emergency situations scenarios are being formed.

At the second stage, optimization calculations of the fuel and energy complex model are carried out, in the framework of which:

- balanced annual version of the statistical information is being debugged;

- the variant for the daily maximum load in energy systems is being debugged;

- abnormal situations are calculated with various options for disconnecting potential vulnerable elements.

At the third stage, the results of assessing the state of critical infrastructures when disconnecting the specified elements are formed:

- the composition of the analyzed calculated states is corrected by excluding the states with a relative deficit acceptable for the country as a whole for at least one resource;

– criteria for the significance of elements for the territories or their groups are determined;

– the ranks and categories of criticality for the analyzed elements are determined;

– a list of critical elements for the fuel and energy complex is formed.

3. An illustrative example of testing the methodology

The methodology was tested on the basis of daily model for the functioning of the fuel and energy complex. Therefore, when choosing critical elements of the fuel and energy complex, three types of daily variants for the model are formed: the average daily balanced option, the maximum load variant for energy facilities, and variant for disconnecting industry-specific critical elements with a maximum workload of facilities. In the last two cases, when deficits arose, model mechanisms of structural redundancy of systems in the form of reservation and diversification of energy resources became active.

This model represents the territory of the country by 79 subjects of the federation. Technologically, model consists of industry subsystems of the energy complex (gas, coal, oil refining (in terms of fuel oil supply), and electricity and heat industries) [4].

A number of experimental calculations were carried out in which, as design conditions, the influence of the disconnection of critical elements of the gas industry [5] on the functioning of the fuel and energy complex as a whole was analyzed. The result of a turn-off of selected gas industry facilities was the under-production of the energy resources considered in the model and, accordingly, their shortage among consumers.

In the whole country, for each switch-off facility in the gas industry, the following situation has developed, Fig. 2.

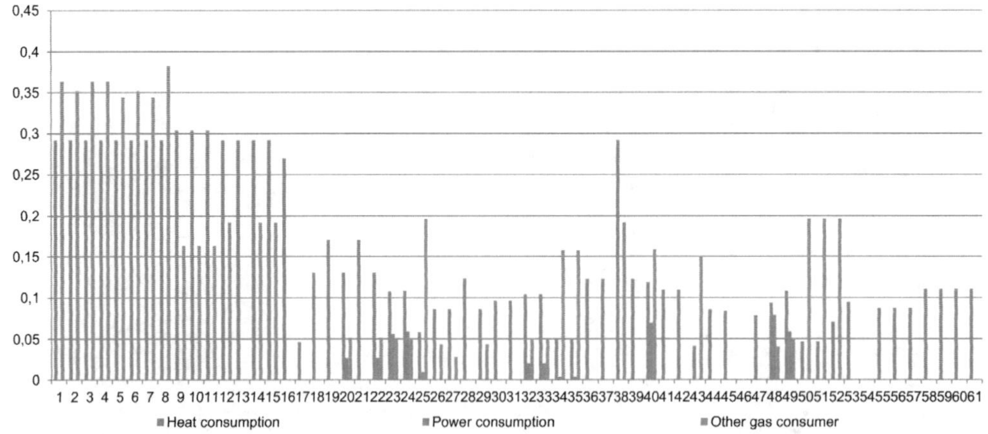

Fig. 2. The relative shortage of energy resources when turning off critical elements of the gas system, the share of units

When determining the coefficients of significance, potential critical elements were discarded, when energy shortages amounted to less than 5% of the needs. As a result, from the initial list of 61 critical elements in the gas industry, 52 elements remained, the disconnection of which led to more significant shortages (Fig. 3).

The analysis of the obtained graphs showed that the ranked list of critical elements of the gas industry almost corresponds to disconnections of these elements during the mutually agreed operation of all energy systems in a single complex (fuel and energy complex) with rare exceptions.

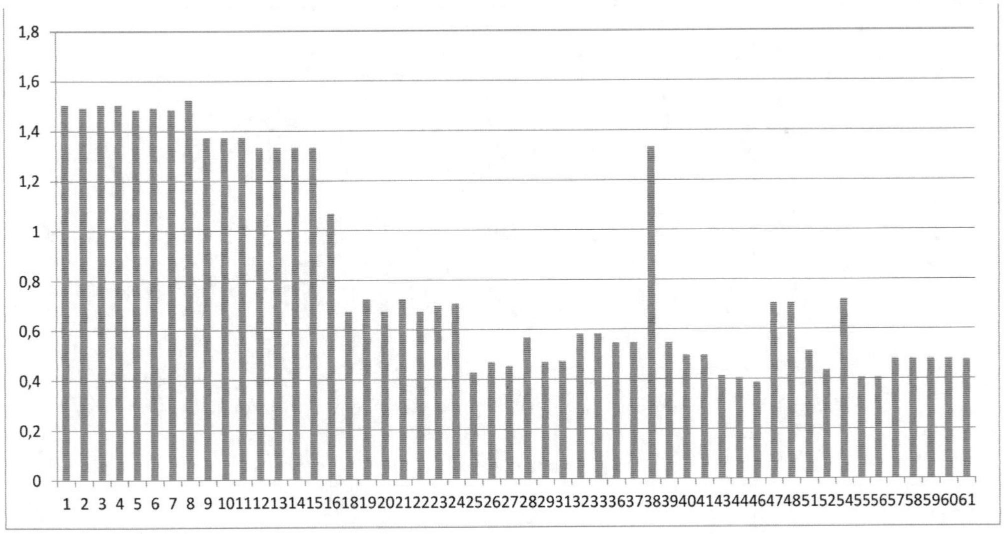

Fig. 3. The coefficients of significance of the critical elements of the gas system

According to the obtained coefficients of significance, three categories of critical elements were identified by their impact on energy supply (by the magnitude of the deficit from their shutdown): 1 category, in which elements causing a total relative deficit of more than 70%, 2 category from 30 to 70%, 3 category - below 30%. The number of critical elements for these groups throughout the country (in federal districts) was distributed as follows (Table 1). A significant deficit was obtained in territories whose fuel and energy balance is focused on the use of natural gas. This is the North Caucasian, Southern and Volga regions, where the number of critical elements of the first category ranged from 17 to 25.

Table. 1. Classification of critical elements of the gas industry by the impact on energy supply to consumers

Federal District	1 category (70-100%)	2 category (30-70%)	3 category (0-30%)
Central	15	3	34
Northwestern	14	27	11
South	17	20	15
North Caucasian	25	17	10
Volga	17	15	20
Ural	6	7	39
Siberian	5	3	44
Far Eastern	3	49	-

Further, critical elements were considered according to the technological criterion: sources (gas production enterprises) (Fig. 4), compressor stations (Fig. 5), sections of main gas pipelines (Fig. 6). The coefficients of significance of these elements differ in the depth of the deficit caused by them in the territories under consideration. A significant deficit causes shutdowns of the compressor stations and individual sections of gas pipelines.

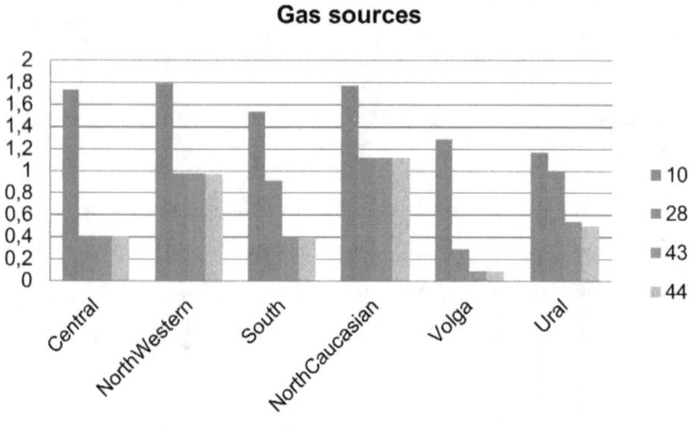

Fig. 4. The coefficients of significance of critical elements of the gas system by district

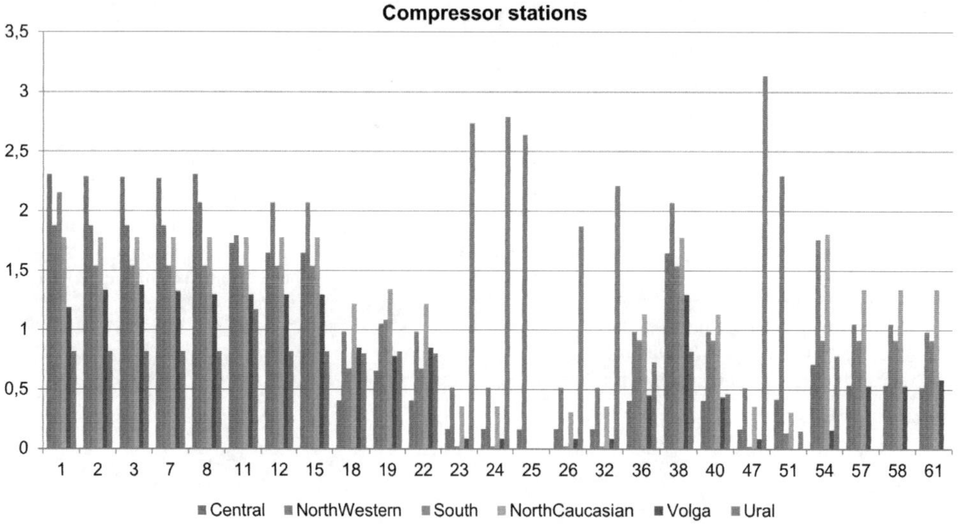

Fig. 5. The coefficients of significance of critical elements of the gas system by district

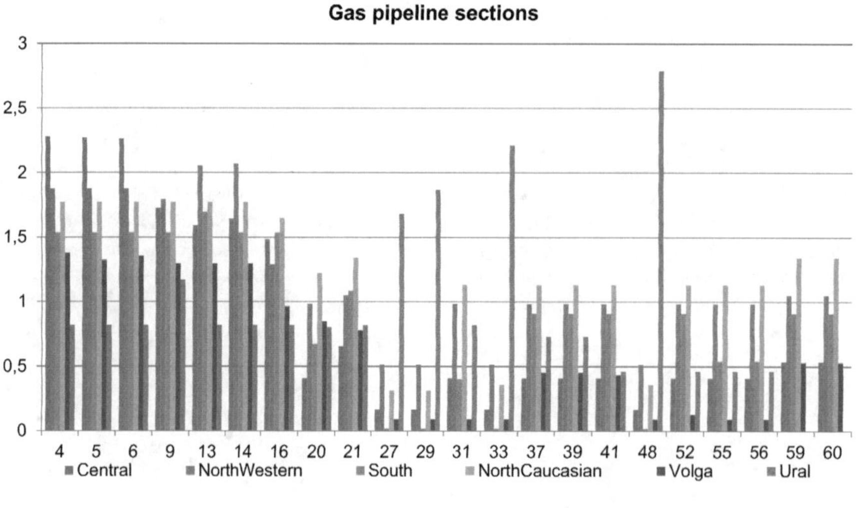

Fig. 6. The coefficients of significance of the critical elements of the gas system by district

An analysis was made of the frequency (repeatability) of the entry of vulnerable elements into the most critical first category in all federal districts. It turned out that these are mainly compressor stations and sections of gas pipelines, the frequency of which is 5 and 6, from the sources - one element, the frequency of which in terms of repeatability across the territory was 4.

This example illustrates the results of experimental studies evaluating the significance of elements of the gas industry for reliable energy supply to consumers in a single complex. Similar studies to identify critical elements and reliable energy supply are carried out in the electricity industry [6, 7]. In the future, it is planned to carry out experimental studies to assess the importance of critical elements of the electric power industry for the operation of the entire fuel and energy complex, as well as combinations of elements of the gas and electric power industries in analyzing the reliability of energy supply to consumers.

4. Conclusion

The paper presents the main provisions of the methodology for determining critical (vulnerable elements) in the fuel and energy complex. This methodology is based on the use of a hierarchical system of models of industries in the fuel and energy sector, on the analysis of the interconnected work of industry critical energy infrastructures in normal functioning and during emergency situations. The results of testing the methodology and developed models for studying the influence of critical elements of the gas industry on the mutually agreed work of the energy industries in a single complex are also presented. The coefficients of significance of the elements under consideration, taking into account the systemic effect, are obtained.

The results were obtained in the framework of the project for state assignment of the ESI SB RA No. AAAA-A17-117030310451-0, some aspects were worked out in the framework of projects supported by the RFBR grant No. 19-07-00351.

References

1. Kondratiev A. Current trends in the study of critical infrastructure in foreign countries // Foreign Military Review No. 1, 2012, P.19-30.
2. S. Rinaldi, J. Peerenboom, and T. Kelly. Idenlifying, Understanding, and Analyzing Critical Infrastructure lnterdependencies. — IEEE Control Systems Magazine — IEEE, December 2001. — Pp. 11-25.
3. Energy security of Russia: problems and solutions / N.I. Pyatkova [et al.], Resp. ed. N.I. Voropay, M.B. Cheltsov; Grew up. Acad. Science, Sib. Department, Institute of Energy Systems named after L.A. Melentieva. - Novosibirsk: Publishing House of the SB RAS, 2011 .-- 198 p.
4. Beresneva N., Pyatkova N. Choosing of the FEC critical objects in model researches of energy security problems and its features // Rudenko International Conference "Methodological problems in reliability study of large energy systems" (RSES 2018). E3S Web Conf. Volume 58, 2018. (DOI 10.1051/e3sconf/ 20185803004)
5. S.V. Vorobev, A.V. Edelev. Application of the method for determining critical elements in networks of technical infrastructures to search for critical objects of the gas transmission network of Russia // Energy Policy, 2018, No. 1, p. 45-51.
6. Iakubovskii D., Komendantova N., Rovenskaya E., Krupenev D., Boyarkin D. Impacts of earthquakes on energy security in the Eurasian economic union: Resilience of the electricity transmission networks in Russia, Kazakhstan, and Kyrgyzstan // Geosciences (Switzerland). Vol.9. №1. ID: 54. 2019. DOI: 10.3390/geosciences9010054
7. Krupenev D. Determination of Critically Objects of Electric Power Systems from the Position of Energy Security // E3S Web of Conferences. Vol.58. ID: 03009. 2018. DOI: 10.1051/e3sconf/20185803009.

https://doi.org/10.1051/e3sconf/201913901018

Factoring in scheduled repairs of generating units when assessing the resource adequacy of electric power systems

Dmitry Krupenev[1,*], *Denis* Boyarkin[1], and *Dmitrii* Iakubovskii[1]

[1] Melentiev Energy Systems Institute of Siberian Branch of the Russian Academy of Sciences, 664033, Irkutsk, Lermontov 130 st..

Abstract. This study presents the algorithms for modeling of scheduled repairs of generating equipment when assessing the resource adequacy of electric power systems by the Monte Carlo method that relies on random events. We present an analysis of the algorithms adopted in the available software and computer systems used to assess the resource adequacy and highlight their shortcomings. We propose an algorithm for modeling of scheduled repairs of generating units that does away with the flaws of the algorithms adopted elsewhere. In the final part of the article, we present the findings of numerical experiments that test the performance of the algorithms. **Keywords**: resource adequacy, electric power system, scheduled repairs, generating unit, Monte Carlo method.

Introduction

Assessment of the reliability of electric power systems (EPS) is an integral procedure for solving many problems of control over the operation and development of the EPS. When planning the development of the EPS, it is necessary to assess the balance reliability to substantiate the regulatory reserve requirements to be met by generating capacity, the structure, and throughput capacity of the power grid. The problem of assessing the EPS resource adequacy can be stated as follows: For given annual hourly profiles of power consumption, the composition, technical and reliability parameters of generating equipment within the reliability zones, the composition, technical and reliability parameters of power lines in the inter-zone connections to determine the resource adequacy for the reliability zones and for the entire system for a one-year period. Reliability parameters of generating equipment should include scheduled repairs.

In the electric power industry, scheduled repairs are a mandatory procedure. There is a regulation on the timing and amount of repairs of power facilities. For most types of the power equipment operated in Russia, this regulation and the rules for taking generating equipment out of service for repairs are presented in [1-3]. The issues of scheduled repairs of nuclear power plants and some other types of power equipment were not addressed there.

Ultimately, reliability of power supply to consumers and reliability of power plants depend on timely carrying out of scheduled repairs. The functional relation is manifested in the increase in the fault rate of power equipment. Repairs of energy equipment require certain costs, but it is known that the damage in case of failure of the unrepaired equipment exceeds the costs.

Since the resource adequacy assessment is carried out for prospective schemes of EPS development, the plans for repairs of generating equipment are oftentimes unknown. Under such circumstances, scheduled repairs of generating equipment are factored in based on custom-built algorithms that simulate real rules for working out repair schedules. An algorithm for modelling of repair schedules should most accurately reflect the actual conditions of repairs of generating equipment. This study covers the algorithms for modelling scheduled repairs of generating equipment employed for balance reliability assessment in the available software and computer systems (SCS), identifies their shortcomings, and proposes a new algorithm that improves the process of modelling of scheduled repairs of generating equipment as part of the assessment of resource adequacy of the EPS.

1 Algorithms for modelling of scheduled repairs

The methodology behind assessing the resource adequacy of the EPS that relies on the Monte Carlo method is made up of the following three main steps [4-6]:

1. The step of generating random states of the EPS.
2. The step of minimizing the power shortage under random states of the EPS.

3. The step of calculating the resource adequacy indicators of the EPS.

At the first step, alongside other calculation procedures, scheduled repairs of generating units are factored in. At the same time, the following aspects of the operation of the EPS should be taken into consideration as well:

- Special considerations for establishing requirements to be complied with by scheduled repairs of NPP, CCGT and GTU equipment;
- Scheduled repairs of generating equipment that operates in the heat and power generation mode outside the heating season;
- Scheduled repairs of hydropower plant (HPP) generating equipment outside the flood season;
- Planned repairs of generating equipment during periods of the seasonal decrease in consumption, taking into account the peculiarities of regional power systems characterized by summer peak power consumption;
- Compatibility with repairs of power lines, as well as other generating and power grid equipment;
- Amount of unscheduled and emergency repairs of generating equipment.

There are several types of scheduled repairs [1]: major repairs, medium repairs, and running repairs. All of them are defined by the scope of work to be performed and the timing of carrying out such scopes of work.

When assessing the resource adequacy of prospective EPS schemes, there may be several options with the information available on scheduled repairs of generating equipment. In case of long-term planning of the power system operation, plans for scheduled repairs of generating equipment are known up to 1 year ahead; these plans are developed by the System Operator when processing applications of generating companies. In such case, one should take generating units out of service as per these plans when assessing resource adequacy. For the period of planning of the operation and development of the EPS up to 3 years ahead, the quantities of taking generating units out of service for scheduled repairs may be known, but there is no known order and sequence of taking them down. For prospective schemes of EPS development for periods exceeding 3 years, the amount of planned repairs and the order of taking generating equipment out of service are not known, and when assessing the resource adequacy, which is usually carried out for a one-year period, it is assumed that the amount of scheduled repairs of generating units is equal to the average amount of scheduled repairs for each unit over its repair cycle, which is presented in [1]. Let us consider various options available for modelling of scheduled repairs of generating units when assessing the resource adequacy of the EPS.

1.1 Factoring in scheduled repairs of generating units based on statistical data

The easiest way to factor in scheduled repairs of generating units when assessing the balance reliability of the EPS is to factor them based on their statistical data of an earlier period. The algorithm behind this process is as follows:

1. Determination of the generation capacity within the reliability zones that was taken out of service for scheduled repairs as averaged over the past period:

$$x_{avg,sch,i,\chi} = \frac{\sum_{n=1}^{N} x_{sch,i,\chi,n}}{N}, \quad i = 1,\dots,I, \chi = 1,\dots,12 \tag{1}$$

where: $x_{sch,i,\chi,n}$ is the actual average amount of planned repairs of generating units within the reliability zones in month χ of year n, MW; where N is the number of years for which statistical data are used; I is the number of reliability zones.

2. Determination of the operating generating capacity within the reliability zones in the design states of the EPS given scheduled repairs of generating units:

$$x_{op,i} = x_{av,i} - x_{em,i} - x_{av,sch,i,\chi} \quad i = 1,\dots,I, \quad \chi = 1,\dots,12 \tag{2}$$

where: $x_{av,i}$ is available capacity of generating units within the reliability zones, MW; $x_{em,i}$ is generating capacity within the reliability zones taken out of service for emergency repairs, MW.

The above approach in not free of the following shortcomings:

1. Lack and unavailability of information on the statistical data on scheduled repairs of generating units.

2. Changing conditions for the operation of EPS in prospective schemes, including the commissioning and decommissioning of generating equipment.

3. Duplication of generating units when factoring in emergency and scheduled repairs.

1.2 Factoring in scheduled repairs of generating units into the load profile

This algorithm has been implemented in the Yantar software and computer system (SCS) [4] and other SCSs. Major and medium repairs are factored in by means of a dip in the load profile for monthly peaks of reliability zones. The algorithm behind such factoring in consists of the following steps:

1. Determination of the amount of medium and major repairs of generating equipment within the reliability zones averaged over a year as measured in MW per day:

$$V_{amr,i} = \sum_{h=1}^{H} x_{h,i} t_{amr,h,i}, i = 1,\dots,I \tag{3}$$

where $x_{h,i}$ is the available capacity of generating unit h in reliability zone i, MW; H is the number of generating units in reliability zone i; $t_{amg,h,i}$ is the average annual time to carry out medium and major repairs of generating unit h in reliability zone i per year, days.

2. Ordering (ranking) of the monthly peak power consumption profile from the minimum to the maximum

value in each of the reliability zones. When ranking, the information on which calendar month a given value belongs to is retained.

3. Determination of the amount of medium and major repairs each month. For this purpose, a loop is introduced to fill in the dip of the ranked monthly peak load profile starting from the minimum load on, proceeding in ascending order until $V_{amr,i}$ is fully included in the annual load profile of the reliability zone. As a result, for each χth month the value of the "amount" of medium and major repairs will be obtained for the reliability zones $V_{amr,i,\chi}$.

4. Determination of the value of generating capacity taken out for medium and major repairs within the reliability zones:

$$x_{amr,i,\chi} = \frac{V_{amr,i,\chi}}{N}, i = 1, \dots, I, \chi = 1, \dots, 12, \tag{4}$$

where: N is the number of days in each month.

5. Determination of the values of the amount of running repairs of generating equipment within the reliability zones averaged over a year as measured by MW a day as per the following formula:

$$V_{rr,i} = \sum_{h=1}^{H} x_{h,i} t_{rr,h,i}, i = 1, \dots, I \tag{5}$$

where $t_{rr,h,i}$ is the average annual time for running repairs of generating unit h in reliability zone i per year, days.

6. Determination of the value of the generating capacity taken out for running repairs in the reliability zones:

$$x_{rr,i} = \frac{V_{rr,i}}{365}, i = 1, \dots, I. \tag{6}$$

7. Generation of the monthly peak values profile with scheduled repairs of generating equipment within the reliability zones factored in:

$$\bar{y}_{sch,i,\chi} = y_{i,\chi} + x_{amr,i,\chi} + x_{rr,i}, \quad i = 1, \dots, I, \chi = 1, \dots, 12, \tag{7}$$

where $\bar{y}_{i,\chi}$ is the peak load in reliability zone in month χ, MW; $x_{amr,i,\chi}$ is the amount of generating capacity taken out of service for major and medium repairs in reliability zone i in month χ, MW; $x_{rr,i}$ is the amount of generating capacity taken out of service for running repairs in reliability zone i, MW.

Figure 1 shows the process of generating the load profile with scheduled repairs of generating equipment within the reliability zones factored in.

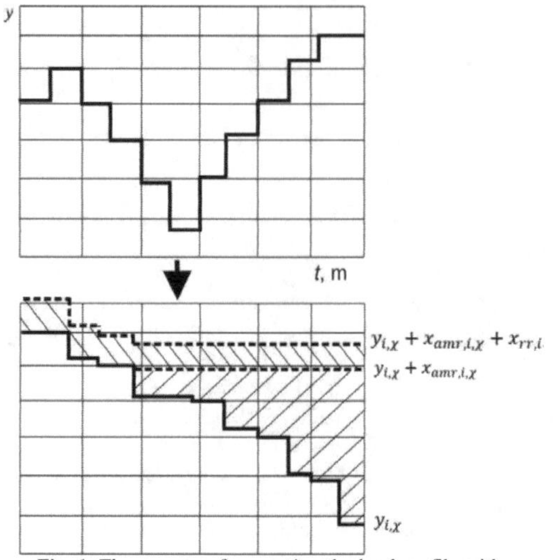

Fig. 1. The process of generating the load profile with scheduled repairs of generating equipment within the reliability zones factored in.

The main drawbacks of this approach to factoring in scheduled repairs are as follows:

1. Duplication of generating units when factoring in emergency and scheduled repairs.

2. "Smoothing out" the timetable of scheduled repairs within the reliability zones with a small number of generating units and / or a minor dip in the load profile. If the profile of monthly load peaks is "smooth" and is free of a pronounced dip, then within the reliability zones with a small number of generating units, the amount of medium and major repairs will be distributed over several months, which distorts the actual situation in the EPS, according to which each generating unit is under scheduled repairs for a certain number of days of a month.

1.3 Factoring in scheduled repairs of generating units when determining the operating capacity of reliability zones

This approach is based on the combination of the two above presented approaches to factoring in scheduled repairs of generating equipment when assessing the EPS resource adequacy. The approach consists of the following steps:

1. Determination of the amount of scheduled repairs in each month of the reference year according to the methodology presented as part of the previous approach (items 1 to 6).

2. Determination of the values of generating capacity within reliability zones that is taken out of service for scheduled repairs each month:

$$x_{scr,i,\chi} = x_{amr,i,\chi} + x_{rr,i} \quad i = 1, \dots, I, \chi = 1, \dots, 12. \tag{8}$$

3. Determination of the operating generating capacity within the reliability zones in design states of the EPS given scheduled repairs of generating units:

$$x_{ope,i} = x_{av,i} - x_{acc,i} - x_{sch,i,\chi}$$
$$i = 1, \dots, I, \qquad \chi = 1, \dots, 12 \tag{9}$$

The approach has the same flaws as the previous one.

1.4 Factoring in scheduled repairs of generating units in the process of generating design states of the EPS

In this algorithm, unlike with those presented earlier, scheduled repairs can be incorporated not only into the load profile of the reliability zone, but also into a larger interconnection, for example, in the case of the Unified Power System of Russia the latter can be interconnected power systems (IPS). In this case, there will be a dip in the load profile with a probability higher than the one obtained when considering an individual reliability zone. Where an isolated system or a system with loose external links is considered, and it has a relatively "smooth" profile of monthly peak loads and a small number of generating units, individual rules for assigning generating equipment for repairs should be applied to this system. One more feature of this algorithm is that scheduled repairs are dealt with as a part of the same computational procedure as emergency repairs which rules out a possibility of their duplication. The algorithm behind this approach consists of the following steps:

1. The first six steps of this algorithm are similar to the first six steps of the second algorithm presented above, with the only difference being that instead of reliability zones it is the parts of the power system that are considered: they may include multiple reliability zones and are generated by experts. As a result, expression (8) determines the values of the generating capacity taken out of service for scheduled repairs in individual parts of power systems.
2. Determination of the relative amount of scheduled repairs within each interconnected power system each month:

$$k_{sch,i,\chi} = \frac{x_{sch,i,\chi}}{x_{av,i}}\ i = 1, \dots, I, \chi = 1, \dots, 12. \tag{10}$$

3. Determination of the share of the time each generating unit is under scheduled repairs relative to the total downtime due to scheduled repairs of all generating units in an individual part of the energy system i:

$$t'_{av,h,i} = \frac{t_{av,h,i}}{\sum_{h=1}^{H} t_{av,h,i}}, i = 1, \dots, I, \tag{11}$$

where: $t_{sch,h,i}$ - average annual time for scheduled repairs of generating unit h in power system's part i per year; H is the number of generating units.
4. Determination of coefficients used to adjust the availability of generating units for scheduled repairs:

$$k_{adj,sch,h,i} = t'_{sch,h,i} \cdot H, h = 1, \dots, H, i = 1, \dots, I. \tag{12}$$

5. Determination of the relative amount of scheduled repairs of generating units within each interconnected power system each month:

$$k_{adj,h,i,\chi} = k_{sch,i,\chi} \cdot k_{ajd,sch,h,i},$$
$$h = 1, \dots, H, i = 1, \dots, I, \chi = 1, \dots, 12. \tag{13}$$

Coefficient $k_{sch,h,i,\chi}$ is used as an additional parameter for playing out the states of the generating equipment, it is to be summed up with its fault rate.

2 Experimental studies of the algorithms under consideration

Experimental studies of the algorithms were carried out based on the three-zone model of the EPS presented in Figure 2.

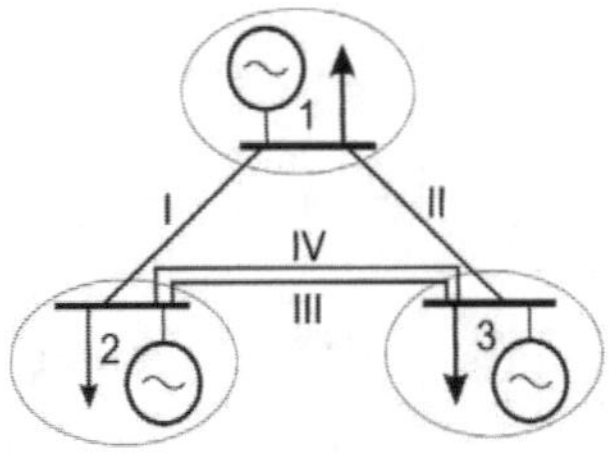

Fig. 2. Schematic diagram of the study EPS.

Table 1 presents the main characteristics of the reliability zones of the study EPS. Reliability zones No. 1 and 2 have a small number of generating units 7 and 6, respectively, while reliability zone No. 3 have 45 generating units to it. The available capacity of the EPS under consideration is 6,050 MW, the peak load of the power system is 5,800 MW, and the capacity margin is 14%. Table 2 presents the main characteristics of inter-zone links. Figure 3 presents profiles of monthly peak loads of reliability zones. As noted above, one of the main challenges when factoring in scheduled repairs by means of the dip of the load profile of the reliability zone is its "smooth" form. In this particular case, the profiles of monthly peaks of the second and third reliability zones are "smooth" relative to the first reliability zone.

Table 1. Characteristics of reliability zones of the study EPS.

Reliability zone No.	Available capacity, MW	Number of units	Generating unit fault rate	Average amount of repairs per year, days			Maximum load, MW
				Maj	Med	Run	
1	50	5	0.05	5		3.6	400
	100	2	0.05	8	3.2	4.8	

2	50	5	0.05	5		3.6	400
	100	1	0.05	8	3.2	4.8	
3	50	15	0.05	5		3.6	4500
	100	15	0.05	8	3.2	4.8	
	200	15	0.05	12.3	6.25	14.5	

Note: Maj - major repairs; Med - medium repairs; Run - running repairs.

Table 2. Description of inter-zone links of the study EPS.

Link	Capacity limit, MW	Length, km	Fault rate, unit fraction /100km	Specific loss factor
1-2	135	400	0.001	0.0001
1-3	135	400	0.001	0.0001
2-3	135	400	0.001	0.0001
2-3	135	400	0.001	0.0001

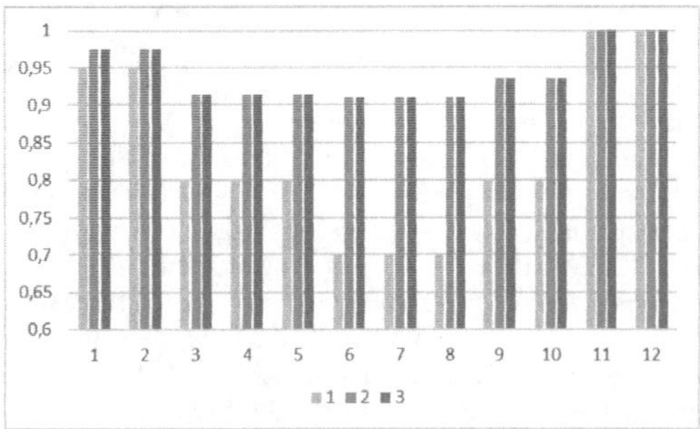

Fig. 3. Schedule of monthly maximums of the load of the reliability zones, the investigated EPS, p.u. (relative to annual maximum load).

Table 3 presents the results of the resource adequacy assessment by the second and fourth algorithms of modeling of scheduled repairs of generating units.

Table 3. Probability values of shortage-free operation when using various algorithms of modeling of scheduled repairs of generating units

Reliability zone number	Probability of shortage-free operation mm	
	Algorithm 2	Algorithm 4
1	0.9991	0.9958
2	0.9965	0.9833
3	0.9963	0.9756
System	0.9951	0.9709

As can be seen from the findings presented in Table 3, the algorithm for modeling of scheduled repairs of generating units affects the probability of shortage-free operation of the EPS. In the case of algorithm No. 4 that we have developed, the probability values for shortage-free operation are lower than those in the case of algorithm No. 2. This is due to the above indicated shortcomings of algorithm No. 2, although algorithm No. 4 can still be improved upon as well in order to provide more satisfactory modeling of scheduled repairs of generating units when assessing the EPS resource adequacy. The study of algorithms was performed based on a test scheme that has zones of reliability with a small number of generating units and a "smooth" profile of monthly peak loads. This is what makes the above difference in the results. The biggest difference in the probability of shortage-free operation is in Zones 2 and 3, as they have a "smoother" profile of monthly peak loads (see Figure 3). When examining the resource adequacy of real-life EPS, similar situations with the load profile and the number of generating units within the reliability zones occur less frequently, so the difference in resource adequacy indicators will be less pronounced.

Conclusion

Dedicated software and computer systems are used to assess the EPS resource adequacy. In these systems, one of the procedures is to factor in scheduled repairs of generating units. The efficacy of an algorithm used for scheduled repairs determines the accuracy of the resource adequacy indicators obtained. On the basis of the analysis of the employed algorithms of factoring in scheduled repairs when assessing the resource adequacy relying on the Monte Carlo method, it is possible to draw a conclusion that all the algorithms have their inherent assumptions affecting the accuracy of the reliability indicators. This study proposes a new algorithm that makes for more accurate assessment of the resource adequacy of electric power systems. In the course of experimental studies, the differences that arise when using the algorithm advocated by the authors and the algorithm used in existing software and computer systems were highlighted.

The research was carried out under State Assignment 17.5.3 (reg. no, AAAAA-A17-117030310450-3).

References

1. Rules for managing the maintenance and repairs of electric power industry facilities. Order of the Ministry of Energy of the Russian Federation dated October 25, 2017, No. 1013. (in Russian)
2. GOST R 57285-2016. Unified electric power system and isolated power systems. Electric power systems. Operations and dispatch control. The procedure of drawing up final reports on the possibility of decommissioning of power plant generating equipment related to dispatching facilities. Standards and requirements. (in Russian)
3. On taking electric power facilities out of service for repairs and their decommissioning. Decree of the Government of the Russian Federation, dated July 26, 2007, N 484. (in Russian)
4. Kovalev G.F., Lebedeva L.M. Reliability of Power Systems. Springer. 2019. 237 p.
5. Billinton R., Li W. Reliability Assessment of Electric Power Systems Using Monte Carlo Methods. Springer. 1994. 361 p.
6. Boyarkin D.A., Krupenev D.S., Iakubovskii D.V. Machine learning in electric power systems adequacy assessment using Monte-Carlo method. Bulletin of the South Ural State University, Series: Mathematical Modelling, Programming and Computer Software. Vol.11. No.4. 2018. P.146-153.

Methodical for the decommissioning of generating units for modernization from the position of ensuring the adequacy of electric power systems

Dmitry Krupenev [1,*], *Gennady* Kovalev [1], and *Ludmila* Lebedeva[1]

[1] Melentiev Energy Systems Institute of Siberian Branch of the Russian Academy of Sciences, 664033, Irkutsk, Lermontov 130 st.

Abstract. The article presents a methodical for determining the procedure for decommissioning for the purpose of modernizing generating equipment while ensuring the highest possible level of adequacy of electric power systems (EPS). The model for assessment of the adequacy of EPS is basis of the methodical. We can choose the most effective option for decommissioning generating units from the standpoint of ensuring the adequacy of EPS using the methodical. The effectiveness of the proposed methodical is shown In the final part of the article.

Introduction

The operational experience of technical equipment, including generating equipment, shows that it wears out and needs to be repaired or replaced. The main requirement is to increase the efficiency of the facility when we carrying out these operations. This can be ensured by the modernization of equipment based on the latest achievements of scientific and technological progress. At the same time, the challenge arises in the electric power systems (EPS) themselves to create conditions that allow for modernization work related to the decommissioning of equipment for quite long periods (up to a year or more). In some cases, a power and energy reserve should be provided for in the EPS to carry out equipment modernization.

The problem of modernization of power equipment for the Russian electric power industry is very urgent. The presence of significant volumes of generating equipment, worn out morally and physically is the reason for this. Thermal power plants (TPP) are very worn out. This equipment reduces the efficiency in the power systems, leads to an increase of the cost of electricity. Currently, a procedure has been organized at the state level for the modernization of obsolete generating equipment at TPPs in the Unified Energy System of Russia [1–3]. At the same time, modernization implies the decommission of generating units for a long period of up to 2 years [2]. This affects the reliability level of the power systems. In [3], criteria for the selection of projects for the modernization of each type of TPP equipment and the rules for the interaction of electricity market entities during selection are presented in detail. The purpose of this article is to develop a methodic for determining the procedure for decommissioning for the purpose of modernizing generating equipment while ensuring the highest possible level of adequacy of EPS. In this case, all the technical characteristics of EPS development schemes necessary for assessing the adequacy [4–5] and the composition of the generating equipment, which must be modernized for a certain period of time, should be known.

1 Methodic for the decommissioning of generating units for modernization from the position of ensuring the adequacy of electric power systems

The initial action to justify the decommissioning of generating equipment for modernization is the formulation of withdrawal criteria. In [1, 3] the following criteria are formulated:

– lifetime;

– demand for power of a generating facility in the power system.

It is also advisable to attribute to the decision-making criteria for the decommissioning of an facility for modernization:

– specific fuel consumption for electricity generation;

– the volumes of unplanned and emergency repairs.

All criteria should be converted into weighting coefficients and integrated into one complex criterion, which should be compared with an economically feasible threshold value. Non-compliance of the actual value of the criterion with the threshold value is the rationale for the decommissioning of equipment for modernization. It is also advisable in the future to develop a system for forecasting the criterion of modernization in order to develop long-term plans for the modernization of generating equipment.

We need the features of the functioning of energy systems should be taken into account when

https://doi.org/10.1051/e3sconf/201913901019

decommissioning equipment for modernization. The main of which are:

– modernization of generating units of thermal power plants should be carried out during the period of the valley of the peak load schedule;

– for heating equipment, it is necessary to provide for the possibility of modernization outside the heating period;

– if there are hydroelectric power plants (HPPs) in the power system, it is necessary to take into account that during the flood period, the HPP can compensate for the retired capacity for modernization.

These features can be comprehensively taken into account when assessing the adequacy of EPS. Adequacy indicators can be signals of determining the best time frame for modernization. We can use indicators such as the probability of a fail-free operation and the mathematical expectation of under-supply of electricity.

The result of modernization of generating equipment is increase of the efficiency and demand of the generating unit in the electricity and capacity market.

The methodic for determining the procedure for decommissioning for the purpose of modernizing generating equipment while ensuring the highest possible level of adequacy of the EPS is a sequence of the following actions:

1. Assessment of the adequacy of the initial plan of the development of the EPS for the period allotted for the modernization of generating equipment, without decommissioning generating units for modernization. A methodic based on the Monte Carlo method is used to assessment of the adequacy. The probability of a fail-free operation and the mathematical expectation of under-supply of electricity is a result for each month of the billing period.

2. Analysis of the values of the probability of fail-free operation in each month and in each EPS (reliability zone) in which the generating equipment selected for modernization is located. Identification of the sequence of months with the highest integral of fail-free probability of a job without deficiency. Moreover, the sequence of months should be equal to the number of months that are required for the modernization of the generating facility with the worst value of the modernization criterion.

3. Assessment of the adequacy of the EPS development option for the period allotted for the modernization of generating equipment, with the decommissioning of the generating unit for modernization.

4. Repeat the second and third stages of this methodic until all generating equipment intended for modernization is taken into account.

According to the results of the compilation of a schedule for modernization of generating equipment, it may turn out that the probability of deficit-free operation in the EPS is low, which does not meet the requirements of electricity consumers for ensuring reliability. In this case, there are several options for solving this problem:

1. If it is technically possible to postpone the modernization dates to later ones, then it is advisable to change the modernization period for some generating facilities.

2. If the forecast schedule for the modernization of generating equipment is relatively "tight" and there is no possibility to postpone the date of modernization, then for the considered EPS, as part of the design for its development, it is necessary to provide an additional reserve of generating capacity to compensate for decommissioning of generating equipment for modernization purposes. The considered excess power is not a part of the power providing the adequacy of the EPS. Additional excess power in large complex EPSs should be evaluated taking into account the place of its placement in the system, the type and parameters of this power, including its reliability.

2 Experimental research

Experimental calculations were carried out according to the scheme of the interconnected power system (IPS) of Siberia. The IPS scheme of Siberia for this research is due to inaccurate source data. The calculations do not pretend to specific recommendations, but are indicative, justifying the proposed methodic. These calculations can be characterized as test ones.

The calculated scheme of Siberia IPS is shown in Figure 1.

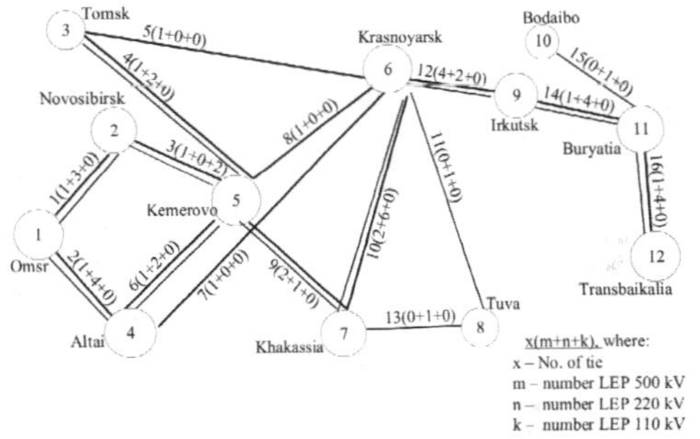

Fig. 1. The calculated scheme of the Siberia IPS.

The main initial data and the results of assessing the adequacy of the initial version of the development of the Siberia IPS (the first step of the methodic for determining the order) are presented in tables 1 and 2.

As can be seen from table 2, in almost all EPS, the values of the probability of deficit-free operation are higher than 0.996, with the exception of those that we call problem power systems. These EPS have their own deficit of generating capacity and insufficient bandwidth of connections with neighboring EPS. The reserve of the system as a whole is 12%, which corresponds to that recommended in [6]. In the initial version, 12% was achieved by reducing the available capacity at hydroelectric power plants.

Table 1. Initial information and the resulting indicators for assessing the adequacy of the initial version (without the output of units for modernization.

No. node	Available power, MW	Absolute maximum of load, MW	Own power reserve	
			MW	% of Pмax
1	1479	1894	-415	-21,91
2	2730	2894	-164	-5,67
3	918	1732	-814	-47
4	1444	1918	-474	-24,71
5	5028	4637	391	8,43
6	11007	7738	3269	42,25
7	4680	2185	2495	114,19
8	39	218	-179	-82,11
9	9670	9526	144	1,51
10	20	90	-70	-77,78
11	898	1029	-131	-12,73
12	1156	1468	-312	-21,25
IPS	39069	35329	4186	12,00

Table 2. Initial information and the resulting indicators for assessing the adequacy of the initial version (without the output of units for modernization.

No. node	Probability of a fail-free operation	The m.e. of under-supply of electricity MW h
1	0,993505	12202,4
2	0,998538	208,2
3	0,998465	442,6
4	0,99853	201,1
5	0,998551	242,1
6	0,998527	307,2
7	0,998848	202,9
8	0,998127	1079,6
9	0,998313	1658,2
10	0,808054	8292,8
11	0,996902	1566,7
12	0,940244	56656,4
IPS	0,782396	83060,1

Table 32 presents the plans for the decommissioning of generating units of the Siberian IPS for modernization for 2023–2025 according to [1–3].

Further, according to the methodic for determining the procedure of decommissioning for the purpose of modernizing generating equipment while ensuring the highest possible level of adequacy of EPS for each year under consideration, a procedure was established for decommissioning generating units. The results were compared with the decommissioning procedure of generating units, presented in table 3.

Table 3. Schedule for the modernization of units at power plants of the Siberia IPS.

No. node	Name of generating facility	Equipment station number	Installed capacity of the unit, MW	Modernization Start Date	Number of months
2023					
5	Tom-Usinsk SDPS	TG-7	200	01.01.2023	18
6	Krasnoyarsk TPP 2	TG-1	110	01.01.2023	12
9	Irkutsk CHP-10	TG-2	150	01.01.2023	10
9	Novo-Irkutsk TPP	TG-3	175	01.01.2023	11
2024					
6	Krasnoyarsk TPP 1	TG-10	87	01.01.2024	6
6	Krasnoyarsk TPP 1	TG-15	70	01.12.2024	11
6	Krasnoyarsk TPP 1	TG-16	70	01.12.2024	11
6	Krasnoyarsk TPP 3	TG-2	110	01.12.2024	12

9	Irkutsk TPP-9	TG-6	60	01.01. 2024	10
9	Irkutsk TPP-10	TG-7	150	01.05. 2024	10
9	Irkutsk TPP -10	TG-8	150	01.01. 2024	10
9	Irkutsk TPP -11	TG-3	50	01.01. 2024	11
2025					
6	Krasnoyarsk TPP 1	TG-15	70	01.01. 2025	10
6	Krasnoyarsk TPP 1	TG-16	70	01.01. 2025	10
6	Krasnoyarsk TPP 3	TG-2	110	01.01. 2025	11
9	Irkutsk TPP -10	TG-7	150	01.01. 2025	2
1	Omsk TPP-4	TG-9	135	01.01. 2025	12
3	TPP CXK	TG-1	60	01.07. 2025	6
3	TPP Seversky Chemical Combine	TG-2	60	01.07. 2025	6
4	Biysk TPP - 1	TG-6	110	01.12. 2025	9
9	Ust-Ilim TPP	TG-3	110	01.05. 2025	9
9	Irkutsk TPP ТЭЦ-10	TG-5	150	01.12. 2025	9
12	Chita TPP-1	TG-4	87	01.11. 2025	7
12	Chita TPP - 1	TG-3	80	01.12. 2025	10

Table 4 shows an example of adjusting the timing of modernization of units for 2024. The column "Date of modernization commencement" in bold represents the modified dates compared to the dates of the commencement of modernization in the regulatory documents presented in Table 3.

Table 4. Adjustment of terms for modernization of generating equipment in 2024.

Name of generating facility	No. Equipment Station	Installed capacity of the unit, MW	Modernization Start Date	Number of months	Upgrade End Date
Krasnoyarsk TPP 1	TG-10	87	**01.04**	6	30.10
Krasnoyarsk TPP 1	TG-15	70	01.12	11	modernization postponed to 25
Krasnoyarsk TPP 1	TG-16	70	01.12	11	
Krasnoyarsk TPP 3	TG-2	110	01.12	12	
Irkutsk TPP -9	TG-6	60	**01.02**	10	30.11
Irkutsk TPP -10	TG-7	150	**01.02**	10	30.11
Irkutsk TPP 10	TG-8	150	**01.03**	10	30.12
Irkutsk TPP -11	TG-3	50	**01.02**	11	30.12

Analyzing the results obtained, we can conclude that the period from April to October with the highest adequacy indicators for each of the considered years (2023–2025) coincides or mainly coincides with the period of the lowest loads. In March and November, the probability of a fail-free operation is at a fairly high level. In December, the situation with this indicator worsens somewhat, but it is still high (0.99 ... –0.999 ...). And in January and February, these indicators are quite low (0.90 ... –0.98 ...). Periods of modernization are adjusted, maintaining their duration. If by duration it is possible to place them in the field of high adequacy, then they are placed there.

The indicative result of adjusting the timing of modernization of generating equipment is the value of the mathematical expectation of under-supply of electricity. Tables 5–10 show the probabilities of fail-free operation and the mathematical expectation of under-supply of electricity received as a result of settlements for 2023–2025 according to plans for modernization of generating equipment and with the adjustment of their terms.

Table 5. The results of the calculations taking into account the modernization of units according to plan in 2023.

No. zone	Probability of a fail-free operation	The m.e. of under-supply of electricity MW h
1	0,987308	26134,1
2	0,996307	605,5
3	0,996170	1345,9
4	0,996311	638,9
5	0,996422	750,4
6	0,99636	840,9
7	0,997362	429,6
8	0,995664	2919,1
9	0,995694	6233,0
10	0,809202	10172,2

11	0,992951	4204,4
12	0,934592	72070,9
IPS	0,781171	126344,8

Table 6. The results of the calculations, taking into account the adjustment of the terms of modernization of units in 2023

No. zone	Probability of a fail-free operation	The m.e. of under-supply of electricity MW h
1	0,989189	21523,4
2	0,997087	487,7
3	0,996997	1156,1
4	0,997065	505,2
5	0,997194	630,5
6	0,997124	707,0
7	0,997895	290,7
8	0,996595	2284,6
9	0,996684	4270,9
10	0,808671	9446,5
11	0,99438	3244,9
12	0,936554	66474,7
IPS	0,781422	111022,2

Table 7. The results of the calculations taking into account the modernization of units according to plan in 2024.

No. zone	Probability of a fail-free operation	The m.e. of under-supply of electricity MW h
1	0.988387	23446.5
2	0.996750	605.1
3	0.996634	1196.8
4	0.996726	574.7
5	0.996824	702.1
6	0.996778	768.4
7	0.997662	304.4
8	0.996219	2531.1
9	0.996252	4962.7
10	0.809111	9659
11	0.993715	3732.1
12	0.935852	69100.1
IPS	0.781349	117583

Table 8. The results of the calculations, taking into account the adjustment of the terms of modernization of units in 2024.

No. zone	Probability of a fail-free operation	The m.e. of under-supply of electricity MW h
1	0.991550	15974.7
2	0.998021	358.7
3	0.997996	781.5
4	0.998014	359.6
5	0.998097	407.7
6	0.998046	438.3
7	0.998503	265.3
8	0.997573	1569.7
9	0.997741	2382.8
10	0.808300	8857.6
11	0.995755	2339.9
12	0.938384	61348.1
IPS	0.781816	95084,0

Table 9. The results of the calculations taking into account the modernization of units according to plan in 2025.

No. zone	Probability of a fail-free operation	The m.e. of under-supply of electricity MW h
1	0,981878	40739,6
2	0,996607	546,0

3	0,996564	1436,9
4	0,996629	526,4
5	0,996744	772,6
6	0,996698	839,2
7	0,997584	377,8
8	0,996031	2416,5
9	0,996194	4753,7
10	0,810172	9725,8
11	0,992625	3730,1
12	0,896632	122255,4
IPS	0,763576	188120,1

Table 10. The results of the calculations, taking into account the adjustment of the terms of modernization of units in 2025.

No. zone	Probability of a fail-free operation	The m.e. of under-supply of electricity MW h
1	0,981670	40561,9
2	0,996718	483,3
3	0,996683	1405,4
4	0,996693	474,0
5	0,996818	780,0
6	0,996840	804,1
7	0,997634	317,2
8	0,996097	2290,3
9	0,996394	4167,8
10	0,809292	9764,8
11	0,992727	3503,4
12	0,925088	79047,4
IPS	0,770604	143599,7

The results of adjusting the timing of modernization of generating equipment showed:

– for 2023 the mathematical expectation of under-supply of electricity of the IPS as a whole, taking into account modernization according to the plan, is 126344.8, and with an adjustment of terms – 111022.2 MW h. That is, the mathematical expectation of under-supply of electricity of IPS decreased by 15322.6 MWh, which is approximately 12%.

– for 2024 the mathematical expectation of under-supply of electricity of the IPS as a whole, taking into account modernization according to plan, is 117583, and with an adjustment of the terms – 94773 MW h. That is, for IPS the mathematical expectation of under-supply of electricity decreased by 22810 MWh, which is approximately 20%.

– for 2025 the mathematical expectation of under-supply of electricity of the IPS as a whole, taking into account modernization according to plan, is 188,120.1, and with an adjustment of the terms – 143,599.7 MW h. That is, the mathematical expectation of under-supply of electricity of the IPS decreased by 44520.4 MW h, which is approximately 23.6%.

Analyzing the probabilities of a fail-free operation, we can conclude that their values correlate with the values of the mathematical expectation of under-supply of electricity and taking into account modernization according to plans is slightly less than in calculations with adjustment of the terms of modernization. An exception in some cases are the most problematic nodes. There are objective reasons for this.

The calculation results for all the years analyzed show that the application of the proposed methodic, based on indicators of adequacy, makes it possible to increase the efficiency of decision-making. As shown above, the increase in efficiency is most evident in terms of the mathematical expectation of under-supply of electricity.

Conclusions

The modernization of electric power facilities makes it possible to increase their efficiency. During modernization, facilities are decommissioned for a certain period of time. The decommissioning of facilities affects the adequacy of the EPS. It is important that the time period is optimally predicted for decommissioning of EPS. At the same time, minimal damage was caused to the EPSs during this period. The article proposes a methodic for determining the procedure for decommissioning for the purpose of modernization of generating equipment. At the same time, the maximum possible level of adequacy of EPSs should be ensured.

Experimental studies have shown the effectiveness of the methodic.

The research was carried out under State Assignment 17.5.3 (reg. no, AAAAA-A17-117030310450-3), with the financial support by the Russian Foundation for Basic Research within the framework of the scientific project No. 18-58-06001.

References

1. "On the selection of projects for the modernization of generating facilities of thermal power plants." Decree of the Government of the Russian Federation of January 25, 2019 No. 43. (in Russia)
2. "The list of generating objects, the capacity of which is supplied under the purchase and sale (supply) agreements of the capacity of modernized generating objects." Order of the Government of the Russian Federation of 02.08.2019 No. 1713-r. (In Russia)
3. "The procedure for the selection of projects for the modernization of generating equipment for thermal power plants" was approved on September 20, 2018 (Minutes No. 20/2018 of the meeting of the Supervisory Board of NP Market Council Association). (In Russia)
4. G. F. Kovalev, L. M. Lebedeva. Reliability of Power Systems. Springer. 237 p. (2019),
5. R. Billinton, W. Li. Reliability Assessment of Electric Power Systems Using Monte Carlo Methods. Springer. 361 p. (1994)
6. Methodical Instructions for designing the development of power systems. Izd-vo NTs ENAS, Moscow. (2003). (In Russia)

https://doi.org/10.1051/e3sconf/201913901020

Method for assessing the voltage stability limit during emergency outages of the line and generators in the power system

Nariman Rahmanov[1], *Victor* Kurbatsky[2], *Huseyngulu* Guliyev[1], *Nikita* Tomin[2], *Famil* Ibrahimov[1]

[1] Azerbaijan Scientific–Research and Designed–Prospecting Institute of Energetics, Baku, Azerbaijan
[2] Melentyev Energy Systems Institute Siberian Branch of the Russian Academy of Sciences, Irkutsk, Russia

Abstract. The paper proposes method and algorithm for obtaining the maximum transmitted power limit in the power system under the voltage stability conditions in case of emergency outages of its elements - power lines and generating sets. The method is based on the analysis of the exaggerated states of the system created by increasing the values of the power sources (generators of power plants). The model of system states in extreme modes is described in the form of "extreme transmitted power – generating power values" relationships. The efficiency of the method is investigated on IEEE test schemes and the real Azerenergy system scheme. A comparison of the proposed method results with the evaluation of the critical conditions on the basis of trained neural network is presented.

1. Introduction

In recent years, the emergency outages of main elements happened in modern power systems – generating units, power lines and transformers, as a result of limiting the possibilities of transit of large powers through transmission networks, lack of generation powers, corrosion of equipment and other factors. During frequent outages of these components, especially in maximum consumption conditions, the power system activity leads it to a state close to the voltage instability boundary, and the subsequent state can lead to a complete "blackout" of its most part.

Taking into account the existing state formed in the power system and frequent emergency outages of its components, there is a need to strengthen the requirements for voltage stability.

Other important factors leading to the formation of voltage stability are the electrical system configurations, the characteristics of generating sets and loads. With the aim of preserving the power system stability in the postemergency modes, subject to the cutoff of the main elements, the shunt capacitances (capacitor banks) and flexible transmission system (FACTS) installed in the "weak" nodes of the network are used [1-4].

The practical implementation of a large number of proposed methods for the determination of the static stability limit of the power grid leads to the analysis and control of successive exaggerating modes of electric networks up to maximum transmitting power. Usually, the exagerration is carried out by means of the control of the power generated by generators or the consumed power of load.

In paper [5] the linear and quadratic estimation of the network load (NL) for the predetermined system parameters, including the generator power values and loads in the nodes, is performed using the sensitivity analysis method. The searching of the node, in which the power value exaggeration from the normal to the critical state is possible, cannot be carried out by means of realization of this approach. The possibility of the power system mode exaggeration with the participation of all sources to the maximum power transmission in the network is considered in the paper.

The method of increase (expansion) of safety limits on the basis of exaggeration by means of the control of the power generated by the sources is proposed in the paper [6]. This method allows for obtaining the maximum generating power at the local point of the network.

The method of mode exaggeration to the optimal critical value using the source model is proposed in the paper under consideration. The proposed methodology allows for the determining the load limit in the form of generation power change dependence. This method makes it possible to reach the load limit of the network with the participation of all generating sources in the mode exagerration process.

2. Methods of system mode exaggeration with the increase of the consumed power of load

The essence of these methods is to identify the share of each source in the increase of the transmitted load and the losses of this power that occurred during transmission from the network. Let's assume that K_{Gi} is an indicator of increase of the generated active power of the generator, and $P_{Gi,0}$ is the basic (nominal) generated power of this source. In this case, the generation power P_{Gi} at the point, following the power increase, will be as follows:

$$P_{Gi} = P_{Gi,0}(1 + K_{Gi}) \quad (1)$$

here for all the generators involved $i = 1,2,...,n$.

The value K_{Gi} of i generator shows the degree of participation in the violation of the voltage stability limit. Analysis of existing methods allowing to establish a list of generators involved in the determination of voltage stability is given below.

Traditional approach. Usually in the system the generation, as in the case of planning in accordance with the existing rotary speed reserve, exceeds the predetermined value [7]. After the load power increase, the i generator power can be represented as follows:

$$P_{Gi} = P_{Gi,0}(1 + K_{Gi}) = P_{Gi,0} + \Delta P_{Gi} \quad (2)$$

$$\sum_{NG} \Delta P_{Gi} = \Delta P_D + \Delta P_{loss} \quad (3)$$

where P_{Gi} – generated power of generator i; $P_{Gi,0}$ – power of generator i during rated load; ΔP_{Gi} – increase of power of generator i; ΔP_D – load power increase; ΔP_{loss} – power loss increase; NG – number of generators.

Approach based on simulation of power distribution optimization. The traditional model of optimal power distribution can be formed taking into account the voltage stability criterion [8]. Minimization of the total costs for the generated power of generating units is as follows:

$$A(P_{Gi}) = \sum_{NG} (a_{Gi}P_{Gi}^2 + b_{Gi}P_{Gi} + c_{Gi}) \quad (4)$$

In this case, the following condition must be met:

$$P_{Gi} - (1+\lambda)P_{Di,0} - \sum_{j=1}^{n} |U_i||U_j|(G_{ij}\cos\delta_{ij} + B_{ij}\sin\delta_{ij}) = 0 \quad (5)$$

$$Q_{Gi} - (1+\lambda)Q_{Di,0} - \sum_{j=1}^{n} |U_i||U_j|(G_{ij}\sin\delta_{ij} - B_{ij}\cos\delta_{ij}) = 0 \quad (6)$$

$$|P_{Gi}|_{\min} \le |P_{Gi}| \le |P_{Gi}|_{\max} \quad (7)$$

$$|U_i|_{\min} \le |U_i| \le |U_i|_{\max} \quad (8)$$

$$S_{ij} = \sqrt{P_{ij}^2 + Q_{ij}^2} \le S_{ij,\max} \quad (9)$$

where a_{Gi}, b_{Gi}, c_{Gi} – the coefficients of the "costs – i unit power" dependence function; P_{Gi}, Q_{Gi} – active and reactive powers of the unit i respectively; $P_{Di,0}, Q_{Di,0}$ – active and reactive powers of the power in the node i respectively; $|U_i|_{\min}, |U_i|_{\max}$ – lower and upper voltage limits in the node i respectively; $|P_{Gi}|_{\min}, |P_{Gi}|_{\max}$ – lower and upper generated voltage limits in the node i respectively; P_{ij}, Q_{ij}, S_{ij} – active, reactive and total powers in ij line respectively.

In this method, the exaggeration value is determined in the form of the difference between the optimal value and the originally defined values.

Method of selection of the boundary voltage value in electrical networks of the power system. This method allows for determining the maximum exaggeration value - load limit as a function of the active powers of generating units. The practical realization of this method can be carried out in the following three stages:

1st stage. The "load limit $P_{H\max}$ - power increase value of each generator" dependence is determined. If the boundary load value is not taken into account as a basic, then it is determined by the $P-U$ curves constructed for the series of corresponding generators P_{Gi}. The $P_{H\max}(P_{Gi})$ graph is then approximated in the form of the following polynomial [9]:

$$P_{HM(j)} = B_{j,0} + B_{j,1}P_{G,j} + ... + B_{j,n}P_{G,j}^n = = B_{j,0} + \sum_{p=1}^{n} B_{j,p}P_{G,j}^p = B_0 + P_{Hm,j} \quad (10)$$

where $P_{HM(j)}$ – polynomial approximating the P_{HM} curve in the case of two generators (basic generator and other generator associated with j node); B_{ij} – coefficients of approximating polynomial; n – number of coefficients; $P_{HM}1$ – if the constant is not taken into account, then it is equal to P_{HM}.

2nd stage. The $P_{H\max}$ surface approximated for the multidimensional case (the case when more than 2 generating units participate in the maximization of the load limit of the network) was based on the division condition. In the case of participation of all units, the $P_{H\max}$ is determined by the following formula:

$$P_{H\max} = B_0 + \sum P_{H\max,j} \quad (11)$$

Equation (11) describes the $P_{H\max}$ surface for all possible generators in the power system. Data for this can be obtained from combinations of all polynomial equations in the form of (10).

3rd stage. The last stage of the proposed method consists of determination of the best vector $P_{G,i}$, which gives a high value of the surface $P_{H\max}$. This value can be determined by means of maximization of the formula (11) within the following constraints:

$$\sum P_{Gj} = 1, \quad 0 \le P_{Gj} \le 1$$

Studies at the first stage, consisting of determination of the dependence of the transmitted power limit $P_{\max}$ on the generated power value of each generator for the cases of the most probable emergency failures of individual elements of the system, are carried out by means of designed industrial program for the existing probability of power distribution. ETAP is used as such a program in the paper [10].

Thus, series of dependences of $P_{\max}^j$ on the power P_{Gi}^j of individual generator i can be obtained for each case of emergency failure "j" during the design experiments:

$$P_{\max}^{j} = F\left(P_{Gi}^{j}\right) = b_0^j + \sum P_{\max} \quad (12)$$

For each case of emergency failure, the global maximum of the quantity $P_{\max}^{j}$ reflecting the impact of the change of powers of all sources on $P_{\max}^{j}$ is determined by the extreme value of the function (12). For this purpose, one of the traditional methods can be used (for example, the Lagrangian method).

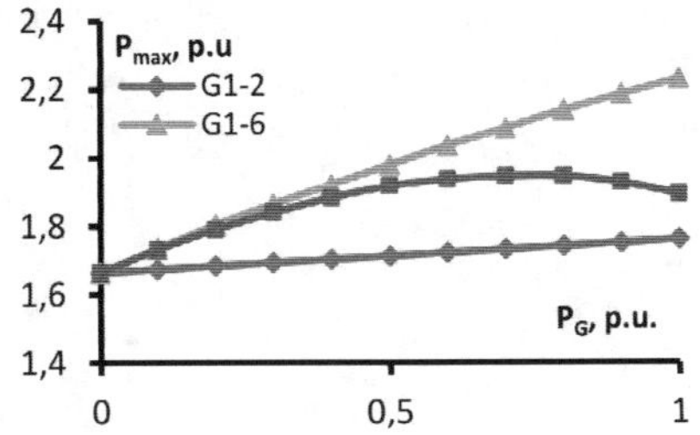

a)

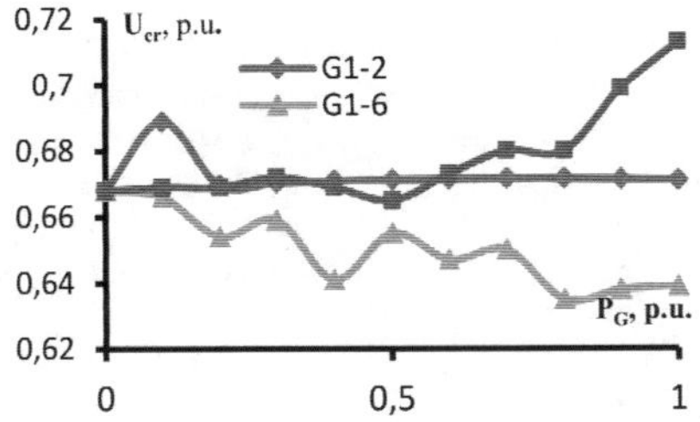

b)

Fig. 1. Curves of transmitted power limit (a) and critical voltage for option of 2 generators allocated for regulation

3. Evaluation of the maximum system stability limit in different states of the network elements and generating unit

In accordance with the proposed method, the limiting state of the system is estimated to reach the maximum of the $P-U$ curve at different values of the power increase and powers of generating sources. The size of the spatial state of the system in coordinates of the quantity of powers is determined by the number of sources selected for the operational dispatching control of the power system. The number of sources involved in the operational control of the 14-node IEEE test scheme is 4. First, the effect on the voltage stability of 2 generators, then 3 generators and finally 4 generators is studied. The powers of sources connected to nodes 1, 2, 3, 8 are 150, 78, 40 and 40 MW, respectively. The total power consumption in the system is 260 MW.

When regulating during the simulation with the participation of two generators, the options of paired generators associated with the following buses were considered: 1 and 2 (G1-2), 1 and 6 (G1-6), 1 and 8 (G1-8). In all studied options the generator connected to bus 1 was accepted as balancing one.

The characteristic of the transmitted power limit change of line 1-2 is presented in the Fig.1,a. These curves were constructed on the basis of calculations for the voltage stability limit during the change of powers of generators 2, 6, 8. It should be noted that the overall effect obtained from all pairs of generators is estimated as follows:

$$P_{G,1} + P_{G,2} = 1; \; P_{G,1} + P_{G,6} = 1; \; P_{G,1} + P_{G,8} = 1; \quad u$$

This suggests that, for example, if 20% is compensated by G2, then the remaining 80% is compensated by balancing source.

The graphs of critical voltage limit changes in the cases of the considered pair generators are represented in the Fig. 1,б.

As can be seen from Fig. 1, during various options of paired generators involved in the power regulation, the boundary parameters ($P_{\max}$, U_{cr}) have different nature of change. Approximating relationships are presented in Table 1.

As can be seen from Fig.1, the maximum transmitted power limit value is formed in generators 6 and 8 respectively during the generation increase. For example, in 8th generator the power increase only in the range of 0.6 (r. u.) can provide the maximum power limit in the network.

Table 1. Coefficients of approximating relationships $\left(P_{\max}\right)_i = f\left(P_{G,i}\right)$

Coefficients	G1-2 (j=2)	G1-6 (j=6)	G1-8 (j=8)
$b_{j,0}$	1,663	1,663	1,663
$b_{j,1}$	0,0984	0,6518	0,7668
$b_{j,2}$	-0,0011	1,3101	-0,9968
$b_{j,3}$	0,000609	-9,5352	2,3964
$b_{j,4}$		27,3342	-4,4472
$b_{j,5}$		-39,3054	3,1079
$b_{j,6}$		27,9063	-1,1972
$b_{j,7}$		-7,7942	

The values of the coefficients of approximating polynomials, which are given in Table 1, represent the curves in Fig.1 in the following form:

$$P_{\max(2)}\left(P_{G,2}\right) = B_0 + \sum_{p=1}^{3} B_{2,p} P_{G,2}^{p}$$

$$P_{\max(6)}\left(P_{G,6}\right) = B_0 + \sum_{p=1}^{7} B_{6,p} P_{G,6}^{p}$$

$$P_{\max(8)}\left(P_{G,8}\right) = B_0 + \sum_{p=1}^{6} B_{8,p} P_{G,8}^{p}$$

These polynomials are used to describe the spatial state surface of the system in the case of 3 and 4 generators involved for the subsequent control of the power balance mode in the system.

The curves of change of power limit and critical voltage in the system for the options of participation of paired generators 1-2, 1-6, 1-8 in the regulation of general load transfer mode in the case of emergency failure of one of the lines (case N-1) are represented in the Fig.2.

In the case of pair generator 1-8, the maximum transmitted power limit reaches $0{,}6P_{\max(n-1)}$ at $P_{G,8}=0{,}65$, and this indicates the decrease of the probable value of the transmitted power limit to $1{,}8P_{\max}/0{,}6P_{\max(n-1)}=3$. Accordingly, during the failure of one of the lines in the system, the indicators, that determine the critical voltage value, deteriorate.

Equations approximating the curves in Fig.2, and their coefficients are presented below (Table 2).

$$P^{(2)}_{\max(n-1)} = B_0 + \sum_{p=1}^{7} B_{2,p} P^{p}_{G,2}$$

$$P^{(6)}_{\max(n-1)} = B_0 + \sum_{p=1}^{6} B_{6,p} P^{p}_{G,6}$$

$$P^{(8)}_{\max(n-1)} = B_0 + \sum_{p=1}^{8} B_{8,p} P^{p}_{G,8}$$

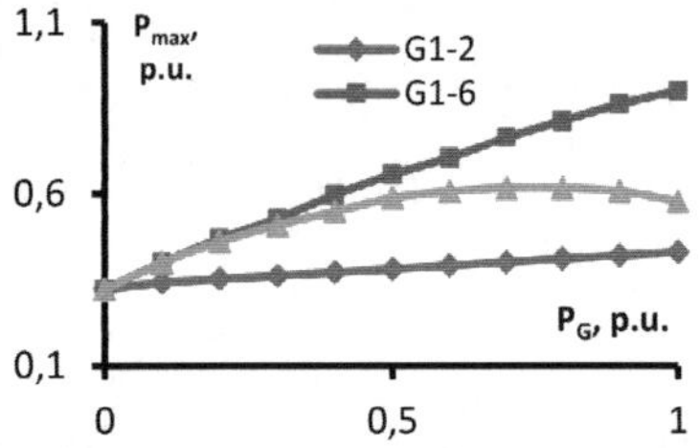

a)

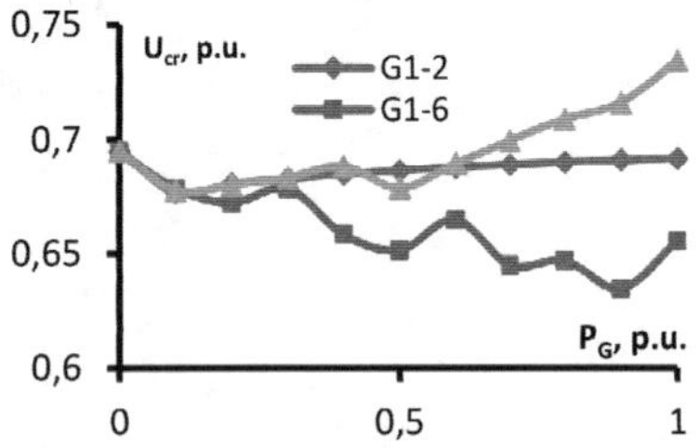

b)

Fig. 2. Change of the limiting state of the system by transmitted power value in the network in the case of regulating source 2 and emergency failure of one element a – transmitted power limit; b-critical voltage

Table 2. The values of coefficients of approximating relationships in the case of regulating source 2 and emergency failure of one element

Coefficients	G1-2 (j=2)	G1-6 (j=6)	G1-8 (j=8)
$b_{j,0}$	0,3228	0,3228	0,3228
$b_{j,1}$	0,3129	0,9175	0,9044
$b_{j,2}$	-1,740	-1,184	-1,1935
$b_{j,3}$	6,9375	6,3475	-1,1513
$b_{j,4}$	-15,19	-11,56	12,9309
$b_{j,5}$	18,461	10,086	-28,649
$b_{j,6}$	-11,69	-3,362	25,9782
$b_{j,7}$	3,0103		-8,5654

Case of three generators. Let's consider the option of three generators placed in nodes 1, 2, 6 (G1-2-6); 1, 2, 8 (G1-2-8) and 1, 6, 8 (G1-6-8) respectively. The transmitted power limit on the network elements can be approximated by combination of the considered options for the case of two generators in all three options, showing the spatial limit state of the system in coordinates. For example, when considering G1-2-6 option, the $P_{\max}$ surface can be obtained by combining curves for generators 2 and 6 as follows:

$$P_{\max} = B_0 + \sum_{p=1}^{3} B_{2,p} P^{p}_{G,2} + \sum_{p=1}^{5} B_{6,p} P^{p}_{G,6} \qquad (13)$$

Similarly, the model can also be identified for options 1-2-8 and 1-6-8:

$$P_{\max} = B_0 + \sum_{p=1}^{3} B_{2,p} P^{p}_{G,2} + \sum_{p=1}^{6} B_{8,p} P^{p}_{G,8} \qquad (14)$$

$$P_{\max} = B_0 + \sum_{p=1}^{5} B_{6,p} P^{p}_{G,6} + \sum_{p=1}^{6} B_{8,p} P^{p}_{G,8} \qquad (15)$$

It is possible to determine the powers of the generators P_G, allowing to estimate the maximum number $P_{\max}$ for each option from the sum total of considered generators, by optimizing the above equations. For example, for G1-2-6 option, the optimization process is described as follows:

$$\max P_{\max} = B_0 + \sum_{p=1}^{3} B_{2,p} P^{p}_{G,2} + \sum_{p=1}^{5} B_{6,p} P^{p}_{G,6} \qquad (16)$$

The conditions are as follows:

$$0 \le P_{G,j} \le 1; \qquad j = 2,6$$

$$P_{G,1} + P_{G,2} + P_{G,6} = 1;$$

Approximated curves and surfaces for the options G1-2-6, G1-2-8, G1-6-8, G1-2-6-8 obtained from equations (13) - (15) are shown in Fig.3. The spatial description, shown in Fig.4, is obtained by optimization and has the maximum coordinates $P_{\max}$ of 0.2 and 0.8, respectively, for the generators 2 and 8.

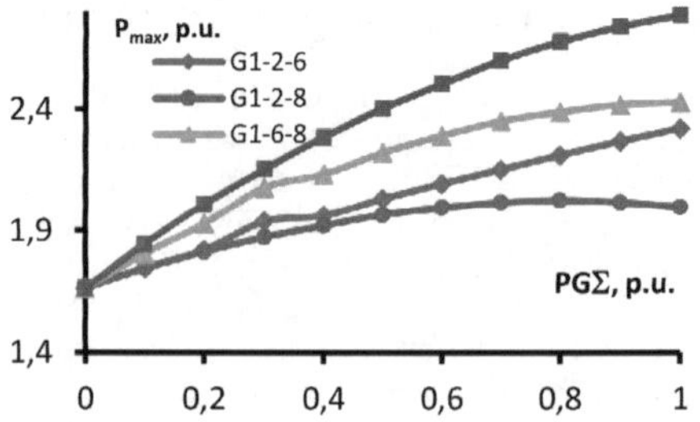

https://doi.org/10.1051/e3sconf/201913901020

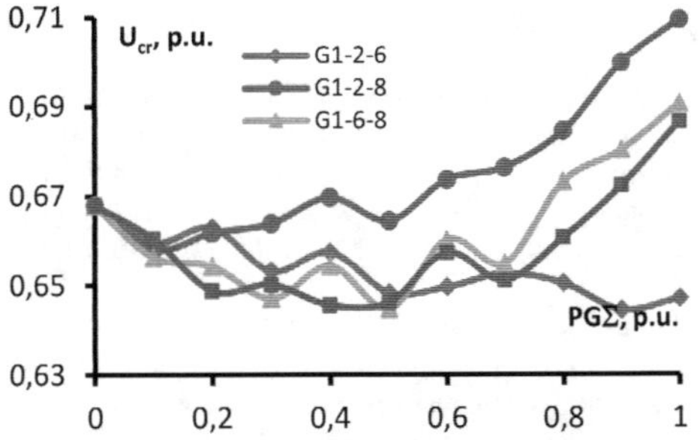

a)

Fig. 3. Curves of power limits (a) and critical voltage (b) in cases of regulating sources 3 and 4

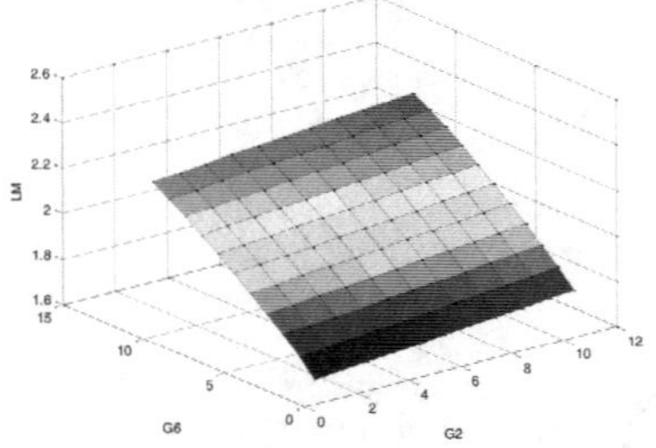

a)

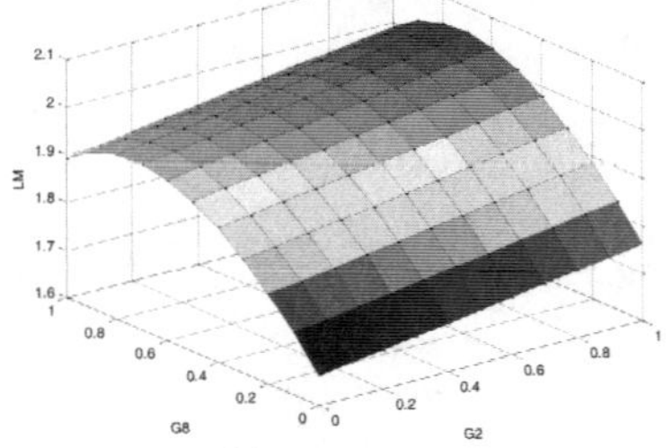

b)

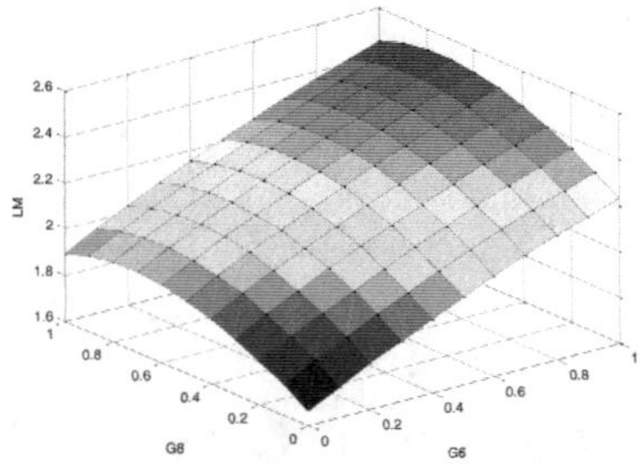

c)

Fig. 4. Surfaces of power limit for different cases a – case G1-2-6; b – case G1-2-8; c – case G1-6-8

This analysis was carried out on the basis of the results of computational experiments for the previously considered schemes and modes of the power system, but taking into account emergency failures of lines and sources. The curves and surfaces of the limit of power transmitted from the network in the case of options with different number of generators involved in power regulation are shown in the Fig.5-7.

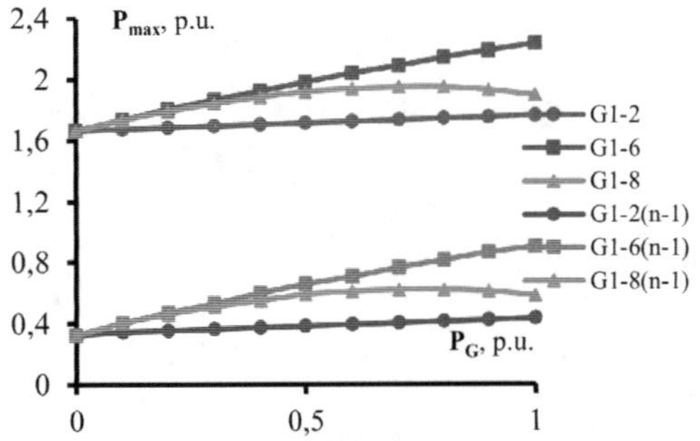

a)

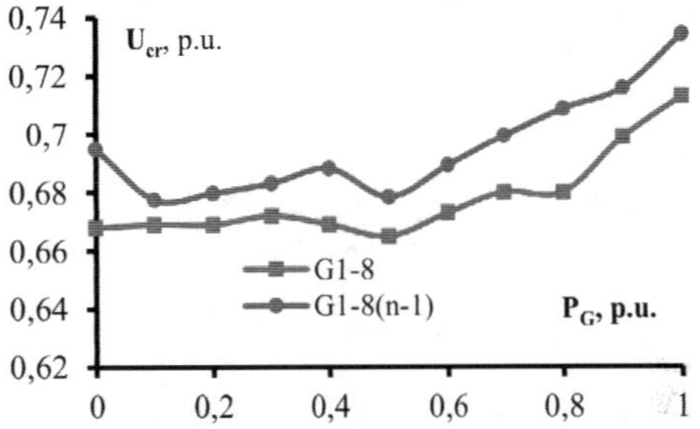

a) б)

Fig. 5. 2 Curves of power limits (a) and critical voltage(b) in cases of regulating source 2 and emergency failure of one element

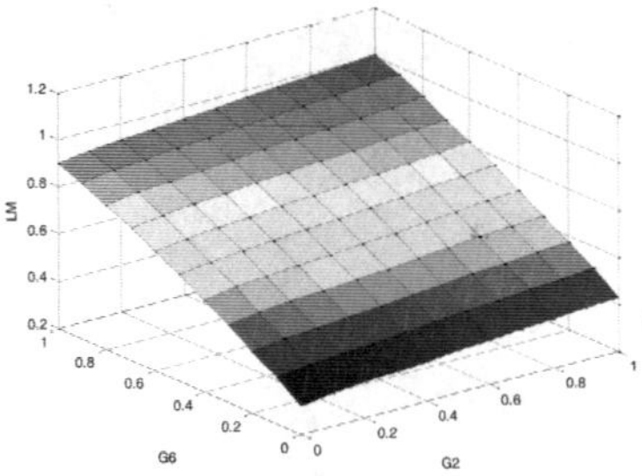

a)

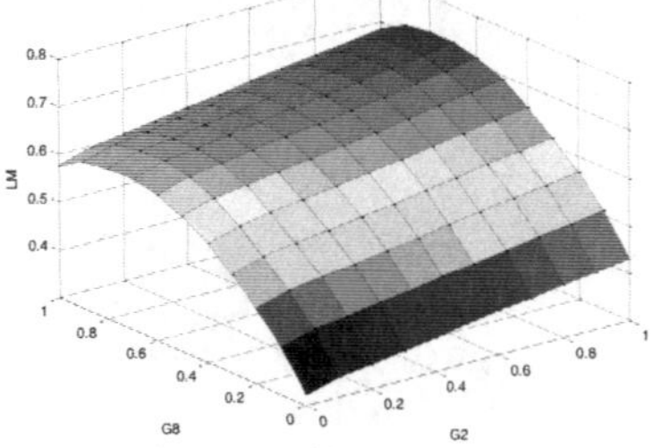

b)

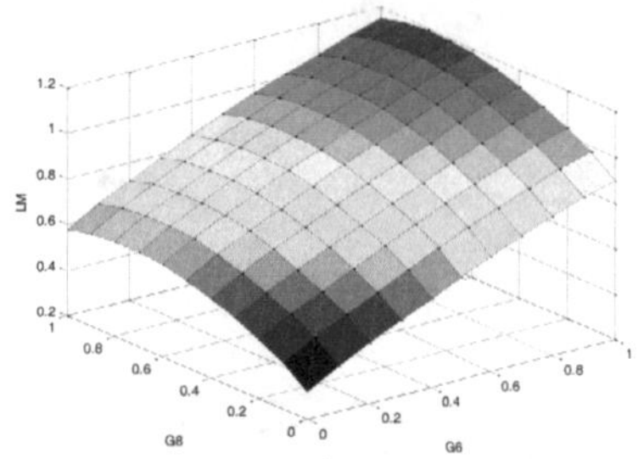

c)

Fig. 6. Surfaces of power limit transferred according to N-1 criterion
a – case G1-2-6; b – case G1-2-8; c – case G1-6-8

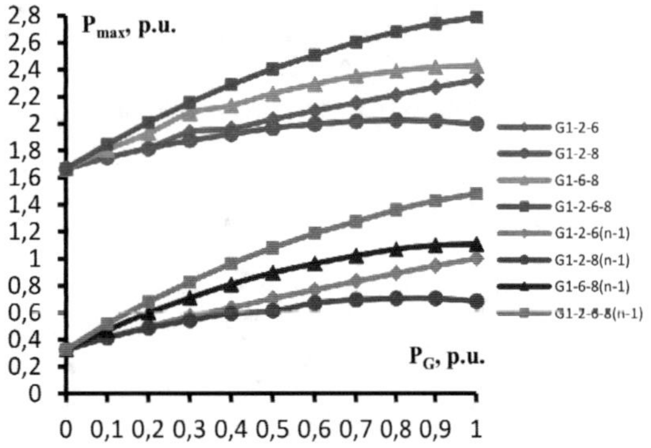

a)

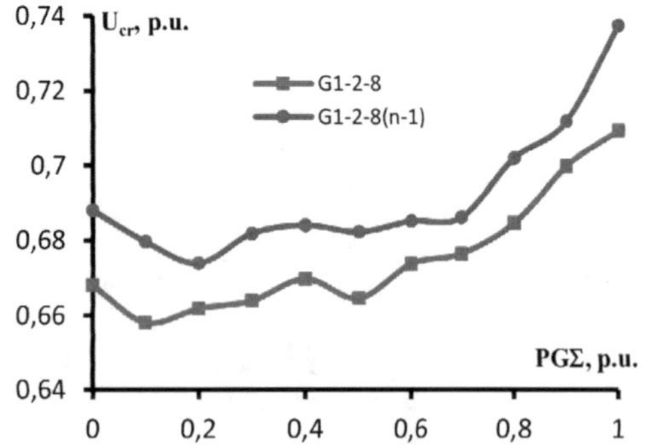

b)

Fig. 7. Curves of power limits (a) and critical voltage (b) in cases of regulating sources 3 and 4 and emergency failure of one element

The results obtained for parameters $U_{cr,i}$ and $P_{\max,i}$ under conditions of different loads of generators and cases of action of one or more sources along with the base node in the system make possible to perform efficient measures for operational control of modes and determination of the most critical modes in different states of the power system.

4. Conclusions

1. The methods of increase of the consumed power of the load and the system mode exaggeration were investigated, these methods allow for determining the number and power of sources involved in the mode exaggeration, but can not be used in multi-machine systems.

2. Calculations were performed on standard schemes of the probable power distribution in cases of outage of lines and generators due to chance causes. On the basis of the calculation results, the critical voltage deviations in U_{kr} nodes and the most critical nodes were determined under the condition of stability of the maximum power flows in the $P_{\max}$ lines. The effects of individual outages on critical quantities $U_{cr,i}$ and $P_{\max,i}$ were determined .

3. Critical quantities $U_{cr,i}$ and $P_{\max,i}$ were determined for the cases of different generator loads and the availability of one or more sources along with the backbone node in the system, as well as the dynamics of change. The obtained results allow for developing rational measures for operational control of modes and determination of the most critical rconditions in various states of the power system.

References

1. A complex of intelligent tools for the prevention of major accidents in power systems / Voropay N.I., Kurbatsky V.G., Tomin N.V. et al. Novosibirsk: Nauka, 2016, 332 p. ISBN 978-5-02-038717-1.

2. A. . Sode – Yome and N. Mithulananthan, "A maximum loading margin method for static voltage stability in Power Systems", IEEE Transactions on Power Systems, vol. 21., No. 2, may 2006. Pp. 799-808.

3. N. Rahmanov., J. Bilbao, N. Tabatabayev, O. Kerimov, R. Rahmanov, "Probabilistic estimation of voltage stability limits in a power system with integrated distribution generation". Proceedings of the 12-th International Conference on Technical and Physical Problems of Electrical Engineering, 7-9 September 2016, Bilbao, Spain, pp. 5-11.

4. R. Shadmesgaran, A. Hashimov, N. Rahmanov "Optimal Location and capacity of parallel FACTS devices in order to improve static voltage stability and power losses reduction using genetic algorithm". Proceedings of the 12th International Conference on Technical and Physical Problems of Electrical Engineering, 7-9 September 2016, Bilbao, Spain, 102-106.

5. S. Grun, V. Dobson, and F. Alvarado "Sensitivity of loading margin to voltage collapse with respect to arbitrary parameters", pp. 362-372, Feb. 1997.

6. L. C. P. Da Silva, Y. Wang, V.F. da Costa, ana W. Xu, "Assesment of generator impact on system power transfer capability using modal participation factors", Proc. Inst. Elect. Eng., Generation, Transm., Distrib. Vol. 149 No-5., pp. 570-664. Sep. 2002.

https://doi.org/10.1051/e3sconf/201913901020

7. R. Wong and R.H. Lasseter, "Redispatching generation to increase power system security margin and support low voltage bus". IEEE Trans. Power Syst. Vol. 15., No -2., pp. 446-501, may 2000.

8. W.D. Rosehart, C.A. Canizares, V.H. Quintana, "Effect of detailed power system models in traditional and voltage stability constrained optimal power flow problems", IEEE Trans. Power Syst., vol. 18, No. 1, pp. 27-35, Feb. 2003.

9. A. Sade – Yome, and N. Mithulananthan, "Maximizing static voltage stability margin in power systems using a new generation pattern", in Proc. Anstralasion Universities Power Engineering Conf. Brisbane, Australia, Sep. 26-29. 2004.

10. ETAP, Power Station 4.0, User Guide. Volume II, Analysis of Modules. Operation Technology, Inc. Registered to ISO 9001, Certification No. A3147, 2002.

Long-term availability of generation capacity in the market-driven electric power industry

Oleg V. Khamisov[1], *Sergei V.* Podkovalnikov[1,*], and *Konstantin A.* Semenov[1]

[1]Melentiev Energy Systems Institute of the Siberian Branch of the Russian Academy of Sciences (ESI SB RAS), 664033 Lermontov str., 130, Irkutsk, Russia

Abstract. The study investigates the issues of long-term availability of generation capacity in a market environment. The expansion of the electric power industry is analyzed, taking into account its one-product and two-product structural organization as well as its splitting into separate generation companies pursuing their own economic interests in an imperfect market. A case study is provided.

1 Introduction

With the introduction of market relations into the electric power industry of Russia, a number of actors have been established that are in charge of the justification of its expansion and the decision-making related thereto. Power companies now operate as if acting primarily in their own economic self-interests, when justifying and making decisions on the expansion of production capacity (though they are constrained by various reference documents, regulations, and rules). Pursuing these interests, especially so under imperfect competition that arises in the power market (PM), does not necessarily yield a system-wide optimal state. Therefore, research and assessment of the effectiveness of the expansion of power systems (PS) under their splitting into individual power companies that operate in an environment of imperfect competition are required to supplement the conventional optimization-driven technical and economic studies employed to justify the expansion of the electric power industry.

To do research, given the unbundling of power systems into separate generation companies (GenCo), and taking into account their interests, we need dedicated mathematical equilibrium models that would enable us to optimize the objective functions of the effectiveness of these companies. Such models were developed abroad [1-4]. In Russia, similar models were developed by the authors of this paper for one-product and two-product types of organization of the power market. They are presented in [5-7] and are not covered here. Unlike the research published abroad, they take into account the technical properties of the power system in sufficient detail.

2 Input data and background

We studied the Interconnected Power System (IPS) of the Center of the European section of the Unified Power System (UPS) of Russia. In this interconnected power system, transmission networks are highly developed, which serves as a certain rationale behind treating it as concentrated and, hence, representing them for the purposes of modelled calculations as a single node. The requirement of a single-node representation of the power system follows from the fact that the equilibrium models, as it stands now, do not allow finding solutions that apply to entire networks.

The above IPS is made up of a wide range of generation companies of both federal (wholesale generation companies: WGC-1,3,4,5,6, RusHydro, Rosenergoatom, INTER RAO UES) and regional (territorial generation companies: TGK-2,3,4,6, MPC [Moscow Power Company]) significance.

The expansion of various types of power plants, as well as the demand for electricity, are in compliance with a number of documents forecasting the expansion of the electric power industry of Russia, including the draft document of the Energy Strategy of Russia to 2035, the up-to-date adjustments of the Scenario Conditions for the Electric Power Industry Expansion to 2030, the Master Plan and the Expansion Program of the UPS of Russia for years 2018 to 2024, and other publications [8-10].

According to the above documents, the power demand and the annual maximum load for the IPS under consideration were set as equal to 422 TWh/year and 69 GW for the assumed target year of 2030.

Table 1 shows the economic indicators of power plants, including investments, fixed operating and fuel costs [10-14]. The number of hours of utilizing the power of hydroelectric power plants (HPP) in each season was assumed to be equal to: 631 (winter), 432 (spring), 761 (summer) and 652 (autumn) hours. The daily number of hours of utilizing the power of the pumped storage power plants (PSPP) was set to 5. The number of equivalent operating days and holydays in each season was assumed as: 64 and 26 (winter), 64 and

27 (spring), 65 and 25 (summer), and 64 and 26 (autumn) days respectively.

The electricity demand function was obtained relying on the method presented in [5]. To this end, the long-term elasticity of demand for electricity was used, which is uncertain. Various sources present different estimates of the long-term elasticity of demand for electricity [11,12]. Based on the analysis of these sources, this elasticity was assumed to be taking the following values: -0,5; -0,7; -0,9. The average of these values is considered as the most likely one.

Table 1. Power plants economic indicators

Indicators	PSPP	Coal Thermal	Gas Thermal	Nuclear	Cogeneration
Capital investment, USD/kW	840	1400	750	2250	1050
Fixed operating costs, %/capital investment	3	4	5	4	3
Fuel costs, cent/kWh	-	2.2-3.2	3.5-4.8	0.7	3.5-4.6

The long-term elasticity of demand for capacity (when constructing the function of demand for capacity in the PS expansion model while taking into account a two-product organizational structure of electric power industry) was assumed to be similar to the above indicated long-term elasticity of demand for electricity.

3 Electric power industry expansion scenarios

Two groups of calculation scenarios were formed for the one-product and two-product electric power industry structures. The one-product organizational structure, although it does not fully capture the actual organization of the modern electric power industry of Russia, is required, first of all, as a baseline to assess the two-product structure against.

Under the first scenario of the first group (S_O1; subscript "o" denotes the one-product organizational structure), the situation of perfect competition is considered when GenCo's, acting in their self-interests, actually maximize the system-wide effectiveness. This scenario serves as a "baseline" to assess the subsequent scenarios.

Under the second scenario (S_O2), the organizational unbundling of the electric power industry split into separate companies is supposed to maximize the objective functions of the performance of GenCo's and takes into account imperfect competition. This captures the trend of horizontal integration, that is GenCo's mergers. As a result, with the advent of larger GenCo's in the Center IPS, their number is declining, thus serving as prerequisites for their strategic behavior.

Under the third scenario (S_O3), the PM entry of a new GenCo in the long-term period is investigated, as well as its effect on the main market parameters (equilibrium price, electricity supply, and new capacity additions).

Under the fourth scenario (S_O4), it is assumed that the long-term expansion of Rosenergoatom, as a state-owned company, will progress in accordance with established plans. To this end, the NPP capacity in the model is not optimized, but is set (fixed) at a certain planned level (for the assumed target year).

Under the first scenario of the second group (S_T1; subscript "T" indicates the two-product organizational structure), the situation of perfect competition on the two-product electricity and capacity market is considered, when GenCo's actually maximize social welfare. This scenario also acts as the baseline (for comparison) for the next scenario.

Under the second scenario of this group (S_T2), we assume organizational unbundling of the electric power industry split into separate companies, maximization of GenCo's objective functions of effectiveness, and the presence of imperfect competition.

4 Electric power industry expansion in a one-product organizational structure

The key findings of a study of the expansion of a one-product organizational structure of the electric power industry with one-product (electricity) trading are shown in Figures 1-4 and Table 2 for the average value of long-term elasticity of demand for electricity (-0,7).

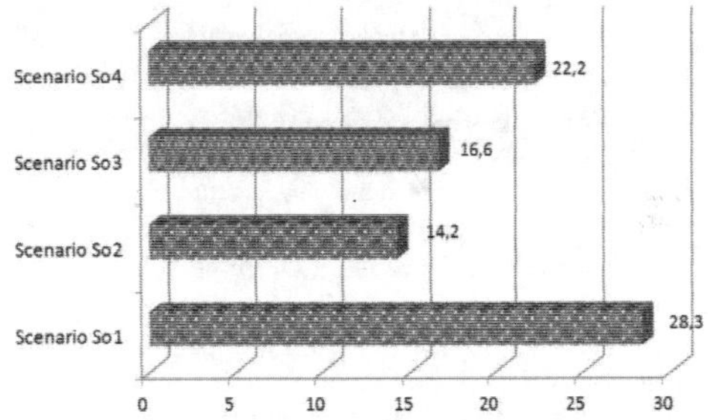

Fig. 1. New generation capacity additions under the one-product organizational structure of the electric power industry, Center IPS, 2030, GW.

As can be seen from Figure 1, under organizational unbundling, independent maximization of GenCo's objective functions, and imperfect competition in the absence of the necessary regulatory influences, new generation capacity additions diminish. Additions are kept to a minimum in the case of GenCo merging (S_O2), accounting for only half of the additions that take place under perfect competition. This is due to the fact that the GenCo merger paves the way for the dominance of aggregated companies, hence enhancing their market power. A new competitor that is established on the basis of new power plants (scenario S_O3), expands its capacity to the maximum. This leads to an increase in total additions if compared to the S_O2 scenario, under which this capacity is only partially added. Under the S_O4

 https://doi.org/10.1051/e3sconf/201913901021

scenario, the maximum additions of generation capacity (over all "market-driven" scenarios) is provided, although they are still somewhat lower than under the S_O1 scenario of perfect competition. This result is due to establishing the requirements on nuclear power plants expansion.

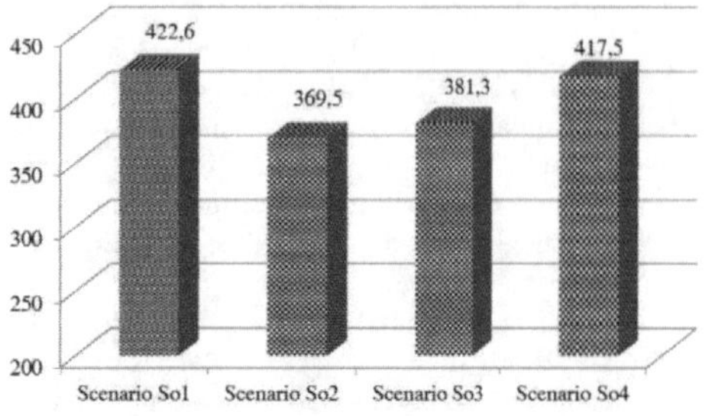

Fig. 2. Electricity supply for a one-product organizational structure of the electric power industry, Center IPS, 2030, TWh/year.

The above strategic GenCo behavior leads to corresponding electricity undersupply by generation companies. Under the S_O2 and S_O3 scenarios, the supply is significantly lower than that under perfect competition (see Figure 2). Under the So4 scenario, the electricity supply is higher if compared to the above two scenarios, although it is still lower than under the reference case. As a result, consumers are forced to cut down their electricity consumption in the long run, moving away from the amount that is optimal social welfare-wise and suffering a corresponding economic loss. This loss is due to an increase in equilibrium electricity prices. As follows from Figure 3, the average annual equilibrium electricity prices for the S_O2 and S_O3 scenarios increase by 14-18% if compared to the S_O1 scenario. It should be noted that under the S_O3 scenario, the price gets slightly lower relative to the S_O2 scenario. This is explained by the entry of a new power market participant under the S_O3 scenario, which somewhat limits the market power of other participants and their negative impact on the level of equilibrium prices. Under the S_O4 scenario, the increase in the equilibrium price is kept to a minimum.

Figure 3 also shows the economic losses of consumers from overstatement of equilibrium prices. As can be seen, these losses are quite large, and in the case of Center IPS reach 4.6-5.5 billion dollar/year by the year 2030. It should be noted, however, that consumer losses under the S_O3 scenario are almost 1 billion dollars lower than those under the S_O2 scenario. As explained above, this is due to the emergence of a new participant of the power market under the S_O3 scenario. Losses of consumers are kept to a minimum under the S_O4 scenario, when NPP capacity additions are regulated/fixed, it, in turn, increases electricity supply by this type of power plants and hence decreases the equilibrium price.

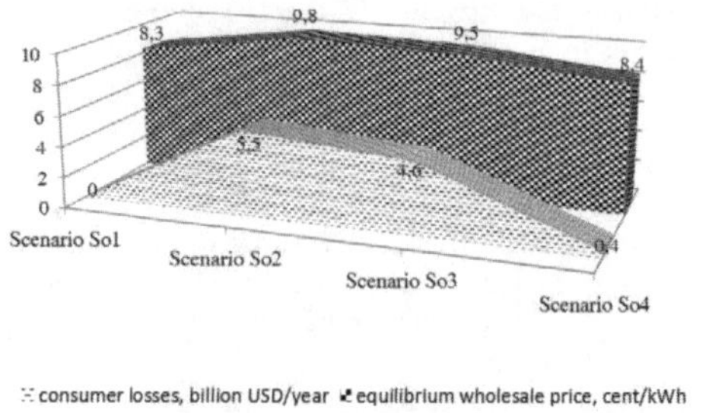

Fig. 3. Equilibrium prices and losses of electricity consumers in a one-product organizational structure of the electric power industry, Center IPS, 2030.

Under the organizational splitting of the electric power industry into GenCo's and independent optimization of their objective functions, not only the generation capacity additions and electricity supply are reduced, but also the mix of electricity generation by various types of power plants changes significantly in some cases. At the same time, the effectiveness of electricity production as a whole decreases.

It should be emphasized that under all considered scenarios and all conditions, both hourly power balances and annual electricity balances were checked and met, alongside ensuring the necessary operating modes of various types of power plants for given daily, weekly, and annual power consumption modes.

5 Electric power industry expansion in a two-product organizational structure

With the introduction of the capacity trading mechanism, along with electricity trading, the new generation capacity additions increase significantly even under imperfect competition (see Figure 4). As can be seen from Figure 4, under imperfect competition, new capacity additions are a mere 1 GW short of those under perfect competition. At the same time, for a one-product organization of the electricity industry, capacity additions are 6 to 14 GW lower under imperfect competition, depending on the conditions under consideration (see Figure 1).

Same time, the aggregate equilibrium prices for a two-product organizational structure are higher than those for a one-product one (see Figures 3 and 4). This is due to the fact that in the one-good (electricity) market, the equilibrium price is formed based on the total cost (including investment) of the marginal facility. When dividing the power market into two markets (those of electricity and capacity), each of them has the price formed based on different marginal facilities.

Under scenario S_T2, given imperfect competition, the price increases relative to that under perfect competition (scenario S_T1). At the same time, if compared to the conditions of imperfect competition and a one-product electric power industry structure (scenarios S_O2, S_O3, S_O4), under scenario S_T2, as was already noted, more new capacity additions are introduced into the power system, thereby increasing their long-term supply availability.

Although, as mentioned above, new generation capacity additions under imperfect competition in the case of a two-product organization of the electric power industry decrease slightly if compared to that under perfect competition, as the calculations show, the mix of these additions noticeably altered. Under scenario S_T2, the share of additions of effective nuclear power plants is significantly reduced, which leads to deterioration of the mix of electricity generation under this scenario.

The mix of electricity generation under perfect competition for scenario S_T1 despite closely resembling the generation mix for scenario S_O1 (one-product structural organization of the electric power industry), is still somewhat different from it. This is due to the fact that there is a transition from optimization of a unified one-product electric power structure to what is actually the optimization of two different structures independently (one of which is for trading electricity, while the other one is for capacity). In this case, a certain loss of effectiveness occurs, as evidenced by the growth of equilibrium aggregate prices under scenario S_T1 as compared to scenario S_O1.

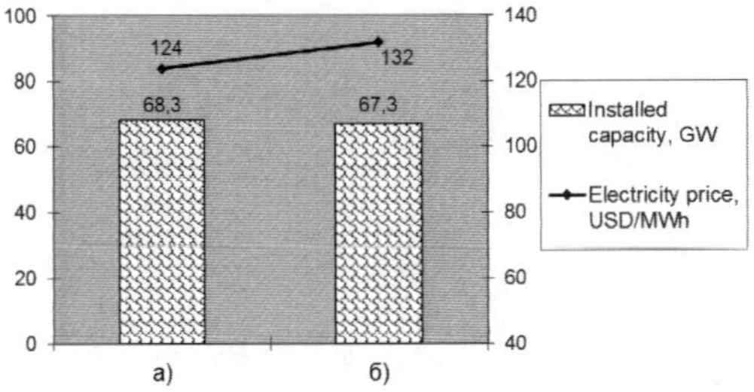

Fig. 4. Installed capacity and aggregate equilibrium price of electricity, Centre IPS, 2030: a) scenario S_T1 of a two-product structural organization of the electric power industry under perfect competition; b) scenario S_T2 of a two-product structural organization of the electric power industry under imperfect competition.

In general, it should be noted that for a two-product PM under imperfect competition, the mix of electricity generation by type of power plants differs from the generation mix under perfect competition to a lesser extent than is the case for a one-product market.

6 Conclusion

1. Given a one-product structure of the electric power industry and the absence of the necessary regulatory influences, new generation capacity additions decrease (if compared to the conditions of maximizing the system-wide effectiveness of the electric power industry). At the same time, the additions are kept to a minimum in the case of GenCo's merging, accounting for only half of the additions that take place under perfect competition. This is due to the strategic behavior of the dominant consolidated companies. This decrease in additions leads to corresponding electricity undersupply. As a result, under imperfect competition, equilibrium prices increase by 14-18% if compared to those under perfect competition. Losses of consumers in the Center IPS due to such overpricing can reach 4.6-5.5 billion dollar/year by the year 2030.
2. Within the two-product electric power industry structure, with the introduction of the capacity market mechanism, the new generation capacity additions significantly increase if compared to the one-product structure even under imperfect competition. Within the framework of this organizational structure, under imperfect competition, capacity additions are a mere 1 GW short of those under perfect competition. Accordingly, the long-term availability of generation capacity increases. However, this triggers an increase in the equilibrium price.
3. The scope of the studies traditionally carried out in Russia to justify the expansion of electric power industry should be extended to address the issue of the organizational unbundling of the electric power industry and self-interests of power companies in order to identify and adjust their behaviour in the long run.

The research was carried out under State Assignment, Project 17.4.2 (reg. no. AAAA-A17-117030310438-1) of the Fundamental Research of Siberian Branch of the Russian Academy of Sciences.

References

1. J. Bushnell, J. Ishii, CSEM, **164** (2007)
2. M. Ventosa, A. Baillo, A. Ramos, The Energy J. **33**, 7 (2005)
3. F. Murphy, Y. Smeers, Oper. Res. **53**, 4 (2005)
4. A.V. Borisenko, S.E. Saukh, Proceedings of Modeling-2010, IPME, 185-194 (2010) (In Russian)
5. S.V. Podkovalnikov, O.V. Khamisov, Izvestija RAN. Energetika, 2, 66-86 (2011) (In Russian)
6. O.V.Khamisov, S.V.Podkovalnikov, Proceedings of 2011 IEEE PowerTech, Trondheim University, 506-512 (2011)
7. S.V. Podkovalnikov, K.A. Semenov, O.V. Khamisov, Izvestija RAN. Energetika, 4, 3-14 (2014) (In Russian)
8. The draft document of the Energy Strategy of Russia to 2035, Electronic publication, (2017). https://minenergo.gov.ru/system/download-pdf/1920/69055 (In Russian)
9. Scenario conditions for the development of the electric power industry to 2030, Electronic publication, (2011). http://www.e-apbe.ru/5years/sc_2012_2030/SC_2012-2030-new.php.html (In Russian)
10. Order of the Energy Ministry of Russia dated 02.28.18 No. 121, Electronic publication, (2018). https://minenergo.gov.ru/view-pdf/11323/82788 (In Russian)

11. L. Gilotte, D. Finon, CIRED, Online reference, (2006). www.centre-cired.fr/IMG/pdf/9_Gilotte-_Finon_inv_pouvoir_marche._pdf.pdf
12. *The power to choose. Demand response in liberalized electricity markets*. IEA, 156 (2003)
13. E.P. Volkov, V.A. Barinov, A.S. Manevich, *Methodology of substantiation and prospects for the development of the electric power industry of Russia* (Energoatomizdat, Moscow, 2010) (In Russian)
14. N.I. Voropai, S.V. Podkovalnikov, V.V. Trufanov, Substantiation of the development of power systems: Methodology, models, methods, and their application (Nauka, Novosibirsk, 2015) (In Russian)

Topical Issues in Corporate Governance Development in the Russia's Electric Power Industry

G.I. Sheveleva[1]*

[1]Melentiev Energy Systems Institute, Siberian Branch of the Russian Academy of Sciences, Irkutsk, Russia

Abstract. Corporate governance in the Russian power companies is developing slowly, and its importance for increasing their investment appeal remains underestimated. In order to identify the main problems hindering such development, the assessment of corporate practices of these companies was carried out. It was performed mainly as per the criteria of the guidelines of Russia's Bank on compliance with the principles of the Corporate Governance Code, as well as those not included in its guidelines. The results obtained were compared to those of the TopCompetence Corporate Governance Center, the Platforma Center for Social Engineering, the Center for Strategic Studies at the MGIMO University, the Corporate Governance Index of the Independent Directors Association and the HSE University. We benchmarked PJSC LUKOIL that ranks among the world's top 10 companies in terms of the total shareholder return as estimated by the Boston Consulting Group. On the basis of a generalization of all the obtained findings we delineated the topical issues of the development of corporate governance. These issues were mainly concerned with the Boards of Directors and were grouped according to the stages of their nomination, election, current operations and performance evaluation. The emphasis was put on the issues related to the adoption of new managerial technologies.

Introduction

In the course of reforming the Russia's electric power industry in 2003-2008, the main task of attracting investments remained unsolved. A non-market driven procedure of capacity contracting (CC) was adopted for financing, as a result of which the reformed power generating companies were indemnified for their investment in new capacity additions. After the expiration of the CCs, another large-scale upgrading program for combined heat and power plants was enacted again. The projects deemed eligible for the program will also be entitled to guaranteed break-even arrangements by imposing higher capacity tariffs on the consumer [1]. The investment appeal of Russian power companies (hereinafter "power companies") remained at a relatively low level. Its improvement depends on the quality of corporate governance, which is considered one of the determining factors in international practices [2].

The main indicator of the quality of corporate governance is the compliance of companies with the Corporate Governance Code (hereinafter Code), developed on the basis of international standards [3]. This has resulted in allocating a high priority to assessment of compliance of corporate governance in power companies with the guidelines of Russia's Bank (hereinafter Bank) on compliance with the Code and the identified criteria of the Code, which were not included in its guidelines.

We performed a comparative analysis of the obtained findings against the evaluation data by analytical centers, for the purposes that included obtaining further information on the quality of corporate governance as based on online surveys of investors [4-8].

PJSC LUKOIL's benchmarking was carried out, mainly to identify corporate governance problems faced by power companies when adopting new technologies.

We studied the 12 public joint-stock companies of the electric power industry that published Annexes "On Compliance with the Principles and Guidelines of the Corporate Governance Code" in their annual reports for the year 2017 (Inter RAO, Enel Russia, RusHydro, OGK-2, Unipro, TGK-1, TGK-2, Mosenergo, Quadra, T Plus, TGK-14, Irkutskenergo) [9-20]. We made use of additional data: those coming from official corporate websites; the Bank's guidelines on compliance with the Code [21]; the author's published research on the subject [22]; the New Economic School [23]; the Russian Institute of Directors [24]; and the Boston Consulting Group total shareholder return rating for the year 2019 [25].

Unlike with the widespread approach that deals only with the largest power companies and thus precludes us from getting an overall view of the quality of corporate governance in the Russia's power industry, we studied virtually all of them.

The results of the study are applicable to improving the efficiency of corporate governance in power companies and to making more informed and balanced investment decisions by investors.

1 Corporate governance practices

The current practices of corporate governance in power companies are identified by the following assessment methods:

- Compliance with the Bank's guidelines on compliance with the Code (a conventional and an in-depth methods),
- Compliance with the Code criteria not included in the Bank's guidelines,
- Compliance with the findings of the analytical centers,
- PJSC LUKOIL's benchmarking.

1.1 Compliance with the Bank's guidelines on adhering to the Code

The assessment of compliance of corporate governance with the Bank's guidelines on compliance with the Code in power companies was mainly based on the data provided by the companies themselves in Annexes "On Compliance with the Principles and Guidelines of the Code of Corporate Governance" in their annual reports for the year 2017. The assessment was performed by a conventional and an in-depth method.

1.1.1 Conventional method

The conventional method is typical of most companies and assesses their compliance with the 79 Level 2 principles of corporate governance of the Code that have three-digit codes assigned to them (for example, "1.1.1." or "1.2.3.", etc.). The status of compliance with these principles ("compliant", "partially compliant", "non-compliant"), when relying on this method, was treated without a further breakdown by the compliance criteria for each of them.

The conventional method of assessment revealed the presence of leading and outsider companies with respect to their compliance with corporate governance principles. Of the 79 principles reviewed, the leading companies (PJSC Inter RAO, PJSC Enel Russia and PJSC RusHydro) complied with the 73, 62 and 59 principles, respectively. PJSC Quadra and PJSC TGC-2 that are the outsider companies that ranked at the bottom of the list complied with 33 and 32 principles respectively. PJSC Inter RAO, PJSC Enel Russia, and PJSC RusHydro proved partially-compliant with 4, 13, and 17 principles, respectively. The corresponding values for PJSC Quadra and PJSC TGK-2 are 27 and 32 principles respectively, thus lagging behind PJSC Inter RAO and scoring 7 to 8 times lower. The disparity between the companies that performed best and worst with respect to the number of principles they failed to comply with was even more pronounced. In particular, the number of principles PJSC Quadra does not comply with exceeds that of PJSC Inter RAO by a factor of 9.6.

These results are more evident when it comes to the share of each compliance status ("compliant", "partially compliant", "non-compliant") in the total of 79 assessed principles: PJSC Inter RAO - 92%, 5.5%, 2.5%; PJSC Enel Russia - 75%, 22%, 3%; PJSC RusHydro - 74.7%, 21.5%, 3.8%. Outsider companies: PJSC Quadra - 41.8%, 34.2%, 24.0%, PJSC TGC-2 - 40.5%, 40.5%, 19.0%. Outsider companies saw a clear increase in the number of principles they "partially comply with" and "do not comply with" to that of the leading companies.

1.1.2 In-depth method

The in-depth method of assessment allowed for a more unbiased view of corporate governance practices in the companies we studied. All 128 criteria for compliance with Level 2 principles of the Code recommended by the Bank were taken into account here. The leading companies identified earlier had the following values of the compliance status with respect to these criteria: PJSC Inter RAO - 91%, 6%, 3%, PJSC Enel Russia - 81%, 13%, 6%, and PJSC RusHydro - 78%, 20%, 2%. The shares of compliance status values have been redistributed, mainly towards an increase in the number of the criteria that were not complied with. This was more typical of the outsider companies: PJSC Quadra - 51%, 22%, 27% and PJSC TGC-2 - 47%, 31%, 23%. However, this method of assessment significantly reduced their lagging behind PJSC Inter RAO in terms of all compliance status values. Moreover, we identified PJSC TGC-1 as having the compliance status values comparable to those of the leading companies, that is 80%, 18%, and 2%, respectively.

We would like to reiterate that the assessment was made on the basis of the data provided by the power companies in their reports "On Compliance with the Principles and Guidelines of the Corporate Governance Code". A thorough analysis of their reliability was not carried out. However, the study revealed unreliable data, which increased the number of the criteria that power companies failed to comply with. In particular, all power companies claimed their compliance with Level 2 principle 5.1.3 that states that "A company has the anti-corruption policy in place". However, only two of them (PJSC Inter RAO and PJSC RusHydro) actually developed, approved, and presented the Fraud and Corruption Prevention Policy in their internal documents.

1.2 Compliance with the Code criteria not included in the Bank's guidelines

Criteria not included in the Bank's guidelines were identified by comparing the guidelines of the Code [3] against those of the Bank [21]. Their compliance was assessed on the basis of the information coming from official websites of power companies. Based on the results of the assessment of compliance with the identified criteria of the Code that were not included in the criteria recommended by the Bank, the number of criteria the power companies fail to comply with increased by 6-10%.

Of all the criteria of the Code that were identified as those not complied with, we highlighted the criteria that the overwhelming majority of power companies failed to meet. They were grouped according to the principles of the Code. More than half of them were directly related to violations of the principles that deals with the Board of Directors. The other criteria that were not met were

 https://doi.org/10.1051/e3sconf/201913901022

related to the principles of remuneration systems, material corporate actions, information disclosure, and shareholder rights. All these criteria were indirectly related to the Boards of Directors, as they belonged to their area of responsibility within the scope of delegated authority.

1.3 Alignment with the findings of analytical centers

The obtained findings were compared to the relevant data of the TopCompetence Center and the Corporate Governance Index of the Independent Directors Association and the HSE University based on the criteria of the British Good Governance Index and turned out to be comparable with respect to the assessment criteria. However, taking into account the data of analytical centers based on online surveys of stakeholders, none of the power companies made it to the top 10 [6,8]. It turned out that these companies were not competitive scoring low on the factors that are deemed crucial for investors that is the Board of Directors, market capitalization, transparency and consistency of the dividend policy.

Based on the Q4 Quarterly Reports for the year 2018, only the leading companies we identified enjoyed average market capitalization (more than 2 billion US dollars): PJSC Inter RAO - 5846.7 million US dollars, PJSC RusHydro - 2977.9 million US dollars. The rest of the power companies (except PJSC Unipro) had low market capitalization. In particular, the outsider companies were valued as follows: PJSC Quadra – 80.8 million US dollars, PJSC TGK-2 – 49.0 million US dollars. Internationally, companies with low market capitalization generally have bad reputations and are risky. In most cases, they are considered to be fraudulent, involved in uncontrolled cash withdrawals to foreign offshore zones, including non-transparent transactions with offshore partners that are oftentimes affiliated with their key shareholders.

The findings of the study of dividend history data of power companies in 2013-2018 showed a relatively low return on their shares, lack of transparency, and an inconsistent dividend policy.

1.4 PJSC LUKOIL's benchmarking

The Russian energy company PJSC LUKOIL was chosen for benchmarking due to the high total shareholder return in the Boston Consulting Group rating published in 2019. Benchmarking was carried out mainly with respect to the adoption of new managerial technologies in the corporate practice, including the use of telecommunication means at the stages of nomination, election, current activities, and performance evaluation of the Boards of Directors. The benchmarking data were taken into account when formulating the problems of corporate governance development faced by power companies.

2 Problems of corporate governance development

On the basis of the generalized data of the performed assessment of corporate governance in power companies we arrived at the topical issues inherent in its development that were mainly related to the Boards of Directors. A strong Board of Directors as it stands in modern corporate practice is the key management body and the main instrument to ensure the leadership of companies in the field of corporate governance. Its role is becoming truly strategic in the face of the new challenges of changing the general paradigm of corporate governance and the development of digital technologies.

The main problems of corporate governance development were grouped according to the "life cycle" stages of the Boards of Directors:
- Nomination.
- Election.
- Current activities.
- Performance evaluation.

2.1 Nomination

• Absence of a list of persons entitled to participate in the General Meeting of Shareholders (GMS) that enables shareholders to do the following: evaluate the balance of power at the upcoming meeting; jointly nominate candidates to the Board of Directors; discuss and agree on possible voting options; and appoint their representative to participate in the GMS.

• Non-transparent procedure for nomination of candidates to the Board of Directors.

• Absence of comprehensive information on the candidates to the Board of Directors approved for voting, which is necessary to get an idea of their personal and professional qualities (information on who nominated them, whether they are free of conflicts of interest, as well as the information on their relations with affiliated persons and major counterparties of the Company, compliance with independence criteria, and other information that may affect the performance of their duties).

• Absence of an online bulletin board for preliminary discussion by shareholders of candidates to the Board of Directors, which would allow taking into account the diversity of opinions of shareholders and ensuring that the Board's composition complies with the requirements of the law, the objectives of the company, and its corporate values. The special importance of such a procedure lies in the discussion of the ethical and behavioral qualities of potential candidates, their ability to work as a team and to make concessions, as well as the identification of advocates of "support groups", bystanders, conventionalists, and "constant" critics who are undesirable as the Board of Directors members.

2.2 Election

• Absence of telecommunication facilities that would ensure remote access of shareholders to participate in the GMS (GMS broadcasting on the Company's website, adoption of video-conferencing).

• Absence of an electronic voting system that would enable filling in the ballot through a personal account on the company's website and ensuring adequate reliability, security, and authentication of persons participating in the GMS.

2.3 Current activities

• Absence of appropriate Committees of the Board of Directors and (or) their poor performance, which compromises the quality of important decisions made at the meetings of the Board of Directors and efficiency of their consideration without proper preparation by these Committees.
• Absence of the team-spirit in the activities of the Board of Directors members when making decisions without comprehensive discussions of the issues under consideration and not being informed by different opinions.
• Meetings tend to be held by correspondence with no telecommunication facilities for the participation of members of the Board of Directors in the discussion of important issues by means of remote access.
• Low competence of members of the Board of Directors in the field of innovation, information, and digital technologies.
• Inadequacy of individual critical decisions and the culture of the decision-making process.
• Lack of monitoring of the implementation of decisions taken.
• Failure to update the composition and content of the body internal corporate documents to meet changed conditions and new challenges of our time.
• Lack of an actual anti-corruption policy and efficient models of interaction with persons interested in reporting such violations. The inefficiency of a traditional confidential communication channel for such complainants available on websites of some of the companies as it generally boils down to an email address or phone number details.

2.4 Performance evaluation

• Lack of a formalized procedure for performance evaluation of the Board of Directors, analytical tools, and the use of the retrieved data.
• Absence of the Regulation "On Performance Evaluation of the Board of Directors" developed and approved in the internal documents of the companies, including its justification; procedure and tools; and the format of public disclosure.
• Non-transparency of the amount and determination of annual remuneration for the members of the Board of Directors.
• Substantial amounts of this remuneration without objective grounds.
• Absence of correlation between the amount of annual remuneration of the Board of Directors members and the performance of power companies, which does not contribute to the convergence of their financial interests and those of shareholders.

Conclusion

In order to enhance the investment appeal of power companies, it is necessary to handle the issues of the quality of corporate governance. For the purpose of identifying such problems, an assessment of their current corporate practices was performed. It stood out from commonly used assessment methods in two ways. First of all, it was the coverage of virtually all companies active in the electric power industry, not just the biggest ones. Secondly, it was the application of all 128 criteria of compliance with the Level 2 principles of the Code recommended by the Bank along with those not included in its guidelines.

A significant number of criteria were identified that were power companies failed to comply with. We delineated the topical issues of corporate governance development based on the criteria related to the Boards of Directors that the overwhelming majority of power companies fail to comply with. The issues were grouped according to the stages of the Boards nomination, election, current activities, and performance evaluation. These issues were clarified and supplemented by the results of their comparison with the data published by analytical centers and the PJSC LUKOIL's benchmarking, mainly in the context of the adoption of new managerial technologies.

It is possible that the identified problems of corporate governance development in electric power companies will be mitigated in the future, in particular, by assessing the criteria for compliance with the G20/OECD corporate governance principles that were not included in the Russian Code.

References

1. The RF Government approves 1.9 trillion ruble spending on the programme of combined heat and power plants upgrading. [Online]. Available: http://www.bigpowernews.ru/print/86618.phtml
2. OECD. Principles of corporate governance. G20/OECD Publishing, Paris (2016). [Online]. Available: http://dx.doi.org/10.1787/9789264252035-ru
3. Corporate Governance Code. Regulations and up-to-date information of the Central Bank of the Russian Federation, Vestnik Banka Rossii, **40** (2014)
4. Transparency in corporate reporting: A report by Transparency International – Russia (2018). [Online]. Available: https://transparency.org.ru/special/trac2018russia/
5. National Corporate Governance Index (2018). [Online]. Available: http://www.topcompetence.ru/Images/Uploads/Publications/National-CG-Index-2018-TopCompetence.pdf

6. *Corporate Governance in Russia: Genre identity crisis and hopes for the future* ("Platforma" Center for Social Engineering, Moscow, 2018). [Online]. Available: http://www.nand.ru/upload/medialibrary/news
7. Zavyalova E.B., Kondratiev V.B., *Corporate governance as a contributing factor to enhancing the investment appeal* (MGIMO University, Moscow, 2014)
8. Corporate Governance Index of Russia (2017). [Online]. Available: http://nand.ru/upload/ux/corporate_governance_index_2017.pdf
9. Report on compliance of JSC TGK-2 with the principles and recommendations of the Corporate Governance Code. [Online]. Available: http://www.tgc-2.ru/upload/iblock/f8d/areport_2016.pdf
10. Report on compliance of PJSC Quadra with the principles and recommendations of the Code of Corporate Governance recommended by the Bank of Russia. [Online]. Available: https://www.quadra.ru/wp-content/uploads/2018/06/gosa2018_AnnualReport.pdf
11. Report on the compliance of PJSC TGC-1 with the principles and recommendations of the Corporate Governance Code. [Online]. Available: http://www.tgc1.ru/fileadmin/ir/reports/annual/2017/godovoi_otchet_tgk-1_2017.pdf
12. Information on compliance of PJSC Enel Russia with the principles and recommendations of the Corporate Governance Code. [Online]. Available: https://www.enelrussia.ru/content/dam/enel-ru/documents/ru/investors/annual/Enel_Annual_report-2017-rus.pdf
13. Report on compliance of PJSC Unipro with the principles and recommendations of the Corporate Governance Code. [Online]. Available: http://www.unipro.energy/shareholders/disclosure/annual_reports/
14. Report on compliance of PJSC OGK-2 with the principles and recommendations of the Corporate Governance Code. [Online]. Available: https://www.ogk2.ru/rus/si/infodisclosure/year/
15. Report on compliance of PJSC TPlus with the principles and recommendations of the Corporate Governance Code. [Online]. Available: http://www.tplusgroup.ru/fileadmin/f/voltgc/ir/disclosure/documents/GO/T_Pljus_PAO_GO_2016.pdf
16. Report on compliance of PJSC Mosenergo with the principles and recommendations of the Corporate Governance Code. [Online]. Available: http://www.mosenergo.ru/d/textpage/ad/173/godovoj-otchet-17.pdf
17. Report on compliance of PJSC TGC-14 with the principles and recommendations of the Corporate Governance Code. [Online]. Available: https://www.tgk-1.com/upload/medialibrary/876/go_2017_dlya_sayta_itog.pdf
18. Report on compliance with the principles and recommendations of the Corporate Governance Code. [Online]. Available: http://www.interrao.ru/upload/doc/Inter_RAO_AR2017_ru_web.pdf
19. Report on compliance with the principles and recommendations of the Corporate Governance Code. [Online]. Available: http://www.rushydro.ru/upload/iblock/90d/Kniga-Prilozhenij-2017-rus.pdf
20. Report on compliance of PJSC Irkutskenergo with the principles and recommendations of the Corporate Governance Code. [Online]. Available: http://www.irkutskenergo.ru/gi/31500
21. Bank of Russia Information Letter No. IN-06-52/8 dated February 17, 2016 [Online]. Available: https://www.garant.ru/products/ipo/prime/doc/71237796/
22. G.I. Sheveleva, Areas for the improvement in corporate governance in the Russian electric power industry, Proceedings of 2017 Tenth International Conference "MLSD", Moscow, 1-4 Oct., IEEE, DOI: 10.1109/MLSD.2017.8109686
23. Guriyev S.M., *Myths of economics. Fallacies and stereotypes disseminated by mass media and politicians* (Mann, Ivanov, and Ferber, Moscow, 2017)
24. I.V. Belikov, Aktsionernoye obshchestvo: voprosy korporativnogo upravleniia **4**, 14 (2008)
25. Value Creators Rankings (2018). [Online]. Available: https://www.bcg.com/ru-ru/publications/2018/interactive-value-creators-rankings-20th-anniversary-edition.aspx

The investigation of frequency response analysis for power transformers winding condition

Artyom Nazarov[1], *Vitaliy* Savelev[2*] and *Dennis* Frogh

[1]AO GK Sistemy i tekhnologii, Department of science, 600014 Vladimir, Lakina st, 8, Russian Federation
[2] AO GK Sistemy i tekhnologii, Department of science, 600014 Vladimir, Lakina st, 8, Russian Federation

Abstract. In modern electrical transmission and distribution systems, power transformers are critical components within the network. In the event that a failure occurs in service, the impact can be far reaching. The majority of mechanical deformation within power transformers is due to short circuit faults. The ageing transformer population increases the likelihood of failure so a reliable diagnostic tools required to determine the remaining life of these assets. Although, frequency response analysis (FRA) has been recently recognized as the most reliable detection tool for mechanical deformation in transformers, in the event of minor fault, the current FRA interpretation approach may not able to detect any variations between healthy and faulty FRA signatures. This paper focuses on FRA tests for power transformer and interpretation of obtained FRA signatures.

One of the most important problems of transformer maintenance is the monitoring system and technical diagnosis development. More than 40 % of transformers have exceeded their design life (25y, established by standard papers [1]). Nowadays, there are several diagnostic techniques for power transformers, for example, vibration analysis, ultrasonic contact fault detection, dielectric loss angle test and partial discharge [2]. The diagnosis of power transformer can be carried out offline, online and partly. Parameters used for online transformer tests are temperature, pressure, leakage currents, vibration amplitude, noise level, velocity vibration and acceleration vibration. They must have properties, such sensitivity, accuracy of measurement, stability, depth of investigation and simplicity of diagnosis. The transformer defects occur due to unsuitable means of transporting short circuit currents, earthquakes, oil ignition in tank etc [3]. Due to CIGRE working group A 2.26 the most common cause of power transformers fault is inter-disk faults [4].

One of the most effective techniques of condition monitoring for power transformers are Sweep Frequency Response Analysis (SFRA) and Impulse Frequency Response Analysis (IFRA) [5]. Lots of investigations are held to improve this technique at the moment, but from 1978 there is no general opinion how to make a conclusion about technical condition [6].The main advantage of this technique is its high sensitivity: even minor local changes of winding elements(turns, disks) result in dramatic changes on diagnosis signature [7]. Different types of deformations lead to changes in FRA signature in different range. Conventional techniques of condition monitoring are not so sensitive. That's why FRA is the best technique for inter-disk fault detection.

Frequency Response Analysis has two directions of use: sweep and impulse [3].

IFRA is provided by low voltage impulse given on winding (The impulse form can be rectangular or conventional or exponential). Current or voltage on bushing terminals or frequency current transformers of other windings is measured. The transition process in windings is recorded with the help of analog-digital converters. After that the response is transformed to frequency area using Fourier algorithm. Finally, transfer function as a ratio of injected and measured signal is founded.

SFRA is used in frequency area directly. In this case the sine 10 V signal from sweep generator is injected on the bushing terminal. It's changes from several Hz to several MHz. The response is measured from other terminals as a function of the variable frequency [3]. Each way has its advantages and disadvantages. For example, SFRA is more sensitive on low frequencies and quite simple maintenance as much as in interpretation results [8]. However the process takes long time.

A foreign power transformer was selected as an object of study. Its characteristics are given in table 1. This transformer was chosen because of its initial data, which were used for modelling.

Table 1. The power transformer parameters

№	Classification	Value
1	Rated power	40 MVA
2	Impedance	13,8 %
3	Primary voltage	66 kV
4	Secondary voltage	11 kV
5	Height of HV winding	1074 mm
6	Height of LV winding	1136 mm
7	Core cross-section diameter	560 mm
8	Insulated core cross-section diameter	579 mm
9	Inner diameter of HV winding	825 mm
10	Inner diameter of LV winding	612 mm
11	Primary turns	1200
12	Secondary turns	200

The parameters of the frequency characteristics with the parameters of the winding could be correlated because the power transformer winding is a multi-frequency oscillatory circuit [9]. It consists of series resistance, series inductance and series capacitances.

A transformer winding equivalent circuit was compiled in the Multisim software package to carry out the experiments (fig.1).Transformer parameters are:

- series inductance and series resistance of HV and LV windings (L_s, R_s);
- shunt capacitance between HV and LV windings (C_{sh});
- series capacitances between HV / LV windings and the earthed tank / core (C_o);
- series capacitances between the HV and LV windings (C_{HL}).

Table 2. Parameters of the equivalent circuit [10]

№	Parameters of the equivalent circuit	HV	LV
1	L_s [μH], R_s [Ω]	8,91;1	8,91; 0,025
2	C_{sh} [pF]	61,196	115,53
3	C_o [pF]	567,96	333,24
4	C_{HL}[pF]	89,283	

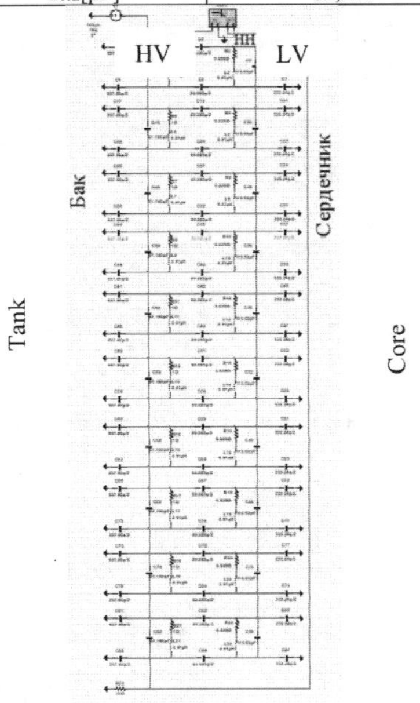

Fig. 1. The equivalent circuit of HV and LV winding

The equivalent circuit of the first HV winding disk is shown in Fig. 2.

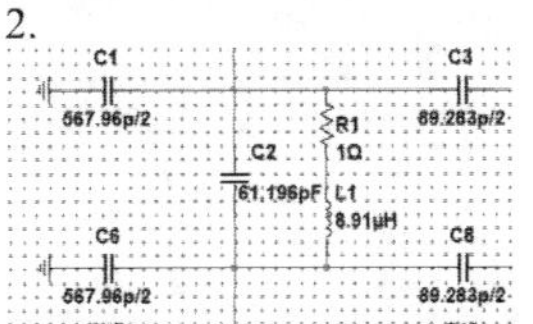

Fig. 2. The equivalent circuit of the first HV winding disk

C_1, C_6 (C_o) – capacitance between first HV winding disk and the earthed tank

C_2 (C_{sh}) – capacitance between first and second HV winding disk.

C_3, C_8 (C_{HL}) – capacitance between first HV winding disk and LV winding.

L_1 (L_s) – inductance of the first HV winding disk.

R_1 (R_s) – resistance of the first HV winding disk.

To conduct tests, the elements implemented in Multisim in the form of blocks were used. They are:

- The generator of sine signal, which allows to change the amplitude, frequency and duration of the signal. The amplitude is10 V.
- BodePlotter, used to get FRA signature.
- Load resistance, which consists of the cable resistance and measuring instrument resistance, etc.

The view of the BodePlotter block is shown in Fig. 3.

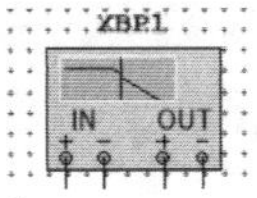

Fig. 3. BodePlotter block

The view of the generator of sine signal block is shown in Fig. 4.

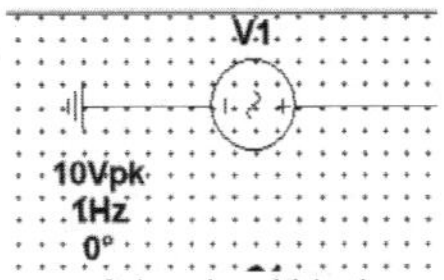

Fig. 4. The generator of sine signal block

The view of load resistance block is shown in Fig. 5

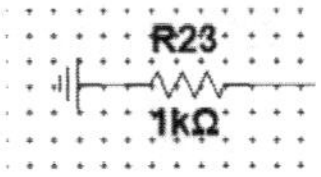

Fig. 5. The load resistance block

During FRA test the transfer function – frequency response was found. This function is the ratio of injected signal on the top of the winding to the measured signal from the winding end. It's a plot, defined by frequency

Each of 10 HV winding disks was short-circuited in this test.

Under the test capacitance C_{sh} (C_2) was enlarged in 100 times. It was made to ensure accurate results. At one plot damaged transformer FRA signature and healthy transformer FRA signature were placed. (fig. 6, fig.7 and fig. 8).

Fig. 6 shows: healthy transformer FRA signature, the first, the second and the third short-circuited winding disk FRA signature. Fig. 7 shows: healthy transformer FRA signature, the fourth, the fith, the sixth and the seventh short-circuited winding disk FRA signature. Fig. 8 shows: healthy transformer FRA signature, the eighth, the nineth and the tenth short-circuited winding disk FRA signature .

 https://doi.org/10.1051/e3sconf/201913901023

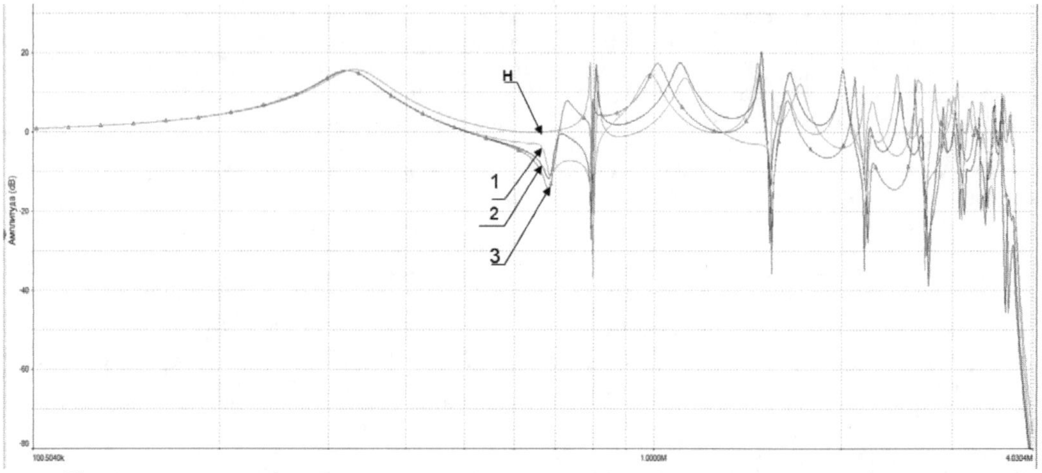

Fig. 6. The plot of transfer function (dB) as a function of frequency. Disk windings short-circuited in series. H – healthy disk windings condition, 1 – first disk winding short-circuited, 2 – second disk winding short-circuited, 3 – third disk winding short-circuited.

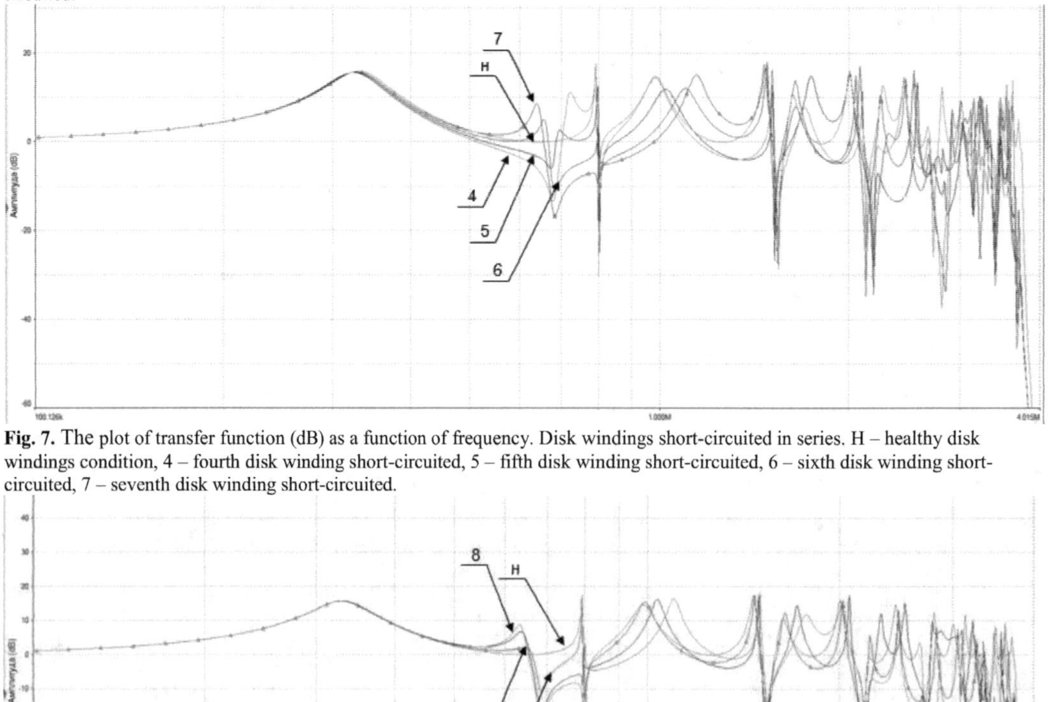

Fig. 7. The plot of transfer function (dB) as a function of frequency. Disk windings short-circuited in series. H – healthy disk windings condition, 4 – fourth disk winding short-circuited, 5 – fifth disk winding short-circuited, 6 – sixth disk winding short-circuited, 7 – seventh disk winding short-circuited.

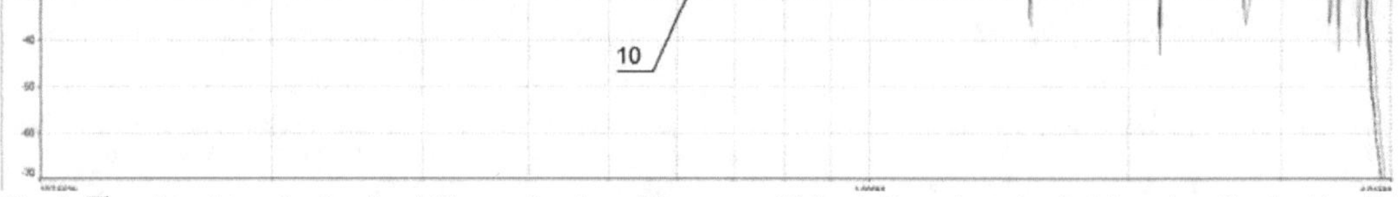

Fig. 8. The plot of transfer function (dB) as a function of frequency. Disk windings short-circuited in series. H – healthy disk windings condition, 8 – eighth disk winding short-circuited, 9 – nineth disk winding short-circuited, 10 – tenth disk winding short-circuite

To compare damaged and healthy FRA signatures it is noticed:

- before 2 MHz changes in curves are noticeable;

The first resonance frequency on heathy winding is about ~320 кГц, and the first antiresonanse is ~800 кГц; When frequency changes from 1,5 MHz to 4 MHz there are lots of resonanses and antiresonanses. When different disk winding short-circuited the next thing should be marked:

- fig.6 shows the growth of resonance amplitude at displacement from 1 disk to 3 disk on ~680 kHz;
- fig. 7 shows the growth both resonance and antiresonanse displacement amplitude at the middle of the winding(from 4 disk to 7 disk);
- The decrease of resonance amplitude is observed at the end of the winding (from 8 to 10 disk);
- The first resonance frequency varies from 500 kHz to 600 kHz, depending of the fault place.

The changes above were observed at ~680 kHz.

The amplitude varies from 18 to -43 dB. After that, the signal is decreasing.

Conclusion

1. FRA could be used as a technique of power transformer winding diagnosis .
2. To explain FRA results the equivalent circuit was made in Multisim. It helps to make an experiments on disk windings
3. To make an experiment the series capacitances, series inductance and series resistance has to be known
4. Using FRA for diagnosis, the algorithm of fault windings determination should be developed.
5. Winding deformation comes to changes in inductance and capacitance, so FRA signature is changing too (number of resonanses, amplitude, location on the curve).
6. The most informative part of FRA signature located from 500 kHz to 600 kHz. The reason is that till 1 kHz there are a lot of electromagnetic noises in power grid and over 2 MHz the response is too much sensitive to all electromagnetic changes.
7. The change of series resistance connected with increase or decrease amplitude of FRA signature.
8. The first antiresonanse frequency moved to the right at the change of fault location from the top to the middle. The amplitude increases.

References

1. M. Y. Lvov Damage analysis of power transformers with voltage of 110 kV and higher – *Elektrichestvo*,2, (2010).
2. O. A. Kuzmin Modern diagnostic methods for power transformers URL:https://energoboard.ru/post/411/
3. Drobyshevsky A.A. Electrodynamic stability of power transformers – *Novosti elektrotekhniki*, №5(65), (2010).
4. CIGRE Working Group A2.26, "Mechanical-condition assessment of transformer windings using frequency response analysis (FRA)" 2007.
5. Rahimpour, Hossein & Mitchell, S.D. &Tusek, Joe. (2016). The application of sweep frequency response analysis for the online monitoring of power transformers. 1-6. 10.1109/AUPEC.2016.7749347.
6. E. P. Dick and C. C. Erven, "Transformer diagnostic testing by fre-quency response analysis," IEEE Trans. Power App. Syst., vol. PAS-97, no. 6, pp. 2144–2153, Nov. 1978.
7. Steven D. Mitchell and James S. Welsh, "Modelling Power Trans-formers to Support the Interpretation of Frequency-Response Analysis", IEEE transactions on power delivery, vol. 26, no. 4, October 2011
8. Belinda J Small and A. Abu-Siada «A New Method for Analysing Transformer Condition Using Frequency Response Analysis» 1-5. 10.1109/PES.2011.6039470.
9. R.G. Ildarkhanov «Diagnosis of defects in transformer windings using low voltage pulses» - *Energetika Tatarstana*, №1 (2009).
10. O. Aljohani, A. Abu-siada, and S. Islam «Impact of Insulating Oil Degradation on the Power Transformer Frequency Response Analysis» IEEE 11th International Conference on the Properties and Applications of Dielectric Materials (ICPADM), 2015

E3S Web of Conferences 139, 01024 (2019) https://doi.org/10.1051/e3sconf/201913901024
RSES 2019

Low-power nuclear power plants as the most important factor of social development, development of remote and remote areas of the country and ensuring its energy security

Vitaliy Savelev[1,*], *Yuri* Saraev[2], *Vera* Bataeva[3]

[1]AO GK Sistemy i tekhnologii, Department of science, 600014 Vladimir, Lakina st, 8, Russian Federation
[2] AO GK Sistemy i tekhnologii, Department of science, 600014 Vladimir, Lakina st, 8, Russian Federation
[3] Ivanovo State Power Engineering University, 153003 Ivanovo, Rabfakovskaya st., 34, Russian Federation

Abstract. The article is devoted to the analysis of specific requirements, problems and tasks requiring solutions for the practical application of low-power nuclear power plants, as well as their impact on the level of energy security of remote and remote regions and the country as a whole.

1 Introduction

Currently, there is a growing interest in small nuclear power due to its special attractiveness in social development and development of remote areas. The total capacity of low-power consumers in Russia is not less than 30% of the total generation. The number of such consumers, especially kilowatt and megawatt capacity, is in the hundreds and thousands, surpassing the number of tens and hundreds of times the number of consumers of large and medium-sized energy.

One of the main tasks to improve the territorial and production structure of the fuel and energy complex, taking into account the need to strengthen the unity of the economic space of Russia, is the development of infrastructure and facilities of the fuel and energy complex of Eastern Siberia, the Arctic zone of the Russian Federation and the Far East [1]. Social development and development of these hard-to-reach and remote areas is of strategic importance for ensuring energy and national security of Russia, as these regions are home to the main explored hydrocarbon deposits and the most important infrastructure facilities of various industries.

Priority energy consumers of these territories are [2]:
- objects related to exploration, development of oil and gas fields: drilling rigs, pumps, shift camps, etc.;
- objects of the Northern sea route: port terminals, Arctic aviation, navigation and hydrographic support, meteorological services, navigation tracking systems, shipbuilding and ship repair plants;
- military facilities: observation bases of the Northern fleet, border outposts, military bases;

Consumers of low power are also large, but local energy consumers, whose energy supply is difficult economically and logistically. The principles of power supply of such consumers differ depending on the specifics of economic development of territories, working conditions and infrastructure. The total capacity of such consumers in the country is estimated at least 30 % of the total generation, and their number is estimated in the hundreds and thousands, exceeding ten times the number of consumers of large and medium-sized energy. Traditional energy supply chains have limited use for them. Therefore, the construction and development of energy supply systems of these territories should take into account their specific features in terms of energy security.

The following features are typical for the energy of the above regions [3]:
- technological isolation and inaccessibility;
- the presence of a large number of dispersed low-power energy sources;
- the disconnection of consumers of energy resources;
- absence of main interregional electric networks;
- multiple links and time restrictions on fuel delivery ("Northern delivery»);
- high cost of production of electric and thermal energy, due to the significant share of the transport component in the final cost of fuel (up to 70 %);
- extremely high electricity tariffs and technical losses in the transportation of energy, as well as the need for subsidies to equalize tariffs;
- reduction in the number and critically low number of inhabitants in the territorial entities;
- capital-intensive maintenance of networks and equipment and high maintenance costs of power lines.

In view of the noted features, the risks of energy security for these territories are realized more vividly, and the consequences of their implementation can be catastrophic, since in this case it is essentially about the survivability of technological processes, about the life and health of people. To ensure the energy security of such territories, there is only one option – to have their own sources of generation.

In Autonomous power supply systems, approximately 96% of electricity generation is accounted for by the most common energy sources for territories with decentralized power supply – diesel power plants [4]. These territories are also suitable to a certain extent for the use of renewable energy generation sources. However, their specificity creates a number of difficulties. In table.1 the main limitations in the application of traditional sources for Autonomous systems are given.

In recent years, a new technology of production, transmission and distribution of electric and thermal energy has been actively developing all over the world. Its innovative platform is distributed generation. In a broad sense, distributed generation is considered to be those objects that are near the final consumption, regardless of who owns them. In a narrower sense, distributed generation (Distributedgeneration) is defined as the construction and operation of electric (thermal) energy sources and, if necessary, distribution networks by consumers for their own needs, as well as for the issuance of surpluses to the General electric or thermal network [5].

Table 2. Restrictions on the use of generation sources.

Source type	Restrictions on the use
Solar power plant	Icing of solar panels and the need for constant snow sweeping
Wind power station	Storm winds up to 36 m/s, frost coating, icing
Diesel power plant	-only Arctic diesel fuel with a solidification temperature of 45°C can be used for storage, preparation and combustion in the Arctic, which makes it very expensive; -the cost of delivery is also high, so the cost of annual "Northern delivery" (diesel fuel, fuel oil, coal) costs 500 billion rubles (this is 20 % of the expenditure part of the budget of the Northern regions); -fuel transportation costs are up to 70 % of the energy tariff; -fuel is delivered only once a year and can exceed the cost of fuel itself, which causes huge fuel costs; -the burned fuel has a negative impact on the environment due to the by-products released into the protected environment; -high wear of generating sources.

In Russia, facilities that are classified as distributed generation are conventionally divided into three categories: block stations, small CHP plants in populated areas and small and medium-sized generation facilities whose owners are not electric power companies.

The growth of interest in small and medium generation is due to:
- exhaustion of a resource of a considerable part of sources of thermal and electric energy of small capacity (DES, boiler rooms, small and micro HPPs, etc.) feeding numerous Autonomous consumers;
- decrease in quality and reliability of power supply from centralized sources and constant growth of tariffs;
- the duration of restoration of power supply after technological failures and natural disasters that have become more frequent in recent years;
- constant growth in oil, gas, coal prices and fuel delivery costs.

Currently, there is a growing interest in small (AFM) and micro (ASMI) nuclear power due to its special importance in social development and development of hard-to-reach and remote areas.

Preliminary estimates show that the total need to cover the deficit of electric and thermal energy due to ACMM for the period up to 2030 is about 20 GW.

AFM is a functionally ready-to-operate module or energy capsule consisting of reactor (RU) and turbine generator sets (TSU).

ASMM reactor plant is an integrated water-water reactor of high self-protection with the following features:
- negative values of the reactivity coefficient of the core in the entire operational range of parameters;
- controlled start of the core from the critical state;
- high level of natural coolant circulation, providing effective cooling and heat removal from the core in case of design and beyond design accidents;
– high heat storage capacity of metal structures and a large mass of coolant in the reactor, providing a relatively slow flow of transients in accidents associated with a violation of the normal heat sink from the core;
- a deep-layered system of barriers to the spread of ionizing radiation and radioactive fission products of uranium into the environment, as well as the implementation of a set of technical and organizational measures to protect these barriers from internal and external influences;
- high level of reliability of security systems;
- the use of passive systems and safety devices that operate on the basis of natural processes that do not require the supply of energy from the outside;
- protection against personnel errors due to the lack of continuous maintenance of the RC and limited access to ITS systems.

Interindustrial creative team, organized on a voluntary basis at the LLC "Cold test Site" of the Academy of Sciences of the Republic of Sakha(Yakutia) as part of M. M. Trevgoda, JSC "SSC RF-FEI", Obninsk, A. I. Chomchoev, JSC IPH an RS (I), Yakutia, A. S. Kosoy, LEMZ, GC Almaz Antey, Moscow., I. A. Syrovatskaya, LLC Praktika, the state Duma developed the draft concept "Development of nuclear power plants of micro capacity - up to 10 MW". (ASMI) as a factor of socio-economic development of remote and remote regions of the country". It gives a list of the main requirements for ASMM and ASMI:
- ensuring the number of hermetic barriers to the spread of radioactive fission products according to the standards and the exclusion of harmful emissions during the operation of the station;
- refusal to use partially activated fuel in ASMM;
- the use of low enriched fuel, which is important from the standpoint of non-proliferation;

- exclusion of nuclear and radiation hazardous works with fresh spent fuel at the site of the plant installation, up to the complete exclusion of works on fuel overload at the station within the resource of its operation;
– for stations of large capacities, from a range of small capacities, for exception of works with fresh spent fuel to use two-position execution of reactor installation with combination in one block of functions of the reactor and storage without opening of the case;
the exception in the construction of the station chemically dangerous environments, and their combinations (structural materials and coolant of the reactor circuit, the working fluid of the Converter, etc.);
- reducing the power voltage of the equipment.;
- reduction of the probability of equipment failure due to improved quality of its manufacture and Assembly in the factory and transportation of ASMI equipment in the form of factory-ready units.

These and other technical solutions, in comparison with traditional high-and medium-power ASMS, should ensure the safety of ASMS at a qualitatively new level.

The presented concept was discussed with the participation of the authors of the report at a workshop of specialists in the field of nuclear energy, held at the initiative of the " International Union of veterans of nuclear energy and industry "(ICAEP) on the development of micro and small nuclear power plants (ASMI).

On it, in particular, it was noted that.
- the existing opinion that small nuclear AFMS are economically unprofitable in comparison with large power plants, as it does not take into account local conditions and the social direction of regional development.
– there is a real danger of Russia lagging behind the United States and Japan, which are working on the introduction of microreactors for civilian use.

It is known from open sources that the American company "Westinghouse" together with the Japanese company "Toshiba" are developing a line of nuclear portable microreactors "eVinci" with a capacity of 200 kW to 15 MW, which can be transported with a single truck;
- in the UK (in Derby) with the support of the government opened a nuclear research centre "NAMRCMidlands" for the development of small modular reactors.
– to develop regional energy supply with the use of ASM required computational design and feasibility study of ASM at the level of the Technical proposal (TP) and Feasibility study (FS) based on multivariate analysis, including social aspects and then comparing the set of parameters of different alternatives to ASM.

The meeting noted the need for:
– contact the state Corporation "Rosatom" of the Russian Federation and the Ministry of economic development, to the formation of the Program of atomic energy development for the period after 2020, the inclusion of the section "Development of nuclear power stations micromodule (ASMI) for a sustainable energy supply of remote regions of the country and their socio-economic development" with the development of the feasibility study for the development of ASM followed to be considered at STC of ROSATOM of the Russian Federation;
- JSC "SSC RF-FEI" to acquaint specialists with the draft Concept and conduct a comprehensive analysis of the demand for ASMI, as well as to develop TP and feasibility study of ASMI for specific regions, followed by consideration of TP and feasibility study at the NTS of JSC "SSC RF-FEI" and " Rosatom»;
- to recommend to administrations of separate regions to consider expediency of construction in them ASMM and to hold public discussion about possibility and expediency of introduction of ASMM at the regional level.

The use of small nuclear power is hampered by the following main reasons:
- historically established priority attention to high-capacity nuclear power plants;
- a large number of specific requirements, different application conditions, technical difficulties, strict requirements for safety and for the duration of the cycle;
- impossibility of direct transfer of experience of development, creation and operation, accumulated since the beginning of development of nuclear power in the field of special nuclear power plants, etc.;
- there are restrictions on the dissemination of information on new ASMM.

2 Conclusion

In this article, the authors have tried to reveal the prospects of using ASMM to ensure energy security, development and development of hard-to-reach and remote regions. Mass implementation of ASMM in hard-to-reach and remote areas will have a positive effect on energy security, however, the lack of real experience in the operation of ASMM in energy systems makes a correct assessment of the impact of threats to energy security at the moment in these areas impossible due to the lack of real data.

To date, the most urgent tasks are:
- ensuring the safety of ASMM taking into account the prospects of their mass introduction;
- analysis of the potential market of ASMM, requirements and conditions of selection of the most suitable types of ASMM for specific conditions of their application;
- exclusion of permanently present personnel during the operation of the station;
- automatic operation of the station while ensuring safety and self-regulation;
- remote monitoring of the station status from the centralized control panel and with the possibility of reliable automatic reactor shutdown in case of emergency situations or unauthorized interference in its operation;
- inclusion of reserve power sources in the equipment of ASMM and provision of automatic transition to them when the reactor is stopped;

 https://doi.org/10.1051/e3sconf/201913901024

- implementation, where possible, and the inclusion of ASMM in the "smart grid" to reduce the cost of heat and electricity;
- mass implementation in remote and remote areas of the country and ensuring its energy security;
- automatic protection against natural disasters (forest fire, flood, earthquake), i.e. its automatic jamming;
- comparison of energy supply options with alternative sources of generation in remote and remote areas of the country by a set of parameters;
- assessment of the impact of commissioning of ASMM on the energy security of the regions and the country as a whole.

References

1. V.V. Bushuev, N.I. Voropai, S.M. Senderov, V.V. Saenko. About the Doctrine of energy security of Russia. Regional economy., No. 2 (30), pp. 40-50 , (2012)
2. N.Yu. Kirsanov. Features of power supply of the Arctic zone of the Russian Federation. Fundamental and applied research in the modern world., No. 21, pp. 169-171, (2018)
3. V.R. Kiushkina. Optimization of local energy of decentralized territories of Northern regions through strengthening of positions of energy security (on the example of the Republic of Sakha (Yakutia))//Internet journal. Science of science., Vol. 9, No. 6, pp. 101-110, (2017)
4. D.A. Soflan, V.R. Kishkina. Problems of reliability in Autonomous power supply systems decentralized energozone. Energy. Innovation in the energy sector. CALS-technologies in the energy sector, Vol. 1, pp. 46-52, (2016)
5. D. Newslos, I.Y. Repin. Tendencies of development of distributed generation Energy, No. 7, pp. 18-25, (2012)

 https://doi.org/10.1051/e3sconf/201913901025

Use of phasor measurements for modeling FACTS at the electric power system state estimation

Irina Kolosok[1], *Elena* Korkina[1,*], and *Aleksandr* Tikhonov[1]

[1]Energy Systems Institute by L.A.Melentiev, Siberian Branch of the Russian Academy of Sciences, Irkutsk, Russia

Abstract. As the Smart Grid is being developed in the Russian Federation, a need in FACTS arises at resolving the issues of increasing the transmission capacity of electric power lines, voltage control, etc. The paper considers the issues of installing phasor measurement units (PMU) at the FACTS for the electric power system state estimation of (EPS SE). The relevance of the paper is explained by the mandatory introduction of PMUs at newly commissioned power engineering facilities (PEF) scheduled for 2020 and farther. PMUs registering voltage and current phasors at FACTS allow increasing the precision of results of the EPS SE problem and improving the electric modes at the considered PEF.

1 Introduction

Smart Grid making stipulates for the development of organization and structural solutions in mode management as well as the upgrade of available grids for voltage regulation, increase the transmission capacity of electric power lines, etc. The actions on EPS upgrade are conducted, in particular, through the installation of controllable (flexible) systems of AC transmission - FACTS. FACTS allows managing not only certain parameters of electrotechnical equipment but a group of parameters in dynamics (U, δ, P and Q).

Transition to Smart Grid is not possible without introducing the wide area monitoring system (WAMS) [1]. PMUs with high precision register the measurements of state parameters. WAMS development in Russia gains its pace: in 2005 only two foreign PMUs were installed while in 2018 at the EP facilities 715 PMU were installed (mostly produced in Russia) [2]. A system operator of the Unified Power System plans the PMU mandatory installation at newly commissioned EP facilities starting from 2020. As it is stated in [3], PMU is the only today's technology allowing for the monitoring of dynamic processes in real time.

Dispatching control practice most frequently solves two problems to form a model of the current state: load flow calculation (LFC) and the EPS SE based on the measurements of SCADA and WAMS [4]. As of today SCADA is a modern software complex collecting and transferring data from the facilities to a central two-way station (CTWS); for remote control of electrotechnical equipment at the facility; for control algorithm implementation; for identifying the emergency state and informing the dispatching staff at the electric power company. The measurement system SCADA, applied today, cannot provide sufficient basis for making a comprehensive Smart Grid that would correspond to all the requirements set to "smart grids", including those relating to FACTS (being its main component) control. It is explained by insufficiently high discretization of transmitted signals as well as their reliability as there are limitations existing for signal transmission from the sensors to CTWS because of the system architecture and algorithms. In addition, one SCADA measuring device registers only a small number of parameters - it is not suitable in case of FACTS.

One of the possible solutions of the problem is applying PMU allowing for registering virtually all FACTS parameters with a high discretization level at continuous data reading. This allows implementing the algorithms of FACTS control at the facilities and transmit the information for forming control actions to the upper level (the dispatching center) taking into account the mode and circuit state of EPS. The tandem use of FACTS and PMU allows preventing, controlling and localize system incidents with minimum aftermath for EPS operation.

Both abroad and in Russia primarily the research were conducted on inclusion the FACTS models in the LFC problem [5-9], and later - in the the EPS SE problem [10-15]. [12] discusses a limited transmission capacity of electric power lines (EPL) and involving FACTS into the SE problem. [13] suggests a state estimation algorithm in power systems where the FACTS and PMUs are located. Russian papers [14, 15] of 2012 suggest a two-stage procedure of local SE for EPL with the account of FACTS the results of which confirm the efficiency of the PMU application at the validation of grid and FACTS parameters. [15] suggests local SE on the basis of STATCOM. The measurements at the EPL ends including a synchronous static compensating device (STATCOM) allow estimating not only the electric state parameters but also the parameters of transmission equivalent circuits that significantly reduces the EPS SE resulting error. Unfortunately, in those years, the number of PMUs was insufficient and the problem was not resolved. Now several FACTS are installed in Russia. Not all of them are equipped with PMU.

In 2016 Melentiev ESI SB of RAS conducted a large-scale work on FACTS modeling in the EPS SE algorithms [16-21]. In this paper the authors present the models of controlled shunt reactors (CSR), phase-shifting transformer (PST) and HVDC included in the SE problem. To increase the grid observability, equipping these models with PMUs is suggested. It is shown that the solution of the SE problem with the inclusion of FACTS equipped with PMUs provides a significant improvement of quality of measurement estimates.

2 Modeling FACTS at solving the problem of EPS state estimation

2.1 The EPS state estimation problem

The EPS SE problem consists in calculating the EPS steady-state conditions based on measurements $\bar{y}$ [2]. State estimation algorithms traditionally deals with SCADA: voltage magnitudes U_i, nodal power injections P_i, Q_i, power flows in lines P_{ij}, Q_{ij}, more rare – nodal and line currents I_i, I_{ij}. So, measurement vector looks like:

$$\bar{y} = (P_i, Q_i, P_{ij}, Q_{ij}, U_i, I_i, I_{ij}) \tag{1}$$

Since WAMS has been created a new generation of measurements started

$$\bar{y} = (\dot{U}_i, \dot{I}_i, \dot{I}_{ij}), \tag{2}$$

with synchronized phasor measurements of nodal voltage $\dot{U}_i = U_{ai} + jU_{ri}$; nodal $\dot{I}_i = I_{ai} + jI_{ri}$ and line currents $\dot{I}_{ij} = I_{aij} + jI_{rij}$.

For solving the EPS SE problem the Test Equation method was developed in MESI SB RAS [4]. Test Equations (TE) are the steady state equations that consist of only measurements $\bar{y}$:

$$w_k(y) = 0 \tag{3}$$

When using the TE method the EPS SE problem concludes in minimization of a function value:

$$J(y) = (\bar{y} - \hat{y})^T R_y^{-1} (\bar{y} - \hat{y}) \tag{4}$$

subject to constraints in the form of steady state equations (3). In (4), R_y - the covariance matrix of measurement errors. At the second stage, based on estimates of basic measurements, estimates of the state vector (complexes of nodal voltages) are calculated then line power flows and nodal power injections are calculated.

2.2 PMU arrangement at EPS facilities

The problem of PMU location is well-known to the researchers and designers occupied with the issues of PMU application in power engineering. To provide for correct operation, all PMU are time-synchronized by means of signal receivers of global positioning systems (GPS, GLONASS). Each variant of PMU installation aimed at some specific purposes:

• EPS facilities of special significance [2],
• the boundary nodes of adjacent EPS at the decomposition of the SE problem [22],
• the nodes improving the properties of the EPS SE problem (criteria redundancy increase; eliminating critical measurements and groups) [23],
• at calculating the linear SE in the areas which can be fully observed by PMUs [1,24],
• the substations with FACTS [25-29, etc.].

Semiconducting technology development in 90-s of the XXth century allowed making turn-off thyristors (GTO and GCT), fast diodes and insulated-gate bipolar transistor (IGBT) operating in the voltage range of 2.5-6kV, with cutoff currents 1,500-4,000A. This breakthrough facilitated the invention of a new type of converters - voltage converters (VC) – due them modern FACTS of the second generation were made (for example, PST and HVDC on the voltage controller basis).

2.3 PMU arrangement on FACTS models at the EPS state estimation

To demonstrate PMU arrangement at the approbation of the PST, CSR, HVDC, the authors use the 500kV grid of the real power system (19 nodes, 28 lines) (Fig.1) [21]. The Northern region of this system locating the largest electric power sources is redundant while the Southern is deficient. The lines 7-8, 8-9 are loaded for a nominal power that leads to significant losses of electric power in them (approximately 50 % of all power losses of that power system).

2.3.1 Modeling a phase-shifting transformer

PST is used to control power flows by changing voltage angles at the line ends (Table 1, Fig. 2,a, line 3-1). The traditional PST structure uses a cascade of two transformers - an regulated transformer T1 connected directly to a bus supplying the line on the part of the power source and a series transformer T2 (Table 1, Fig. 2,b) which is connected in series to overhead line by means of a secondary winding [21]. By means of regulating voltage at T1 the authors change the voltage magnitude and angle at T2 thus changing the power flows transferred in the line.

To model a booster transformer T2 with the capacity of 630 MVA, the authors include the node 20 and line 3-20 in the network while to model a reducing transformer T1 and PN2 - the node 21 and the line 3-21. To initially set PST and further control its operation as well as to increase its observability, it is suggested installing PMU in the node to which a series transformer is connected, i.e., in the node 3.

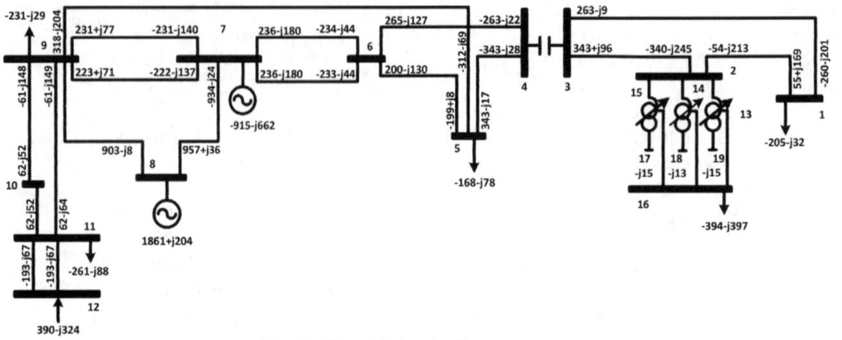

Fig. 1. 19-nodal circuit of power system

As it is clear from Table 1, PMU installation increase the observability of the network while the measurement estimates in the node 3 and incident lines become more precise. In the basic mode along the line 3-1 an "excess" active power is transmitted. It exceeds the load required in the node 1, while along the line 3-2 the power insufficient for load supply in the node 2. Along the line 1-2 an additional (reversive) power flows from the node 1 to the node 2. When inclusing Phase-Shifting Transformer the active power flow in the line 3-2 increased from 343 MW to 397 MW while in the line 3-1 it decreased from 262 MW to 208 MW. A total power in the node 3 virtually did not change (-606 MW) while the power flow towards the node 1 is virtually equal to zero.

Table 1. Phase-shifting transformer modeling

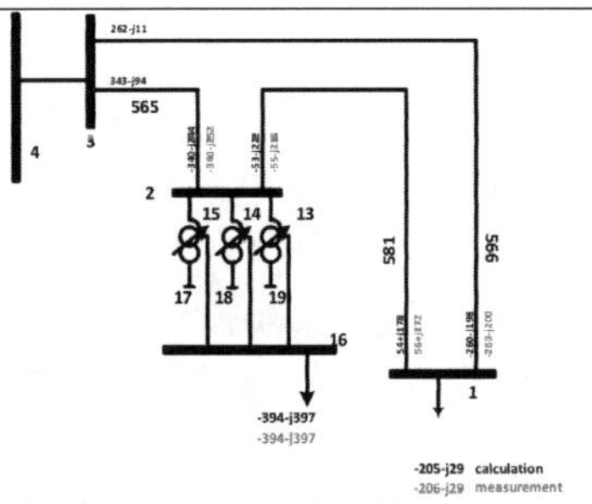

Fig. 2, a. Original grid fragment.

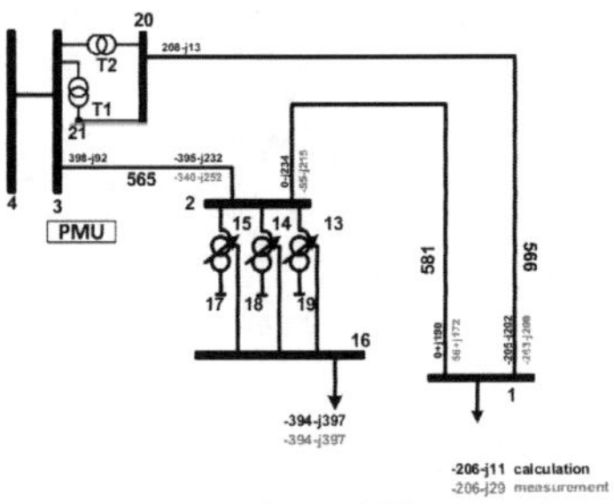

Fig. 2, b. PMU and Phase-Shifting Transformer

Original grid fragment											
	U_1	δ_1	U_2	P_{3-1} *without FACTS*		U_3	δ_3			P_{3-2}	P_{3-4}
Measurements	503	-	498	-	-	-		-	-	-	-
Calculations	503 .7	-6.9	495 .5	262	-	516.8	-3 .2	-	-	343	-606.9
$\Delta = \delta_3 - \delta_1 = 3.7°$ $P_{1-2\_revers} = 56$											
Grid with a phase-shifting transformer equipped with a phasor measurements (pm)											
	U_1	δ_1	U_2		P_{20-1} *with FACTS*	U_3 *pm*	δ_3 *pm*	P_{3-20} *pm*	P_{3-21} *pm*	P_{3-2} *pm*	P_{3-4} *pm*
Measurements	503	-	498	-	-	518 .8	-3 .1	76 .4	132 .8	397 .3	-605 .9
Calculations	504 .3	-7 .7	499	-	208	518 .6	-3 .1	76 .4	132 .8	397 .3	-606 .2
$\Delta_{PSD\&pm} = \delta_3 - \delta_1 = 4.6°$ $P_{1-2\_revers} = 0.55$											

2.3.2 HVDC modeling in the EPS SE problem

HVDC are applied for a coordinated operation of AC and DC power electric grids; for the grids with various electric current frequency; for improving the transmission capacity of grid components containing "weak" links [31]. HVDC can be presented by two voltage sources [8]. Each voltage source is connected to the AC system via a transformer. HVDC is modeled at the substation in the node 12 (Table 2, Fig.3,a) to which one connects two branches modeling the transformer 12-112 and the rectifier 112-1112. To connect an inverter to the circuit, the authors add the node 122 and two lines modeling the transformer 22-122 and the inverter 122-1122. For HVDC the authors introduce an additional equation describing the process of active power transfer via HVDC, i.e. the active power equality $P_{rectifier}$, consumed by the rectifier and $P_{inverter}$ provided by the inverter:

$$P_{rectifier} + P_{inverter} = 0 \quad (5)$$

The equation (5) is considered as a test equation at the solution of EPS SE.

Table 2. HVDC modeling

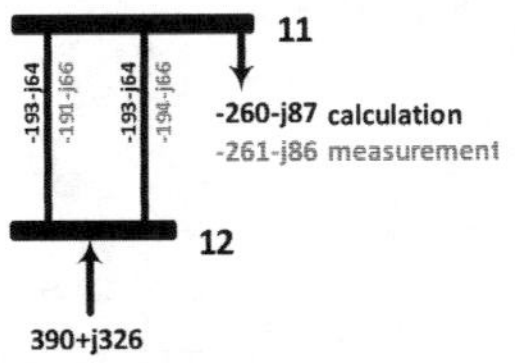

Fig.3, a. Original network fragment

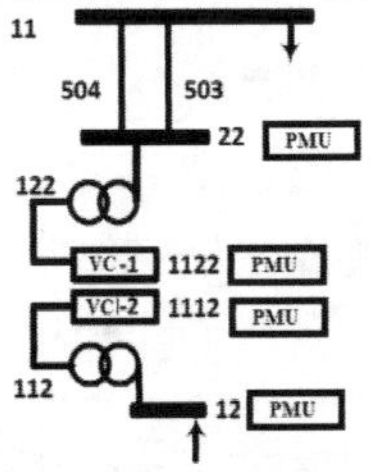

Fig.3, b. PMU and HVDC on the basis of voltage converter

	U_{12}	δ_{12}	P_{1112}	Q_{1112}	P_{1122}	Q_{1122}	U_{1112}	δ_{1112}	U_{1122}	δ_{1122}	U_{22}	δ_{22}
Original network												
Meas.	512	-	-391	85	-	-	-	-	-	-	512	-
Calc.	512	0	-391	85	392 .8	-187 .6	11	-10 .8	9 .3	25	511 .8	11 .8
Network with a HVDC on the basis of voltage converter equipped with PMU												
	U_{12}	δ_{12}	P_{1112}	Q_{1112}	P_{1122}	Q_{1122}	U_{1112}	δ_{1112}	U_{1122}	δ_{1122}	U_{22}	δ_{22}
Meas.	511 .9	1 .8	-391	166	391	-188	9 .3	-9	9 .3	25	511 .9	11 .8
Calc.	511 .9	1 .8	-391	166	391	-188	9 .3	-9	9 .3	25	511 .9	11 .8

At making the model of HVDC on the voltage converter basis undesirable problems of measurement lack occur. In practice PMUs are installed at voltage converters that is why the nodes 1112 and 1122 were selected for this purpose. In addition, two PMU are installed in the nodes 12 and 22 (Table 2, Fig.3,b). As it is clear from Table 2, the obtained estimates of voltages are equal to the measurement of voltages in the nodes where PMUs are installed and, due to more precise voltage, the authors specified the reactive power estimate.

Table 3 shows that the condition (5) at the transfer of active power on HVDC is satisfied ($P_{1112} + P_{1122} = 0$).

Table 3. HVDC "cut-in" conditions

Rectifier PN2 emits active power in the grid and gets reactive power $U_{1112} > U_{112}$ (11.4 kV >11.0 kV)
$P_{1112} = -391 kW; \quad Q_{1112} = 166 MVar.$
Inverter PN1 receives active power from the grid and puts out reactive power $U_{1122} < U_{122}$ (9.3kV <10.0kV).
$P_{1122} = 391 kW; \quad Q_{1122} = -190 MVar.$

2.3.3 Modeling a controlled shunt reactor (CSR).

Some time ago two uncontrolled shunt reactors (180MVar each) (Table 4, Fig.4,a) were installed at the 500kV substation (node11). They did not allow smooth compensation of excess reactive power occurring at the substation due to traction load at the side of 110kV of that substation and the availability of system connecting overhead lines of 500kV with the power system of neighbour area. To provide smooth load control, in 2012 one of SR was replaced by a controlled shunt reactor (CSR). The CSR, providing a smooth control of inductance, allows maintaining a required voltage level in the overhead line node connection with a large charge capacity. At the use in the EPS SE problem of the CSR model with the control angle the CSR is modeled by a variable conductivity in the installation node the value of which is defined during calculation. The node voltage is registered. Depending on the obtained conductivity value, one can define the thyristor ignition angle. To improve the CSR observability, one should install PMU at the bus where CSR is installed (node 111) (Table 4, Fig.4,b).

3 Conclusion

Modern means for measuring the parameters of electrical power system modes (PMU, digital measurement devices) and electrical power management units (FACTS, energy accumulators, etc.) significantly increase the observability and controlability of electrical power systems. The solution of the EPS SE can be more precise and efficient due to the application of the precise measurements of WAMS.

https://doi.org/10.1051/e3sconf/201913901025

Table 4. Modeling a controlled shunt reactor (CSR).

Fig.4, a. Original circuit fragment.

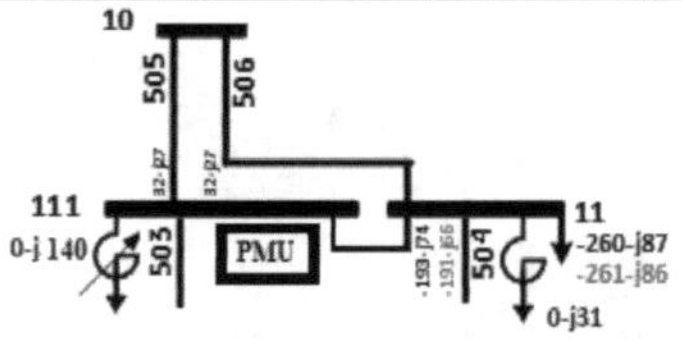

Fig.4, b. PMU and CSR

Original circuit											
	U_{10}	δ_{10}	U_{11}	δ_{11}	P_{11}	Q_{11}	U_{111}	δ_{111}	P_{111}	Q_{111}	b_{111}
Meas.	-	-	516	-	-261	-86	-	-	-	-	-
Calc.	514.9	1.8	514.5	1.8	-260.7	-87	-	-	-	-	-
Circuit with a controlled shunt reactor equipped with PMU											
Meas.	-	1.8	516	1.8	-261	-91	516	1.8	-	-	522
Calc.	516.4	1.8	516	1.8	- 260 .4	-91	516	1.8	0	-140	522

The PMU provides for the precise setting of FACTS functioning and subsequent monitoring. The calculations conducted for the section of the 500 kV grid of the Irkutsk energy system proved the correctness of the results obtained. This makes it possible to adopt the data for the electric power system state estimation in real life.

The research was carried out under State Assignment, Project III.17.4.2 (reg. № AAAA-A17-117030310438-1) of the Fundamental Research of Siberian Branch of the Russian Academy of Sciences.

References

1. A. Phadke, J. Thorp. *Synchronized phasor measurements and their applications.* (Springer, NewYork, 2008)
2. A.V. Zhukov Experience of applying WAMS technology for problems of monitoring and control in power engineering. Report at the meeting of Non-commercial Partnership Science&Technical Council of Uniform Power Grid. 10.10.2018
3. Prospects of phasor measurement application// http://digitalsubstation.com/blog/2018/05/07/perspektivy-primeneniya-sinhronizirovannyh-vektornyh-izmerenij/
4. A. Gamm, A. Glazunova, Yu. Grishin, I. Kolosok, E. Korkina. Electricity, **6** (2009).
5. N.Hingorani, L.Gyugyi. *Understanding FACTS concepts and technology of flexible AC transmission systems* (IEEE Press, 2000)
6. E. Acha, C.R. Fuerte-Esquivel, H. Ambriz-Perez, C. Angeles-Camacho. *FACTS. Modelling and Simulation in Power Networks.* (John Willey&Sons, England, 2004)
7. V.Kochkin. News of Electrotechnics, **4(46),** (2007)
8. N.Voropai, A.Osak. Development of Equipment and Control Systems for Large Power Engineering Systems. Science&Technology Report, 2008-0-2.7-31-01-007. Irkutsk, 2009
9. P.Yerokhin, V.Neuimin et al. News of Science&Technical Center of Uniform Power Grid, **66(1)**, 2012
10. T. Okon, K. Wilkosz. PowerTech'2011, Norway.
11. E. Zamora-Cárdenas, B.Alcaide-Moreno, C.Fuerte-Esquivel. EPSR, **81** (2011)
12. J. Kumar, N. Kumar. Intern. Conf. on Recent Advances in Management, Science, Technology, Education and Legal Issues, 2019.
13. V. Presada, C. Cristea, M. Eremia, L. Toma. UPEC'2014. doi: 10.1109/UPEC.2014.6934712.
14. P.Bartolomey, S.Eroshenko, E.Lebedev, A.Suvorov. The 3d Intern. Science&Technology Conf., **2** (2012)
15. P.Bartolomey, S.Eroshenko, E.Lebedev, A.Suvorov. IEEE PES ISGT, Germany, 2012.
16. I.Kolosok, A.Tikhonov. Methodological problems in reliability study of large energy system. Minsk, **66** (2015)
17. I.Kolosok, A.Tikhonov. Industrial Power Engineering, **10** (2015)
18. I.Kolosok, A.Tikhonov, E.Korkina. Methodological problems in reliability study of large energy system. Syktyvkar, **67** (2016)
19. A.Tikhonov. *Thesis* (Melentiev ESI SB RAS, Irkutsk, 2017)
20. I.Kolosok, A.Tikhonov, E.Korkina, A.Mahnitko, A.Gavrilovs. doi: 10.1109 /RTUCON.2018.8659809.
21. I.Kolosok, A.Tikhonov. Methodological problems in reliability study of large energy system. Irkutsk, **69** (2018)
22. I.Kolosok, E.Korkina, A.Paltsev. PowerTech'2011, Norway.
23. A. Gamm, A. Glazunova, I. Kolosok, E. Korkina. DRPT'08, China.
24. I.Kolosok, E.Korkina, E.Buchinsky. PSCC-2014, Poland.
25. Yang Guangya, M.Gordon, A.Nielsen, J. Østergaard. Future Energy Systems Workshop, 2009.
26. B.Singh, N.Sharma, A.Tiwari, K.Verma, S.Singh. MultiCraft Intern. Journ. of Engineering, Science and Technology, **3(3)** 2011
27. M.Dalali, H.Kargar. PSPC2015, Tehran.
28. M. Khan and A.Sh. Siddiqui Perspectives in Science, **8** (2016)
29. E.Zamora-Cárdenas,C.Fuerte-Esquivel, A.Pizano-Martínez, H.Estrada-García, EPSR, **133** (2016).
30. Edris R. A., Bake M. H. et al. IEEE Trans. on Power Delivery, **12(4)** (1997)
31. HVDC. http://www.ntc-power.ru/innovative_projects/insert_dc/ (access date: 27.02.2018).

Assessment of risks due to cyberattacks on information and communication infrastructure of cyber-physical system in the electric power system control[1]

Irina Kolosok[1,2], *Liudmila* Gurina[1,*]

[1]Melentiev Energy Systems Institute SB RAS, Irkutsk, Russia
[2]Irkutsk National Research Technical University, Irkutsk, Irkutsk region, Russia

Abstract. The paper proposes an approach to risk assessment in the electric power system control to identify possible failures in the electric power system functioning caused by cyberattacks on SCADA systems and WAMS. There is a large number of information, hardware and software factors, which increase the automated dispatching control system vulnerability to cyberattacks. The study shows the influence of unforeseen cyber-incidents in the information and communication infrastructure on electric power systems through false control actions on their physical part. The effectiveness of the developed risk assessment approach is demonstrated by its performance under uncertainty.

1 Introduction

The digital transformation of the energy industry is aimed at creating an intelligent energy system (IES) with cyber-physical infrastructure based on digital (information) models [1]. The wide-scale use of information and communication technologies and the adoption of synchronized vector measurement technologies increase the controllability and observability of electric power systems (EPSs), thus providing their reliable and efficient operation [2,3]. Apart from the undoubted advantages of these changes, the number of cybersecurity vulnerabilities is increasing due to the growing scale and complexity of EPSs, the information interaction between the EPS information-communication and physical subsystems, the possibility of hidden threat occurrence when using hardware and software, and a great number of tools and devices for measuring, transmitting and processing information at all levels of hierarchy of the automated dispatching control systems of EPS. The integration of new information-communication technologies with the existing ones creates new cyber threats to the EPS control. Cyberattacks at the information and communication level can lead to control errors, power supply interruptions, delayed system restoration, and other adverse consequences. The need to provide and maintain reliable operation of the EPS requires an analysis and assessment of risks in the EPS control, which are created by possible cyberattacks on the measuring, information transmitting and processing devices, which constitute the SCADA systems and WAMS. Such cyberattacks can lead to generation of incorrect control actions and cause EPS malfunction.

In this regard, the paper proposes an algorithm for assessing risks in the EPS control due to cyberattacks on SCADA systems and WAMS, given the preservation of their cybersecurity properties. The development of the algorithm rested on the theory of fuzzy sets and probability theory.

The effectiveness of the approach developed for risk assessment is confirmed by its verification under the conditions of uncertainty caused by the lack of statistics on cyberattacks on SCADA systems and WAMS.

2 Cyber-physical electric power system and its cyber-vulnerability

The electric power system development based on the adoption of new information and communication technologies and digitalization of the electric power industry has led to the creation of a cyber-physical system that consists of information - communication and physical subsystems.

The efficiency of dispatching control of complex EPSs is provided by multiple-stream constantly updated and reliable data. The specific parameters and state variables of EPS are measured, processed, transmitted, analyzed and visualized by SCADA systems and WAMS, which are part of the information-communication system using modern technical means, which also produces, transmits and implements the necessary control actions.

[1] This study is supported by the Siberian Branch of the Russian Academy of Sciences (Project III.17.4.2) of the Federal Program of Scientific Research (№ AAAA-A17-117030310438-1).

* Corresponding author: gurina@isem.irk.ru

Even in the present-day power systems, the physical and information-communication subsystems become comparable in complexity and responsibility in terms of normal functioning of electric power systems and increasing requirements of consumers for the power supply reliability and the quality of electricity supplied to them (in the future power systems this trend will persist to an even greater extent). In the context of EPS digitalization, it is increasingly necessary to consider complex integrated cyber-physical electric power systems that are characterized by new properties and face exacerbating reliability problems [4].

The SCADA and WAMS data are used to perform the control and monitoring functions with the aid of hardware and software of their components [5].

The data acquisition system transfers the RTU and PMU measurements and the information about the positions of switching devices (switches, disconnectors) to the real-time data processing system. The transmission system provides data transfer to control centers, interaction, and coordination of actions between operators in various control centers. The information processing system, in addition to data processing, also performs the EPS state estimation, which should provide a reliable picture of the system state based on measurements, pseudo-measurements, and information on the network topology. The graphical interface displays the results of state estimation and analysis of unforeseen situations. It also shows alert messages and provides access to other information about the system in real time.

Cyber incidents can occur in any of the described components of the information-communication infrastructure and cause a failure of control functions. In [6], the authors analyze and describe possible IES vulnerabilities.

The SCADA systems and WAMS are the most vulnerable to cyberattacks in the information-communication infrastructure. Since the control system affects the EPS through control actions or its output data, the consequences caused by cyber threat events in these systems pose the greatest danger to the EPS operation. This is because the control of EPS is based on the information from the SCADA systems and WAMS.

To reduce the dangerous consequences of cyberattacks for EPS, it is important to maintain the following cybersecurity properties of SCADA systems and WAMS [7]:

1. Timeliness of information. The SCADA systems and WAMS are rigid real-time systems [8, 9]. Transfer, receiving, and processing of data, as well as generation of control actions, are time-critical and must be done on time. This property means that any information should correspond to the current state of EPS.
2. Availability means that any component of the SCADA system and WAMS should be ready for use when needed.
3. Integrity requires that the data to be collected, transmitted, displayed, and stored in SCADA systems and WAMS be authentic and intact disregarding the unauthorized interference.
4. Cybersecurity is the ability of a system to constrain the local impact of cyberattacks, identify and delay the stream of corrupted data within the area vulnerable to cyberattack, without further transfer and use of these data in the control of a physical subsystem, in order not to cause emergencies leading to the development of system blackouts.
5. Confidentiality means that an unauthorized person should not have any access to the information related to a specific SCADA system and WAMS.

3 The effect of the information-communication infrastructure failures on the EPS operation

The reliable operation of EPS depends on the applied information and communication technologies in the automated control system. The power system itself becomes more vulnerable to information failures and cyber incidents in the information-communication infrastructure. Since the physical and information-communication systems are closely interconnected, the incompleteness and inaccuracy of information due to cyberattacks on SCADA systems and WAMS can lead to the generation and implementation of incorrect control actions and the development of emergencies in EPS. For example, false data injection attacks can lead to an incorrect EPS state estimation, which will negatively affect the correctness of control actions. Moreover, attackers, if not detected, can change the network topology and system state estimation. Table 1 shows the examples of possible malfunctions of the physical subsystem because of cyberattacks on the information-communication system, and their impact on the EPS control functions.

In order to prevent an EPS control failure, it is required to develop measures (techniques, tools, methods) to preserve the cyber-resilience properties of the system during cyberattacks. Solving this problem calls for an analysis and assessment of the factors influencing the control system, where the methods of risk theory are an important tool.

The assessment of risks during EPS control focuses on the information provided by SCADA systems and WAMS, given their cybersecurity properties.

4 An algorithm of risk assessment during EPS control based on the information factors

Threat and risk are the determining factors for cybersecurity. Threat is a potential cause that can lead to negative consequences for and damage to the system. Risk is the possibility of an undesirable outcome as a result of an incident, event or disturbance, as determined by its probability and the associated consequences. In fact, risk is a combination of the probability of threat event occurrence and its consequences (damage) with respect to a protected asset (resource). The consequences are determined by the level of effect [10, 11].

Table 1. Effects of cyberattacks on control functions and related consequences for EPS.

Threats	Information-communication system	Automated dispatching control system	Physical system
False data injection	SCADA (RTU), WAMS (PMU, PDC), EMS, DMS, state estimation software.	Generation of incorrect control actions. Loss of frequency and voltage control. Loss of observability. Incorrect dispatching commands. False characteristics of failures will affect the distribution and transmission operations.	Stability loss. Short circuits (false line overload, sagging and overheating of wires). Large fluctuations in the system dynamics: - tripping of additional lines; - shutdown of generators; - load shedding. Blackout.
Time Synchronization Attacks (spoofing attacks, etc.)	WAMS (PMU, communication channels between PMU and PDC, GPS), state estimation software.	False visualization of the current state, which leads to erroneous control and protection actions. Loss of frequency and voltage control. False information regarding the presence and location of a malfunction. A mismatch between the commands to switch off / operate smart devices.	Power outage, with a subsequent cascading chain of emergencies in the system. Loss of stability.
Denial of Service attacks (DoS jamming attacks, etc.)	SCADA (RTU), WAMS (PMU, PDC), EMS, DMS, state estimation software.	Delay in control. Generation of incorrect control actions. Loss of frequency and voltage control. Blocking of a control signal. Loss of observability.	Loss of stability. Increase in the system restoration time. Generation-load imbalance. Failure to disable/operate smart devices.
Replay, Distribution DoS attacks	SCADA (RTU), WAMS (PMU, PDC), state estimation software.	Control delay. Loss of observability. Violation of frequency and voltage control.	Loss of stability.
Coordinated Attacks	SCADA (RTU), WAMS (PMU), state estimation software.	All of the above.	Loss of stability. Cascade emergencies in the system.
Malicious software (Backdoor, Virus, worms, Trojan horse)	SCADA (HMI), WAMS.	Incorrect generation of control actions. Wrong operation of hardware and software. Loss of voltage and frequency control.	Unwanted tripping/operation of smart devices. Loss of stability. Voltage collapse. Blackout.

Information, software, and hardware, etc. can be considered as an asset. SCADA systems and WAMS are complex assets that include the above components. The possibility of vulnerability occurrence in the system assets is the main threat to cybersecurity.

The authors of [7] developed an algorithm for assessing the cybersecurity risks of information-communication infrastructure in the case of cyberattacks based on the theory of fuzzy sets. In this paper, the algorithm has been further developed and has undergone some changes in the semantic description of risk factors associated with the specific features of EPS control.

The risk assessment during the EPS control considers SCADA systems and WAMS as an asset. Loss of control is the damage caused by the cyber threat, which entails adverse consequences for the EPS operation.

The algorithm consists of two stages:

1. Assessment of risk during control for each cyber threat event;
2. Assessment of the resulting risk.

4.1 Risk assessment during control for each cyber threat event

The algorithm is based on the theory of fuzzy sets. To this end, input and output linguistic variables are defined.

The input linguistic variables are the intentions, capabilities, and purposes of an attacker; the vulnerability of the information-communication infrastructure; and the consequences determined by the level of impact on the information-communication infrastructure. The output linguistic variables (the probability of the threat initiation, the probability of a threat event, the probability of the threat event occurrence) are determined by the developed rules of the Mamdani Fuzzy Inference System (FIS), taking into account the recommendations [12].

A term set $\{VL, L, M, H, VH\}$ is defined (Table 2) and its semantic description is given for each risk factor [7]. A range of changes in the membership function in the segment $[0;1]$ is specified.

Table 2. A term set of risk factor.

Notation	Value	Range of membership function change
VL	Very low	[0;0,4]
L	Low	[0,05;0,2]
M	Moderate	[0,21;0,79]
H	High	[0,8;0,95]
CH	Critically high	[0,96;1]

Table 3 presents a semantic description of risk levels in EPS control.

Table 3. Risk levels.

Level	Description
Very low risk	The threat event is expected to have a slight adverse effect on the EPS control.
Low risk	A threatening event may have a limited adverse effect on the EPS control; the consequences for functioning are local in nature.
Moderate risk	A hazardous event can have a serious adverse effect on the EPS control.
High risk	A threatening event can have a serious or catastrophic adverse effect on the EPS operation.
Critically high risk	A threat event can have numerous serious or catastrophic adverse effects on EPS operation.

A hierarchical fuzzy system (Fig. 1) has been proposed for risk assessment during the EPS control, in which four systems of fuzzy logical inference $FIS1$, $FIS2$, $FIS3$, $FIS4$ are embedded. The factors such as the capabilities, intentions, and purposes of the attacker are used to assess the probability of threat initiation. Combinations of factors, such as the attacker capabilities and the information and communication infrastructure vulnerabilities, were used to assess the probability of a threat event as a result of adverse effect. Combination of these probabilities was used to determine the total probability of the threat event occurrence. The combinations of the probability of threats and the levels of impact (consequences) on SCADA systems and WAMS determine the risk estimate during the EPS control (Fig. 2).

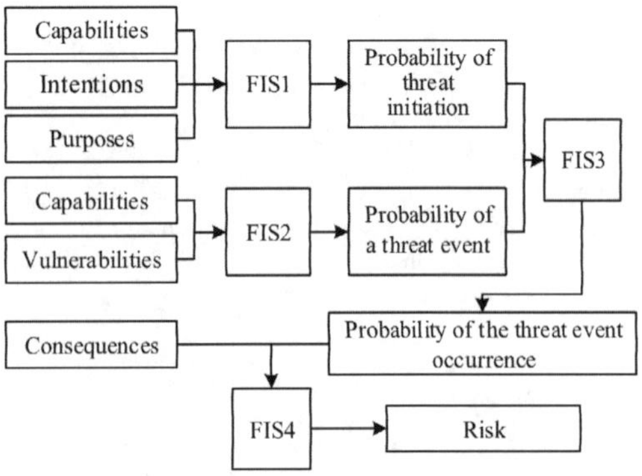

Fig. 1. Hierarchical fuzzy system of risk assessment during EPS control.

A matrix has been obtained to assess risk during EPS control (Fig. 2).

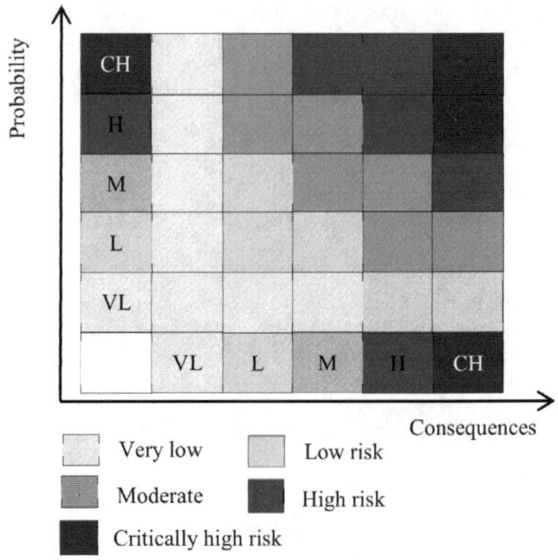

Fig. 2. Matrix of risk assessment during EPS control.

4.2 Assessment of a resulting risk

An error in the EPS control can be caused by several cyberattacks. Events of i -x cyber-threats ($i = \overline{1,n}$) are joint independent events (they may appear simultaneously and the probability of one threat event occurrence does not change depending on the other threat event occurrence [13]). In this case, the resulting risk estimate during control R_c at n cyberattacks is calculated according to:

$$R_c = R_{c1} + R_{c2} + \ldots + R_{cn} - R_{c1}R_{c2} - \ldots - \\ - R_{n-1}R_n + R_{c1}R_{c2}R_{c3} + \ldots + R_{cn-2}R_{cn-1}R_{cn} - \\ - \ldots + (-1)^{n-1} R_{c1}R_{c2} .. R_{c3} \ldots \quad , \tag{1}$$

where risk under the i -th cyberattack R_{ci} is calculated according to the method presented in the previous section (Figs.1,2).

The obtained cyberattack-related risk assessments can be useful in choosing the models and methods of processing SCADA and WAMS data streams in real time, which could provide the necessary completeness and reliability of the information [14, 15] used for EPS control.

5 Case study

To assess the risk during EPS control, we considered a jamming attack (a type of DOS attack) and a spoofing attack (a type of time synchronization attack) [16] on WAMS.

Table 4 shows the values of the input linguistic variables for each cyberattack.

Table 4. Input linguistic variables.

	Jamming-attack	Spoofing-attack
Capabilities	0.75	0.97
Intentions	0.76	0.92
Purposes	0.6	0.7
Vulnerabilities	0.85	0.92
Effects	0.92	0.93

The values of the output linguistic variables are obtained according to the rules of fuzzy inference of systems $FIS1$, $FIS2$, $FIS3$, $FIS4$ (Table 5).

Table 5. Input linguistic variables.

	Jamming-attack	Spoofing-attack
Probability of threat initiation	0.63	0.81
Probability of threat event	0.76	0.81
Probability of threat event occurrence	0.69	0.75
Risk (see Fig. 3)	0.67 – the moderate level	0.72 – the moderate level

The resulting risk estimate $R_c = 0.91$, obtained according to (1), indicates a high level of risk and allows us to draw the conclusions (which are confirmed by the calculations) that successful joint jamming and spoofing attacks can cause far greater damage to the power system and cause more serious consequences for its operation (Table 1) than each of the cyberattacks individually.

Conclusion

The proposed algorithm for risk assessment during EPS control in case of cyberattacks makes it possible to assess the consequences for the EPS functioning. It can also be useful to conduct further studies aimed at improving the quality of the information data streams required for monitoring and control, and to develop measures to enhance the cybersecurity of the entire EPS infrastructure.

References

1. N.Voropai, M. Gubko, S. Kovalev, L. Massel, D. Novikov, A. Raikov, S. Senderov, V. Stennikov, Control Sciences, **1**, 2-14, (2019). doi: 10.25728/pu.2019.1.1
2. Remote control at substations: 330 kV Gubkin substation and 500 kV Shchelokov substation. Available at: http://digitalsubstation.com/blog/2018/06/27/teleupravlenie-na-podstantsiyah-ps-330-kv-gubkin-i-ps-500-kv-shhyolokov/
3. A. Zhukov, D. Dubinin, Releyshchik, **3**, 22-29, (2013)
4. N. Voropai, I. Kolosok, E. Korkina, A. Osak, Energy Policy, **5**, 53-61, (2018)
5. Yu. Ivanov, A. Cherepov, D. Dubinin, E. Satsuk, Energetik, **3**, 8-12, (2016)
6. B. Papkov, *Methodological problems in the reliability study of large energy systems. Research and provision of the reliability of energy systems,* **68**, 441–451, (2017)
7. I. Kolosok, L. Gurina, Information and Mathematical Technologies in Science and Management, **2 (14)**, 40-51, (2019). DOI: 10.25729/2413-0133-2019-2-04
8. B. Ayuev, P. Erokhin, Yu. Kulikov, *Technologies for controlling the operating conditions of energy systems of the 21st century: Collected papers of All-Russian Scientific and practical. conf.,* 83-92 (Novosibirsk: Publishing House of NSTU, 2006).
9. STO 59012820.35.240.50.004-2011. Dispatching control systems in the electric power industry. System of Supervisory Control And Data Acquisition (SCADA) in dispatching control. Available at: https://www.so-ups.ru
10. B. Papkov, A. Kulikov, *Cyber-threats and cyberattacks in the electric power industry*, (Nizhny Novgorod: NRU RANEPA, 2017)
11. A. Dorofeev, A. Markov, Voprosy Kiberbezopasnosti, **1(2)**, 67-73, (2014)
12. National Institute of Standards and Technology NIST Special Publication 800-30 rev. 1 (Sep. 2012), Guide for Conducting Risk Assessments. Available at: National Institute of Standards and Technology
13. E. Wentzel. *Probability Theory: Textbook for universities,* (5th ed. M.: Higher School, 1998)
14. I. Kolosok, L. Gurina, Voprosy Kiberbezopasnosti, **3 (27)**, 63-69, (2018) . doi: 1021681 / 2311-3456-2018-3-63-6915
15. I. Kolosok, L. Gurina, *2018 International Conference on Industrial Engineering, Applications and Manufacturing (ICIEAM)*, (Moscow, Russia, 2018). doi: 10.1109/ICIEAM.2018.8728768
16. K. Gai, M. Qiu, Z. Ming, H. Zhao, L. Qiu, IEEE Transactions on Smart Grid, **8 (5)**, 2431-2439, (2017). doi: 10.1109/TSG.2017.2664043

Best overall dimensions of synchronous generator with permanent magnets for small power wind plants and micro hydropower plants

Nurali Pirmatov[1,*], *Shaxnoza* Tosheva[1], and *Sherzod* Toshev[2]

[1]Tashkent State Technical University named after I. Karimova, 100095, Tashkent, Uzbekistan
[2]Joint stock company Uzbekgidroenergokurulish, 100077, Tashkent, Uzbekistan

Abstract. The overall dimensions of a synchronous generator with permanent magnets are determined by its main dimensions - the length l_p and active part of the rotor with a diameter D_p , so the larger the air gap, the smaller the magnetic induction in it and vice versa. The force of gravity, in turn, will also depend on the air gap. The structurally incorporated unevenness of the air gap leads to an increase in the attractive force in the place of the smallest gap value, which leads to the appearance of a bending moment. As a result, the disc skews and the load on the bearings increases .

Introduction

Renewable energy is the energy received as a result of natural processes, which are replenished faster than they are consumed. Common sources of renewable energy include solar, wind, geothermal and some types of biomass.

More than 80% of global energy consumption is fossil resources, which are depleted faster than they can be recovered. Due to growing concerns about carbon emissions and uncertainty in fossil fuel supplies, interest in clean and renewable energy is constantly growing [1].

Discussion

The generator is the most important element of the electrical equipment of the wind turbine and micro-hydroelectric power plant. The advantages of generators with permanent magnets include high reliability, ease of construction and maintenance associated with the absence of sliding contacts and rotating winding, autonomy since no constant current is required for excitation, less heating due to the absence of excitation losses [2].

In a low-speed synchronous generator, electromagnetic processes occur that affect the performance of the structure.

The setting parameter of the electromagnetic processes of the generator is magnetic induction, knowing which, you can find the force with which the rotor is attracted to the stator, and an important argument will be the size of the working air gap. The larger the air gap, the smaller the magnetic induction in it and vice-versa. The force of gravity, in turn, will also depend on the air gap. Structurally, the unevenness of the air gap leads to an increase in the attractive force at the place of the smallest value of the gap, which leads to the appearance of a bending moment. As a result, there is a skew of disks and the load on bearings increases [3].

A generator with a permanent magnet rotor can be built according to various schemes, differing from each other by the joint arrangement of windings and magnets. Alternating polarity magnets are located on the rotor of the generator. Windings with alternating winding direction are located on the generator stator. If the rotor and stator are coaxial disks, then this type of generator is called axial or disk.

The overall dimensions of the SGPM are determined by its main dimensions - the length l_p [mm] and the active part of the rotor with a diameter D_p [mm] [4, 5]:

$$D_p^2 \cdot l_p = \frac{C_A \cdot P_{est}}{n},$$
$$C_A = \frac{6{,}1 \cdot 10^7}{\alpha_i \cdot k_\Phi \cdot k_O \cdot A \cdot B_\delta}, \qquad (1)$$

where P_{est} - estimated power of SGPM, [W];

C_A - Arnold's machine constant;

α_i - estimated coefficient of pole overlap;

k_Φ - coefficient of the shape of the curve of the field;

k_O - winding coefficient;

A - linear load, [A / cm];

B_δ - the maximum value of induction in the air gap at rated load [T];

n - rotor speed [r / min];

The estimated power is determined by the formula:

https://doi.org/10.1051/e3sconf/201913901027

$$P_{est} = \frac{\kappa_e \cdot P_н}{\cos \varphi} \qquad (2)$$

Where $k_e = \dfrac{E_i}{U_н}$ - coefficient characterizing the EMF of the anchor winding.

SGPMs are characterized by the highest energy efficiency, they are characterized by the minimum possible internal losses.

$$\Delta P = \Delta P_{copper} + \Delta P_{steel} + \Delta P_{extra} \qquad (3)$$

Comparative results

It should be noted that the ratio of power losses in high-frequency and low-frequency SGPMs differ from each other. In classic low-frequency machines, the main share is electric (in copper) and magnetic (in steel) losses and additional losses associated with rotor friction against air and losses in bearings account for only 5-10% of the total power loss [4,6,7].

On the contrary, in high-frequency EM, the proportion of additional losses increases significantly. In particular, low-speed high-torque synchronous generators with a fixed armature and rotating magnets and with additional transverse windings can be used as wind generators, high-frequency electric generators and in stand-alone power plants [5,8,9,10].

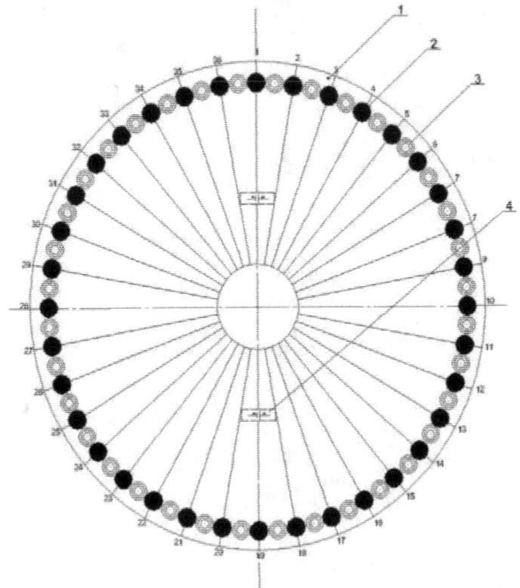

Fig. 1. Rotor: 1- rotor disk, 2-permanent magnet, 3-additional transverse windings, 4-battery

A low-speed permanent magnet electric generator contains a rotor (Fig. 1.) in the form of two flat disks, a stator (Fig. 2.) is placed between the rotor disks and is made in the form of a disk connected to a fixed case, the anchor winding is located on the disk along the radii - reel magnets with alternating poles and additional transverse windings mounted on the side parts of the rotor in an amount of 36 to 360 on each disk.

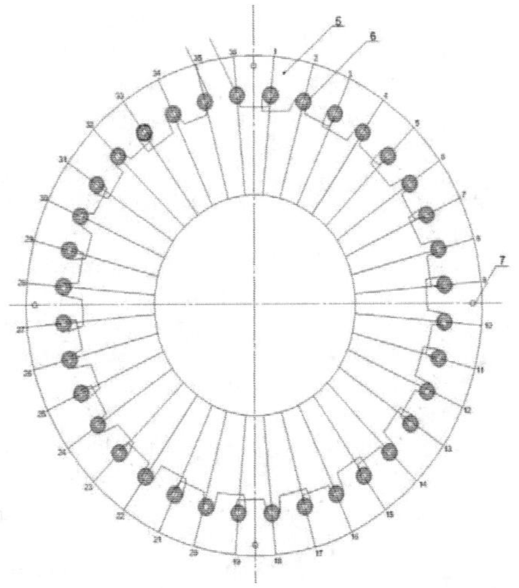

Fig. 2. Stator: 5-stator disk, 6-stator winding, 7- places for fixing

The implementation of 36 to 360 alternating permanent magnets on each rotor disk and the installation of an additional transverse winding ensures the efficiency of the proposed low-speed permanent magnet electric generator Fig. 3.

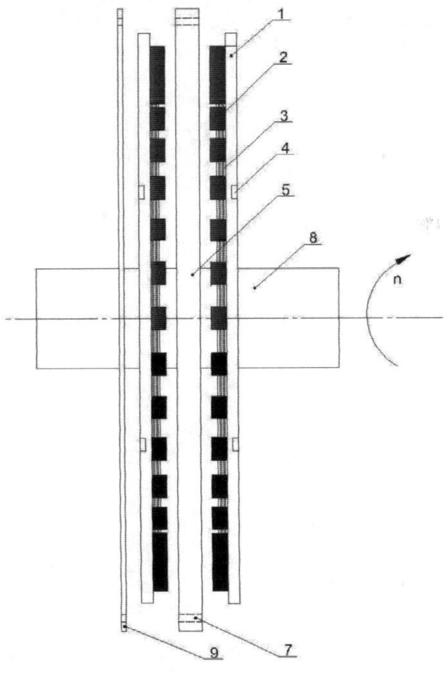

Fig. 3. General view of the generator: 8-shaft, 9-body

In the known electric generator, the number of constantly alternating magnets is in the range from 40 to 120, while the possibility of rotation of the rotor is from 50 to 60 rpm, which does not allow the use of this electric generator in various farms, agricultural enterprises on ravine rivers with a low flow rate, while to ensure the rotation of the rotor 50-60 rpm, a step-up reducer is required, which complicates and increases the cost of the electric generator. Power supply to consumers due to the distance from the power system and low power requires high costs. The great length and branching of power lines, wear and tear of electric networks cause significant losses of electricity. As a result, there is an increase in electricity costs, which generally reduces the efficiency of production at enterprises and worsens the conditions of comfortable living of the population. In the further development strategy of the Republic of Uzbekistan for the years 2017-2021 was determined "...to reduce energy intensity and resource intensity of the economy, wide introduction in manufacture of energy-saving technologies, increased use of renewable sources of energy, increasing productivity in sectors of the economy...".

The task, which is aimed at solving, is to increase the efficiency of the generator. The technical result is an increase in power, an increase in EMF, a decrease in rpm and rated speed, as well as a decrease in electromagnetic torque.

Conclusion

The proposals presented in this paper open up the prospect of further research on the extremely important problem of estimating the power loss in high-frequency and low-frequency SGPMs. In classical low-frequency machines, the main share is electric (in copper) and magnetic (in steel) losses, and additional losses associated with rotor friction against air and losses in bearings account for only 5-10% of the total power loss.

References

1. A.Mohamed. Ph.D Disser., **175** (2014)
2. Krivtsov, V.S., Oleinikov, A.M., Yakovlev, A.I. Inexhaustible Energy, Pr. Textbook. **400** (2003)
3. Balagurov V.A. Design of special AC electric machines. - M .: Higher. School, **272** (1982)
4. Balagurov V.A., Galteev F.F. Permanent magnet electric generators. - M.: Energoatomizdat, **280** (1988)
5. Pirmatov N.B., Toshev Sh.E., Tosheva Sh.N., Egamov A. Low-speed generator of a new type used to generate electric energy based on renewable energy sources. // International scientific conference. **123-124** (2018)
6. K.S. Fediy, A.L. Vstovsky. Optimization of the operating modes of electrical systems. Interuniversity collection of scientific papers. - Krasnoyarsk CPC KSTU. **56-61** (2006)
7. Booth D.A. Contactless electric machines: textbook. YES. Booth M .: Higher school, **416**(1990)
8. Pirmatov N.B., Toshev Sh.E., Haydarov S.D., Tosheva Sh.N. European Conference on Innovations in Technical and Natural Sciences.:,15th International scien. Conf.. **105-110** (2017)
9. Toshev Sh.E., Pirmatov N.B., Haydarov S.D., Duvlonov J.N., Yakubova D.K. Europ. Scien. R.:Aust, Vienna,**11-12**, 134-136 (2016)
10. Akhmatov M. G. Synchronous machines. Special course. - Moscow: higher school, **136** (1984)

E3S Web of Conferences 139, 01028 (2019) https://doi.org/10.1051/e3sconf/201913901028
RSES 2019

Towards digital energy (experience of digital technology implementation by Russian companies)

Yulia Zhilkina[1,*], *Dmitry* Vodennikov[2]

[1] Candidate of economic sciences, Assets management Department, Federal Grid Company of Unified Energy System, Moscow, Russia
[2] Chairman of the board - Chief Engineer, Federal Grid Company of Unified Energy System, Moscow, Russia

Abstract. Today, digitalization is increasingly "capturing" the world economy. The inclusion of Russia in this trend is a matter of domestic companies maintaining their positions in the international arena. In the future, the creation of our own high-tech solutions will allow our country not only to win a competitive race in traditional markets, but also to reach new borders and take a leading position in the global market of complex systems and services of intellectual energy.

1 Introduction

Today, digitalization is increasingly "capturing" the world economy. The inclusion of Russia in this trend is a matter of domestic companies maintaining their positions in the international arena. In the future, the creation of our own high-tech solutions will allow our country not only to win a competitive race in traditional markets, but also to reach new borders and take a leading position in the global market the complex systems and intellectual energy services.

Innovative development is a driver of increasing efficiency in the electric power industry, as well as reliability, quality and economy energy supply to consumers on the basis modern technologies with transformation of them into the intelligent core of technological infrastructure the electric power industry.

The priority direction of innovative development of PJSC FGC UES is the introduction digital technologies and platforms. In modern conditions, the use digital data is a key factor of production in all spheres the economy, a condition of competitiveness and economic development of the country, ensuring the growth of the quality of life of people. The transition to digital mode of operation is a response to external technological challenges and future structural changes in the energy system.

2 Major trends of energy complex development

Within framework concepts of clean (green, low-carbon) sustainable development Global energy transformation through introduction new technologies (including digital technologies) is taking place. Major trends in the energy sector include:

- Creation of new and modernization of existing transport infrastructure (gas pipelines, oil pipelines, high-capacity product pipelines, laid including in hard-to-reach and remote regions, sea and land terminals for fuel transhipment, new generation networks, other facilities);
- Modernization of the thermal generation segment (increase of efficiency of boilers and turbines of gas and coal-fired power plants/CHPP, reduction of fuel consumption, co-generation, switching of coal-fired power plants to gas and biomass, change of quality composition of used gas by inclusion of hydrogen, other substances contributing to reduction of harmful emissions into the environment, increase of heating value of energy carrier and other methods);
- Expansion of RE-capacity in electric and thermal power (using water, wind, solar, Earth heat, biomass, household and industrial waste);
- Modernization of the nuclear industry through the introduction of technologies that reduce production and operational risks and create small-capacity power units (except for individual countries that do not develop this direction for social, economic and other reasons);
- Introduction of energy-efficient and resource-saving technologies in all sectors of the energy sector;
- Construction of air, cable and combined long-distance and super-long-distance transmission lines with alternating and direct current;
- Introduction of technologies to capture and absorb carbon dioxide, sulphur and other harmful substances, creation of carbon dioxide storage facilities;
- Creation of active and adaptive networks ("smart") the distributive systems uniting [1] industrial facilities of generation of various power and type (by the form the energy carrier and the used technologies of transformation of a substance/resource to energy), [2] the centralized energy stores, [3] electric transport, [4] household power stations, electric systems;
- Partial decentralization of energy supply systems.

Developed States control the situation in world energy direct the development of economic and other ways: through public opinion formation, standards development, indicators systems, recommendations to politicians, i.e. "push" society to make concrete decisions. In 2018, International Economic Forum using the Energy Transition Index (see table 1) presented for the first time rating of 114 States ranked, which on several dozen indicators, assesses the level of readiness of national energy to integrate into a future safe, sustainable, affordable and universal energy system, which creation is also a global challenge.

Table 1. Energy Transition Index-2018 index of the certain countries of the world

Rating	Country	Index
1	Sweden	75,8
2	Norway	75,0
3	Switzerland	72,9
4	Finland	72,4
5	Denmark	72,4
54	Armenia	55,3
70	**Russia**	**50,9**
75	Kazakhstan	49,7
112	Kyrgyzstan	39,3

2.1 New technologies and energy.

Digital introduction (as well as space and biotechnology) into the energy sector began more than half a century ago and continues to be carried out as a matter of priority, following the industries of the military-industrial complex, steadily, at all technological stages and "floors" (During field exploration, production, transportation, storage, processing, energy distribution, waste management, energy trade). In order to optimize production, increase reliability and availability of safety of power supply (see Table 2).

Natural-like technologies, as "core" of the next technological order, make it possible to improve the efficiency solving multifaceted problems of sustainable "clean" development, including: (1) resource consumption optimization , (2) anthropogenic influence on the environment reduction, (3) efficiency increase of fuel and energy industries, (4) predictability increase demand and supply for energy, (5) consumers and producers market coverage expansion, "energy" poverty reduction, (6) living standards of the population increase, and even (7) the cost reduction of unit energy.

Digital technologies are transforming not only the sphere of production, but also the sector of transportation/distribution of electricity, the development of which is going towards the creation of active-adaptive ("smart") electricity networks aimed at solving the key problems sustainable development, including:

- Combine into a common/unified energy system generating objects, which differ in power and type of energy carrier used, physically located at short, long and ultra-long distances;
- Provide continuous monitoring of equipment condition, flows, power accumulation and distribution;
- To expand scope interaction market participants (consumers with suppliers and among themselves in various combinations), to implement principle of "sending excess to the network - obtaining energy in case of shortage."

Table 2. Digital Applications in Global Energy 2018

Activity Type	Type of work/technology	Leading country
Geological exploration	Aerospace survey of the Earth 's surface, construction of virtual models of subsoil and mining schemes (geological-mathematical modeling)	USA, Russia
Drilling wells	With significant distance from vertical and high length (about 15 km), self-propelled, non-navigational, offshore drilling platforms	USA, France, Russia
Production of liquid and gaseous hydrocarbons	Hydraulic fracturing, production technologies in Arctic, offshore and deep oceans, underwater deserted production complexes	USA, Canada, Norway, France, Russia, China
Preparation for and transportation of oil and gas	Cleaning, conversion to different phases, transhipment and transportation by land and sea	Russia, USA, Canada, Qatar, Australia, China
Oil and gas refining	Release nano - and biopolymers, biotechnological, anticorrosive, crystal, fire-resistant materials, materials for transformation of energy, biosensors, pharmaceutical products	USA, Japan, Germany, Great Britain, France, China, India, Brazil, Russia
Coal production and processing	Gasification, flexible robotic systems based on artificial intelligence, deserted production, unmanned transport (mines, cuts, etc.), nano- and biotechnology of coal and waste processing	United Kingdom, Japan, United States, Germany, Czech Republic, Russia, South Africa
Atomic sector	Computer design and product lifecycle management, security systems, flexible robotic systems based on artificial intelligence, deserted production	Russia, USA, France, Great Britain, Germany
Power industry	Active adaptive networks, DC power transmission, superconducting-based AC transmission	USA, European Union, Republic of Korea, China, India, Russia
Production of	On land, in the oceans and seas	China,

gas hydrates		Japan, USA, Canada
Renewable energy	Solar, geothermal, offshore and terrestrial wind power, biofuel technologies	USA, Brazil, Germany, Denmark, Norway, Spain, Iceland, Russia, China, Japan
Activity Type	Type of work/technology	Leading country
Geological exploration	Aerospace survey of the Earth 's surface, construction of virtual models of subsoil and mining schemes (geological-mathematical modeling)	USA, Russia

The countries of EU, USA, Japan, China, Republic of Korea, and Australia demonstrate the greatest success in the construction of "intellectual" networks. From a technical point of view, smart networks cannot yet be considered breakthrough solutions, given the fact that their creation and development are based on known principles and technologies.

In other words, in the modern network complex there is a process of deep modernization. More efficient, reliable and safe equipment (often on a new element base) is introduced, while the network structures created continue to serve as "closing" technologies for the current technological order. The policy of comprehensive digitalization is aligned with the global policy of combating climate change, the genesis of which is not defined by science. The main beneficiaries of these processes are large and ultra-large TNCs with high scientific and technological potential, developed production base and trade and sales infrastructure.

2.2 Interpretation of the concept of "digitalization"

Digital economy - the system of the economic, social and cultural relations based on use of digital information and communication technologies.

The Digital Energy Development Center insists on its own unique content having in terms of "digital energy" and "digitalization". In digital energy the main concept is "economy." It should be noted that the terms "digital energy" and "digitalization" have emerged in the context of digital economy formation processes, and it makes sense to consider them only in this connection. It follows from the many definitions of the digital economy that economic activity, commercial transactions and professional interactions based on new principles through the use of information and communication technologies are a special subject.

The essence of digital energy is the re-assembly and development of a set of industrial and economic relations in the industry based on digital approaches and means. In sum, in the phrase "digital economy" (and thus "digital energy"), the defining word is "economy," and the adjective "digital" only indicates the means to achieve the goal.

The concept of "digital economy" was best defined in 1995 by computer scientist Nicholas Negroponte [1]. He used the following metaphor to represent the essence of this concept: "Transition from the movement of atoms to the movements of bits" and presented the concepts of weight, raw materials and transport - shortcomings of the past, putting them as opposed to the concept of the absence of the weight of goods, virtually.

According to the European Community (EU), the digital economy is the result of the transformational effects of new general-purpose technologies in the field of information and communication [2]. This has affected all sectors of the economy and social activities, e.g. retail, transport, financial services, manufacturing, education, health care, media, etc.

Digitalization usually also means a wider use of data. Modern substations already provide a huge amount of data, and by purposefully adding new devices and sensors to the substation automation system (ideally designed with an open architecture) the amount of data may still be increased once this seems to be justified by the benefits achieved.

In this context, one highly relevant application area of substation data is asset management support. The acquisition and intelligent analysis of data (including pattern recognition) allows giving early warnings to induce preventive maintenance actions assures the increased availability of important assets like transformers or switchgears and supports the definition of an effective investment and maintenance strategy.

When it came to the use of computing digitization replaced computerization, computers and information technology to address selected economic challenges. [3]

2.3 Main documents

President of the Russian Federation Vladimir Putin signed a Decree "On National Goals and Strategic Tasks for the Development of the Russian Federation for the Period up to 2024" May 07, 2018. The Decree sets priorities for the digital transformation of the country's economy, including specifically the energy sectors, in an extremely concentrated manner.

Digital transformation (digitalization) in energy is primarily the creation of new business models, services and markets based on the capabilities of the digital economy Digitalization is not only a fashionable technological process, but also a much needed process! "Figure" in energy should give the same qualitative information leap as the arrival of IT-technologies in due time The advent of computer networks has increased the efficiency of all production processes and brought the world to a new level of communication Energy is one of the last areas not yet digitized It is obvious that such modernization requires considerable costs However, it in turn will have positive effects: it will open up opportunities for new equipment and new approaches that will allow to provide consumers with electricity services in a more comfortable mode Intelligent energy will fundamentally change the speed, reliability and cost of work - and this is the saving of both labor and money.

2.4 Objectives of digital transformation of electric power industry

The main challenge of digital energy is eliminate dramatically rising costs of integrating distributed energy and market transactions. The goal of digital transformation is to change the logic of processes and move the company to risk-based management based on digital adoption and big data analytics.

Digital Transformation Objectives: (1) Adaptability of the company to new challenges and challenges; (2) Improvement of power supply reliability characteristics of consumers; (3) Improving the efficiency of the company; (4) Improved availability of grid infrastructure; (5) Development of human resources and new competencies; (6) Diversification of the company's business through additional services.

Basic principles of digital transformation are:

Within the company: - ensuring observability of network objects and modes of their operation; technological and corporate processes management automation; - automated risk-oriented management application principles; - construction digital CIM-model according to a single industry standard and information interaction with all counterparties (Networks, consumers and other entities of the electric power industry); - integration and integration of various IT systems at hierarchical levels; - integration of network information (technological and corporate) systems;

With regard to regulated activities of the company: - ensuring reduction of electricity losses; - optimization of operating and capital expenditures; - reduction of technological connection time; - improvement of reliability of consumers 'electricity supply; - increase of openness and transparency of the company's activities; - containment of tariff growth rate;

To the loop of interaction with other entities: - creation of a publicly available, reliable, transparent and verifiable system of intelligent commercial accounting of electricity; - creation of infrastructure for simple and effective interaction with consumers (load management, distributed generation, "requests"); - creation opportunity for automation contract relations (smart contracts) in terms of provision of transmission services, technological connection, etc.;

With regard to the development of new unregulated services, the development of a fundamentally new infrastructure for an affordable, efficient and flexible process of electricity exchange among all market stakeholders with minimal transaction costs.

The potential benefits of applying digitalization for substation asset management are numerous: The efficiency of operational maintenance of primary assets and patch management of secondary assets can be significantly improved. Investments into new primary assets can be made at the right point of time, considering their actual heath status and their strategic relevance. Unplanned downtimes can be avoided, with a positive effect on grid availability and asset productivity. Decisions can be made more consciously and faster.

The digital transformation will increase the energy security of the regions of the country by creating new infrastructure opportunities and ensuring a new standard of living for the population thanks to new standards of service. The systemic effects of digital technologies introduction to be increased reliability of electric power supply and reduction of the duration of power supply interruptions are expected, possibility to optimize energy consumption at the level of end-users, increase efficiency of asset use, reduce power losses, accessibility for renewable distributed energy. Table 3 shows the main effects of the Concept implementation.

Table 3. Key Effects from Concept Implementation

Subject	Effects
State	-Ensuring energy independence and infrastructural security of economic development. -Advance modernization of the underlying infrastructure company. -Radical improvement of the quality and accessibility of transmission and technological connection services, development of competitive markets for related services. -Containing tariff growth.
Company	- Getting savings in the implementation of investment programs, and, accordingly, the possibility of ahead of the development of network infrastructure while increasing the profitability of the company's business. Ensure that infrastructure is prepared to develop new challenges, improve sour quality and reliability of consumer energy supply. - Increase the speed and quality of decision-making at all levels of the company's management. -Reducing losses by timely detection of uncontracted and unaccounted electricity consumption. Reducing the cost of ongoing equipment maintenance (OPEX) - moving from planned repairs to repairs as of the state. -Optimizing the logistics of equipment delivery. -Improving staff competence.
Consumers	- Quality and electricity transmission accessibility improving and technology connectivity services. - participate in regulating one's own consumption. - Tariff growth deterrence.

3 Digital technologies by Russian energy companies: practical implementation

The resource for increasing the efficiency and quality of FGC's power grid infrastructure is found in the use of digital technologies.

Main priority is the development of a network of low-maintenance digital facilities with a high degree of reliability. In addition, over the past ten years, the Company has been introducing elements of digital substation technology on the basis of the international standard IEC 61850. By 2025, we will build 32 high and ultra-high voltage facilities with integrated digital solutions. At the same time, all substations will be equipped with digital communication systems capable of remote control from unified centers.

However, digitisation is not the only task of innovative development. In 2017, two of FGC's projects received the national status. The first of them is focused on increased energy efficiency of substations, while the second relates to the use of high-temperature superconductors. This technology minimises energy losses during transport and the subject of scientific research in many countries. Last year FGC UES completed testing the world's largest superconducting cable line. It will be put into operation in 2020 in St. Petersburg.

In 2018, in particular, the connections between the power systems of the Northwest and the Centre have been strengthened, previously isolated Yakutia has become part of Russia's Unified Energy System thanks to the 750-kilometre power transit, and the Tobol substation was commissioned - the first ultra-high voltage facility to integrate a set of digital solutions.

PJSC Rosseti	PJSC "FGC UES"
- Intelligent metering devices	-Remote control of substation equipment
- "Smart Networks"	-Digital substations
-Digital substations	

Currently, when creating digital substations, the Federal Grid Company seeks to solve two tasks: reliability improvement and cost reduction. The former is achieved through a continuous monitoring of the component integrity and operability. The overall facility cost is determined by the expenditures on its design, construction and maintenance. Today, if a digital substation is erected from scratch, it turns out more expensive than a conventional substation. However, if some of the maintenance costs can be avoided, and the substation equipment is inspected only if needed (rather than scheduled), the digital substation will eventually be cheaper.

Instead of an electrician inspecting the secondary circuits, we would be forced to employ a well-paid cyber security expert and a system administrator, who would repeatedly take care of the complex information systems. Such systems would fail during the operation because the server's operational lifetime is 10 years and the substation is in service for around 40 years. All costs associated with the IT infrastructure modernization would be included in the tariff, so we could face a modern and innovative power industry with huge tariffs for electricity and a minimum number of consumers. The Federal Grid Company is working on the creation of digital substations and considers them quite promising. However, if we fail to justify the costs for cyber security solutions by showing evidence they do improve the reliability of the electrical facility operation, power engineers will once again have to resort to the electromechanical relays manufactured at "Cheboksary Electromechanical Factory" (which have been in service for more than 40 years without any failures).

Key Digitalization Effects:

• Reducing the duration of power outages and the average frequency of technological failures (SAIDI / SAIFI) by 5% by 2024;

• Increasing the level of technical condition of production assets of the electric power industry for facilities by 5% by 2024 without increasing the cost of maintaining the technical condition;

• A 20% reduction in accident rate at electric power facilities related to the technical condition of production assets by 2024.

Digitalization in the energy sector is necessary because it will ensure uninterrupted power supply, preventing accidents in advance. Digital technologies are set to transform grid operations. Increasingly, drones will carry out inspections, replacing workers in the field, while sensors enable real-time network monitoring. Predictive maintenance, based on machine learning and artificial intelligence, will reduce power outages and improve companies' investment and maintenance decisions. And robotic process automation will speed up back-office functions. Sensor, data collection, and storage costs are one-tenth what they were ten years ago, and the resulting proliferation of data and advanced analytics will also boost efficiency. But while digital transformation of the grid is an opportunity, it also adds complexity and involves fundamental changes in operating models.

References

1. A. Dobrynin, K. Chernykh, V. Kupriyanovsky, P. Kupriyanovsky, S. Sinyagov Digital economy - various ways to the effective application of technologies // International Journal of Open Information Technologies ISSN: 2307-8162. vol. 4. (2016)
2. OECD Digital Economy Outlook 2015, OECD July 15, 2015 OECD Publishing [Electronic resource] URL: https://read.oecd-ilibrary.org/science-and-technology/oecd-digital-economy-outlook-2015_9789264232440- en # page26 appeal date: 07/19/2019

3. Yu. Zhilkina Electricity development: vertical integration or further liberalization of the industry? // Bulletin of KGEU, No. 2 (38). (2018)
4. P. Livinsky Digitalization of networks will become the basis for future transformation processes in the electric power industry // Website of PJSC Rosseti [Electronic resource]. URL: https://www.rosseti.ru/press/news/index.php?ELEMENT_ID=34079 (accessed: 07/03/2019).
5. Digital transition to the electric power industry of Russia // Expert-analytical report, ed. Knyaginina V.N., D.V. Kholkina. Moscow: Publishing House of the Center for Strategic Research Foundation Innovation Development Project Centre. (2017)

Analysis of the emergency control and relay protection structures approached from the point of view of EPS reliability and survivability by taking into account cybersecurity threats

Alexey Osak[1,*], *Daniil* Panasetsky [1], and *Elena* Buzina[1]

[1]ESI SB RAS, 664033 Lermontov str., 130, Irkutsk, Russia

Abstract. Cyber threats pose an increasing threat to energy objects. It is essential to ensure the cybersecurity of automatic control systems, such as relay protection devices (RP), devices of regime control (RC) and emergency control (EC), automated control systems. At the same time, the issues of cybersecurity include not only the problem of hacker attacks, but also the whole complex of problems relating to adequate functioning of cybernetic systems in the power industry. The authors propose a methodical approach to the analysis of the structure of automatic means of regime and emergency control in terms of their impact on the reliability and survivability of power systems (EPS), taking into account the known threats to cybersecurity.

Introduction

Over the past decade, there have been active discussions on the topic of digital substations and the implementation of solutions for the power industry based on IEC 61850, these discussions take place against the background of the process of introducing these technologies to power objects. Participating in these discussions, the authors repeatedly noted the presence of the problem of cybersecurity in the mass introduction of digital substation technologies. In recent years, Russia has begun to open the active discussion about total digitalization and digital transformation of the power industry while also taking practical steps towards said digitalization [1-4]. In these conditions, the urgency of the problem of cybersecurity has significantly increased.

Against the background of rapidly changing external conditions, cyber threats pose an increasing threat to energy objects [5]. It is of great importance to ensure the cybersecurity of automatic control systems, such as relay protection devices, devices of regime and emergency control, automated control systems [6]. At the same time, the issues of cybersecurity include not only the problems of hacker attacks, but also the whole complex of problems of adequate functioning of cybernetic systems in the power industry. It is important to pay attention to the influence of reliability and cybersecurity of digital subsystems on the overall reliability of power objects, EPS and their associations [7].

1. Reliability of digital control systems

Effective and adequate operational and emergency control is one of the factors determining the reliability of EPS. It is known that the operation of the EPS is possible only with appropriate continuous control, both over individual electrical installations and the EPS as a whole.

On the one hand, digital technologies make it possible to create complex and flexible algorithms for operational-dispatch and emergency control, which, combined with a new generation of primary equipment with high performance and monitoring and control capabilities, increases the overall reliability of the EPS. On the other hand, digital technologies and microprocessor technology are characterized by the possibility of a relatively simple change in functionality by reprogramming, which, when properly used, allows to improve technologies and control algorithms without replacing equipment, but also becomes the basis for new types of threats to the EPS – threats to cybersecurity. The versatility of communication networks and microprocessor devices allows them to solve any information problems, both useful and obviously malicious functions in the process of cyberattacks, which could not be said about traditional devices, especially on an electromechanical basis. Therefore, block diagrams and the composition of software and hardware do not characterize the functionality of the control system (because similar software and hardware can create completely different control systems), especially in the process of a cyber attack, when the functionality of the devices may even change.

The subject of cybersecurity has become extremely popular in recent years, but in most publications and regulations on cybersecurity of power objects, the emphasis is on unauthorized, intentional and malicious actions of certain persons who seek to gain access to information, resources and means of the attacked party through cyberspace. No matter how the software and hardware that perform application and communication functions at power facilities are improved in resistance to cyberattacks, and no matter what additional special technical means are used to protect against cyberattacks, all this does not solve the problem of the human factor [9].

While the aspect of protection against external threats is important, it is not the only one, just as the general concept of the term security is not limited to the state of protection against external threats only. An equally important aspect is to ensure protection against internal threats, which include flaws and errors in the software. By considering only external threats, one overlooks the shortcomings of the design and development of modern automatic and automated control systems.

Cyberthreats [10] are executions not of the specified (required) functions, but of unintended functions, which can be interpreted as a partial or complete failure of the control system of the power object. Possible threats (disturbing factors) for electric power objects are named below [11]:

- internal threat:
  - undetected errors in algorithms and software, which result in information and control systems of the power object operating according to the wrong algorithm;
  - errors of operational personnel of the power object, which lead to incorrect changes in the mode of operation of the devices, to disabling the protection systems of external communication channels, to replacing the software with a non-project version, to infection with viruses, etc.
- external threat:
  - malicious software defects (Spyware) embedded in the software of microprocessor devices for the purpose of controlled system failure or unauthorized access to them;
  - cyberattacks from the outside, through external digital communication channels of the power object, by intercepting telemechanics and telecontrol channels, general corporate control channels or embedding malicious software code into control systems (virus infection).

A comprehensive approach for common analysis of the structure of RP, RC and EC systems from the perspective of cybersecurity is proposed.

2. Approach for general assessment of reliability of modern RP, RC and EC systems from the cybersecurity perspective

The assessment of reliability of any system is connected with an assessment of probability and scales of consequences of failures of elements of system. The following types of failures can be distinguished for automatic control systems in the electric power industry (RP, RC and EC) [12]:

- the failure of the hardware of the computing system;
- interface and I/O failures;
- errors in the applied algorithms;
- bugs in the software;
- errors in settings.

Considering the reliability of modern automatic control systems, it can be noted that the hardware of the microprocessor computer system has a high internal reliability. The failures encountered in practice are mainly due to a complete failure of functioning, and not to errors in logical or arithmetic calculations. Accordingly, the required reliability of the microprocessor system is provided by duplication of devices.

Failures of interface and I/O include failures of any adjacent connections: circuits and I/O modules of analog and discrete signals, external digital communications. On objects with classical DC operational circuits, analog current circuits and voltage circuits, all the problems of cable connections are manifested. Unlike electromechanical devices, microprocessor terminals and controllers require intermediate circuit interface modules. The application of IEC 61850 has its own problems of digital communications, which are potentially vulnerable to cyber attacks from the outside.

Errors in applied algorithms can be divided into two groups:

- Errors in logic;
- Errors in principle decisions when some state-mode situations, certain emergencies or equipment failures are not taken into account.

Software errors can be divided into two groups:

- Errors (bugs) made in both system and application software.
- Errors resulting from inconsistencies between designed algorithms and real software. Quite often, designed algorithms are schematic drawings made in a universal graphics editor such as Autocad or Visio. This schematic drawing is agreed and approved at the stage of design or working documentation, but it is not a program code in itself. And the problem here is that the software of the terminals RP, RC and EC in the source code is not sent and, accordingly, is not checked by either the customer or other matching organizations. The quality of this software depends on the personal qualities of programmers and the quality of the organization of internal processes of software development and testing within the manufacturer. This is all opaque, and accordingly, the actual quality of the software is quite random and unpredictable.

Errors in settings can be divided into two groups:

- Errors in settings, from incorrect recording of parameters in paper forms to incorrect setting of parameters in terminals. To exclude this factor rechecks and tests are necessary, but the human factor can lead to the omission of these errors.
- Errors caused by misunderstandings between manufacturers, designers, operating organizations, no matter how improved standards and requirements

are, there is always the influence of the human factor.

Modern computers and controllers, as well as their programming tools, contain hundreds of thousands or even millions of lines of program code and a corresponding number of machine code commands. And these lines actually contain errors (bugs), which is confirmed by systematic software updates of household and corporate devices. The problem is compounded by the fact that public devices are more widespread, which due to the mass leads to the detection of more errors, and industrial software has a small number of installation, which greatly increases the risk of undetected errors.

Considering the finished product as a "black box", by conducting external tests, even theoretically it is impossible to detect all possible errors. In practice, in accordance with the current standards, a very limited number of test experiments is required, which amounts to tens or hundreds, which is several orders of magnitude less than the number of lines of program code, which, in fact, is checked by testing the "black box".

External "black box" tests are not enough to get a reliable hardware and software product. Modern programming methodologies indicate that for the development of correctly working (error-free) software it is necessary to use a whole range of organizational, methodological and technical means, starting with the selection of personnel and improvement of their skills. Testing should be performed starting with an automatic or automated check of the operability of each program procedure (function), even if it contains only a few lines of code. But all this significantly increases the initial cost of the software product. And at the level of competitive procedures for the purchase of equipment, where the price criterion is decisive, the excess cost will not contribute to the choice in favour of the quality software product.

It is important to note that in modern conditions, organizational and managerial forms of limited liability are widely used, starting with the fact that these are all joint-stock companies and limited liability companies (the situation is identical in all countries of the world). The liability of specialists is limited by labour legislation. All this is subject to financial and time constraints. The result of all this is the fact that absolutely error-free software simply does not exist, if we talk about products that solve complex problems.

Summarizing the above, it can be noted that any microprocessor-based automatic control device can operate incorrectly, both in terms of failure, and in terms of false or excessive operation. The complexity of the structure of modern devises and systems, and the growth of the number of modern devices of RP, RC and EC, leads to the fact that many failures are difficult to predict.

It is proposed to consider the failures of automatic control systems of different types in the design and planning of power systems. At a minimum, it is necessary to take into account the possibility of false operation of any device in the form of issuing the largest or most adverse control action. Any terminal RP or EC due to a certain failure can simultaneously issue all connected control actions, even if their combination is not provided by the application algorithm, because such a failure may be caused by an error in the software. In this regard, it is proposed to systematize and expand the types of failures of RP, RC and EC devices that need to be worked out when designing and planning the development of power systems.

A clear illustration of the problem of cybersecurity in the presence of internal threats is the accident in the United Power System of Russia in 27.06.2017 [12], where there were only two initiating failures: the first, a false issue of the maximum load shedding stage, the second – an error in accounting for the power direction sign. During this accident there was no short circuit or damage to the primary power equipment, but the volume of disconnected generation was about 7 GW, disconnected load about 4 GW. According to the results of the investigation of this accident, changes were made to the algorithms of the EC of the 500 kV transmission of Bratsk – Irkutsk in order to detect and prevent a false operation. It is important to note that the problem of false alarm was not raised when performing design on this complex Automatic Stability Control System, including not raised by the operating organizations and branches of JSC "SO UPS". In the future, it is necessary to take into account the experience of this accident (and others like it), and to raise these issues and find solutions to prevent false alarms and minimize the potential volumes of falsely issued excessive operation for load shutdown, generation and change of the state of network elements early at the design stage.

It is possible to improve business processes in companies developing hardware and software, you can take into account the experience and reputation of manufacturers and engineering companies. But as soon as cybersecurity issues move into the sphere of geopolitical rivalries, these methods of improving quality will not be enough.

3. The problem of targeted external cyberattacks

In the future, in the era of total digitalization, the situation may be aggravated by the fact that cyberattacks or other negative ways of affecting the digital infrastructure of critical infrastructure objects and systems, which include the power systems, will become elements of geopolitical and military confrontation, which is already publicly spoken about by senior officials of various countries of the world.

When targeted external cyberattacks are launched by foreign countries or large corporations, significant resources, both financial and human, are allocated to their implementation. The qualification of attacking hackers can be significantly higher than the qualification of most specialists in the power industry. If a cyberattack is blocked by technical means, it is possible to bribe, blackmail or deceive specialists at power objects, specialists of engineering companies or enterprises producing technical means for the power industry. In the context of the comprehensive use of smartphones, smart gadgets, social networks and other tools of digital communications, the task of bribery, blackmail or deception of specialists is greatly simplified if it is done by representatives of the special services of foreign countries. For these purposes, they have access to fairly complete information about the specialist, his family, interests, Hobbies, friends and so on. Contacts and information about close family members are available, including their current location, audio, photos, and videos. And here it is important to note that we are talking about the impact of foreign intelligence services on ordinary professionals, not on intelligence officers. Ordinary employees do not give the legally and morally binding oath at the workplace, do not have special training, etc. So if not this, so another specialist, if not this, so another object will succumb to bribery, blackmail or deception, respectively, open access, disable protection, etc. Therefore, the probability of a successful attack of this sort is almost 100%. In this case, the preparatory stage will be invisible from the side of the energy object itself, i.e. the attack is likely to be unexpected.

These are typical problems of any defence, because the attacking side can concentrate all efforts on one area, attract the best specialists, allocate large funds for the attack. And not knowing place and time attacks, protection and defence will have provide on all objects and systems, that causes dispersion forces and funds, and as a consequence, natural lack of these forces and funds, in camping on champion personnel in place full-scale attacks.

Therefore, it can be concluded that if the hostile impact on the digital component of critical infrastructure facilities and systems become elements of geopolitical and military confrontation, then all echelons of cyber defence will be overcome in the point of a full-scale complex attack. The exception here can be only a few objects, which even in normal conditions apply super-strength in the field of cybersecurity. Accordingly, we are no longer talking about repelling an attack on a typical power object, we are talking about minimizing the consequences and damage after a successful cyber attack. In this case, the magnitude of the expected damage in comparison with the cost of a massive cyber attack will be the criterion of whether or not a particular critical infrastructure will be attacked. Accordingly, measures to reduce possible damage will be an effective means of preventing cyberattacks [8].

As a result of a cyberattack of this kind, for one reason, you can get the following negative consequences:

- Simultaneous failure of a large number of digital devices RP, RC and EC of one manufacturer at one power object or a group of power objects located in the same information space that has a physical connection to the Internet or public access channels. However, logical defenses such as firewalls or routing can be disabled as part of this attack.
- Simultaneous failure of a large number of intra-and inter-site digital networks (channels) located in the same information space that has a physical connection to the Internet or public access channels.
- Simultaneous access to a large number of digital devices RP, RC and EC of one manufacturer at one power object or a group of power objects located in the same information space that has a physical connection to the Internet or public access channels. Use this access to change the settings and algorithms of operation or for remote control, including to create an emergency situation.

It is important to note that the principles of short-range and long-range redundancy in relay protection, as well as the principles of several echelons of emergency control do not imply simultaneous and mass failure of a large number of protections and automatics. Accordingly, there may be unrecoverable short circuit, operation of the equipment in overload mode and other emergencies that can lead to damage to the primary equipment.

Another dangerous consequence of a cyberattack of this kind is the long recovery time of the EPS. Given the complete dependence of all spheres of public life and the economy on the availability of electricity, the disruption of electricity supply to a large number of consumers at the same time with a long recovery time is already catastrophic.

Thus, universalism of digital solutions, unification of digital interfaces, hardware platforms, operating systems, availability of centralized administration tools, common information space at the physical level, significantly increase the likelihood of large-scale cyberattacks, as they increase the potential damage from a successful cyberattack. Accordingly, the heterogeneity of solutions, their incompatibility, lack of integration into a single information space, reduce the likelihood of large-scale cyberattacks, because the potential damage from a successful cyber attack is limited due to the limited number of devices and systems that can be subjected to this kind of attack.

Therefore, when building automatic control systems in the power industry in the era of total digitalization, it is necessary to adhere to the layered principle, where the systems of the last tier must be either isolated or minimally integrated into digital control systems. If an expensive cyber attack, requiring the participation of unique specialists-hackers, can not lead to significant damage, and will not lead to a significant increase in the recovery time of EPS after an accident caused by a cyber attack, then the feasibility of such an attack becomes far from obvious in a geopolitical or military confrontation.

Conclusion

The paper shows that cyber threats in the power industry should be understood not only as cyber attacks in the form of hacker activities, but also the whole complex of possible failures of the cybernetic control system, without which the power system is unable to function.

The main types of failures of automatic control systems in the electric industry (RP, RC, EC) are analyzed and recommendations on the account of these failures in the design and planning of power systems development are given. Situations with potentially possible cyberattacks initiated by foreign countries or large corporations in the framework of geopolitical confrontation are considered separately.

The work was carried out within the project III.17.4.2 (No. AAAA-A17-117030310438-1) of the fundamental research program of the Siberian Branch of the Russian Academy of Sciences.

References

1. A.L. Teksler, "Power industry digitalization: from process automation to the digital transformation of the industry", Energy Policy. № 5. 2018. p. 3-6. (in Russian)
2. D.V. Holkin, I.S. Chausov, "Digital transition in Russian power engineering: in search of meaning", Energy Policy. № 5. 2018. p. 7-16. (in Russian)
3. Report "Digital Transition in the Electric Power Industry of Russia", Ed. V.N. Knyaginin, D.V. Holkin. CSR, 2017. URL: https://www.csr.ru/wp-content/uploads/2017/09/Doklad_energetika-Web.pdf (in Russian)
4. The concept of "Digital transformation 2030", PJSC "ROSSETI", Moscow. 2018. 31p. URL: https://www.rosseti.ru/investment/Kontseptsiya_Tsifrovaya_transformatsiya_2030.pdf (in Russian)
5. Massel' L.V., Voropay N.I., Senderov S.M., Massel' A.G., "Cyber Danger as One of the Strategic Threats to Russia"s Energy Security" Cybersecurity issues, 2016, №4 (17), p. 2-10. DOI: 10.21681/2311-3456-2016-4-2-10. (in Russian)
6. Kulikov A. L., Osokin V. L., Papkov B. V., Shilova T. V. "The extension of the concept «reliability» in modern electric power industry", Bulletin NGIEI. 2018. № 3 (82). p. 88–98. (in Russian)
7. A.B. Osak, A.I. Shalaginov, D.A. Panasetsky, E.Ya. Buzina, "The impact of cybersecurity of power objectson the reliability of the EPS", Methodic problems of studies of reliability of large power systems, Eds. N.I. Voropai, Yu.Ya. Chukreev, Syktyvkar: LLC "Komi Republican Printing House", 2016, p. 377-385. (in Russian)
8. A.B. Osak, A.I. Shalaginov, D.A. Panasetsky, E.Ya. Buzina, "Cybersecurity of power objects as a factor of EPS reliability", Methodic problems of studies of reliability of large power systems, Eds. N.I. Voropai, M.A. Korotkevich, A.A. Mikhalevich. – Minsk: Belarusian National Technical University, 2015, p. 258-264. (in Russian)
9. A.B. Osak, A.I. Shalaginov, D.A. Panasetsky, E.Ya. Buzina, "Human factor in ensuring cybersecurity of power objects", Book of reports of international conference "Modern tendencies of System Development of Relay Protection and Automation of Power System", Sochi. 1-5 June 2015. (in Russian)
10. A.S. Alpeev, "Terminology of security: cybersecurity, information security", Cybersecurity issues, 2014, №5(8), p. 39-42. (in Russian)
11. A.P. Doukhvalov, "Cyber attacks on critical facilities – the probable cause of the accident", Cybersecurity issues, №3(4), 2014, p. 50-53. (in Russian)
12. A.B. Osak, A.I. Shalaginov, D.A. Panasetsky, E.Ya. Buzina, " Reliability of emergency control and relay protection from the position of cybersecurity", Methodic problems of studies of reliability of large power systems, Eds. N.I. Voropai, ESI SB RASm 2018, Vol.2. p. 99-108. (in Russian)

E3S Web of Conferences **139**, 01030 (2019) https://doi.org/10.1051/e3sconf/201913901030
RSES 2019

Overvoltage in the free phase of the stator winding in case of asymmetric short circuit implicit pole synchronous generator biaxial excitation

Nurali Pirmatov[1,*], *Sherzod* Toshev[1]

[1]Tashkent State Technical University named after I. Karimova, 100095, Tashkent, Uzbekistan

Joint stock company Uzbekgidroenergokurulish, 100077, Tashkent, Uzbekistan

Abstract. Reviewed and analyzed issues associated with overvoltage in the free phase winding of the stator with asymmetric short-circuit implicitly pole synchronous generator biaxial excitation, and compared the results of experimental measurements of the phase currents in the study of a mathematical model of the synchronous generator biaxial excitation (MM SGBD) and physical model SGBD and the SG at steady-state single-phase short circuit.

Introduction

In order to conduct a comparative analysis of the research results of the developed mathematical model of synchronous generator biaxial excitation (SGBD) with experimental data, two physical models were adopted: an implicit pole synchronous generator (SG) and SGBD, converted from MTO-12-4 type induction motors, with a power of 2, 2 kW, voltage 220/380 V (Fig. 1).

An important place in the development strategy of the electric power industry is occupied by generators that produce electrical energy. The problems of managing transients of synchronous generators, primarily in terms of optimizing the processes of the generator, from the standpoint of energy-saving and minimizing costs, is an important task.

An effective way to solve the above problem is to convert mechanical energy into electrical synchronous generators biaxial excitation (SGBD) and asynchronous synchronous generators (ASTG). In this regard, in recent years there has been a growing interest of engineers and researchers to these machines, which can fully solve the tasks.

Fig. 1. General view of a synchronous biaxial excitation generator

Of great importance is the study of asymmetric modes of SGBD with the network. At the same time, there are a number of specific issues of joint work of the SRS with the electric network.

Discussion

Studies of the mathematical model and the physical model of SGBD, as well as SG in short circuit modes, were carried out.

With a single-phase short circuit:

$$U_a = 0, \quad i_b = i_c = 0 \tag{1}$$

Then we write the equations of flux linkage of the stator winding in the form [2,3,4]:

$$\left.\begin{aligned} \frac{d\psi_d}{dt} - \omega_r \cdot \psi_q - (2r_c + 3r_k) \cdot i_d = 0 \\ \frac{d\psi_q}{dt} + \omega_r \cdot \psi_d - (2r_c + 3r_k) \cdot i_q = 0 \end{aligned}\right\} \tag{2}$$

The equations of flux linkages of the stator chain are as follows [5,6,7]:

$$\left.\begin{aligned} &\psi_d = X_{ad} \cdot (1 + \cos 2\gamma_r) \cdot (i_{fd} + i_{1d}) - \\ &- X_{ad} \cdot \sin 2\gamma_r (i_{fq} + i_{1q}) - 2(X_d + X_k) \cdot i_d \\ &\psi_q = -X_{ad} \cdot \sin 2\gamma_r \cdot (i_{fd} + i_{1d}) - \\ &- X_{aq} \cdot (1 - \cos 2\gamma_r) \cdot (i_{fq} + i_{1q}) - 2(X_d + X_k) \cdot i_q \end{aligned}\right\} \tag{3}$$

We write the flux linkage in the single-phase short circuit mode in matrix form [8,9]:

$$\begin{vmatrix} -2(X_d+X_k) & 0 & X_{ad}(1+\cos 2\gamma_r) & -X_{ad}\sin 2\gamma_r & X_{ad}(1+\cos 2\gamma_r) & -X_{ad}\sin 2\gamma_r \\ 0 & -2(X_d+X_k) & -X_{ad}\sin 2\gamma_r & -X_{aq}(1-\cos 2\gamma_r) & -X_{ad}\sin 2\gamma_r & -X_{aq}(1-\cos 2\gamma_r) \\ X_{ad} & 0 & X_{fd} & 0 & X_{ad} & 0 \\ 0 & X_{aq} & 0 & X_{fq} & 0 & X_{aq} \\ X_{ad} & 0 & X_{ad} & 0 & X_{1d} & 0 \\ 0 & X_{aq} & 0 & X_{aq} & 0 & X_{1q} \end{vmatrix} \times \begin{vmatrix} i_d \\ i_q \\ i_{fd} \\ i_{fq} \\ i_{1d} \\ i_{1q} \end{vmatrix} = \begin{vmatrix} \Psi_d \\ \Psi_q \\ \Psi_{fd} \\ \Psi_{fq} \\ \Psi_{1d} \\ \Psi_{1q} \end{vmatrix}$$

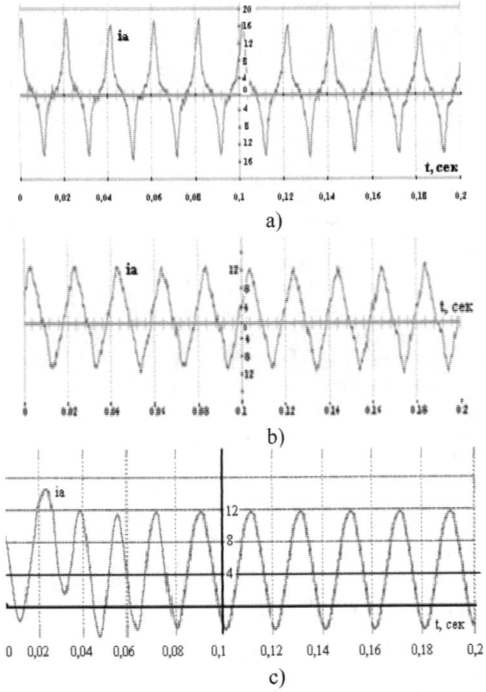

Fig. 2. Oscillogram of stator currents (a) SG ia = 17.3 p.u. and (b) SGBD ia = 12.1 p.u., c) MM SGBD ia = 13 p.u. at single-phase short circuit

As is known, in the asymmetric short circuit mode, overvoltage occurs in the free phase of the stator winding of the SGBD. This is due to electrical and magnetic asymmetry in the rotor circuit of the machine.

For an asymmetric short circuit, the SGBD coefficients in the differential equations become time-dependent, which leads to an increase in the voltage in the free phase of the stator winding for a certain ratio of circuit parameters. In this case, the magnetic field created by the introduced transverse excitation winding prevents the voltage increase (Fig. 3), [1].

The initial value of the voltage in the free phase of the stator winding depends on the scattering magnetic flux, and not on the total total magnetic flux. The subsequent change in voltage depends on the constant time of the field winding and the stator. To avoid overvoltage, the direction of the magnetic flux of the transverse excitation winding should be selected so that when the main magnetic field of the stator is exposed, the magnetic field of scattering is minimal.

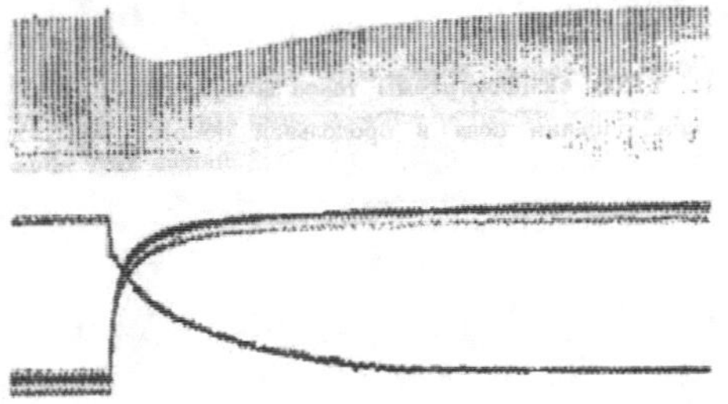

Fig. 3. Oscillogram of EMF (1), longitudinal current i_{fd} (3) and current transverse i_{fq} (2) of the field winding

The increase in the scattering of the stator magnetic flux depends on the free currents in the rotor windings. The additional magnetic flux created by these free currents is oriented in the same direction as the magnetic fluxes of the field winding, which prevents the passage of the magnetic flux of the stator. The larger the drop in the magnetic voltage in the rotor winding (scattering of the magnetic field energy), the lower the EMF induced in the free phase of the stator winding and decrease into the flux [7,8,9]. Limitations for single-phase short circuit [10]:

$$\left.\begin{array}{l} U_a = 0 \\ i_b = 0 \\ i_c = 0 \end{array}\right\} \quad (4)$$

We determine the voltage of phases B and C from the expression:

$$U_b = \frac{d\psi_b}{dt} - r_c i_b \qquad U_c = \frac{d\psi_c}{dt} - r_c i_c \quad (5)$$

$$U_b = \frac{d\psi_b}{dt} \qquad U_c = \frac{d\psi_c}{dt} \quad (6)$$

here

$$\psi_b = \psi_d \cos(\gamma - 120^0) - \psi_q \sin(\gamma - 120^0) + \psi_0 \qquad \psi_c = \psi_d \cos(\gamma + 120^0) - \psi_q \sin(\gamma + 120^0) + \psi_0 \quad (7)$$

here γ_0- the angle between the phase magnetic axis of phase A and the axis of the rotor. For single phase short circuit:

$$U_b = \left(\frac{d\psi_d}{dt} - \psi_q\right)\cos(\gamma - 120^0) - \left(\frac{d\psi_q}{dt} + \psi_d\right)\sin(\gamma - 120^0) \quad (9)$$

$$U_c = \left(\frac{d\psi_d}{dt} - \psi_q\right)\cos(\gamma + 120^0) - \left(\frac{d\psi_q}{dt} + \psi_d\right)\sin(\gamma + 120^0)$$

ψ_d, ψ_q - according to equation (4) for a single-phase short circuit mode, respectively.

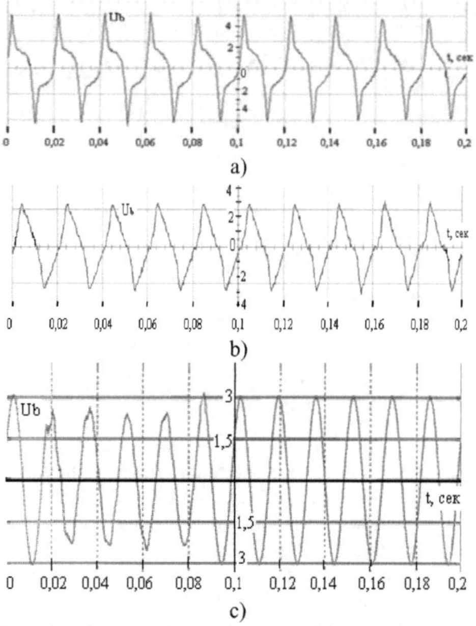

Fig. 4. Oscillogram of the overvoltage of the free phase of the stator (a) SG (U = 30 V (1 pu), Ub. = 142 V (4.73 pu)) and (b) SGBD (U = 30 V (1 p.u.) Ub = 85.2 V (2.84 pu), (c) MM SGBD (U = 30 V (1 pu), Ub = 90 V (3 pu) for single-phase short circuit

Comparative results

In the table. Figures 1 and 2 present the results of experimental measurements of phase currents and overvoltage in the free phase of the stator winding, in which the mathematical model under study of the synchronous generator biaxial excitation (MM SGBD) corresponds to the physical model of SGBD and SG with a steady state single-phase short circuit.

Comparisons of the results of experimental measurements of SG and calculation of the stator phase winding current on the mathematical model of SGBD and for single-phase short-circuit are given in Table 1.

Table 1.

	Stator phase currents				
	Steady state		Short Circuit		Multiplicity
	Ampere	pu	Ampere	pu	
SG	7,0	1,0	121,18	17,312	17,312
SGBD	6,9	0,98	83,5	12,1	12,1
MM SGBD	7,0	1,0	91	13	13

From tables 1 and 2 it follows that:
- by turning on the transverse excitation winding of a synchronous biaxial excitation generator, a reduction in current and overvoltage in modes with a single-phase short circuit is achieved. generators almost doubled;
- the currents of a single-phase short circuit in the experimental SGDV installation and in the solutions of MM SGBD are the same. These statements are confirmed by the taken oscillograms of short-circuit currents during experimental studies of SG and SGBD, which are shown in Fig. 2.

Table 2. Comparison table of overvoltages in modes at single-phase short-circuit generators

	Voltage				
	Steady state		Free phase overvoltages		Multiplicity
	B	o.e.	B	o.e.	
SG	30	1,0	142	4,73	4,73
SGBD	30	1,0	85,2	2,84	2,84
MM SGBD	30	1,0	90	3	3

- overvoltage on the free phase of the stator winding during a single-phase short-circuit in the experimental installation of SGDV and in the solutions of MM SGBD are the same. These statements are confirmed by the recorded waveforms of the overvoltage in the free phase of the stator winding during experimental studies of SG and SGBD, which are shown in Fig. 4.

Conclusion

The proposals outlined in this paper offer the prospect of further research is an extremely important problem - the estimation of the surge in the free phase winding of the stator with asymmetric short-circuit implicitly pole synchronous generator biaxial excitation.

References

1. Akhmatov M. G. Synchronous machines. Special course. - Moscow: higher school, **136** (1984)
2. Khaidarov S.Zh., Toshev Sh.E. - Bulletin of TSTUniversity. **1**, 68-74 (2016)
3. Chernykh I.V. DMK Press; St. Petersburg: Peter, **288** (2008)
4. Allaev K.R., Khaidarov S.D. Izv. Academy of Scien. of Uzb., a series of tech. // Science, **5**, 23-26 (1986)
5. Pirmatov N.B., Salimov D.S., Toshev Sh.E. — J. of Prob. of Energy and Resour. S. **1-2**, 163-165 (2012)
6. Toshev Sh.E., Taniev M., Egamberdiev D.Kh. - The Repub. of scien. and tech. conf. - Fargana, **182-184** (2016)
7. Pirmatov N.B., Toshev Sh.E. - Bulletin of the Almaty Ins. of Ener. and Communi. **2(2)**, 40-44 (2008)
8. Pirmatov N.B., Toshev Sh.E., Haydarov S.D., Tosheva Sh.N. Europ. Conf. on Inn. in Tech. and Natu. Scien.:,15th Int. scien. conf.. **105-110** (2017)
9. Toshev Sh.E., Pirmatov N.B., Haydarov S.D., Duvlonov J.N., Yakubova D.K. Europ. Scien. R.:Aust, Vienna,**11-12**, 134-136 (2016)
10. Pirmatov N.B., Toshev Sh.E. – Int. Scien. and tech. conf. Collection of works. Tashkent.: **175-177** (2006)

E3S Web of Conferences **139**, 01031 (2019) https://doi.org/10.1051/e3sconf/201913901031
RSES 2019

Power System Optimization in Terms of Uncertainty of Initial Information

Tulkin Gayibov[1,*], *Bekzo*d Pulatov[1], *Sherxo*n Latipov[1], and *Gulnaz* Turmanova[2]

[1]Tashkent State Technical University named after Islam Karimov, Uzbekistan
[2]Karakalpak State University, Uzbekistan

Abstract. In the article the problem of power system optimization under partial uncertainty of initial information is considered. The possibilities of using the methods of determinization to solve the problem of power system optimization taking into account the regime and technological constraints in the form of equality and inequality are researched. In basis of performed calculation studies using the example of the test scheme of power system, it was found that the optimization algorithm by the method of determinization can be effectively used for this proposes.

1 Introduction

The problem of optimization of power system development and functioning is a difficult problem of nonlinear mathematical programming with many limiting conditions in the form of equality and inequality imposed on adjustable parameters and on some of their functions. These problems are also characterized, like similar problem for any large system, by the presence of various kinds of initial information. Currently, the initial data used to solve such problems can be divided into four types: deterministic, probabilistic, fuzzy, and interval.

The deterministic data gives a complete picture of the state of object, which meets the requirements of this problem. The last three types of initial data are uncertain data. With a probabilistic initial data, the laws of distribution of probabilistic quantities in known ranges of their possible values are usually set. Fuzzy initial data suggests their task by formulated linguistic distribution of possible parameter values. In the case of interval initial information, the intervals of possible values of the initial parameter (or state) are usually known, and no laws are known about their possible values (or states) in this interval.

Currently, the methods and algorithms for solving power system optimization problems in the conditions of deterministic initial data are sufficiently developed [1-7]. The solution of the problem in conditions of probabilistic initial data is carried out by probabilistic-statistical methods. They provide for the use of the mathematical expectation functions of the objective function and the constraint functions of probabilistic quantities. In this case, the calculation algorithms do not differ much from the algorithms for deterministic initial data. The optimization problems in the conditions of fuzzy initial data are solved by methods of the theory of fuzzy sets [8, 9]. In this paper, we consider the problem of Power system optimization in the conditions of interval (partial) uncertainty of the initial parameter. In the existing literature, there are no an advanced algorithms for optimization of the development and functioning of power system, taking into account all the constraints in conditions of partial uncertainty of the initial information. In [10–12], the efficient algorithm of power system optimization in terms of partial uncertainty of initial information is described. The algorithm is based on the sequential solution of deterministic problem for the values of initial parameter selected in a given interval and select the best plan for various criteria. In this case, preference is given to the criteria of minimax and minmax risk. At the same time, in these works, the questions of taking into account of constraints in the form of inequality remains opened. The algorithms proposed in [13] for taking into account inequality constraints during optimization under the conditions of probability and partial uncertainty of the initial information do not completely overcome the noted difficulties. In some cases, the determination of a conditionally optimal plan, where all the specified constrains are satisfied for all the possible values of the initial parameter in a given range, requires additional complex calculations. In [14–16] an algorithm of optimization under conditions of interval uncertainty of the initial parameter by the method of determinization, taking into account the constraints is described. This paper presents the results of study of possibility of using this algorithm to solve problems of optimization of a development and functioning of power system.

2 The optimization algorithm

We consider the possibilities of using the algorithm of optimization in terms of uncertainty by determinization method for solving the problem of power system development optimization, which is formulated as follows:

minimize the function of estimated costs

$$3(a_1, a_2, ..., a_m, x_1, x_2, ..., x_n) \rightarrow min \quad (1)$$

taking into account

- simple constraints imposed on adjustable parameters

$$x_i^{min} \leq x_i \leq x_i^{max}, \qquad i = \overline{1, n}; \quad (2)$$

- functional constraints in the form of equality and inequality imposed on some functions of adjustable parameters

$$f_j(a_1, a_2, ..., a_m, x_1, x_2, ..., x_n) = \bar{f}_j, \quad j = \overline{1, N}; \quad (3)$$

$$g_l(a_1, a_2, ..., a_m, x_1, x_2, ..., x_n) \leq \bar{g}_l, \quad l = \overline{1, L}, \quad (4)$$

where $\bar{f}_j, \bar{g}_l$ are specified values of the constraint functions.

At partial uncertainty of initial parameters, when their minimum and maximum possible values are known, all functions, in particular, the objective function and the constraint functions, are also partially indefinite. Then the problem described above is reduced to minimizing the functions of the estimated costs determined at the minimum and maximum possible values of the initial partially indefinite parameters, as in [15]

$$3(a_1^{min}, a_2^{min}, ..., a_m^{min}, x_1, x_2, ..., x_n) \rightarrow min, \quad (1a)$$

$$3(a_1^{max}, a_2^{max}, ..., a_m^{max}, x_1, x_2, ..., x_n) \rightarrow min \quad (1б)$$

taking into account the constraints (2) and functional constraints in the form of equality and inequality imposed on functions determined by the minimum and maximum possible values of the initial parameters

$$f_j(a_1^{min}, a_2^{min}, ..., a_m^{min}, x_1, x_2, ..., x_n) = \bar{f}_j^{min}, \quad j = \overline{1, N}; \quad (3a)$$

$$f_j(a_1^{max}, a_2^{max}, ..., a_m^{max}, x_1, x_2, ..., x_n) = \bar{f}_j^{max}, \quad j = \overline{1, N}; \quad (3б)$$

$$g_l(a_1^{min}, a_2^{min}, ..., a_m^{min}, x_1, x_2, ..., x_n) \leq \bar{g}_l^{min}, \quad l = \overline{1, L}; \quad (4a)$$

$$g_l(a_1^{max}, a_2^{max}, ..., a_m^{max}, x_1, x_2, ..., x_n) \leq \bar{g}_l^{max}, \quad l = \overline{1, L}. \quad (4б)$$

Solution of the obtained optimization problem (1a) - (4b) reduces to solving the equivalent two deterministic problems. In the first problem, the objective function is represented in the form (1a), and the constraints consist of (2), (3a) - (4b). In the second problem, the objective function form is (1b), and the constraints are (2), (3a) - (4b). As a result of solving these two deterministic problems, the interval values of the sought-for controlled parameters are determined. In this case, the lower boundary values of the intervals are obtained as a result of solving the first problem, and the upper boundary values are obtained as a result of solving the second problem. For the final solution of the original problem, one point is taken at which all the given constraints are satisfied. The coordinates of this point are determined by the values of the parameters found in the obtained intervals. The value of the objective function at the point of optimal solution is taken as an interval, the boundaries of which are determined by the results of solving the first and second deterministic problems.

3 The results of calculation experiments

To study the effectiveness of the described algorithm, we consider its application for solving the problem of power system mode optimization using the example of the test circuit (Fig. 1), given in [4] for given interval (partially undefined) values of active power of load nodes 2, 3, and 4. Minimum and maximum possible values of loads in these nodes are given in table 1. The active power of node 5 is deterministic and $P_5 = 500$ MW.

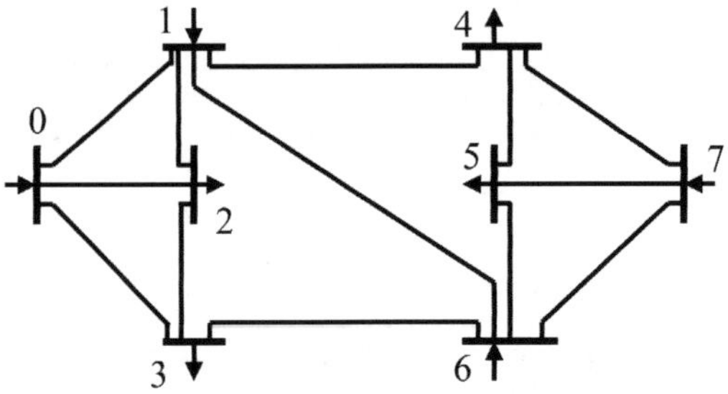

Fig. 1. The scheme of the power system.

Table 1. Minimum and maximum limit values of active loads.

The number of node, i	P_i^{min}, MW	P_i^{max}, MW
2	380	420
3	570	630
4	190	210

In four nodes 0, 1, 6 and 7 there are thermal power plants (TPPs) with oil equivalent consumption characteristics, presented in the form

$$B_i = c_i + d_i P_i + e_i P_i^2, \quad t.o.e./h.,$$

where *t.o.e./h.*- tonne of oil equivalent per hour.

The values of coefficients of oil equivalent consumption characteristics of thermal power plants are given in table 2. The minimum and maximum permissible capacities of all the TPPs are the same and, MW

$$100 \text{ MBт} < P_i < 1000 \text{ MBт}, \quad i = 0, 1, 6, 7.$$

In three power transmission lines (TPLs) 6-3, 6-5 and 0-3 the active power flow are controlled, MW:

$$P_{6-3} \leq 480, \quad P_{6-5} \leq 70, \quad P_{0-3} \leq 100.$$

Power flows through a controlled TPLs are determined by linearized formulas using the node power distribution coefficients:

$$P_l = \sum_{i \in \Gamma} C_{li}^{\Gamma} P_i^{\Gamma} - \sum_{j \in H} C_{lj}^{H} P_j^{H} + P_{l0},$$

Table 2. The coefficients of oil equivalent consumption characteristics of TPPs

The node number with TPP	c_i	d_i	e_i
0	35	0,14	0,00035
1	42	0,14	0,00042
6	21	0,105	0,00028
7	28	0,175	0,000175

where C_{li}^{Γ}, C_{lj}^{H} are the power distribution coefficients of the i-th generating and j-th load nodes by PTL l (table. 3); P_i^{Γ}, P_j^{H} are active powers of the i-th generating and j-th load nodes: P_{l0} is a free member of the linearized expression of power flow over the PTL l.

Table 3. Power distribution coefficients of nodes.

PTL	Nodes						
	1	2	3	4	5	6	7
6-3	0,2536	-0,0713	-0,172	0,4106	0,489	0,6343	0,4986
6-5	-0,0701	0,02025	0,0484	-0,281	-0,4223	0,1029	-0,2986
0-3	-0,14	-0,1735	-0,267	-0,1744	-0,1917	-0,224	-0,1939

Thus, in the problem under consideration, functional constraints in the form of both equality and inequality are linear. The right-hand sides of functional constraints in the form of equality and inequality at interval node loads are determined based on interval arithmetic operations. As a result, we have the following interval values for the right-hand sides of the constraint functions in the form of equality on power balance and inequality on power flow in the controlled PTLs 6-3, 6-5, and 0-3, respectively:

[1640; 1760], [664,168; 685,592], [-164,877; -155,543], [-273,554; -247,106].

In the problem under consideration, the constraint in the form of equality is a condition for the active power balance in power system, therefore, the left-hand sides of the constraints in the form (3a) and (3b) are the same. When translating the components associated with the interval loads of the nodes in constraints in the form of inequality on power flows in the controlled NPLs, the left-hand sides of the constraints (4a) and (4b) also become the same. In this case, in each deterministic problem, one constraint in the form of equality takes part. In the first problem, the right-hand side of the constraint is equal to the lower boundary value, and in the second, to the upper boundary value of the total load of power system specified in the interval form.

Since the left-hand sides of the constraints in the form of inequality obtained from (4a) and (4b) are the same, one of these constraints is involved in both deterministic problems, the right-hand side of which is less than the second. In this case, the second constraint with a large right-hand side turns into a passive restriction.

Thus, the first deterministic optimization problem, for the considered example, according to the described algorithm has the following form:

The objective function

$$B = B_0 + B_1 + B_6 + B_7 = 126 + 0{,}14P_0 + 0{,}00035P_0^2 + 0{,}14P_1 + \\ + 0{,}00042P_1^2 + 0{,}105P_6 + 0{,}00028P_6^2 + 0{,}175P_7 + \\ + 0{,}000175P_7^2 \to min$$

Constraints

$$P_0 + P_1 + P_6 + P_7 = 1640\,,$$

$$0{,}2536P_1 + 0{,}6343P_6 + 0{,}4986P_7 \le 664{,}168\,,$$

$$-0{,}0701P_1 + 0{,}1029P_6 - 0{,}2986P_7 \le -164{,}877\,,$$

$$-0{,}14P_1 - 0{,}224P_6 - 0{,}1939P_7 \le -273{,}554\,,$$

$$100 < P_i < 1000, \quad i = 0, 1, 6, 7\,.$$

The second deterministic problem is represented as follows:

The objective function

$$B = B_0 + B_1 + B_6 + B_7 = 126 + 0{,}14P_0 + 0{,}00035P_0^2 + 0{,}14P_1 + \\ + 0{,}00042P_1^2 + 0{,}105P_6 + 0{,}00028P_6^2 + 0{,}175P_7 + \\ + 0{,}000175P_7^2 \to min$$

Constraints

$$P_0 + P_1 + P_6 + P_7 = 1760\,,$$

$$0{,}2536P_1 + 0{,}6343P_6 + 0{,}4986P_7 \le 664{,}168\,,$$

$$-0{,}0701P_1 + 0{,}1029P_6 - 0{,}2986P_7 \le -164{,}877\,,$$

$$-0{,}14P_1 - 0{,}224P_6 - 0{,}1939P_7 \le -273{,}554\,,$$

$$100 < P_i < 1000, \quad i = 0, 1, 6, 7\,.$$

It should be noted that the first and second deterministic problems, for the considered example, differ from each other only by the right-hand side of the constraint in the form of equality.

In table 4 the results of solving the first and second deterministic problems are shown.

According to the results given in table 4, we determine the interval values of the optimal powers of the settlement plants and the total of oil equivalent consumption:

P_0 = [127,720; 247,701] MW, P_1 = [547,769; 547,715] MW, P_6 = [326,838; 326,673] MW, P_7 = [637,673; 637,910] MW, B = [599,280; 631,857] t.o.e./h.

Table 4. The Results of solving the deterministic optimization problems.

TPPs load, MW				Power flow on controlled TPPs, MW			The total fuel equivalent consumption, t.o.e./h.
P_0	P_1	P_6	P_7	P_{6-3}	P_{6-5}	P_{0-3}	B
The results of solving the first deterministic problem							
127,720	547,769	326,838	637,673	480,00	39,70	100,00	599,280
The results of solving the second deterministic problem							
247,701	547,715	326,673	637,910	480,00	39,62	100,00	631,857

For the final solution of the problem, according to the above marked rules, we take any point whose coordinates are determined by the powers of the plants in the received intervals. In this case, the lower and upper boundary values of power plant power intervals at nodes 1, 6, and 7 were the same. And the power of the balancing plant can be taken as the average value in the obtained interval. Thus, we take the optimal power of plants as
$P_0 = 187$ MW, $P_1 = 547{,}75$ MW, $P_6 = 326{,}75$ MW, $P_7 = 637{,}75$ MW.
The minimum total oil equivalent consumption in power system has an interval value: B = [599,280; 631,857] t.o.e./h.

4 Conclusion

The problems of power electric system development and functioning optimization taking into account simple and functional constraints in the form of equality and inequality in conditions of partial (interval) uncertainty of initial parameters, can be effectively solved by the method of determinization.

In power system mode optimization on active power and presenting the expression of power flow over a controlled PTL using linear distribution coefficients, solving the problem described by the method of determinization is much easier by halving the number of functional constraints in the form of equality and inequality compared to the classical optimization algorithm in basis of determinization method.

References

1. *Methods for power system mode optimizaion.* V.M. Gornstein, B.P. Miroshnichenko, A.V. Ponomarev and others; Edited by V.M.Gornstein (Energy, Moscow, 1981).
2. Krumm L.A. *Methods of optimization and control of electric power systems* (Nauka, Novosibirsk, 1980).
3. *Automation of dispatch control in the electric power systems* / Edited by Yu.N. Rudenko and V.A. Semenov (Publishing MPEI, Moscow, 2000).
4. Fazylov H.F., Nasyrov T.Kh. *Steady state mode of power systems and its optimization* (Moliya, Tashkent, 1999).
5. Nasirov T.Kh., Gayibov T.Sh. *Theoretical foundations of power system mode optimization* (Fan va texnologiya, Tashkent, 2014).
6. Gayibov T.Sh., Reimov K.M., Proc. of the Seventh All-Russian Scientific and Technical Conference with international participation. Energy: Management, quality and energy efficiency. Blagoveshchensk, (2013).
7. Gayibov T.Sh, Latipov Sh.Sh. *Taking into Account of Functional Constraints in Optimization of Modes of Power Systems by Genetic Algorithms*. Engineering, **11 (4)**, (2019).
8. Articles written on the occasion of the 50th anniversary of fuzzy set theory // Didier Dubois, Henri Prade. Rapport Interne IRIT RR-215-11-FR. December (2015).
9. M. Valdma, M. Keel, H. Tammoja, J. Shuvalova. *Oil Shale*, vol. 24, **2**, Special, (2007).
10. Chaoyi Peng, Shunbo Lei, Yunhe Hou, Felix Wu. *CSEE Journal of Power and Energy Systems*, vol. 1, March **1**, (2015).
11. Valdma M, Keel M, Liik O, Tammoja H. *Proc. of IEEE Bologna PowerTech*, Bologna, Italy, June 23-26, Bologna, Italy. Paper **252**, (2003),
12. Arzamastsev D.A. and other. *ACS and optimization of power system modes*: (Higher. school, Moscow, 1983).
13. Gayibov T.Sh., Reimov K.M. *Problems of energy and resource saving (special issue)*. - Tashkent – **4**, (2013).
14. V.I. Levin. *Automatic Control and Computer Sciences*, Vol. 46, **4**, (2012).
15. V.I. Levin. *XII All-Russian Meeting on Management Problems of the VSPU-2014.* June 6-19, (2014).
16. Nemkova EA. *Abstract of diss. for Degrees Ph.D.*, Penza, (2014).

Features of Estimation of Structural Reliability of Power Supply Systems

Boris Papkov[1*] and *Vladimir* Osokin[1]

[1]Nizhny Novgorod State Engineering and Economic University, Chair of Electrification and Automation, 606340 Knyaginino, Oktyabrskaya str. h .22A, Russia

Abstract. Development and operation of modern power supply systems of consumers which are the part of distributed generation (DG) systems with their communication with electric power system (EPS) and renewable energy sources (RES) demand changes in estimation of their reliability indicators on the basis of corresponding mathematical apparatus. It is shown that DG systems refer to structures with several overlapping zones of action, which determines their peculiarities, which are represented by some integral characteristic - efficiency of operation, which characterizes the degree of expediency of use of the DG system in different modes. The approach based on evaluation of effectiveness of such systems is offered. Examples of calculation of relatively simple structures of DG systems performing several tasks simultaneously are given.
Keywords: reliability, structure, efficiency, distributed generation, mode

1 Status of issue

A specific feature of distributed generation (DG) systems is the complexity of their energy, technological and organizational structure. On the one hand, it allows to solve a complex of problems of ensuring reliable and effective power supply of consumers, and, on the other hand, to provide stable operation in case of violation of the working capacity of separate generating installations (GI) and even their groups. It is known that for such systems there is no generally accepted concept of "failure", because changes in the structure of DG system due to failures of separate GI and their connections lead, as a rule, only to decrease in its functioning since the boundaries between serviceable and faulty, as well as between operational, partially operational and non-operational states are blurred and often conditional [1]. This is due to some redundancy of the structure, presence of reserve GI (overlapping zones), switching capabilities and connections, peculiarities of relay protection (RP) and emergency automation (EA), means of correction modes (reactive power compensation and voltage regulation), possible errors of operational personnel.

Since DG systems are represented by structures with several overlapping zones of action, this determines the features of their work, which can be assessed by some integral characteristic – efficiency of functioning – the degree of expediency of using the DG system in different modes. However, issues related to the peculiarities of calculating the efficiency of systems with overlapping zones have not been fully investigated [2-5, 8, 9], which makes the topic relevant and the practical implementation of the overlapping zones method necessary.

Even formally in the state of full operability the DG system may not perform all the functions required of it. This is due to unfavorable circumstances (uneven distribution of power by renewable energy sources (RES), overload with increased consumer demand, unauthorized external influences (cyber attacks)) [7].

It should be taken into account that the behavior of autonomously operating GI of DG systems differs significantly from their behavior in the system. In the latter case, the features of hierarchical structure, modes and switching capabilities of network may appear; interdependence of failures of individual elements of DG system; interdependence (sequence) of their recoveries, etc. This leads to the need to take into account the fact that the behavior of single K element of x_k is determined by the background of some or even all other elements [4]:

$$x_k(t+dt) = f_k[x_k(0,t),\ x_1(0,t), \dots, x_n(0,t)]\,,$$

where f_k – function, determined by the specific features of the system.

Therefore, the state of the system is determined by the n-dimensional random value

$$X(t) = [x_1(t), x_2(t), \dots, x_n(t)],$$

and the trajectory of its behavior in the time interval $(t,\ t+\theta)$ n-dimensional random function

$$X(t,\theta) = [x_1(t,\theta\,), x_2(t\theta), \dots, x_n(t\theta)].$$

Therefore, there is a need to expand the concept of "reliability" [7] and the transition to performance indicators of DG system when it performs certain tasks in specific conditions [2-5]. Efficiency indicator is a measure that quantifies the quality of functions performed by DG system, that is, a measure of usefulness of DG system operation in a certain situation while providing power supply to consumers in specific operating modes of the GI.

2 DG system with overlapping zones of action

Since any electric power system (EPS), including DG, belongs to the class of large systems with several levels of functioning, has a complex structure, multifunctionality, redundancy, it cannot be represented by single classical reliability indicators. Failures of its individual elements (in this case the GI and the links between them) do not lead, as a rule, to the failure of the system, but only reduce the efficiency of its functioning.

In the simplest case [8] efficiency is the probability of performing an autonomous GI task for uninterrupted power supply to consumers. The efficiency coefficient E in this case additionally acquires the meaning of probability that the implementation of this task will not be disrupted due to the impossibility of implementing some modes (failures). We consider that N_Σ is a set of all possible operating modes of GI implemented to provide power supply to consumers; N is a set of actually existing modes. However, some additional permissible modes are possible, forming a plurality of N_0 together with N. Then $N \in N_0 \in N_\Sigma$ (fig. 1).

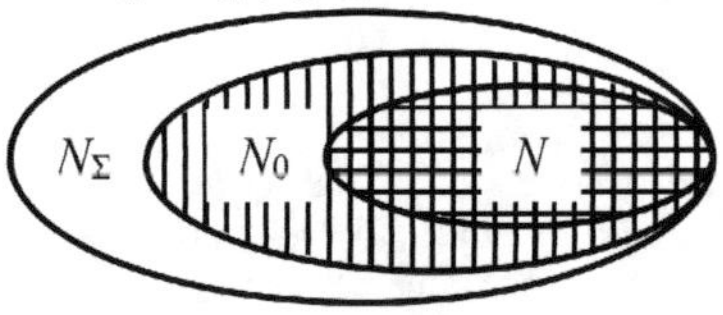

Fig. 1. Possible ratio of GI modes.

Efficiency here is the probability p that the mode n belongs to the set N, that is, $E = p\{n \in N\}$. Similarly is $E_0 = p\{n \in N_0\}$. The probability that the mode of operation of GI simultaneously belongs to two sets N and N_0 will be

$$p\{n \in N, n \in N_0\} = p\{n \in N\},$$

because $N \subset N_0$ (fig. 1).

According to the probability multiplication theorem we get

$$p\{n \in N, n \in N_0\} = p\{n \in N_0\} \cdot p\{n \in N | n \in N_0\}.$$

The efficiency coefficient is defined as

$$E = \frac{p\{n \in N\}}{p\{n \in N_0\}} = \frac{p\{n \in N, n \in N_0\}}{p\{n \in N_0\}} = p\{n \in N | n \in N_0\}. \quad (1)$$

The given expression (1) has probability that power supply to consumers will be carried out in any mode of GI operation.

Considering the difficulty of obtaining estimates of efficiency of DG systems, it is necessary to develop and improve the corresponding mathematical apparatus. In general, the efficiency of the DG $E_{syst}(t)$ is come down to calculation according to the formula offered in [2]

$$E_{syst}(t) = \sum_{k=1}^{2^n} p_{X_k}(t) E_{X_k}, \quad (2)$$

where $X = (x_1, x_2, \ldots, x_n)$ is system condition of DG; n is number of system's elements, being in two conditions ($x_i = 1$ – efficient, $x_i = 0$ – failure); $p_{X_k}(t)$ – probability that the system is X_k condition at time t; E_{X_k} – efficiency of system being in X_k condition.

When n is large, the calculation of efficiency according to (2) is quite complicated. Therefore, we consider the approach offered in [2, 3, 5]. It is based on the fact that in some territory there are several GI united in the system, providing power supply to consumers not only directly connected to "their" GI, but also (with a sufficient reserve of power and network capacity) other, geographically close and located in the overlapping zones of adjacent GI.

If in the system of n elements the i element with probability p_i is in good condition, the results of its operation apply to the areaS_i. Coverage zones of two, three, etc. elements can intersect, forming area $S_{i,j}, S_{i,j,k}, \ldots, S_{i,j,k,\ldots,n}$ with corresponding elements of influence. In general case such zones can be 2^n. Scope of the whole system is represented by the union of all zones of action of elements $S = \bigcup_{i=1}^{n} S_i$ [2].

Suppose that conditional power district representing system of DG and consisting of four GI (power plants), each of which provides operation of consumers of electric power of "their" zone. $\Omega = \bigcup_{i=1}^{4} Z_i$. There are areas in each zone, which intersect adjacent two or more zones. Consequently, system with overlapping zones of action is formed.

The part of zone that does not intersect with others is designated as S_i, $(i = \overline{1,4})$, and the areas of intersection are $S_{i,j}$, $S_{i,j,k}$ $(i = j = k = \overline{1,4})$. All selected zones do not intersect and define the district S of GI operation of investigated DG system (Fig. 2).

The efficiency coefficients of selected areas are denoted by W with corresponding indexes. If, for example, GI whose main consumers are located in zone Z_2 fails, the efficiency coefficients of areas covered by this zone will change. Instead of E_{12}, E_{23}, E_{24}, E_{234} we get E_1, E_3, E_4, E_{34}. Naturally, in each practical problem the efficiency coefficients for intersection areas are determined individually.

Such division of energy district of DG system, consisting of n GI can significantly reduce the number of selected areas relative to the value of 2^n, that is, simplifies the calculation of the efficiency coefficient of DG system.

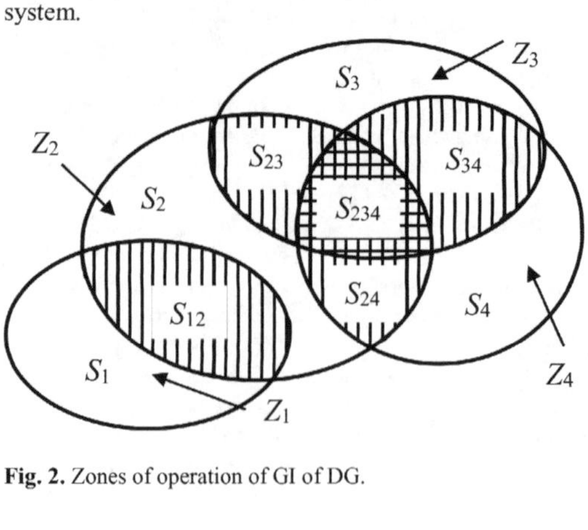

Fig. 2. Zones of operation of GI of DG.

Its calculation includes calculation of efficiency coefficients of each areas and their summation taking into account

- power output from generating units;
- switching capabilities of power supply schemes;
- network mode settings;
- composition, features of operation and installations of RP and EA systems;
- power consumption and technological features of consumers located in corresponding zone;
- other indicators characterizing the transition of DG system to restrictive regime.

The component of efficiency coefficient e_i in resulting sum for S_i area without S overlapping with other zones is defined as

$$e_i = S_i E_i p_i,$$

where p_i is probability of GI failure in Z_i area with power supply failure of consumers in S_i. zone.

The component for the overlap area of Z_i and Z_j zones is equal to

$$e_{ij} = S_{ij}\left(E_{ij} p_i p_j + E_i p_i q_j + E_j q_i p_j\right).$$

The component formed by the overlap of three zones Z_i, Z_j and Z_k

$$e_{ijk} = S_{ijk}(E_{ijk} p_i p_j p_k + E_{ij} p_i p_j q_k + E_{ik} p_i q_j p_k + E_{jk} q_i p_j p_k + E_i p_i q_j q_k + E_j q_i p_j q_k + E_k q_i q_j p_k)\ .$$

Estimation of efficiency coefficients of each area with more zones is done in the same way.

Note that efficiency coefficient determines the average level of quality of the system, due to the reliability of its elements. It can be dimensionless (probability of execution of requirements of consumer) and have the dimension (the amount of generated power, the amount of consumption, profit, damage, etc.) In case of GI failure in Z_i zone losses (damage) are determined by consequences of power supply failure of consumers located in S_i area. In other areas of overlap with other zones, there will (or may) be some decrease in the efficiency of functioning of consumers (reduced productivity, changes in the parameters of the technological process and (or) the quality of products) located in these zones [1]. The damage Y_i in zone S_i can be estimated by expression

$$Y_i = S_i E_i + S_{ij}\left(E_{ij} - E_i\right) + S_{ijk}\left(E_{ijk} - E_i\right) + \ldots\ .$$

An approximate estimate of the magnitude of total damage Y_Σ based on the probability p_i of failure GI in different zones of DG systems will be

$$Y_\Sigma = \sum_{i=1}^{n} p_i Y_i.$$

3 Bases of estimation of efficiency of DG systems with overlapping zones of action

Consider a system of DG consisting of two identical GI, each of which can operate in two modes: x_1 и x_2.

If both modes $x_1 \cup x_2$ are possible then the effectiveness of system of DG – $E = S_2$; if $\bar{x}_1 \cup x_2$ or $x_1 \cup \bar{x}_2$, then $E = S_1$, (if $\bar{x}_i$, $i = 1{,}2$ mode fails) and $S_1 < S_2$; if $\bar{x}_1 \cup \bar{x}_2$, then $E = 0$ [9].

The state of DG system is characterized by vector with components

$$\left\{x_1^{(1)} x_2^{(1)} x_1^{(2)} x_2^{(2)}\right\}, \quad (3)$$

where the upper index is the number GI, lower index - corresponding mode.

Each component (3) takes two values: x_i^jj - power supply mode of consumers is performed; $\bar{x}_i^j$ - the mode is impracticable.

When $E = 0$ we have:

$$\left\{\bar{x}_1^{(1)} \bar{x}_2^{(1)} \bar{x}_1^{(2)} \bar{x}_2^{(2)}\right\}.$$

When $E = S_1$ two modes can be: *a*) arbitrary, but only one component of vector (3) takes the value x_i^j, and other – $\bar{x}_i^j$; *b*) DG system state is characterized by one of vectors

$$\left\{\bar{x}_1^{(1)} x_2^{(1)} \bar{x}_1^{(2)} x_2^{(2)}\right\},\ \left\{x_1^{(1)} \bar{x}_2^{(1)} x_1^{(2)} \bar{x}_2^{(2)}\right\}. \quad (4)$$

When $E = S_2$ two modes are also possible: *a*) at least three components of vector (3) take the value x_i^j (one of GI is completely serviceable); *b*) DG system state is characterized by one of vectors

$$\left\{\bar{x}_1^{(1)} x_2^{(1)} x_1^{(2)} \bar{x}_2^{(2)}\right\},\ \left\{x_1^{(1)} \bar{x}_2^{(1)} \bar{x}_1^{(2)} x_2^{(2)}\right\}. \quad (5)$$

Comparison of expressions (4) and (5) shows that in both cases the system consists of two GI operating in one of possible modes; the number of impossible modes is also the same, but efficiency indicators are different. The difference is determined by the nature of conditions for power supply of consumers. According to (4) the DG system is fully operational at the expense of two partially operational GI.

First of all the estimation of effectiveness of DG systems should be based on fundamental knowledge about their structural, technological and organizational peculiarities, principles of operation of the main and control elements, relay protection systems and automation, which are part of them [10, 11]. The order of even the preliminary estimation of the effectiveness of DG systems is determined by the following sequence:

- obtaining information on main circuit-mode characteristics and technical and economic indicators of operation of DG system;
- possible tasks of DG system operation are investigated;
- DG system operation conditions are determined in parallel with EPS, independently and in combined mode;
- estimated the expected frequency of repetition of tasks and modes of DG system operation;
- functional diagram of the system is drawn up;
- DG system is divided into separate elements (zones, regions);
- quantitative measure of functioning quality of acceptable for this system of DG is chosen;
- reliability indicators of elements characterizing probability of state of each element at different moments of time are calculated;

- probabilities of average (if necessary permissible and/or limit) states of DG system are calculated on the basis of probabilities of states of separate elements;
- efficiency indicators of possible states of DG system are evaluated.

If for each state of DG system it is possible to estimate the "instant" value of its functioning efficiency (output effect), the efficiency coefficient is defined as the average value of the output effect for all states of the system [2].

4 Examples of indicative estimation of the effectiveness of DG systems

We illustrate the method of calculating the efficiency coefficient of DG system on simple examples in which the traditional calculation of reliability indicators is quite complex.

1). ***Non-intersecting coverage zones of GI in DG system.*** Each of two identical GI in DG system provides power supply to consumers in its zone with probabilityp $= 0,9$. The coefficient of readiness of each GI will take as $K_g = 0,95$. The coefficient of interval readiness $K_g^*(t, t + t_0)$ is introduced in order that the system was efficient not only in any required timepoint, but also it would be able to work smoothly during the required time interval$t + t_0$. In general case $p^*(t) \neq p$ is because p is probability of no-failure operation, since condition of full working capacity, and $p^*(t)$ is probability of no-failure operation, since one of possible intermediate conditions of working capacity. For simplicity, assume that $t \to \infty$ and $p^*(t) = p = 0,9$. Then

$$K_g^*(t) = K_g p^*(t).$$

In case of failure of one of GI DG system splits into two independent subsystems, each of which can provide only half the load of consumers. The efficiency coefficient of one GI is defined as

$$e_{gi}(t) = 0,5K_g p = 0,5 \cdot 0,95 \cdot 0,9 = 0,4275.$$

Therefore, the considered system of DG provides the efficiency coefficient equal to $E_\Sigma = 2e_{gi} = 0,855$.

2). ***Point-by-point estimation of operation efficiency of DG system in case of the simplest GI.*** The power supply system (DG) consists of two identical power plants (GI): *Sa* and *Sb*, providing power to consumers with the maximum power consumption $P_\Sigma = 36$ MW. The *Sa* station operates in the power output mode in the range from $P = 0$ to $P = 22$ MW, and the *Sb* station – from $P = 14$ to $P = 36$ MW. The power distribution zones of these stations are shown in Fig. 3.

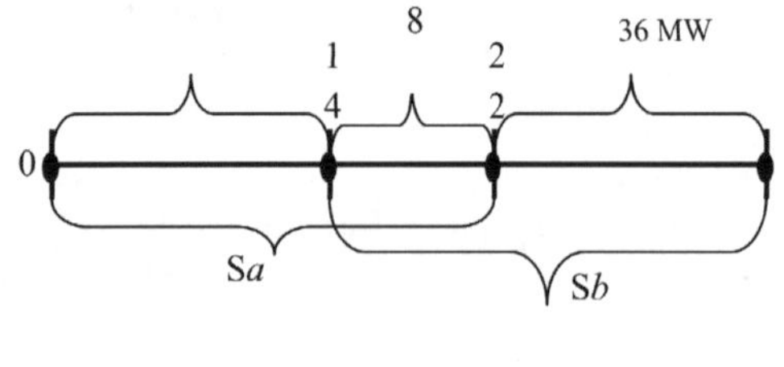

Fig. 3. Zones of power distribution of GI.

It is required to find the probability of providing the load of consumers appearing at any time in the range $0 \leq P \leq 36$ MW, if the reliability of providing the necessary power by these stations at this moment: $R_a = R_b = 0,95$. Thus probability of reliable power supply of the consumers which are in a zone of action of each of GI – $R_1 = R_2 = 0,9$, and in zone of simultaneous action of both stations (in a zone of overlapping)

$$\mathcal{P} = 1 - (1 - R_1)^2 = 0,99.$$

Suppose that the load of consumers at given time is P_0 MW. The probability that the DG system will be able to provide it at this point in time

$$\mathcal{P}_0 = R_a \cdot R_b = 0,95 \cdot 0,95 \approx 0,9.$$

At the interval from $P = 14$ to $P = 22$ MW (overlap zone – $P_п = 8$ MW), the load of the consumer is provided by both stations, and within zones (from $P = 0$ to $P = 14$ and from $P = 22$ to $P = 36$ MW) – by one GI.

The efficiency index E_0 of DG system state at given load P_0MW is determined by weighted average

$$E_0 = \frac{P_{S_a \cap S_b}}{P_\Sigma}\mathcal{P} + \frac{P_{S_a-(S_a \cap S_b)} + P_{S_b-(S_a \cap S_b)}}{P_\Sigma}\mathcal{P}_0$$
$$= \frac{8}{36} \cdot 0,99 + \frac{14 + 14}{36} \cdot 0,9 = 0,92.$$

Note that "weight" here is states of DG system, and not its separate elements (GI). In the same DG system, "weight" of an element essentially depends on the state for which it is considered. If the failure of *i*-element occurred for all working elements, it is obvious that it is accompanied by consequences different from the case when this *i*-element failed after the failure of one *j* $(i \neq j)$or several elements of this system. However, for DG systems, in many cases, an element is more important than another in one state of the system proves to be more important in another state of the system.

The probability of system state when the load limits $0 < P_a < 22$ MW has one station *Sa*, and in the range $22 < P < 36$ MW – none,

$$\mathcal{P}_a = R_a(1 - R_a) = 0,95 \cdot 0,05) \approx 0,05.$$

The efficiency index E_a of DG system state at a load of P_a MW will be determined as

$$E_a = \frac{22}{36} \cdot 0,9 + \frac{14}{36} \cdot 0 = 0,55.$$

Since in given example the states of *Sa* and *Sa* are absolutely identical, so we get the resulting efficiency of operation of analyzed DG system:

$$E_\Sigma = \mathcal{P}_0 E_0 + 2\mathcal{P}_a E_a = 0,9 \cdot 0,92 + 2 \cdot 0,05 \cdot 0,55 = 0,883.$$

3). ***Estimation of technical efficiency of DG system at a certain time interval.*** The power supply system (GI) consists of two identical power plants (GI). In case of operability of both stations the load provided by them is P MW. In case of failure of one of them, the value of provided load is reduced to the value of $0,3P$. In case of failure of two stations, the power supply of consumers is disrupted completely. The failures of GI are independent events. The probability of failure-free operation of each GI obeys the exponential distribution law

$$p(t) = e^{-\frac{t}{T}} = e^{-\lambda t}.$$

Considering DG systems of this type it is necessary to take into account that the GI will not give the required power to consumers when stopped as a result of preventive maintenance, as well as the impact of any external factors (weather conditions, if the composition of the GI is determined by RES). However, in order to simplify the problem, we assume that the distribution law remains exponential.

If we take the duration of the calculation period $\tau = t$ equal to the average uptime of each GI $t = T = 1$ in a year, the probability of uptime GI will be

$$p = e^{-\frac{t}{T}} = e^{-1} = 0{,}368.$$

Indicators of technical efficiency of DG system states are determined by multiplication of transmitted power on time of operation:

1) technical efficiency of DG system operation when both stations are operable during considered period τ

$$E_0 = P\tau$$

2) technical efficiency of DG system operation, when the first GI failed at the moment of t_i

$$E_i(t_i) = 0{,}3P\tau + 0{,}7Pt_i \quad t_i < \tau; \quad i = a, b;$$

3) technical efficiency of DG system operation, when the first GI failed at the moment of t_i, and the second one – at the moment of t_j

$$E_{ij}(t_i, t_j) = 0{,}3Pt_j + 0{,}7Pt_i \quad t_i < t_j < \tau,$$

The resulting estimation of efficiency of analyzed DG system according to the theory set in [4, 5] is got according to formula:

$$E = p^2P\tau + 2p\{0{,}3P\tau(1-p) + 0{,}7P\tau[1 - p(1+\lambda\tau)]\} +$$
$$+2\{0{,}3P\tau[0{,}75 - p(1+\lambda\tau) + 0{,}25p^2(1+2\lambda\tau)] +$$
$$+0{,}7P\tau[0{,}25 - p + (0{,}75 + 0{,}5\lambda\tau)p^2]\} =$$
$$= 0{,}368^2P\tau + 2\cdot 0{,}368\{0{,}3P\tau(1-0{,}368) + 0{,}7P\tau[1 - 0{,}368(1+1)]\} +$$
$$+2\{0{,}3P\tau[0{,}75 - 0{,}368(1+1) + 0{,}25 \cdot 0{,}368^2(1+2)] +$$
$$+0{,}7P\tau[0{,}25 - 0{,}368 + (0{,}75 + 0{,}5)0{,}368^2]\} =$$
$$= (0{,}135 + 0{,}28 + 0{,}138)P\tau = 0{,}553P\tau.$$

The analyzed DG system can be considered as a system with additive efficiency indicators [4, 5], since each GI of this system contributes its independent share of c_k to the overall output effect. Then the effectiveness of system will be determined according to following expressions

$$E_0 = P\tau\mathcal{P}_0 = P\tau[1 - (1-p)^2] = P\tau[1 - (1-0{,}368)^2] = 0{,}6P\tau;$$

$$E_i(t_i) = 0{,}3P\tau\mathcal{P}_0 + 0{,}7Pt_i(1 - e^{-\frac{t_i}{\tau}});$$

$$E_{ij}(t_i, t_j) = 0{,}3Pt_j\left(1 - e^{-\frac{t_j}{\tau}}\right) + 0{,}7Pt_i\left(1 - e^{-\frac{t_i}{\tau}}\right);$$

$$E(t) = \sum_{k=1}^{n} E_k c_k = E_0 c_0 + E_i(t_i)c_i + E_{i,j}(t_{i,j})c_{i,j}, \quad \sum_{k=1}^{n} c_k = 1,$$

where c_k is the coefficient of ratio participation of each GI operating in DG system.

4 Conclusion

Despite the fact that the reliability of the elements is increasing, the rate of this growth lags behind the rate of increase in the complexity of the system. The nature of functions performed by DG system requires its high reliability, but a large number of elements, their links and modes of operation can adversely affect the efficiency of the system as a whole.

The absence of practical methods for estimation of effectiveness of DG systems, characterized by a variable structure that can be changed at random and intersecting GI zones of operation, is extremely relevant and requires solving these problems, meeting the requirements of competitiveness in the world market.

The effectiveness of operation of DG system must be assessed by a specialist familiar with the system as a whole, who knows the requirements for it, its structure, modes, methods and means of control and protection.

References

1. B.V. Papkov, A.L. Kulikov *Theory Systems and System Analysis for electric power engineers* (Publishing house Urait, 2016)
2. I.A. Ushakov *Course of the theory of reliability of systems* (Dropha, 2008.)
3. K. Rainshke, I.A. Ushakov *Reliability evaluation of systems using graphs*(Radio и communication, 1988)
4. I.A. Ushakov Effi. of fun. of com. Sys. *About reliability of complex technical systems* (Soviet radio, 1966)
5. I.A. Ushakov *Research methods of efficiency of functioning of technical systems* (Publ.house Zhnanie, 1976)
6. A.L. Kulikov, V.L. Osokin, B.V. Papkov Prob. and pec. of dist. elec., NGIEI, **11 (90)** (2018)
7. B.V. Papkov Vuln. and resil. of elec. facil. *Research and ensuring the reliability of energy systems,* **68** (Irkutsk: ISEM SO RAN, 2017)
8. E.V. Dzirkal *Specifying and verifying reliability requirements for complex products* (Radio и communication, 1981)
9. V.V. Plyamovatyi To the theor of "surv" of comp. syst. *Main Issues of Theory and Practice of Reliability* (Soviet radio, 1980)
10. Yu.E. Gurevich, P.V. Ilyushin *Features of calculations of modes in energy district with distributed generation Nizhny Novgorod* (Nizhny Novgorod Institute of Management, Branch of Ranepa, 2018)
11. P.V. Ilyushin, A.L. Kulikov *Automatic control of normal and emergency modes of energy districts with distributed generation* (Nizhny Novgorod Institute of Management, Branch of Ranepa, 2019)

https://doi.org/10.1051/e3sconf/201913901033

Method of calculation of system reliability on the basis of construction of the logical function of the system

Dilshod Okhunov[1*], *Mamatjon* Okhunov[2], *Mukaddas* Akbarova[2]

[1]Fergana Branch of the Tashkent State University of Information Technologies named after Muhammad al-Xorazmiy. Fergana, Uzbekistan
[2]Fergana polytechnical institute. Fergana, Uzbekistan

Abstract. The article considers methods of calculating the reliability of systems using the logical functions of the system on the basis of the structural scheme of the system. And also the appliance of this method to calculate the probability of hacking the system based on the logical function of it.

1. Introduction

As it's known, the reliability of the system is the ability of it to perform a set of specified algorithms for processing information with a specified reliability or the property of the system to maintain its own operability at some point in time in the specified operating conditions.

Hacking system, unlawful penetration into it, in the vast majority of cases over time should lead from a low-noticeable to an impressive increase in the number of failures in the system, due to (even minor) fluctuations in the operating modes of the system [1].

This assertion is quite obvious if illegal penetration is carried out, for example, using special electromagnetic channels, which immediately leads to the passage beyond the permissible limits of such parameters as the current strength at the penetration node or voltage that are included in the number of necessarily controlled parameters of the system. It is well known that a high level of reliability (0.99999) requires absolutely accurate finding of any controlled parameter within a certain tolerance, which is controlled by the object management system [2,3,4]. Failure is a fundamental concept of reliability theory - a sudden or gradual (metrological) hacking into a system will necessarily react to this by increasing the failure rate λ, equal to the average number of failures per unit time. But the reliability of the system ($P= e^{-\lambda t}$), as can be seen, with increasing λ, will decrease.

2. Methodology

It is known, that calculating the reliability of a system has been used very successfully for a long time, based on the construction of a structural diagram of the system (block diagram) appropriate to the system itself [5,6,7]. According to this, the so-called logical function of working capacity corresponding to this scheme is calculated, from which, after a number of mathematical operations (by using logical symbols ∧, ∨, -), the reliability of the system is calculated.

The heart of our calculation methodology is the representation of our system by a block diagram, but as a parameter of each block, we will not consider the value of its reliability($P_{b/o}$), we will analyze the probability of hacking this block(P_{break}), which is naturally calculated in advance. The resulting logical "hacking" function is processed completely analogously to the logical function of working ability, widely used in reliability theory.

We will write down an algorithm of sequence of operations for calculating the reliability of a complex cybernetic system:

1. To formulate verbally the condition of the operability of the system, product, device.
2. Based on the verbal formulation, write down the logical function of working capacity.
3. Using a set of allowed postulates that correspond to the algebra of logic, try to convert, simplify the logical function F, bring it to a form that can not be simplified, or, in other words, minimized.
4. In F replace logical operations with their corresponding arithmetic, connected by signs «+,-, * ...».
5. In the arithmetic function FA, replace the simple events, connected by arithmetic signs, with their probabilities.
6. Using the formula for FA, calculate the probability of working capacity interests us, by replacing the corresponding probabilities with their numerical estimates.

This table of postulates is used in the calculation of complex systems.

1. $a \vee b = b \vee a$
2. $ab = ba$
3. $a \vee (b \vee c) = (a \vee b) \vee c$
4. $a \vee (b \wedge c) = (a \vee b)(a \vee c)$
5. $a(b \vee c) = ab \vee ac$
6. $a \vee \bar{a} = 1; a \vee 0 = a, a \wedge 0 = 0$
7. $a * 1 = a, a \vee 1 = 1$ *the absorption law*
8. $a * \bar{a} = 0$
9. $a \vee \bar{a}b = a \vee b$
10. $a \vee a = a$
11. $a * a = a$
12. $\overline{(a \vee b)} = \bar{a}\bar{b}$
13. $\overline{(ab)} = \bar{a} \vee \bar{b}$
14. $ab \vee a\bar{b} = a$
15. $a(a \vee b) = a$
16. $F_l(a, b, c, \ldots k) = aF_{l1}(1, b, c, \ldots k) \vee \bar{a}F_{l2}(0, b, c, \ldots k)$

Each postulate can be associated with a certain structural scheme (and vice versa), for example:

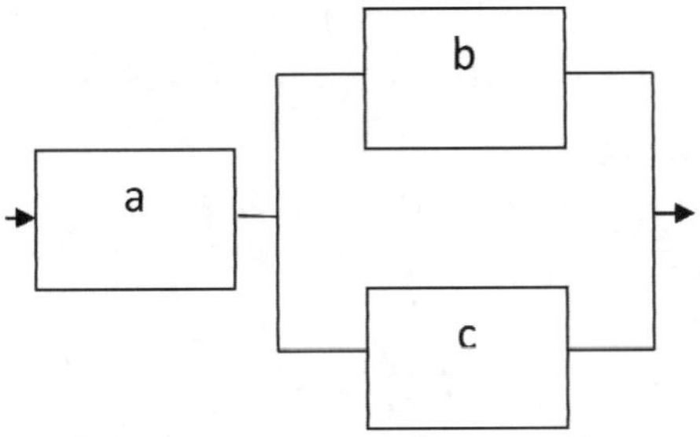

Fig. 1. Structural scheme example

In this case, postulate № 5 $a(b \vee c) = ab \vee ac$ describes the condition of its operability.

It is clear that the system is operable, that means it transmits information, if a and b, a and c or all three elements are operable, that's why this logical expression can be put into correspondence with this logical function:

$$F_Л = ab \vee ac \vee abc = ab \vee ac = a(b \vee c)$$

We have reduced logical function F_l to an elementary, minimal form, from which we can make the transition to the arithmetic function F_A, using the standard set of arithmetic postulates:

$$a \vee b = a + b - ab$$
$$a \wedge b = ab$$
$$\bar{a} = 1 - a$$

3. Experimental results

We need to define the probability of an operational state of the system if it is known that the reliability of the elements "a" and "b" are: P_a =0,8 and P_b =0,8.

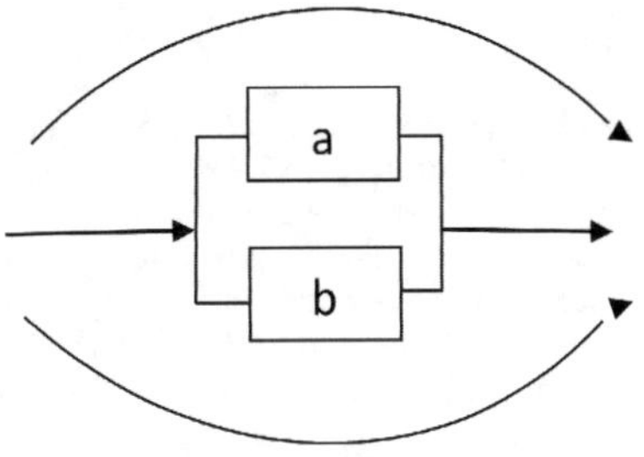

Fig. 2. The structural scheme at P_a =0,8 and P_b =0,8.

The first way to solve:

$$F_l = a \vee b \vee ab$$
$$P_l = a\{1 \vee b \vee 1 * b\} \vee \bar{a}\{0 \vee b \vee 0 * b\} =>$$
$$=> a * 1 \vee \bar{a} * b => a \vee ab => a \vee b$$

Pass to the arithmetic logical function:

$$F_a = a + b - ab = P_a + P_b - P_a * P_b =$$
$$= 0{,}8 + 0{,}8 - 0{,}8 * 0{,}8 = 1{,}6 - 0{,}64 = 0{,}96$$

Reliability (probability of failure-free operation) is 0.96

The second way to solve:

$$F_l = a \vee b \vee ab => a(1 \vee b) \vee b(1 \vee a) => a * 1 \vee b * 1 => a \vee b$$

Passing to the arithmetic logical function F_a, we obtain

$$P_{b/o} = P_a + P_b - P_a * P_b \cong 0{,}96$$

We calculate the reliability of the following scheme, if $P_a = P_b = P_c = 0{,}8$.

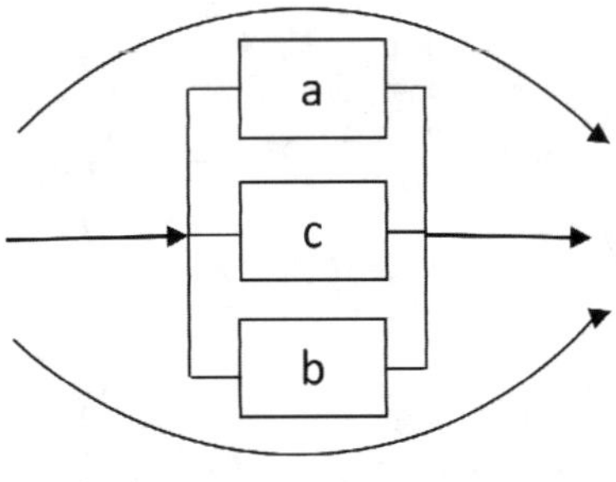

Fig. 3. The structural scheme at $P_a = P_b = P_c = 0{,}8$

It is easy to see that

$F_l = a \vee b \vee c \vee ab \vee ac \vee bc \vee abc$

Using the postulate 16:

$F_l(a, b, c) = aF_1(1, b, c) \vee \bar{a}F_2(0, b, c)$, *we get:*

$F_l(a) = a * 1 \vee \bar{a}(b \vee c) => a \vee b \vee c$

We have obtained an elementary, primitive formula for a logical function. Turning to an arithmetic logic function

$$F_a = a + b + c - ab - bc - ac + abc$$

substituting $a \to P_a = 0{,}8, b \to P_b = 0{,}8, c \to P_c = 0{,}8$, arrive

$$P_{b/o} = 0{,}992.$$

 https://doi.org/10.1051/e3sconf/201913901033

4. Conclusion

Obviously, this principle of calculating reliability can be applied to the calculation of the probability of hacking of a certain system, if this system is represented as some structural scheme containing its channels of unauthorized access [8]. Consider the problem of calculating the probability of hacking the system based on the logical function of the system. This task is entirely in the class of tasks to model the threats of impact on all sorts of information sources: modeling of ways of penetration to the place of accumulation of information (for example, the head office); simulation of optical leakage channels (windows, doors, television embedded device); simulation of acoustic leakage channels; simulation of electronic leakage channels. This list can be continued further. However, the most diverse options are available for both the type of leakage channel (optical, electronic, etc.) and (in the general case of a random number) of simultaneously cracked channels [9]. If we try to clarify the probability of hacking the entire system, then this approach (from "private to general") is difficult to do, because the model of the entire system in these methods of hacking is absent from the outset, and the entire problem of estimating the probability of breaking (as well as the probability of hacking individual channels) to the class of weakly formable problems [10, 11].

However, if we look at the estimation of the probability of hacking the system, remaining within the framework of the system approach considered above, which can be implemented as some structural scheme of the system (where separate channels of illegal intrusion into the system act as separate blocks), then the problem of a general, the whole system can be solved with varying degrees of accuracy, depending on the correctness of the block diagram itself, composed of channels of hacking. This method consists in using a standard structural scheme of the system and compiling a logical function of operability, with its further standard processing. In this case, the reliability of each block of the structural diagram must be replaced with the probability of unauthorized penetration through this block (Pi_{break}). Further it is necessary to follow strictly on items (1-6) of the algorithm given above.

In conclusion, it should be noted that using this technique for calculating the logical function of the structural scheme, it is possible to calculate the reliability of more complex structural schemes - for example, schemes of a logical bridge, a triangle, etc.

References

1. Barsukov. V.S. *Safety: technologies, means, services.* Moscov, Kudishch-image Publ., (2001).
2. A.M. Polovko. *The basis of the theory of reliability.* - Moscov, “Nauka” Publ., (1964).
3. Ryabinin I.A. *Reliability and safety of structurally-difficult systems.* SPb.: publishing house of the St.-Petersburg university, (2007).
4. Ryabinin I. A. *Reliability of Engineering Systems. Principles and Analysis.* - Moscov, Mir Publ., (1976).
5. Polovko A.M., Gurov S.V. *Of the basis of the theory of reliability.* - SPb.:BHV- Petersburg, (2006).
6. Polovko A.M., Gurov S.V. *Of the Basis of the theory of reliability.* A practical work. - SPb.:BHV-Petersburg, (2006).
7. Ryabinin I. A. *Logiko-likelihood methods of research of reliability of structurally-difficult systems.* Moscov, (1981).
8. Sovetov B.J. Yakovlev S.A. *Modelling of systems.* Moscov, Higher school Publ., 1989.
9. Titarenko G. A, Braga V.V., Vdovenko L.A. *Information technologies of management.* Moscov, UNITI Pabl., (2005).
10. Grinberg A.S., Gorbachev H.H, Teplykova A.A *Protection of information resources of the government.* Moscov, UNITI Pabl., (2003).
11. Grinberg A.S., Gorbachev H.H, Bondarenko A.S. *Information of technology of management.* Moscov, UNITI Pabl., (2004).

https://doi.org/10.1051/e3sconf/201913901034

Device for conversion of equalizing current at the site of the traction ac network

Abdurauf Safarov[1], *Khurshid* Sattarov[2,3], and *Sarvar* Jumaboyev[1]

[1]Tashkent institute of railway engineering, Department of Power Supply of Railways, Adilkhodjayev – 1, Tashkent, 100167, Uzbekistan
[2]Tashkent university of information technologies, Department of Power Supply Systems, A.Timur, 108, Tashkent, 100200, Uzbekistan
[3]Tashkent state technical university, Department of Power Supply, Universitet, 2, Tashkent, 100097, Uzbekistan

Abstract. The article proposes the design of a device for converting the equalizing current using a compensating winding, which meets the requirements for accuracy and reliability on the part of various automated control systems, control and measurement of currents.

Introduction

In connection with the aggravation of the problem of saving fuel and energy resources due to the constant increase in prices for them, the issues of further introduction of modern energy-efficient and energy-saving technologies are of particular importance [1].

In particular, one of the main problems in railway transport is the problem of reducing the losses of electric energy in the traction power supply system of electrified Railways, which are the largest consumers of electricity in the industry.

The formation of the indicator "loss of electricity" begins with the energy consumption for traction of trains received by the network (meters of traction substations on the primary side of traction transformer substation) and released to the consumers (counters of the electric rolling stock). Thus actual values of losses of the electric power which size depends on reliability of the account of the electric power consumption are defined. For example, in [2] given the accounting data on the electricity consumption for traction by results of work on the Railways of Russia.

In accordance with these reporting data, the range of changes in the relative unbalance of electricity reception for traction trains by meters of traction substations and consumption by meters of electric rolling stock was 2.1...27.9%. The difference between the actual and calculated values of losses of electric energy is caused by unsatisfactory technical and organizational condition of systems of the account of electric energy on traction substations and the electric moving structure.

Statement of a problem

Determination of structure and actual losses of the electric power, to allocate the factors causing these losses for the purpose of development and the organization of actions for reduction of losses of electric energy in a traction network.

The concept of the problem decision

An electrified AC railway is an asymmetrical non-linear consumer with an uneven load. It is an extended receiver of electric energy, and its traction substations can't be powered from one point of connection to the external power supply system [3,4].

Traction substations are powered by various nodes of one or more power systems. The voltages at the points of connection of traction substations to power systems are different in modulus and phase, which causes the flow of power flows in the traction power supply system between adjacent substations even in the absence of traction load. Connection of systems of external and traction power supply on power transit is accepted to estimate an equalizing current in a traction network [5].

Equalizing current is the current determined by the parameters and mode of operation of external and traction power supply systems. The traction network between two adjacent traction substations in its electrical essence is a small resistance current line connected in parallel to the power line of the external power supply system at long distances. In accordance with Kirchhoff's first law, part of the load current flowing through the power lines of the external power supply system branches off into the traction network (a system of parallel wires). For the traction network, this is a harmful current that causes additional losses of electrical energy, heating of wires, reducing the capacity of trains in this inter-substation zone. The presence of transit of electrical energy in the traction network complicates the system of accounting of electricity for traction of trains.

On Railways seek to reduce the value of the equalizing current. Therefore, the problem of its determination for the purpose of subsequent minimization is acute [6].

Losses of electric energy in the traction network, caused by equalizing currents depend on power supply and partitioning merpttazinol areas and are determined by the nature of the traction load, parameters, elements and modes of operation of power lines.

Under operating conditions, these factors change, and therefore it is important to be able to quickly assess the loss of electrical energy in the traction network for various power supply circuits of the inter-substation zone and choose a power supply scheme to minimize energy losses.

Equalizing currents cause additional losses of electricity in the contact network regardless of whether there are trains on the inter-station zone or not. In the absence of a load on the zone, equalizing currents can flow, causing energy losses in the contact network, which are recorded by the meters of substations, but are not measured by the meters of electric locomotives. Therefore, the more time there are no trains in the zone, the greater will be the difference in the readings of the energy meters of substations and electric locomotives.

To estimate this difference, in addition to the value of the equalizing current, it is necessary to know the dimensions of the train movement. Disadvantages of the mode of operation of the traction power supply system of electrified AC Railways are amplified by the imperfect system of accounting for electric energy at traction substations and electric rolling stock.

A significant drawback in the accounting of electricity for traction trains at traction substations is the lack of accounting for electricity on the feeders of the contact network, which prevents the study of the balance of received and distributed electric energy [7,8].

The lack of control of the balance of received and distributed electricity for the traction of trains occurs for the following reasons:

- the absence of metering devices on the feeders of the contact network does not allow to control the distribution of electricity on the supply arms;
–not defined zone with an increased energy consumption for traction of trains with the aim of developing joint solutions services E and T to reduce specific energy consumption for traction of trains substation in these areas.

The electric power accounting on the electric rolling stock of alternating current is carried out now by counters of the electric power spent on traction of trains and returned at regenerative braking. They make it possible to determine the specific cost of electricity for the transportation of goods in operation.

Since cash settlement currently on the counters of electric locomotives do not produce, so today, the energy metering on the electric rolling stock is made by induction and electronic electricity meters with lower accuracy class of energy measurement.

Due to the absence of electric voltage transformers, voltage all meters are powered from winding their own needs at the secondary side of traction transformer locomotive, and the circuit from the primary winding of the traction transformer locomotive. However, there is no proportional relationship between changes in the voltage in the contact network and the voltage on the winding of the transformer's own needs.

Therefore, to estimate the specific energy consumption for traction trains for different AC locomotives have to take the average transformation coefficients. For example, switching at an angle of 40...45° rectifier electric locomotive caused by a large current load of the engine, and because of voltage losses in the dispersion reactance of the secondary windings of traction transformer voltage drop in the winding their own needs differs from the reduction of voltage on the pantograph of the electric locomotive 10 to 20% in magnitude and 3...4° in phase [7].

As a result, the error of electricity metering on the electric rolling stock is not less than 10 %. These errors of the account of the electric power entail emergence of a certain share of conditional losses in system of traction power supply.

Thus, today there is a need to develop a comprehensive system of scientific, technical and organizational measures in the form of an automated system of control and accounting of electricity (ASCAE), taking into account the above factors. This will identify areas with increased losses of electrical energy, develop measures to reduce them, reduce the total energy consumption in the traction power supply system and improve the technical and economic performance of the traction power supply system. For a complete picture of the electric energy consumption for all elements of the traction power supply system, it is necessary to take into account the electric power consumption on the feeders of the contact network and on the electric rolling stock of the phase [5,9].

In addition to continuous monitoring and accounting of electricity, the introduction of ASCAE in the traction power supply system will allow to assess the status of power supply devices, to conduct current control of electricity losses and quality indicators of the received electricity, as well as to identify irrational electricity consumption.

The creation and implementation of such a system on electric rolling stock will allow:

- to control reliability of the account of electric energy on counters of the electric rolling stock and receiving information after processing of route sheets of drivers;
- to provide technical and organizational improvement of the account of the electric power on the electric moving structure;
- control the consumption of electric power by electric locomotives from neighboring roads.

The main objectives for the implementation of ASCAE on electric rolling stock are:

- development of technical solutions for its implementation and on its basis the organization of current control of electricity consumption by electric locomotives on the traction of trains;
- compliance with the legal provision of technical means of ASCAE, which will be possible with the use of a current measuring Converter both on the 25 kV side and on the 10 kV side of an electric locomotive traction transformer and modern electricity metering devices;

- implementation by means of technical means of binding of power consumption of the electric rolling stock to the served zone (owing to the fact that the electric locomotive is the mobile unit and moves on a site of the railway).

In this paper, the authors propose a design of the device to convert the surge current to meet the requirements of AMR both at 25 kV and 10 kV traction transformer locomotive.

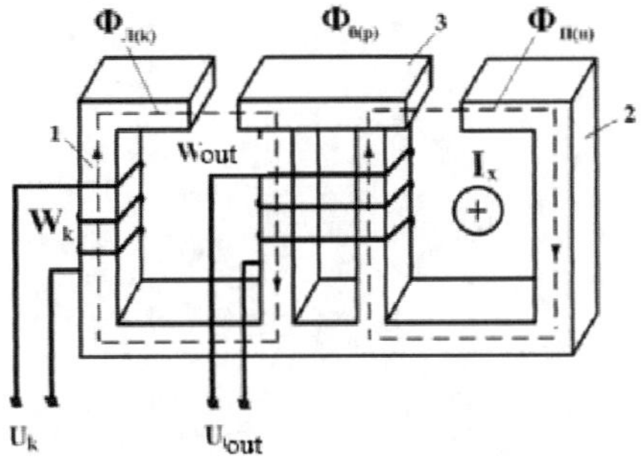

Fig.1. The design of the device to convert the compensating current

The device for converting the equalizing current (Fig.1) is built on the basis of the design [10] for the case of transformation of alternating current and consists of the extreme rods 1 and 2, the magnetic core rod with a cutout 3, the conductive bus c measured current I_x, a compensating winding wound W_k on the core 1, an output winding W_{out}, encompassing a Central stem with the neck 3.

The principle of operation of the Converter is as follows. When the measuring current I_x flows through the bus, as well as when the compensating winding W_k is connected to an alternating voltage source, a compensating current is formed (Fig. not shown), and in rods 1 and 2 there are magnetic currents proportional to the corresponding currents.

With the same direction of currents in the tires, the magnetic flows $\Phi_{L(k)}$ and $\Phi_{R(i)}$ are closed along the extreme rods 1 and 2 and through the Central rod 3 in opposite directions. The output winding W_{out} is permeated by magnetic flows $\Phi_{L(k)}$ and $\Phi_{R(i)}$ counter direction. If the bus currents are not equal to each other, then the resulting flow $\Phi_{0(p)}$ is the flow difference, $\Phi_{L(k)}$ and $\Phi_{R(i)}$, and the corresponding voltage at the output of the output winding is determined by the derivative of the flow difference $\Phi_{L(k)}$ and $\Phi_{R(i)}$ in time:

$$U_{\text{out}} = -W_{\text{out}} \frac{d(\Phi_{\text{L(k)}} - \Phi_{\text{R(i)}})}{dt} = -W_{\text{out}} \frac{d\Phi_{0(\text{p})}}{dt}$$

where is W_{out} - the number of turns of the output winding.

Conclusion

Thus, the device for converting the equalizing current using the compensating winding allows you to convert current to AC voltage with high accuracy and reliability, which expands the scope of its application in various automated control systems, control and measurement of currents.

References

1. O merax po dalneysimu vnedreniyu sovremennix energoeffektivnix I energosberegayushix texnologiy. Postonovlenie Prezidenta Respubliki Uzbekistan ot 23.08.2017g. № PP-3238.
2. Molin N.I. Potentsial energosberejeniya v sisteme tyagovogo elektrosnabjeniya peremennogo toka jeleznix dorog// Materiali Vserossiyskoy nauchno-texnicheskoy konferensi s mejdunarodnim uchastiem: V 2t. T.1/ Otv. red. V.P. Surov. – Krasnoyarsk: Grotesk, 2005g., s. 39-46.
3. Markvardt K.G. Elektrosnabjenie elektrifistirovannix jeleznix dorog. M.: Transport, 1982. 528 s.
4. Tuygunova A.G., Bardushko V.D. Analiz uravnitelnix tokov potyagovim setyam na osnove rschetno-eksperimentalnix dannix ob urovnyax naptyajeniya i ix fazax v uzlax podklyucheniya tyagovix podatantsiy. IrGUPS, Sovremennie texnologii. Sistemniy analiz. Modelirovaniya. №2 (50), 2016.
5. Kashtanov A.S. Povishenie effektivnosti kontrolya elektropotrebleniya i usloviy soglasovaniya system vneshnego i tyagovogo elektrosnabjeniya po dannim ASKUE. Diss. na soisk. kand. tex. nauk, - Omsk, 2007g., s.164.
6. Osipov V.A. Zakonomernosti protekaniya uravniteknix tokov v elektrotyagovoy seti i metodi ix minimizatsii. Diss. na soisk. kand. tex. nauk, - Rostov-na-Donu – 2000g., s.174.
7. Chermisin V.T. Sistema meropriyatiy po snijeniyu poter elektricheskoy energii v tyagovoy seti ot uravniteknix tokov/ V.T. Chermisin, V.A. Kvashuk// Jeleznodorojniy transport TsNIITEI MPS, – 2000. – Vip. 3., s. 9–21.
8. Cheremisin V.T., Kvashuk V.A. Kontrol uravnitelnix tokov, metodika otsenki poter v tyagvoy seti peremennogo toka i meropriyatiya po ix snijeniyu//Sb. tr. Vtorogo Mejdunar. sipoziuma ENSB–KE–EMS–EME–2000. M.: MIIT, 2000.
9. Chermisin V.T. Texniko-ekonomicheskiy aspect vnedreniya ASKUE na jeleznodorojnom transporte/ V.T. Cheremisin, M.M. Nikiforov, A.L. Kashtanov// Jeleznodorojniy transport. Moskva, 2007, №3, s. 51-53.
10. A.s. №1019503 (SU). Ustroystvo dlya preobrazovaniya postoyannogo toka v peremenniy/ Zaripov M.F., Axrorov N.A., Safarov A.M., Petrova I.Yu.// Opublikovano "Byulleten izobreteniy"-1983, №19.

 https://doi.org/10.1051/e3sconf/201913901035

Automated control of distributed generation enabled power districts

Pavel Ilyushin[1*#]

[1]Nizhny Novgorod State Technical University n.a. R.E. Alekseev, Minin str., 24, Nizhny Novgorod, Russia

Abstract. The paper presents the results of integrating distributed generation (DG) facilities in the distribution grids and internal grids of industrial enterprises. It proves such locations require many more additional relay protection and automation devices, including emergency controls; the cost of such systems is on par with that of a DG facility. The paper presents the prerequisites for equipping DG-enabled power districts with routine and emergency controls (REC). It substantiates the use of phasor measurement units as REC sensors in low- and medium-voltage grids. Presented herein are the fundamental principles behind, and functionality of, REC. The paper proves feasible using offline-computed parameters to implement a REC system for further transition to online software-based computing of steady states and transients.

1 Introduction

Over the past decade, advancements in distributed generation were mainly based on commissioning DG facilities based on gas-turbine, gas-piston, and diesel generator sets (GS).

Russia plans to launch 5,278.3 MW of installed capacity provided by solar and wind farms by 2024; this will require novel approaches to parametric control [1].

Integrating too many DG facilities in a low- or medium-voltage grid will cause the possible operating situations to become too diverse, making visual identification and manual control of their parameters impossible while also complicating parametric control due to significantly higher dimensionality.

Despite the experience of implementing multiple effectively running DG facilities that operate as designed, negative experience is also quite substantial. It is associated with the complex technical problems that emerge in operation and have financial implications for the owners.

To enable a DG facility to efficiently run in parallel with the energy system, relay protection and automation (RPA) in the adjacent grid must be redesigned when designing the power-delivery diagram. A need may arise to use too numerous additional RPA devices. Such project may require replacing switching units (SU) and instrument transformers, as well as installing voltage cabinets, ACRs, emergency event recorders, communication lines, etc. Such extras may run as high as the cost of the DG facility itself, making its construction cost-ineffective.

A REC can control the parameters of the internal grids in DG-enabled power districts whether running in parallel with the energy system or islanded, thus requiring less extra systems for more cost-effective operation.

2 Prerequisites for Implementation power district REC systems

2.1 DG-enabled power districts have quite specific operating parameters

What makes them specific is the low mechanical constants of GS inertia as well as the slow loading of ICE-based GS. This results in significant deviations in the operating parameters when exposed to external disturbances, including loading/load-shedding, while emergencies progress faster. Proper REC operation requires fast yet accurate measurements of operating parameters [2-3].

In islanded operation, such power districts may exhibit significant deviations in electricity quality from the rated values due to large-scale use of power electronics, non-linear loads, and stochastic output from renewable-energy facilities. Operating parameters must be measured accurately and reliably in the context of deviating quality and various distortions.

2.2 Algorithms of energy automation units are not suitable for DG-enabled power districts

Connecting DG GS alters the operating situation of the grid and affects the running automation units. Such units are required to run per the algorithm and to trigger when encountering a specified setpoint; faults are not

https://doi.org/10.1051/e3sconf/201913901035

acceptable, neither are false-positives. However, the effects of a connected DG GS may hinder the ability to meet these requirements.

Automations in operation are usually not capable of recognizing the location-specific operating parameters, not to mention any algorithmic adaptation to such parameters or reconfiguring their own trigger setpoints to tailor them to the actual equipment conditions. REC must use enhanced linear and sectional automatic transfer switches (ATS) [4], under-frequency load shedding (UFLS) [5], pole slip protection (PSP) [6], automatic under-voltage protection (AUVP) [7], and automatic overload protection (AOP) [8] algorithms.

2.3 DG-enabled power districts must be islanded to supply power to consumers

Chances are high of a system-wide accident that might cause outage of DG GS or make necessary islanding the critical consumers. Better islanding systems (IS) are required to prevent significant damage.

Multi-parameter islanding system (MIS) algorithms used by REC must enable islanding a DG-enabled power district from the energy system in case of an accident, which poses high risk of inadequate power supply, to prevent GS outage, critical consumer disconnections, or voltage collapse. Location-specific MIS algorithm and configuration (starters and stoppers, interventions, their scope and points) must be specifically computed to match the operating parameters of the external grid, power district, and load [9].

2.4 Centralized DG control algorithms to synchronize the power district with the energy system

One of REC functions is to synchronize an islanded DG-enabled power district with the energy system. Automatic synchronization enables the power district to restore normal external and internal power supply after an emergency is contained and the operating parameters are normalized. Solving this problem requires special switching unit controls to be in place in the internal and external grids as well as special parametric controls for all DG GS that can adjust the parameters of, or reconfigure the automatic frequency and excitation controls (AFC, AEC).

2.5 Implementation of basic in-district REC algorithms, no communication with the energy system automations

Aside from local REC algorithms such as UFLS and LS that are run by single units, a power district can use coordinating algorithms (ATS, PSP, AUVP, and AOP), or centralized algorithms (group active/reactive power controls, or GAPC/GRPC) that do not need to interact with the outer automations.

The coordinating and the centralized REC algorithms require a communication network capable of sufficiently fast data transmission. Given that DG-enabled power districts are usually quite small (a few dozens square kilometers at max), the above-mentioned REC algorithms can be implemented to have sufficient performance.

The today's trend is to simultaneously implement local automatic control systems (ACS) within industrial power districts. Barely coordinated U-based controls by visual assessment of a limited set of local parameters does not seem efficient, while non-coordinated effects of local ACS aggrieves the consequences of emergency disturbances. REC uses coordinated U-based controls within district grids, which stabilizes ACS operations in any operating situation.

2.6 PMU as basic REC sensors

Cutting-edge communications and PMUs open up ample opportunities for REC design by providing reliable phasor measurements in the context of electromagnetic (electromechanical) transients, which requires faster data transmission and greater phasor measurement range in case of emergency. Current and voltage phasors for a connection can be used to calculate the required parameters; if measurements are taken for all connections, such phasors enable district-wide parametrization.

To avoid duplicating voltage phasor measurements, each connection must use separate current PMUs (measurement winding and relay winding of measuring CTs) and voltage PMUs (measuring VTs).

3 Transients in islanded operation

When DG-enabled power districts are islanded, their transients caused by a sudden power shortage may differ from those in larger energy systems if:

– The power district is dominated by generator sets with relatively low mechanical constants of inertia (T_J): multi-shaft gas-turbine units (T_J up to 2–3 seconds) and gas-piston units or diesel-generator units (T_J about 1 s),

– The load and the operating situation may contribute to a voltage collapse.

Let us analyze the calculated transients that occur when islanding causes a power shortage. The initial power shortage d is defined as the difference between the available capacity of the running GS (110% of P_{nom}) and the load corresponding to a rated frequency and a normal voltage.

Let us find the available time (T_{avl}) as a function of GS parameters; the available time is the time during which the REC must record the occurrence of a power shortage, generate control actions (CA) to shed a load, and take these actions. This paper presents similar transients in steam turbine GS powered systems for comparison.

Table 1 enumerates the main factors that may affect LS performance, where T_{avl} values are given for d = 56%.

Table 2 compares transients caused by power shortage for the case an islanded power district is only covered by a group of similar GS of different types.

Table 1. Main factors that may affect LS performance.

T_J, s	The droop regulation GS, %	The time constant of the load GS, s	The setpoint of the frequency protection Hz	s	T_{avl}, s
7	4,5	1,5	46,5	1	1,584
3	4,5	1,5	46,5	1	1,266
1	4,5	1,5	46,5	1	1,110
1	0,1	1,5	46,5	1	1,115
1	0,1	20	46,5	1	1,112
1	0,1	20	47,5	1	1,422
1	0,1	20	47,5	2	2,422

Table 2. Technical parameters GS.

№ graph	Type GS	T_J, s	The setpoint of the frequency protection GS, s
1	Steam turbine	7	46,5 Hz (1 s)
2	Gas-turbine multi-shaft	3	47,5 Hz (20 s), 45,0 Hz (0 s)
3a	Gas-piston (Wärtsilä)	1	47,5 Hz (2 s)
3b	Gas-piston (GE Jenbacher)	1	49,0 Hz (0,2 s)

Figure 1*a* compares T_{avl} values for different GS types. All examples found herein imply a single LS CA, where the disconnected load equals the initial power shortage, whereas the time to disconnect the load (T_{LS}) is counted from the transient onset as a value below T_{avl}.

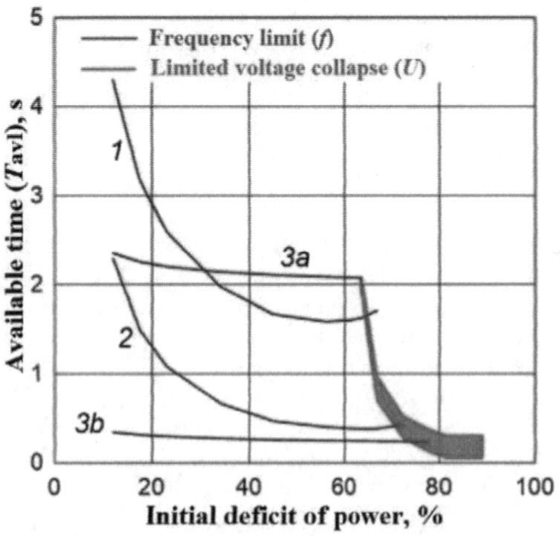

Fig. 1*a*. Dependence graph $T_{avl} = \varphi(d)$ for different types of GS.

Figure 1*a* shows the scatter of T_{avl} that is limited by the probable voltage collapse in the power district; what makes it probable is the varying parameters of the asynchronous motors (AM) in the load. At relative low initial power shortage (Figure 1*a*), transients cause an insignificant drop in frequency. We'll refer to it as an *f-transient*, see Figure 1*b* for an example (GTU, d = 34%, T_{avl} = 0.67s along Curve 2, T_{LS} = 0.5s).

For *f-transients,* the effective LS CA must prevent too significant drop in f and restore the frequency to a normal value; to that end, REC uses conventional UFLS algorithms.

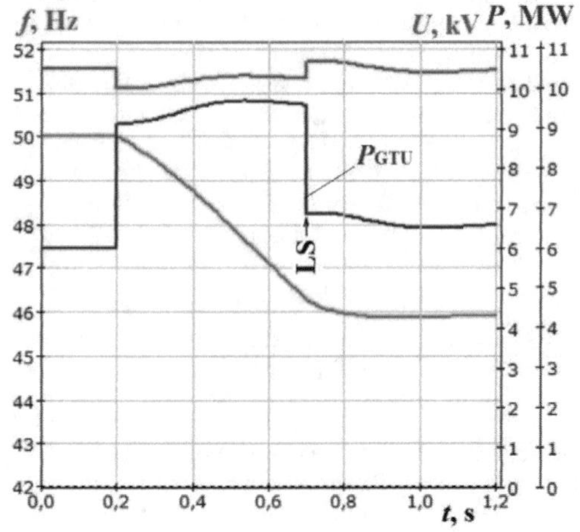

Fig. 1*b*. Graph *f-transient*, d = 34%, T_{avl} = 0.67s, T_{LS} = 0.5s.

At low T_J, islanded GS have significantly higher frequency reduction rates even if external disturbances do not change. In this case, the voltage collapse may occur so fast that conventional UFLS devices won't be able to prevent it. It's also effective to use additional automatic load shedding (AALS) that takes place in case of emergency power shortage before frequency starts dropping.

Higher UFLS performance is also necessary when RPA disconnects GS or critical consumers before UFLS can take a control action.

Figure 2*a* shows a transient at d = 72.5%, where T_{avl} = 0.42s, T_{LS} = 0.3s. In this case, the initial shortage d is close to the region of high voltage collapse probability, see Figure 1*a*, a situation associated with significant voltage drop (*fU-transient*).

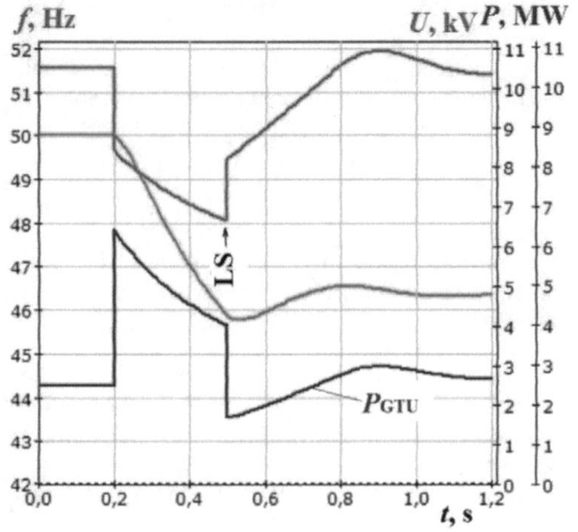

Fig. 2*a*. Graph *fU-transient*, d = 72.5%, T_{avl} = 0.42s, T_{LS} = 0.3s.

In case of significant initial deficit of power (Figure 1*a*, *d* > 65%), the shortage itself is the cause of voltage collapse. The transient and the voltage drop rate are the same as in case of a three-phase short-circuit at some distance from the GS. Implications for consumers in terms of maximum permissible fault-clearing time or time to take LS CA are identical. Engines can return to normal operation if the voltage dip has not caused an AM speed reduction that would render impossible their self-start.

An *fU-transient* might cause active power shortage when a power district is islanded even if no SC occurs, or when a GS is disconnected in islanded operation, which causes the AMs to consume more reactive power as *f* is falling; this situation requires a CA.

It's not that important how long *U* is below critical if the AMs account for 10% to 20% of the load. However, in case of average load, especially in industrial settings, T_{LS} is pretty limited. A transient that features a deep voltage dip (a *U-transient*) will not damage the consumers too significantly if T_{LS} is close to zero, see Figure 2*b*.

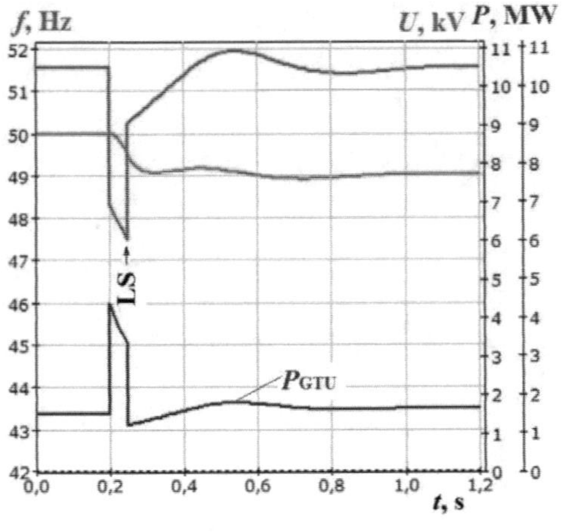

Fig. 2*b*. Graph *U-transient*, *d* = 83.5%, T_{LS} = 0.05s.

In case of significant power shortage, a voltage collapse occurs before frequency drops significantly, which means UFLS systems do not have time to respond. Deep and fast voltage dips cause significant load shedding, which restores the active power balance and normalizes the *f* values in the power district. Conventional UFLS is not suitable for such operations, which is why REC uses an enhanced UFLS algorithm.

The value T_{avl} depends on T_J of GS in islanded power districts, on the RPA setpoints, technological protection and automation setpoints, as well as on the maximum permissible outage duration for critical consumers.

4 REC basics

REC must have a hierarchy comprising local, coordinating, and centralized algorithmic levels.

All the REC algorithms rely on the parametric changes communicated by the PMUs; PMU placement is optimized when designing the system with due account of hardware and software redundancy requirements.

REC algorithms are implemented as functions of a single hardware and software redundancy-enabled platform.

The REC uses non-terminal solutions based on Russian industrial computers and specialized Russian-developed software that uses IEC 61850 communication protocols.

Initially, REC uses offline-computed parameters to implement a REC system for further transition to online software-based computing of steady states and transients. Using offline computing requires preliminary simulations of all possible operating situations for all the types and scopes of CA the REC must be capable of.

5 REC functions

No REC can be designed without monitoring the electric parameters in any operating situation; as such, it must feature a communication bus to transmit data from REC sensors and units as well as to convey control actions to the controlled units.

The REC as implemented here mainly functions to:

– Control DG GS load-shedding by active power given specific control ranges and technological limitations. This function uses GAPC among other things. provide secondary active power control in islanded operation. Balance the active power in an islanded power district, which uses enhanced UFLS and and AALS algorithms,

– Control DG GS load-shedding and SVC by reactive power given specific control ranges and technological limitations. This function uses GRPC among other things. provide secondary reactive power control in islanded operation, prepare the circuitry to launching high-power AM or AM groups. balance the reactive power in an islanded power district, which uses enhanced AUVP algorithms,

– Automatically maintain the specified magnitude and direction of active/reactive power flow within the power district and at the points of its connection to the energy system,

– Control the previous parameters (PP) to be used by the REC: the power district generation power *P*, *Q*; the district-to-energy system power flow *P*, *Q*; the in-district consumption *P*, *Q*, as well as the SU locations within the district,

– Implement MIS to successfully island a DG-enabled power district (preventively on the basis of its parameters; for repairs without an SC; for routine operation or repairs with an initiating SC),

– Automatically synchronize a DG-enabled power district to the energy system when commanded so by the dispatcher, e.g. after the district is reconnected to the energy system or after the system parameters are stabilized,

– Prevent unnecessary GS disconnections in case of significant f and U deviations by using an energy storage system (ESS) that is independently controlled by active/reactive power,

– Identify and prevent the desynchronization of GS within the district, or desynchronization within the district grids, as well as at the district-to-system PTL by means of an adaptive PSP algorithm,

– Prevent non-permissible equipment overloads within the district grid (PTL and power transformers) by means of enhanced AOP algorithms,

– Stabilize the dynamics of engine loads by controlling the dynamic voltage dip compensators,

– Control the district-wide electricity consumption to prevent emergencies, to maintain normal operation, or in a power-saving mode [10],

– Enable automatic cold and hot starting, which begins with EPS initiation,

– Implement enhanced linear and sectional ATS algorithms. Prevent ATS when load-shedding the energy district or when the district is connected to the energy system by multiple PTLs,

– Monitor the available capacity reserves by predicting the short-term renewable generation,

– Control wind and solar farms connected via frequency-controlled inverter converters by imposing absolute and relative limits on the power output, as well as by setting the power ascension and descension rates),

– Integrating the system in the district grid and controlling the microgeneration facilities by plug-and-play automatics (a future work).

Conclusion

Routine and emergency controls have specific features, are hierarchical, and fall into local, coordinating, and centralized control levels.

REC uses small-size PMUs as sensors, which enables it to calculate the connection-specific and district-wide parameters on the current and voltage phasor basis.

The REC must use non-terminal solutions based on Russian industrial computers and specialized software (mainly Russian) that uses IEC 61850 communication protocols.

Russian-made automatics must be designed to freely integrate new microgeneration facilities in power-district grids on a plug-and-play basis.

REC enhancements will require Russian-developed software to compute steady states and transients online.

Bibliography

1. P. Ilyushin, P. Berezovskiy, Energetik, **3**, 12-18 (2019).
2. M. Sharygin, A. Kulikov, Power Plant, **2**, 32-39 (2018).
3. A. Kulikov, P. Ilyushin, Electricity, **5**, 4-11 (2019).
4. P. Ilyushin, Relay protection and automation, **4**, 28-36 (2017).
5. P. Ilyushin, Relay protection and automation, **3**, 20-27 (2018).
6. P. Ilyushin, A. Mokeev, V. Narovlyanskii, Power Plant, **1**, 52-59 (2019).
7. P. Ilyushin, A. Kulikov, Relay protection and automation, **1**, 55-65 (2019).
8. P. Ilyushin, Relay protection and automation, **3**, 44-49 (2014).
9. P. Ilyushin, Relay protection and automation, **2**, 12-24 (2018).
10. M. Sharygin, A. Kulikov, *Protection and automation of power supply to active industrial consumers: monograph* (RANKHiGS, Nizhny Novgorod, 2017).

Modeling of reliability of power supply systems with autonomous energy source

M.S. Ershov[1], *R.B.* Jalilov[2*], *S.P.* Shayimova[2], *A.I.* Choriyev[2], *M.R.* Charieva[2]

[1]Russian State University of Oil and Gas (National Research University) named after I.M. Gubkin, Moscow, Russia
[2]Bukhara Engineering-Technological Institute

Abstract. The methodology for calculating the reliability of power supply systems with autonomous and mixed energy sources has been considered. Autonomous sources are widely used in industrial power supply systems as primary, backup or emergency sources. The functioning and interchangeability of these systems has features that affect the reliability of power supply. The proposed methodology is based on the method of semi-Markov random processes and can be used for comparative analysis of the reliability of options for power supply systems. The methodology has been brought to software implementation and is accompanied by an example of calculating the reliability of the power supply system for stationary marine oil production platforms.

Introduction

Objects of production, transportation and refining of oil and gas are complex continuous production, which put increased requirements for the reliability of power supply [1-4]. Autonomous power plants of own needs (PPPON) based on diesel or gas turbine units are used to power consumers of electricity in oil and gas industries. Depending on the number and type of PPPON units used, the conditions for their parallel operation with centralized power supply sources, various options for power supply systems with autonomous power supplies can be used. For a comparative analysis of the reliability of options for power supply systems, a technology based on the mathematical method of semi-Markov random processes is proposed [5-8].

The methodology for modeling and calculating the reliability of power supply system

Includes the following steps [2, 6, 7]:

- Formation of the space and graph of the state of the system;
- Mathematical description of the space and graph of the state of the system;
- Determination of indicators of system reliability.

Formation of the space and graph of the state of the system.

When performing this stage, for power supply systems it is necessary to take into account the working, standby and repair conditions of power supplies and system connections, i.e. power lines, connecting distribution devices of 6 (10) kV technological installations to which generators of autonomous power sources are connected. Along with structural redundancy, characterized by a reserve of generating capacities and backup links, power supply systems also have a temporary reserve [4, 7, 10]. Temporary redundancy consists in the fact that consumers allow a short interruption in power supply. A break in the power supply during the operation of the automatic switch-on reserve (ASOR) does not stop the process and not considered as failure. Along with failures of elements (sources, lines) and failures in the operation of the reserve input, it is necessary to take into account the possibility of system failures, which may be due to interruptions in the supply of fuel to generating electrical units and the failure of relay protection of power supply systems

Failures of system elements occur with the intensity λ_i, 1/year; and recovery occurs with the intensity μ_i, 1/year; $(i=1,...n)$.. n is the number of the main elements (sources and lines) of the system. Failure and restoration intensity of elements are inverse to their average time between failures and average times of recovery, respectively.

The possibility of an unsuccessful ABP is characterized by probability q_{ABP}. The probability of a successful ABP is characterized by probability $(1-q_{ABP})$. The probabilities of events make it possible to "sift" the flow of failures and recoveries [9, 11]. In case of unsuccessful ABP, the operator carries out the input of the generating unit or the turning on of system communication manually. Manual entry of the power plant unit into operation is carried out with intensity μ_{ap} .Manual connection of communication or sectional switch is carried out with intensity μ_{cp} .

System failures are characterized by the intensity of complete λ_S or the intensity of partial λ_{SS} failures and recoveries μ_S or μ_{SS}, respectively. The rates of system failures associated with the probability q_{P3} of failure of the protection are determined by the product of the rate of accidents in the network of the system λ_C and the specified probability.

The state and transition graph is formed based on the analysis of possible states and includes a finite set of states $\mathbf{X} = \{x_1, x_2, x_3, ...x_n\}$, which is divided into subsets of functional $\mathbf{X}_{PC}$ and non-functional $\mathbf{X}_{HPC}$ states. In the formation of states, it is necessary to take into account the main states and neglect the secondary ones. It is accepted that the combination of independent failures of more than three main elements of the system is impossible [3, 5]. A subset of non-functional states can be divided into a number of levels, ranked by the energy deficit in the system.

Mathematical description of the space and graph of state of the system.

Initially, the process is described by the matrix $\{\lambda_{ij}\}$of intensities of transitions of the system from one state to another. Based on the intensity matrix, the vector $\{a_i\}$ of the average residence time of the system in states and the matrix $\{p_{ij}\}$ of transition probabilities between states are determined. The element a_i of the vector of the average residence time in the states is defined as the reciprocal of the sum of the intensities of the transitions leaving this state. The element of the transition probability matrix is defined as the ratio of the intensity of the transition from state x_i to state x_j to the sum of the intensities of all transitions out the state x_i .

Based on the original matrix $\{p_{ij}\}$, the vector $\{\pi_i\}$ of stationary probabilities of the process is defined. Stationary probabilities π_i are defined from the system of equations [1]

$$\pi_i = \sum_{i,j}^{n} p_{ij}\pi_j \tag{1}$$

and normalization conditions

$$\sum_{i=1}^{n} \pi_i = 1 \tag{2}$$

The Gauss method is used to solve the system of equations.

Determination of system reliability indicators

Calculation of the reliability of power supply to consumers comes to the determination of reliability indicators. A set of reliability indicators includes:

T_H, year – average operating time for failures;

T_B, hour – average recovery time;

K_Γ – availability coefficient;

F, 1/year – average failure rate;

ΔW , kWh/year – average annual undersupply of electricity.

Operating time for failure of system T_H is defined as the average residence time of a process in a subset of functional states $\mathbf{X}_{PC}$. For calculation of T_H following equation is used

$$T_H = \frac{\sum \pi_k a_k}{\sum \pi_k \sum p_{kl}}, \quad (x_k \in \mathbf{X}_{PC},\ x_l \in \mathbf{X}_{HPC}), \tag{3}$$

where x_k - functional states; x_l - non-functional states.

The average recovery time is defined as the average residence time of the process in a subset of non-functional states $\mathbf{X}_{HPC}$. For calculation of T_B following equation is used

$$T_B = \frac{\sum \pi_l \cdot a_l}{\sum \pi_l \sum p_{lk}}, \quad (x_k \in \mathbf{X}_{PC},\ x_l \in \mathbf{X}_{HPC}), \tag{4}$$

The system availability coefficient is defined as the stationary probability of a process being in a subset of operational states $\mathbf{X}_{PC}$. For calculation of K_Γ following equation is used

$$K_\Gamma = \frac{T_H}{T_H + T_B} = \frac{\sum \pi_k a_k}{\sum \pi_k a_k + \sum \pi_l a_l}, \tag{5}$$

The average system failure rate is determined from the relation

$$F = \frac{K_\Gamma}{T_H} = \frac{\sum \pi_k \sum p_{kl}}{\sum \pi_k a_k + \sum \pi_l a_l}, \tag{6}$$

https://doi.org/10.1051/e3sconf/201913901036

The indicators defined by formulas (3) - (6) can be differentiated according to the levels of power shortage in non-working conditions.

The average annual undersupply of electricity at the ith level of power shortage in power supply systems is carried out based on calculated reliability indicators according to the formula

$$\Delta W_i = \Delta P_i \cdot F_i \cdot T_{Bi}, \quad (7)$$

where ΔP_i – power shortage level, kW; F_i –rate of failure leading to a given level of power shortage, failure/year; T_{Bi} , hour. – average recovery time after failure with a given power shortage.

During scheduled preventive repairs (SPR) of the main elements of the system, i.e. units of power plants and system connections, the level of redundancy of the system decreases, which reduces its reliability during the specified period. The evolution of the system during SPR corresponds to its own graph of states and transitions, which allows to calculate the reliability indicators of the system in the corresponding period. The resulting reliability indicator of the system is calculated as a weighted average, taking into account the value of the corresponding indicator and the duration (hour) of the system during the year in normal and repair situations.

Reliability calculation algorithms are implemented in the RELIABILITY program, using which calculations were performed in the example below.

An example of calculating the reliability indicators of the power supply system of a platform of compressor station of the marine oil production

The compressor station of power plant consists of three units with a rated power of 2800 kW each. In normal mode, two units operate at a combined tire system, loaded at 50%; the third unit is in reserve. In addition to the power plant units, a backup source for the compressor station is communication with the power supply system of the central platform (CPU). Communication throughput is 1600 kW.

In case of failure of the power plant unit (intensity λ_1), it turns off and the standby unit automatically turns on. In case of unsuccessful ABP, part of the load is disconnected by the automatic unloading system. Unsuccessful ABP is characterized by probability q_1, and successful ABP by probability $q_3 = 1 - q_1$. After an unsuccessful ABP, the backup unit can be manually entered into operation. If there are no units in the reserve (the third unit is under repair), then if one of the two workers fails, communication with the central system is automatically turned on. The failure in automatically turn-on the communication may be due to the failure of the ABP and the non-functional state of communication at the required time.

An inoperative state of communication is determined by probability q_2 , an operational state is determined by probability $q_4 = 1 - q_2$. Communication may also fail in the course of work (intensity λ_2). During operation, system failures are possible. The intensity of these failures is $q_5 \lambda_3$. The process of system recovery is determined by the intensities: μ_1 - recovery of units; μ_2 - recovery of communication with the CPU; μ_3 - recovery after system failures; μ_4 - manual input into the operation of the unit; μ_5 - operational switching in the system.

The graph of states and transitions of the power supply system of the compressor station in the normal mode is shown in Fig. 1. The states of the system in the normal mode are shown in Table 1. Fig. 2. shows the data input interface.

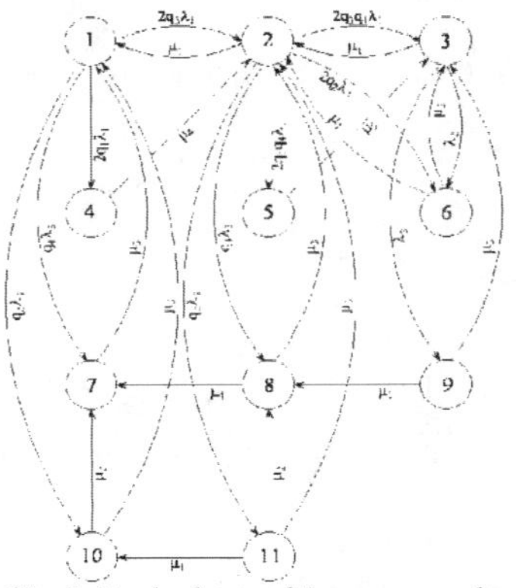

Fig. 1. Graph of state of the power supply system of the compressor station

Table 1. Identification of the state of the power supply system of the compressor station

No. of states	The number of units in modes				Connection with CPU	Communication state	Deficit, kW
	Operation	Backup	Repair	Repair waiting			
1	2	1	0	0	0	-	0
2	2	0	1	0	0	-	0
3	1	0	1	1	1	1	0
4	1	1	0	1	0	-	400
5	1	0	1	1	0	1	400
6	1	0	1	1	0	0	400
7	0	0	0, S	0	1	1	1600
8	0	0	1, S	0	1	1	1600
9	0	0	1, S	1	1	1	1600
10	0	0	0, S	0	0	0	3200
11	0	0	1, S	0	0	0	3200

Note. S - systemic fault; communication with the CPU: 1 - enabled, 0 - disabled; communication state: 1 - functional, 0 – non-functional

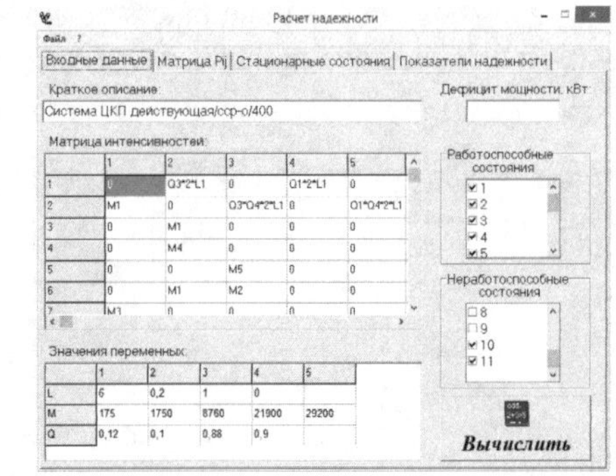

Fig. 2. Presentation form of program input data

Fig. 3. shows a fragment of the report of calculation results

Russian State University of Oil and Gas named after I.M. Gubkin
Reliability calculation program for power supply systems
Name: Compression station power supply system
Power shortage: 1400 kW
Input data:

L(1) = 6   M(1) = 175   Q(1) = 0,12
L(2) = 0,2   M(2) = 1750   Q(2) = 0,1
L(3) = 1   M(3) = 8760   Q(3) = 0,88
M(4) = 21900   Q(4) = 0,9
M(3) = 29200
Non-functional state: 7 8 9 10 11
Average residence time and final state probabilities:
A[1] = 0.07692   п[1] = 0.441948
A[2] = 0.00532   п[2] = 0.438208
A[3] = 0.00568   п[3] = 0.0279
A[4] = 5E-5   п[4] = 0.048954
A[5] = 3E-5   п[5] = 0.003021
A[6] = 0.00052   п[6] = 0.002829
A[7] = 0.00011   п[7] = 0.031205
A[8] = 0.00011   п[8] = 0.002139
A[9] = 0.00011   п[9] = 0.000158
A[10] = 0.0001   п[10] = 0.003403
A[11] = 9E-5   п[11] = 0.000233
Reliability indicators:
Operating time for failure Tн = 1.000128 year
Average recovery time Tв = 0.99864 hour
Availability coefficient Kг = 0.999886
Average failure rate W = 0.999758 failure/year
Expected energy deficit dW = 1397.757743 kWh/year

Fig. 3. Fragment of listing of the output report of the program

Conclusion

A method for calculating the reliability of power supply systems with autonomous power supplies is proposed, which is developed based on analysis of the features of power supply for oil and gas industry objects and allowing more reasonably choosing options for power supply of industrial complexes during their design and reconstruction. The technology was created on the basis of semi-Markov random processes, brought to a software implementation and is effective for rapid assessment of the reliability of systems with autonomous power sources.

References

1. Menshov B.G., Ershov M.S. Reliability of power supply systems of compressor stations. – M.: Nedra, 1995.
2. Ershov M.S., Zyong Hoang Hai. The methodology for calculating the reliability of SES complexes of stationary marine oil production platforms // Automation, telemechanization and communication in the oil industry, 2006, No. 9, p. 35-38.
3. Reliability of energy systems and their equipment. Reference: in 4 vol. Vol. 2. Reliability of electric power systems / Ed. M.N. Rudenko. - M .: Energoatomizdat, 2000.
4. Sukharev M.G., Ershov M.S. The reliability model of a duplicated system with limitations of the backup element // Reliability and quality control, 1985, No. 3, p. 18-22
5. Reliability of technical systems. Handbook / Ed. I.A. Ushakov. - M .: Radio and communications, 1985.
6. Polovko A.M., Gurov S.V. Fundamentals of reliability theory. - 2nd ed., reviewed and added - SPb .: BHV-Petersburg, 2006.
7. Credentser B.P. Prediction of the reliability of systems with temporary redundancy. - Kiev: Naukova dumka, 1988.
8. Billinton R., Allan R., Reliability evaluation of power systems, N.Y.: Plenum Press, 1987.
9. P. Wang, R. Billinton, "Reliability cost/worth assessment of distribution systems incorporating time-varying weather conditions and restoration resources", IEEE Trans on Power Delivery, vol. 17, no. 1, 2002, pp. 260-265.
10. Jalilov R.B. Structural redundancy to increase the reliability of power supply to consumers. // Electrics, 2007, No.6, p. 27-29.
11. Jalilov RB, Sitdikov R.A. Graph-analytical method for studying the reliability of electric power systems. // Problems of energy and resource saving, Tashkent, 2017. No.1-2, pp. 23-29.

 https://doi.org/10.1051/e3sconf/201913901037

Calculation of reliability indicators of power supply systems of consumers

S.T. Latipov[1*], *G.N.* Aslanova[1], *L.A.* Nematov[1], *A.A.* Akhmedov[1], *M.R.* Charieva[1]

[1]Bukhara Engineering-Technological Institute

Abstract. The article covers the study of methods for calculating the reliability of schemes of power supply for consumers at the present stage. It analyzes the existing methods for calculating the reliability of various schemes of power supply systems for consumers. Methods for determining the reliability indicators of redundant systems are reviewed. Examples are given

Introduction

A comprehensive assessment of the reliability of electric power systems (EPS) [1-5, 8-10] includes a joint review of electrical networks (EN), power plants and substations. A lot of publications in Russia [1-5, 8-10] and abroad [6, 7] cover the methods of its analysis.

As it is known [3-5], in accordance with accepted terminology, the reliability of power systems (power associations) is understood as their ability to fulfill their main function – continuous power supply of consumers with electricity of the required quality. With a certain convention in the general complex concept of reliability of power systems, functional concepts can be distinguished by particular concepts related to the main structural components of a power system: reliability of a generation system — the property of power plants to maintain the required power balance at a standard frequency value; reliability of the main electric network - the property to stably transmit power from parts of the power system with excess power to parts of the power system with a deficit of power; reliability of the distribution network - the property of this network is to provide continuous power to the load nodes.

In the situational aspect, two particular concepts can be distinguished [9,10]: reliability in the steady state (according to the terminology adopted in the USA, – adequacy) is the balance reliability, i.e. the ability to balance power and electric energy (EE) with the standard quality of EE; Reliability in the transition process (Security) is the mode reliability, i.e. the ability of individual structural parts and the energy system as a whole to withstand violations of the mode and provide power to consumers.

The reliability of power supply to consumers depends on the reliability of the power system at all its levels, i.e. in power generation, transmission and distribution systems. The reliability of stations, substations, and energy transmission and distribution systems depends on many deterministic and stochastic parameters and, above all, on the structure, as well as the operational safety and recoverability of the main equipment [1-3, 5, 10].

The task of assessment the reliability of a particular electricity supply option arises at the decision-making stage in the design of new and development of existing systems. The initial information in this case is the scheme and retrospective data on the reliability of its main elements and automation elements. Calculations of the circuit reliability of generating facilities, switchgears, and transmission and distribution systems of electric power have their own specific features, therefore specific methods are oriented to a certain level - a specific class of power system facilities. At the same time, the basis of these calculations consists from general formulas that allow to determine the probabilities, frequencies, and average durations of simultaneous downtime of several elements when their failures coincide, or when a failure of one element is imposed on a scheduled repair of another. These formulas are obtained based on general analytical models [1-8, 10].

In the reliability of the electric power industry, two components can be conditionally distinguished [1-3, 5]:
1) the first one, due to the structure of the system, i.e. the composition of the elements, their interconnections, throughputs, without a quantitative consideration of the operational features of the functioning of the elements - structural reliability;
2) the second one, due to the features of the modes in the system, their limitations on the throughput of the elements when changing the structural system in various states - functional reliability.

Such a subdivision is conditional, because the interconnection of these components is obvious, but it allows simplifying the tasks of assessing the reliability of networks of complex structure.

Currently, reliability requirements are accounted in accordance with regulatory guidelines in directive materials for the design and operation of power systems. The complexity of modern energy facilities and the processes of their functioning, the insufficient accuracy

* Corresponding author: zhalilov.rashid@mail.ru

https://doi.org/10.1051/e3sconf/201913901037

of the source information, increasing with the increase in the lead-time for the formation of decisions, the need to develop mass solutions determines the feasibility of the predominant application of the normative approach.

An example of the feasibility of using standards in the development of mass solutions is the establishment of standards for the technological design of typical circuits for switchgear of power plants and substations
The analysis showed that at present it is recommended to distinguish between two methods of normalizing reliability [2-6, 9, 10]:

♦ establishment of normative quantitative values of indicators characterizing individual properties of reliability;

♦ "indirect" standardization of reliability levels based on regulatory requirements for the energy system (UES, UES), its individual elements, control systems, means of ensuring reliability, etc.

Direct normalization of reliability indicators is limited by the fact that the work on substantiating the choice of these indicators is not completed and the tested composition of these indicators is not available. At the present stage, "indirect" rationing has gained predominant use.

For non-redundant systems, it is not possible to recovery failed elements during operation; they must be disabled. At the same time, reducing the time of switching and recovering failed elements increases the availability of the system, but practically does not affect its reliability.

If the system has redundancy, then recovery is an effective way to increase reliability. By duplicating the functions of the recovered elements and reducing the recovery time, an arbitrarily high degree of system reliability can be achieved.

Determination of system reliability indicators

The features of redundant systems with recovery is advisable to consider on the example of two mutually redundant elements 1 and 2, for example, parallel power lines, mutually redundant power transformers in substations, etc. [3, 4, 8, 9].

Parallel connection of lines and other circuits, as well as parallel operation of several units is modeled with a structure with parallel connection of elements. This can be considered a system consisting of n elements or pieces of equipment, if r elements are necessary for normal operation, and n-r elements are backup. The system failure occurs when m elements fail, i.e. while the number of backup elements is the number of failed, the system does not fail. Thus, the failure condition is $-1 = n - r$.

Consequently, the probability of a system failure is defined as the probability of coincidence of failures $(n - r + 1)$ or m elements. If the failures of one element are independent of the failure of other elements, then the formulas for estimating the probability of failure-free operation or the probability of system failure can be obtained by forming the sum of the products of the probabilities $Q_i(t)$ and $P_i(t)$ based on the product of n binomials

$$[P_i(t) + Q_i(t)] = 1:$$

$$\prod_{i=1}^{n}[P_i(t) + Q_i(t)] = P_1(t)P_2(t)\dots P_n(t) + Q_1(t)P_2(t)\dots P_n(t) + Q_1(t)Q_2(t)\dots Q_n(t) \quad (1)$$

Choosing terms with the value $Q_i(t) \geq m$, from this sum, we can form an expression for the probability of failure from them. And choosing the terms with the value $Q_i(t) < m$, we can form an expression from them for the probability of failure-free operation of the system with parallel connection of elements.

Under the condition of equal reliability of elements:

$$Q(t) = \sum_{k=m}^{n} C_n^k P^{n-k}(t) Q^k(t); \quad (2)$$

$$P(t) = \sum_{k=r}^{n} C_n^k P^k(t) Q^{n-k}(t) \quad (3)$$

where $C_n^k = n!/[(n-k)!\,k!]$.

A system with parallel connection of elements is a redundant system, i.e. failure of one or more elements does not cause system failure. Redundancy is permanent if all elements are in operation and the system does not fail until a certain number of them fail. Redundancy by replacement is a redundancy in which redundant elements are turned on only after automatic shutdown of failed elements. In the energy sector, numerous ABP devices, as well as constantly rotating and hidden reserve of generators, transformers and electric motors carry out redundancy by replacement.

At present, in Russia and abroad, there has already planned a transition to the next generation of digital devices for relay protection and automation (DDRPA) with the integration of the functions of relay protection, measurement and commercial metering of energy efficiency, regulation and control of electrical devices within a single information complex [11]. In almost all countries of the world, the concept of smart electric power grid (Smart Grid) has been adopted as a technological platform for the future electric power industry. The use of intelligent EES technologies (FACTS, power storage devices, power electronics, artificial intelligence systems, information and communication technologies, modern methods of control theory) will fundamentally change the properties of future intelligent EES [13]. This in turn necessitates the feasibility of developing research methods and improving the reliability of consumer EPS. At the same time, it is necessary to take into account the reliability of the technological object and to characterize the events, conditions and processes that determine the conditions for the functioning and development of energy supply systems (ESS). Reliability analysis of the power supply system should be closely related to the study of the reliability of the units and the entire production functioning as part of the technological facility, depending on the properties of the power plant, instrumentation and automation equipment, process automation (components of industrial control systems), as well as technological redundancy [10. 11].

https://doi.org/10.1051/e3sconf/201913901037

Based on the foregoing, our research is based on probabilistic methods of calculation using stochastic modeling [10-12].

The probability of a fail-free operation of a system with redundancy is determined by the reliability of not only the elements themselves, but also of circuit breakers, which, when constant redundancy, must disconnect the failed element, and when redundant by replacement, also include the redundant one. If the entire system fails when the disconnecting device fails to disconnect, the probability of failure-free operation of the system with constant redundancy will be

$$P_c = P_k P_{o.c}, \quad (4)$$

where $P_{o.c}$ – probability of a failure of operation when disconnecting a failed element; P_k – the probability of failure of a group of elements with a multiplicity of redundancy k.

When redundancy by replacement, the probability of system failure is determined by the formula for the total probability:

$$Q_c = (S|A_1A_2)P(A_1)P(A_2) + Q(S|\overline{A_1}A_2)Q(A_1)P(A_2) +$$
$$+Q(S|A_1\overline{A_2})P(A_1)Q(A_2) + Q(S|\overline{A_1}\,\overline{A_2})Q(A_1)Q(A_2), \quad (5)$$

where $Q(S|A_1A_2)$- conditional probability of system failure in the absence of device failures; $Q(S|\overline{A_1}A_2)$ – the same when disconnecting a failed element; $Q(S|A_1\overline{A_2})$ – the same with the failure to turn on the redundancy; $Q(S|\overline{A_1}\,\overline{A_2})$ – the same if the failure to disconnect coincides with the failure to turn on; $P(A_1)$, $Q(A_1)$ – probability of failure abcence and failure to turn-off; (A_2), $Q(A_2)$ – probability of failure abcence and failure to turn-on.

Examples of calculating the reliability indicators of a power supply system

Example 1. Consumers of their own needs are supplied from two independent sources according to the circuit shown in Fig. 1.

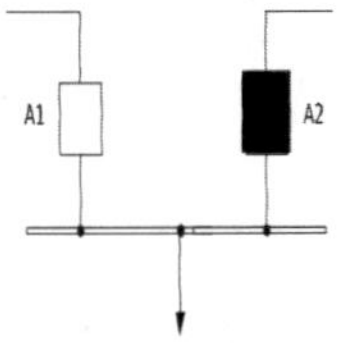

Fig. 1. Supply scheme for consumers of their own needs to two sections

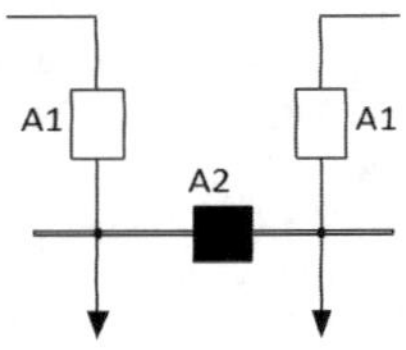

Fig. 2. Connecting consumers

In accordance with the conditions of example примера $Q(A_1) - 0.05$; $P(A_1) = 0.95$, one source is turned on constantly, the other is turned on by the action, ABP. The probability of failure of the source during the time between scheduled stops of the unit is 0.90. The probability of finding a backup source in working condition at any time between scheduled stops is 0.99. The probability of failure in the shutdown is 0.05 and in the inclusion of 0.01. To determine the probability of failure-free operation of the power supply system during the time between scheduled turn-off, taking into account the possibility of using a backup source, considering that the probability of its failure during the recovery of a working source can be neglected.
$Q(A_2) = 0.01$. $P(A_2) = 0.99$. $Q(S|\overline{A_1}A_2) = 1$;
$Q(S|A_1\overline{A_2}) = 1$.

The probability of a system failure in the absence of device failures is determined by the product of the probability of a working source failure and the probability of backup emergency downtime:
$Q(S|A_1A_2) = (1 - 0.9)(1 - 0.99) = 0.1 \cdot 0.01 = 0.001$.

On equation (5) we have

$Q_c = 0.001 \cdot 0.99 \cdot 0.95 + 1 \cdot 0.05 \cdot 0.99 + 1 \cdot 0.95 \cdot 0.01 + 1 \cdot 0.05 \cdot 0.01 = 0.06044$

Probability to be determined

$$P_c = 1 - 0.06044 = 0.939\ 56.$$

Example 2. Consumers of their own needs can be connected to two different sections, powered by two independent sources according to the circuit shown in Fig. 2. The probability of emergency downtime, source failure, and hardware failure are as in the previous example. The difference between the circuits lies in the operation of the switches: the sectional switch provides the ABP of the section left without power due to the neighboring section and its source. We determine the probability of uninterrupted power supply to any of the consumers of this system. According to the terms of the example:

$Q(A_1) = 2 \cdot 0.05 = 0.1$; $P(A_1) = 0.9$; $Q(A_2) = 0.01$; $P(A_2) = 0.99$. Each of these consumers may be connected to one of the sections with a probability of 0.5, therefore $Q(S|\overline{A_1}A_2) = 0.5$; $Q(S|\overline{A_1}\,\overline{A_2}) = 0.5$; $Q(S|A_1\overline{A_2}) = 0.5$.
In the absence of equipment failures, a system failure occurs when the failure of one of the sources coincides with the emergency downtime of another, i.e.
$Q(S|A_1A_2) = 2 \cdot (1 - 0.9)(1 - 0.99) = 0.002$. On equation (5) we have

$Q_c = 0.002 \cdot 0.9 \cdot 0.99 + 1 \cdot 0.5 \cdot 0.1 \cdot 0.99 + 0.5 \cdot 0.9 \cdot 0.01 = 0.056\ 28.$

Probability to be determined

$$P_c = 1 - 0.056\,28 = 0.943\,72\,.$$

Example 3. In a case when a consumer is used a complete automatic transformer substation in a PSS, developed by the author [11,12]. We determine the probability of uninterrupted power supply to any of the consumers of this system

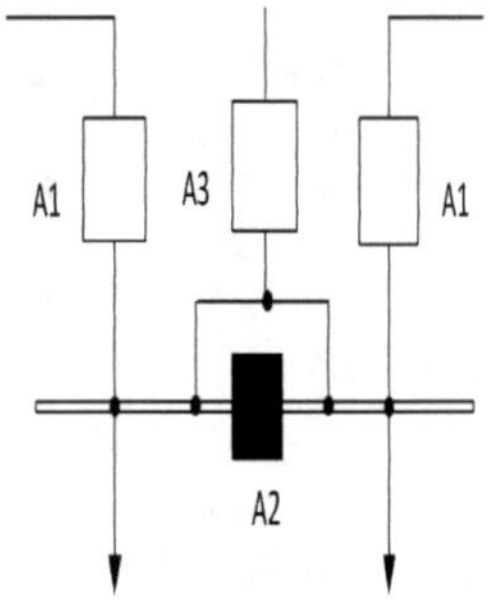

Fig. 3. Consumer supply according to the scheme from ABP to CB and implementation of the 3rd independent source with automatic reverse transition.

In the absence of failures of the switches of the two main inputs, the failure of the system occurs when the failure of one of the sources coincides with the emergency downtime of the other, i.e.

$$Q(S|A_1A_2) = 3 \cdot (1 - 0.85)(1 - 0.99) = 0.0045\,.$$

Conditional probability of failure of the 3rd independent source

$$Q(S|A_1A_2\overline{A_3}) - = 2 \cdot (0.5 \cdot 0.9) \cdot 0.01 - 3 \cdot 0.01 = 0.039$$

On equation (5) we have

$$Q_c = 0.0045 \cdot 0.85 \cdot 0.99 + 0.5 \cdot 0.1 \cdot 0.99 + 0.5 \cdot 0.9 \cdot 0.01 + 0.5 \cdot 0.9 \cdot 0.01 = 0.0037\,8\,7 + 0.0495 + 0.0045 + 0.0045 - 0.009 = 0.062\,29 - 0.039 = 0.023\,29.$$

Probability to be determined

$$P_c = 1 - 0.023\,29 = 0.976\,71.$$

Comparative results

The calculation results and their comparison are presented in the table.

Table. Comparison of reliability indicators of options for power supply schemes

Estimated indicators	Existing scheme options		Designed from ABP to CB and implementation of the 3rd independent source with automatic reverse transition
	with ABP on one of the inputs	with ABP on CB	
$Q(A_1)$	0.05	0.1	0.15
$P(A_1)$	0.95	0.9	0.85
$Q(A_2)$	0.01	0.01	0.01
$P(A_2)$	0.99	0.99	0.99
$Q(S\|\overline{A_1}A_2)$	1	0.5	0.5
$Q(S\|\overline{A_1}\,\overline{A_2})$	1	0.5	0.5
$Q(S\|A_1\overline{A_2})$	1	0.5	0.5
$Q(S\|A_1A_2)$	0.001	0.002	0.003
$Q(S\|A_1A_2A_3)$	-	-	0.039
Q_c	0.060 44	0.056 28	0.023 29
P_c	0.939 56	0.943 72	0.976 71

The use of a circuit with constant redundancy and ABP on a sectional switch increases the probability of continuous power supply to consumers. In addition, partitioning reduces the probability of a complete loss of power to all consumers. Without sectioning, failure of the switch of any of the connections leads to complete extinction, and during sectioning, to the disconnection of only one of the sections, i.e. to discharge half of consumers. The application of the ABP scheme at the CB and the implementation of the 3rd independent source with automatic reverse switching to supply from the main inputs is the most maneuverable and even more increases the probability of uninterrupted power supply to consumers. The probability of failure of the power supply system during the time between scheduled stops, taking into account the possibility of using a backup source, is high and the probability of its failure during the restoration of the working source is not high.

Conclusion

1. One of the main ways to improve the efficiency of methods and programs for assessing the reliability of ESS is to automatically select significant and ordered failures of not only power equipment, but also DDRPA devices, instrumentation and automation equipment, technological automation (components of industrial control systems), as well as devices and complexes of technological redundancy of consumers.

2. To improve the methods of analyzing the reliability of EPS, it is necessary to develop methods, algorithms, and software for the comprehensive assessment of the reliability of electric power plants and power systems, distribution and transmitting electric power systems, devices of central distribution and

automation systems, DDRPA and TM, as well as assessing the level of safety and survivability of EPS.

References

1. Voropay N.I., Domyshev A.V., Nepomnyashchy V.A. Models and methods for research the operational reliability of electric power systems // Reliability of energy systems: Problems, models and methods for solving them. Novosibirsk: Nauka, 2014, p. 57-73.
2. V.A. Ovseychuk, V.A. Nepomnyashchy. Rationing of reliability and quality of power supply to consumers // Innovation and Expertise. 2016. Issue 1/16. p.178-185. [Electronic resource] .URL: http: // www. energystrategy. com / projects / energy_21.htm (date of application: 04/20/2019).
3. Kitushin. Reliability of energy systems. Part 1. Theoretical basis: Textbook.-Novosibirsk: Publishing House of NSTU.-2003.p. 256.
4. Guk B.Yu. The theory of reliability in the electric power industry. - Moscow: Energoatomizdat. Leningrad branch, 1990 . p.207.
5. GOST 27.301-95. Reliability in technology. Reliability calculation. Key Points. - Moscow: Publishing House of Standards, 1995.
6. Endreni, J. Modeling in reliability calculations in electric power systems: Translated from English / Ed. Yu.N. Rudenko. M .: Energoatomizdat, 1983.
7. Billinton R., Allan R. N. Reliability evalition of Power Systems. –New York, 1984.
8. Menshov B.G., Ershov M.S. Reliability of power supply systems of compressor stations. - M .: Nedra, 1995.
9. Savina N.V. The theory of reliability in the electric power industry. Amur State University - Blagoveshchensk, 2007. p.228.
10. Jalilov R.B., Sitdikov R.A. A systematic approach to ensuring the reliability of operation of electric power systems // Problems of energy and resource saving, Tashkent, 2016. No. 3-4, p. 127-132.
11. Jalilov R.B. A new approach to the creation of an intelligent automated substation // Industrial Energy, 2008. No. 4, p.15-18.
12. Jalilov RB Complete automatic transformer substation. Patent of Republic of Uzbekistan for useful model No. FAP 01303. 04/27/2018.
13. Voropai N.I., Efimov D.N., Kurbatsky V.G., Panasetskiy D.A., Tomin N.V. Algorithms of survivability and self-recovery of intelligent electric power systems // [Electronic resource]. URL: http: // www.energystrategy.ru/ projects / -energy_21.htm (date of application: 04/20/2019).

 https://doi.org/10.1051/e3sconf/201913901038

An adaptive single-phase reclosing of line in the scheme with intermediate system

German Samorodov[1], *Tatyana* Krasil'nikova[1], and *Saiyod* Jononaev[1,*]

[1]Novosibirsk State Technical University, Automated Electrical Power Systems Department, 630073, Novosibirsk, Russian Federation

Abstract. Providing the reliability of adaptive single-phase reclosing (ASPR) in line equipped with any number of shunt reactors (SR), including the case of their absence, are considered and analyzed in this paper. The essence of the proposed work is that in order to provide high reliability of ASPR, is carry out short-term (for 0.1 seconds) one-sided switching on of the emergency phase from the side of the intermediate system, is measuring the current value of the current of this phase at the switched end and comparing it with the specified setpoint. If the current is higher than the specified setpoint, then a short or arc fault is detected in the phase and then it gives a command to disconnecting the line, otherwise it is not a short circuit in the phase and it is successfully will turned on to work on the other side of the line.
Keywords: adaptive single-phase reclosing, weak intermediate system, input resistance argument, controlled current in the emergency phase

The dead time of the SPR, was recommended on the basis of experimental data, is selected from the conditions for quenching the secondary arc and subsequent deionization of the arc gap. However, the true arc extinction time is often much less than that selected on the basis of the experimental curve. This circumstance reduces the effectiveness of SPR due to the overestimated effects of emergency automation. On the other hand, with stable single-phase faults (SPF) or a protracted secondary arc, unsuccessful SPR are not excluded. According to data of statistics, 15-30% of SPR may be unsuccessful, which increases the number of effects of short-circuit currents on power and switching equipment.

By the creation of powerful generating units and HV and EHV trunk lines, the problem of the successful elimination of the most probable single-phase faults has become more escalated. Really, an unsuccessful SPR worsens the operating conditions of circuit breakers that are at the limit of it interrupting capacity, on the other hand, the overestimated dead time of SPR on trunk lines requires an increase in the volume of emergency measures to provide the dynamic stability of power systems, which, in turn, leads to increased damage from the fault.

However, if you find a way to determine the time of extinction of the feed arc, then you can minimize the dead time of SPR for this accident associated with short circuit on the line. To achieve this goal, it is necessary to fix certain signs of the fact of self-destruction of the damage in the disconnected phase or the persistent nature of the damage.

Devices that determine the dead time of SPR in accordance with the true time of the existence of the feed arc are called adaptive devices for controlling the dead time of SPR, and SPR that satisfies this condition is called adaptive SPR. In the literature [1, 2, 3–11], it is proposed to control a number of operational parameters to determine the fact of the extinction of the feed arc and to decide on re-inclusion, such as:

- voltage module at the ends of the disconnected phase;
- voltage argument, measured in relation to a synchronously rotating vector;
- natural frequencies as part of the voltage of the disconnected phase;
- high-frequency components of the transition process during the burning of the feed arc;
- aperiodic current component in the neutral of the shunt reactor.

In the domestic literature [1, 2, 3–7], preference is given to monitoring the voltage of the disconnected phase to establish the fact of the extinction of the feed arc. If the voltage at the beginning of the emergency phase is selected as the controlled parameter, then it is necessary to compare the calculated situations shown in (Fig. 1).

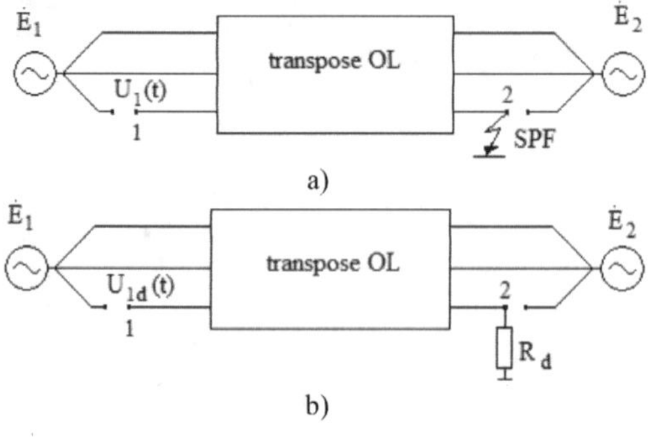

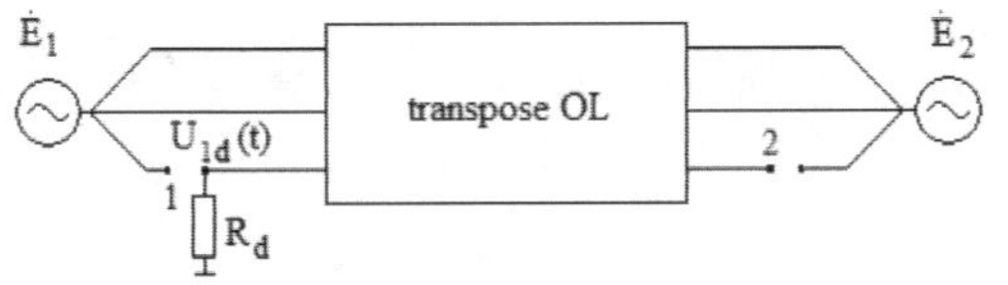

c)

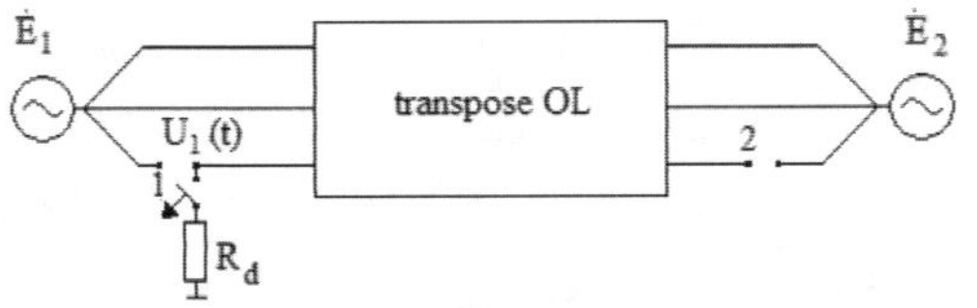

d)

Fig. 1. Comparable design situations with ASPR: a - stable SPF at the remote end of the overhead line (OL); b – non-missing secondary arc at the remote end of the overhead line; c - non-missing secondary arc at the controlled end of the overhead line; d - successful extinction of the secondary arc.

The developed ASPR reduce the likelihood of a phase being included at the final stage of ASPR on an sustained single-phase fault, but they do not completely exclude such a situation. So, for example, if one of the phases in the span to a tree is blocked, the arc will go out after the emergency phase is turned off, the gap strength will be restored, and ASPR will give a signal to turn on the phase. However, its inclusion will lead to repeated overlapping of the gap, since there is a hidden weak spot on the line that cannot be detected by using known methods of ASPR.

The method for implementing ASPR considered below, which is intended for use in intersystem connections that include a weak intermediate system (Fig. 2), is simple and allows to identify with a high degree of reliability all possible cases noted in Fig. 1. By an intermediate system we mean a weak system whose power is much less than the power of the connected systems.

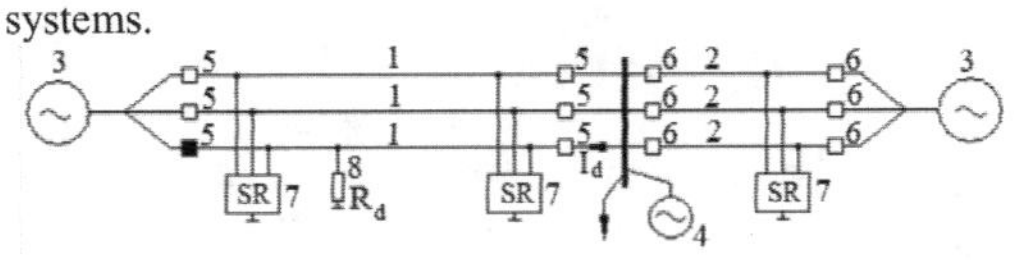

Fig. 2. ASPR of power lines in the presence of a weak intermediate system

The essence of the proposed solution is that after a two-sided disconnection of the damaged phase of the line for a selected dead time, a short-term (within 0.1 second) unilateral inclusion of the emergency phase from the side of the intermediate system 4 is carried out, the current value of the current of this phase at the switched on end is measured and compared with the specified setpoint.

The device contains phase lines 1 and 2 connected to systems 3 and 4 using switches 5 and 6 for any number of shunt reactors 7 on the lines. Any phase 1 or 2 may cause damage 8.

The method is as follows. A two-way disconnected emergency phase can be in one of three states. When the damage resistance is Rd ≈ 10 Ohm, an stable single-phase short circuit occurs, at Rd ≈ 500 Ohm, an arc fault occurs and when Rd → ∞, there is no phase damage.

However, in the latter case, as noted above, there is no full guarantee that adaptive ASPR will be successful. Therefore, the two-sided disconnected emergency phase 1 is turned on one-sided from the side of the intermediate system 4 using the switch 5, the current value of the current of this phase at the switched end is measured and compared with the specified setpoint. If the measured current is higher than the specified setpoint, then there is a short or arc fault in the phase, and a command to disconnect the line is given. Otherwise, there is no short circuit in the phase, and it successfully turns on at the other hand.

The specified current setting is selected from the condition of the detuning from the highest current, determined at the one-sided end of the phase in the absence of damage.

$$I_s \ge I_{max} \cdot K_r \cdot K_s$$

where K_r = 1,1 - reliability coefficient, K_s = 1,25 - sensitivity coefficient.

Switching on from the side of the intermediate system 4, the power of which, as a rule, is significantly less than the power of the adjacent systems 3, is characterized by low short-circuit currents if the fault is maintained in phase and therefore does not pose a risk to the equipment.

The technical result is to provide a high degree of reliability of ASPR of power lines with any number of SR on the line. In addition, the method guarantees reliable identification of arc damage that cannot be detected on a two-sided emergency phase.

The diagram in Fig. 2 depicts the successful one-way switching on of the emergency phase of the line from the side of the intermediate substation at Rd → ∞. The mode parameters necessary for identifying the success of one-way switching are found according to an algorithm recorded in phase coordinates.

As a controlled value, you can use either the current from the side of the intermediate system, or the argument of the input impedance from the side of the intermediate system $\arg Z(\delta)$.

Figure 3 shows the dependence of the effective value of the monitored current of the emergency phase when it is turned on unilaterally from the side of the intermediate system from the angle.

The influence of the angle on the controlled current is insignificant, but the number of reactors connected to the line must be taken into account.

The number of reactors connected to the line has a smaller effect on the controlled parameter if the input impedance argument for the emergency phase from the side of the intermediate system is selected as the controlled parameter (Fig. 4).

https://doi.org/10.1051/e3sconf/201913901038

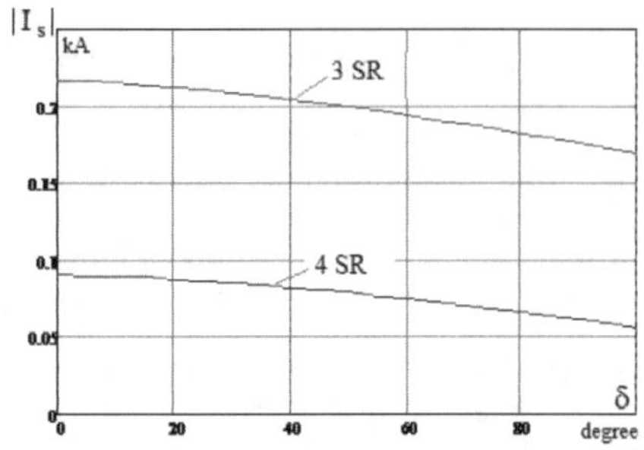

Fig. 3. Controlled current in the emergency phase from the side of the intermediate system with one-way switching on the overhead line

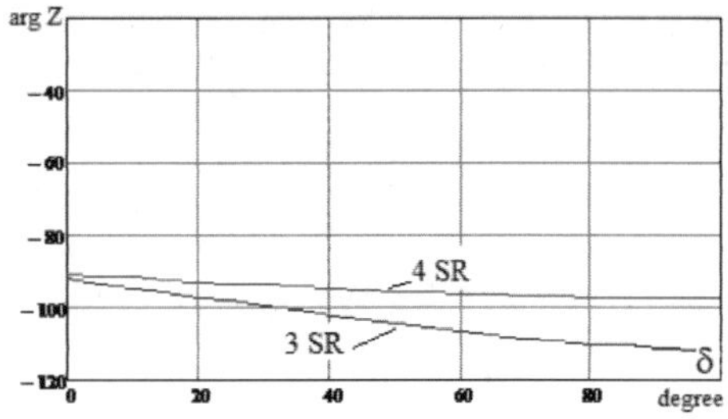

Fig. 4. Argument of input impedance for the emergency phase from the side of the intermediate system with one-way switching on the overhead line

The diagram in Fig. 2 depicts the one-way switching on of the emergency phase to a stable single-phase fault (at Rd ≈ 10 Ohms) or to an arc fault (at Rd ≈ 500 Om) of the line from the side of the intermediate substation.

Either the line current from system 4 or the input impedance argument from the side of the intermediate system is also used as a controlled quantity $Z_d(\delta, R_d)$.

Fig. 5 shows the dependences of the controlled current on the side of the intermediate system during an arc fault on the side of the open end of the overhead line

A significant difference in the currents at the beginning of the line in the modes of successful unilateral switching on of the emergency phase and its unilateral switching on the arc fault allows reliable identification of the outcome of the successful extinction of the secondary arc.

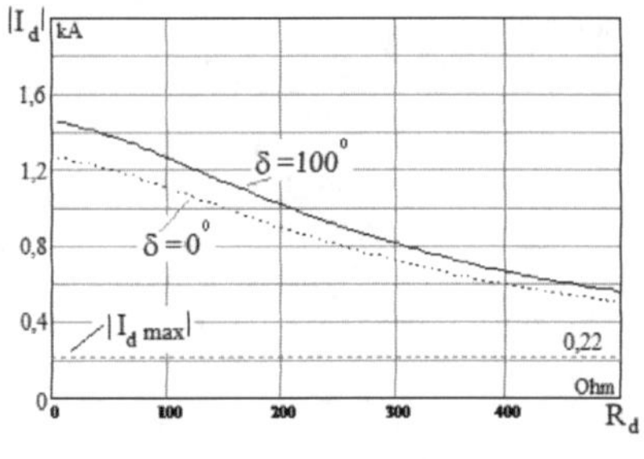

Fig. 5. Controlled current in the emergency phase from the side of the intermediate system depending on the arc resistance on a one-way switched on 500 kV overhead line

The identification of a successful ASPR is more efficient by the argument of the input resistance relative to the buses of the controlled phase at the intermediate substation (Fig. 6).

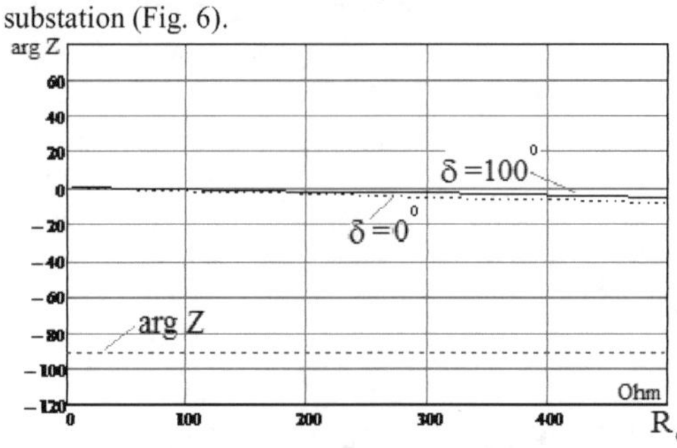

Fig. 6. Argument of the input impedance for the emergency phase from the side of the intermediate system depending on the arc resistance on a one-way switched on 500 kV overhead

As the analysis showed, for EHV lines in the range of lengths of 100-500 km, depending on the number of SR on the line, the specified setpoint is no more than 0.25 kA for lines 500 kV and 0.55 kA for lines 750 kV, while possible the current range at the switched-on end of the emergency phase in the presence of a fault or arc is at least 0.8 kA for 500 kV lines and 1.2 kA for 750 kV lines, which is more than 3 times higher than the setting for 500 kV lines and more than 2 times the setting for lines 750 kV [8 –11]. Thus, there is a reliable guarantee of identification of damage.

Let us further consider how permissible a one-way switching on from the side of the intermediate system is under the condition that the single-phase fault current affects the adjacent system. It follows from the table that the current of a single-phase short circuit on the side of this system, when turned on unilaterally from the side of the intermediate system, depending on the location of the single-phase short circuit on the overhead line, is in the range of 3-4 kA. And this means that according to the table, these currents are approximately 3 times less than the single-phase short-circuit current in the initial circuit, i.e. not dangerous to the adjacent system.

Table. Normal mode and single-phase short-circuit currents

Normal mode current in system 3, kA	Single-phase short current in system 3 for the initial circuit, kA	Single-phase short current in system 3 at unilateral inclusion from the side of the intermediate system, kA
2	11	3 – 4

Conclusion

A new effective way of implementing ASPR based on a short-term one-side connection of the fault phase from the side of the intermediate system is proposed. Connecting from the side of an intermediate system, which has power significantly less than the power of adjacent systems, is characterized by low short-circuit currents if a single-phase short circuit is maintained by the phase and therefore does not pose a risk to equipment. The technical result is to provide a high of reliability of ASPR with any number of shunt reactors on the line. In addition, the method guarantees reliable identification of arc damage which cannot be detected on a two-sided disconnected faulted phase.

References

1. Aleksandrov G.N. The design of EHV transmission lines. Leningrad, Energoatomizdat Publ., 1993. 560 p.
2. D'iakov A.F. The electric networks of EHV and UHV UES of Russia. Theoretical and practical issues. Moscow, NTF "Energoprogress" Corporation "EEEK" Publ., in 3 vols., 2012. 696 p.
3. A. s. 1056341 USSR. MKI3 H02H 3/06. A device for monitoring the state of a phase of a power line with shunt reactors disconnected in the cycle of single-phase automatic reclosing / M.L. Levinstein, M.I. Khoroshev, A.F. Kalinichenko, Yu. I. Lyskov // Discovery. Inventions 1983 No. 46.
4. A. s. 1092642 USSR, MKM3 H02H 3/06. A method for controlling the disappearance of a short circuit on a disconnected power line and a device for its implementation / V. A. Ilyin, B. And Kovalev, M. And Khoroshev, Yu.I. Lyskov // Discoveries. Inventions 1984. No. 18.
5. A. s. 632023 USSR. MKI3 H02H 3/06. The method of single-phase automatic reclosing of the power line / V. N. Sobolev, V.F. Lachugin, Yu. I. Lyskov, V.G. Kagan // Discovery. Inventions 1978, No. 41.
6. A. s. 740108 USSR, MKM3 H02H 3/06 11/00. A method for monitoring the state of a power line phase disconnected in a single-phase automatic reclosing of a power line with shunt and compensation reactors / N. N. Belyakov, V. S. Pashkes, K. V. Hoscian // Discovery. Inventions 1981.№26.
7. S.P. Websper, A.T, Johns, R.K. Aggarwal. And R.W. Dunn «An investigation into breaker reclosure strategy for adaptive single pole autoreclosing», IEE Proc.- Gener. Transm. Distrib., vol.142, no.6, pp.601-607, Nov.1994.
8. S. Ahn, Ch.Kim, R.K. Aggarwal. and A.T, Johns, «An alternative approach to adaptive single pole autoreclosing in high voltage transmission systems based on variable dead time control», IEEE Trans. Power Delivery, vol.16, no.4, pp.676 – 686, Oct. 2001.
9. Krasil'nikova T.G., Dzhononaev S.G. Comparative analysis of ways to eliminate unstable single-phase short circuits in non-transposed lines of EHV and UHV// Electricity . 2017.no. 11, pp. 22–29.
10. Krasil'nikova T.G., Dzhononaev S.G. Effect of duration of pause SPAR on the throughput of the transmission in condition for dynamic stability.// Scientific Bulletin of Novosibirsk state technical University. 2017. no. 2 (67), pp.163–176.
11. Dzhononayev S.G. Elimination of unstable Single-Phase Short-Circuits in transposed lines in the SPAR cycle using automatic phase shunting // Scientific Bulletin of Novosibirsk state technical University. 2018. no. 3 (72), pp.157–174.

Controlled switching circuits based on non-linear resistive elements

R.Ch. Karimov [1*], *M.K.* Bobojanov [1], *A.N.* Rasulov [1], *E.G.* Usmanov [1]

[1]Tashkent State Technical University named after Islam Karimov, 100095, Uzbekistan, Tashkent, University St. 2A

Abstract. Based on the results of the analysis of resistive circuits, a new circuit of an optoelectronic contactless voltage relay is developed in the article and its application in the power supply system is considered. The possibility of controlling the operating modes of various installations for amplifying electrical signals using optothyristors is presented and methods for their inclusion in electrical circuits are given. Optoelectronic contactless voltage relays were tested under laboratory conditions to obtain input-output characteristics and capacitance voltage waveforms. Based on the improvement of the proposed circuit, a circuit of an optoelectronic contactless voltage relay is developed and provides a high degree of reliability. The proposed voltage relay circuit is simulated using the Electronics Workbench and MATLAB R2014a programs, the results of which show that the changes in the shape of the output voltage curve are close to a sinusoid and coincide with the results obtained analytically.

Currently, power non-contact semiconductor equipment is widely used, which has significantly expanded the scope of semiconductor technology in a number of areas of electrical engineering and has provided qualitatively new results, which made it possible to master a new class of electrical devices.

Based on the foregoing, there is a need for the development of a non-contact voltage relay with a simple and reliable circuit that allows switching large currents of loads. For this purpose, in this paper we consider simple nonlinear resistive circuits [1-2].

High-quality devices developed on the basis of nonlinear dynamic circuits are widely used in various fields of automation, radio electronics, computer technology and power supply systems. The non-linear electric circuit under study consists of a diode connected in series to the *VD* network, active resistance R_1 and capacitance C and connected in parallel to this resistance circuit R_2 (Fig.1) [3-5].

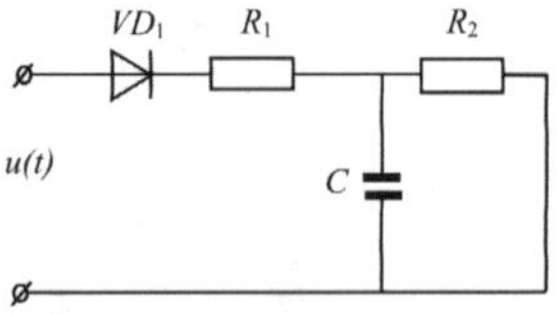

Fig.1. Scheme of a nonlinear electric circuit

To study the proposed nonlinear electric circuit, we use the classical method.

The voltage of the circuit will change according to a sinusoidal law:

$$u = U_m \sin \omega t \quad (1)$$

We accept that the angle of inclusion of the circuit in the network is zero. Consider the transients for each diode opening. In this case, the voltage across the capacitance is described by the following equation [3, 5]:

$$u_C = U_m \cdot \frac{R_2}{R_1 + R_2}\left(1 - e^{-\frac{R_1+R_2}{R_1 \cdot R_2 \cdot C}t}\right) \quad (2)$$

here, U_m is the maximum value of the mains voltage; R_1, R_2 and C circuit elements.

We take the characteristic of the diode ideal and assume that $U_{input} = U_m \sin \omega t$. Then, from the moment $t=0$ to t_1, the diode is open and when the condition $0 \leq t \leq t_1$ is fulfilled, the circuit equation has the following form [3, 5]:

$$U_m \sin \omega t = R_1 i + u_c \quad (3)$$

given that, we have $i = C\frac{du_c}{dt}$,

$$U_m \sin \omega t = R_1 (C\frac{du_C}{dt} + \frac{u_C}{R_2}) + u_C \quad (4)$$

From equation (4) we define $\frac{du_C}{dt}$ and from here we get:

$$\frac{du_C}{dt} = \frac{1}{R_1 C}\left(U_m \sin \omega t - u_C\left(1 + \frac{R_1}{R_2}\right)\right) \quad (5)$$

The differential equation of a nonlinear electric circuit described by expression (5) (Fig.1) is solved by the Lagrange method [5-8].

For an approximate solution of the differential equation of the chain, expression (5) is multiplied by $\frac{1}{R_1 C}$ [5, 7, 9]:

$$\frac{du_C}{dt}=\frac{U_m \sin\omega t}{R_1 C}-u_C\left(\frac{1+\frac{R_1}{R_2}}{R_1 C}\right) \tag{6}$$

By changing the equation we get:

$$\frac{du_C}{dt}+u_C\left(\frac{1+\frac{R_1}{R_2}}{R_1 C}\right)=\frac{U_m}{R_1 C}\sin\omega t \tag{7}$$

Equating the left side of differential equation (7) to zero, we obtain [7]:

$$\frac{du_C}{dt}+u_C\left(\frac{1+\frac{R_1}{R_2}}{R_1 C}\right)=0$$

From here:

$$\frac{du_C}{dt}=-u_C\left(\frac{1+\frac{R_1}{R_2}}{R_1 C}\right) \tag{8}$$

We multiply equation (8) by $\frac{dt}{u_C}$, integrate:

$$\int\frac{du_C}{u_C}=-\int\frac{R_1+R_2}{R_1 R_2 C}dt \tag{9}$$

The natural logarithm of this equation is calculated as follows:

$$\ln u_C=-\left(\frac{R_1+R_2}{R_1 R_2 C}\right)t \tag{10}$$

From here:

$$u_C=A(t)\cdot e^{-\left(\frac{R_1+R_2}{R_1 R_2 C}\right)t} \tag{11}$$

Substituting this expression into formula (8) we obtain [7].

$$\frac{du_C}{dt}=A(t)^{'}\cdot e^{-\left(\frac{R_1+R_2}{R_1 R_2 C}\right)t}-A(t)\cdot\left(\frac{R_1+R_2}{R_1 R_2 C}\right)\cdot e^{-\left(\frac{R_1+R_2}{R_1 R_2 C}\right)t} \tag{12}$$

Inserting the introduced changes in equation (7), we solve:

$$A^{'}(t)\cdot e^{-\left(\frac{R_1+R_2}{R_1 R_2 C}\right)t}-A(t)\cdot\left(\frac{R_1+R_2}{R_1 R_2 C}\right)\cdot e^{-\left(\frac{R_1+R_2}{R_1 R_2 C}\right)t}+A(t)\cdot\left(\frac{R_1+R_2}{R_1 R_2 C}\right)\cdot e^{-\left(\frac{R_1+R_2}{R_1 R_2 C}\right)t}=\frac{U_m}{R_1 C}\sin\omega t \tag{13}$$

After reductions, we get the following [7]:

$$A^{'}(t)=\frac{U_m}{R_1 C}\sin\omega t\cdot\left(\frac{R_1+R_2}{R_1 R_2 C}\right)dt \tag{14}$$

we denote $\frac{R_1+R_2}{R_1 R_2 C}=a$.

We integrate equation (14) [7]:

$$A(t)=\frac{U_m}{R_1 C}\int\sin\omega t\cdot e^{at}dt \tag{15}$$

We integrate the results of equation (15) by parts:

$$I=\int\sin\omega t\cdot e^{at}dt=\left|\begin{matrix}\sin\omega t=U & dv=e^{at}dt\\ \omega\cos\omega t dt=du & v=\frac{1}{a}\cdot e^{at}\end{matrix}\right|=\frac{1}{a}\cdot e^{at}\sin\omega t-\int\frac{\omega}{a}\cdot e^{at}\cos\omega t dt=$$
$$=\left|\begin{matrix}\cos\omega t=U & dv=e^{at}dt\\ -\omega\sin\omega t dt=du & v=\frac{1}{a}\cdot e^{at}\end{matrix}\right|=\frac{1}{a}\cdot e^{at}\sin\omega t-\frac{\omega}{a^2}\cdot e^{at}\cos\omega t-\frac{\omega^2}{a^2}\int e^{at}\cdot\sin\omega t dt \tag{16}$$

here, $I=\sin\omega t\cdot e^{at}dt$, the following notation is accepted: $\sin\omega t=U$; $e^{at}dt=dv$; $\int udv=uv-\int vdu$, we use the separate integration formula [7].

Simplify equation (16):

$$\int\sin\omega t\cdot e^{at}dt=\frac{1}{a}e^{at}\sin\omega t-\frac{\omega}{a^2}e^{at}\cos\omega t-\frac{\omega^2}{a^2}\int e^{at}\sin\omega t dt$$

$$\int\sin\omega t\cdot e^{at}dt\left(1+\frac{\omega^2}{a^2}\right)=e^{at}\left(\frac{1}{a}\sin\omega t-\frac{\omega}{a^2}\cos\omega t\right)$$

$$\int\sin\omega t\cdot e^{at}dt=\frac{a^2}{a^2+\omega^2}\cdot e^{at}\left(\frac{1}{a}\sin\omega t-\frac{\omega}{a^2}\cos\omega t\right)$$

$$\int\sin\omega t\cdot e^{at}dt=\frac{1}{a^2+\omega^2}\cdot(a\cdot\sin\omega t-\omega\cdot\cos\omega t)e^{at}+const \tag{17}$$

We introduce a constant in equation (17), taking into account (11) and (16) after some transformations we get [5, 7]:

$$u_C=\frac{U_m}{R_1 C}\cdot\left(\frac{1}{\left(\frac{R_1+R_2}{R_1 R_2 C}\right)^2+\omega^2}\right)\left(\frac{R_1+R_2}{R_1 R_2 C}\sin\omega t-\omega\cos\omega t\right)+const\cdot e^{-\left(\frac{R_1+R_2}{R_1 R_2 C}\right)t} \tag{18}$$

Figure 2 shows the graphs of the voltage change across the capacitance determined using formula (7).

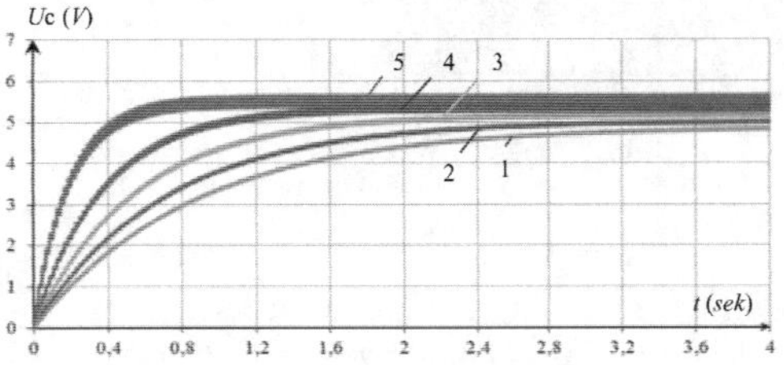

Fig.2. The voltage variation curves at the capacitance $U_c=f(t)$ with constant C and R_2 and with changing parameters R_1

Figure 2 shows the curves of the voltage change at the capacitance at: network voltage U_m=18 V, capacitance C=1 μF, resistance R_2=3,0 kOhm, the integration step h=0,001, and when the resistance R_1 changes (1 –0,3 kOhm ; 2 – 0,5 kOhm; 3 – 0,8 kOhm; 4 – 1,0 kOhm; 5 – 1,2 kOhm).

To verify the reliability of the simulation results of the considered circuit, computer simulation software was used [7-10].

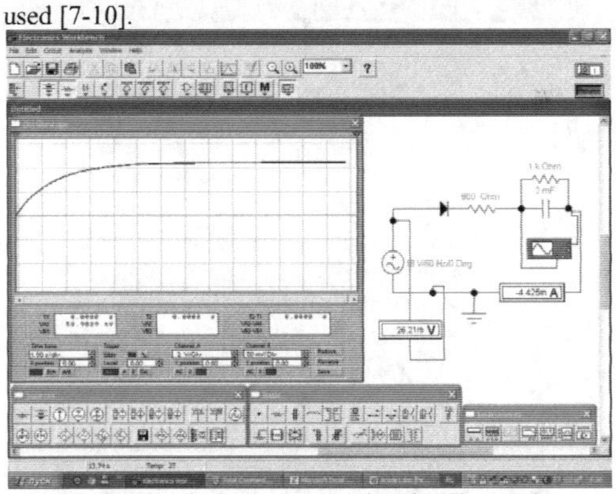

Fig.3. Simulation process using the Electronics Workbench software

Figure 3 shows a simulated graph of the voltage variation across the capacitance $U_c=f(t)$, considered in Fig.1 non-linear circuit using Electronics Workbench [6-7, 9-10, 12].

The graph of changes in the capacitance voltage $U_c=f(t)$ is shown in Fig.4 [13].

In the analysis time, by the Simulink block of the MATLAB R2014a program, a nonlinear electrical circuit (Fig.1), the process progresses over time, the results show that this program makes it possible to conduct more data for the analysis of the processes of voltage change in time [11, 14].

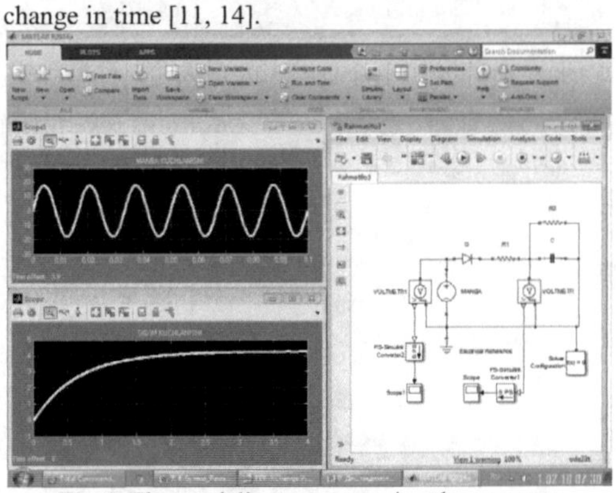

Fig.4. The modeling process using the program MATLAB R2014a

Figure 5 shows a comparison of the results obtained by mathematical modeling methods and programs: *a* - Electronics Workbench; *b* - MATLAB R2014a and *c* – analytical research method.

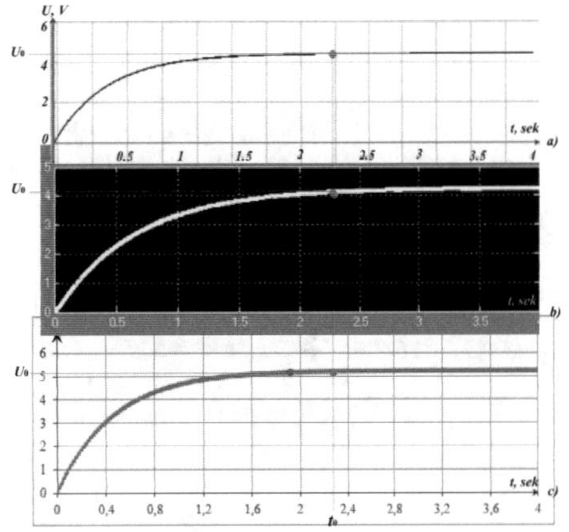

Fig.5. Comparison of the results of analysis of a nonlinear electric circuit

An analysis of the results of the graph of the voltage across the capacitance shows that the difference between the Lagrange methods and modeling using the Electronics Workbench and MATLAB programs is 1.5-2%. This is explained by the input of a constant value when solving the circuit under consideration by the Lagrange method. The results obtained by modeling a nonlinear electric circuit and the obtained analytical methods give the same results and coincide with the theoretical curves [6].

The shape of the load current can be influenced by controlling the opening angle of the thyristor. When the phase shift between the load current and the control signal is equal to zero, the curve of the load current will be in the form of a sinusoid. Thus, in order to ensure a sinusoidal shape of the current and voltage curves on the load, it is necessary to apply a signal to open the thyristor when the current passes through zero [4, 15].

Compared to electromagnetic installations, optoelectronic relays differ in reliability, durability and resistance to transient and frequency characteristics, improved technical and economic indicators and speed, makes it possible to use transformer windings and relay protection installations without switching special devices [3, 16-17].

The implementation of the optoelectronic contactless voltage relay using a thyristor and an optocoupler improves weight and dimensions. Figure 6 shows a circuit diagram of an optoelectronic contactless voltage relay with a sinusoidal waveform of a voltage across a load [4].

Consider the operation of an optoelectronic contactless voltage relay. The non-contact voltage relay consists of a diode bridge VD_2-VD_5 connected to a power source in series with the load R_{load}, into the diagonal of which the thyristor VT_2 is connected. The relay is also equipped with a capacitor *C*, a thyristor VT_1, three resistors R_1, R_2, R_3, a thyristor optocoupler *VU* and a diode VD_1. The control electrode of the first thyristor VT_2 is connected through a resistor R_3 to the capacitor plate *C* and the cathode of the thyristor optocoupler *VU*. The second lining of the capacitor *C* is connected to the cathodes of the thyristor VT_2, the cathode of the second thyristor VT_1 and the second terminal of the supply network. The anode of the thyristor VT_1 is connected to the cathode of the diode of the optocoupler *VU*, the anode of which is connected to the output of the resistor R_2. The second output of the resistor R_2 is connected to the anode of the thyristor of the optocoupler *VU* and the first output of the resistor R_1 also has connections to the first terminal of the supply network. The second output of the resistor R_1 is connected to the anode of the diode VD_1, the cathode of which is connected to the control electrode of the thyristor VT_1 [15-16].

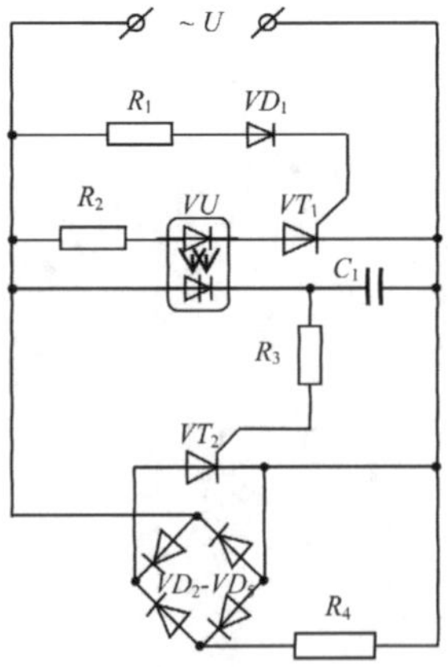

Fig.6. Schematic diagram of the optoelectronic contactless voltage relay

Consider the operation of an optoelectronic contactless voltage relay. When the unlocking signal at the control electrode reaches a value sufficient to unlock the thyristor, the thyristor VT_1 opens with an opening angle of 90^0 and connects the diode of the *VU* optocoupler to the network through a resistor. This leads to the opening of the thyristor of the optocoupler *VU* and the flow of current through the diode of the optocoupler, thereby leading to the opening of the thyristor of the optocoupler *VU*, and turns on the capacitor *C*. Since a constant signal is supplied from the capacitor *C* through the resistor R_3 to the control electrode of the thyristor VT_2, it remains open, and a sinusoidal current flows through the load R_4. The triggering of thyristors VT_1 is controlled by changing the parameter of the resistor R_1 [16-17].

This optoelectronic contactless voltage relay has been tested in the laboratory of the Department of Power Supply of the Energy Department of the Tashkent State Technical University named after Islam Karimov. In this case, thyristors *KU202I*, *KU201I*, respectively, were used as thyristors VT_1, VT_2, *D226B*, as diodes VD_1, as resistors R_1, R_2, R_3, R_4, respectively, resistors of 820 Ohms, 15 kOhm, 3.8 kOhm, 16 kOhm , capacitor *C* with a capacitance of 1 μF as capacitance *C*, *AOU103V* thyristor optocoupler, VD_2-VD_5 diode bridge - *KTs402E* used as optocouplers. Experimental studies have shown that the load R_4 was included in the network at a voltage of 18 V. The shape of the load voltage curve is sinusoidal [16-17].

Figure 7 shows the input-output voltage characteristic of an optothyristor non-contact voltage relay and it.

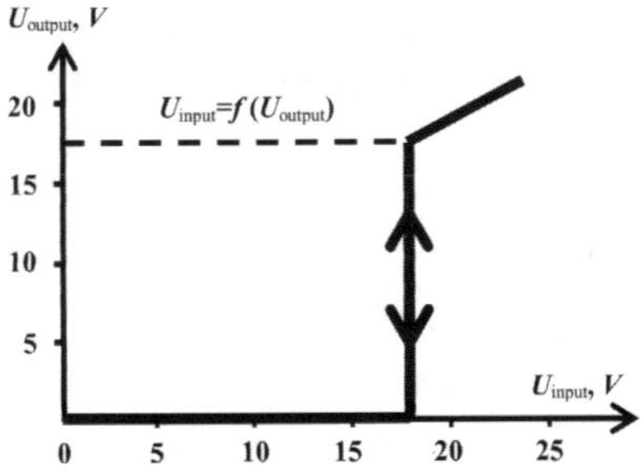

Fig.7. The characteristic "input-output" voltage optoelectronic proximity voltage relay

Using the MATLAB R2014a program, the circuit of the optoelectronic contactless voltage relay is shown (Fig.6), the results of which are shown in Fig.8 [11, 16-17].

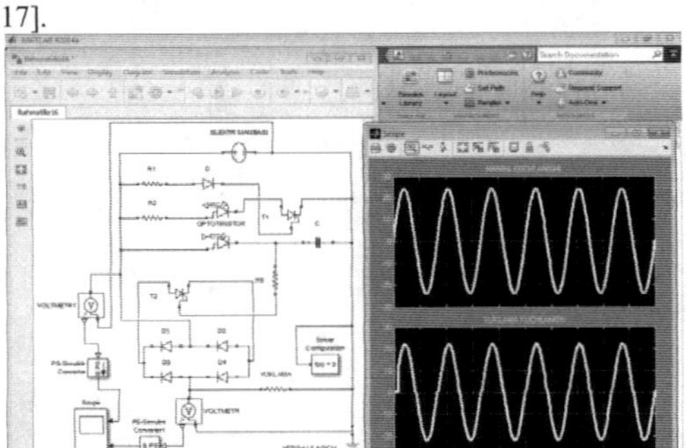

Fig.8. Modeling of an optoelectronic contactless voltage relay

Thus, the optoelectronic contactless voltage relay has a sinusoidal shape at the output [16-18].

In the laboratory of the Department of Power Supply, experimental studies were conducted using a digital oscilloscope of the LeCroy WaveRunner 64 Xi-A type (USA) and the voltage amplitude value was taken.

Experimental waveforms are shown, the input voltage variation curves within the range of 176÷241 V, while the output voltage varies within the range of 217÷224 V. The results of the experiments show that the proposed voltage stabilizer, depending on the change in the voltage of the network, provides voltage stability at the load, i.e. within ±2% of the nominal [19-20].

Figure 9-10 shows the oscillograms of the input-output voltage, the stabilizer for active load.

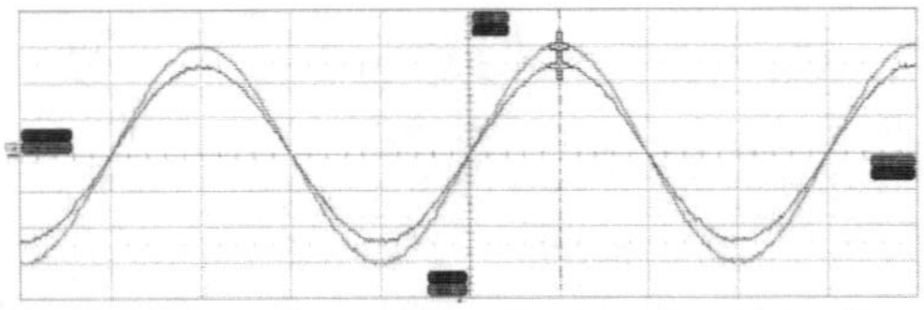

Fig.9. The waveform of the voltage input-output voltage stabilizer for active load: U_{input}=175 V; U_{output}=214 V

Figure 9 shows the oscillogram of the voltage change at U_{input}=175 V (deviation from the nominal -20,4%) and load voltage U_{output}=214 V (deviation from the nominal -2,72%), the input-output voltage difference is 39 V.

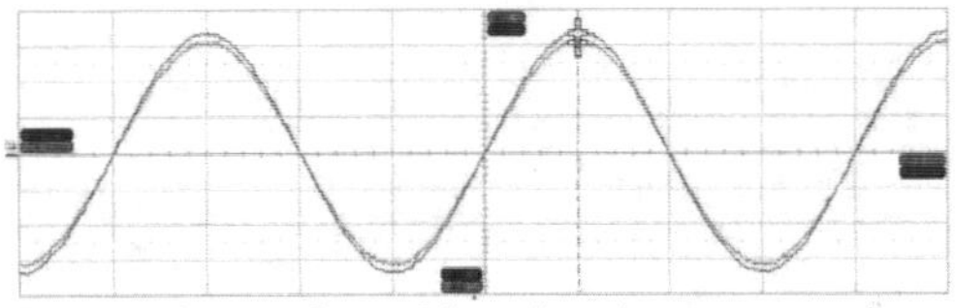

Fig.10. The waveform of the voltage input-output voltage regulator for an active load: U_{input}=241 V; U_{output}=224 V

Figure 10 shows the oscillogram of the voltage change at U_{input}=241 V (deviation from the nominal value of 8,71%) and the voltage at the load U_{output}=224 V (deviation from the nominal value of 1,78%), the input-output voltage difference is 17 V.

The results of the analysis of the voltage waveform show that, when the input voltage varies from -20,4% to +8,71%, the output of the voltage regulator changes from -2,72% to +1,78%. Thus, the developed prototype voltage stabilizer meets the requirements of the state standard [19-20].

Thus, the goal and objectives set in this scientific work are obtained, such an optoelectronic contactless voltage relay has been developed; for use in the switching control system of the magnetizing windings of a boost transformer [16].

The tested prototype of the voltage stabilizer provides a deviation of the voltage at the load within the permissible range of ±5% from the nominal one, which leads to an improvement in the quality of electricity.

Conclusion

To analyze the dynamic processes of a nonlinear electric circuit, a solution to the differential equation is developed.

For the analysis of a nonlinear electric circuit, it is recommended to use a numerical method for solving the differential equation of state of a circuit. Using computer technology, this circuit was studied vertically experimentally and it was found that the time to establish the voltage on the capacitance depends on the value of the parameters of the diode-capacitive circuit.

Based on the improvement of the proposed circuit, a circuit of an optoelectronic contactless voltage relay is developed that eliminates false alarms and provides a high degree of reliability.

The simulation of the proposed voltage relay circuits is performed using the MATLAB R2014a program, the results of which show that changes in the shape of the output voltage curve are close to a sinusoid and coincide with the results obtained analytically.

Experimentally, using the LeCroy WaveRunner 64 Xi-A type oscilloscope, the proposed voltage stabilizer was tested and it was determined that the use of this stabilizer reduces the starting torque of the active-inductive load by 0,2 sec.

References

1. ES Filippov, Textbook, Energy, 504 (1985).
2. MK Bobojanov, RCh Karimov, Bulletin of TSTU, **4(101)**, 53-57 (2017).
3. MK Bobojanov, EG Usmanov, EH Abduraimov, RCh Karimov, JESR, **1-2**, 210-212 (2018).
4. AN Rasulov, RCh Karimov, JEE, **4**, 174-178 (2015).
5. RCh Karimov, Collection of materials of the IV international scientific conference "Modern Trends in Technical Sciences", 33-35 (2015).
6. EG Usmanov, AN Rasulov, RCh Karimov, DGU 05406 (2018).
7. RCh Karimov, EG Usmanov, AN Rasulov, MR Ruzinazarov, DGU 06134 (2019).
8. WJ Cunningham, Textbook, Gosenergoizdat, 456 (1962).
9. VP Sigorsky, AI Petrenko, Textbook, Publishing. 2nd, Soviet Radio,. 608 (1976).
10. II Aliev, Textbook House of Radio, 112 (2003).
11. VV Teryokhin, Textbook, Kuzbassvuzizdat, 376 (2004).
12. http://studentlib.com, http://radio.bobrodobro.ru/10215
13. VS Scherbakov, AA Ruppel, VA Glushets, Textbook, Publishing House, 160 (2003).
14. AM Namestnikov, Methodological guidelines, UISTU, 72 (2004).
15. MM Fayziev, NA Kurbonov, IN Karimov, JYS, **6**, 215-219 (2016).
16. EG Usmanov, EKh Abduraimov, RCh Karimov, Patent of Uzbekistan. IAP 05122 (2015).
17. EKh Abduraimov, RCh Karimov, INNOVATION-2016, 128-129 (2016).
18. AX Sulliev, RCh Karimov, JTREI, **4**, 149-154 (2018).
19. Interstate standard GOST 32144-2013.
20. The state standard of Uz. is O'zDSt 1044:2003.

The support vector machine application in the implementation of multidimensional relay protection

A.L. Kulikov[1] , *D.I.* Bezdushniy[1,*], *M.V.* Sharygin[1] and *V.Yu.* Osokin[1]

[1]Nizhniy Novgorod State Technical University, Russia

Abstract. Power engineering digital transformation, the use of different intelligent electronic devices (IEDs), high-speed communication protocols provide extensive opportunities for relay protection and automation systems modernization of power utilities. One of the most promising avenues of power engineering development is design of new protection devices, whose principles are based on the elements of artificial intelligence and machine learning. The article discusses the features of the application of one of the most common machine learning algorithms, the support vector machine, by the example of constructing a three-dimensional fault detector, which would serve to increase a transmission line stepped protection selectivity. The proposed fault detector has high recognition ability and ease of technical implementation as part of the protection IED.

1 Introduction

Modern electric power systems are technically complicated complexes consisting of a large number of interconnected elements. At the same time, both producers and consumers of electricity impose increasingly stringent requirements for reliability of power supply. Relay protection plays an important role in the process of power engineering intellectualization. Modern relay terminals, in addition to the protection functions, quite often perform the functions of control, recording, oscillography, and a number of others. The exchange of information with other devices located in and outside the substation is provided by means of data acquisition and transmission systems [1]. In particular, the implementation of the IEC 61850 standard assumes the existence of a single intra substation process bus, not only for the relay protection devices, but also for other automation equipment [2]. This allows digital devices to access a large amount of information about the protected object in real time. Unfortunately, the basic algorithms of relay protection fundamentally unchanged over the past decades, and in fact they are digital analogues of their electromechanical predecessors [3, 4].

One of the most promising approach to the relay protection and automation organization is based on the multiple simulation and statistical processing of the simulation results [4 - 9]. In particular, it is advisable to use machine learning to implement the recognition functions of electric network states [10 - 12]. The use of this approach to power utilities protection potentially has the following advantages:

- features that help to make a decision on the protection operation can be individual for each object and be selected according to the condition of maximum information value. Since machine learning algorithms, as a rule, are multidimensional [3, 6, 7], it can be any number of features for a single protection limited only by the performance of a digital device that runs this algorithm;
- the choice of the parameters of the relay operating is connected with learning of the recognition algorithm via sample data characterizing the behavior of the protected object in normal and emergency states. Such data can be obtained both by the simulation modeling results and directly during operation. This approach allows protection to adapt to the operating conditions in real time (self-study).

Machine learning algorithms can be used not only as part of relay decision-making procedures, but also as a part of fault detectors that complement conventional types of relay protection and improve their main characteristics (speed, selectivity, sensitivity). Below is an example of the development of special fault detector for distance relay. It serves for the transmission line protection and it is based on one of the simplest and most illustrative but at the same time effective [11, 12] machine learning methods [6, 7] - support vector machine.

2 Principles of the support vector machine application

Problems solved by the machine learning can be divided into two main classes: classification and regression problems. The regression problem is to restore the form of some function $F(x)$ according to the available empirical information about the function values $F_1...F_n$ and the arguments of the function $x_1...x_n$. The classification problem is to categorize some feature vector x as one of the classes $Y_1...Y_m$, based on a training sample consisting of a set of vectors $x_1...x_n$, which classes are known. The relay protection object is very close to the classification problem and can be formulated in machine learning terms. In this case, the subject of classification is the vector consisting of the parameters of currents and voltages calculated by measuring elements of protection device and the classes are the operating states of the protected object (normal operating state, short circuit in the protected zone, short circuit outside the protected zone, etc.).

The simplest form of the support vector machine is a two alternative classification method, the idea of which is to form a hyperplane in the feature space and separate elements related to different classes. Furthermore, a hyperplane should be as far as possible from representatives of both classes.

Suppose there is a training sample $\mathbf{x}_1 ... \mathbf{x}_N$, which consists of N vectors located in the M-dimensional feature space; then for each element of the training sample, there is a class label $y_1 ... y_N$ such that if the i-th element belongs to one class (we denote it by class α), then $y_i = 1$ and if the i-th element belongs to another class (we denote it by class β), then $y_i = -1$. Further, if the training sample is linearly separable, it follows that there is a hyperplane such that all the points on one side belong to the class α and all the points on the other side belong to the class β. Thus, we have the hyperplane equation

$$Q(\mathbf{a}) = \mathbf{w}^{\mathrm{T}} \cdot \mathbf{a} + b = 0, \tag{1}$$

where $\mathbf{w}$, b are coefficients that define the hyperplane; $\mathbf{a}$ is a vector describing an arbitrary point in the feature space where the hyperplane is constructed.

If the hyperplane really separates the training sample points, then we get

$$\forall i \in 1...N,\ y_i \cdot \left(\mathbf{w}^{\mathrm{T}} \cdot \mathbf{x}_i + b\right) > 0. \tag{2}$$

In the general case, there is a hyperplane such that it satisfies expression (2) and it is not unique. The Fig. 1 shows the training sample points belonging to two classes and two options for the location of the separating hyperplane **a** and **b**. Since the considered example corresponds to two-dimensional space (Fig. 1), we see that the hyperplanes degenerate into straight lines. Despite the fact that both straight lines properly divide groups of points between themselves, it is clear from Fig. 1 that **a** is more suitable for the classification of the analyzed data because of providing a wide bandwidth between representatives of different classes. The width of this band is called the margin. The SVM method makes it possible to find a hyperplane that provides the maximum margin between classes.

Proven [11], that the equation of the hyperplane that provides the maximum margin can be obtained by finding a conditional minimum

$$\begin{cases} \min_{\mathbf{w}_m, b_m} \frac{1}{2} \mathbf{w}_m^{\mathrm{T}} \cdot \mathbf{w}_m, \\ \forall i \in 1...N,\ y_i \cdot \left(\mathbf{w}_m^{\mathrm{T}} \cdot \mathbf{x}_i + b_m\right) \geq 1, \end{cases} \tag{3}$$

where $\mathbf{w}_m$, b_m are coefficients that define the hyperplane such that ensures the maximum margin.

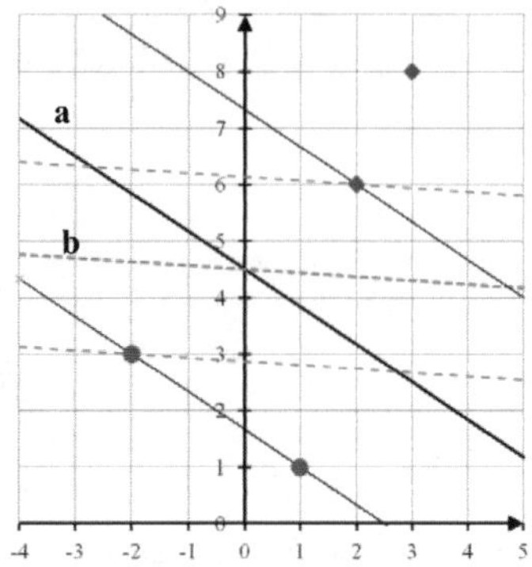

Fig. 1. Examples of separating hyperplanes.

However, in most practical cases, the training sample is not linearly separable. In other words, there is no hyperplane that satisfies a system of inequalities (2). For the SVM algorithm to be used in these cases, it is necessary to be modified in such a way as an incorrect classification of training sample objects is allowed. For the optimization objective, since there is a penalty for a total classification error, we have

$$\begin{cases} \min_{\mathbf{w}_m, b_m, \xi_1..\xi_N} \frac{1}{2} \mathbf{w}_m^{\mathrm{T}} \cdot \mathbf{w}_m + C \cdot \sum_{i=1}^{N} \xi_i \\ \forall i \in 1...N,\ y_i \cdot \left(\mathbf{w}^{\mathrm{T}} \cdot \mathbf{x}_i + b\right) \geq 1 - \xi_i\ a, \\ \forall i \in 1...N, \xi_i \geq 0 \end{cases} \tag{4}$$

where ξ_i is a slack variable; C is an error penalty.

In practice, the dual form of above optimization problem is widely used. Besides the fact that it gives the similar answer to the initial problem, it can be solved by faster iterative methods. In these terms, the optimization is performed with respect to the variables $\lambda_1...\lambda_n$, corresponding to Lagrange multipliers. We say that the dual SVM problem for a linearly inseparable sample is based on the primal problem [7] and write

$$\begin{cases} \min_{\lambda_1...\lambda_n} \left(\sum_{i=1}^{n} \lambda_i - \frac{1}{2} \cdot \left(\sum_{i=1}^{n} \sum_{j=1}^{n} \lambda_i \lambda_j y_i y_j \mathbf{x}_i^{\mathrm{T}} \mathbf{x}_j \right) \right) \\ \forall i \in 1...N,\ 0 \leq \lambda_i \leq C, \\ \forall i \in 1...N,\ \lambda_i y_i = 0. \end{cases} \tag{5}$$

https://doi.org/10.1051/e3sconf/201913901040

It is necessary to note that quadratic programming problem (5) can be solved by means of one of the well-known methods [for example, 9].

Since the solution gives the optimal Lagrange multipliers $\lambda_1 \ldots \lambda_n$ (5), this means that we can recover the equation of the separating hyperplane

$$\mathbf{w} = \sum_{i=1}^{n} \lambda_i y_i \mathbf{x}_i, \qquad b = \frac{1}{y_s} - \mathbf{w} \cdot \mathbf{x}_s, \tag{6}$$

where s is the index number of the feature vector such that $0 < \lambda_s < C$.

Thus, using the coefficients $\lambda_1 \ldots \lambda_n$, we get the following classification algorithm

$$a(\mathbf{x}) = \operatorname{sign}\left(\sum_{i=1}^{n} \lambda_i y_i \mathbf{x}_i^{\mathrm{T}} \mathbf{x} - b \right). \tag{7}$$

Let us remark that for the correct operation of the classifier described by the expression (7), it is sufficient to sum over those i such that $\lambda_i \neq 0$. In other words, addition combines only support vectors. Note, that the number of vectors is considerably smaller than the total size of the training sample.

3 The support vector machine application in relay protection problems

Let us consider an example of using the SVM algorithm to increase the recognition ability of protection devices. Suppose a distance relay device is installed on a transmission of the electric grid section (Fig. 2). Suppose the primary zone of the relay protects the entire line ω1 and the backup zone is meant for protection of the adjacent lines ω_2 and ω_3. Besides, distance relay have to operate faults on transmission line ω_2 and ω_3 by separate zones, because it is necessary to provide required sensitivity and speed of a distance relay. Note also that zones have different sets and time delays and are denoted by "Zone ω_2" and "Zone ω_3". Now we implement the recognition algorithm for the distance relay such that it distinguishes faults on the lines ω_2 and ω_3.

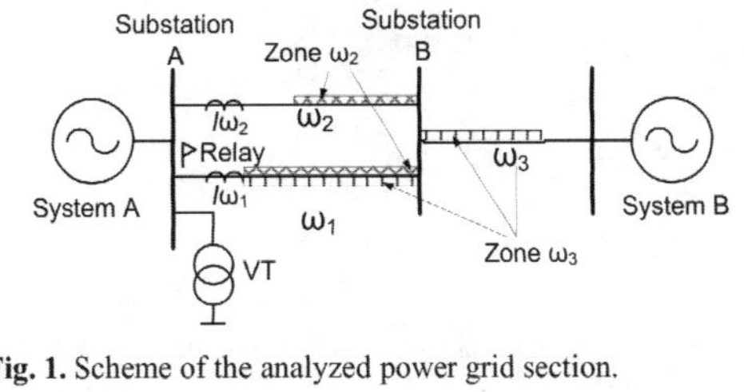

Fig. 1. Scheme of the analyzed power grid section.

Repeatedly simulating faults on the lines ω_2 and ω_3 in accordance with Monte Carlo method [10], we obtain the mode parameters (currents and voltages) at the protective device location in the form of complex values.

An analysis of the obtained model data shows that on the complex plane of active and reactive resistances, many short circuits on the lines ω_2 and ω_3 of different distances and with different values of the transition resistance will be located as shown in Figure 3.

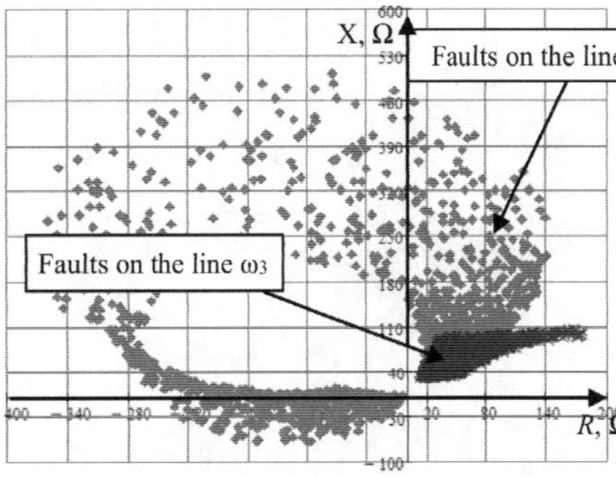

Fig. 2. The set of complex impedances in cases of different fault distances on the lines ω_2 and ω3.

According to Fig.3, practically all faults on line ω_3 are characterized by similar to line ω_2 fault resistances such that they are located in the "overlap" area, therefore, that they cannot be definitely identified. It is obvious that it is not possible to select the distance relay zone characteristics such that one of them ensures relay operation only at the line ω_2 faults and the other only at the line ω_3.

Let us use the support vector machines and design additional fault detector for operating at faults on the line ω_3 and blocking at faults on the line ω_2. Fig.4 shows the logic of the cooperative functioning of zone ω_2 and ω_3 fault detector and additional one.

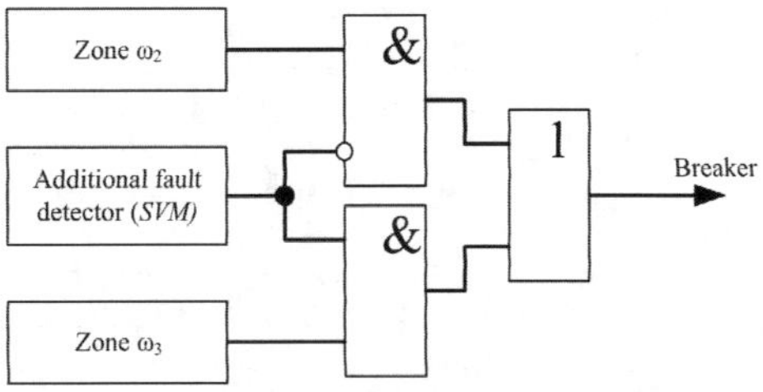

Fig. 3. The logic of the cooperative functioning of zone ω_2 and ω3 fault detector and additional one based on SVM.

Suppose that resistance (R), reactance (X), and the parallel line current (Iω_2), as shown in Fig.2, are used as information features for implementation of an additional fault detector; then combine that features in a single three-dimensional space. The areas of existence of feature values in the resulting space are shown in Fig.5. Since the additional fault detector functions properly in conjunction with above zones (Fig.4), we see that it is possible to limit data points that are used as a training set by doing experiments such that at least one of these zones operates.

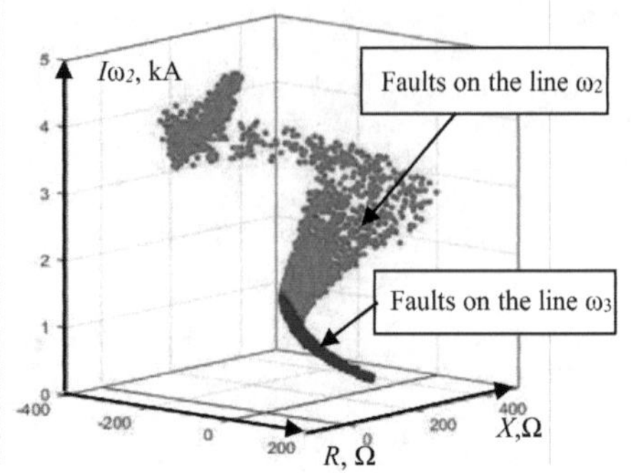

Fig. 4. The area of existence of fault points on the lines ω_2 and ω_3 in a three-dimensional space.

Such approach will increase the accuracy of the algorithm whereas learning time will be reduced.

Thus, the training sample consists of the data points we get in experiments such that faults occur in hatched areas of the line ω3 (shown in Fig.2), which we denote by α-states, and hatched areas of the line ω_2 we denote by β –states. Fig.6 shows the training data points in the feature space.

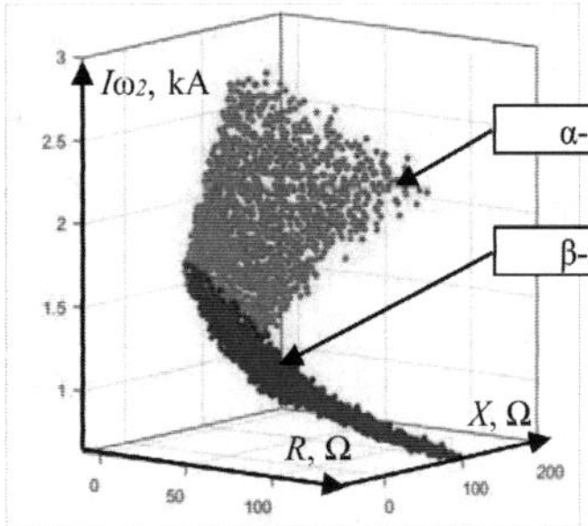

Fig. 5. The points in the training set

Let us find out how each of the selected features is effective for the classification problem solving individually. Fig.7 shows distributions of conditional probability density functions of the selected features in α- и β-states.

Based on each feature, we implement three one-dimensional classifiers. The classifiers will choose one of two hypotheses: "The fault on the line ω_2" and "The fault on the line ω_3" depending on the value of the corresponding feature. The decisive rule for the one-dimensional classifiers will be implemented according to the maximum likelihood estimation [for example, 14]. Each of the classifiers will choose the hypothesis which allows the obtaining feature value to be more likely.

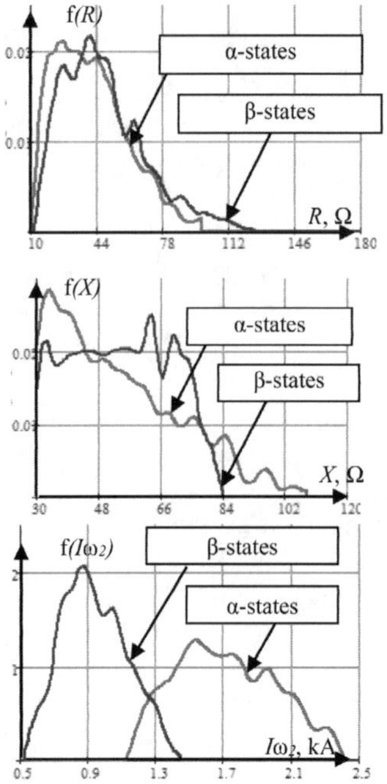

Fig. 6. Conditional probability density distribution functions of classification features.

Table 1 shows what error probability is provided by each of the features when the feature is used in the one-dimensional classifier. The probability distributions show that such features as resistance (R) and reactance (X) used alone provide probability of correct recognition almost the same as choosing the answer "at random".

Table 1. Error probabilities of one-dimensional classifiers.

R	X	$I\omega_2$
40,32%	41,21 %	4,79 %

Let us implement the "learning" procedure of the SVM algorithm with a linear kernel function [8] and the coefficient C = 100 on a previously obtained sample. The separating plane obtained as the result of the learning procedure is shown in the Fig.8.

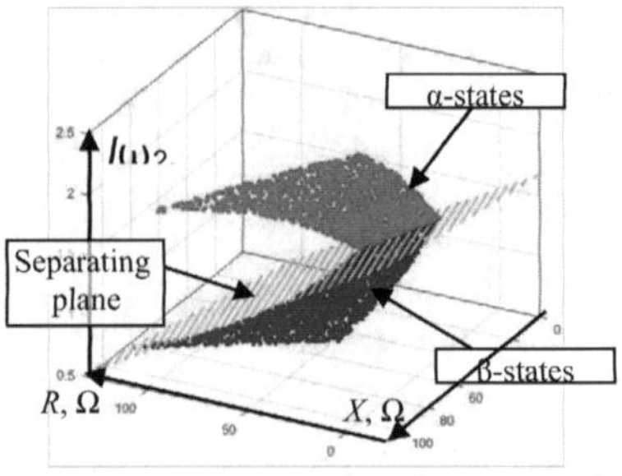

Fig. 7. Separating plane obtained as a result of the support vector machines.

Thus, the error percentage of the designed SVM-based recognition algorithm does not exceed 0.03% which is much less than the most effective one-dimensional classifier can provide. Therefore, the considered example confirms that the combination of information features into a single multidimensional feature space allows to achieve significant benefits in the task of relay protection.

4 Conclusion

The digitalization of power engineering and the development of modern information technologies made it possible to apply a fundamentally new approach to the relay protection designing, which based on simulation and machine learning.

The application of the support vector machine is promising in relay protection problems, both in the formation of new protection algorithms and in the use of SVM as an additional mean for increase the selectivity and speed of existing types of protection.

The combination of information features in a multidimensional feature space let us to implement a classifier such that it has a greater probability of the correct recognition than the use of information features.

This work was supported by the Ministry of Science and Higher Education of the Russian Federation (agreement No. 075-15-2019-1209 of May 31, 2019. Unique project identifier RFMEFI57717X0244).

References

1. I.N. Lizunov, A.N. Vasev, R.Sh. Misbakhov, News of Universities. Energetics problems. J. **1-2**, 52 (2017)
2. M. Adamiak, D. Baigent, R. Mackiewicz, Prot. Control J. **1**, 61 (2009)
3. E.M. Schneerson, *Digital relay protection* (Energoatomizdat, Moscow, 2007)
4. M.V. Sharygin, A.L. Kulikov, *Protection and automation of power supply systems with active industrial consumers* (NIU RANEPA, Nizhny Novgorod, 2017)
5. A.L. Kulikov, M.V. Sharygin, Electrical engineering J. **2**, 58 (2019)
6. D.I. Bezdushniy, A.L. Kulikov, Relay protection and automation J. **1**, 20 (2019)
7. M.V. Sharygin, A.L. Kulikov, Power stations J. **2**, 32 (2018)
8. H. Van Trees, *Detection, Estimation, and Modulation Theory: Detection, Estimation, and Linear Modulation Theory* (John Wiley & Sons, Inc., New York, 2001)
9. E. Gose, *Pattern recognition and image analysis* (Prentice Hall PTR, Upper Saddle River, 1997)
10. C.M. Bishop, *Pattern Recognition and Machine Learning* (Springer, Berlin, 2006)
11. L. Wang, *Support Vector Machines: Theory and Applications* (Springer, Berlin, 2005)
12. L. Bottou, C-J. Lin, Large Scale Kernel Machines J. **1**, 1 (2007)
13. S.M. Ermakov, *Monte Carlo Method in Computational Mathematics: Introductory Course* (St. Petersburg, 2009)
14. A.G. Zyuko, D.D. Klovsky, M.V. Nazarov *Signal theory* (Radio and communication, Moscow, 1986).

https://doi.org/10.1051/e3sconf/201913901041

Determining single-phase earth fault location with applying short-term double earth fault

A.L. Kulikov[1], *V.Ju.* Osokin[1*], and *D.I.* Bezdushniy[1]

[1]NSTU n.a. R.E. Alekseev, Nizhniy Novgorod, Russian Federation

Abstract. The single-phase earth faults location are the predominant type of damage in distribution networks of 6-35 kV. The problem of remotely determining the fault location during single-phase earth faults has not definitely accepted practical and accurate solution. The article proposes and substantiates the intellectual methods of determining the fault location with using artificial introduction of short-term double earth faults. In the future by recorded oscillograms of currents and voltages, an accurate calculation of the distance to the damage is assumed.

1 Introduction

Single-phase earth fault is the predominant type of damage in medium voltage distribution electrical networks [1-6]. The task of determining the fault location does not have no unambiguous solution [8-18] due to the specificity of electromagnetic transients under single-phase earth faults, caused primarily by the neutral point grounding [1, 7]

The long-term operation of the network with single-phase earth faults leads to increase accident rate and probability of the development of single-phase earth faults into multiphase short circuits, which could in turn cause significant economic damage.

Fast and accurate detection of fault location is one of the main condition for eliminating the emergency and restoring the normal operation of the electrical network.

Previously the proposed methods of synthetic introduction of two-phase earth faults [19-22] for determining the location of single-phase earth faults are prospective in used, but, unfortunately, have not yet found practical implementation in operating organizations, including due to low accuracy.

2 The fault location algorithm for single-phase earth faults

Short-term connection of ballast resistance to the reserve cell of the bus section of the 6-35 kV switchgear is proposed for solving the problem of determining the location. A circuit breaker installed in the standby cubicles switchgear with a phase drive, whose the main contacts are closed between themselves and connected to the ground through ballast resistance.

After occurrence the single-phase earth fault, one of the undamaged phases of the circuit breaker is closed with ballast resistance, ensuring that a short-circuit current of a limited magnitude and time flows through the ballast resistance, thereby simulating a double earth fault [16, 17]. After registering current flowing through the ballast resistance, the circuit breaker is turned off and network operation mode is restored.

Figure 1 shows an equivalent circuit describing the current flow in the synthetic introduction of double earth fault, and the following designations are accepted: $\underline{E}_{ph1}$, $\underline{E}_{ph2}$, $\underline{E}_{ph3}$ – is the equivalent of system electromotive force; $\underline{Z}_s$ – equivalent resistance of system; $\underline{z}_L$ – line resistivity (Ohm/km); $\underline{z}_m$ – resistivity of mutual induction (Ohm/km); $\underline{Z}_{load}$ – equivalent load resistance; R_t – transient resistance in the place of single-phase earth faults; l - the actual distance to point of ground fault; R_b – ballast resistance; L – length of the line (km).

The use of the scheme shown in Figure 1, in phase coordinates [7], allows obtaining expressions to determine distances to the single-phase earth fault.

 https://doi.org/10.1051/e3sconf/201913901041

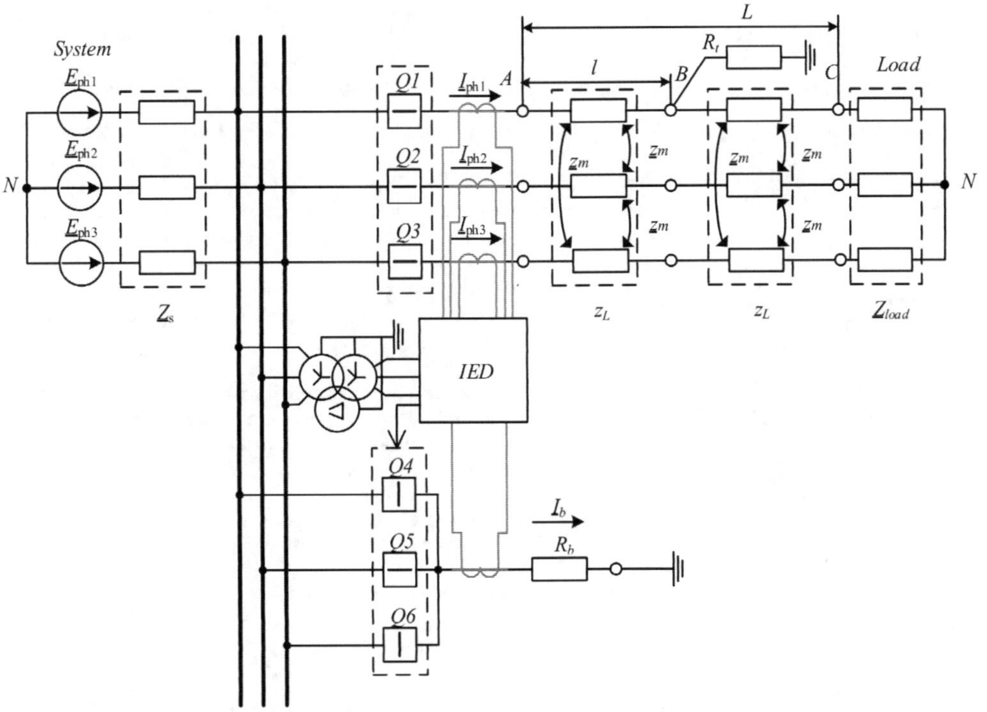

Fig. 1. The scheme for replacing the network in the single-phase earth fault mode with synthetic introduction of double earth fault

The equivalent circuit shown in figure 1 is converted to the form shown in figure 2 for the convenience of solving the scheme relatively to unknown parameters

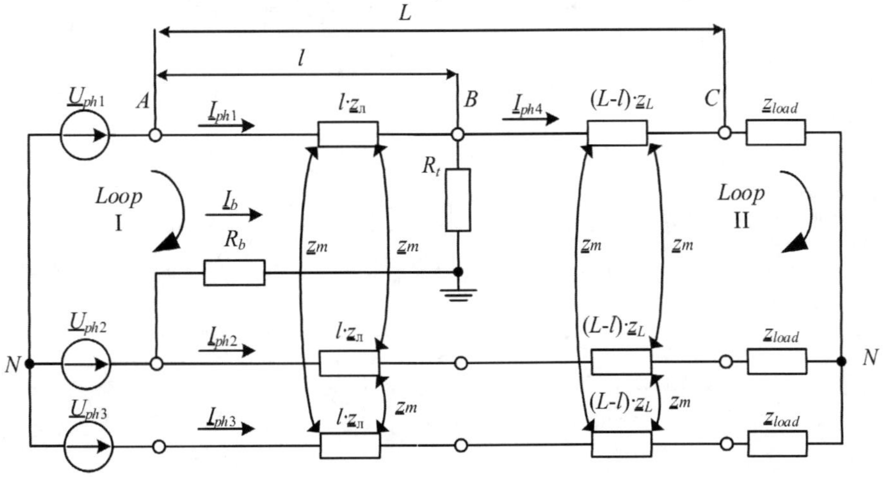

Fig. 2. The scheme for replacing the network in in double earth fault mode

Write Kirchhoff's voltage law for the first and second loops (figure 2), after specifying their current directions on the circuit.

From loop 1 of figure 2, and moving in a clockwise direction as indicated, gives the following expression:

$$\underline{I}_{ph1}\cdot \underline{z}_L \cdot l + \left(\underline{I}_{ph2}+\underline{I}_{ph3}\right)\cdot \underline{z}_m \cdot l - \underline{I}_b \cdot \left(R_b + R_t\right) = \underline{U}_{ph1} - \underline{U}_{ph2} \quad (1)$$

The expression for the second loop is as follows

$$\begin{aligned} &\underline{I}_{ph4}\cdot(L-l)\cdot\underline{z}_L+\underline{I}_{ph4}\cdot\underline{z}_{load}+ \\ &+(\underline{I}_{ph2}+\underline{I}_{ph3})\cdot(L-l)\cdot\underline{z}_m- \\ &-\underline{I}_{ph2}\cdot(\underline{z}_{load}+\underline{z}_L)-\underline{I}_{\phi3}\cdot\underline{z}_m\cdot L- \\ &-\underline{I}_{\phi1}\cdot\underline{z}_m\cdot l-\underline{I}_{\phi4}\cdot\underline{z}_m\cdot(L-l)+ \\ &+\underline{I}_{б}\cdot(R_б+R_n)=0 \end{aligned} \quad (2)$$

Having expressed from the equation for the first circuit (R_b+R_t) and substituting in equation (2), we obtain the equation for determining the distance to the point of damage.

$$l=\frac{(\underline{I}_{ph1}+\underline{I}_b-\underline{I}_{ph2})\cdot(\underline{z}_L\cdot L-\underline{z}_m\cdot L+\underline{z}_{load})}{\underline{I}_b\cdot(\underline{z}_L-\underline{z}_m)}-\frac{(\underline{U}_{ph1}+\underline{U}_{ph2})}{\underline{I}_b\cdot(\underline{z}_L-\underline{z}_m)} \quad (3)$$

In general terms, the calculation expressions look like this:

$$l=\frac{(\underline{I}_{dam.ph.}+\underline{I}_b-\underline{I}_{ph.b})\cdot(\underline{z}_L\cdot L-\underline{z}_m\cdot L+\underline{z}_{load})}{\underline{I}_b\cdot(\underline{z}_L-\underline{z}_m)}-\frac{(\underline{U}_{dam.ph.}+\underline{U}_{ph.b})}{\underline{I}_b\cdot(\underline{z}_L-\underline{z}_m)} \quad (4)$$

where, $\underline{I}_{dam.ph}$ – the current of the damaged phase on which there was single-phase earth fault; $\underline{I}_b$ – the current flowing through the ballast resistance; $\underline{I}_{ph.b}$ – the current of the phase closed on the ballast resistance; $\underline{U}_{dam.ph}$ – the voltage of the damaged phase on which the fault occurred; $\underline{U}_{ph.b}$ – the voltage of the phase closed on the ballast resistance.

It should be noted that during the formation of the final ratio (4) substitution and compensation of the sum (R_b+R_t) were realized, as a result of which the calculation of the distance to the single-phase earth faults location does not depend on the R_t. This circumstance is extremely important, since in networks with isolated and compensated neutral, almost all single-phase earth faults are followed by time-varying transient resistance. On the other hand, the independence of l (expression (4)) from R_b emphasizes that the introduction of ballast resistance is necessary only to provide the required range of current and voltage fixation in case of short-term double earth fault.

In the equivalent circuit (figure 2) and obtained the expression (4) is not taken into account the influence of the active conductance and capacitance conductance of the line. In order to avoid possible distortion of calculations from capacitive conductivity of power lines, it is proposed to consider only the actual part of expression (4), and use of correction ratios, in further calculations, will reduce the influence of active conductivity on accuracy characteristics of obtained results.

In order to use expression (4), it is necessary to know the information about the parameters of the electric network that can be obtained by the operating organizations:

- from design or reference data ($\underline{z}_L$, $\underline{z}_m$, L), taking into account the design of the power line;
- by analyzing data from operational measuring complexes, automated metering systems of electric energy, SCADA systems or by requesting dispatch services of consumers to obtain information about the load of the power line $\underline{z}_{load}$.

3 Results of the modelling

The accuracy of the obtained calculation algorithm (4) was estimated by simulating the damaged section of the electrical network in the Matlab software package in accordance with the scheme shown in Figure 3.

The considered scheme has the following parameters:

- voltage of 35 kV network;
- length of the lines $Line_1$=20 km; $Line_2$=15 km; $Line_3$=10 km;
- specific resistance of the phase: $\underline{z}_L$=0.825+j0.6949 Ohms/km;
-specific resistance of mutual induction: $\underline{z}_m$=0.0485+j0.3517 Ohms/km;
- transient resistances R_t in the places of closures are determined by a random quantity distributed with uniform law in the range from 0 to 100 Ohms;
- load power consumption: $\underline{S}_{load}$ = 45+j18.6 MBA that corresponds to resistance $\underline{z}_{load}$ = 16.242 + j6.713 Ohms;
- ballast resistance R_b =100 Ohms.

 https://doi.org/10.1051/e3sconf/201913901041

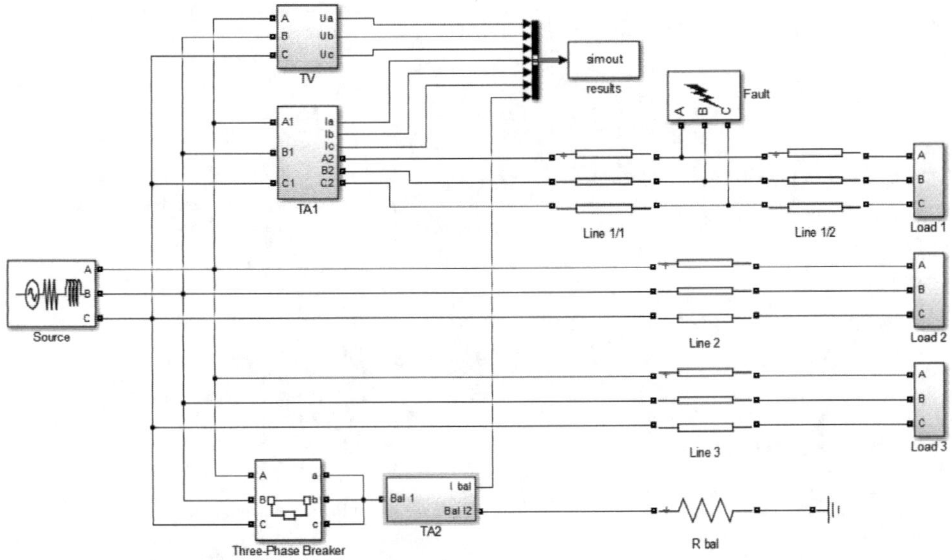

Fig. 3. Simulation modelling of the electrical network section in the mode of double earth fault

Network modeling was carried out in two stages. At the first stage, a model of power line with an inductive coupling between the phases was formed, and the transient resistance during single-phase earth faults was set randomly from a given range (R_t = 0…100 Ohm). As a result of simulation experiments, the dependence (Figure 4) of the calculated distances on the actual (i.e., set during the simulation) was constructed and the maximum error in estimating the distance to the single-phase earth faults location was determined.

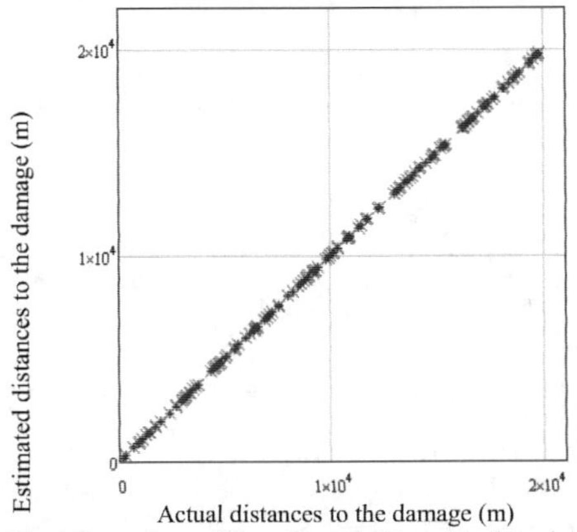

Fig. 4. Dependence of the estimated distances to the points of damage from the actual, in the absence of line conductivity

Analysis of the obtained results showed that due to the creation of a short-term artificial double earth fault and the use of a calculated algorithm appropriating to expression (4), it is possible to determine the distance to the single-phase earth faults location with an error not exceeding 1% of the line length. The maximum error of the calculated distance from the actual distance (Figure 4) was about 20 meters.

However, in practice, it is impossible to neglect existence of conductivity of the line, therefore, at the second stage of modeling the power line with the distributed parameters having capacitive conductivity $3.04 \cdot 10^{-9}$ Sm/m was imitated.

According to the results of simulation experiments, the maximum error of calculations by expression (4) was 12 %, and the maximum deviation of the calculated distances from the actual ones respectively was 1.5 km (figure 5A). The introduction of the correction ratio taking into account capacitive conductivity and the obtained dependence (Figure 5a), allows reducing the error of determination of distance to single-phase earth faults to the value of 1%, which according to the results of simulation is about 100 meters (Figure 5b)

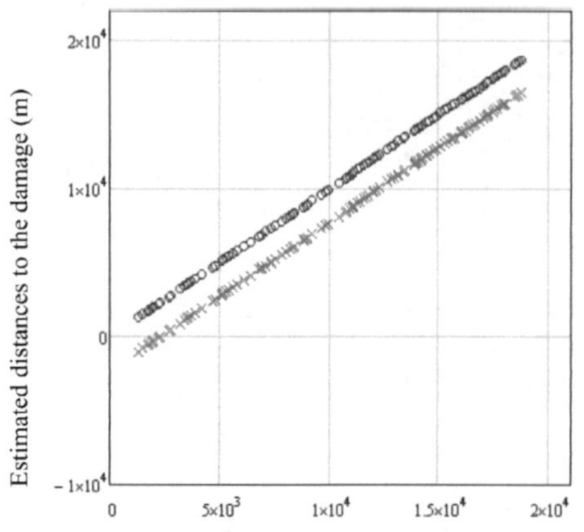

a)

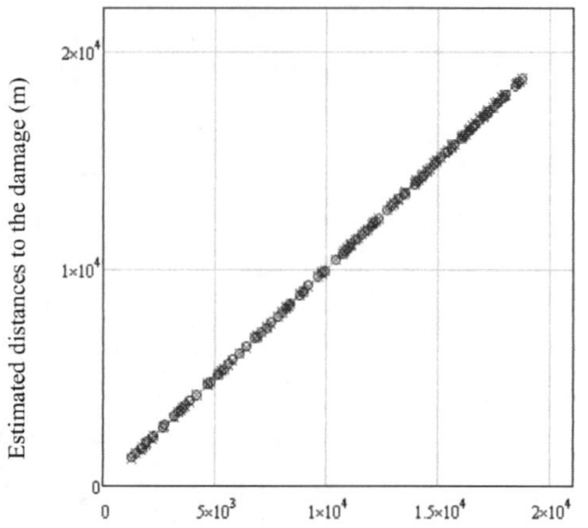

b)

Fig. 5. Dependence of the estimated distances to the points of damage from the actual, in the account of line conductivity

Thus, there is obtained fast and accurate algorithm for determining of single-phase earth faults on power transmission lines 6-35 kV, which will allow not only to reduce the time spent on elimination of damage and recovery of normal operation of the network, but also to significantly reduce the probability of damage transition to the short circuit, which is disconnected by the standard action of relay protection, and, as a result, to reduce the number of short-term power supply violations of consumers. The accuracy of the proposed algorithm in cable and cable-air networks can be reduced due to the presence of high-frequency discharge components of the transition current of the single-phase earth fault, as well as a significant influence of the total capacitive current of the network [1].

It is important to note that solving the problem of distance detection of the single-phase earth faults location in 6-35 kV distribution networks is one of the important tasks related to the implementation of the concept digital transformation accepted in 2018. The proposed algorithms can be widely used in digital complexes of relay protection and automation systems as part of digital distribution networks and digital substations.

4 Conclusions

1. The proposed algorithms allow to accurately determine the single-phase earth faults location in electric networks 6-35 kV with isolated or compensated neutral through the use of simulation, accumulation of statistical information and compensation of errors in real calculation on the simulation results.

2. An important advantage of the developed algorithms is the complete elimination of the influence of the transient resistance on the accuracy of determination of the single-phase earth fault location, and the introduction of a correction factor taking into account the uniform distribution of conductivity along the length of the line allows to minimize the corresponding calculation error.

3. The proposed algorithms for determining the single-phase earth faults location due to short-term double earth faults can be implemented in modern relay protection devices and correspond to the concept of creating smart electrical networks (Smart Grid).

The reported study was funded by RFBR, project number 19-38-90144.

References

1 V.A. Shuin, A.V. Gusenkov, *Zashchity ot zamykaniy na zemlyu v elektricheskikh setyakh 6-10 kV, M.: NTF Energopro-gress, Bibliotechka elektrotekhnika*, P.104 (2001)

2 A. L. Kulikov, M. Sh. Misrikhanov, A. A. Petrukhin, *Opredelenie mest povrezhdeniy LEP 6 – 35 kV metodami aktivnogo zondirovaniya, – M.: Energoatomizdat*, P.162 (2009)

3 E.A.Vorob'eva, D.I. Gandzhaev, G.A. Filatova, V.A. Shuin, *Informatsionnye parametry elektricheskikh velichin perekhod-nogo protsessa dlya opredeleniya mesta zamykaniya na zemlyu v raspredelitel'nykh kabel'nykh setyakh napryazheniem 6-10 kV, Vestnik IGEU*, **I.2.**, pp.34-42 (2017)

4 E.F. Khakimzyanov, R.G. Mustafin, A.I. Fedotov, *Opredelenie rasstoyaniy do mest dvoynykh zamykaniy na zemlyu na linii elektroperedachi raspredelitel'noy seti srednego napryazheniya, Izvestiya vuzov, Problemy energetiki*, **I.3-4**, pp.132-137 (2015)

5 A.N. Visyashchev, *Diagnostika sostoyaniya vozdushnykh liniy elektroperedachi 10-110 kV v normal'-nykh i avariynykh rezhimakh, IrGTU, Uchebnoe posobie,* P. 270 (2012_

 https://doi.org/10.1051/e3sconf/201913901041

6 Geoinformatsionnaya sistema OMP 6-35 kV. URL: www.relematika.ru (data obrashcheniya: 17.07.2017) www.relematika.ru/produkty/6-35_kv/sistema_omp_6-35_kv.

7 B.V. Papkov, *Toki korotkogo zamykaniya v elektricheskikh sistemakh, Ucheb. Posobie NGTU,* P.275 (2005)

8 Yu. S. Belyakov, *Aktual'nye voprosy opredeleniya mest povrezhdeniya vozdushnykh liniy elektroperedachi, Bibliotechka energetika, prilozhenie k zhurnalu «Energetik»,* **11**, P.80 (2010)

9 A.L. Kulikov, M.D. Obalin, P.A. Kolobanov, *Kompleksnye algo-ritmy OMP LEP na baze statisticheskikh metodov, Energetik*, **1** (2012)

10 A. L. Kulikov, M. D. Obalin, P. A. Kolobanov, *Analiz i povyshenie tochnosti pri opredelenii mesta povrezhdeniya liniy elektroperedachi, Izvestiya vuzov. Elektromekhanika*, **5**, pp.57-62 (2013)

11 H. Diaz, M. López, *Fault location techniques for electrical distribution networks: a literature survey, in Proc. the Fifth IASTED International Conference*, pp. 311-318 (2005)

12 M. Lehtonen, *Novel techniques for fault location in distribution networks, in Proc. Power Quality and Supply Reliability Conference*, pp. 1-6 (2008)

13 A.L. Kulikov, V.V. Anan'ev, *Adaptivnoe volnovoe opredelenie mesta povrezhdeniya linii ehlektroperedach, Vestnik IGEU,* **l.4**, pp.21-25 (2014)

14 V.F Lachugin., D.I. Panfilov, A.N. Smirnov, *Realizatsiya volnovogo metoda opredeleniya mesta povrezhdeniya na liniyakh elektroperedachi s ispol'zovaniem statisticheskikh metodov analiza dannykh, Izvestiya Rossiyskoy akademii nauk, Energetika*, **6**, pp.137-146 (2013)

15 A.L. Kulikov., M. D. Obalin, V.Ju. Osokin, T.R. Sharafeev, *Primenenie imitacionnogo modelirovaniya LEP 6–35 kV dlya povysheniya tochnosti opredeleniya rasstoyaniya do dvojnyh zamykanij na zemlyu, Vestnik IGEU,* **1**, pp.40-49 (2018)

16 A.L. Kulikov., V.Ju. Osokin, M. D. Obalin, *Povyshenie tochnosti algoritmov opredeleniya mesta povrezhdeniya LEP 6–35 kv pri dvojnyh zamykaniyah na zemlyu s pomoshch'yu vvedeniya iteracionnyh procedur, Elektroenergiya, peredacha i raspredelenie*, **1**, pp.54-59 (2019)

17 A.L. Kulikov., V.Ju. Osokin, M. D. Obalin, *Recovery time reduction of a damaged 6–35 kV transmission line after double earth fault applying the iteration fault location method, E3S Web of Conferences, Rudenko International Conference «Methodological problems in reliability study of large energy systems»* (2019)

18 A.L. Kulikov., V.Ju. Osokin, M. D. Obalin, *Improving Accuracy the Fault Location on Transmission Line 6–35 kV for Double Earth Fault, 2018 ICIEAM*

19 A.L. Kulikov., V.Ju. Osokin, A. A. Loscutov, *Sposob opredeleniya mesta i rasstoyaniya do mesta odnofaznogo zamykaniya na zemlyu v elektricheskih setyah 6–35 kV s izolirovannoj ili kompensirovannoj nejtral'yu*, Patent RF, no. 2685746, IPC G01R31/08, (2019)

20 A.L. Kulikov., V.Ju. Osokin, A. A. Loscutov, *Sposob opredeleniya mesta i rasstoyaniya do mesta odnofaznogo zamykaniya na zemlyu v elektricheskih setyah 6–35 kV s izolirovannoj ili kompensirovannoj nejtral'yu*, Patent RF, no. 2685747, IPC G01R31/08, (2019)

21 V.A. Fastunov, *Sposob opredeleniya mesta i rasstoyaniya do mesta odnofaznogo zamykaniya na zemlyu v elektricheskih setyah 6–35 kV s izolirovannoj ili kompensirovannoj nejtral'yu,* Patent RF, no. 2293342, IPC G01R31/08, (2007)

22 V.N. Pupynin., V.H. Nguen, *Ustrojstvo dlya opredeleniya rasstoyaniya do mesta odnofaznogo zamykaniya na zemlyu v setyah 6–35 kV elektricheskih sistem s izolirovannoj ili kompensirovannoj nejtral'yu*, Patent RF, no. 2096795, IPC G01R31/08, (1997)

Adaptation of power lines fault location methods to voltage characteristics deviations

Mixail Obalin[1], and *Georgij* Majstrenko[2,*]

[1] NNSTU, Power Engineering, Electricity Supply and Power Electronics Department, 603950 Nizhny Novgorod, Russia
[2] NNSEEU, Electrification and Automation Department, 606340 Knyaginino, Russia

Abstract. The influence of voltage characteristics deviations, acceptable in normal operation of power grid, at power lines emergency state fault location methods precision is considered. The error distributions of emergency state fault location methods across power line length under frequency deviation, single rapid voltage changing and supply voltage unbalance variation are received. The way of emergency state fault location adaptation to voltage characteristics deviations is offered.

Introduction

Ensuring reliable supply of electricity to consumers is one of the main goals of organization, existence and development of power systems. As part of this issue, it is necessary to quickly and correctly locate the faults. Modern digital computing equipment used by relay protection and fault location devices is sensitive to voltage characteristics deviations [1]. In practice, voltage characteristics for networks of different voltage levels and purpose can deviate from ideal ones and lie both within [2] and outside [3] the limits.

The existing directions of research in this area can be separated into two groups:

1) the use of voltage characteristics as an argument of the fault location method [4];

2) development of fault location methods adapted to voltage characteristics deviations [5].

The approach implemented in [4] involves permanent analysis of voltage characteristics over the entire area of the observed network, using synchronized measurements. In case of violation of electric power quality caused by fault, iterative algorithm of multi-terminal fault location [6] is implemented (argument of fault location formula is vector of voltage drops at observed points). It is obvious that such an approach involves high requirements for network infrastructure and, consequently, a significant amount of investment.

The approach implemented in [5] eliminates the possible influence of supply voltage unbalance due to the calculation in phase coordinates. Iterative algorithm of one-terminal fault location with target function as reactive component of fault resistance is implemented. The disadvantage is the introduction of an additional error associated with the representation of the entire load concentrated at the most electrically distant point.

Thus, the described fault location methods are either expensive to implement or have a certain methodological error. Research in general tends to focus on excluding or compensating for the influence of some single voltage characteristic.

In this work, the way of adaptation of fault location methods to deviations of voltage characteristics, existing in normal mode, is investigated. It can be implemented both one-terminally and multi-terminally. The adaptation method is not a source of additional methodological error. In the case of known deviations of voltage characteristics, it is possible to use the adaptation algorithm to compensate for the influence of any of them.

The study of normative documentation on voltage characteristics issues [7] and previous studies [8] have shown the expediency of studying the effect on the accuracy of emergency state fault location of the following voltage characteristics long-term existing in normal network operation: frequency deviation, single rapid voltage changing, supply voltage unbalance. According to [7], simulated deviations of voltage characteristics were within the limits specified in Table 1.

Table 1. Range of voltage characteristics deviations

№	Voltage characteristic	Range
1	Frequency deviation, %	± 5
2	Single rapid voltage changing	± 6
3	Supply voltage unbalance, %	0-4

1. Fault location methods and power line model

For the study, a group of one-terminal impedance based emergency state fault location methods was selected.

The calculation formulas and the decryption of the values used are shown in Figure 1 and Table 2.

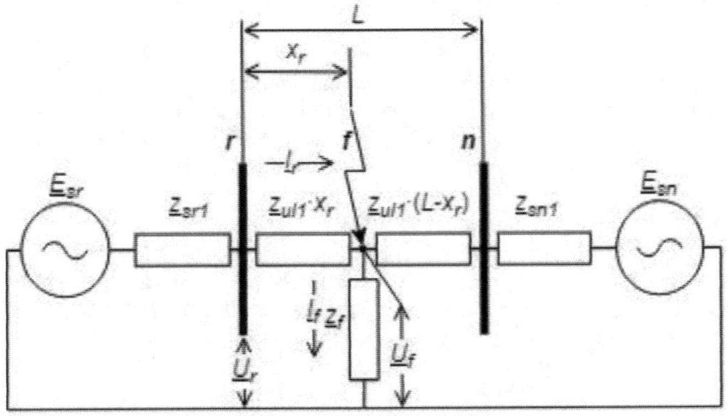

Fig. 1. Diagram of faulted line in post fault mode

Table 2. Information about fault location methods

№	Algorithm	Formulas and explanations
1	Takagi [9]	$$x_e = \frac{\mathrm{Im}\left(\underline{U}_r \cdot \Delta\hat{\underline{I}}_r\right)}{\mathrm{Im}\left(\underline{I}_r \cdot \Delta\hat{\underline{I}}_r \cdot \underline{z}_{ul1}\right)} \quad (1)$$, where $\underline{U}_r$ voltage at measuring point; $\underline{I}_r$ – current at measuring point; $\Delta\hat{\underline{I}}_r$ – complex conjugate pure fault current at measuring point; $\underline{z}_{ul1}$ – positive sequence line impedance per 1 km ; x_e fault location evaluation from measuring point.
2	Ericsson [10]	$$x_e = \frac{B_1 - \sqrt{B_1^2 - 4 \cdot B_2 \cdot B_0}}{2 \cdot B_2} \cdot L \quad (2)$$, where $B_2 = \mathrm{Im}\left(\hat{\underline{k}}_3\right)$; $B_1 = \mathrm{Im}\left(\hat{\underline{k}}_3 \cdot \underline{k}_1\right)$; $B_0 = \mathrm{Im}\left(\hat{\underline{k}}_3 \cdot \underline{k}_2\right)$; $\underline{k}_1 = \frac{\underline{U}_r}{\underline{I}_r \cdot \underline{z}_{ul1}} + 1 + \frac{\underline{z}_{sn1}}{\underline{z}_{ul1}}$; $\underline{k}_2 = \frac{\underline{U}_r}{\underline{I}_r \cdot \underline{z}_{ul1}} \cdot \left(1 + \frac{\underline{z}_{sn1}}{\underline{z}_{ul1}}\right)$; $\hat{\underline{k}}_3$ complex conjugate of $\underline{k}_3 = \frac{\Delta\underline{I}_r}{\underline{I}_r \cdot \underline{z}_{ul1}} \cdot \left(1 + \frac{\underline{z}_{sn1} + \underline{z}_{sr1}}{\underline{z}_{ul1}}\right)$; $\underline{z}_{sn1}$ positive sequence impedance of system behind measuring point; $\underline{z}_{sr1}$ positive sequence impedance of remote system.
3	Ankamma [11]	$$x_e = \frac{\mathrm{Im}\left(\underline{U}_r \cdot \hat{\underline{I}}\right)}{\mathrm{Im}\left(\underline{I}_r \cdot \hat{\underline{I}} \cdot \underline{z}_{ul1}\right)} \quad (3)$$, where $\hat{\underline{I}}$ - complex conjugate of $\underline{I} = \underline{a}_{f1} \Delta\underline{I}_{r1} + \underline{a}_{f2} \Delta\underline{I}_{r2}$; $\underline{a}_{f1}, \underline{a}_{f2}$ – weighting coefficients; $\Delta\underline{I}_{r1}$, $\Delta\underline{I}_{r2}$ – pure fault positive and negative currents at measuring point, respectively.
4	Wiszniewski [12]	$$x_e = \frac{X_r}{x_{ul1}} - \frac{R_r \cdot \tan(\varphi_l) - X_r}{x_{ul1} \cdot \left(\frac{a}{b} \cdot \tan(\varphi_l) - 1\right)} \quad (4)$$, where $R_r = \mathrm{Re}\left(\frac{\underline{U}_r}{\underline{I}_r}\right)$; $X_r = \mathrm{Im}\left(\frac{\underline{U}_r}{\underline{I}_r}\right)$; $\varphi_l = \arg\left(\underline{z}_{ul1}\right)$; $a = \mathrm{Re}\left(\frac{\Delta\underline{I}_r}{\underline{I}_r \cdot e^{j\cdot\gamma}}\right)$; $b = \mathrm{Im}\left(\frac{\Delta\underline{I}_r}{\underline{I}_r \cdot e^{j\cdot\gamma}}\right)$; $\gamma = \arg\left(\frac{\underline{z}_{ul1} \cdot L - X_r \cdot \cot(\varphi_l + j) + \underline{z}_{sn1}}{\underline{z}_{sr1} + \underline{z}_{ul1} \cdot L + \underline{z}_{sn1}}\right)$.
5	Arzhannikov [13]	$$x_e = \frac{\left(X_r - R_r \cdot \tan(\alpha - \beta)\right) \cdot \tan(\varphi_l)}{\left(\tan(\varphi_l) - \tan(\alpha - \beta)\right) \cdot x_{ul1}} \quad (5)$$, where $\alpha = \arg\left(\frac{\underline{I}'_f}{\underline{I}_r}\right)$; $\beta = \arg\left(\frac{\underline{I}'_f}{\underline{I}_r}\right)$; $\underline{I}'_f$ current, calculating by values at measuring point, with phase close to phase of fault current

The power line model was created using the SimPowerSystems section blocks of the Simulink library of the MATLAB software package. In order to solve the system of equations of the model in the state space, the Fixed-step discrete solver with a sampling period of 32 samples per period of industrial frequency (sampling frequency - 1600 Hz) was selected, which showed sufficient accuracy in performing similar tasks [14]. The power supply source was set by the Three-Phase Programmable Voltage Source, deviations of voltage characteristics were simulated by changing parameters of this unit. The Three-Phase Series RLC Branch block modeled little active resistance to avoid parallel connection of the voltage source and capacitive branches of the following blocks. The resistance of the positive and zero sequence system was modeled by the Three-Phase PI Section Line unit of unit length; Power line sections - Three-Phase PI Section Line; Fault - Three-Phase Fault; Load - Three-Phase Series RLC Load. The Three-Phase V-I Measurement, From, to Workspace, and Scope blocks are designed to visualize and export data to the MATLAB workspace. The powergui unit is required for the operation of the SimPowerSystems units.

Parameters of the simulated system are given in Table 3. The scheme of the simulation model of the power line in Simulink is shown in Fig. 2.

Table 3. Data for simulation

Parameter	Value
Nominal voltage, kV	110
Neutral grounding	Solid
Positive/zero sequence system impedance, Ohm	3,900+j9,900/ 3,200+j13,300
Positive/zero sequence line impedance per 1 km, Ohm/km	0,118+j0,142/ 0,268+j1,260
Line length, km	40
Loading power, MVA	14+j26

All types of faults (phase-to-ground, phase-to-phase, phase-t-phase-to-ground, three-phase) were imitated.

Complex amplitudes from instantaneous values were obtained by quadrature filters of industrial frequency using discrete Fourier transform of one period by sliding window method [15].

The analyzed time of the model operation was 20 full periods of industrial frequency - 0.4 s: 10 periods were for normal mode of operation; 10 for emergency.

2. Analysis of fault location error distribution along power line length under voltage characteristics deviations variations

Relative fault location error is determined by formula (6):

$$\varepsilon = \frac{x_e - x_r}{L} \cdot 100\%, \qquad (6)$$

, where x_r real distance to fault; L line length (Fig.1).

The fault location error distributions over the line length during variation of voltage characteristics deviations on the example of phase A phase-to-ground fault for the Ericsson method are shown in Figure 3. Similar distributions have been obtained for all studied fault location methods and types of faults.

Analysis of Figure 3 shows that frequency deviation has the greatest impact on the accuracy of the emergency state fault location. For a single rapid voltage changing and supply voltage unbalance, the error of the fault location during variation of voltage characteristics deviations is explained by the methodological error of the implemented algorithms. The noted patterns are qualitatively repeated for all investigated fault location methods in any types of fault.

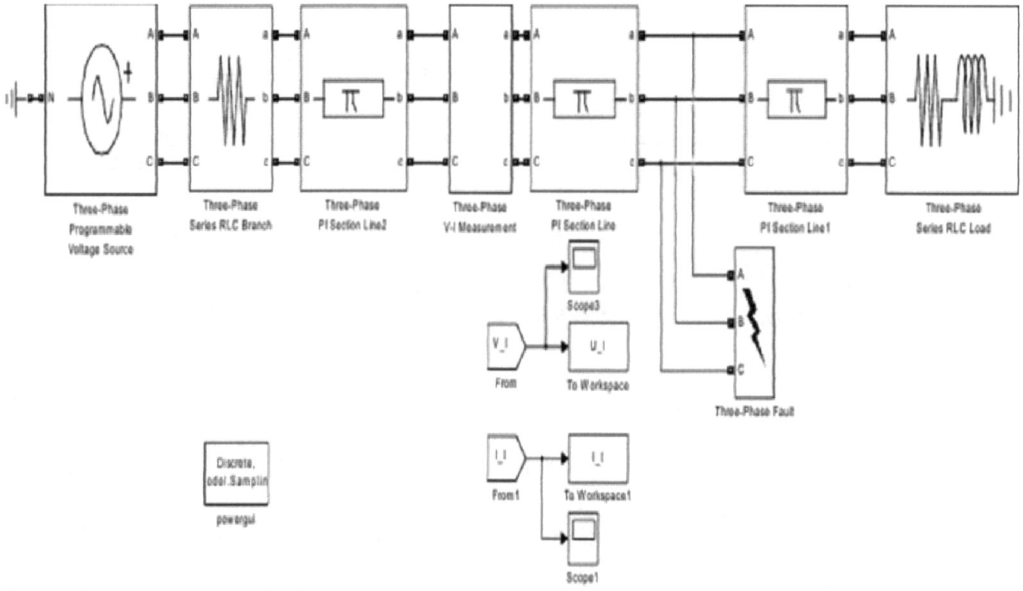

Fig. 2 Scheme of imitation model in Simulink

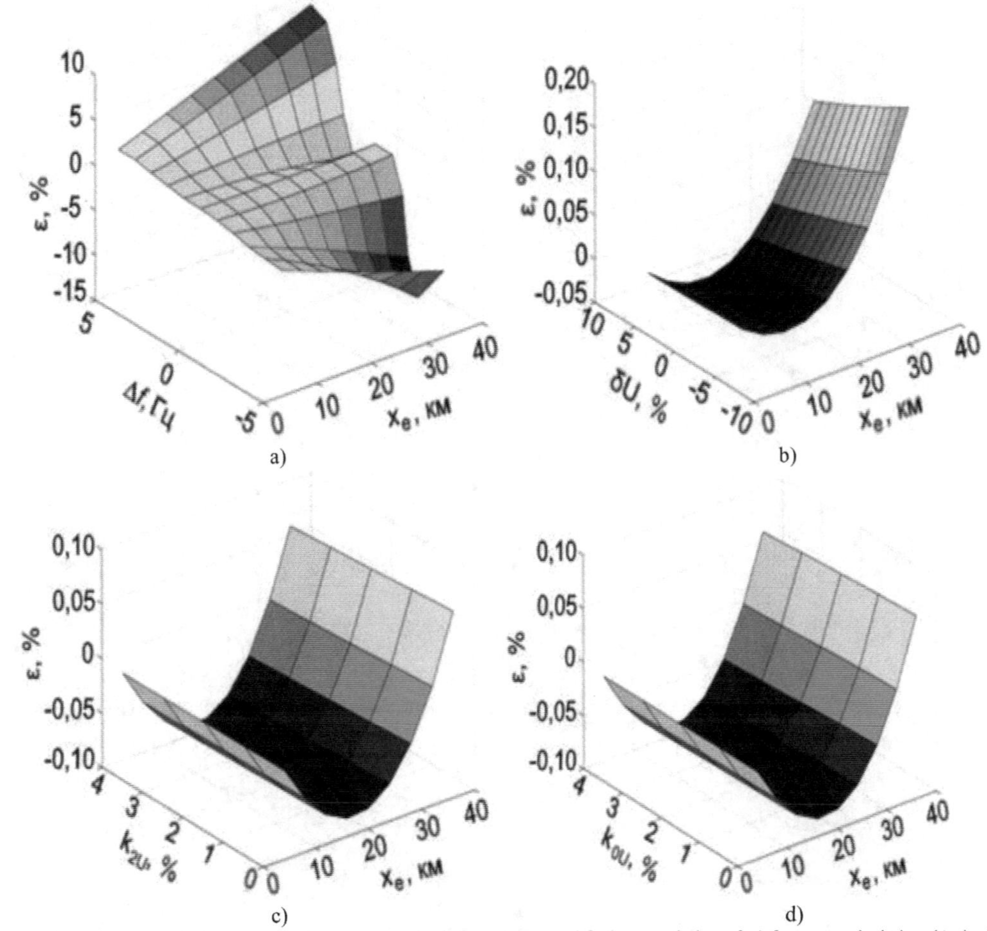

Fig. 3 Distribution of relative error of Ericsson method phase-to-ground faults at variation of: a) frequency deviation; b) single rapid voltage changing; c) negative sequence supply voltage unbalance; d) zero sequence supply voltage unbalance

3. The way of adaptation of emergency state fault location algorithms to voltage characteristics deviations.

Figure 4 shows the flow chart of the algorithm implementing the way of adaptation of emergency state fault location to voltage characteristics deviation - the algorithm of emergency state fault location adaptation to voltage characteristics deviations.

In case of fault, current and voltage oscillograms are recorded and complex amplitudes of this margins are obtained by quadrature filters of industrial frequency. Further, recorded instantaneous values and obtained estimates of complex amplitudes of currents and voltages are used to check presence of deviations of voltage characteristics (for example, by algorithms proposed in [4]). If there is no deviation of voltage characrteristics, the result of fault location will be values obtained from non-adaptive formulas (1-5). In case of voltage characteristics deviation, the value of the nearest error is determined based on the results of simulation database of fault location methods error distributions along the length of the power line at voltage characteristics deviations. The result of fault location will be the value obtained by the adaptive formula (7).

$$x_f = x_e - \frac{\varepsilon_e \cdot L}{100}, \quad (7)$$

, where x_f – fault location evaluation from measuring point after adaptation, ε_e – nearest error in falt location errors database, corresponding to found voltage characteristic deviation and x_e .

Figure 5 shows the results of the proposed method of adaptation in case of random deviations of different voltage characteristics for phase-to-phase faults. As can be seen from the results for the simulated example, the application of the proposed adaptation method not only significantly reduces the error of the emergency state fault location methods from the deviation of the voltage characteristics, but also compensates for their own errors.

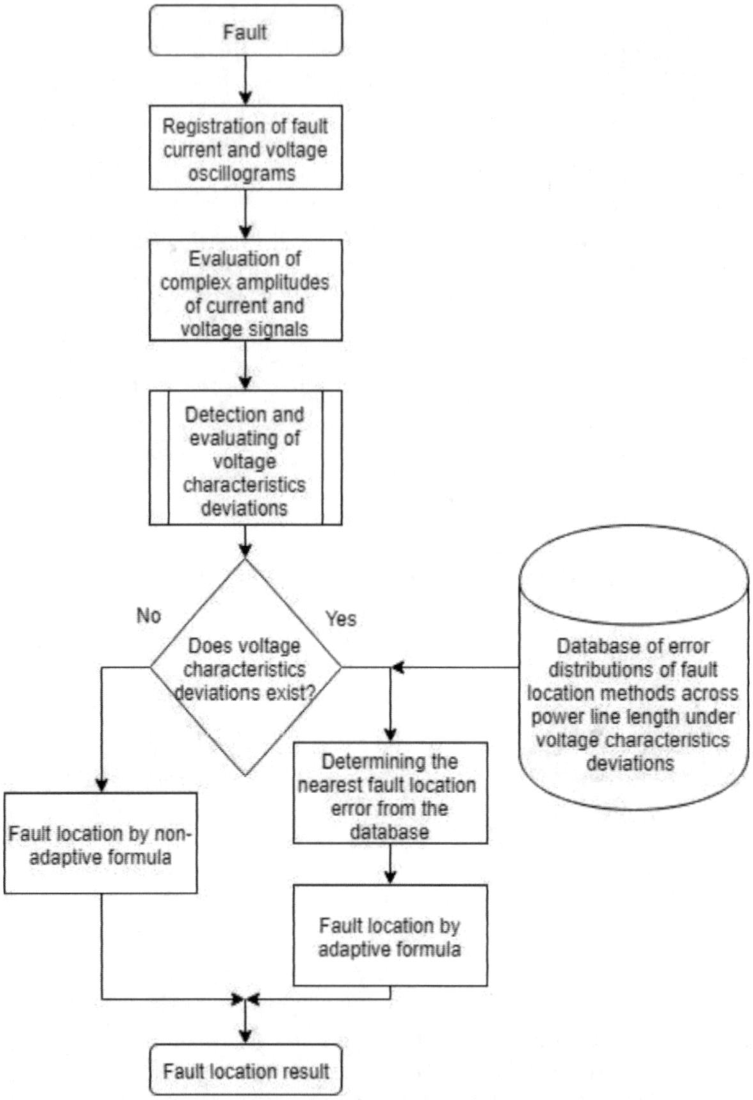

Fig. 4 – The algorithm of emergency state fault location adaptation to voltage characteristics deviations

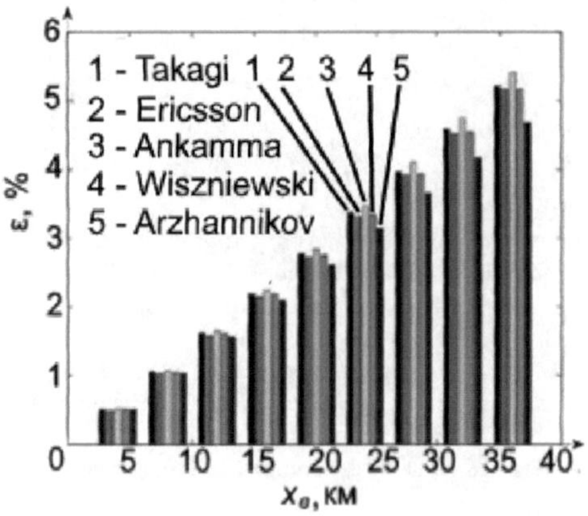

a)

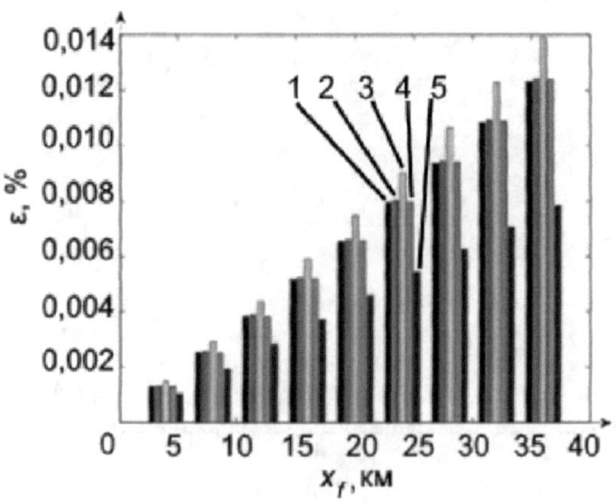

e)

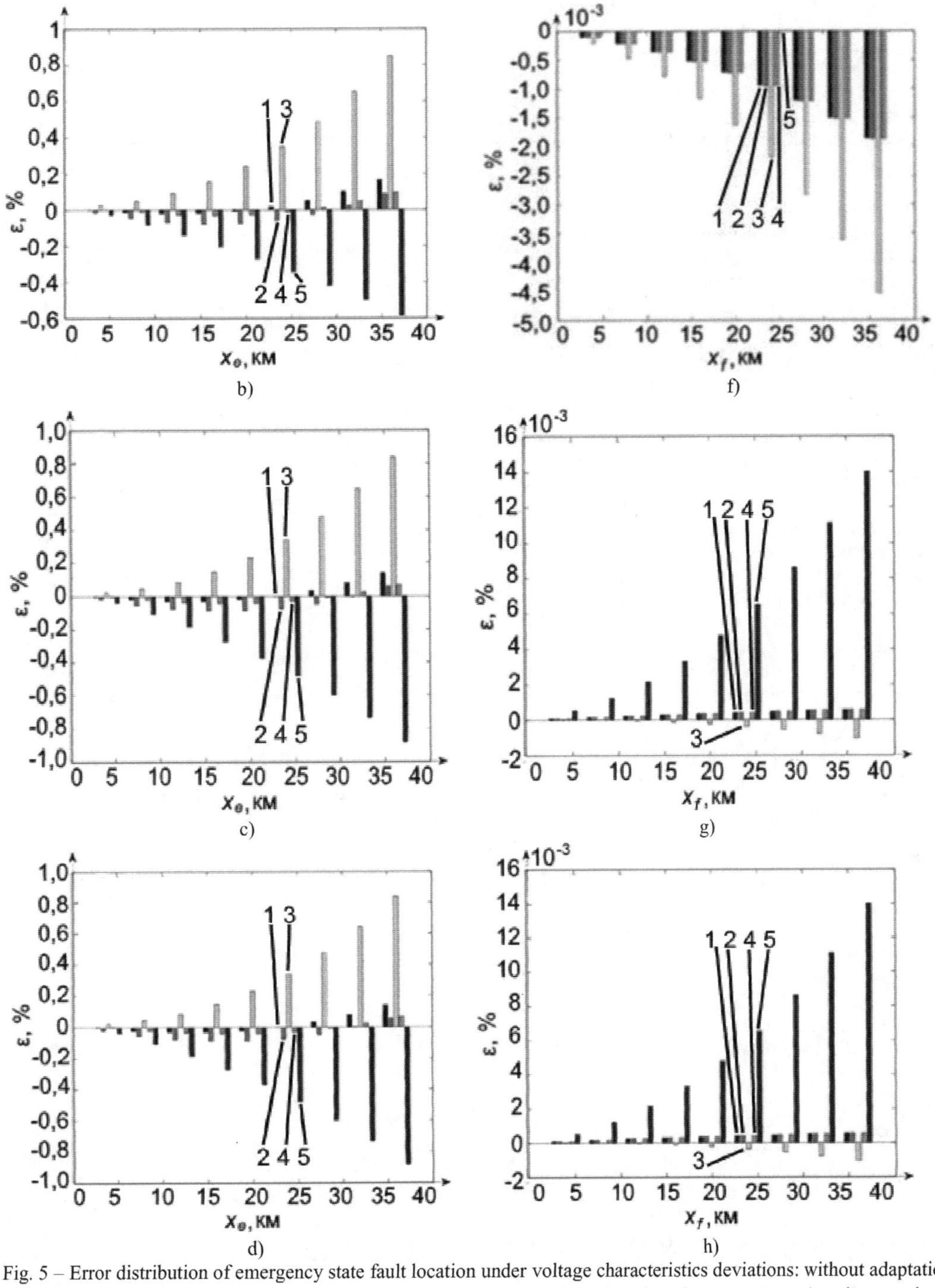

Fig. 5 – Error distribution of emergency state fault location under voltage characteristics deviations: without adaptation: a) under frequency deviation $\Delta f = 3{,}14$ Гц; b) under single rapid voltage changing $\delta U = -2{,}66\%$; c) under negative sequence supply voltage unbalance $k_{2U} = 3{,}83\%$; d) under zero sequence supply voltage unbalance $k_{0U} = 3{,}71\%$; with adaptation: e) under frequency deviation $\Delta f = 3{,}14$ Гц; f) under single rapid voltage changing $\delta U = -2{,}66\%$; g) under negative sequence supply voltage unbalance $k_{2U} = 3{,}83\%$; h) under zero sequence supply voltage unbalance $k_{0U} = 3{,}71\%$;

Conclusion

Proposed way allows to correct both methodological error of fault location and additional error introduced into calculation by voltage characteristics deviations. Of scientific and practical interest is the extension of the studies carried out to the combinations of voltage characteristics deviations characteristic of specific groups of electric users existing in the normal operation of network. Further development of the study will be useful in correcting the results of multi-terminal fault location under voltage characteristics deviations.

References

1. Shneerson J. M. Digital Relay Protection, Moscow: (Energoatomizdat, Moscow, 2007)
2. Boyarskaya N. P., Dovgun V. P., Temerbaev S. A., Shahmatov S. N. Bull. of KrasGAU, **3**, 169-181 pp., (2012)
3. Nevretdinov Y. M., Fastij G. P., Yaroshevich V.V. Karpov A. S. Bull. of MGTU, **1**, 67-76, (2014)
4. Bíscaro A.A.P., Pereira R.A.F., Kezunovic M., Mantovani J.R.S. IEEE Trans. Power Deliv, **31 (2)**, 428-436, (2016)
5. Morales-Espana G., Morga-Flores J., Vargas-Torres H. IEEE/PES Transmission and Distribution Conference and Exposition: Latin America, 803–809, (2010)
6. Pereira R.A.F., da Silva L.G.W., Kezunovic M., Mantovani J.R.S. IEEE Trans. Power Deliv, **24(2)**, 852-862, (2009)
7. GOST 32144-2013. Electric energy. Electromagnetic compatibility of technical equipment. Power quality limitsin the public power supply systems. 2014–07–01. Mosccow: FGUP «STANDARTINFORM», 2013. 18 p.
8. Majstrenko G. V. , Kulikov A. L., Papkov B. V., Obalin M. D. Bull. of AGAU, **4**, 71-81, (2019)
9. Takagi T. Yamakoshi Y., Yamaura M., Kondow R., Matsushima T. IEEE Trans. Power Appar. Syst., **PAS-101**, 2892-2898, (1982)
10. Eriksson L., Saha M.M., Rockefeller G.D. IEEE Trans. Power Appar. Syst., **PAS-104 (2)**, 423-435, (1985)
11. Ankamma Rao, J, Bizuayehu B. Internat. Journ. of Engin. Res. & Techn. (IJERT), **4 (02)**, 492-495, (2015)
12. Wiszniewski A. IEE Proc. C Gener. Transm. Distrib, **130**, 311-314, (1983)
13. Arzhannikov E. A. News of higher education institutions. Power industry, 2, 42-45, (1984)
14. Kulikov A. L., Majstrenko G. V., Obalin M. D., Papkov B. V. Bull. of NNSEEU, **96**, 61-77, (2019)
15. Bhide S.R. *Digital power system protection,* (PHI Learning ,Dehli:, 2014)

E3S Web of Conferences 139, 01043 (2019) https://doi.org/10.1051/e3sconf/201913901043
RSES 2019

About accidents cyclicality and mathematical model of failure rate in main electrical grids of power systems

Irek Galiaskarov [1], *Misrikhan* Misrihkanov[2], *Vladimir* Ryabchenko[3*]

[1]JSC "CEMC UES", 4, Belovezhskaya st., Moscow, Russia
[2]National research university "MPEI", Institute of Electrical Power Engineering, 14 Krasnokazarmennaya St, Moscow, Russia
[3]JSC "R&D Center FGC UES", Building 3, 22 Kashirskoye Highway, Moscow, Russia

Abstract. The accidents cyclicality of 500 kVoverhead lines of a large region over an extended time period was studied. Significant fluctuations in the values of their failure rate (failure frequency) under the impact of natural and socio-economic factors are revealed. The wavelet spectrums of the failure frequency were constructed, which revealed several historical periods of accidences. It is proposed to consider the failure rate as the output signal of a dynamic system with many difficultly formalized inputs. The stability of this dynamic system is evaluated. The ways of formalizing a multidimensional dynamic model of accidence in main electrical grids of power systems are outlined.

1 Introduction

It is known that the causes of major accidents in electrical grids with massive damage to overhead lines (OHL), often called extreme climatic conditions. These include elevated ice and rime coatings, hurricanes, natural fires, etc.

Human activity also has a comparable impact on accident rate - unauthorized exposure to overhead line elements and their poor-quality operation: collision of vehicles with towers, trapping wires by hoisting mechanisms, untimely detection defects, etc. Thus, in the general case, the accident rate of overhead lines randomly depends on the impact of natural and social (socio-economic) factors.

In the Table 1 as an example, the structure of stable (i.e. unrecoverable by the autoreclosing), failure of 500 kV overhead lines is given with a total length of about 8.5 thousand km of the central European part of the country for the period 2011–2018. From table 1 it follows that the social (paragraphs 1, 4) and natural influences (paragraphs 2, 3, 5) are approximately equally affect the accident rate of overhead lines.

Earlier in [1], it was noted that the available oscillation periods accidents in electrical grids are close to (quasi) eleven solar activity cycle.The latter definitely affects the change in natural and climatic characteristics (there is an opinion that cycles solar activity have an effect on social phenomena). However, this assumption in [1] was considered as a hypothesis, requiring verification and targeted research. Accident cycles in electric networks were considered in [1] on example of statistical data on unrecoverable failures of 500 kV overhead lines also the central European part of the country for the period 1974-2001. (specific damage was given, 1 / (year 100 km), or rather, the average failure rate or the failure frequency ω). Therefore represented keen interest to extend this time period and clarify the hypothesis of cyclical accidents in the main electrical grids of power systems.

Table 1. Organizational causes of failures for the period 2011–2018.

№	Failure cause	Failure number	
		pcs	%
1	Failure to comply with deadlines, failure to perform the required amount of maintenance or repair of equipment and devices, including:	34	12.8
1.1	Untimely identification and elimination defects (broken or untwining wires and ground wires, the destruction of the set of insulators)	17	6.4
1.2	Other interruptions	17	6.4
2	Birds intervention	5	1.9
3	Exceeding the parameters of the impact of natural disasters regarding the conditions of the project	8	3.0
4	The intervention of unauthorized persons and organizations not involved in the technological process	91	34.2
5	The effects of recurring natural disasters, including:	121	45.5
5.1	Ice and rime coatings	17	6.4
5.2	Atmospheric overvoltage (thunder)	57	21.4
5.3	Natural fires	17	6.4
5.4	Other effects of adverse natural events (falling trees)	30	11.3
6	Undetected reasons	7	2.6
7	Total	266	100.0

https://doi.org/10.1051/e3sconf/201913901043

2 About accidents cyclicality

Archives for all technological violations of OHL 500 kV of the considered region were raised for the subsequent 2002–2018 In fig. 1 The frequency of unrecoverable OHL failures of 500 kV over 45 years in the interval from 1974 to 2018 is presented with overlaying four peaks held eleven-year cycles of solar activity (from the 21st to the 24th, the last cycle according to the reported data had two bursts: 2013 and 2014); the peak of the 25th cycle is expected in 2023.

As can be seen from fig. 1, the values of the failure frequency are oscillatory in nature, varying in a wide range: from 0.11 1 (year 100 km) to 1980 and 2018 up to 0.86 1/(year 100 km) in 1998 (Historical note: in 1980, the summer Olympic Games were held in our country, in 2018 - soccer world cup, and in 1998 - was declared default).

In years of the highest solar activity (Fig. 1), there is a slight increase in the OHL failure frequency. However, in between practically of each cycle, bursts of OHL accidence are found that exceed those for the peaks of solar activity. For example - this is earlier mentioned 1998 or 2010 with its most severe fires in the European part of the country as a result of an abnormally hot summer.

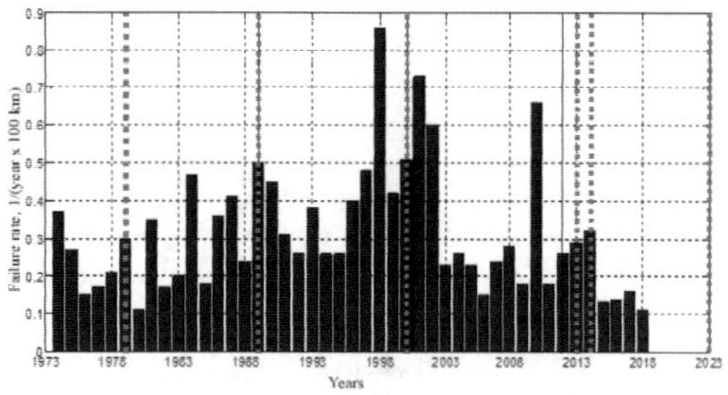

Fig. 1 - Values of the failure rate of 500 kV overhead lines for the period 1974–2018 with solar activity cycles (dashed)

Fig. 1 is an amplitude-time representation parameter (signal). But, as you know, no less significant information hiding in its frequency domain. For this purpose, the graph in Fig. 1 was subjected to the so-called wavelet analysis [2], which allowed the time-frequency separation of the process in Fig. 1.

Figure 2 shows the wavelet spectrum of the failure rate with the Haar mother wavelet. By the way, almost identical given in fig. 2 wavelet spectra obtained using as mother wavelets of Daubechies (db), Koifman (coif), simlet (sym), biorthogonal spline wavelet (bior) and reverse biorthogonal spline wavelet (rbio).

In the behavior of the failure rate in Fig. 2 distinguishable three time periods from the standpoint of the intensity (disturbance) of its frequency spectrum: from 1974 to the beginning of the 1990s; since the early 1990s before early 2000s and since the beginning of the 2000s. Until now. The first period has an unperturbed, relatively uniform, calm frequency spectrum. The second cycle is the spectral "storm". Finally the third period - stabilization and calming of the frequency spectrum; some the surge in intensity is apparently associated with the events of 2010 (see above). It must be assumed that Fig. 2 reflected events that occurred in our country. In particular, the second period is the time of decay once a single country and the collapse of its economy.

Based on the analysis of events in Fig. 1 and 2 does not seem possible to confirm the attractive hypothesis expressed in [1], that the available periods of fluctuations of accidence in electrical networks close to the eleven-year cycle of solar activity. This is too simple an explanation of the processes in the issue of reliability.

Cause-and-effect links are much more complicated. Their resulting exposure is determined by multifactorial and difficult to formalize addition of environmental influences and socio-economic relationship. However, regardless of this, the failure rate under consideration is not a set of fixed values that depend, for example, on the material of the towers or the nominal line voltage, and a dynamic process with a change in characteristic time periods generated by some dynamic system.

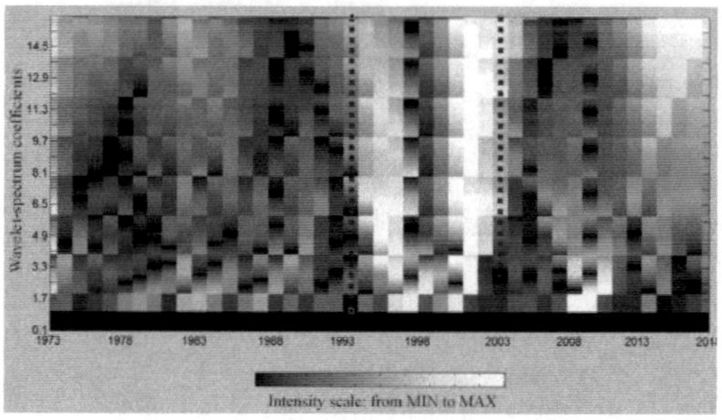

Fig. 2 - Wavelet spectrum of the failure rate of a 500 kV overhead line

As is known, mathematical models of dynamical systems are conventionally divided into two disjoint classes: autonomous and input-output [3]. In the first case, the output signals of the system are generated by internal state transitions under the action of nonzero initial conditions. Moreover, all non-formalized external influences of this system are converted into equivalent initial conditions [3]. In the second case, the system output signals are a homomorphic (not one-to-one) transformation of the corresponding input signals.

In our case, we formulate a particular problem: which dynamic system reproduces the sequence of output signals with the smallest possible deviations from the observed series in the form of failure rate (Fig. 1)?

In this case, an identifiable model should be suitable for analysis of its stability. (Not strictly speaking the steady movement of a dynamic system is determined by two trends that depend on the physical nature of the process: conservative, seeking to compensate for external influences, resisting these influences and leading, as a rule, to oscillatory motion, and dissipative, dissipating the energy of these oscillations.) We formulate the calculated conditions for the model of the dynamic system under study:

 https://doi.org/10.1051/e3sconf/201913901043

– It belongs to the class of discrete dynamic systems [3], since changes in the failure rate occur discretely, with a step of "integration" of one year;
– It is positive (positive, non-negative), i.e. generates only positive values of the output signal, since the parameter of the failure rate is at least a non-negative value [4];
– it can be represented in the state space, i.e. is described a set of physical or abstract variables characterizing the behavior of the system in the future, provided that the state at the present moment of time is known [3];
– It possesses the property of non-stationarity, as its parameters change in time, from step to step;
– It considered as an autonomous system, as in this case there is no additional information that could be taken as input;
– It's limited to the set of data analyzed, starting in 2002 – see the wavelet spectrum in Fig. 2. Otherwise, the identification results will be significantly affected by the so-called "tails" data relating to previous historical periods [5].

We can realize Identification of the model for changing the failure rate per 2002–2018 in the class of positive, discrete, non-stationary, autonomous dynamical systems in the state space, on the base of transformations of the Hankel matrix (a square matrix with all diagonals perpendicular to the main equal elements formed from a data array) using the least squares method [5]; we obtain the following discrete equations:

$$\mathbf{x}(t+1) = \boldsymbol{A}(t)\mathbf{x}(t),\ \omega(t) = \mathbf{c}^{\mathrm{T}}\mathbf{x}(t) \quad (1)$$

Where $\mathbf{x}(t)$ - state vector, $\omega(t)$ - system's output signal (failure rate), $\mathbf{c}^{\mathrm{T}} = [1 \quad 0 \quad 0]$.

$$\mathbf{x}(1) = \begin{bmatrix} \omega_{2002} \\ \omega_{2003} \\ \omega_{2004} \end{bmatrix},\ \mathbf{x}(2) = \begin{bmatrix} \omega_{2003} \\ \omega_{2004} \\ \omega_{2005} \end{bmatrix}, \ldots,\ \mathbf{x}(N) = \begin{bmatrix} \omega_{2016} \\ \omega_{2017} \\ \omega_{2018} \end{bmatrix}$$

- values of the state vector at the corresponding step of "integration"; t = 1,2,...,N. N is the interval of "integration" (in this case, N = 15);

$$\boldsymbol{A}(1) = \begin{bmatrix} 0 & 1 & 0 \\ 0 & 0 & 1 \\ 0,4655 & 0,0621 & 0,2404 \end{bmatrix}, \quad (2)$$

$$\mathrm{eig}\,\boldsymbol{A}(1) = \{-0,33264 \pm j0,6439;\ 0,8933\}$$

$$\boldsymbol{A}(N) = \begin{bmatrix} 0 & 1 & 0 \\ 0 & 0 & 1 \\ 0,4125 & 0,2326 & 0,1548 \end{bmatrix}, \quad (3)$$

$$\mathrm{eig}\,\boldsymbol{A}(N) = \{-0,3773 \pm j0,5579;\ 0,9093\}$$

- matrices of eigen (free) dynamics in the steps of "integration" and the corresponding sets of eigenvalues (poles).

The distribution of poles (eigA eigenvalues) of the identified failure rate model (1) - (3) for the period 2002– 2018 years presented in fig. 3.

Since for the asymptotic stability of a discrete dynamical system it's necessary and sufficient that the location of its poles on the complex plane C inside the unit circle, centered at the beginning of C, then from the diagram in Fig. 3 that the identified model is asymptotically stable [3] or, in other words - Shurov model. In this case, the margin (radius, degree) of stability of the indicated model of the failure rate is shown in Fig. 4.

We recall that in the case of a discrete dynamic system (1), the stability margin is calculated as follows:

$$\rho_{st} = |\mathrm{Re}\lambda_{max}| < 1 \quad (4)$$

The stability margin is an extremely important characteristic of a dynamic system and characterizes the attenuation rate of transients in it. Moreover, the lower ρ_{st}, the closer the transitional oscillation in the system to aperiodic and rapidly decaying processes.

However, the presence of one or another margin of stability in a dynamic system is not a guarantee that sufficiently small perturbations will not lead to a loss of stability. Therefore, we consider the ε-spectrum [6] in as a method for analyzing robustness of stability (rudeness – freedom from influence of disturbances) of the dynamic model of the failure rate. This method operates with complex numbers z_i that make up the set of the following form (ε-spectrum):

$$\mathrm{eig}_{\varepsilon}(\boldsymbol{A}) = \{z_i : \sigma_{\min}(z_i\boldsymbol{E} - \boldsymbol{A}) \le \varepsilon\} \quad (5)$$

In formula (5), $\sigma_{\min}$ is the minimum singular number of the matrix $z_i\boldsymbol{E} - \boldsymbol{A}$. ε is the given positive number (tolerance), E is the identity matrix. In other words, the ε-spectrum is the set of complex numbers z such that all the matrices $z\boldsymbol{E} - \boldsymbol{A}$ have minimal singular numbers that do not exceed some finite number (tolerance) ε that characterizes these matrices as "close to degenerate" or "practically degenerate"

Stability analysis of the failure rate model for period 2002–2018 based on the ε-spectrum (Fig. 5) gives an estimate of a small degree of robustness, lying in the range $(0,75...1,25)\times10^{-1}$. In other words, the stability of this model is subject to significant the influence of relatively small perturbations of the elements of matrix A.

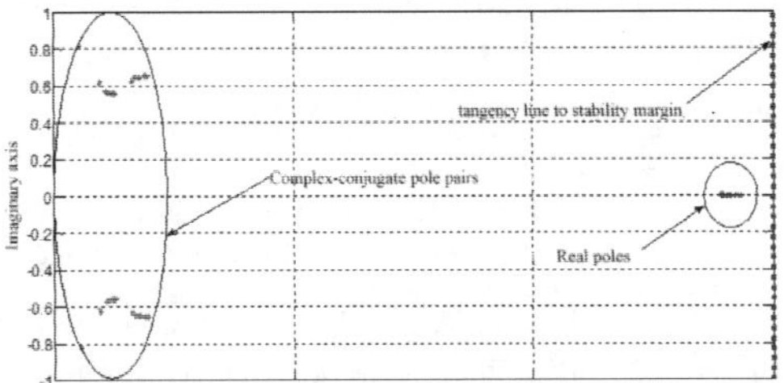

Fig. 3 - Poles of the model of the failure rate of 500 kV overhead lines for the period 2002–2018 years in the state space

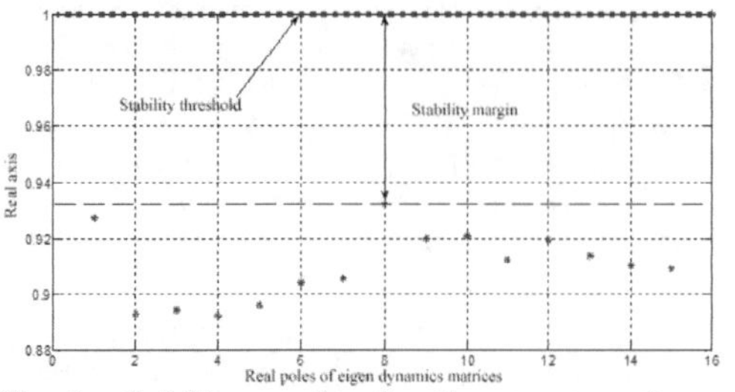

Fig. 4 - Stability margin of a discrete non-stationary dynamic failure rate model

This estimate is obtained by numerically analyzing regions on the complex plane in which the poles (eigenvalues) of all matrices A of the model of the dynamical system (1) obtained at relatively close but different initial approximations are localized.

In this sense, the ε-spectrum method is regular with respect to the used strategy of "weighting" (increasing the intensity of disturbances), i.e. guarantees the independence of the obtained stability margin estimates from the choice of initial approximations and the "weighting" strategy.

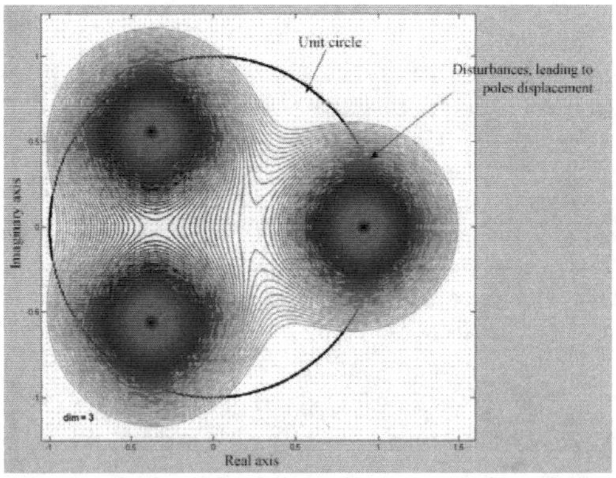

Fig. 5 - Evaluation of the robustness of the stability margin of the failure flow parameter model ε-spectrum based (two-dimensional looking)

Thus, the dynamic model of the 500 kV OHL failure rate is stable over the considered time period, but is subject to the influence of relatively small perturbations. We relate this theoretical assumption to current practice of operation of main electrical grids.

Figure 6 shows the age characteristics of the OHL of 500 kV in the region under consideration. As you can see, about 40% of OHL of 500 kV is older than 50 years, and lines "younger" 30 years, i.e. built after the events of the 90s last century - less than 20%.

According to [7, 8], there is a noticeable increase in the accident rate of the deterioration character of overhead lines of all classes stress due to mass aging the country. However, the evidence on fig. 1 (if you do not take into account the anomalous 2010) indicates on the contrary: the reliability of the 500 kV OHL is continuously improving despite the fact that there is almost no new construction or reconstruction of them. A possible explanation for this paradox is as follows.

An analysis of the reliability characteristics of the overhead lines of the Unified National Power Grid revealed [9] that over the past 30 years, the average duration of scheduled overhead lines repair in the main electrical grids of power systems has grown from 12–17 to 95–149 hours, i.e. almost 10 times.

At the same time, the main part of deliberate outages is associated with repairs or maintenance of the line itself, and not with its reconstruction or other external reasons.

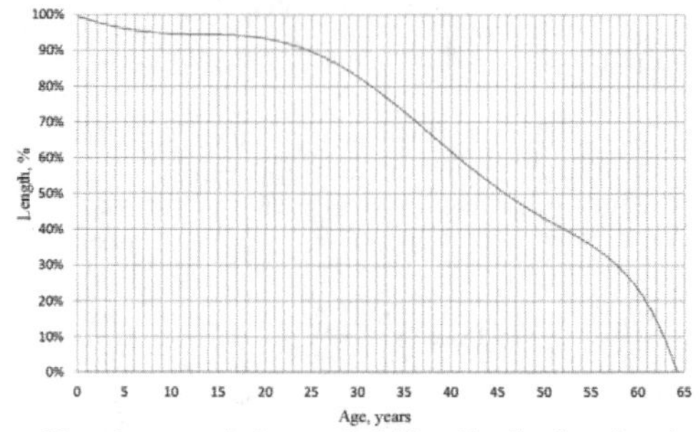

Fig. 6 - cumulative probability distribution function of age characteristics of 500 kV OHL

In other words, maintaining the operability of morally and physically worn out grid elements is provided not by reconstruction, but by repairs. It was on this model of management that domestic energy systems were forced to work in the 90s of the last century in the conditions of the collapse of the economy.

Thus, in spite of the favorable reliability characteristics of the 500 kV overhead lines in Fig. 1, it is hardly possible to speak of a favorable prognosis of their accident rate.

3 Conclusions

1. Overhead lines of the Unified National Power Grid form the backbone of energy systems, while being the most damaged elements. Therefore, accident risks are traditionally given close attention.
2. Over the past decades, significant fluctuations in the failure rate of high-voltage lines of 500 kV under the influence of natural and socio-economic factors were revealed: from 0.11 to 0.86 1/(year 100 km). Moreover, the previously stated hypothesis is not confirmed that the available periods of accident fluctuations in electric networks are close to the eleven-year cycle of solar activity.
3. Using wavelet analysis, the presence in the frequency spectrum of the failure rate of a 500 kV overhead line of three distinct time periods, approximately corresponding to the socio-economic changes in the country, is shown. This once again confirms that the parameter under consideration is determined by the multifactorial and difficult to formalize addition of environmental influences and socio-economic relations.

4. Based on the oscillatory process of changing the failure rate of a 500 kV overhead lines, this parameter is described as an output signal in the class of positive discrete non-stationary autonomous dynamic systems in the state space.
5. The asymptotic stability of this dynamical system is investigated. The system was steady. However, the robustness of the model of the failure rate of a 500 kV overhead lines turned out to be insufficiently favorable in terms of reliability. That is, its stability is significantly affected by relatively small perturbations.
6. The results of mathematical modeling and the corresponding conclusions regarding insufficient robustness are indirectly confirmed by an analysis of the age structure of the 500 kV OHL and change in their reliability characteristics in terms of repair and maintenance services in the direction of their deterioration.

References

1. V. Skopintsev *The quality of electric power systems: reliability, safety, efficiency, survivability* (Energoatomizdat, 2009)
2. S Malla *Wavelets in Signal Processing* (Mir, 2005)
3. T. Kailath *Linear Systems* (Prentice Hall, Englewood Cliffs, NJ. 1980)
4. W. Haddad, V. Chellaboina, Q. Hui *Nonnegative and compartmental dynamical systems* (Princeton University Press, 2010)
5. Van Overschee P., de Moor B.L. *Subspace identification for linear systems: Theory, Implementation, Applications* (Springer Science and Business Media, 2012)
6. L.N. Trefethen, M. Embree *Spectra and pseudospectra*. (Princeton Univ. Press, 2005)
7. R. Kaverina, F Kogan, L. Yakovlev, Improving the reliability of overhead lines 35–750 kV. General issues of OHL condition. Novosti Electrotehiki, **№ 4(46)** (2007)
8. R. Kaverina, F Kogan, L. Yakovlev Improving the reliability of overhead lines 35–750 kV. Reliability of wires and lightning protection cables. Novosti Electrotehiki, **№ 5(47)** (2007)
9. A. Abdurahmanov, S. Glushkin, I. Protasenko, A. Shuntov About reliability characteristics of OHL in electrical power grids. Electrichestvo, **№ 8**, p. 12–17 (2018)

Currents and Voltages Induced by Electric Field in Two Converging Single-wire Overhead Transmission Lines

Kirill Zimin[1], *Nina* Rubtsova[2], *Andrey* Tokarskskiy[1]

[1] Joint-Stock Company "Research and Development Center of the Federal Grid Company of Unified Energy System", Kashirskoe shosse, 22, bld.3, Moscow, Russia
[2] Federal State Budgetary Scientific Institution "Izmerov Research Institute of Occupational Health", prospect Budennogo,31, Moscow, Russia

Abstract. The calculations of induced by electric field voltage and current for *i* line grounded and ungrounded ends depending on *p* - line convergence part length, a_0 minimal distance between convergence lines, and Θ - the angle of lines convergence. Limit values of *p*, a_0 and Θ are determined, one of which disturbance allows to neglect the electric effect of converging transmission lines.

Introduction

Personnel working on overhead transmission lines (OTL) under induced voltage health maintenance requires these voltages limit values compliance. These requirements compliance ensure is possible by use the developed algorithms for calculation the current and voltage distribution along the disconnected grounded single-wire (single-phase) *i* line, induced by electric field (EF) of operating single-wire (single-phase) *k* line, converging with the *i* line under Θ angle.

1 Partial capacitance between convergent single-wire *i* and *k* lines

Consider first two parallel single-wire *i* and *k* OTLs located from each other at *a* distance and at $h_i = h$ and $h_k = H$ height above the ground (Figure 1, 1a). Figure 1b shows the scheme of *i* and *k* single-wire OTLs connection between lines and with the ground by partial capacities.

C_{ii0}, C_{kk0} и C_{ki0} specific partial capacities are determined by Maxwell's formulas first group potential factors use (1).

$$\left.\begin{aligned} &\alpha_{11} = \alpha_{kk} = \frac{1}{2\pi\varepsilon_0}\ln\frac{2H}{R}, \quad \alpha_{22} = \alpha_{ii} = \frac{1}{2\pi\varepsilon_0}\ln\frac{2h}{r}, \\ &\alpha_{12} = \alpha_{21} = \alpha_{ki} = \alpha_{ik} = \frac{1}{2\pi\varepsilon_0}\ln\frac{\sqrt{a^2+(H+h)^2}}{\sqrt{a^2+(H-h)^2}}, \end{aligned}\right\} \quad (1)$$

where: *H* and *h* are the heights above the ground; *R* and *r* are *k* and *i* line wires radiuses, respectively.

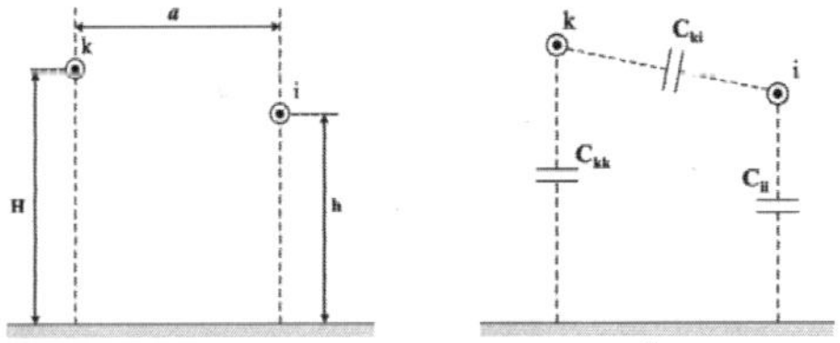

a b

Fig. 1. *I* and *k* single-wire OTLs layout (a) and connection between lines and with the ground by partial capacities (b)

Potential factors α matrix has the form (2):

$$\alpha = \begin{bmatrix} \alpha_{11} & \alpha_{12} \\ \alpha_{21} & \alpha_{22} \end{bmatrix}. \quad (2)$$

Matrix α inversion leads to obtain the Maxwell's formulas second group potential factors β matrix (3):

$$\beta = \begin{bmatrix} \beta_{11} & \beta_{12} \\ \beta_{21} & \beta_{22} \end{bmatrix} = \begin{bmatrix} \alpha_{11} & \alpha_{12} \\ \alpha_{21} & \alpha_{22} \end{bmatrix}^{-1} = \alpha^{-1}. \quad (3)$$

Potential factors have dimension is meter per farad [m/F], and capacity factors dimension is farad per meter [F/m].

Consider lines *k* and *i*, converging at Θ angle (Figure 2).

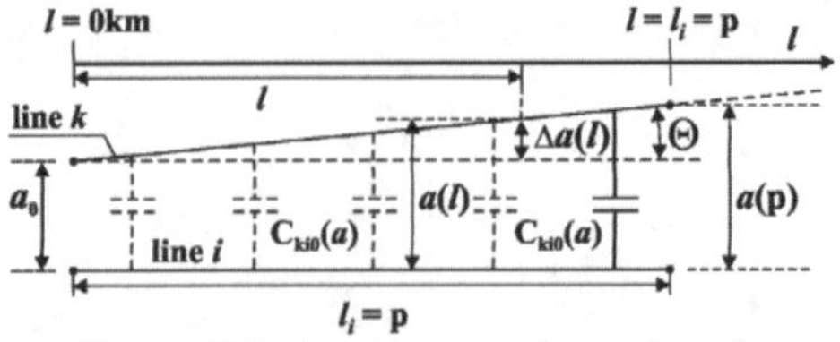

Fig. 2. OTLs *k* and *i*, converging at Θ angle

We know: *i* line length l_i = p, the distance between *k* and *i* lines at the beginning a_0 and at the end *a*(p) of *i* line, and so at the and a(p) of line we know k и i Θ converging angle, because:

$$\Theta = arctg\left(\frac{a(p)-a_0}{p}\right) = arctg\frac{\Delta a(p)}{p}.$$

Let's count, that a_0 = 5 m and a(p) = 100 m. Lines k and i parameters are: H = 19 m, h = 17.5 m, R = r = 0.014 m. Partial capacity values between k and i lines are determined by the expression [1, 2] $C_{ki0} = -\beta_{ki} = -\beta_{12} = -\beta_{21}$, then on expressions (1) ÷ (3) we obtain $C_{ki0}(a)$ values with a change from 5 m to 100 m in 5 m increments, as it is shown in table 1.

It should be noted that to obtain the values of partial capacities between phases and lighting wires of converging working and disconnected three- or more-phase power lines, $C_{ki0}(a)$ value calculation must be carried out taking into account all of these OTL phases and lighting wires. $C_{ki0}(a)$ values can be calculated for case from one to six three-phase OTL by computer program "OTL EMF" [3].

Table 1. Capacity $C_{ki0}(a)$ values depending on the distance a, obtained by the expressions (1) ÷ (3)

a, m	$C_{ki0}(a)$, $\times10^{-12}$ F/m	a, m	$C_{ki0}(a)$, $\times10^{-12}$ F/m	a, m	$C_{ki0}(a)$, $\times10^{-12}$ F/м	a, m	$C_{ki0}(a)$, $\times10^{-12}$ F/m
5	1.8704	30	0.4081	55	0.1636	80	0.08477
10	1.2197	35	0.3304	60	0.1412	85	0.07386
15	0.8774	40	0.2718	65	0.1229	90	0.06826
20	0.6614	45	0.2258	70	0.1078	95	0.06172
25	0.5137	50	0.1916	75	0.09529	100	0.05606

To represent $C_{ki0}(a)$ capacity value changes from a distance in the graphical form, fill in the MathCAD the matrix named Z, containing one row and c = 20 columns, C_{ki0} values from the table. 1. Next, we transform it into column matrix and, choosing the boundaries and the step of ψ_j argument changing we obtain Z_j curve with C_{ki0} values linear interpolation. To smooth Z_j curve, create ZZ(y) function and set the limits and the step of argument a changing (4):

$$\left.\begin{array}{l} \text{ORIGIN} := 1 \quad Z := (1.8704\ \ 1.2197\ \ ......\ \ 0.06172\ \ 0.05606) \\ Z := Z^T \cdot 10^{-12} \quad c := 20 \quad \psi_j := j \cdot 5 \\ ZZ(a) := \text{interp}(\text{cspline}(\psi, Z), \psi, Z, a) \quad a := 5, 5.1..100 \end{array}\right\} \quad (4)$$

ZZ(a) function allows carrying out numerical mathematical operations with it, but it is not analytically determined. Define an analytical expression for $C_{ki0}(a)$ capacity value.

For this purpose, in MathCAD, using the expressions (1) and (2) at $\varepsilon_0 \equiv \varepsilon$, we carry out the operation of matrix α analytical inversion analytical equal sign (→) using:

$$\alpha^{-1} \to \frac{-2\pi\varepsilon}{\ln\left[\frac{\sqrt{a^2+H^2+2Hh+h^2}}{\sqrt{a^2+H^2-2Hh+h^2}}\right]^2 - \ln\frac{2H}{R}\cdot\ln\frac{2h}{r}} \times$$

$$\times \begin{bmatrix} \ln\frac{2h}{r} & -\ln\left[\frac{\sqrt{a^2+H^2+2Hh+h^2}}{\sqrt{a^2+H^2-2Hh+h^2}}\right] \\ -\ln\left[\frac{\sqrt{a^2+H^2+2Hh+h^2}}{\sqrt{a^2+H^2-2Hh+h^2}}\right] & \ln\frac{2H}{R} \end{bmatrix}.$$

Then analytically obtained $C_{ki0}^{An}(a)$ capacity value can be written (5):

$$C_{ki0}^{An}(a) = -\frac{2\pi\varepsilon\ln\left[\frac{\sqrt{a^2+H^2+2Hh+h^2}}{\sqrt{a^2+H^2-2Hh+h^2}}\right]}{\ln\left[\frac{\sqrt{a^2+H^2+2Hh+h^2}}{\sqrt{a^2+H^2-2Hh+h^2}}\right]^2 - \ln\frac{2H}{R}\cdot\ln\frac{2h}{r}} \quad (5)$$

$C_{ki0}^{An}(a)$ function disadvantage is complexity, and complexity degree increases dramatically with increasing number of converging lines.

Will hold a rough interpolation (approximation) ZZ(a) curve function (6) using:

$$C_{ki0}^{Int}(a) = d\cdot e^{b\cdot(a-5)} + \frac{d}{1-v\cdot(a-5)} \quad (6)$$

Let us take follows values of d, b and v indexes: d = ZZ(a = 5)/2 = $0.935\cdot10^{-12}$, b = -0.6 и v = -0.15. Figure 3a shows ZZ(a) and $C_{ki0}^{An}(a)$ functions distribution curves, and Figure 3b shows ZZ(a) and $C_{ki0}^{Int}(a)$ distribution under a change from от 5 m to 100 m.

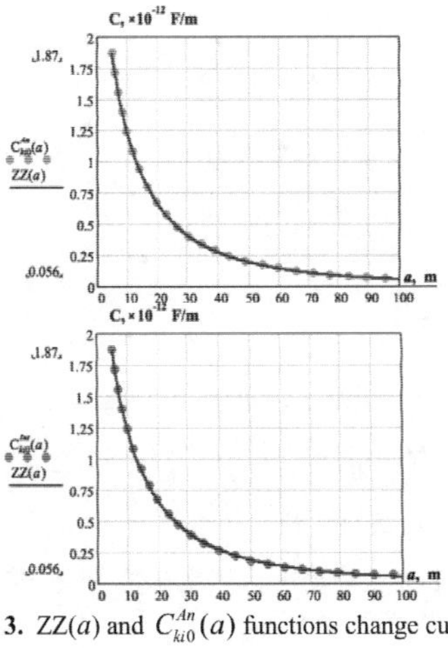

Fig. 3. ZZ(a) and $C_{ki0}^{An}(a)$ functions change curves (a), as well as ZZ(a) and $C_{ki0}^{Int}(a)$ (b)

The coincidence of ZZ(a), $C_{ki0}^{An}(a)$ and $C_{ki0}^{Int}(a)$ curves is good.

Thus, when calculating transverse voltage induced on the disconnected and grounded single-wire *i* line by converging single-wire *k* line electric field, partial capacity between these lines, depending on the distance between them, can be described by $C_{ki0}(a)$ = ZZ(*a*), $C_{ki0}^{An}(a)$ and $C_{ki0}^{Int}(a)$ functions.

2 Currents and voltages distribution along grounded at one end *i* line converging to *k* line

OTL *k* l_k length is in idle mode under voltage $\dot{U}_k = \dot{E}_k$ = 127 kV, and *i* line $l_i < l_k$ length is disconnected and grounded at one end remoted from *k* line substation (in this case, the left end) (Figure 4).

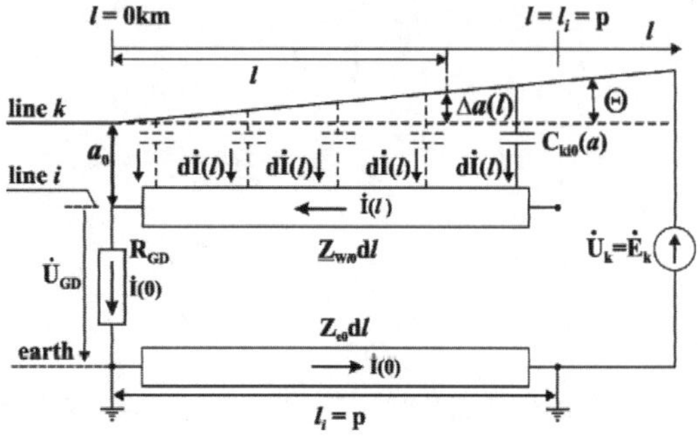

Fig. 4. OTLs converging at Θ angle: operating *k* line and *i* line disconnected and grounded at one end remoted from substation

Descriptions in Figure 4:
$\underline{Z}_{\mathrm{W}i0} = R_{\mathrm{W}i0} + \mathrm{j\,Im}\left[\underline{Z}_{ii\ 0}^{(Car)}\right]$ - is *i*, line specific active-inductive resistance (Ohm/km); $Z_{e0} = \mathrm{Re}\left[\underline{Z}_{ii\ 0}^{(Car)}\right]$ - is earth resistivity (Ohm/km); R_{GD} - is grounding device (GD) resistance (Ohm); $\underline{Z}_{ii\ 0}^{(Car)}$ - is own *i* line wire specific inductive resistance by Carson (Ohm/km) [4-6]:

$$\underline{Z}_{ii\ 0}^{(Car)} = \frac{j\omega\mu_0 10^3}{2\pi}\left(\ln\frac{\sqrt{2}\delta_e}{r_{Wi}} - j\frac{\pi}{4} + \frac{4}{3}\frac{h_{Wi}}{\delta_e}(1+j) - 0{,}0772\right).$$

Where: $\omega = 2\pi f$ – angular frequency, f = 50 Hz; μ_0 = $4\pi\cdot10^{-7}$ H/m – permeability of vacuum; r_{Wi} – *i* line wire radius, m; $\delta_e = \sqrt{2\rho_e/\omega\mu_0}$ – penetration depth, m; $\rho_з$ – earth resistivity, Ohm·m; h_{Wi} - *i* wire height above the earth. For case ρ_e = 100 Ohm·m $\underline{Z}_{ii\ 0}^{(Car)}$ = 0.0473+j0.716 Ohm/km, wire resistivity $R_{\mathrm{W}i0}$ = 0.074 Ohm/km, R_{GD} = 10 Ohm.

The distance *a*(*l*) between *k* and *i* OTLs change (Figures 2 and 5) is described by the expression (7):

$$a(l) = a_0 + \Delta a(l) = a_0 + l\cdot tg\Theta\,. \quad (7)$$

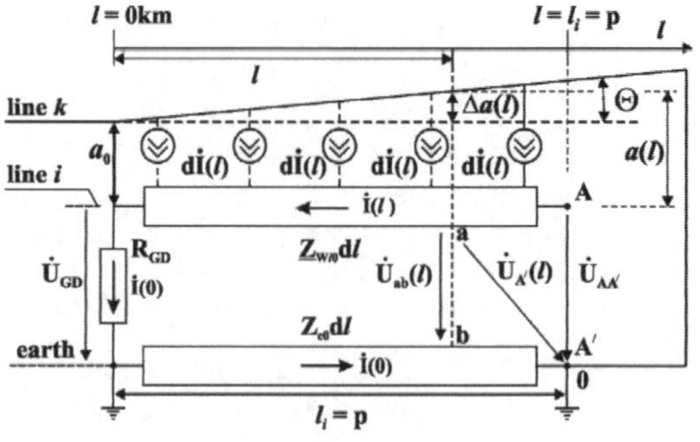

Fig. 5. Scheme of *k* and *i* OTLs converging at Θ angle with electromotive force and specific capacitance replacement at elementary current source

Substituting in $C_{ki0}(a)$ expression as well as expressions (4) ÷ (6) instead of variable *a* *a*(*l*) function, get the equations for $C_{ki0}(l)$ = ZZ(*a*(*l*)), $C_{ki0}^{An}(l)$ and $C_{ki0}^{Int}(l)$.

Let us replace electromotive force $\dot{E}_k$ and specific capacitance $j\omega C_{ki0}(l)dl$ with elementary current source $d\dot{I}(l)$ (Figure 5): $d\dot{I}(l) = j\omega C_{ki0}(l)\dot{E}_k dl$.

Because $d\dot{I}(l) = j\omega C_{ki0}(l)\dot{E}_k dl$, the current in *i* line find by expressions in graphic, analytical and interpolation forms:

$$\dot{I}(l) = \int_l^{l_i=p} d\dot{I}(l) = j\omega\dot{E}_k\int_l^{l_i=p} ZZ(a(l))dl =$$
$$= j\omega\dot{E}_k\int_l^{l_i=p} C(l)_{ki0}^{An}dl = j\omega\dot{E}_k\int_l^{l_i=p} C(l)_{ki0}^{Int}dl\,.$$

Voltage $\dot{U}_{A'}(l)$ value to point A′, which potential is assumed to be 0, and voltage $\dot{U}_{ab}(l)$ value along *i* line find by equations:

$$\dot{U}_{A'}(l) = \dot{I}(0)(R_{GD} + Z_{30}\,\mathrm{p}) + \underline{Z}_{\mathrm{W}i0}\int_0^l \dot{I}(l)dl,$$

$$\dot{U}_{ab}(l) = \dot{I}(0)(R_{GD} + Z_{30}\,l) + \underline{Z}_{\mathrm{W}i0}\int_0^l \dot{I}(l)dl.$$

Figure 6 shows current module $\dot{I}(l)$ distribution along *i* line length l_i = p = 100 km, calculated of capacities: $ZZ(a(l))$, $C(l)_{ki0}^{Int}$ and $C(l)_{ki0}^{An}$ at a_0 = 5 m, Δ*a*(p) = 100 m and angle Θ = 0.0573° (a), and of parallel *i* and *k* lines at a_0 = 5 m, Δ*a*(p) = 0 m and angle Θ = 0° (b).

Angle Θ of lines converging increase from 0° to 0.0573° leads to current at line *i* grounded end decreased by 5.7 from 7.46 A to 1.31 A.

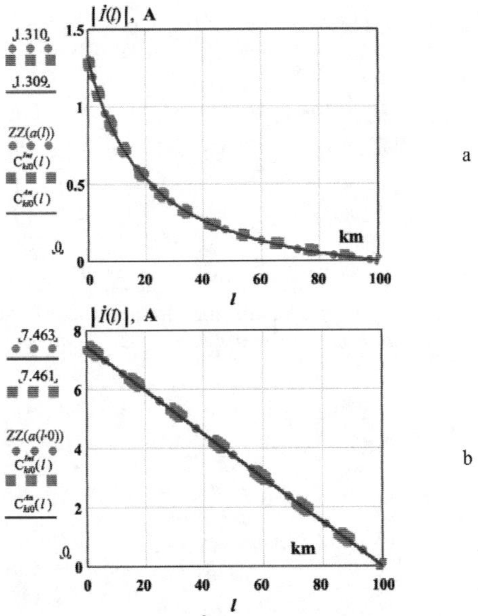

Fig. 6. Current module $\dot{I}(l)$ distribution of $ZZ(a(l))$, $C(l)_{ki0}^{Int}$ and $C(l)_{ki0}^{An}$ at p = 100 km, a_0 = 5 m, Δa(p) = 100 m: (a) converging *k* and *i* lines; (b) at p = 100 km, a_0 = 5 m, Δa(p) = 0 m parallel *k* and *i* lines

Figure 7 shows the distribution of voltages $\dot{U}_{A'}(l)$ and $\dot{U}_{ab}(l)$ modules (a) and arguments (b) of $ZZ(a(l))$, $C(l)_{ki0}^{Int}$ and $C(l)_{ki0}^{An}$ at a_0 = 5 m, Δa(p) = 100 m, p = 100 km.

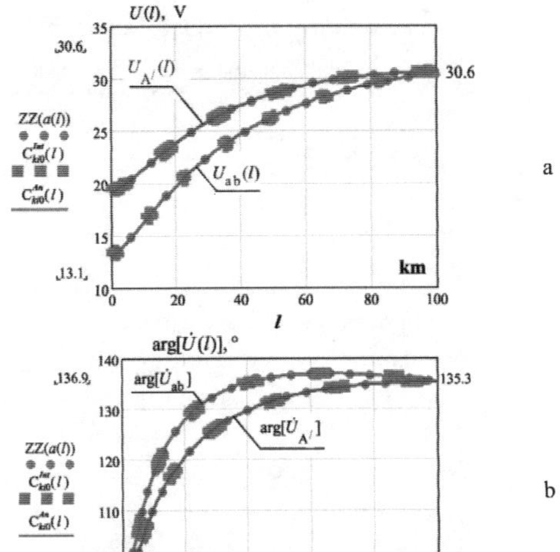

Fig. 7. $\dot{U}_{A'}(l)$ and $\dot{U}_{ab}(l)$ voltages modules (a) and arguments (b) distribution for $ZZ(a(l))$, $C(l)_{ki0}^{Int}$ иand $C(l)_{ki0}^{An}$ under a_0 = 5 m, Δa(p) = 100 m, p = 100 km

Consider the case of *i* line grounded at near to *k* line right end (Figure 8).

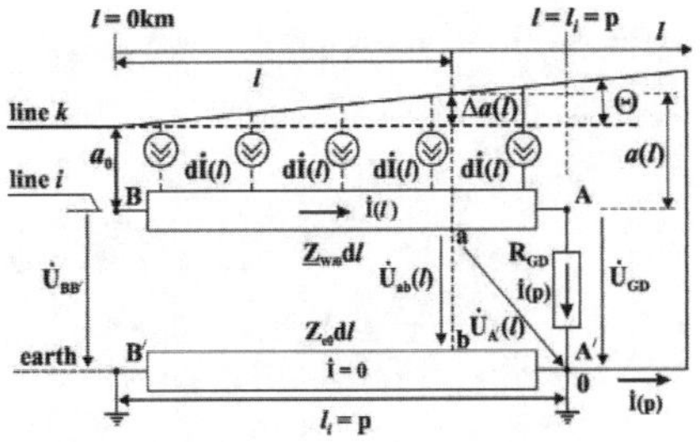

Fig. 8. The scheme of converging at Θ angle *k* and *i* OTLs with *i* line grounded at right end

$\dot{I}(l)$ current in graphic, analytical and interpolation forms are presented in (8):

$$\left.\begin{aligned}\dot{I}(l) &= \int_0^l d\dot{I}(l) = j\omega\dot{E}_k\int_0^l ZZ(a(l))dl = \\ &= j\omega\dot{E}_k\int_0^l C(l)_{ki0}^{An}dl = j\omega\dot{E}_k\int_0^l C(l)_{ki0}^{Int}dl.\end{aligned}\right\} \quad (8)$$

Since there is no current in Z_{e0} resistance, the voltage is $\dot{U}_{ab}(l) = \dot{U}_{A'}(l)$:

$$\dot{U}_{ab}(l) = \dot{U}_{A'}(l) = \underline{Z}_{Wi0}\int_l^{p}\dot{I}(l)dl + \dot{I}(\mathrm{p})R_{GD}. \quad (9)$$

Figure 9 shows $\dot{I}(l)$ current module distribution along grounded at left end *i* line l_i = p = 100 km length, calculated for $ZZ(a(l))$, $C(l)_{ki0}^{Int}$ and $C(l)_{ki0}^{An}$ capacities: at a_0 = 5 m, Δa(p) = 100 m and Θ = 0.0573° angle (a); as well as for parallel *i* and *k* lines at a_0 = 5 m, Δa(p) = 0 m and Θ = 0° angle (b).

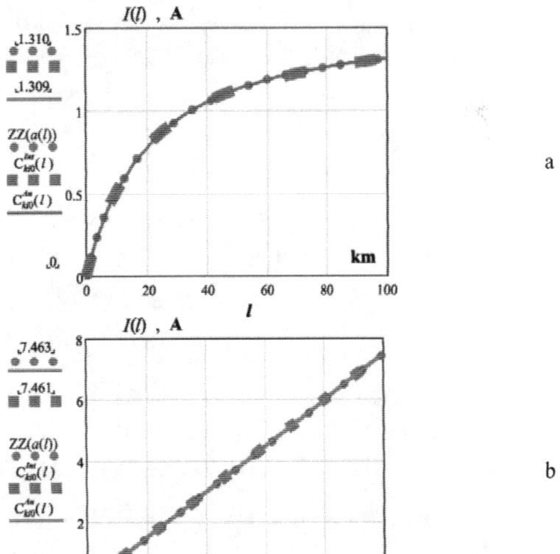

Fig. 9. Current module $\dot{I}(l)$ distribution for $ZZ(a(l))$, $C(l)_{ki0}^{Int}$ and $C(l)_{ki0}^{An}$: (a) converging *i* and *k* lines at p = 100 km, a_0 = 5 m, Δa(p) = 100 m; (b) parallel lines at p = 100 km, a_0 = 5 m, Δa(p) = 0 m

 https://doi.org/10.1051/e3sconf/201913901044

Figure 10 shows voltages $\dot{U}_{A'}(l) = \dot{U}_{ab}(l)$ modules (a) and arguments (b) distribution for $ZZ(a(l))$, $C(l)_{ki0}^{Int}$ and $C(l)_{ki0}^{An}$ at a_0 = 5 m, Δa(p) = 100 m, p = 100 km.

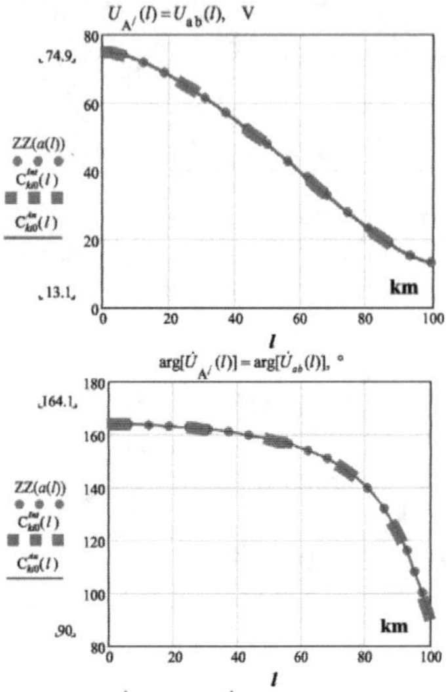

Fig. 10. Voltages $\dot{U}_{A'}(l) = \dot{U}_{ab}(l)$ modules (a) and arguments (b) distribution for $ZZ(a(l))$, $C(l)_{ki0}^{Int}$ and $C(l)_{ki0}^{An}$ at a_0 = 5 m, Δa(p) = 100 m, p = 100 km

Voltage module $\dot{U}_{ab}(l) = \dot{U}_{ab}(0)$ at the ungrounded l line end, when it is grounded at far from k line end конце (Figure 8), exceed 2.4 times voltage module $\dot{U}_{ab}(l) = \dot{U}_{ab}(\mathrm{p})$ value at ungrounded i line end (Figure 7a) when it is grounded at near to with k line end (Figure 5). This is mathematically explained by larger area under the curve $\dot{I}(l)$ in Figure 9a, than in Figure 7a, calculated by the integral $\int \dot{I}(l)dl$, included in the expression of voltage $\dot{U}_{A'}(l)$ и $\dot{U}_{ab}(l)$. The physical explanation is that in case of i line grounded at far from k end induced current $\dot{I}(l)$ module from the beginning to the end of i line is higher than in case of i line grounded at near k line end.

3 Currents and voltages of grounded at one end *i* line depending on length *l*i=p, angle of convergence Θ and minimal distance *a*0 between *k* and *i* lines

Consider the case of i line grounding at far from k line right end (Figure 8), when induced voltage in ungrounded end of i line is most significant.

In expression (7) a function variables will be a_0, l and Θ: $a(a_0,l,\Theta)$. Partial capacity between lines we present in analytical form: $C_{ki0}(a) = C_{ki0}^{An}(a)$. Substituting $a(a_0,l,\Theta)$ function in expression (5) instead of variable a get $C_{ki0}^{An}(a_0,l,\Theta)$. Similarly, by introducing additional variables into formulas (7) and (8), we obtain equations for current and voltage: $\dot{I}(a_0,l,\Theta)$ and $\dot{U}_{ab}(a_0,l,\Theta)$.

Take the angle Θ = 0.0573°. Figure 11 shows the distribution of current modules $\dot{I}(a_0, l = \mathrm{p}, \Theta)$ (a) and voltage modules $\dot{U}_{A'}(a_0, l = \mathrm{p}, \Theta) = \dot{U}_{ab}(a_0, l = \mathrm{p}, \Theta)$ (b), for case of grounded (l = p) i line, as well as Figure 12 shows voltage $\dot{U}_{A'}(a_0, l = 0, \Theta) = \dot{U}_{ab}(a_0, l = 0, \Theta)$ module values for case of ungrounded (l = 0) i line end l_i = p = 100, 50 и 10 km length with minimum distance a_0 between k and i lines change from 5 to 100 m.

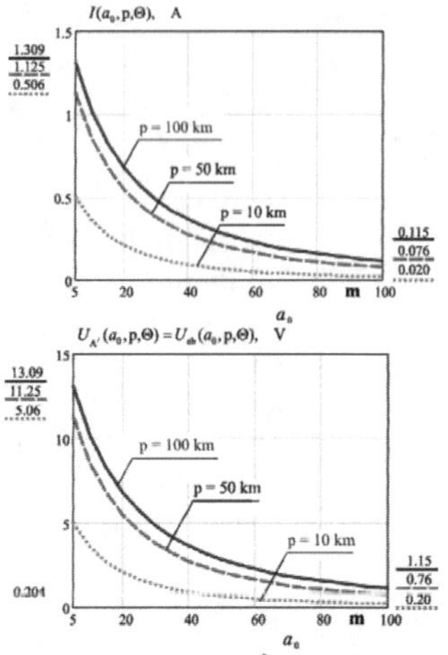

Fig. 11. Current module value $\dot{I}(a_0, \mathrm{p}, \Theta)$ distribution (a), voltage module value $\dot{U}_{A'}(a_0, \mathrm{p}, \Theta)$ distribution (b) for case of grounded i line (l = p) when p = 100, 50 и 10 km at a_0 change from 5 to 100 m

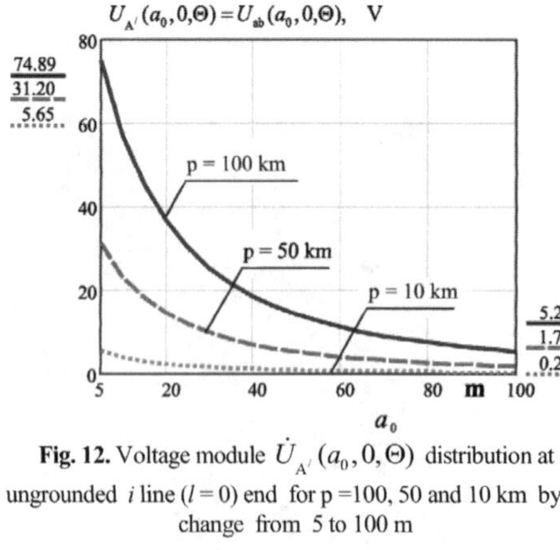

Fig. 12. Voltage module $\dot{U}_{A'}(a_0, 0, \Theta)$ distribution at ungrounded i line (l = 0) end for p =100, 50 and 10 km by a_0 change from 5 to 100 m

Current module values in ungrounded end as well as voltage module values at both i line ends decrease rapidly for all line' p length values with a_0 increase.

Take the minimal distance a_0 = 50 m. Figure 13 shows current module values $\dot{I}(a_0 = 5, l = \mathrm{p}, \Theta)$ (a), and voltage module values $\dot{U}_{A'}(a_0 = 50, l = \mathrm{p}, \Theta) = \dot{U}_{ab}(a_0 = 50, l = \mathrm{p}, \Theta)$ (b) distributions for i line (l = p) l_i = p = 100, 50 and 10 km length grounded case with Θ angle change from 0° to 1°. Figure 14 shows voltage module $\dot{U}_{A'}(a_0 = 50, l = 0, \Theta) = \dot{U}_{ab}(a_0 = 50, l = 0, \Theta)$ values distribution for the same *i* line (l = 0) ungrounded at the end.

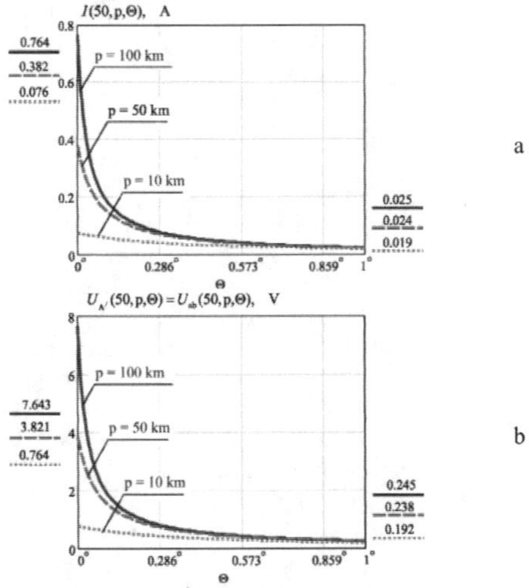

Fig. 13. Current module $\dot{I}(50, \mathrm{p}, \Theta)$ value (a); voltage module $\dot{U}_{A'}(50, \mathrm{p}, \Theta) = \dot{U}_{ab}(50, \mathrm{p}, \Theta)$ (b), distribution for *i* line (l_i = p) l_i = p = 100, 50 and 10 km length grounded case with Θ angle change from 0° to 1°

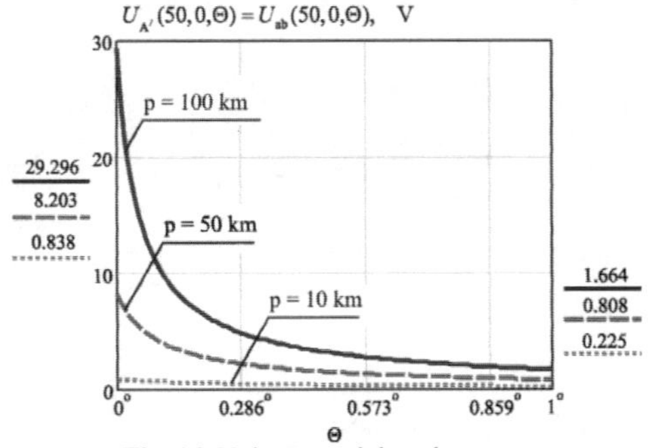

Fig. 14. Voltage module values $\dot{U}_{A'}(50, 0, \Theta) = \dot{U}_{ab}(50, 0, \Theta)$ distribution at ungrounded end of *i* line (l = 0) with l_i = p = 100, 50 and 10 km length under a_0 = 50 m and Θ angle change from 0° to 1°

Angle Θ increase from 0° to 1° at a_0 = 50 m leads to the same sharp decrease in current module values $\dot{I}(50, \mathrm{p}, \Theta)$, and in voltage module values $\dot{U}_{A'}(50, \mathrm{p}, \Theta) = \dot{U}_{ab}(50, \mathrm{p}, \Theta)$ in grounded *i* line end: for p = 100 km by 31 times, for p = 50 km by 16 times, and for p = 10 km by 4 times. Voltage module values $\dot{U}_{A'}(50, 0, \Theta) = \dot{U}_{ab}(50, 0, \Theta)$ at *i* line ungrounded left end reduced by less: for p = 100 km by 17.6 times, for p = 50 km by 10.2 times and for p = 10 km by 3.7 times.

Current and value changes curve have are similar but more expressed for the same conditions and minimal distance a_0 = 5 m (Figures 15 and 16).

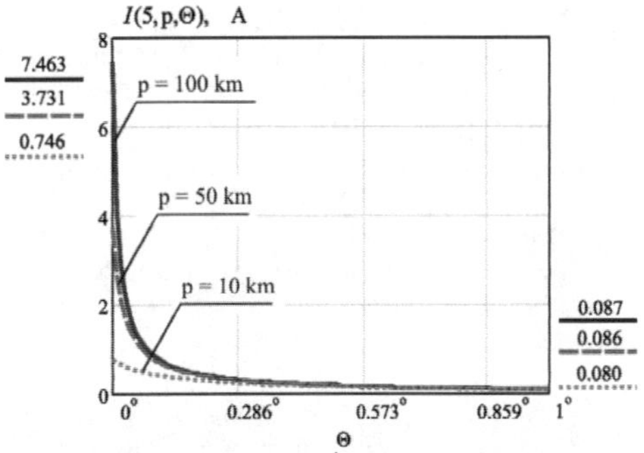

Fig. 15. Current module $\dot{I}(5, \mathrm{p}, \Theta)$ distribution in grounded end (l = p) of *i* line with l_i = p = 100, 50 и 10 km under при a_0 = 5 m and Θ angle change from 0° to 1°

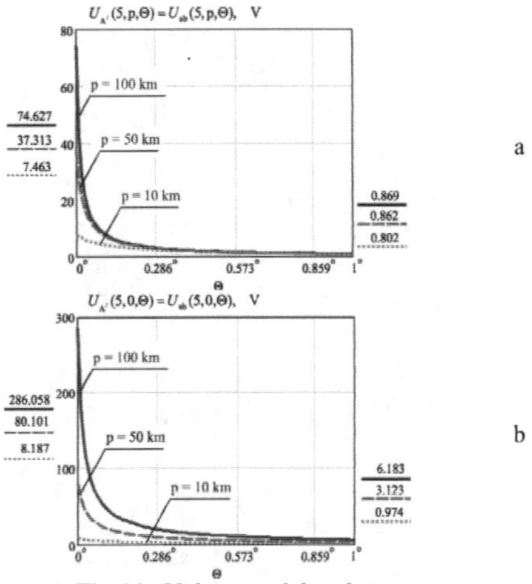

Fig. 16. . Voltage module values $\dot{U}_{A'}(5, \mathrm{p}, \Theta) = \dot{U}_{ab}(5, \mathrm{p}, \Theta)$ distribution at the ends of *i* line with length l_i = p = 100, 50 and 10 km under a_0 = 5 m and Θ angle change from 0° to 1°: (a) grounded (l = p) and $\dot{U}_{A'}(5, 0, \Theta) = \dot{U}_{ab}(5, 0, \Theta)$ line end; (b) ungrounded (l = 0)

Θ angle increase from 0° to 1° under a_0 = 5 m leads to the same sharp decrease the values of current module $\dot{I}(5, \mathrm{p}, \Theta)$ and voltage module $\dot{U}_{A'}(50, \mathrm{p}, \Theta) = \dot{U}_{ab}(5, \mathrm{p}, \Theta)$ of *i* line grounded end: for p = 100 km by 85.8 times, for p = 50 km by 43.3 times and for p = 10 км by 9.3 times.

Herewith voltage module $\dot{U}_{A'}(5, 0, \Theta) = \dot{U}_{ab}(5, 0, \Theta)$ values at ungrounded left end of *i* line reduce by less: for p = 100 km by 46.3 times, for p = 50 km by 25.6 times and for p = 10 km by 8.4 times.

Conclusion

Calculation results synthesis shows that for partial capacity values between phases and cables of converging under voltage and unconnected three-phase lines determination $C_{ki0}(a)$ calculation must be carrying out taking into account the phases and cables of all, both converging and parallel lines of the considered part.

https://doi.org/10.1051/e3sconf/201913901044

Transfer from single-phase to three-phase designs, both under voltage and unconnected OTL leads to $C_{ki0}(a)$ capacity value decrease from $1.87{\cdot}10^{-12}$ F/m (Table 1) to $1.43{\cdot}10^{-12}$ F/m under H = 19 m, h = 17.5 m and a = 5 m, that in result leads to induced current and voltage values in disconnected line decrease.

Given algorithms with superposition method use allow to determine the values of converging three-phase OTL induced voltage that lead to personnel' work reliability and safety elevation.

U_{ab} transverse voltage induced in i unconnected line by electric field of converging with it k line under voltage can be neglected in case of violation from the values of any tested parameters presented in table 2 as well as in [7]: the length of i line converging part l_i = p, minimal distance a_0 between converging lines, convergence angle Θ.

Table 2. Tested parameters values

Parameter value for U_{ab} voltage possibility of neglect	U_{ab} voltage as well as another parameters
- The length of i line converging part $l_i = p < 1$ km	$\dot{U}_{A'} \approx 2$ V at ungrounded end, even under a_0 = 5 m and converging angle Θ = 0.0573° (Figure 12)
- Minimal distance between converging lines $a_0 \geq 100$ m	Highest voltage value $\dot{U}_{A'}$ = 5.3 V at ungrounded end, even under l_i = p = 100 km and converging angle Θ = 0.0573° (Figure 12)
- Converging angle Θ ≥ 5° (with reserve)	$\dot{U}_{A'}$ = 6.2 V at ungrounded end, even under a_0 = 5 m, l_i = p = 100 km and Θ = 1° (Figure 16b)

Reference

1. Korovkin N.V., Selina E.E., Chechurin V.L. Theoretical foundations of electrical engineering. – S-Pb.: Piter, 2006. – 512 p. (in Russian)
2. Neiman L.R., Demirchjan K.S. Theoretical foundations of electrical engineering. V.2. – L.: Power. 1967. – 408 p. (in Russian).
3. Misrikhanov M.Sh., Iostson Yu.A., Rubtsova N.B., Tokarskiy A.Yu. Computer program "Electromagnetic parameters of overhead transmission lines № 2006613744, 27.10.2006 (in Russian).
4. Kostenko M.V., Perelman L.S., Shkarin Yu.P. Wave processes and electrical disturbances in multi-wire high-voltage lines. Energy. M., 1973.- 272 p. (in Russian)
5. Tsitsikjan G.N. Electromagnetic compatibility in power industry. St-Petersburg, ELMOR., 2007- 184 p. (in Russian)
6. Guidelines for the calculation of short-circuit currents and the selection of electrical equipment. G153-34.0-20.527-98. Energia. M, (1998). - 83 p. (in Russian)
7. A.Yu. Tokarskiy, N.B.Rubtsova Distribution of voltage induced by parallel transmission line along dead and grounded line. Safety in the Technosphere, 2015, 4 (June-July), 32-38 (in Russian)

https://doi.org/10.1051/e3sconf/201913901045

About reliability characteristics of overhead electrical grids 6-10 kV elements

Abdula Abdurakhmanov[1], *Sergey* Glushkin[1], and *Andrey* Shuntov[1]

[1]all - Moscow Power Engineering Institute, Department of Electric Power Systems, Krasnokazarmennaya 14, Moscow, 111250 Russia.

Abstract. In the paper, failure outage statistics in 6–10 kV overhead electric grids of one of the largest electric grid companies in the European part of the country were analyzed to clarify the reliability characteristics of their elements. Failure intensity and mean restoration time for covered and uninsulated overhead lines, low-oil and vacuum circuit breaker cells, power transformers and secondary electrical grid elements were obtained. The obtained reliability characteristics of elements should be used in the tasks of technical justification and selection 6–10 kV overhead electrical grid preferred structure.

1 Problem description

Problem of power systems reliability performance existed before and continue to be relevant. One of its aspects is traditionally a research of reliability characteristics of electrical grids. At the same time, close attention is paid to the indicators characterizing their failure outages: failure intensity ω and mean restoration time T_r.

In the USSR there was centralized collection and processing of statistical data on the failure outages of power system elements. Its latest results are dated to the period 1977-1982 [1], which presents the reliability characteristics of elements of 6–10 kV overhead electric grid: overhead lines (OHL) with uninsulated wires, 6–10/0.4 kV power transformers, and low-oil circuit breaker cells.

In plenty of publications for OHL, an extremely wide range of failure intensity is given. So, in paper [1], ω=7.64 1/(year per 100 km), in [3] - 2–20 1/(year per 100 km), and in [2] - 25 1/(year per 100 km) and even more [4]. In addition, covered overhead lines, (i.e. with wires having a protective insulating outer sheath) apply more and more in the country's medium voltage grids. Compared with uninsulated wires they have certain advantages in terms of reliability, for example, in [5] it is noted that the covered OHL are protected from overlapping, have high mechanical strength, practically do not form icy-hoarfrost deposits and others. In [6] there was an assumption that the use of covered OHL will reduce the failure intensity of elecrical lines by 2.4 times. The widespread use of covered overhead lines in distributing power systems dates back to beginning 2000-s and now this type of line occupies an increasingly significant share.

For power transformers, published failure intensity are also very different. For example, in [1] values of 0.008-0.016 1/year are indicated, and in [2] - 0.035 1/year. However, the studies presented were carried out almost 40 years ago.

For low-oil circuit breakers, the values of ω=0,009 1/year and T_r=20 h [1], obtained in the 80s of the last century, are similar to the reliability characteristics that are still given in various publications. For example, in [7] values ω–0,01 1/year is presented, including 0,003 1/year - fault type of short circuit and 0,007 - fault type of break circuit.

However, vacuum circuit breakers become more widely used over the last decade in comparison low-oil ones. For vacuum circuit breakers, the expert value ω=0.004 1/year is given in [8]. The same publication contains a reference that, according to leading world manufacturers, the mean operating time to failure of a modern vacuum circuit breaker is 2000 years, whence ω=0.0005 1/year.

Note that in the analysis of circuit breaker failure models consider not a only circuit breaker, but a connection failure, i.e. circuit breaker with all related equipment - disconnectors, measuring transformers, connecting buses, relay protection and automation devices, insulators and bushings [9].

Thus, the presented information on the reliability characteristic of the elements of 6–10 kV overhead electric grids is outdated or contradictory in a number of positions, and the reliability characteristic of elements of 6–10 kV grids, such as fuses, measuring transformers and etc. are almost not presented in the literature.

Therefore, the authors set themselves the task of analyzing failure outage statistics in 6–10 kV overhead electrical grids of one of the largest electric grid companies in the European part of the country and clarifying their reliability characteristic.

2 Reliability characteristics of 6–10 kV covered and uninsulated overhead lines

Statistical sample for uninsulated OHL and covered OHL for 2015–2017 shown in table 1. As can be seen in the table 1 every year about 1000 km of covered OHL are put into operation and to date their share in the region has reached about one third of total length.

In the table 2 shows the total number of failure outages and their total duration for covered and uninsulated OHL from table 1. Based on the data in table 1 and 2, the values of failure intensity and mean restoration time are calculated - see table 3.

Table 1. The total length of the electrical grid in the region

Year	The total length of the electrical grid 6-10 kV, km		
	uninsulated OHL	covered OHL	Total
2015	20 033	8 067	28 100
2016	19 437	8 901	28 338
2017	18 708	10 177	28 885

Table 2. Overhead lines outage

Year	uninsulated OHL		covered OHL	
	Number of outages, ea	Duration of outages, hrs	Number of outages, ea	Duration of outages, hrs
2015	557	1038,4	25	86,5
2016	438	836,7	29	96,5
2017	345	862,5	31	189,5
Total	1340	2737,6	85	372,5

As can be seen from the table 3, the overhead line failure intensity was only 2.3 1/(year per 100 km). In the region, abnormal ice rains preceding the period under review caused numerous failure outages in electric grids of 6-10 kV and long-term restrictions on the power supply of consumers. Further, technical measures were organized, such as accelerated reconstruction of grids, widespread and uncompromising clearing of «no tree» zone OHL, replacing uninsulated OHL with covered OHL, which made it possible to achieve shown in table 3 characteristics.

It should be noted those 3-5 days each year, during which a quarter or more of all overhead line outage occur due to abnormal weather conditions (strong winds, ice rains, etc.).

Table 3. Reliability characteristics of overhead lines

Year	uninsulated OHL		covered OHL	
	ω, 1/(year per 100 km)	T_r, hrs	ω, 1/(year per 100 km)	T_r, hrs
2015	2,78	1,86	0,31	3,60
2016	2,25	1,91	0,33	3,33
2017	1,84	2,50	0,30	6,11
2015–2017	2,30	2,04	0,31	4,38

The failure intensity of covered OHL are approximately an order of magnitude more reliable (7.4 times) than uninsulated OHL: 0.31 compared with 2.3 1/(year 100 km) (table 3). A significant decrease in the failure intensity of covered OHL is partially offset by an increase in their mean restoration time. Obviously, the multiplied ωT_r is the expected average annual outage of line: 2.3·2.04=4.7 hours per 100 km for uninsulated OHL and 0.31·4.38=1.36 hours per 100 km for covered OHL, so we can conclude that the covered OHL have 4.7/1.36≈3.5 times more favorable characteristics.

Additionally, in the table 4 shows the organizational structure of the total failures of uninsulated OHL and covered OHL for the period 2015–2017 (classification of failures are taken from [10]). Data analysis table 4 shows that the failure structure of uninsulated OHL and covered OHL proved to be expected. So, the percentage of failures related with the intervention of unauthorized persons and organizations and adverse natural phenomena (tree contact), reaches in both cases 70–80%. In addition, the share of covered OHL failures associated with atmospheric overvoltages is several times higher than that for uninsulated OHL, as was mentioned earlier in [5].

Table 4. Overhead lines failure structure

Failure cause	uninsulated OHL		covered OHL	
	Number of outages, ea/%	ω, 1/(year per 100 km)	Number of outages, ea/%	ω, 1/(year per 100 km)
Bad-quality OHL maintenance	97/7,2	0,167	6/7,1	0,022
Intervention of unauthorized person or organization	192/14,3	0,330	20/23,5	0,074
Birds	69/5,2	0,119	–	–
Natural disasters including	888	1,526	54	0,199
– Icy-hoarfrost deposits;	10/0,8	0,017	–	–
– Wind speed load (wires overlapping);	103/7,7	0,177	–	–
– Atmospheric overvoltage (lightning);	27/2,0	0,046	7/8,2	0,026
– Tree contact	748/55,8	1,286	47/55,3	0,173
Installation defects	35/2,6	0,060	5/5,9	0,018
Unclassified cause	59/4,4	0,101	–	–
Total	1340/100	2,30	85/100	0,31

https://doi.org/10.1051/e3sconf/201913901045

3 Reliability characteristics of 6–10 kV low-oil and vacuum circuit breaker cells

The failure outage statistics for the cells of low-oil and vacuum circuit breakers 6–10 kV at 611 substations 35–220/.../6–10 kV in the region for the period 2016–2018 were analyzed. The size of the statistical sample, see table 5.

As can be seen from the table 5, the share of low-oil circuit breakers is more than half of the installed equipment fleet, despite the fact that for many years they have not been used in new construction or reconstruction.

In the table 6 shows the total number of failure outage and their total duration for the circuit breaker cells from table 5. Based on the data in table 5 and 6, the values of the failure intensity and the mean restoration time of the circuit breakers are calculated - see table. 7.

Table 5. Number of circuit breaker cells 6–10 kV at substations

Year	Low-oil	Vacuum
2016	11 500	7063
2017	11 282	7281
2018	10 929	7634

Table 6. Circuit breaker cells outages

Year	Low-oil circuit breakers		Vacuum circuit breakers	
	Number of outages, ea	Duration of outages, hrs	Number of outages, ea	Duration of outages, hrs
2016	26	102,7	17	63,4
2017	15	35,4	3	15,7
2018	34	265,1	5	7,5
Total	75	403,2	25	86,6

Table 7 shows that the failure intensity and the mean restoration time of the vacuum circuit breaker cells are about two times less than the low-oil. One of the indirect cause for this is connected that the number of components and moving parts of a vacuum circuit breaker is half that of a low-oil one.

Table 7. Reliability characteristics of circuit breaker cells

Year	Low-oil		Vacuum	
	ω, 1/year	T_r, hrs	ω, 1/year	T_r, hrs
2016	0,0023	3,9	0,0024	3,7
2017	0,0013	2,4	0,0004	5,2
2018	0,0031	7,8	0,0007	1,5
2016–2018	0,0022	5,4	0,0011	3,5

Note that the operational logs that record failure outages usually indicate the time to restore the initial condition of the circuit, but not actual time for repair of the circuit breaker cell. The reason of it is that cells with withdrawable elements are mainly installed at substation. Therefore, reserve withdrawable circuit breaker cell are used to quickly restore initial state of the circuit in case of a serious failure of circuit breaker.

Table 8 shows the failure structure of circuit breaker cells at substations. During the period under review, only two cases (2016 and 2018) of loss of airtightness of the vacuum circuit breaker arc chamber were recorded. The low-oil circuit breaker drive shaft failure share of about 20% of all failures. In general, the failure structure of low-oil circuit breaker cells in table 8 and in [8] are comparable. For vacuum circuit breaker cells, the main cause of failures was malfunctions of the control circuits, and for low-oil ones - insulation breakdown.

A comparison of the data from table 7 and the paper [1,7,8] shows that the obtained reliability characteristics of the circuit breaker cells become more optimistic. It can be assumed that this is because substation cells were analyzed. The operational specificity of substation's circuit breaker cells is the relatively insignificant (several times a year) number of switching and the presence of duty personnel at substation. Perhaps we should expect a deterioration in the reliability characteristics for circuit breaker cells in the auxiliary systems of 6 kV power plants, where technological conditions require much more frequent (once or twice a day [8]) switching.

Table 8. Circuit breaker cells failure structure

Failure cause	Constructional part failure, ea./%	
	Low-oil	Vacuum
Arc chamber	10/13,3	2/8,0
Drive shaft	17/22,7	0/0
Control circuits	19/25,4	13/52,0
Insulators and bushings	25/33,3	9/36,0
Bus bars	4/5,3	1/4,0
Total	75/100	25/100

For circuit breakers installed in prefabricated 6-10 kV distribution substation should be expected higher mean restoration time. Firstly, this failure takes extra time to detect it and personnel to arrive at the failure substation. Secondly, the design of the cells does not include withdrawable elements. However, a document, published more than 30 years ago, already described that in the scheme of medium voltage electrical grid, the construction of a 6-10 kV distribution substation should be used only in the present of feasibility studies [11].

Thus, vacuum circuit breaker cells 6–10 kV are approximately two times more reliable than outdated low-oil circuit breaker cells in terms of failure intensity and mean restoration time.

4 Reliability characteristics of 6–10 kV power transformer

The failure outage statistics on power transformers 6–10 kV in the distribution electrical grids of the region for 2015–2017 were processed. The size of the statistical sample, see table 9. Accepted designations in table 9: capacity - total installed capacity of transformers).

Note that the installed capacity of transformers in city cable grids in this region is three times higher than that in the overhead electric grids of suburban areas, although the number of transformers in both cases is comparable. Moreover, in overhead electric grid installed capacity of transformers 6–10 kV three to four times exceed the maximum of their actual load, possibly due to the crisis that occurred in the country (the period of the 90s of the last century).

In the table 10 shows the total number of failure outage and their total duration for power transformers from the table 9.

Based on the data table 9 and 10, the values of failure intensity and mean restoration time of the considered power transformers are calculated - see table 11.

Table 9. Number of 6-10 kV transformers substation and power transformers

Year	Cable electrical grids			Overhead electrical grids		
	Transformer substation, ea	Power transformers, ea	Total capacity, MVA	Transformer substation, ea	Power transformers, ea	Total capacity, MVA
2015	15379	28247	18621	19645	23037	5854
2016	15494	28415	18960	21424	24681	6479
2017	15623	28683	19350	22709	25174	6886

Table 10. Power transformers outages

Year	Cable electrical grids		Overhead electrical grids	
	Number of outages, ea	Duration of outages, hrs	Number of outages, ea	Duration of outages, hrs
2015	32	331,4	6	25,8
2016	29	473,1	8	39,2
2017	31	538,85	5	30,1
Total	92	1343,35	19	95,1

From the table 11 it follows that the failure intensity of power transformers in cable electrical grids is about four times higher than in overhead electrical grids. Apparently, this is due to two reasons associated with operation on urban areas: the obviously higher load of power transformers and their installation in enclosed spaces, which worsens cooling conditions.

Table 11. Reliability characteristics of 6–10 kV power transformers

Year	Cable electrical grids		Overhead electrical grids	
	ω, 1/year	T_r, hrs	ω, 1/year	T_r, hrs
2016	0,00113	10,4	0,00026	4,3
2017	0,00102	16,3	0,00032	4,9
2018	0,00102	17,4	0,00019	6,0
2016–2018	0,00106	14,6	0,00026	5,0

Comparing reliability characteristic from table 11 with previously published data, a significant improvement in the reliability of power transformers is noticeable: the failure intensity decreased from 0.008 (see above) to 0.001 1/year (Table 11), which is a favorable factor.

The failure structure of power transformers 6–10 kV is given in table 12. As can be seen from it, the failure share of transformers associated with internal failures reaches approximately 70%, which must be taken into account during their operation.

Table 12. Power transformers 6–10 kV failure structure

Type of failure	Number of failures, ea/%
Internal equipment failure (short-circuit between phases or winding coil, breakdown in insulation, breakdown in tap-controller)	76/68,5%
External equipment failure (breakdown of bushings, short-circuit cable bus-bar)	29/26,1%
Breakdown in transformer tower or transformer tank, oil leak	6/5,4%

5 Reliability characteristics of 6–10 kV secondary electrical grid elements

In the table 13 shows the reliability characteristics of measuring transformers, fuses (on three phases) and disconnectors. Analysis of the data from the table 3, 7, 11 and 13 leads to the conclusion that the main part of the failures of 6–10 kV overhead electric grids consist of overhead line failures. Failures of other network elements can be neglected in reliability assessment

Table 13. Reliability characteristics of 6–10 kV secondary electrical grid elements

Electrical grid element	Number of outages, ea				The number of observed elements	ω, 1/year
	2015	2016	2017	2015–2017		
Current transformer	7	7	6	20	474711	$0{,}42\cdot10^{-4}$
Voltage transformer	1	1	2	4	158237	$0{,}25\cdot10^{-4}$
Fuse	13	11	5	29	474711	$0{,}61\cdot10^{-4}$
Disconnector	6	2	1	9	158237	$0{,}57\cdot10^{-4}$

 https://doi.org/10.1051/e3sconf/201913901045

6 Conclusion

1. Based on the processing of representative statistical sample, the reliability characteristics of uninsulated and covered OHL in 6-10 kV electrical grids are obtained. It turned out, that the failure intensity of covered OHL has significantly more favorable reliability characteristics compared to uninsulated ones.

2. Vacuum circuit breaker cells 6–10 kV are more reliable than outdated low-oil ones: about two times from the terms of the failure intensity and mean restoration time. At the same time, the actual values of the reliability characteristic obtained at this time interval turned out to be more favorable compared to previously published ones.

3. The reliability characteristics of 6–10 kV power transformers turned out to be much more favorable than previously published values. However, it cannot be ruled out that the equipment fleet and its operating conditions have been updated over the past 40 years since the last reliability studies.

4. The failure intensity of secondary electrical grid elements (measuring transformers, disconnectors, fuses) is significantly lower than overhead lines, circuit breaker cells and power transformers described above. Therefore, in a first approximation, they can be neglected in reliability assessment.

5. The obtained reliability characteristics of the 6–10 kV overhead electric grid elements should be used in the tasks of technical justification and selection their preferred structure.

References

1. *Guidance on the application of reliability characteristics of power system elements and the operation of power units with steam turbine units* (SPO Soyuztekhenergo Publ., 1985)

2. V. Pruss, V. Teslenko *Improving the reliability of rural electric grids* (Energoatomizdat, 1989)

3. L. Rybakov, Z. Ivanova, *Failure forecasting and planning of the reserve and spare elements, apparatus and equipment of 10 kV distribution grids*. Bulletin of the Chuvash University, **1**, 104 (2015)

4. L. Doletskaya, V. Kavchenkov, R. Solopov *Evaluation of the effectiveness of methods to improve the reliability of distribution electric grids*. NAUKOVEDENIE, **6**, 1 (2015)

5. *Advantages of self-supporting covered wires 6–35 kV. Ways to protect overhead lines from lightning surges.* Electrical engineering news. **3**, (2002). [Electronic resource]. URL: http: // www.news.elteh.ru/arh/2002/15/09.php.

6. V. Gulkov. *10 (6) kV overhead lines with insulated wires. PhD candidate's abstract.* (1997)

7. Voropay N.I. *Reliability of power supply systems.*
(Nauka, 2015).

8. A. Nazarychev. *Analysis of the main advantages of using vacuum circuit breakers*. Energoekspert, **4-5**, 58, (2007)

9. P. Grudinsky, S. Mandrykin, M. Ulitsky *Technical operation of the main electrical equipment of power stations and substations.* (Energia, 1974)

10. Order of the Ministry of Energy of Russia dated 02.03.2010 No. 90 (as amended on 07.27.2017) "*On approval of the form of the act on the investigation of the causes of accidents in the electric power industry and the procedure for filling it out.*

11. *Instructions for the design of urban and rural electric networks.* (Ministry of Energy of the USSR, 1984)

Overheard electrical grids 6-20 kV partition

Abdula Abdurakhmanov, *Sergey* Glushkin, and *Andrey* Shuntov[a]

[a]all - Moscow Power Engineering Institute, Department of Electric Power Systems, Krasnokazarmennaya 14, Moscow, 111250 Russia.

Abstract. The paper analyzes the technical principles of construction and the typical structure of overhead electric grids 6-20 kV. Reliability calculations were performed for 18 segments of 6-10 kV electrical grids of one of the largest electric grid companies in the European part of the country. Shown, when assessing the reliability of overhead electric grids, it is permissible to take into account only reliability parameters of electric transmission lines. It is permissible to neglect failures of other grid elements in a first approximation (power transformers, circuit breaker e.t.c.)

1 Problem description

In publications [1, 2], the authors analyzed failure outage statistics in 6–10 kV overhead electric grids of one of the largest electric grid companies in the European part of the country to clarify the reliability characteristics of grid elements.

Thus, the failure intensity for uninsulated overhead lines (OHL) is ω_{OHL} = 2.3 1/(year per 100 km) and for covered OHL $\omega_{coverOHL}$ = 0.31 1/(year 100 km). A noticeable decrease in the failure intensity of covered overhead lines is partially offset by an increase in the mean restoration time for them, which is approximately two times higher than for uninsulated OHL. Therefore, the multiplication ωT_r is the expected average annual outage of line: 2.3·2.04=4.7 (hrs per 100 km) for uninsulated OHL and 0.31·4.38=1.36 (hrs per 100 km) for covered OHL, i.e. from these positions covered OHL 4.7/1.36≈3.5 times have more favorable characteristic.

Reliability characteristics were also obtained for power transformers 10(6)/0.4 kV, low-oil and vacuum circuit breaker cells and other grid elements (disconnectors, measuring transformers, etc.). Failure intensity obtained for uninsulated and covered OHL and these values differed by more an order of magnitude from previously known values, published 30–40 years ago: 10–20–25 1/(year per 100 km) and even more [3, 4]. Note that the technical principles of medium voltage networks were justified using those overestimated failure intensity - see, for example, [5].

Earlier in document [6] reliability characteristic of power supply for consumers were established. For consumers of the 2nd category in [6], the standard failure intensity (power outages up to four hours) was approximately 2.3 1/year and for consumers of the 3rd category with a duration of power outages up to 24 hours - 3.0 1/year.

Based on these characteristics, the basic technical principles of 6-20 kV overhead grids and the installation site of sectional devices were formed. According to [6], the total length of the line section (including branches) should not exceed 12 km and be limited to an automatic switching device; also sectionalizers should be installed on branches longer than 2.5 km. Modern regulatory documents [7] partially repeat the previously developed technical principles. For example, if a branch length from overhead feeder line is more than 1.5 km, then installation of a switching apparatus is required. Document [7] also describes the installation requirements of sectionalizers and automatic switching device using vacuum circuit breakers in the overhead feeder line.

Therefore, the authors set themselves the task of justifying the preferred structure of 6-10 kV overhead grids, taking into account the reliability characteristics obtained in the present time interval.

2 Typical configuration of an overhead electrical grids of 6-10 kV

As it is known, a typical configuration of an overhead electrical grid of 6-10 kV (Fig. 1) is a loop circuit with branches. The grid powered from two geographically remote substations of 35...220/10(6) kV, also the network is partitioned by sectional devices (reclosers).

Line branches with transformer substations 10(6)/0.4 kV are connected to the main feeder line using disconnector or recloser installed on the branch (recloser is installed on branches longer than 1.5 km). The grid is partitioned by automatic switching device at the point of current separation with the ability to automatic transfer switch. And this is done not to improve reliability, but to reduce power and electricity losses.

In accordance with [8], the system average interruption frequency index (SAIFI) for consumers is defined as SAIFI= $(\sum^{I} N_i)/N_t$, where $\sum N_i$ – total number of customer experienced interruption, N_t – total number

https://doi.org/10.1051/e3sconf/201913901046

of customers served, I –number of failure outage for the period under review. However, if there is one partition point in the grid, any failure in line section from substation to sectional device (including branches from main feeder line) will lead to a power outage for all consumers in this section of line. Thus, the value of SAIFI and ω in this configuration of electrical grid will be equal to each other.

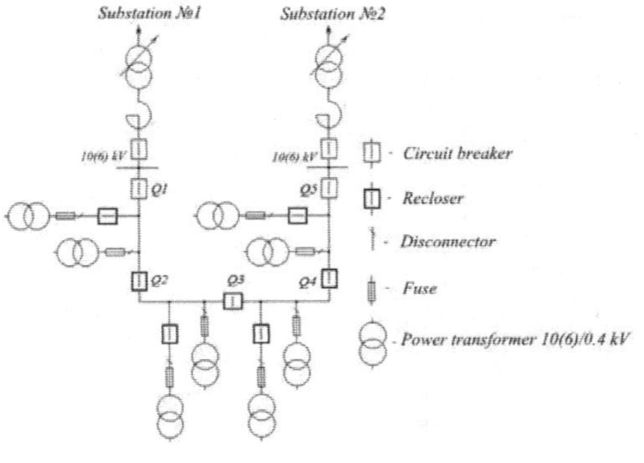

Fig.1. Typical configuration of an overhead electrical grid

Table 1 shows examples of reported data on SAIFI values in EU countries [9] (excluding exceptional events - natural disasters, major system power outage, etc.). These values should be treated with some caution; in different countries, SAIFI calculations use a non-uniform set of source data. Another example is the 2006 summary report [10] for US regions showing SAIFI values of 1.13–1.59 1/year.

3 Reliability calculations

Reliability calculations were performed based on that integrated parameters of 6–10 kV grids obtained in [11]. In [11] were analyzed 18 regions of one of the largest electric grid companies in the European part of the country. Among them were fragments, both adjacent to a big city, and 100-150 km distant from it, where predominantly installed single-transformer substations of pole-mounted type (S_{rated}=160÷250 kV·A) and kiosk type (S_{rated}=400÷1000 kV·A) were installed. The load density was determined from the assumption that transformer substations 10/0.4 kV covers an area with a radius of not more than 0.5 km. In fact, the entire overhead system is made from a wire with a cross section of 70-mm^2, with the exception of two fragments, where cross section of 120-mm^2 wire was partially used on the feeder line.

Table 1. SAIFI reporting data for EU medium voltage electrical grids

Country	2006	2007	2008	2009	2010	2011	2012	2013	2014	2015	2016
Germany	0,43	0,31	0,30	0,27	0,24	0,29	0,25	0,46	0,33	0,78	0,49
Greece	–	–	–	–	–	–	1,65	1,50	1,60	1,44	1,42
Denmark	–	–	–	–	0,30	0,27	0,30	0,24	0,23	0,26	0,27
Ireland	–	–	–	–	–	–	–	–	–	0,90	0,80
Italy	1,87	1,71	1,58	1,61	1,47	1,37	1,38	1,28	1,27	1,41	1,20
Latvia	–	–	–	–	–	–	–	2,45	2,12	1,88	2,00
Lithuania	0,51	0,81	0,68	0,52	0,51	0,64	0,63	0,58	0,52	0,48	0,51
Luxembourg	–	–	–	–	–	–	–	0,27	0,22	0,22	0,15
Netherlands	–	–	–	–	–	–	–	0,20	0,20	0,18	0,17
Norway	–	–	–	–	–	–	1,58	2,30	1,59	1,43	1,27
Portugal	2,66	1,98	2,28	2,68	3,02	1,88	1,57	1,70	1,84	1,64	1,68
Romania	–	–	–	–	–	–	–	6,29	4,80	5,16	6,10
Slovenia	–	–	1,80	1,50	1,40	1,63	2,16	1,59	1,89	1,45	1,21
Czech	–	–	1,45	1,82	1,95	1,93	1,87	1,78	1,78	1,57	1,50
Switzerland	–	–	–	–	0,17	0,20	0,22	0,19	0,14	0,17	0,12
Sweden	–	–	–	–	–	–	0,94	1,03	1,02	0,90	0,99

Table 2 shows the parameters of grid segments. Notations defined in the table. 2: P_{max} and P_{min} - the largest load of a winter and summer day according to the measurement results; l_{sum} - the total line length of the segment; l_{feeder} - the length of feeder line section between substations, excluding branches; σ_{load} - load density; n_S - the number of substations 10(6)/0.4 kV in a grid segment.

Data analysis of table 2 allows us to establish some statistical regularities in the grid construction. For example, the length of the feeder lines between substations varies in a narrow range from 12.9 to 41.3 km with an average value of 28 km, which indicates a relatively uniform distribution of the substations in the area. The load density varies over a wide range from 8.2 to 157.3 kW/km^2 (average value of 76.1 kW/km^2), i.e. more than an order of magnitude. Moreover, the range of average distances between transformer substations varies from 0.76 to 2.33 km with a total average of 1.14 km, which indicates a relatively uniform distribution of transformer substations across the area. Note that the installed power of the transformer substation 10(6)/0.4 kV is 2–9 times (average value 3.6) higher than the actual maximum load, which is somewhat irrational and requires additional research.

Table 2. Integral parameters of electrical grid segments

Segment number	P_{max}/P_{min} kW/kW	l_{sum} km	l_{feeder} км	σ_{load} кВт/км^2	n_S ea
1	3138/1642	40,7	19,8	129,1	48
2	4751/3168	52,4	19,4	154,3	64
3	7358/5816	74,1	32,6	154,3	87
4	3609/1929	70,1	21,0	83,2	71
5	6646/6072	87,6	43,0	157,3	110
6	1781/989	46,9	12,9	57,3	54
7	1840/844	42,5	22,5	45,5	44
8	2486/1244	36,2	22,0	66,8	38
9	2992/1673	62,4	35,8	55,0	47
10	3101/2436	75,7	38,3	45,1	67
11	2716/1338	26,5	13,8	110,4	20
12	2818/1934	54,3	29,6	72,3	60
13	1640/1102	64,6	31,0	39,8	57
14	1193/618	61,8	41,3	24,8	46
15	956/349	50,0	30,9	26,3	33
16	297/78	35,0	30,5	8,2	15
17	3023/1872	42,3	20,3	117,6	56
18	1484/796	107,0	38,5	21,7	68
Average	2879/1833	57,2	28,0	76,1	54

Based on grid configuration and data from table 2 the reliability characteristics of grid segments were calculated in table 3 using the reliability characteristics of:

Only uninsulated OHL (covered OHL);

Uninsulated OHL (covered OHL) and all other elements of the grid.

It was considered that a switching device in the middle of a feeder section partitioned the grid segment. Notations defined in the table 3: ω_{OHL} – failure intensity of a grid segment, using the reliability characteristics of only OHL, $\omega_{OHL\Sigma}$ – same, but using the reliability characteristics of OHL and other grid elements; $\omega_{coverOHL}$, $\omega_{coverOHL\Sigma}$, – similar parameters for covered OHL.

Table 3 shows that the increase in failure intensity of a grid segment, when taking into account all other grid elements, (i.e., in addition to uninsulated OHL) is about 6% (average values of ω_{OHL}=0.66 1/year and $\omega_{OHL\Sigma}$=0.7 1/year), which is not important. For covered OHL this increase is much larger - about 30% ($\omega_{coverOHL}$=0.09 1/year and $\omega_{coverOHL\Sigma}$=0.13 1/year, moreover 27% of this increase is the failure of 10(6)/0.4 kV power transformer. Thus, the value of failure intensity obtained are relatively small. Therefore, in assessing the reliability of medium voltage electrical grids, it is sufficient to take into account the reliability characteristics of only uninsulated OHL (covered OHL) as a first approximation.

Table 3. Reliability characteristics of grid segments

Segment number	ω_{OHL}, 1/year	$\omega_{OHL\Sigma}$, 1/year	$\omega_{coverOHL}$, 1/year	$\omega_{coverOHL\Sigma}$, 1/year
1	0,47	0,51	0,06	0,10
2	0,60	0,65	0,08	0,13
3	0,85	0,91	0,11	0,17
4	0,81	0,86	0,11	0,16
5	1,01	1,09	0,14	0,22
6	0,54	0,58	0,07	0,11
7	0,49	0,52	0,07	0,10
8	0,42	0,45	0,06	0,09
9	0,72	0,75	0,10	0,13
10	0,87	0,92	0,12	0,17
11	0,30	0,32	0,04	0,06
12	0,62	0,66	0,08	0,12
13	0,74	0,78	0,10	0,14
14	0,71	0,74	0,10	0,13
15	0,58	0,60	0,08	0,10
16	0,40	0,41	0,05	0,06
17	0,49	0,53	0,07	0,11
18	1,23	1,28	0,17	0,22
Average	0,66	0,70	0,09	0,13

Moreover, power transformers in overhead electric grids 6-10 kV in country are usually protected by base fuses. The above conclusion can be used for electrical grids in which fuses are excluded [12]. This application, in our opinion, is progressive and may be useful in the reconstruction of grids with isolated transformer neutral to a neutral grounded through a low-ohm resistor (see [13]).

It follows from the table 3 that the use of covered OHL significantly increases the reliability of power supply to consumers ($\omega_{coverOHL}$=0.05–0.17 1/year). Taking into account the data of the table 1 it is enough to have only one point, partitioned by an automatic switching device (recloser) at the current separation point (in the typical configuration in Fig. 1). For additional partitioning, for example, when connecting consumers of the 2nd category, for power grid partition it is enough to use disconnectors or load break switches.

A similar approach can be used for uninsulated overhead lines. Here ω_{OHL} varies in the range 0.4–1.23 1/year. This range of reliability characteristics is present in the practice of a number of industrialized countries (see above). For comparison, in existing grid segments 6-10 kV from table 2, the actual number of automatic switching device (reclosers) is in the range of 9–16, which seems redundant.

4 Conclusion

1. When assessing the reliability of overhead electric grids, it is permissible to take into account only reliability parameters of electric transmission lines. It is permissible to neglect failures of other grid elements in a first approximation.

2. The use of covered OHL significantly increases the power supply reliability to consumers. In a typical configuration of the electric grid, it is enough to have one point, partitioned by a switch (recloser) at the point

of current separation. A similar approach can be used with some limitations for uninsulated overhead lines.

3. The previously established regulatory documents and their corresponding requirements for ensuring the reliability of medium voltage grids turned out to be noticeably worse than the obtained modern reliability characteristics from Table 3. The values obtained are comparable with the level of reliability of power supply to consumers in industrialized countries.

References

1. A. Abdurakhmanov, S. Glushkin, A. Shuntov *About reliability characteristics of uninsulated and covered overhead lines 6–10 kV*. EP. Transmission and Distribution, **1**, 84 (2019)
2. A. Abdurakhmanov, S. Glushkin, A. Shuntov *Reliability characteristics of overhead electrical grids 6-10 kV elements*. ISEM SB RAS (to be published)
3. V. Pruss, V. Teslenko *Improving the reliability of rural electric grids* (Energoatomizdat, 1989)
4. R. Fedosenko *Operational reliability of agricultural grids*. (Energy, 1977)
5. B. Maksimov, V. Vorotnitsky *Evaluation of the effectiveness of automatic sectioning of distribution networks of 6-10 kV using reclosers in order to increase the reliability of power supply to consumers*. EE, **10**, 7 (2005)
6. *Guidance on the design of agricultural electricity. Guidelines for ensuring the design of regulatory levels of reliability of power supply to agricultural consumers*. (Ministry of Energy of the USSR, 1986)
7. Regulation of PJSC «Rosseti». *On the Unified Technical Policy in the Electric Grid Complex* (2017)
8. Order of the Ministry of Energy of Russia dated November 29, 2016 No. 1256 *On approval of the Methodological guidelines for calculating the level of reliability and quality of delivered goods and services for an organization managing a single national electric grid and territorial grid organizations*.
9. *Benchmarking Report 6.1 on the Continuity of Electricity and Gas Supply*. (CEER, 2016) URL: http:// www.ceer.eu/documents/104400/-/-/963153e6-2f42-78eb-22a4-06f1552dd34c.
10. *Tracking the Reliability of the U.S. Electric Power System: An Assessment of Publicly Available Information Reported to State Public Utility Commissions*. (2008) URL: http://www.smartgrid.gov/files/ eto_oct_2008.pdf.
11. A. Mayorov, K. Osintsev, A.Shuntov *On the use of a rated voltage of 20 kV in overhead electrical grids*. Electricity, **9**, 4 (2018)
12. D. Shamanov, S. Sokolov *Distribution networks of Finland. Features of circuit solutions*. Electrical Engineering News, **6** (2005)
13. A. Mayorov, K. Osintsev, A. Shuntov *On the selection of a single-phase earth fault current in a network with low-resistance resistive neutral grounding*. Electricity, **9**, 34 (2017)

E3S Web of Conferences **139**, 01047 (2019) https://doi.org/10.1051/e3sconf/201913901047
RSES 2019

Substantiation model calculation diagram UES of Russia to take balance reliability

Mikhail Chukreyev[1]

[1]Federal Research Center "Komi Scientific Center of the Ural Branch Russian Academy of Sciences", ISE and EPN, 167000, Russia

Abstract. The principles formation the model design scheme UES of Russia are considered to justify the balance reliability in relation to modern conditions of infrastructure changes and intellectualization of energy systems. The definition of territorial zones equal reliability UES of Russia is based on topology information of electrical connections in the UES of Russia and a list of controlled sections in it. Studies on the possible aggregation the model design scheme for spectrum of source information available to researchers are presented.

Introduction

Ensuring balance reliability is associated with the determination and placement of a power reserve in the territorial zones (TK) of electric power system (EPS) and the Unified EPS (UES) of Russia in particular. This task has always been given due attention in planning development electric power industry for a short and long term. Some stagnation was observed in 90s of the past and the beginning of this century. Since 2010, the attitude towards ensuring balance sheet reliability at the planning stage has changed. Since this year, on the instructions the Russian Ministry of Energy[a], JSC "UES of UES" together with PJSC "FGC UES" began annually carrying out the work "Scheme and Program for Development UES of Russia for a seven-year period".

The solution to the problem of providing the balance task is iterative. At each iteration, the problem of evaluating indicators of balance reliability (IBR) is solved, and depending on their difference from standard values, reserve funds change in territorial zones (TK) and levels of maximum permissible power flows (MPF). The process ends with the practical coincidence of the calculated for a particular backup option IBR with standard values. Thus, the main block in the justification the means to ensure balance sheet reliability is the block assessment of IBR.

The following factors and random events affect the values indicators of balance reliability in a complex EPS, and therefore the values standard power reserve:

– A model design scheme UES of Russia with allocated TZ and their connections;

– the available capacities of individual TZ and MPF stocks of links between them;

– structure generating capacities of TZ;

– scheduled equipment repairs;

– schedules changes in TZ loads by year and day;

– irregular load fluctuations and errors in forecasting consumer demand;

– decrease in the generating capacity of TZ and MPF reserves due to accidental damage to power plant units and power lines.

It is completely obvious that all of the factors listed above are taken into account when assessing IBR and are given in sufficient detail in monographs [1–3]. At the same time, the question of formation a model settlement scheme UES of Russia with the allocation the territorial zones and the links connecting them is still debatable and not completely worked out. Below, we consider the issues of creating a model design scheme for tasks of balance reliability, taking into account the modern realities of the functioning UES of Russia and existing power balances.

1 Prerequisites for the justification model design scheme UES of Russia for the task of substantiating balance reliability

In substantiating the balance sheet reliability of EPS development options at all times, a number of various assumptions were made that made it possible to simplify the solution of the problem. This is a representation of the power consumption mode, and a representation operation of generating equipment and much more [2]. The simplifications introduced in the long-term planning naturally relate to issues of aggregation electrical connections of the main generating and network facilities. It is clear that the introduced simplifications that have a certain impact on the decisions taken to manage the development of EPSs should have a certain justification. Therefore, the definition of territorial zones

[a] Rules for development and approval schemes and programs for prospective development electric power industry, approved by Decree the Government of Russian Federation, October 17, 2009 No. 823.

must comply with a number of requirements of a technical (technological) and economic nature.

The model design scheme UES of Russia should be presented in the form of a graph, the vertices which are zones equal reliability, and the edges are system-forming bonds. Zones equal reliability (hereinafter the territorial zone - a concentrated system[b]) are understood as a set of nodes included in the zone, network restrictions between which do not affect the IBR of the zone in question. Such a definition to some extent equates the concept of reliability zones and zones of free power flow (ZFPF).

In the current conditions introduction of market relations in the Russian electric power industry, the main goal of creating a model calculation scheme for the country's UES is to justify reservation tools for holding auctions for commercial power take-offs (CPT), power supply contracts (PSC) and other market procedures. CPT is focused on the selection of generating capacity in accordance with its demand from the power balances developed in the work of the S&D UES. Demand for capacity in addition to the maximum load and export / import includes the value of the normative reserve capacity.

The use values of normative reserve of power, given in the methodological recommendations (MR) for the design and development energy systems when conducting CPT, [2], is currently not entirely correct. These MRs were developed by Institute Energosetproekt OJSC in the mid-90s of the last century, but were approved by the Ministry of Energy of Russia only in 2003. It should be noted that, at the stage of their development, retrospective information of the 80-90s of the last century was used, for which the situation was rather tense with the provision of generating capacity reserves territorial zones UES of the country. The task of substantiating the power reserves was solved in conjunction with the optimization MPF of the backbone links. At the same time, significant funds were required in the construction of new generating sources and system-forming links for the implementation of recommendations to achieve the required level of balance reliability or a minimum of reduced costs.

At present, it has become possible to obtain more detailed and reliable initial information for solving the problem of evaluating IBR and the means to ensure it [2], both by the modes of power consumption and probabilistically determined parameters about random outputs generating equipment for unscheduled repairs. The task of substantiating the power reserve today does not envisage the creation new generating sources and power lines (they are prescribed for the seven-year period in the work of the S&D UES). In 2012, on instructions from JSC UES UES and JSC Institute Energosetproekt, new guidelines[c] were developed that partially took these factors into account, but they were not approved by the Russian Ministry of Energy. In 2018, the Russian Ministry of Energy initiated work on the creation of the "Procedure for determining the size regulatory reserve of generating capacities in the UES of Russia and isolated power systems" [d].

The most difficult task in the formation of an acceptable model design scheme UES of Russia, in addition to highlighting the boundaries of TZ, is the task of determining the MPF power links between them or MPF power in controlled sections. In this case, the model should not be excessively large. This follows from two important circumstances. The first is the difficulty of solving the problem of justifying the power reserve, consisting in the multiple solution problem of estimating the power supply circuit requiring playing a set of random system conditions caused by unplanned findings in the repair of generating and network equipment and load changes due to errors in its forecasting caused by the temperature factor. The second is to obtain the necessary range of source information that affects the for the selected territorial zones. Such information is provided only in the context of territorial zones in the form of integrated energy systems and partially regional ones in the work of the S&D UES.

2 Principles for determining territorial zones of equal reliability

In modern conditions, they are based on the available information on the topology of electrical connections in the UES of Russia presented the work of the S&D UES and the list of controlled sections between and within the UES. The latter information, unlike the map-scheme, is not given in open publications and can be provided, for example, by SO UES JSC. The validity of the use of MPF of controlled sections obtained during the operational management of operating conditions under operating conditions for the formation of technical specifications for tasks of balance reliability is not considered in this article. At the same time, certain assumptions about their use should be noted. They are calculated for a specific mode with a given composition and power of generating equipment, active and reactive load levels, weighting trajectories and many other parameters. In the process of assessing IBR, both the composition of the generating equipment and the levels of power consumption are randomly played. Therefore, for these states, the MPF stocks of controlled sections will not always correspond to reality.

The currently relevant controlled sections are independent of each other, since they were obtained by weighting the mode for different electrical modes of the updated design scheme UES of Russia. This suggests that the application of any laws of transformation of the regime parameters in the problem of circuit equivalentization and obtaining power flows from individual section connections based on data on flows obtained for another section is fundamentally impossible.

[b] A power system is called a concentrated one, inside of which there are no restrictions on the transfer of power through power lines from one zone to another.

[c] Guidelines for designing development of energy systems / Institute Energosetproekt OJSC, 2012.

[d] The task of NP Market Council JSC to perform work of the same name.

 https://doi.org/10.1051/e3sconf/201913901047

Given the independence of the controlled sections, the algorithm for allocating TZ of equal reliability is quite simple. It consists in combining in one TZ those nodes that do not have any specified restrictions on power transmission. Thus, many TZ is obtained that can be used in the formation model design scheme. In fig. 1, for some fragment of the EPS, three TZ and two bonds between them (I and II) are distinguished. It can be seen that inside the allocated TZ, there are no restrictions on the transmission of electricity between the nodes entering it.

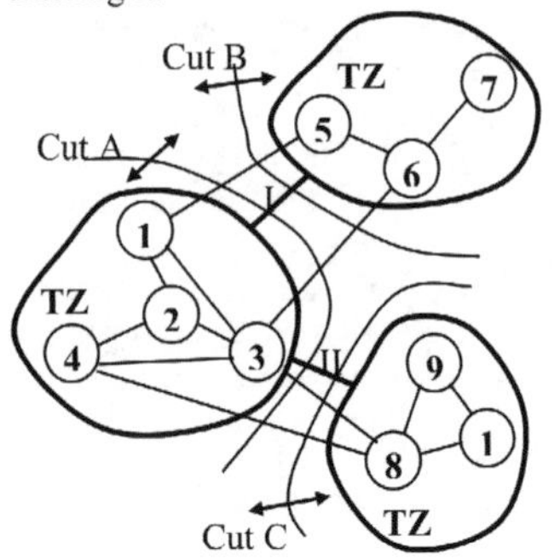

Fig. 1 Fragment EPS scheme

The most difficult is the question of determining the power MPF for the formed and physically non-existent connections I and II between the selected TZ. There are conflicting questions about the choice of the power MPF values in the forward and reverse directions (a double-headed arrow in Fig. 1) for bonds I and II - from the MPF values of sections A and B and sections A and C, respectively. At the same time, it is impossible to simplify based on the laws of electrical engineering, due to the independence of the cross sections. The simplest method, which does not lead to any additional errors, is to take into account, at each random state, the restrictions on power transfer not by links (I and II), but directly by controlled sections A, B and C (see the next section).

After completion the process of formation design scheme, it is necessary to reduce the number of territorial zones to a certain minimum (in foreign and domestic practice to 100). This can be achieved by considering the impact of one controlled section on another from the standpoint of ensuring balance reliability. Such an approach requires detailing the source information for all the territorial zones identified at the first stage.

For example, consider the Ural. If we consider all 190 controlled sections in this UES, then according to the above algorithm, 62 TZ and more than 100 connections between them (most often physically non-existent) can be formed. Specialists of SO UES, JSC, by combining reliability zones, both on a territorial basis and taking into account their interconnectedness, allowed reducing the number of TZ to five (Fig. 2). These TZ have a pronounced territorial division and are interconnected by powerful lines with a voltage of 500 kV. The equivalent bonds formed in this case are determined by 1–3 real lines (in Fig. 2, the links between zones 1001-1004, 1002-1003, 1004-1002, 1002-1006 and 1006-1003) and in only one case more than 3 (Fig. 2 links between zones 1001-1002). Considering this scheme (Fig. 2) from the standpoint of using controlled sections, it can be seen that any of the equivalent bonds is well replaced by a set of sections defining it. Moreover, inside the allocated TZ, there are no restrictions on the transmission of electricity between the nodes entering it. Similarly, relationships with other UESs are described.

3 The ability to account for controlled sections in the model of the design scheme for the task of ensuring balance reliability

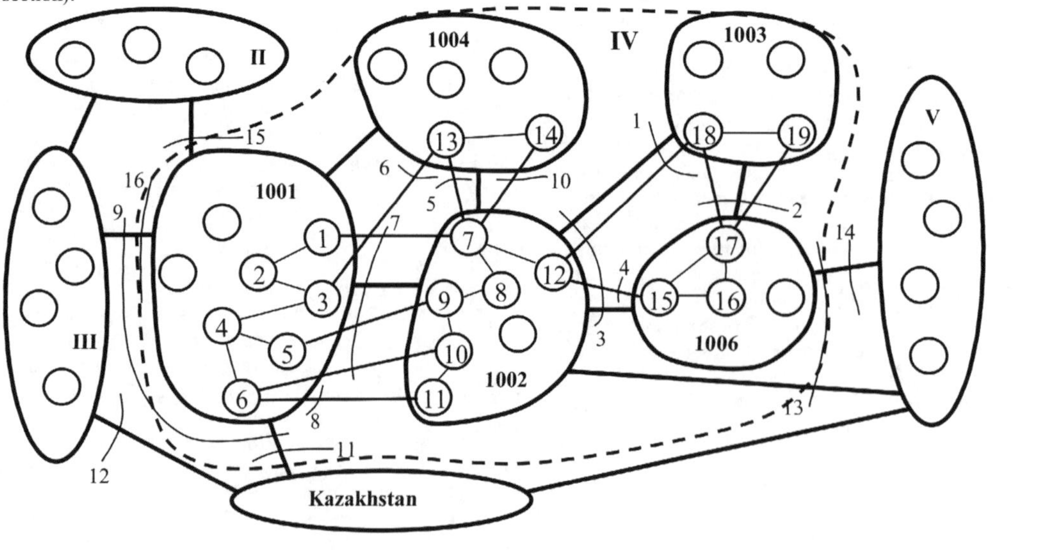

Fig. 2 Ural scheme and its relationship with neighboring UES.
II - Center; III - Middle Volga; IV - Urals; V - Siberia; 1001-1006 - The zones of reliability Urals UES allocated by SO UES.

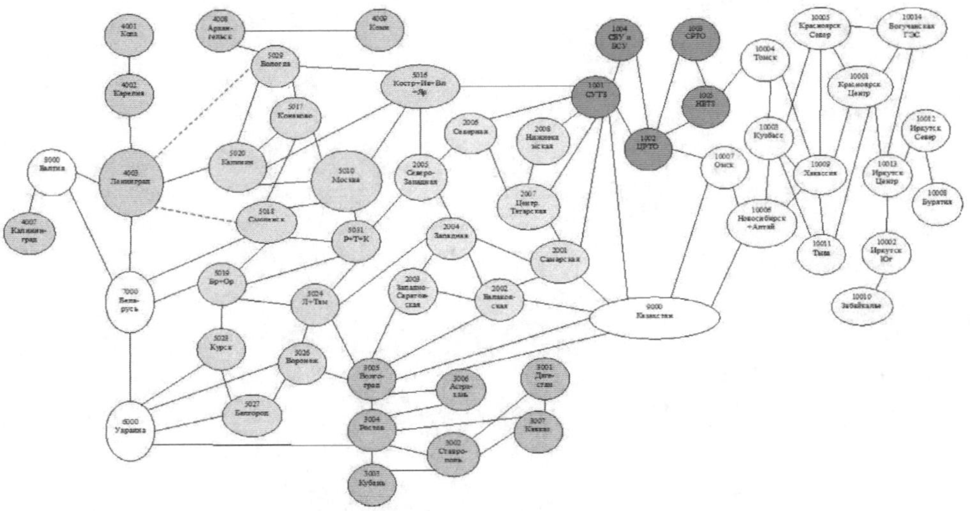

Fig. 3. The model calculation scheme balance sheet reliability UES of Russia without the East.
1001 – 1: 1002 – 2; 1003 – 3; 1004 – 4; 1005 – 5- Ural; 2001 – 6: 2002 – 7; 2003 – 8; 2004 – 9; 2005 – 10: 2006 – 11; 2007 – 12; 2008 – 13 – Middle Volga;
3001 – 14: 3002 – 15; 3003 – 16; 3004 – 17; 3005 – 18: 3006 – 19; 3007 – 20 – South;
4001 – 21: 4002 – 22; 4003 – 23; 4007 – 24; 4008 – 25; 4009 – 26- North-West;

Under the conditions when there is information only on the restrictions of power flow over individual sections of the model of the design scheme of the EPS (Fig. 1), the most acceptable way to solve the problem of assessing a randomly generated state of the system by generating power and load is to take into account precisely these restrictions. With this in mind, the following is a correction of the random state estimation model (power deficit distribution unit) in the EPS, for the case of using the maximum allowable flows in controlled sections as restrictions on power flows between territorial zones. It is clear that the objective function (1) and restrictions (2) - (5) should remain unchanged:

$$\sum_{j=1}^{n} c_j^H(\overline{P}_j^H - P_j^H) \rightarrow min, \qquad (1)$$

with constraints:

$$P_j^H - P_j^{г} + \sum_{l(i,j)\in V} P_{l(i,j)}^L = 0, \quad j = 1, 2, .., n \qquad (2)$$

$$0 \le P_j^H \le \overline{P}_j^H, \quad j = 1, 2, \dots, n; \qquad (3)$$

$$0 \le P_j^{г} \le \overline{P}_j^{г} \quad j = 1, 2, \dots, n; \qquad (4)$$

$$\underline{P}_{l(i,j)}^L \le P_{l(i,j)}^L \le \overline{P}_{l(i,j)}^L, \quad l(i,j) = 1, 2, \dots, m, \qquad (5)$$

where P_j^H, $\overline{P}_j^H$, $P_j^{г}$, $\overline{P}_j^{г}$ – correspondingly covered demand and load demand capacity, used and available generating capacities of j TZ; $P_{l(i,j)}^L$, $\overline{P}_{l(i,j)}^L$, $\underline{P}_{l(i,j)}^L$ – communication power flow and its limiting values in the forward (from the i-th TZ to the j-th) and reverse directions; c_j^H – the coefficients reflecting the cost indicators of the restriction of consumers of the j-th TZ are taken equal to unity; V – many TZ pairs with interconnected system-forming; $n(m)$ – the number of TZ (backbone links) EPS.

To take into account the cross sections, correction of constraints (5) is necessary by changing the power MPF levels for the bonds included in a given cross section to infinitely large values. In addition, it is necessary to introduce additional restrictions on controlled sections in the form of:

$$\underline{P}_k \le \sum_{l\in U} P_l^L \le \overline{P}_k, \quad k = 1, 2, \dots, K, \qquad (6)$$

where P_l^L – calculated power transfer via l-th connection; $\overline{P}_k$, $\underline{P}_k$ – power flow limit values for the k-th section in the forward and reverse directions; U – the set of l-bonds included in the k-th section; K – number of sections considered.

4 Comparison of balance sheet reliability indicators for schemes of various dimensions

In 2011-2012. On the instructions of JSC SO UES, STC UES OJSC, a model of the design scheme of the UES of Russia was developed, which includes 56 territorial zones and 102 system-forming links between them (Fig. 3). For this scheme, MPF capacities between the allocated territorial zones were obtained. The source of information was the materials of the report on work of the S&D Russia UES for 2012-2019.

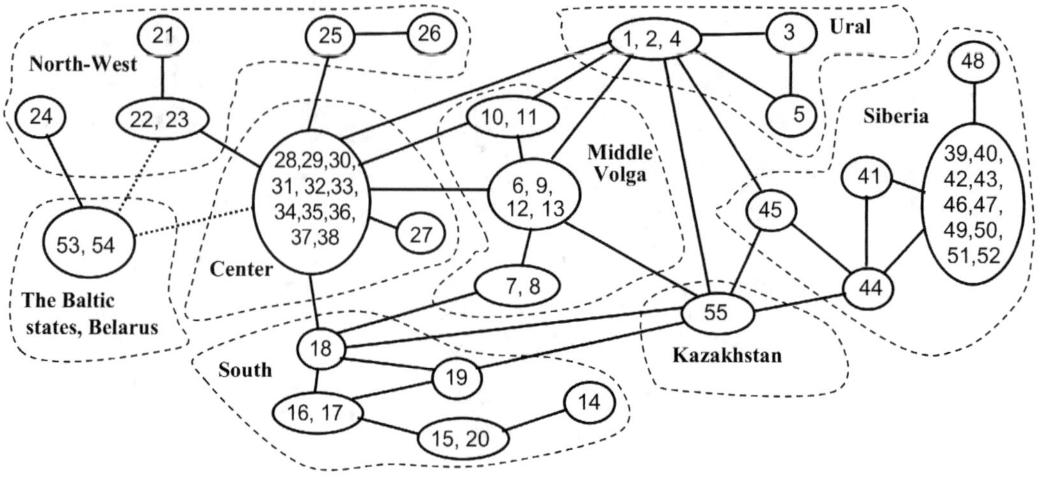

Fig. 4. The UES model of Russia, presented form zones of free power flow (ZFPF).

It is quite obvious that for a really existing composition of generating equipment, their accident rate, and the predicted maximum load values, such sufficiently large design schemes are of certain interest from the standpoint of identifying critical reliability zones and "weak" connections between them. The justification of the means of ensuring balance reliability - operational power reserves for such schemes is fraught with significant difficulties, both of a computational nature and, to a greater extent, of information.

Based on the model of the design scheme of the UES of Russia, presented in Fig. 3, two more were developed. One was prepared in accordance with the recommendations of the MR [4] (in Fig. 3 the numbers of TZ included in the UES are shown in the caption), the other, in accordance with the regulatory document[e] for solving the CPT problem in the UES of Russia (Fig. 4).

In the table 1 presents the results distribution operational capacity reserve UES, which are purely research-based. Information stuffing on the composition of the generating equipment, its accident rates, the MPF levels of the power of the backbone links were presented by SO UES JSC. The levels of energy consumption of TZ are taken from the work of the S&D Russia UES. Forms of changes in the daily load schedule and forecasting errors are presented by SO UES. Important is the constancy of information for all the presented calculations with different representations of the model of the calculation scheme of the UES of Russia. The results speak for themselves - the differences in the TZ reservation levels are very slight. The decrease in the operational power reserve with a decrease in the number of considered territorial zones from 56 to 9 amounted to only 300 MW, or 0.19% with respect to the irregular maximum load (Table 3). The explanation for this is quite simple - inside the UES (Fig. 3), the MPF levels of the communication power practically do not limit the transfer of power from one node to another. The very same situation is observed when instead of MPF is taken into account by the constraints of the flow restriction over the controlled sections. We should not have expected another, due to the lack of information and the conditional formation of MPF over controlled sections on the basis of data on MPF link.

[e] Order of the Ministry of Energy of Russia dated 04/06/2009, No. 99 (as amended on 12/28/2016) "On approval procedure for determining zones free flow of electric energy (power)" (Registered in the Ministry Justice of Russia on June 29, 2009 No. 14153).

Table 1 - the Distribution operational reserve for UES with a different presentation scheme UES of Russia (the work of the S&D UES of Russia for 2014)

	Maximum load, MW	Operational reserve, MW /%		
		56 TZ, the full scheme UES of Russia (Fig. 3)	8 TZ (NorthWest + Kazakhstan, information in Fig. 3,)	TZ (fig. 4)
UES of Russia without East	157670	14101/ 8.943	13801/ 8.753	14076/ 8.928
Ural	37349	1703/ 4.56	1953/ 5.23	1703/ 4.56
Middle Volga	18088	343/ 1.9	293/ 1.62	293/ 1.62
South	15357	637/ 4.15	587/ 3.82	612/ 3.99
North-west	15731	2044/ 12.99	1944/ 12.36	2094/ 12.99
Center	38556	6158/ 15.97	6108/ 15.84	6158/ 15.97
Siberia	32589	3216/ 9.87	2916/ 8.95	3216/ 9.87

5 Conclusion

Justification of the operational component of the normative reserve of power to ensure balance reliability in the conditions of the power market and its commercial selection in modern conditions allows the use of the model of the design scheme of the UES of Russia in the form of integrated power systems with the possible fragmentation of some of them into two or three. This greatly simplifies the process of obtaining initial information from the work "Scheme and development programs of the UES of Russia for a seven-year period" and the computational process with acceptable errors in obtaining the result.

References

1. Chukreev Yu.Ya. Models ensuring reliability of electric power systems. Syktyvkar: Komi Scientific Center, Ural Branch of the Russian Academy of Sciences, 1995.176 p.
2. Chukreev Yu.Ya., Chukreev M.Yu. Models assessing balance sheet reliability indicators for managing development of electric power systems. Syktyvkar: Komi Scientific Center, Ural Branch of the Russian Academy of Sciences, 2014, 207 p.
3. Kovalev G.F., Lebedeva L.M. Reliability of power systems. Novosibirsk: Nauka, 2015, 224 p.
4. Guidelines for designing development of energy systems. (Approved by Order of the Ministry of Energy of Russia dated June 30, 2003, No. 281). M .: Ministry of Energy of the Russian Federation, CO 153-34.20.118-2003.

Adaptive multivariable relay protection of reconfigurable distribution networks

Mikhail Sharygin[1], *Vladimir* Vukolov[1], *Anton* Petrov[2,*]

[1]R. E. Alekseev Nizhnii Novgorod State Technical University, 603950, Minin Str. 24, Nizhnii Novgorod, Russia
[2]JSC NIPOM, advanced development department, 603140, Lenin Ave., 20, Nizhnii Novgorod, Russia

Abstract. Version of relay protection construction of relative selectivity is proposed, which allows increasing its technical perfection and reliability in microgrids with distributed generation sources. This is done by increasing the dimension of the measurement space along arbitrary axes, introducing new methods of recognizing emergency modes, as well as automating the selection of response parameters. Hardware implementation of protection is described. Recommendations for calculating the settings of multidimensional protection and an example of calculation are given. Protection coordination is carried out by using graphical-analytical methods in automatic mode. Automation of the calculation, elimination of the human factor will make it possible to apply the method for adaptation of protection in the conditions of dynamic change of microgrid topology.

1 Introduction

In electric power industry, micro-network is a medium or low voltage network, which includes sources of power, distributed generation and which is capable of operating both in isolated mode and in parallel with the combined power system.

Application of distributed generation and possibility to dynamically changing topology of the electrical network negatively affects sensitivity of relay protection devices, which control condition of the network [1, 2]. Solution to this problem can be an adaptive protection system based on digital technologies and modern communications according to IEC 61850 standard. Such protection devices can detect faults both in the external network, ensuring the micro-network's switching to autonomous operation, and faults in the internal network isolating its least possible part to eliminate the fault [3, 4].

Usually, when a relay protection system is developed, distribution network is divided into local zones covering its individual parts. Connection of a new distributed generation source to the network leads to the need to adjust the protection device settings to ensure required selectivity level, speed and sensitivity at autonomous modes and in parallel operation with the external electrical network. Increase in distributed generation volume within the micro-network makes parameterization of its protection system significantly more complex, and the protection system adaptation to existing operating conditions can be carried out both continuously and discretely, by switching pre-defined groups of settings. In any case, correct protection system operation and fine adjustment is possible only under condition of continuous synchronized measurement of various parameters of the micro-network mode in its nodes with following processing of the information received in a specialized device that controls entire network and is able to process large amounts of information. In case of the micro-network fault, such a device identifies the fault type and location and isolates the faulted part from the rest of the network and cuts-off power generation within this part. At that, the remaining parts can continue normal operation.

2 Proposed concept for protection of micro-networks with distributed generation sources

Perspective structure of a relay protection system in micro-networks with distributed generation sources involves two-level protection scheme with mutual redundancy. For the first level, it is proposed to use high-speed differential-logic protection [5] using IEC 61850 data protocol to exchange information between distributed intelligent electronic devices (IED), taking into consideration the electrical network topology and operating modes. The second level involves application of multiparameter protection devices based on statistical principles [6, 7, 8]. In case of a failure of the information communication network, the multiparameter protection device ensures operation of the facility similarly to common protection devices with relative selectivity, while having, by virtue of its principle of action, greater sensitivity.

Use of IEC 61850 protocols allows continuous adjusting and adapting the IED to the existing modes of the micro-network operation using a centralized or decentralized device for automatic calculation of protection triggering parameters.

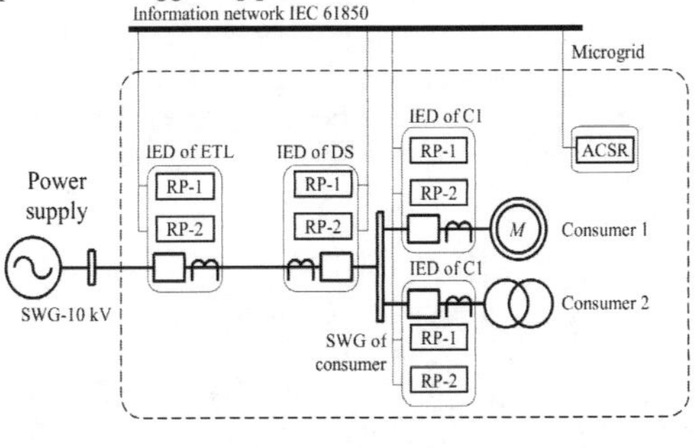

Fig. 1. Application of new protection algorithms for micro-network: ETL – electrical transmission line; DS – input disconnecting switch; C1, C2 - consumers; RP-1 – first level protection device; RP-2 – second level protection device; SWG – switchgear; ASCR – device for automatic calculation of protection setpoints

The proposed concept allows:

- increasing sensitivity and speed of protection of micro-networks with distributed generation sources;
- increasing stability of the relay protection system to failures of information network, measuring transducers or individual devices;
- ensuring joint operation of the protection devices using different algorithms and operating principles implemented as IED of both centralized and decentralized design.

3 Perspective principle of emergency mode recognition

It is reasonable to adapt statistical methods of detection theory as a theoretical basis for development of electric network mode recognition algorithms.

IED, by continuous analysis of parameters of the power network protected element mode, recognizes its current condition as normal (hypothesis H_0) or emergency (hypothesis H_1) operation mode. Criteria values for the facility condition classification are defined so as to minimize average risk R of the protection triggering, in the two possible IED modes they are defined by:

$$R = C_{00} \cdot P_0 \cdot (1 - P_F) + C_{10} \cdot P_0 \cdot P_F + C_{11} \cdot P_1 \cdot (1 - P_M) + + C_{01} \cdot P_1 \cdot P_M , \quad (1)$$

C_{00} – losses at correct choice of H_0 hypothesis; C_{01} – losses at incorrect choice of H_0 hypothesis; C_{01} – losses at incorrect choice of H_1 hypothesis; C_{11} – losses at correct choice of H_1 hypothesis; P_0, P_1 – a priori probability of H_0 и H_1 hypotheses correctness; P_F – conditional probability of false de-energizing; P_M – conditional probability of undetected alarm mode.

Conditional probability P_F is the probability of event when the protection device attributes current condition of the facility to H_1 hypothesis while the actual condition matches H_0 hypothesis. Conditional probability P_M, on the contrary, is the probability of attributing by the protection device the facility condition to H_0 hypothesis while the actual condition matches H_1 hypothesis:

$$P_F = P(H_1|H_0) = \int_{\Psi_1} p_{H_0}(\Phi|H_0) d\Phi ; \quad (2)$$

$$P_M = P(H_0|H_1) = \int_{\Psi_0} p_{H_1}(\Phi|H_1) d\Phi , \quad (3)$$

Φ – vector of monitored mode parameters; $p_{H_0}(\Phi|H_0)$ – density function of observed parameters under the conditions of the existence of the normal mode H_0; $p_{H_1}(\Phi|H_1)$ – density function of observed parameters under the conditions of the existence of the emergency mode H_1; Ψ_0 – part of the observation space corresponding to permissible condition of the protected facility which does not require to be disconnected from the electrical network; Ψ_1 – part of the observation space corresponding to non-permissible (emergency) condition of the protected facility.

The density distribution functions of the regime parameters are obtained by conducting a sufficiently large number of statistical experiments on mathematical models of the electric network. For this purpose, in addition to the information necessary for the calculation of traditional relay protections, information on the probabilistic laws of mode distribution is required.

It is obvious that total observation space Ψ is the sum of:

$$\Psi = \Psi_0 + \Psi_1 . \quad (4)$$

To increase sensitivity and recognizing ability of the relay protection device, dimension of the monitored space Ψ may be increased, and this will be more effective when using parameters that change maximally in emergency mode [6, 7]. Thus, different combination of the monitored parameters gives different probabilities of mode recognition.

Features of up-to-date IED make it possible to perform versatile analysis of power network modes by means of universal multidimensional measurement-starting controls that use the most informative monitoring parameters of the network mode.

Bayes criterion according to which the hypothesis is chosen (recognition of the protected facility mode) is [9]:

$$\begin{cases} H_0, & \text{if} \quad \Lambda(\Phi) > \eta; \\ H_1, & \text{if} \quad \Lambda(\Phi) \le \eta, \end{cases} \quad (5)$$

where η is threshold of likelihood ratio criterion [10]; $\Lambda(\Phi)$ is likelihood ratio:

$$\Lambda(\Phi) = \frac{p_{H_1}(\Phi|H_1)}{p_{H_0}(\Phi|H_0)} . \quad (6)$$

Offset of the element protection device from all normal modes, in which it must not trigger, gives a value of conditional probability $P_F = 0$ which makes it possible to trend the likelihood ratio criterion threshold to infinity: $\eta \to \infty$ [11].

Recognition of emergency modes of the electric network protected element based on the Bayesian method will allow:

- reducing the risk of erroneous triggering of the relay protection and automatic equipment in the course of decision-making;

https://doi.org/10.1051/e3sconf/201913901048

- implementing relay protection and automatic equipment algorithms that take into account actual statistic data of normal and emergency modes of the protected network element;
- developing various new types of automation equipment for both emergency and normal modes.

Let us compare probabilities of short-circuit detection by a supply line protection device at monitoring of various sets of parameters characterizing operating modes of the electric network element which shown is the scheme, in Fig. 2. To do this, based on the results of simulation in different observation spaces Ψ, let us draw a curve of probability density function for following characteristic facility modes: normal; short circuit fault in the supply line; short circuit fault in the switchgear of consumer; short circuit fault in the consumer 1 short circuit fault in the consumer 2.

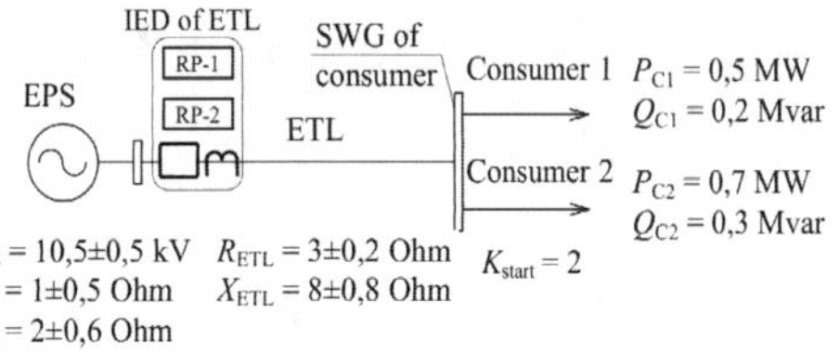

Fig. 2. Scheme for mode simulating

Example of the probability density functions for different modes, obtained based on mathematical modeling, in the observation space that includes only one monitored parameter (current) is shown in Fig. 3. Curves of probability density functions for various modes in multidimensional observation spaces can be drawn in a similar way and they depend on several parameters.

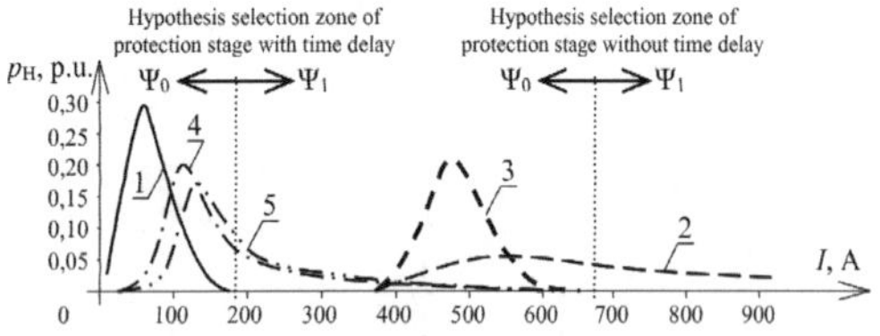

Fig. 3. Function of the facility mode conditional probability density for one-dimensional observation space: 1 – normal; 2 – short-circuit in the supply line; 3 – short circuit fault in the SWG of consumer; 4 – short circuit fault in the consumer 1; 5 – short circuit fault in the consumer 2

Due to variability of characteristics of the factors affecting the protected facility condition, areas of its normal and emergency modes often overlap which does not allow protection to unmistakably assess its condition. Requirements of selectivity and inadmissibility of protection operation in normal mode allow setting the protection setpoints by comparing only the areas of different modes existence without taking into account numerical values of the conditional probability density. Example of such areas comparison for two-dimensional space (current, voltage) is shown in Fig. 4.

Probability of emergency condition recognition can be defined by integrating the function of conditional probability density for emergency conditions $p_{H_1}(\Phi|H_1)$ over the space of monitored parameters limited by the protection device setpoints (Ψ_1). The values obtained for the protections monitored different sets of parameters are given in Table 1.

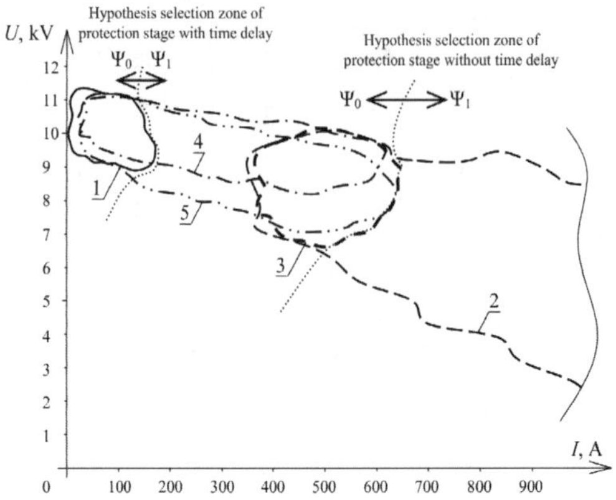

Fig. 4. Mode existence areas in two-dimensional observation space

Table 1. Comparison of probabilities of modes recognition by protection devices

Recognition coefficients	Monitored parameters			
	I	*I, U*	*P, Q*	*I, U, φ*
Probability of short circuit fault recognition in a supply electrical transmission line by an instantaneous action stage	0,597	0,806	0,795	0,898
Probability of short circuit fault recognition in a supply electrical transmission line by a stage with time delay	1	1	1	1
Probability of short circuit fault recognition in the consumer 1 by a stage with time delay	0,255	0,322	0,937	0,955
Probability of short circuit fault recognition in the consumer 2 by a stage with time delay	0,367	0,461	0,950	0,958
Total probability of short-circuit fault recognition in reserve zones by a stage with time delay	0,311	0,392	0,944	0,957

4 Hardware implementation of the proposed protection principles

In the course of the research, a technical task for development of intelligent electronic device with a sufficient for the mentioned functions computing power was prepared. The device prototype presented by the research and production enterprise is shown in Fig. 5. The device feature consists in its ability to function at computing platform of any manufacturer that, depending on mathematical calculation complexity, allows using various productivity microprocessors of both domestic and foreign production without significant structural changes.

Integration into information networks and implementation of the concept for protection of micro-networks with distributed generation sources is carried out by means of fully-fledged use of IEC 61850 data transfer protocol.

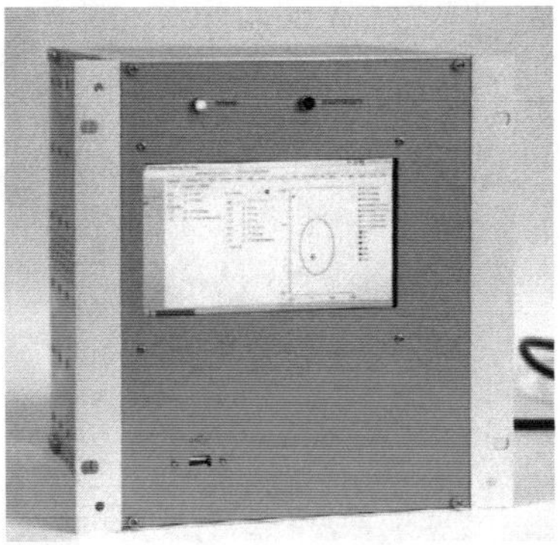

Fig. 5. View of the intelligent electronic device

5 Software package for automatic calculation of the protection setpoints

It is reasonable to carry out configuring and adapting of relay protection and automatic system elements parameters for current operating conditions of the micro-network, including its topology, using technology of automatic calculation and matching of the relay protection. It implemented in a centralized device or a set of decentralized devices that receive information about the micro-network condition using IEC 61850 network communication standard.

Components of the software layout implementing the automatic calculation technology are shown in Figure 6.

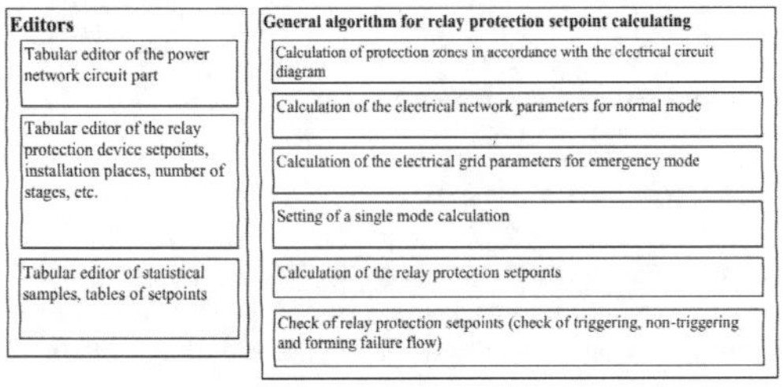

Fig. 6. Components of the software system for automatic calculation and matching triggering parameters of the network relay protection devices

Conclusions

1. The concept of relay protection system in micro-networks with distributed generation sources involving use of two-level relay protection scheme, elements of which use IEC 61850 protocol to exchange of information among themselves, is proposed.
2. Bayesian principle, which applies probabilities of protected element possible conditions, was used as a basis of the method for detection of protected element emergency modes.
3. Probabilities of faults detected by protection devices, which monitor different sets of the facility parameters, were compared.
4. An intelligent electronic device which implements the concept of micro-network protection involving a dynamically changing topology and generation, as well as a software package for automatic calculation of current protection setpoints on the basis of this concept, was developed.

This work was supported by the Ministry of Science and Higher Education of the Russian Federation (agreement No. 075-15-2019-1209 of May 31, 2019. Unique project identifier RFMEFI57717X0244).

References

1. O. A. Onisova, EEPIR J., **5**, 88 (2018).
2. O. A. Onisova, Energetik J., **1**, 17 (2015).
3. P. V. Ilyushin, RZA J, **2**, 12 (2018).
4. TerzijaY. Proceedings of the IEEE., **1(99)**, 80 (2011).
5. A. L. Kulikov, M. V. Sharygin, El. St., **3**, 37 (2018).
6. A. L. Kulikov, M. V. Sharygin, Izv. Ac. N. Energetika, **6**, 38 (2018).
7. V. I Nagay, Izv. VUZ. Elektromechanika, **6(60)**, 84 (2017).
8. A. L. Kulikov, M. V. Sharygin, Elektrichestvo, **7**, 20 (2017).
9. A. L. Kulikov, M. V. Sharygin, Elektrichestvo, **9**, 24 (2017).
10. P. Anil Kumar, IJACEEE, **1(1)**, 19 (2013).
11. G. S. Nudelman, O. A. Onisova, EEPIR J., **4**, 106 (2014).

Increase of Transient Stability Level of Gas Turbine Power Plant Using FACTS

Sergey Solodyankin[1,*], *Andrey* Pazderin[2]

[1]Branch of JSC "SO UPS" UDC Ural, Yekaterinburg, Sverdlovsk Region, Russia
[2]Ural Energy Institute of the Ural Federal University named after the first President of Russia B. N. Yeltsin, Yekaterinburg, Sverdlovsk Region, Russia

Abstract. The article is devoted to the development of the mathematical models of modern devices of flexible alternating current transmission systems (FACTS) when calculating the modes and stability of power systems and to the analysis of influence of the specified devices on transient stability of the generators. The considered scheme contains the generators with the gas turbine drive that have electromechanical parameters providing lower level of transient stability compared to units of higher power rating, which in some cases requires implementation of measures for transient stability enhancement. As examples of FACTS the following devices have been considered: compensating device based on voltage-sourced converter (STATCOM), static synchronous series compensator (SSSC) and the unified power flow controller (UPFC). The known examples of mathematical models of FACTS devices vary in complexity. For a preliminary assessment of the effectiveness of the FACTS devices, it is proposed to use simplified models that adequately reflect their impact on transients. The use of models made it possible to establish a positive impact of the devices on transient stability of generating equipment in case of short circuits in the electric network. The important conclusion here is that the use of the UPFC device based on two converters (with a corresponding increase in cost) compared to one converter device (STATCOM or SSSC) slightly increases the level of transient stability and the limit time of short circuit disconnection. The proposed method of simulating the FACTS devices is suitable for numerical calculations of transient processes in electric power systems, in particular, to estimate the impact on the transient stability level of the parallel operation of power plants in case of disturbances.

1 Introduction

The aim of the article is to describe the mathematical modeling technique of the devices of flexible alternating current transmission systems and the analysis of their efficiency in terms of influence on the level of transient stability of parallel operation of the power plant equipped with generators with deteriorated electromechanical parameters. The devices of the regime regulation based on the power electronics allow to solve problems with providing required indicators of stability and reliability of electrical power systems and also decrease total costs of the electrical energy' transfer because the new network construction is not necessary [1-6]. These devices are used to increase the indicators of controllability and capability of electricity transmissions and define properties of flexible alternating current transmission systems. Electro-technical network equipment of the FACTS is capable to solve problems with the operation modes optimization in electrical power systems by several criteria at once: transmission capability, level of technological losses, stability, redistribution of the power flows, quality of the electric energy, etc. [7-8]. Besides, implementation of the FACTS allows to reduce or eliminate the action adjustments from emergency control systems in a number of circuit-mode situations.

Due to the development of the FACTS technologies, it is relevant to develop the methods of assessing the effectiveness of such devices for further promotion in the United Power System of Russia.

2 Calculation scheme

As a design scheme, in which a gas turbine power plant of relatively low capacity is connected by a power transmission line to a powerful load, the voltage at the connection point can be considered constant (Figure 1). A compensating device of parallel/series action, so-called unified power flow controller is in the middle of the line. The device consists of two voltage-sourced converters connected by a direct current link through which the converters can exchange active power; each of the converters generates independently reactive power.

Such scheme is rather widespread in electrical supply systems of oil and gas extraction complexes. The output of the power plant is 100 MW (4 machines on 25 MW, cosφ=0.9). Electromechanical parameters of generators:

* Corresponding author: solodsal@mail.ru

the synchronous, transient and subtransient reactances are respectively equal to xd = 1.98 p.u., x'd = 0.3 p.u., x"d = 0.216 p.u. The Value of the mechanical inertial constant of the unit: Tj = 2.3 s. These circumstances cause reduced level of the transient stability in short circuits. The connection with the receiving end of the power system is performed by double-circuit power transmission line of nominal voltage 110 kV with length of 60 km (made by steel-aluminum wire 240 mm^2). Different types of multi-phase short circuits (SC) at the beginning of the transmission line (the bus with U1) are used as disturbances [9]. The aim of this article was the analysis of the FACTS devices influence on the electromechanical transition processes at disturbance and definition of limit time of fault clearing at different structure of the compensating device.

It was assumed that the power of each of the UPFC converters was 0.225 p.u. (25 MVA), and the power S of the power plant was taken as the basis power.

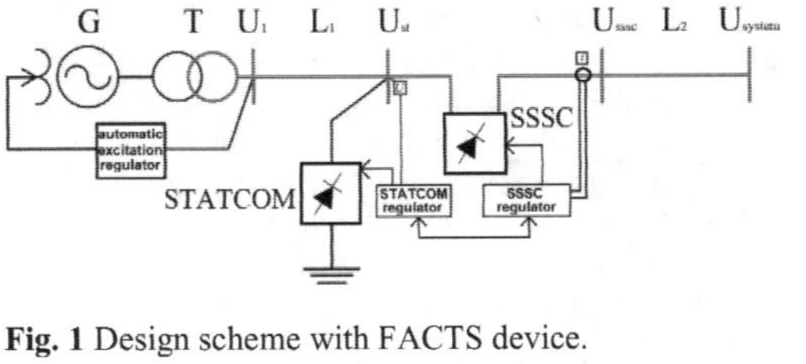

Fig. 1 Design scheme with FACTS device.

3 Modeling of FACTS devices

3.1 Mathematical model of STATCOM

The following dynamic model as a system of algebraic differential equations is introduces in order to carry out calculations of electromechanical transient processes of electric power system with a device of STATCOM type.

The input variable is the voltage on the high voltage buses of network at the connection point with U_{st} set by the projections of U_q and U_d on the coordinate axes of q and d combined with a rotor of the equivalent generator.

The output power of STATCOM S_{st} is set by in basic data.

The increment of the electromotive force (EMF) STATCOM ΔE_{st} is calculated on the basis of the following transfer function:

$$\Delta E_{st} = \frac{K_{st} \cdot (U_{st} - U_{set})}{1 + pT_v} \quad (1)$$

where U_{set} is the set value of voltage setpoint;

K_{st} is the coefficient of regulation of the voltage deviation channel (K_{st} = -5...-100 units of EMF STATCOM/unit of voltage);

T_v is the time constant of voltage deviation channel (T_v = 0.02 s).

Since the simulation uses a common coordinate system q, d associated with the mathematical model of the generator, the coefficients of the EMF increment distribution along the q and d axes are calculated:

$$K_q = abs\ (U_q)/U_{st}$$
$$K_d = abs\ (U_d)/U_{st} \quad (2)$$

Components of STATCOM currents along the axes q and d are calculated in accordance with relations linking components of its EMF and voltages at its connection point:

$$I_{qst} + jI_{dst} = E_{qst} + jE_{dst} - U_q - jU_d \cdot jb_{st} \quad (3)$$

it follows that

$$I_{qst} = -b_{st} \cdot E_{dst} - U_d \quad (4)$$
$$I_{dst} = b_{st} \cdot E_{qst} - U_q \quad (5)$$

Expressions in parentheses represent projections of increments of EMF STATCOM respectively on d and q coordinate axes, so q, d - components of currents are calculated from expressions:

$$I_{qst} = -b_{st} \cdot \Delta E_{st} \cdot K_d \cdot sign\ U_d \quad (6)$$
$$I_{dst} = b_{st} \cdot \Delta E_{st} \cdot K_q \cdot sign\ U_q \quad (7)$$

The value of the EMF increment is limited in such a way that the module of current STATCOM equal to the $\Delta E_{st} \cdot b_{st}$ cannot exceed the S_{st} device power size set in relative units.

3.2 Mathematical model of SSSC

The mathematical model similar to model STATCOM is developed for the SSSC. It is supposed that the device can independently operate as the static synchronous series compensator and also as a part of the unified power flow controller (UPFC) that includes STATCOM exchanging active power with SSSC (according to regulation parameters of the last). Power links of two devices are presented in the set of constraints.

Input variable is the value of current I_m, proceeding on the transformer which is switched on in series in a transmission line in device connection point, set by the projections I_q и I_d on coordinate axes of the general coordinate system q and d.

Inductive reactance of SSSC transformer is taken in relative units equal to x_{sssc} = 0.02... 0.04 p.u.

The power of the SSSC S_{sssc} is set in the basic data (usually taken equal to the power rating of the STATCOM).

The EMF control signal of SSSC is calculated from the known current deviation using the following transfer function:

$$\Delta I_m = \frac{K_{0i}(I_m - I_{set})}{1 + pT_i} \quad (8)$$

where I_{set} is a setting value on current;

K_{0i} is a coefficient of regulation of the current deviation channel (K_{0i} = 10...100 units of EMF SSSC/unit of current);

T_i is a time constant of the current deviation channel (T_i = 0.01 s).

The projections of SSSC EMFs on axes q and d are calculated with transfer functions using additional signal of generator rotor speed deviation s from the synchronous speed (time constants T_{sssc} и $T_{0\omega}$ are 0.01 s):

$$\Delta E_{qsssc} = \frac{\Delta I_m \cdot x_{sssc} \cdot I_{dreg}}{1 + pT_{sssc}} + \frac{k_{0\omega} \cdot s}{1 + pT_{0\omega}} \quad (9)$$

$$\Delta E_{dsssc} = \frac{-\Delta I_m \cdot x_{sssc} \cdot I_{qreg}}{1 + pT_{sssc}} + \frac{k_{0\omega} \cdot s}{1 + pT_{0\omega}} \quad (10)$$

Where I_{dreg}, I_{qreg} are the components of SSSC current, determined in accordance with the control angle ρ of (when SSSC is a part of UPFC, the angle ρ is generally different from zero):

$$I_{dreg} = I_q \cdot \sin \rho \ + I_d \cdot \cos \rho \quad (11)$$

$$I_{qreg} = I_q \cdot cos \ \rho \ - I_d \cdot sin \ \rho \quad (12)$$

When SSSC operates as a part of UPFC, active power exchange with the network depends on STATCOM capabilities in accordance with limitations determined by the power rating of the devices. Active current components I_{qsssc}, I_{dsssc} of SSSC (needs for calculating the power exchange between voltage-source converters) are calculated according to the value of the measured active power of SSSC P_{sssc_1}. Formally equating the value of measured reactive power of SSSC Q_{sssc_1} to zero we can obtain the following relations that determine d, q - components of the SSSC current corresponding to the active power:

$$I_{qsssc} = \frac{P_{sssc_1}}{\frac{\Delta U_{dsssc}{}^2}{\Delta U_{qsssc}} + \Delta U_{qsssc}} \quad (13)$$

$$I_{dsssc} = I_{qsssc} \cdot \frac{\Delta U_{dsssc}}{\Delta U_{qsssc}} \quad (14)$$

Due to the active power exchange of STATCOM with the electrical power network and through the DC link with SSSC, additional components are entered into equations (6) and (7):

$$I_{qst} = -\, b_{st} \cdot \Delta E_{st} \cdot K_d \cdot sign \ U_d \ + I_{qsssc}, \quad (15)$$

$$I_{dst} = b_{st} \cdot \Delta E_{st} \cdot K_q \cdot sign \ U_q \ + I_{dsssc}. \quad (16)$$

The additional voltage increments of SSSC are calculated from obtained components of EMF (formulas (9), (10)) taking into account limitations determined by the active power exchange with STATCOM in accordance with its own reactive power loading. Voltage increments introduced by the SSSC are limited so that they do not exceed preset values of the converter power rating.

The restriction system is constructed so that SSSC has priority over STATCOM as active power supply with SSSC reduces reactive power generation of STATCOM, using for the voltage regulation in connection point.

The simulation results, shown in the next section of this article, had been obtained using the Dymola software. The correctness of the devices simulation performed within the framework of the article was confirmed by comparing transients in normative disturbances with processes calculated in Matlab/Simulink (SimPowerSystems Blockset).

4 Obtained influence transient processes results and limit time of fault clearing

The fault clearing limit time was determined in case of double-phase-to-ground short circuit (SC) and three-phase SC at the beginning of the power transmission line.

The impact of single operating mode of devices (STATCOM/SSSC) and the total composition of UPFC was assessed.

In all cases it was assumed that the generator is equipped with a fast-excitation control system (digital model automatic excitation regulator of strong action based on semiconductors was used), which provided double excitation forcing at the initial stage of the process and subsequent continuous adjustment of voltage regulation, which caused damping of rotor oscillations. The limit times results of SC clearing calculation are summarized in Tables 1 and 2.

The cut-off time limit was determined at the first angle fluctuation.

Table 1. Limit value of time of double-phase-to-ground SC disconnection.

Calculation option	Time SC, s
Without additional devices	0.1592
STATCOM	0.1635
SSSC	0.1667
UPFC	0.1710

Table 2. Limit value of time of three-phase SC disconnection.

Calculation option	Time SC, s
Without additional devices	0,1110
STATCOM	0,1130
SSSC	0,1138
UPFC	0,1155

When analyzing the results given in Tables 1 and 2, it is necessary to keep in mind that the increment of rotor angle value in the short circuit has a quadratic dependence on the disturbance duration. The data of Tables 1 and 2 show that in order to ensure stability with considered disturbances, it is reasonable to use relay protections and switching devices, which ensure the time of SC clearing at the level of 0.1 s.

In determining the limit times of the SC disconnection, rather small values of the control coefficients of the voltage deviation channel of STATCOM (K_{st} = -10) and the current deviation channel of SSSC (K_{0i} = 10) were used, which provide satisfactory stabilization of the specified values during transient processes.

Despite the relatively small increase in the SC cut-off time limit, the effect of the FACTS devices on the transition process is positive. To illustrate this conclusion, Figures 2 and 3 compare processes caused by a double-phase-to-ground SC without and with FACTS devices.

Figure 2 compares curves illustrating the change of power transmission angle and electromagnetic power of the generator in case of double-phase-to-ground SC. We can observe significant decrease in the amplitude value of the angle δ_Σ in the first fluctuation.

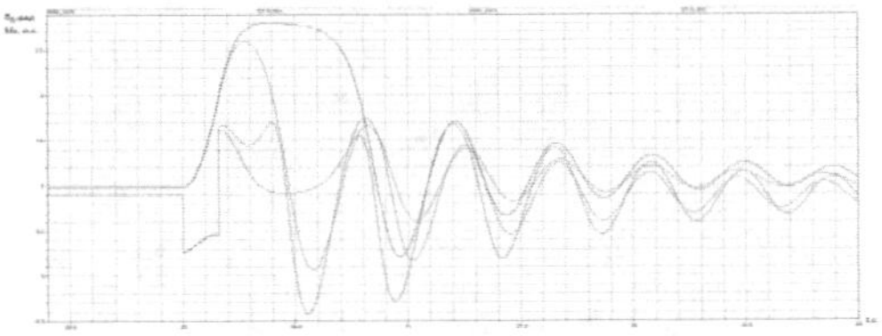

Fig. 2 Change of the angle and electromagnetic power of the generator at double-phase-to-ground SC with duration of short circuit Δt_{SC} = 0,1592 (without STATCOM - blue and red curves, with STATCOM - green and violet curves respectively)

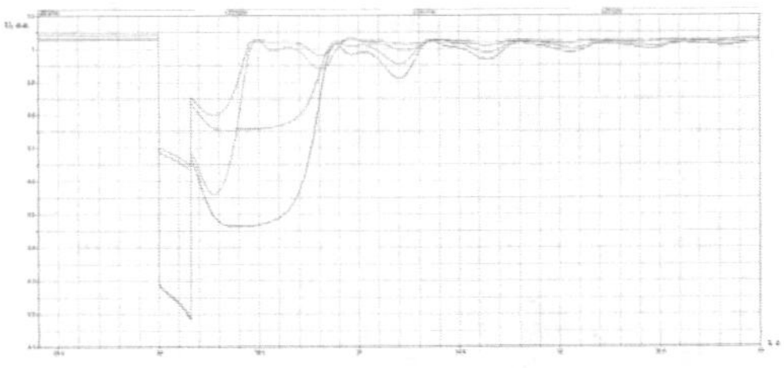

Fig. 3 Change of voltage at the beginning of the line L1 (U_1 - blue and green curves) and at the installation point of STATCOM (U_{st} - red and purple curves); blue and red curves are when STATCOM is switched off, green and violet curves are when STATCOM is switched on.

An important factor affecting the process is the change of the voltage value at the midpoint of the power transmission line [10-13]. As shown in Figure 3, without STATCOM the process is characterized by a long-term reduction of voltage values U_1 and U_{st}.

It should be noted that the main variables (transmission angle, electromagnetic generator power, U_1 и U_{st} voltage) change slightly in comparison when SSSC is used and when the full structure of UPFC is switched on. Therefore the specific structure of the device should be considered with technical and economic comparison, considering that UPFC cost (consists of two voltage-source converters) exceeds twice the cost of STATCOM [14]. Obtained considerations are also confirmed by the relatively small dependence of SC disconnection limit times on the structure of the compensating device.

Conclusions

The method of mathematical modeling of fast-acting devices of the FACTS based on presentation of them by regulated sources of EMF, related by the system of mutual restrictions has been developed. Such device models were verified by comparing transients in disturbances with those calculated in the Matlab/Simulink complex. The presentation of devices in the form of sources of controlled EMF allows to preliminary assess the efficiency of their application in order to increase the level of transient stability.

The influence of the FACTS devices on the limit time of short circuit disconnection has been considered and it has been noted that the use of a device with one converter (STATCOM or SSSC) is more effective. Application of UPFC device based on two converters (with corresponding increase of cost) in comparison with one converter device (STATCOM or SSSC) slightly increases the level of transient stability and the limit time of short circuit disconnection.

References

1. V.I. Kochkin, Y.G. Shakaryan, *Application of the FACTS in power systems* (TORUS PRESS, Moscow, 2011)
2. A.V. Pazderin, S.A. Solodyankin, Power plants, **5**, pp. 38 – 41 (2009)
3. A.V. Pazderin, S.A. Solodyankin, Power plants, **11**, pp. 34 – 39 (2012)
4. A. S. Lyamov, A.V. Pazderin, S.A. Solodyankin, News of STC UES, **1(76)**, pp. 77 – 86 (2017)
5. N. G. Hingorani, L. Gyugyi, *Understanding FACTS* (IEEE Press, N.Y., 2000)
6. M. Eremia, *Advanced Solutions in Power Systems* (IEEE Press, Hoboken, 2016)
7. K.R. Padiyar, *FACTS Controllers in Power Transmission and Distribution* (New Age International Publishers, New Delhi, 2007)
8. X.P. Zhang, C. Rehtanz, B.C. Pal, *Flexible AC Transmission Systems: Modeling and Control* (Springer-Verlag Berlin Heidelberg, 2006)
9. *Methodological guidelines on the stability of power systems* (Approved by the Ministry of Energy of Russia, 2018)
10. B.N. Abramovich, Y.A. Sychev, Industrial energy, **9**, pp. 18-22 (2016)
11. B.N. Abramovich, Y.A. Sychev, Note of Mining institute, **217**, pp. 132-139 (2016)
12. B.N. Abramovich, Y.A. Sychev, Mountain information and analytical bulletin (scientific and technical magazine), **S7**, pp. 737-744 (2015)
13. B.N. Abramovich, Y.A. Sychev, Oil Industry, **9**, pp. 120-123 (2016)
14. Ch.Kiran Kumar, M.Sudheer Kumar, V.SriramBabu, S.Nagulmeera, IOSR-JEEE, **4**, Issue 6, pp. 53-61 (2013)

Load-flow model of a multi-energy system

Dmitry Efimov[1*], and *Dmitry* Bykov[1]

[1]Melentiev Energy Systems Institute, Electric Power System Department, 664033 Irkutsk, Russia

Abstract. Current trends in the energy development includes the creation of multi-energy systems (MES) with several energy carriers that are planned, designed and operated with essential coordination of their subsystems. When managing the development and functioning of such systems, optimization problems should be solved to minimize the cost of production, transmission and distribution of energy resources. Traditionally, the means of such minimization are the redistribution of the loads of generation sources, consumers and energy storage devices. The topology of MES if different for different energy carriers, and topologies are combined through energy hubs – nodes in which energy is converted from one type to another. Such transformation can simply be considered as the consumption of one type of energy with the simultaneous generation of another type. This provides brand new opportunities for optimizing modes specific to a MES. Essential part of the search of optimal solution is the calculation of the load flow – that is, the determination of nodes' and connections' mode parameters of the system subject to balance equations (Kirchhoff's circuit laws). The load flow equations act as constraints in optimization problems. This paper presents an attempt to form a single model for calculating the load flow of different energy carriers in a MES.

1 Introduction

In electric power, gas and heat supply systems, steady-state mode calculations (load flow calculations) in the network are carried out at all time and territorial levels of dispatch control, as well as in design and research organizations. Their purpose is to verify the admissibility of the planned regime in the short, long and long-term planning and solving a number of other issues. In addition, the methods and algorithms for calculating the steady-state mode of the electric network are the basis of methods for more complex calculations, in particular, methods for optimizing the network modes, comprehensive optimization of the power systems' mode by all variables, stability calculations, etc. [1].

Traditionally, energy (electric, gas, heat supply) systems are designed and operated separately. Meanwhile, the interdependence between these systems is becoming increasingly apparent, for example, power generation is increasingly dependent on the gas supply system. In addition, the composition of the problems of managing the operating modes is in many respects similar for the production / transmission / distribution / consumption systems of one or another energy resource. In particular, when managing the modes of electric power, gas and heat supply systems, such problems are solved as state estimation (forming a model of the current mode), steady state calculation, operational adjustment of optimal operating modes within an acceptable area, analysis of possible emergency situations, a comprehensive assessment of the reliability of networks, operational control of parameters and admissibility of the current regime, assessment of reliability indicators of the current mode, assessment of mode's technical and economic indicators, the adjustment of modes and networks by the conditions of reliability, the restoration work plan formation after accidents etc.

Note that all of these problems require load flow equations to be calculated. The mathematical formulation of the problem is concretized at the stage of clarifying the temporal level of mode control taking into account external conditions, the actual state of the system, the availability of the necessary initial data, and other factors.

One of the possible ways to solve the problem of rational distribution of energy resources is the creation of multi-energy systems. The purpose of such systems is the combined production, transmission, conversion and storage of several types of energy resources. Accordingly, a whole range of control and coordination problems is associated with this, both at design and operation stages.

Currently, the joint control of the modes of generation of electric and thermal energy is being held at the level of cogeneration sources - cogeneration plants. At the same time, the issues of controlling the regimes of trigeneration and, especially, quadrogeneration plants need serious theoretical and methodological study. To an even greater extent, this applies to the management of the functioning modes of

* Corresponding author: d.n.efimov@mail.ru

MES. Meanwhile, a joint analysis of, for example, gas and electric infrastructure would make it possible to evaluate their productivity, the economically optimal load on the gas and electricity supply network, satisfying the physical restrictions on the regime parameters while satisfying demand [2].

The load flow problems of various energy resources are also similar from the calculating point of view. Quantitative characteristics of the energy carrier (substance) flow or energy are determined in all problems, which is the speed of their movement in space or the speed of their conversion from one type to another. The substance flow (more exactly, the flow rate) is measured in units of mass or volume per unit of time, and the energy flow (electric power) in units of energy per unit of time, respectively. Systems of nonlinear equations are compiled in all problems in one of two (interchangeable) forms – in the form of a balance of flows in nodes or in the form of a balance of potential changes in the contours of a scheme. The mathematical methods being used in this case are also almost identical.

The difference caused by the different physical nature of the considered energy resources can be traced only in the dependences between the flow along the edge of the graph of the design scheme and the change in potential in the resistance of this edge. These dependencies generally are nonlinear and that is reason why the system being solved is nonlinear.

2 Optimization problems in a multi-energy system

The purpose of applying methods for calculating optimal modes is primarily to find acceptable, i.e. satisfying the conditions of reliability of power supply and proper quality of energy and modes. Of no less importance is the choice of the most profitable modes, since this allows ensuring savings by reducing fuel consumption and network losses for practically no additional costs [1].

Satisfying the needs of the end consumer in different types of energy can be ensured from different primary energy carriers through a chain (sequence) of their transfer and conversion from one type to another. Such variability in the ways of providing energy supply to consumers along with the possibilities of storing energy grants the opportunities for solving problems of energy supply optimization.

The problem of estimating the optimal mode is to find the optimal values of all parameters characterizing its affordability and efficiency. Comprehensive optimization of the regime is required for all variables, which involves the selection of optimal values of the controlled parameters, which, when optimized, are considered as independent variables [1].

The mode optimality criteria is the minimum of total energy costs (or total operating costs) while meeting the requirements of energy and energy carriers demand and observing all technological restrictions for a given period of time [2, 3].

Variable constraints include:

- Equality constraints corresponding to the initial data of the load flow problem;
- Equality constraints that coincide with the equations of the mathematical model of the load flow problem;
- Bilateral inequalities as the results of solving the load flow problem.

From a math point of view, the problem of multi-energy optimal load flow is considered as a non-convex nonlinear bounded optimization problem.

3 Math model for load flow calculating

The math formulation of the load flow problem is based on a topological model in the graph form, which makes it possible to formalize the process of obtaining the set of equations of the model in a generalized ordered form. This representation allows the use of methods and algorithms of graph theory [4].

In load flow calculations, energy sources can be associated with both nodes and edges of a system graph. In the first case, the sources are represented by nodal injections, and balance equations are compiled in accordance with Kirchhoff's first law. In the second case, the sources appear to be active potential difference and the balance equations of the Kirchhoff 's second law are used. In particular, in hydraulic load flow problems, nodal injections and “active” pressure differences at edges are traditionally taken as input, and resource flows and “passive” pressure differences at edges are considered as variable (calculated) values [5].

The structure of the studied system is described by a directed graph containing N nodes and L edges. The incidence matrix (connection matrix) of this digraph can be expressed as

$$\mathbb{A} = \; A_{n\ell} \; n = \overline{1..N}, \ell = \overline{1..L} \qquad (1)$$

Elements of this matrix are equal to one if the edge l leaves the node n, -1 — if it enters into it, 0 — if the edge is not incident to this node. The incidence matrix fully describes the configuration of the graph.

Each l edge of the digraph is associated with its resistance (system parameter) $z_{nk}^{(\ell)}$. Here and below n and k are nodes incident to l edge and double lower index nk means the orientation of the edge from the node n to the node k. Upper and lower indices will be skipped further in case of insignificance for the current study.

The vector of edges' resistance is

$$\mathbb{Z} = \; z_{nk}^{(\ell)} \; \ell = \overline{1..L} \qquad (2)$$

Together with the matrix $\mathbb{A}$, this vector completely determines the topology of the system (that is, along with the configuration, also the "weight" of the edges).

It is necessary in the traditional method of load flow calculation to set a number of mode's parameters. The problem comes down then to determining the remaining mode's parameters in nodes and edges. The load flow equations are considered as constraints in optimization problems. Since optimization requires one or more degrees of freedom, this means that one or more parameters that were fixed in the load flow problem are "released".

The system mode parameters matched to each node n of the digraph:

- Injection of the transported resource flow $\mathcal{J}_n$, which matches with the nodal flow rate of the substance (volume or mass of the transported liquid or gas per unit time) or injection of electric current (or electric power). Injection is considered to be positive when importing a resource into the system and negative when exporting it from the system. Thus, a positive injection matches to the generation of a resource, and a negative injection corresponds to its consumption;
- Nodal potential difference V_n, corresponding to nodal pressure (or its square) in hydraulic networks and nodal voltage - in electric ones.
- The system mode parameters matched to each edge l of the digraph:
- The flow of transported resource $\mathcal{J}_{nk}^{(\ell)}$, which matches with the flow rate of the substance in hydraulic networks, and with current or power in electric networks. According with the Kirchhoff's law, the sum of the flows along the edges incident to the node n is equal to the injection of this node: $_k\mathcal{J}_{nk} = \mathcal{J}_n$;
- "Passive" potential difference $\Delta V_{nk}^{(\ell)}$, which matches with the pressure difference or a voltage drop in the edge's resistance: $\Delta V_{nk}^{(\ell)} = V_n - V_k$;
- "Active" potential difference $E_{nk}^{(\ell)}$ due to the potential source on the edge (which matches to the pressure difference on the edge during the operation of pumping units or electromotive force).

All these parameters (resistance of edges, nodal injections and potentials, flows and potential differences on edges), in the general case, can be considered as complex quantities.

The vectors of nodal injections and potentials for nodes can be expressed as:

$$\mathbb{J} = \mathcal{J}_n \; n = \overline{1..N}^{\;T} \quad (3a)$$

$$\mathbb{V} = V_n \; n = \overline{1..N}^{\;T} \quad (3b)$$

and vectors of flows, passive and active changes in potential – for edges:

$$\mathbb{J}^{edge} = \mathcal{J}_{nk}^{(\ell)} \; \ell = \overline{1..L}^{\;T} \quad (4a)$$

$$\Delta\mathbb{V} = \Delta V_{nk}^{(\ell)} \; \ell = \overline{1..L}^{\;T} \quad (4b)$$

$$\mathbb{E} = E_{nk}^{(\ell)} \; \ell = \overline{1..L}^{\;T} \quad (4c)$$

Taking into account the introduced notation, the mathematical formulation of the load flow problem can be described as

$$\mathbb{A} \cdot \mathbb{J}^{дуг} = \mathbb{J} \quad (5)$$

$$\Delta\mathbb{V} - \mathbb{A}^T \cdot \mathbb{V} = \mathbb{E} \quad (6)$$

$$f^{(\ell)} \; z_{nk}^{(\ell)}, \mathcal{J}_{nk}^{(\ell)}, V_n, V_k \; = 0 \quad (7)$$

Equation (5) expresses the flows' balance (the Kirchhoff's law) at all nodes. Equation (6) determines the balance between the potential difference at the final nodes ($\mathbb{A}^T\mathbb{V}$ $_k$) and the algebraic sum of the passive and active potential differences ($\Delta V_{nk}^{(\ell)}$ and $E_{nk}^{(\ell)}$) of arcs. Equation (7) expresses the dependence between the flow J_{nk}^{ℓ} and the passive potential difference $\Delta V_{nk}^{(\ell)}$ in the edge's resistance l (such dependencies should be determined for all $\ell = \overline{1..L}$).

Equation (7) is nonlinear in the general case - with the exception of electrical calculations (where the flow matches with electric current), as well as hydraulic calculations during laminar motion of the substance. In these two cases, the dependence between the flow and the passive potential difference in the edge's resistance is linear and can be expressed as:

$$\mathcal{J}_{nk}^{\ell} = \Delta V_{nk}^{(\ell)} \;\; z_{nk}^{\ell} \quad (7a)$$

Quadratic dependence between the friction pressure loss and edge flow (turbulent motion of the substance) is the most common dependence in hydraulic calculations of piping systems through which an incompressible medium is transported and can be described as:

$$\mathcal{J}_{nk}^{\ell} \; \mathcal{J}_{nk}^{\ell} \; = \Delta V_{nk}^{(\ell)} \;\; z_{nk}^{\ell} \quad (7b)$$

The passive potential difference is an increasing continuous flow function, taking values from -∞ to + ∞ and equal to zero at zero flow.

In electric power calculations (when the resource flow matches with complex electric power) equation (7) takes the form:

$$J_{nk}^{\ell} = V_n \cdot \Delta V_{nk}^{\ell} \;\; z_{nk}^{\ell} \quad (7c)$$

The dot above the letter symbol indicates the complexity of the indicated quantity, and the sign indicates the conjugation of the complex quantity.

4 Energy hub model as a static characteristic of nodal injection

Network models of different energy carriers are interfaced through energy hubs, which are energy converters of networks. The energy hub is the link between production, transmission and distribution systems. It exchanges energy with sources, consumers, and networks through hybrid input and output ports. A typical energy hub consumes various types of energy at its input ports (connected, for example, to electric, gas supply and heating networks), and produces some types of energy (such as thermal and electric) at the output ports.

In electric load flow calculations, nodal injections are often represented by functions of variable mode parameters. For example, the static characteristic of the of node n injection as a function of its voltage can be represented as a quadratic dependence with constant set coefficients (some of them may be equal zero):

$$\mathcal{J}_n = a_n V_n^2 + b_n V_n + c_n \tag{8}$$

The terms on the right-hand side represent the power of a fixed-conductance shunt, a fixed current source, and a fixed-power source (shunt) respectively.

A model of an energy hub can be formed in the same way. The transformation of the type of energy carrier (converting gas to steam or electricity, steam to electricity, diesel fuel to electricity, electricity to heat, etc.) can be simplified as the consumption of one type of energy carrier with the simultaneous generation of another type (taking into account conversion losses). The terms that are the functions of the nodal injection of another resource are added then to the static characteristic of the injection of the system node corresponding to a certain type of resource.

For example, if the node n of the system is a hub containing a gas turbine unit (GTU), then for a positive nodal injection in the electrical part of the hub (electric power generation) addition ($\Delta\mathcal{J}_n^{electr}$) can be represented as a function of negative injection (resource consumption) in its gas part:

$$\Delta\mathcal{J}_n^{electr} = \lambda_n \cdot \mathcal{J}_n^{gas} \ q_n \tag{9}$$

where $\mathcal{J}_n^{gas}$ is an injection at the node n of the gas network, q_n is a specific gas consumption for the production of a unit of electricity, λ_n is a fraction of injection $\mathcal{J}_n^{gas}$ used to generate electricity. At the same time, the term is added to the static characteristic of the injection of node n of the gas network and can be expressed as:

$$\Delta\mathcal{J}_n^{gas} = -\lambda_n \cdot \mathcal{J}_n^{gas} \tag{10}$$

There are may be more terms in the static characteristics, since the total injection of the electric part of the hub can consist of both the above generation and consumption (for example, for the thermal energy production).

The fraction of the resource injection that is converted to another type of resource indicated in the equations presented here will play the role of an adjustable parameter (independent variable) in optimization problems of the multi-energy system. So, for a gas network, gas turbine is considered as a negative injection (gas consumer), the value of which depends on the mode of the electric network. On the contrary, gas turbine units seem to be a positive injection (source of electric power) in the electric network.

The simplest model of energy or energy resource storage is a positive or negative injection, depending on its operating condition (output or consumption of a resource).

5 Regarding load flow calculation methods

Sources (resource flow and active potential change) can be located both in nodes and on the edges of the graph in equivalent scheme of a multi-energy system, as can be seen from Section 3. This arrangement determines the form of the balance equations of the mathematical model. At the same time, the placement of potential sources in nodes can lead to an unjustified increase in the number of nodes of the design scheme.

On the other hand, since in this paper we use the concept of energy hubs (as common nodes for networks with different energy carriers), we have to put up with this. For example, if we consider a gas pumping unit (GPU) not only as a source of active pressure, but also as a consumer (for example, electricity), then it will have to be placed in the graph node of the gas supply scheme. However, if the GPU is powered by a local source, it is advisable to place it on the corresponding edge of the graph.

Assuming that in equivalent schemes both methods of source arrangement would have to be used, the method of load flow calculation should be combined. This method combines different locations of sources in a single mathematical model [5]. The method is based on the representation of the vectors of the regime parameters and the incidence matrix in block form:

$$\mathbb{J} = \begin{matrix}\mathbb{J}_{\mathcal{D}}\\ \mathbb{J}_{\mathcal{V}}\end{matrix}, \quad \mathbb{V} = \begin{matrix}\mathbb{V}_{\mathcal{V}}\\ \mathbb{V}_{\mathcal{D}}\end{matrix}, \quad \mathbb{E} = \begin{matrix}\mathbb{E}_{\mathcal{D}}\\ \mathbb{E}_{\mathcal{V}}\end{matrix} \tag{11a}$$

$$\Delta\mathbb{V} = \begin{matrix}\Delta\mathbb{V}_{\mathcal{V}}\\ \Delta\mathbb{V}_{\mathcal{D}}\end{matrix}, \quad \mathbb{J}^{edge} = \begin{matrix}\mathbb{J}_{\mathcal{V}}^{edge}\\ \mathbb{J}_{\mathcal{D}}^{edge}\end{matrix} \tag{11b}$$

$$\mathbb{A} = \begin{matrix}\mathbb{A}_{11} & \mathbb{A}_{12}\\ \mathbb{A}_{21} & \mathbb{A}_{22}\end{matrix} \tag{11c}$$

where the subscripts D and V represents, respectively, the given and sought quantities. The equations (5)-(6) can be expressed as

$$\begin{matrix}\mathbb{A}_{11} & \mathbb{A}_{12}\\ \mathbb{A}_{21} & \mathbb{A}_{22}\end{matrix} \ \begin{matrix}\mathbb{J}_{\mathcal{V}}^{edge}\\ \mathbb{J}_{\mathcal{D}}^{edge}\end{matrix} = \begin{matrix}\mathbb{J}_{\mathcal{D}}\\ \mathbb{J}_{\mathcal{V}}\end{matrix} \tag{12a}$$

$$\begin{matrix}\Delta\mathbb{V}_{\mathcal{V}}\\ \Delta\mathbb{V}_{\mathcal{D}}\end{matrix} - \begin{matrix}\mathbb{A}_{11} & \mathbb{A}_{12}\\ \mathbb{A}_{21} & \mathbb{A}_{22}\end{matrix}^{T} \begin{matrix}\mathbb{V}_{\mathcal{V}}\\ \mathbb{V}_{\mathcal{D}}\end{matrix} = \begin{matrix}\mathbb{E}_{\mathcal{D}}\\ \mathbb{E}_{\mathcal{V}}\end{matrix} \tag{12b}$$

and further, after block multiplication:

$$\mathbb{A}_{11}\mathbb{J}_{\mathcal{V}}^{edge} = \mathbb{J}_{\mathcal{D}} - \mathbb{A}_{12}\mathbb{J}_{\mathcal{D}}^{edge}, \tag{13a}$$

$$\mathbb{A}_{21}\mathbb{J}_{\mathcal{V}}^{edge} = \mathbb{J}_{\mathcal{V}} - \mathbb{A}_{22}\mathbb{J}_{\mathcal{D}}^{edge}, \tag{13b}$$

$$\Delta\mathbb{V}_{\mathcal{V}} - \mathbb{A}_{11}{}^{T}\mathbb{V}_{\mathcal{V}} = \mathbb{E}_{\mathcal{D}} + \mathbb{A}_{21}{}^{T}\mathbb{V}_{\mathcal{D}} \tag{13c}$$

$$\mathbb{A}_{12}{}^{T}\mathbb{V}_{\mathcal{V}} + \mathbb{E}_{\mathcal{V}} = \Delta\mathbb{V}_{\mathcal{D}} - \mathbb{A}_{22}{}^{T}\mathbb{V}_{\mathcal{D}} \tag{13d}$$

These four matrix equations, along with a set of coupling equations (7), represent a load flow model for an equivalent scheme with a part of the sources are located on edges and the other part – on nodes.

6 Conclusions

The function of the energy system (regardless of which energy carrier is implied) is the power supply of consumers. A multi-energy system is a system with different and, at the same time, interchangeable energy carriers. So, the task can be set to find a rational or optimal (by some criteria) combination of energy carriers used.

The introduction of multi-energy systems will improve the whole range of performance indicators of the energy system, the main of which are optimization of consumer load schedules, optimization of energy production modes and network infrastructure, improving energy quality and reliability of energy supply.

The infrastructures (networks) of transmission and distribution of a multi-energy system are different both in topology (which is obvious due to the different physical nature of the transmitted resource), and network configuration. These infrastructures are interfaced between themselves in nodes common to them (energy hubs), where the transformation of types of energy from one type to another takes place. Along with this, hubs are capable of producing, consuming and storing various types of energy carriers.

Depending on the specific optimization problems, the load flow calculations of different energy carriers can be integrated at the levels of:

- mathematical model (system of nonlinear equations),
- steps (iterations) of each calculation,
- calculation results (in the general case, iteratively).

The load flow calculation of energy carriers (the calculation of the steady state based on Kirchhoff's laws) is the basic one for any statement of the problem. Methods, algorithms and software for such calculations are developed for specific energy carriers, but still not developed for combination of them.

This paper focuses on the first of these coupling levels. A complex load flow model in gas, heat and electricity systems is proposed. The model is a system of nonlinear algebraic equations, including flow balance equations at all nodes of the structural graph of the energy system (corresponding to the Kirchhoff's law), balance equations of the potential difference at the end nodes and potential changes on the edges, as well as the equations of the connections between the flow of the medium (or energy) by the graph edge and the potential difference of the vertices incident to it.

It is proposed to simulate energy hubs in the load flow calculations of MES in a simplified way of sets of interdependent static characteristics of parts of the hub related to different energy resources.

The work was carried out within the framework of the scientific project of SB RAS III.17.4.1 "Theoretical foundations for creating the integrated intelligent energy systems and control of them" (No. AAAA-A17-117030310432-9).

References

1. V.M. Gornshtejn, B.P. Miroshnichenko, A.V. Ponomarev et al. *Methods for optimization of energy systems conditions,* Energy, 336 (1981). (in Russian)
2. M. Chaudry, N. Jenkins, G. Strbac, *Multi-time period combined gas and electricity network optimisation,* Electric power systems research, **78(7)**, 1265-1279 (2008).
3. M.G. Sukharev, R.V. Samoilov, *Analysis and control of stationary and non-stationary conditions of gas transportation*, Russia state university of oil and gas named after I.M. Gubkin, 399 (2016). (in Russian)
4. L.G. Galperin, *Mathematic modelling*: [electronic resource], Ural federal university named after the first president of Russia B.N. Eltsyn» (2015), https://media.ls.urfu.ru/589/1606/3819/4513/, https://media.ls.urfu.ru/589/1606/3819/4516/ (in Russian)
5. S.P. Epifanov, V.I. Zorkaltsev, *Load flow problem in non-classical statement*, Siberian magazine of industrial mathematics, **13(4)**, 15–24 (2010). (in Russian)

Stability of intelligent energy system and intelligent control methods

Nikolai Voropai, *Dmitry* Efimov*, *Victor* Kurbatsky, and *Nikita* Tomin

Melentiev Energy Systems Institute, Electric Power System Department, 664033 Irkutsk, Russia

Abstract. In modern power systems, a variety of both objects and the tools of control is expected to be much larger than before. As a result, the dynamic properties of these systems are complicated, and the issues of maintaining stability come to the fore. The paper provides a brief overview of the types of stability, including those that, until recently, were considered local in the electric power systems of Russia. It is shown that in today's conditions the violation of these types of stability affects the operation of the electric power system as a whole. Therefore, the coordination of control of both normal and emergency modes of the systems takes on a special role and should become more intelligent. In this regard, a brief overview of machine learning developments of control agents at different levels of the control hierarchy is presented.

1 Stability and instability of the power system

Electric power system (EPS) stability is usually defined as the ability of a system to return to its original or close state after a disturbance. Any system has a "natural" margin of stability, and during its functioning tends to a state of equilibrium - that is, a balance (zero sum) of the forces (in the broad sense) at any point (both at nodes and at any points on the links). Mathematically, this is expressed by the equation of the first Kirchhoff law. However, in the process of its functioning, the system is always subjected to external and internal disturbances. With strong perturbation, the "natural" resource of stability may not be sufficient. In this case, to bring the system to an equilibrium state, admissible by the values of its state parameters, control actions are required, which should be provided by devices and control systems.

Traditionally, in the EPSs of Russia (and earlier, the USSR), under the stability of the system was understood, mainly and primarily, angle stability. Other types of stability have historically been regarded as having a local (non-systemic) character. The traditional principles of controlling the EPS modes were based on the use of the regulatory effect of the load and the frequency characteristics of the generation. It is precisely due to these effects that a "natural" resource of stability was provided.

However, over the past decades, the structure and operation modes of Russian EPSs have undergone significant transformations in all segments – generation, consumption, and the electric grid.

At present and in the near future, the use of power electronics, rectifier-inverter devices, photovoltaic installations, storage devices and other similar devices significantly reduces the regulatory effect of the load in voltage and frequency and the frequency regulatory effect of generation and significantly change the properties and controllability of EPS. Connection of distributed generation units to a distribution electric network radically changes its properties, creating stability problems, forming the need for significant development and fundamental reconstruction of relay protection and automation systems at this level [1].

In addition, the increase in power consumption with the dispersal of generating sources and consumers throughout the territory leads to an increase in the density of transmission and distribution electric networks [1]. A dense network (or, equivalently, a concentrated node) is an electric network that has very strong cross sections that do not limit the exchange of power between its parts. Moreover, if the instability in the angle (instability of the parallel operation of the generators) is characteristic of the connection or cross section between two parts of the EPS (including between concentrated nodes of the EPS), then in a dense network the types of instability are much more diverse (see Table 1). It is in a dense network that such types of stability as voltage, frequency and thermal stability acquire a system-wide nature.

Based on the classical diagram of the stability structure of P. Kundur [2] and taking into account the analysis of the types of stability according to [3], Fig. 1 presents a simplified structure of cause-effect relationships between stability disorders of different types and the corresponding states of the EPS. Violations of stability are closely intertwined and difficultly connected by causal relationships with each other and with the phenomena (events) in the EPS. Some of these

* Corresponding author: d.n.efimov@mail.ru

phenomena are local in nature (for example, an increase in voltage that is dangerous for equipment), while others affect the fundamentals of the operation of EPS, the stability of the entire EPS or a significant part of it.

Table 1. Types of instability characteristic of a dense network.

Violations of stability by type	Causes and manifestations of instability
By frequency	Separation of the dense network from the rest of the EPS with an emergency power imbalance exceeding the regulatory capabilities of the generating sources and the load of the separated part.
By voltage (voltage reduction)	• severe short circuit in a dense network or a slow decrease in voltage due to a shortage of reactive power; • asynchronous mode in section, to which a dense network adjoins as an intermediate load node or additional generation; • synchronous swings in a section close to the unit with the load.
By angle	• "density loss" – that is loss of a number of network elements with a significant weakening of some of its sections compared to other sections; • a heavy protracted short circuit near the power plant tires, which can cause a rapid violation of the dynamic stability of the station generators relative to the rest of the EPS.
By heating (violation of thermal stability)	Overloading the line with current creates heating wires. The most common cause of current overload is the disconnection of parallel or shunt lines. For a power system, a cascade process of overloading and shutting off several shunting lines is especially dangerous.

Adapted from [2], [3]

Note to the table. Like stability in angle, it is also advisable to subdivide voltage stability into state stability at small perturbations (static, Long-Term) and transition stability between states at large perturbations (dynamic, Short-Term).

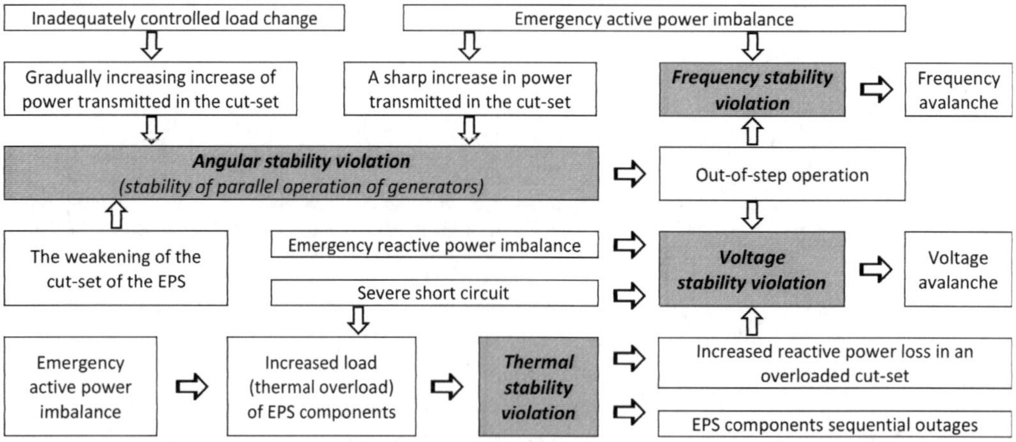

Fig. 1. Violations of stability of various types and related phenomena.

Note. Frequency instability or voltage instability can lead to shutdown of all thermal power plants, the first - due to the shutdown of auxiliary engines, the second - sometimes for the same reason and more often - due to overload of generators with reactive power.

Any of the mentioned phenomena can develop into another, even more dangerous for EPS, spread in one form or another to other parts of EPS, and ultimately lead to its catastrophic collapse [3].

2 Hierarchy of control and hierarchy of intelligence

In the EPS, the control systems (operational dispatch and automatic, normal and emergency modes) are organized according to the hierarchical principle: control of the energy object (generating, consuming, network) - control of the energy district (district power system) - control of the interconnected power system - control of a unified energy system.

If the control in a certain area is performed locally, then the control action performed in the interests of the section of the given region can cause the section to be overloaded in another region of this EPS. This drawback of local control can be eliminated by adding it to the coordination system (the simplest is hierarchical) in order to take into account the real multi-node nature of the network, covering all areas and sections of EPS [3].

Intelligence (or its absence) is manifested in the control of the system, aimed at ensuring the functioning of the system in the field of admissible values of its state parameters. More strictly, the intelligence of the unified can be defined as the intelligence of its devices and control systems, and the ability of higher levels of control to coordinate and adjust the actions of lower levels.

In this case, it is advisable to consider devices and systems that are part of the control hierarchy as agents and use existing and intensively developing approaches to agent control and multi-agent systems. An "agent" (in general terms) is usually defined as a certain entity that (regardless of its physical nature) may possess some or all of the following properties [4]: adaptability (learning ability), autonomy (ability to conduct independent purposeful actions), communicative ability (ability to communicate with other agents), collaboration (ability to collective behavior - that is, coordinated interaction with other agents to achieve common goals), ability to reason (possess of partial knowledge and / or inference mechanisms).

In relation to the power system, the use of agents listed in Table 2 was proposed in [5]. Moreover, the control levels (types) are based on the multilevel organization of complex control systems for a dynamic system proposed in [6], which was used in the development of the IES Concept [7] and detailed in relation to the IES in [5]. Agents, depending on the level of intelligence, take pre-fixed actions, or formed actions with a fixed goal, or carry out adjustment (revision) of goals for lower levels.

Table 2. Intelligent control hierarchy and classification of intelligent energy system agents.

Control level	Control method	Agent type
1,2. Rational response	*Program control* (control of a preset trajectory of the control object)	Unidirectional machine
	Position control (control of the specified state of the control object)	Multidirectional machine
3. Stereotypic response	*Robust control* (synthesis of control of good quality when changing the parameters of the control object or external disturbances)	Multidirectional automatic machine of limited competence
4. Combinatorial response and planning	*Adaptive control* (synthesis of control systems that change the laws of control under the conditions of clause 3)	Software module
5. Strategic planning	*Intelligent control* (a control system with built-in artificial intelligence functions without a goal-setting function)	Programmable model
6. System planning and control	*Intellectual control* (control system with built-in artificial intelligence functions that implement the goal-setting functions)	Control software module
7. Strategic planning based on a systems approach	*System coordination*	The set of managing software systems

Notes to the table:

1. Starting with positional control (level 2), feedback from the control object is taken into account, and intelligent control (level 6) takes into account the influence of the external environment.
2. The key difference between intelligent control (level 6) from intelligent (level 5) is the presence of goal-setting.
3. Starting from level 4, the agent must assess the state of the groups of managed objects, from level 6 - the state of the managed system, at level 7 - also evaluate the possible consequences of decisions.
4. Agents of level 5 and above must have the ability to learn.

According to [7], the existing automated technological control system of the Unified national electric grid of Russia is responsible for the implemented automatic control functions at levels 1-2 and partially at levels 3-4. The active and adaptive control system currently being created for the implemented automated and automatic control functions will correspond to levels 1-6.

It should be noted that the hierarchy of intellectual levels of agents from Table 2 does not necessarily coincide with the hierarchy of EPS control given at the beginning of this section. On the contrary, agents of different intellectual levels can function at almost any level of this hierarchy. Moreover, depending on the presence and orientation of inter-agent relationships, a wide variety of architectures of a multi-agent system is possible (from vertically integrated to fully distributed).

It is clear that the larger the range of control tasks to be solved at a lower level, the less information will have to be exchanged with a higher one (respectively, the speed of information exchange will increase) and the lower will be the load of computing power of a higher level (accordingly, the speed of decision-making on it will increase) At the same time, the more control functions can be delegated from a higher level of energy system control to a lower one, the more "intelligent" a lower level should be.

Thus, the presence of highly intelligent agents is desirable at any level of the control hierarchy. The degree of intelligence of the agent is determined by the variety of information with which the agent is able to work. Increasing the intellectual level of the agent can be carried out in the process of its (agent) functioning through machine learning.

The next section presents a brief overview of ways to creating intelligent agents using machine learning.

3. Algorithms and models of machine learning in the tasks of intelligent power systems control

The development of artificial intelligence (AI) methods has made it possible to significantly accelerate and automate the solution of a whole range of problems in the control of electric power systems (EPS) [8].

One of the actively developed areas is the application and implementation of machine learning (ML) technology, including methods for constructing algorithms that can be trained, such as artificial neural networks (ANNs), reference vector machines, decision trees, etc.

The use of various types of training models of ML: supervised learning, unsupervised learning, reinforcement learning, deep learning, etc., allowed to create separate adaptive, machine learning software modules for regulation and control of both individual components of the EPS, and its mode.

Their main advantages are speed, high adaptability, the ability to approximate nonlinear functions and the presence of a certain kind of machine intelligence, which allows to develop the most autonomous systems capable of independent decision making based on experience and original properties for generalization.

Table 3. Application of various machine learning technologies in the control of intelligent EPS modes

Controlled mode of EPS	**Control tasks**	**Artificial Intelligence Technologies and References**				
		RL	**SL**	**DL**	**DRL**	**USL**
Normal mode	1. Renewable energy control	[9]	[10]	[10]		
	2. Isolated power system and micro-grid control				[11]	
	3. Voltage and reactive power regulation	[14]	[12]			[13]
	4. Energy consumption control			[15]		
Emergency mode	5. Power system blackout prevention	[16]				
	6. Peak load and overload control	[17]				
	7. Control in transition modes	[18]	[19], [20]			
Recovery mode	8. Restoration of the power system	[21]				

Abbreviations: **RL** – Reinforcement Learning; **SL** – Supervised Learning; **DL** – Deep Learning; **DRL** – Deep Reinforcement Learning; **USL** – Unsupervised Learning.

Table 3 presents some of the tasks of control the EPS modes that are successfully solved on the basis of the Ministry of Defense, including in combination with other AI technologies.

Studies show that agents trained offline based on the methods of reinforcement learning, successfully control the individual components of the EPS in the control of normal and emergency conditions.

For example, in [22], the global generator excitation regulator WASCCO, implemented on the basis of reinforcement learning for damping electromechanical vibrations and controlling the transient voltage, allows the excitation system to "learn" when interacting with EPS and to effectively predict its future states.

Tests have shown that this approach gives better damping and a transition characteristic than traditional system stabilizers and allows damping interregional mode oscillations better than a regulator based on heuristic programming, acting independently.

Good results were also obtained when Q-agents were a dynamic brake, a thyristor-controlled series capacitor, synchronous generators, individual or aggregated loads, etc., to implement optimal control strategies [23]. Moreover, the use of reinforcement learning showed good results in the whole spectrum of tasks of operational dispatch and emergency control.

Recent developments indicate that the use of promising technology of deep machine learning allows the development of fully autonomous intelligent control systems. For example, Siemens is already applying deep ANNs in various projects: improving the operation of SIPROTEC multifunctional relay protection devices, optimizing the operation of gas turbines to reduce emissions of toxic nitrogen oxides, intelligently adjusting the position of the wind turbine rotors depending on the wind direction [24].

In the general case, deep machine learning is a part of a wider family of ML methods of teaching representations, where the feature vectors are located simultaneously on many levels.

As a rule, we are talking about special ANNs with many hidden layers of neurons (levels), which are based on special building blocks, for example, restricted Boltzmann machine, which allow for network pre-training, training each layer separately.

However, on the basis of the combination of reinforcement learning methods and deep ANNs, technologies have been created that have become breakthroughs of recent times in the field of AI, such as AlphaGo, Atari Deep Q-learning and others. These developments have shown that such models are able to solve problems better than the person himself (an expert in the subject area) or completely displace the classic, traditional algorithm for solving the problem. According to some experts, this particular technology will allow in the near future to create a full-fledged machine intelligence similar to human [25].

Conclusion

Historically, in the power systems of the USSR, and then Russia, the angle stability was considered as the main type of system stability, and the principles of EPS operation control were based on the use of the regulatory effect of the load and the frequency characteristics of generation. Due to these effects, traditional EPSs had a "natural" resource of stability, and control systems acted when parameters of EPS condition exceeded certain boundaries.

However, the new characteristics of consumers, storages and the generators of future EPSs significantly change the properties and controllability of systems, and the increase in power consumption with the distribution of generating sources and consumers across the territory leads to an increase in the density of transmission and distribution electric networks. Therefore, traditionally "local" types of stability (in terms of voltage, frequency, and heating) acquire a system-wide nature in the EPSs of

Russia, which previously was typical only for "western" EPSs with a dense electric network.

Already now there are, and in the future will be aggravated, new problems associated with the need to strengthen coordination of control of EPS conditions at various levels, improve control efficiency, ensure the reliability of control systems themselves. Therefore, the traditional principles of control of EPS conditions need to be substantially modified and developed.

The intelligence of a power system can be defined as the intelligence of its control devices and systems, and the ability of higher levels of control to coordinate and adjust the actions of lower levels. In this case, it is advisable to consider devices and systems that are part of the control hierarchy as agents, and use existing and intensively developing approaches to agent control and multi-agent systems.

The presence of highly intelligent agents is desirable at any level of the control hierarchy. Increasing the intellectual level of an agent can be carried out in the process of its functioning through machine learning. Currently, various machine learning technologies are being intensively developed for use in the control of intelligent EPS modes, including both normal mode control and emergency control, as well as system recovery control after accidents. Recent developments indicate that the use of deep machine learning technology allows the development of fully autonomous intelligent control systems that have already found application at power facilities.

The work was carried out within the framework of the scientific project of SB RAS III.17.4.2. "'Theory and methods of substantiating the development and control of intelligent electric power systems"' (No. AAAA-A17-117030310438-1).

References

1. Voropai N.I., Osak A.B. Electric power systems of the future // Energy policy. - 2014. - No. 5. - pp. 60-63. (in Russian)
2. Kundur P.C. Power System Stability and Control. McGraw-Hill, 1994.
3. Gluskin I.Z., Iofiev B.I. (with the participation of Meklin A.A., Chekalovets L.N.). Emergency automation in power systems. – M.: "Znack". 2009. -556 p. (in Russian)
4. Pospelov D.A. Multi-agent systems: present and future // Information Technologies and Computing Systems. - 1998. - No. 1. (in Russian)
5. General requirements for the creation of an intelligent energy system with an active adaptive network - IES AAN / Working group of FGC UES, Team Leader V.V. Dorofeev - M., January 2012. - 88 p. (in Russian)
6. Vasiliev S.N., Zherlov A.K., Fedosov E.A., Fedunov B.E. Intelligent control of dynamic systems. - M.: Physical and mathematical literature, 2000. - 352 p. - ISBN 5-9221-0050-5. (in Russian)
7. The concept of the intellectual electric power system of Russia with an active and adaptive network / Ed. Fortov V.E. and Makarov A.A. - M.: NTC FGC UES, 2012. - 235 p. (in Russian)
8. Negnevitsky M. Artificial intelligence: a guide to intelligent systems. 3rd edn, Addison Wesley, Harlow, England, 2011. 504 p.
9. Tang Y., He H., Wen J., Liu J. Power system stability control for a wind farm based on adaptive dynamic programming // IEEE Trans. Smart Grid. 2015. Vol. 6. 2015. P.166–177
10. Liu Y. Machine Learning for Wind Power Prediction. MCS Thesis. University of New Brunswick, Canada. 2016. 88 p.
11. Francois-Lavet V. et al. Deep Reinforcement Learning Solutions for Energy Microgrids Management // In European Workshop on Reinforcement Learning. 2016.
12. Zhukov A., Tomin N., Sidorov D., Kurbatsky V., Panasetsky D. On-Line Power Systems Security Assessment Using Data Stream Random Forest Algorithm / In: Zelinka I., Vasant P., Duy V., Dao T. (eds) Innovative Computing, Optimization and Its Applications. Studies in Computational Intelligence. Vol 741. 2018. Springer.
13. Tomin N., Negnevitsky M., Rehtanz Ch. Preventing Large-Scale Emergencies in Modern Power Systems: AI Approach // Journal of Advanced Computational Intelligence and Intelligent Informatics. 2014. Vol. 18. No.5. P. 714-727.
14. Xu Y., Zhang W., Liu W., Ferrese F. Multiagent-based reinforcement learning for optimal reactive power dispatch // IEEE Trans. Syst., Man, Cyber. 2012. Vol. 42. P. 1742–1751.
15. Mocanu E., Nguyen P.H., Gibescu M. Deep Learning for Power System Data Analysis, In Big Data Application in Power Systems. ed. by Reza Arghandeh and Yuxun Zhou, Elsevier, 2018. P. 125-158.
16. Belkacemi R., Abdulrasheed Babalola A., Zarrabian S. Real-Time Cascading Failures Prevention Through MAS Algorithm and Immune System Reinforcement Learning // Electric Power Components and Systems. 2017. Vol. 45. No. 5. P. 505-519.
17. Yousefian R., Kamalasadan K. Design and Real-Time Implementation of Optimal Power System Wide-Area System-Centric Controller Based on Temporal Difference Learning // IEEE Trans. on Industry Applications. 2015. Vol.1.
18. Zarabbian S., Belkacemi R., Babalola A.A. Reinforcement learning approach for congestion management and cascading failure prevention with experimental application // Elec. Power Syst. Research. 2016. Vol. 141. P. 179–190.
19. Glavic M., Ernst D., and Wehenkel L. A reinforcement learning based discrete supplementary control for power system transient stability

enhancement // Engineering Intelligent Systems for Electrical Engineering and Communications. 2005. Vol. 13. P. 81–88.

20. Kaci A. et al. Synchrophasor Data Baselining and Mining for Online Monitoring of Dynamic Security Limits // IEEE Trans. on Power Systems. 2014. Vol. 29, No. 6. P. 2681-2695.
21. Ye D., Zhang M., Sutanto D. A hybrid multiagent framework with Q-learning for power grid systems restoration // IEEE Trans. Power Syst. 2011. Vol. 26. P. 2434–2441.
22. Liu C. et al. Dynamic security assessment of western Danish power system based on ensemble decision trees // in Proc. of the 12th IET International Conference on Developments in Power System Protection. Copenhagen, 2014.
23. Ernst D., Wehenkel L., Glavic M. Power systems stability control: Reinforcement learning framework // IEEE Trans. Power Syst. 2004. Vol. 19. P. 427–435.
24. Kogan I., Boehme K., Herrmann H-J. The introduction of advanced machine learning technologies in the protection relay // Sat. Proceedings of the International Scientific and Technical Conference and Exhibition "Relay Protection and Automation of Energy Systems 2017". St. Petersburg. Russia. 2017.
25. Irpan A. Deep Reinforcement Learning Doesn't Work Yet [Electronic Document] // Sorta Insightful. 2018. Access Mode: https://www.alexirpan.com/2018/02/14/rl-hard.html

E3S Web of Conferences **139**, 01052 (2019) https://doi.org/10.1051/e3sconf/201913901052
RSES 2019

Placement of shunt reactors in high-voltage network using fuzzy constraints

Arif Hashimov[1], *Huseyngulu* Guliyev[1,*], *Aytek* Babayeva[1]

[1]Azerbaijan Scientific–Research and Designed–Prospecting Institute of Energetics, Baku, Azerbaijan

Abstract. In recent years, controlled shunt reactors (CSR) relevant to the class of FACTS facilities have been widely used to control voltage modes and reactive power flows in the high-voltage electrical network. The selection of location, as well as the definition of the law of CSR control in the conditions of stochastic variability of the operation mode of high-voltage power transmission, are associated with numerous technical and economic factors. At the same time, such constraint conditions as ease of use, performance efficiency, purpose and location in the system, as well as the period of commissioning should be taken into account. In the proposed procedure these factors are considered as fuzzy constraints. The procedure of CSR placement in the 330 kV electrical network of Azerenergy system for control of reactive power flows taking into account the mentioned fuzzy constraints is proposed. The obtained simulation results confirm the advantage of the proposed procedure.

1. Introduction

In the modern period in the power systems of CIS and foreign countries, great importance is attached to the creation of controlled or flexible power lines, which are part of 'smart" networks (Smart Grid) with FACTS facilities [1]. For the optimal control of the modes of such power systems, there is a need for highly efficient means of control of the flows of both active and reactive powers.

To control the voltage and reactive power regimes, in addition to the generators, synchronous and static compensators, switching reactors and capacitor banks, new facilities - controlled shunt reactors (CSR) - have been widely used in recent decades [2-6]. The economic analyses have shown that additional energy losses are at such a high level that despite the availability of expensive equipment, the CSR installation is self-supporting for a period of less than 5 years [7]. There is still the problem of elimination of the surplus reactive power generated in the minimum load modes in most power grids. The main reason for this redundancy is that the charging power in 330 kV lines is higher than the reactive power loss in them, and this can cause the voltage level increase to a level dangerous for the line insulation.

Traditional methods and means used in the modern period to eliminate surplus reactive power are not effective enough and should be replaced by more modern technological means.From this point of view, 330 kV CSR is preferable. Thus, the "reactive power loss –charging power" relation in the networks is not constant, it changes. Therefore, to ensure the balance of reactive powers, the CSR power must be controled in a wide range [8].

During the selection of power and the CSR installation location, as well as changes in the operating mode of power transmission, the definition of the control mode is associated with numerous economic and technical factors.So, the installation location and the CSR characteristics affect the energy losses in the entire power transmission, the mode stability, ensuring voltage regulation within the predetermined limits during various power transmissions, the formation of excess voltages in individual elements of the power transmission. At the same time, other factors should also be taken into account during the CSR selection, such as the convenience of the CSR installation place in the given point of the network from the point of view of operation, operating efficiency, technical and economic indicators. So, for different networks these or other factors are an acceptable option for the selection and placement of compensating devices.

In the conditions of numerous influencing factors, the choice of compensating means, determination of the optimal option taking into account the specified practical cases leads to the solution of the multi-purpose problem. Thus, the solution of this problem without the use of modern mathematical technologies is extremely complicated.

In the present paper the problems of selection and placement of 330 kV controlled shunt reactors on the basis of studies conducted on real perspective scheme of the power system are considered.

2. CSR placement in 330 kv nodes of power grid

https://doi.org/10.1051/e3sconf/201913901052

A special procedure can be used for the reactor placement in the power system. In order to determine the criteria for selection of the most efficient CSR installation locations, it is necessary to analyze their impact on two important indicators of the power system mode. It is known that such indicators are the values of absolute and relative reduction of voltage levels at various points of the network before and after the reactor installation, as well as the values of losses in the network. Calculations should be carried out for the most severe minimum load conditions, so that the voltage levels at the observed points of the network reach the maximum possible value. Obviously, in this mode the reactor power should be maximal. Therefore, during the comparative calculations, the CSR power is assumed to be equal to the reactor rated power for all nodes.

Placement of a single CSR at separate substations will impact differently on the average voltage level in 330 kV nodes of power system and the total loss level in networks.Obviously, the CSR installation at any substation will lower the voltage level both at this station (most of all) and at other substations. Therefore, the average undervoltage can be accepted as the main technical efficiency indicator of the reactor installation. Another important indicator is the power loss reduction in networks. It should be noted that during the installation of a single CSR, in contrast to the voltage, the power loss can both increase and decrease.

Taking into account the above-mentioned, as a special technical efficiency indicator of the reactor installation, the mean absolute δU_{or}, mean relative $\delta\overline{U}_{or}$ voltage reduction and, accordingly, the absolute δP_{Σ} and relative $\delta\overline{P}_{\Sigma}$ total power loss reduction can be accepted. These quantities can be determined by performing multivariate calculations with the alternate CSR placement at different substations. In addition to the above, for a comprehensive assessment of the technical and economic efficiency of the CSR application, the resulting efficiency indicator $E_{ef,\Sigma}$ was proposed, which is expressed as follows [2]:

$$E_{ef,\Sigma} = \delta\overline{U}_{or} \cdot \delta\overline{P}_{\Sigma} \quad (1)$$

It is possible to advance an idea of comparative efficiency of the CSR installation at different points of the network in accordance with the value of this indicator.It should be noted that in the case when the CSR placement has the same effect on the average voltage level (it always reduces), the reactor loss level is double affected.

Table 1. Efficiency indicator values for 330 kV nodes of power grid

Node No.	Node name	Voltage, kV		Total losses in the networ k, MW	Absolute and relative decrease of total losses		Average absolute and relative voltage		Efficiency factor,
		Bus voltage before the SR connection	Average voltage after SR connection		MW	%	kV	%	E_{ef}
39	Absheron 330	344,38	334,23	14,5	0,7	4,61	3,60	1,07	4,912
201	Janub PP	346,53	334,66	14,6	0,6	3,94	3,17	0,94	3,701
101	Yashma 330	342,18	333,96	14,9	0,3	1,97	3,87	1,15	2,261
601	Mini HPP	339,44	335,55	14,7	0,5	3,29	2,41	0,71	2,349
651	AzES330	339,23	335,55	14,7	0,5	3,29	2,28	0,68	2,219
400	Goranboy SG	338,76	334,91	14,9	0,3	1,97	2,92	0,86	1,704
333	Agdjabedi330	342,64	333,78	15,0	0,2	1,32	4,05	1,19	1,577
280	Imishli 330	344,27	333,86	15,0	0,2	1,32	3,97	1,18	1,548
801	Khachmaz330	343,43	333,99	15,0	0,2	1,32	3,84	1,14	1,497
411	Shamkir HPP	331,01	336,13	15,4	-0,2	-1,32	1,70	0,50	-0,663
401	Gandja330	329,82	335,79	15,5	-0,3	-1,97	2,04	0,60	-1,190
456	Samukh 330	328,74	335,98	15,6	-0,4	-2,63	1,85	0,55	-1,443
457	GAZ 330	328,28	335,89	15,8	-0,6	-3,95	1,94	0,58	-2,269
502	Agstafa 330	330,89	335,25	15,9	-0,7	-4,61	2,58	0,76	-3,516

Thus, in this case the loss may both increase (useful effect) and decrease (useless effect). Obviously, in this case, the comparison and alternation of substations according to the $E_{ef,\Sigma}$ indicator gives meaning to its positive values. In other words, certain places for the CSR installation should be selected among the nodes with the $E_{ef,\Sigma} > 0$ condition. In addition, other factors should be taken into account, especially the periods of commissioning of substations, the availability of place for CSR installation, possibility of reactive power flows from neighboring power systems.

The values of the special and resultant efficiency indicators for the 330 kV nodes of the power system (initial loss of 15.2 MW) are given in Table 1. The nodes are arranged in a sequence of decreasing values of $E_{ef,\Sigma}$ indicator.

As can be seen from the table, the $E_{ef,\Sigma}$ value is positive only for 9 nodes of the considered 14 ones, and the reactor installation locations should be selected between the nodes just with this value $E_{ef,\Sigma} > 0$. In this case, in addition to the condition $E_{ef,\Sigma} > 0$, as noted above, other factors should be taken into account (the periods of commissioning of substations, the availability of places for the SR installation, the technical possibility for the electric schematic diagram of switchgear, etc.).

3. Taking into account of constraints during sr placement

The other 5 factors influencing the selection of SR installation place were taken into account in the form of fuzzy constraints: the period of commissioning of substation; the period of the substation operation; the availability of the installation place; the possibility of an electrical wiring diagram; the substation place in the system. For linguistic variables the membership functions of Gauss, type Z and S were accepted.

Gauss membership function:

$$\mu_{ki}(x) = \exp\left(\frac{-(x_i - m_{ki})}{2\sigma_{ki}^{\ 2}}\right), \quad i = \overline{1,n} \quad k = \overline{1,m}$$

where m – coordinate of the maximum; σ – concentration ratio.

zmf and *smf* – membership function

$$\mu_{ki}(x) = \begin{cases} 1, & x_i \le a_{ki} \\ nonlinearapproximation, & a_{ki} < x_i < b_{ki} \\ 0, & x_i \ge b_{ki} \end{cases} \qquad (6)$$

where m – coordinate of the maximum; σ – concentration ratio; a,d – fuzzy set carrier; b,c – fuzzy set kernel, $\mu_{A,i}(x): X_i \to [0,1]$.

After determination of forms of fuzzy implication and membership functions, the output signals were formed on the basis of fuzzy approximation between the input and output vectors.

$$\mu_{A,i}(x): X_i \to [0,1]$$

The membership functions of input and output variables and their terms are shown in Fig.1, and their parameters – in Table 2.

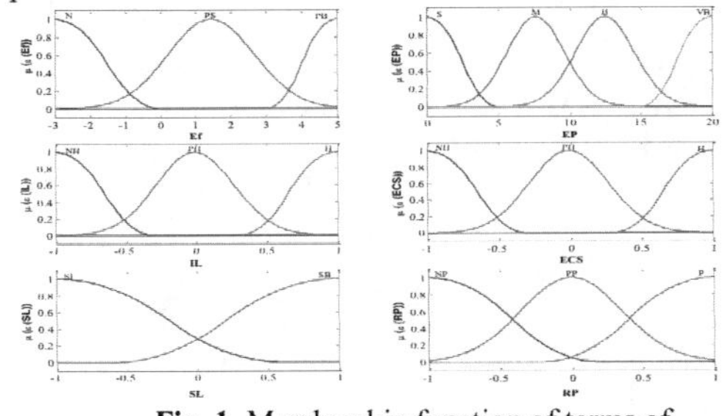

Fig. 1. Membership function of terms of linguistic variables

Table 2. Terms of linguistic variables, membership functions and their parameters

Term-subsets	Membership functions	Parameters
Efficiency factor, EF		
Negative (N)	Zmf	[-3 0]
Positive small (PS)	Gaussmf	[1,23 1,44]
Positive big (PB)	Smf	[3 5]
Commissioning period, EP		
Small (S)	Zmf	[0 5]
Mean (M)	Gaussmf	[2 7,62]
Big (B)	Gaussmf	[2,07 12,47]
Very big (VB)	Smf	[14,97 20]
Installation places, IL		
No (NH)	Zmf	[-1 -0,3]
Частично есть (PH)	Gaussmf	[0,277 -0,009]
Есть (H)	Smf	[0,3 1]
Connection diagram, ECS		
Нет (NH)	Zmf	[-1 -0,3]
Partially available (PH)	Gaussmf	[0,277 -0,009]
Available (H)	Smf	[0,3 1]
Places in the system, SL		
Backbone, SI	Zmf	[-1 0,607]
Between systems, SB	Smf	[-0,6 1]
Reactor installation, RP		
Don't install (MP)	Zmf	[-1 0,18]
Partially possible (PP)	Gaussmf	[0,35 0]
To install (P)	Smf	[-0.2 1]

Fuzzy output mechanism consisting of 65 rules synthesized based on the Mamdani algorithm is shown in Fig.2.

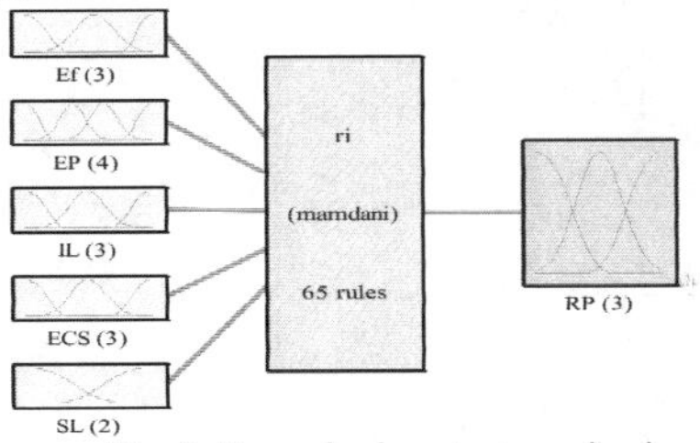

Fig. 2. Fuzzy logic output mechanism

The decision-making procedure for one option according to the given rules is shown in Fig.3.

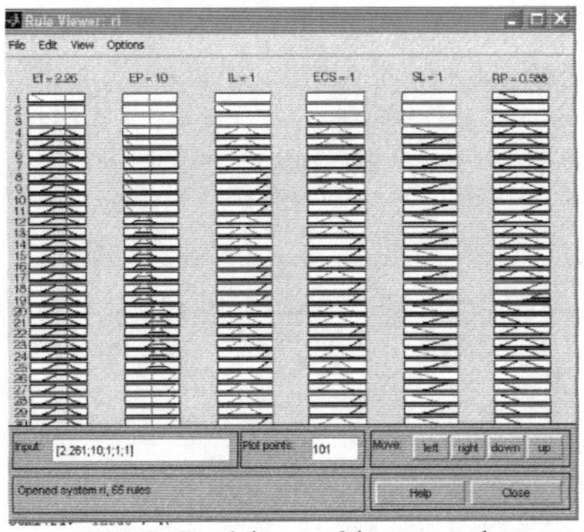

Fig. 3. Decision-making procedure segment

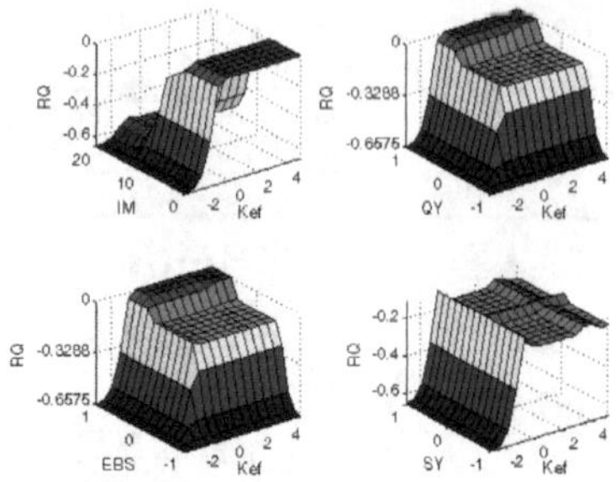

Fig. 4. Fuzzy surface dependences

The "indicated surfaces - output variables" relationships (SR installation node) taking into account fuzzy constraints are presented in Fig.4. Table 3 presents the results of the adjustment of the priority nodes according to this procedure.

Table 3. Efficiency factor values for 330 kV nodes

Node No.	Node name	Voltage, kV		Total losses in network, MW	Absolute and relative decrease of total losses		Average absolute and relative voltage decrease		Efficiency factor, E_{ef}	Reactor installation RP
		Bus voltage before SR connection	Average voltage after SR connection		MW	%	kV	%		
400	Goranboy SG	338,76	334,91	14,9	0,3	1,97	2,92	0,86	1,704	0,518
201	Janub PP	346,53	334,66	14,6	0,6	3,94	3,17	0,94	3,701	0,508
101	Yashma 330	342,18	333,96	14,9	0,3	1,97	3,87	1,15	2,261	0,172
280	Imishli 330	344,27	333,86	15,0	0,2	1,32	3,97	1,18	1,548	0,015
801	Khachmaz 330	343,43	333,99	15,0	0,2	1,32	3,84	1,14	1,497	0,014
333	Agdjabedi 330	342,64	333,78	15,0	0,2	1,32	4,05	1,19	1,577	0,013
456	Samukh 330	328,74	335,98	15,6	-0,4	-2,63	1,85	0,55	-1,443	-0,434
411	Shamkir HPP	331,01	336,13	15,4	-0,2	-1,32	1,70	0,50	-0,663	-0,536
601	Mini HPP	339,44	335,55	14,7	0,5	3,29	2,41	0,71	2,349	-0,551
651	AzES 330	339,23	335,55	14,7	0,5	3,29	2,28	0,68	2,219	-0,551
401	Gandja 330	329,82	335,79	15,5	-0,3	-1,97	2,04	0,60	-1,190	-0,558
457	GAZ 330	328,28	335,89	15,8	-0,6	-3,95	1,94	0,58	-2,269	-0,631
502	Agstafa 330	330,89	335,25	15,9	-0,7	-4,61	2,58	0,76	-3,516	-0,658
39	Absheron 330	344,38	334,23	14,5	0,7	4,61	3,60	1,07	4,912	-0,658

As can be seen, taking into account these factors, the priority nodes are Goranboy SG, Janūb ES, Yashma 330 kV, Imishli SS and Khachmaz 330 kV. Thus, if we take into account the situation in the system and the calculation results, then initiallly the Yashma 330 kV and 330 kV SG Goranboy nodes can be accepted. So, for both nodes the condition $E_{ef,\Sigma} > 0$ is met and in addition, due to the ability to connect the reactor to busbar at the Yashma 330 kV Substation and one circuit of one-and-a-half scheme on 330 kV SG Goranboy to a free node, and due to the availability of appropriate places for the reactor placement, the schemes turn out efficient and reliable.

As can be seen from the Table 1, the 330 kV Goranboy ES buses and 330 kV Imishli nodes can be considered as pretenders for the third node for the reactor connection in the future.

4. Simulation results

To determine the voltage levels in the nodes of the power system, the corresponding mode calculations for the maximum and minimum load should be carried out. The voltage profiles of some characteristic load nodes with voltage of 330 and 500 kV based on calculations performed for the maximum and minimum load modes in real perspective schemes of the power system are shown in Fig.5.

It should be noted that the maximum load regime of the power system was formed according to the data obtained from "Perspective development scheme" Department of "AzRandDSIPE"JSC. And minimum load mode was adopted 0.3 Pmax (Pmax - the maximum active load of the power system).

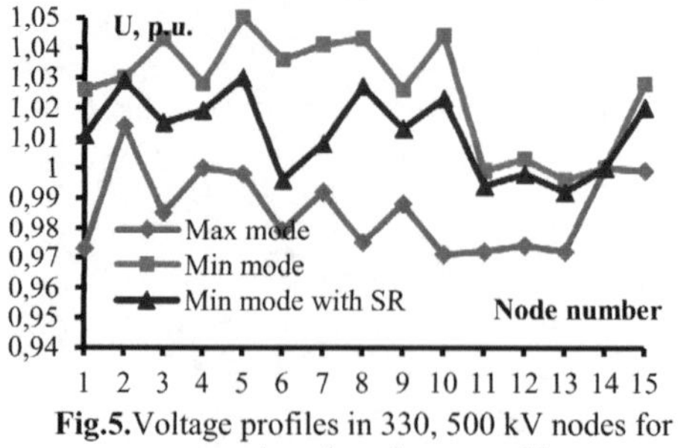

Fig.5. Voltage profiles in 330, 500 kV nodes for normal modes of power grid

As can be seen from the figure, in the maximum load mode (Py=5613.9 MW, Qt=3355.2 MVAr) the voltage in 500 kV nodes varies within the (1.0-1.014)Unom interval, in 330 kV nodes within the (0.972-0.999)Unom interval, and in the minimum mode (Py=1684.2 MW, Qt=1006.6 MVAr) the voltage in 500 kV nodes varies within the (1.0-1.03)Unom interval, in 330 kV nodes within the (0.996-1.05)Unom interval (Table 3). Thus, the voltage in the maximum and minimum modes is within normal limits. In some nodes, the voltage is set to the upper limit.

Taking into account the above, the calculation was repeated for the minimum load mode with the connection of a shunt reactor with a capacity of 180 MVAr to the node Goranboy 330 kV, and the reactor with a capacity of 100 MVAr to the node Yashma 330 kV. As can be seen, in the case of the reactor connection, the voltage profiles in the minimum load mode improve and change within the range(0.992-1.03)of Unom around the nominal value.

Table 4. Voltage variation limits in 500 and 330 kV nodes

Modes	Load		Voltage variation interval	
-	P, MW	Q, MVAr	500 kV	330 kV
Maximal mode	5613,9	3355,2	(1,0-1,014)U_{nom}	(0,972-0,999)U_{nom}
Minimal mode	1684,2	1006,6	(1,0-1,03)U_{nom}	(0,996-1,05)U_{nom}
Minimal mode after reactor connection	1684,2	1006,6		(0,992-1,03)U_{nom}

At the next stage the calculations were performed with the simulation of emergency modes according to the criteria N-1 and N-2. The cases of disconnection of lines Absheron-1, Absheron-8, Agstafa-4, Goranboy-3, Shimal ES according to criterion N-1, and disconnections of lines Shimal ES and Absheron-1 according to criterion N-2 were considered.

Voltage profiles in the nodes based on the results of calculations performed according to the N-1 criterion are presented in Fig.6.

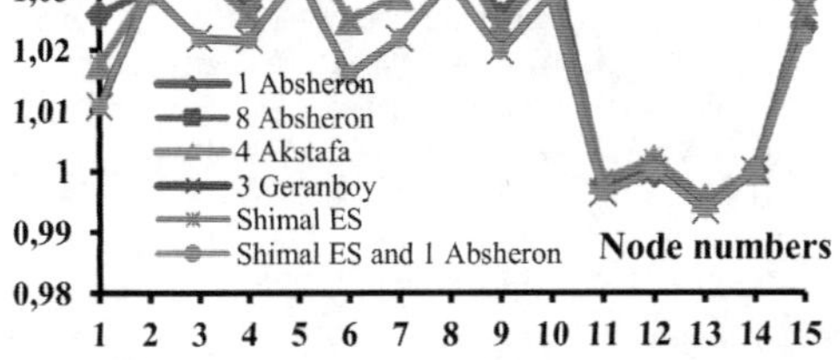

Fig. 6. Voltage profiles for emergency modes

As can be seen in Fig.6, model calculations in emergency modes of power system based on the N-1 criterion indicate the voltage location at the nodes within acceptable limits. The voltage in some nodes, for example, the voltage in the 5th node (Janub ES) during the disconnection of Absheron-8, Agstafa-4 and Goranboy-3 lines is in the upper limit. The same conclusion can be drawn with respect to the Imishli node. Taking this into account, mode calculations were carried out for the case of connection of shunt reactors to certain nodes in some of the considered emergency modes (disconnection of the Agstafa-4 and Goranboy-3 lines).

5. Conclusions

1. In order to control the reactive power flows in the power system, the procedure of selection and placement of CSR devices in 330 kV nodes with fuzzy constraints is proposed.

2. On the basis of the developed procedure, priority nodes for the CSR placement in the Azerenergy system were identified. The mode calculations with the CSR placement in these nodes are carried out. The results confirmed a significant improvement of the voltage profile at the nodes.

References

1. Dolgopolov A.G., Condratenko D.V., Ucolov S.V., Postolati V.M. Reactoare controlate shuntate pentru retele electrice. Problemele Energeticii Regionale, 3(17), 2011, c.1-21.

2. Dolgopolov A.G. Bias-controlled shunt reactors. Operation speed issues // News of Electrical Engineering, No.4(64), 2010, Electronic resource.

3. Bryantsev M.A., Lurie A.I., Spiridonov D.Y. Results of introduction of bias-controlled shunt reactors in 110-500 kV networks // Electro, No.3, 2006, pp.25-31.

4. Kochkin V.I., Nechaev O. P. Application of static reactive power compensators in electrical networks of power systems and enterprises.-M., Publishing house of SC ENAS, 2002, pp.248.

5. Hashimov A.M., Guliyev H.B., Babayeva A.R. IJTPE Journal International Journal on technical and physical problems of engineering (IJTPE), Issue 30, Vol. 9, No. 1, 2017, pp.18-22.

6. Hashimov A.M., Guliyev H.B., Babaeva A.R. Selection and placement of compensating devices of the excess reactive power in high-voltage electrical networks // Problems of Power Engineering, No.1, 2019, pp.3-12.

7. Controlled bias electric reactors, ed. by Bryantsev A. M., ZNAK, 2010, p.288.

8. Hashimov A.M., Guliev H.B., Babayeva A.R. Multi-pole fuzzy logic controller to improve the control of shunt reactors // Electricity, No. 6, 2019, pp. 26-32.

https://doi.org/10.1051/e3sconf/201913901053

Operational reliability of low-voltage asynchronous motors at asymmetrical supply voltage

Victoria Romanova[1]*, *Sergey* Khromov[1]

[1]Trans-Baikal State University, Department of Electric Power Generation and Electrical Engineering, Russia, Chita, 30, Aleksandro-Zavodskaya Str.

Abstract. Issues of operating modes of 4A series asynchronous squirrel-cage motors and AI serried motors at asymmetrical supply voltage for different values of motor shaft load are considered on the basis of simulation using Matlab software package and Simulink package. In the simulation, currents in asynchronous motor phases were evaluated as a function of K_{2U}, K_Z, and based on these results, permissible operation parameters were evaluated for induction motors. The results obtained are the basis for technical solutions aimed at improving operational reliability of asynchronous motors, they will allow taking into account permissible operating parameters of asynchronous motors not only in the design, but also in operation in power supply systems of various facilities.

1 Introduction

Reliability is the most important technical and economic indicator of asynchronous motors (AM) quality which defines ability to operate reliably with unchangeable technical characteristics during a given period of time under certain operating conditions [1,2]. However, in actual electric motor operation conditions, due to various factors, there is a significant deviation from normal operating modes. The vast majority of the reason for this is poor voltage quality in the networks, in particular, asymmetry of supply voltage. Supply voltage asymmetry has a negative impact on AM, reduces operational reliability and increases risk of motor failures. Thus, service life of asynchronous motors operating under full load is reduced by 2 times at a voltage asymmetry of 4%. When the voltage is unbalanced by 5%, available power of the motor is reduced by 5-10% [3,4].

In this paper, operation modes of 4A series squirrel-cage AM and AI series motors at asymmetrical supply voltage with different values of the motor shaft load were studied to identify AM operating modes.

Long-term trouble-free operation of AM is achieved as a result of implementation of measures aimed at maintaining technical and operational characteristics within the limits necessary for their trouble-free operation.

2 Materials and methods of the study

The study of AM operating modes can be effective if computer simulation with virtual model implemented in a visual and effective simulation tool Simulink of interactive programming environment Matlab for T-shaped induction motor equivalent circuit [5,6] are used.

Simulation scheme for studying operation modes of asynchronous squirrel-cage motor, created by means of simulation in the Matlab/Simulink environment, is shown in (Fig.1). Using developed computer model, modes of squirrel-cage AM were studied taking into account the following:

– value of the motor shaft load varied from 0 to 120 % with 20 % increments at different K_{2U} values,

– value of reverse sequence voltage asymmetry coefficient was from 0 to 5 % with 1% increments.

Study of series 4A and AI AM with a nominal motor speed n=1500 rpm, with a significant power spread from 5.5 kW to 200 kW, was carried out. As an example, asynchronous squirrel-cage motors of 4A series (4A132S4Y3 with Pn = 7.5 kW; 4A250S4Y3 with Pn = 75 kW; 4A315M4Y3 with Pn = 200 kW) and of AI series (AIR132S4 with Pn = 7.5 kW; AIR250S4 with Pn = 75 kW; AIR355M6 with Pn = 200 kW) were selected [7,8].

https://doi.org/10.1051/e3sconf/201913901053

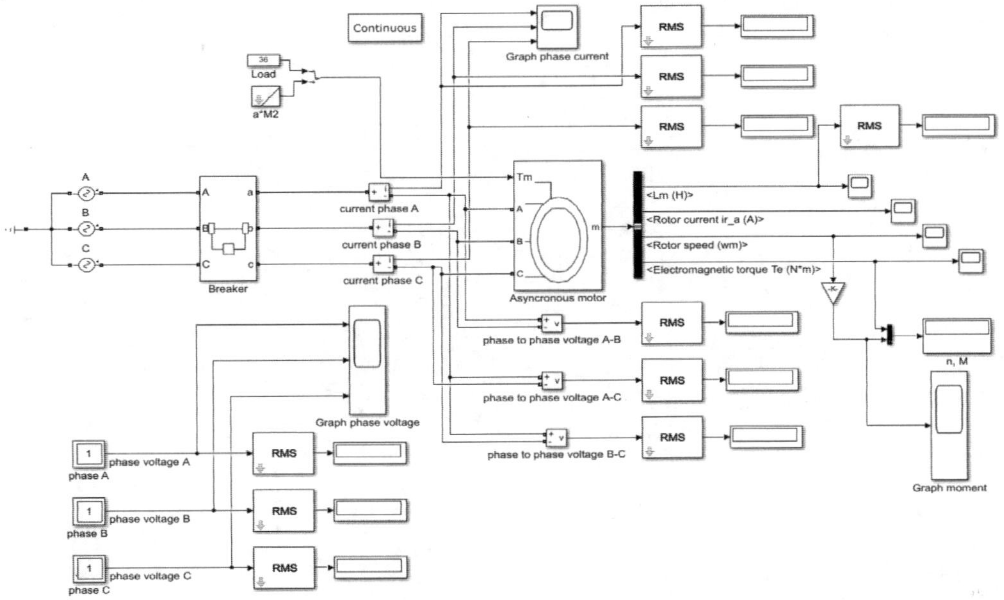

Fig.1. Simulation scheme implemented in Matlab/Simulink environment.

On the basis of the carried-out study, for each phase of the AM under consideration, dependences of currents versus K_{2U} were plotted for various values of motor shaft loads. Main types of current change in "A" phase for series 4A AM, AI with Pn = 7.5; 75; 200 kW, at different values of K_{2U} and different values of the motor shaft load are shown in (Fig.2-4). It should be noted that the control value of permissible overload current equal to 1.1 of the nominal current is taken into account when performing the studies [9,10].

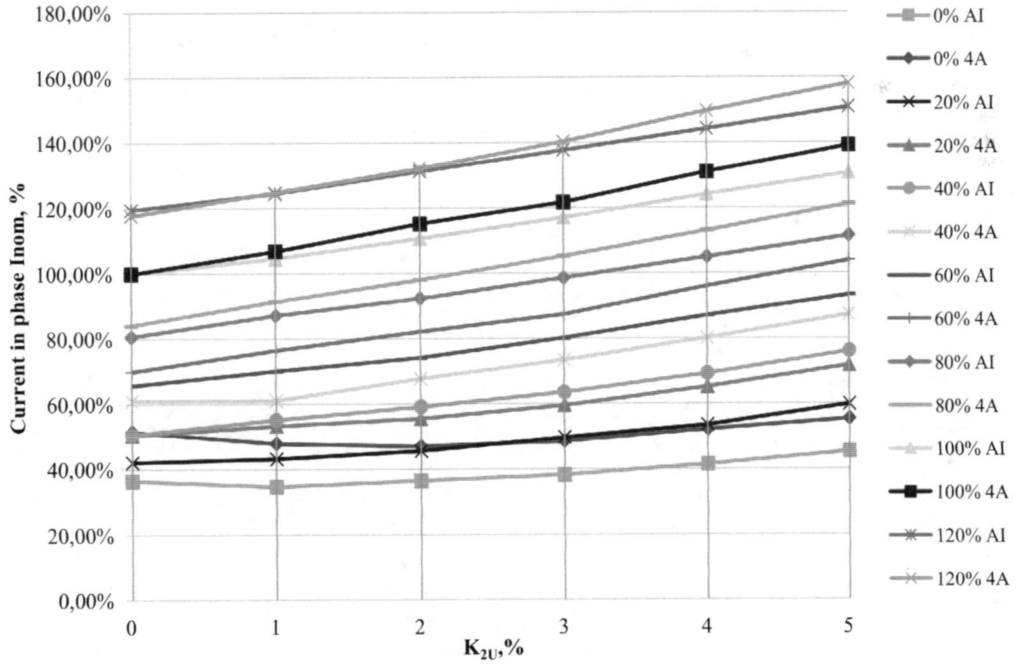

Fig.2. Curve of current in phase "A" AM series 4A, AI with Pn = 7.5 kW from K_{2U} under different permanent load factors.

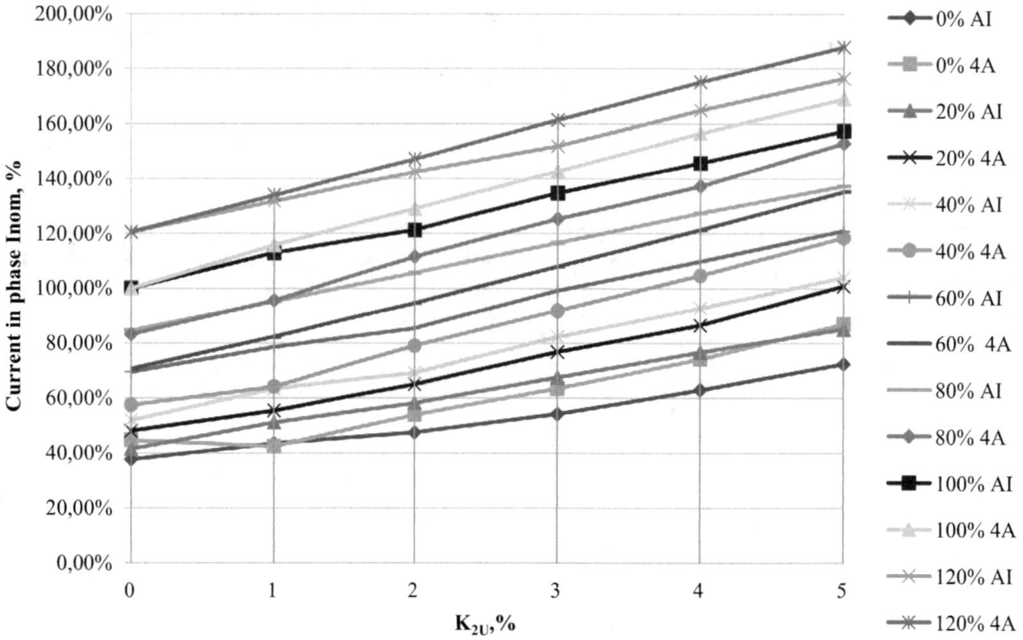

Fig.3. Curve of current in phase "A" AM series 4A, AI with Pn = 75 kW from K_{2U} under different permanent load factors.

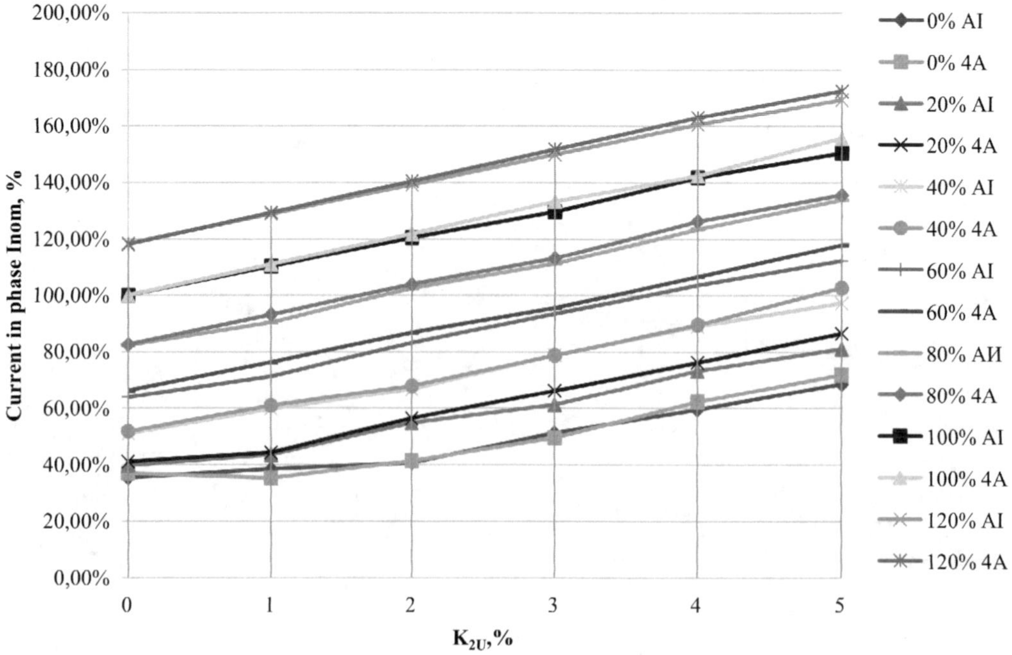

Fig.4. Curve of current in phase "A" AM series 4A, AI with Pn = 200 kW from K_{2U} under different permanent load factors.

3 Results of the study and discussion

On the basis of the results obtained in the course of computer simulation and using linear approximation of the data, analytical dependence of the AM current in the most loaded phase depending on the factors - reverse sequence voltage asymmetry coefficient (K_{2U}), load factor (Kz), type of series and the electric motor power - was derived.

The AM current dependence was defined by formula:

– for 4A series motor:

$$I\,\%\text{ of } I\text{ nom} = ((-0.0002P + 0.163)\,K_{2U} + (0.8126\,Kz + 0.081))*100, \%$$

– for AI series motor:

$$I\,\%\text{ of } I\text{ nom} = ((-0.0003P + 0.1535)\,K_{2U} + (0.8126\,Kz + 0.081))*100, \%$$

The conducted study allowed establishing the following regularities:

1) at supply voltage asymmetry, following indicators influence permissible AM parameters: power, load factor (Kz) and coefficient of negative sequence voltage asymmetry (K_{2U}),

2) permissible AM operation parameters (at normal AM operation mode) are identified based on overload capacity.

Table 1. Permissible values of load factor AM series 4A, AI at different values K_{2U}.

Permissible Kz AM series 4A, AI at different K_{2U}	Motor power capacity, kW		
	7.5	75	200
Kz 4A, of K_{2U} = 0%	110%	110%	110%
Kz AI, of K_{2U} = 0%	110%	110%	110%
Kz 4A, of K_{2U} = 1%	90%	92%	95%
Kz AI, of K_{2U} = 1%	91%	93%	96%
Kz 4A, of K_{2U} = 2%	70%	74%	80%
Kz AI, of K_{2U} = 2%	73%	76%	82%
Kz 4A, of K_{2U} = 3%	50%	55%	65%
Kz AI, of K_{2U} = 3%	54%	59%	68%
Kz 4A, of K_{2U} = 4%	31%	37%	49%
Kz AI, of K_{2U} = 4%	35%	42%	54%
Kz 4A, of K_{2U} = 5%	11%	19%	34%
Kz AI, of K_{2U} =5%	16%	25%	40%

3) the permissible value of K_{2U} is determined, when the engine is fully loaded

Table 2. Permissible values of coefficient asymmetry on the reverse sequence of AM series 4A, AI.

Valid K_{2U} AM series 4A, AI	Motor power capacity, kW		
	7.5	75	200
K_{2U} 4A	1.28	1.39	1.68
K_{2U} AI	1.36	1.58	2.21

Thus, permissible values of K_{2U} are different not only for different AM series, but they also vary within the motor series, and they directly dependent on the motor power.

In accordance with standards GOST 32144–2013, permissible AM operating parameters are:

a) at K_{2U} from 0% to 2%, range of normal operating modes of 4A series AM is Kz from 0 % to 70%, range of normal operating modes of AI series AM is Kz from 0 % to 80%,

b) at K_{2U} from 2% to 4%, range of normal operating modes of 4A series AM is Kz from 0 % to 31%, the range of normal operating modes of AI series AM is Kz from 0 % to 35%.

In the case of long-term operation, normal AM operation at K_{2U} = 2% is possible with reduced load:

- for 4A series motor with Pn = 7.5 kW - to 70%, for AI series motor with Pn = 7.5 kW - to 73%,
- for 4A series motor with Pn = 75 kW - to 74%, for AI series motor with Pn = 75 kW - to 77%,
- for 4A series motor with Pn = 200 kW - to 80%, for AI series motor with Pn = 200 kW - to 82%.

4) permissible K_{2U} range (in the case of AM normal operation) at full load of the motor is defined as follows:

- for 4A series motor with Pn = 7.5 kW - allowable K_{2U} range is from 0 to 1.28%, for AI series motor with Pn = 7.5 kW - allowable K_{2U} range is from 0 to 1.36%,
- for 4A series motor with Pn = 75 kW - allowable K_{2U} range is from 0 to 1.39%, for AI series motor with Pn = 75 кВт - allowable K_{2U} range is from 0 to 1.58%,
- for 4A series motor with Pn = 200 kW - allowable K_{2U} range is from 0 to 1.68%, for AI series motor with Pn = 200 kW - allowable K_{2U} range is from 0 to 2.21%.

As a result of the study on influence of main factors on AM operation mode, it should be noted:

1) regulation of K_{2U} should be performed based on the permissible operating modes of motors and their commercially available models. It was found that, at long-term or repetitive voltage asymmetry, technical balancing means with output asymmetry of not more than 1.3% should be provided for AM power supply sources even if K_{2U} value complies with requirements of GOST 32144 – 2013 [11],

2) necessity of optimal solution to change AM load or application of input technical balancing means in

power supply system is defined primarily by following factors: technical and economic analysis of the solution feasibility; nature of asymmetry in the network; features of the production process in the enterprise. Based on the obtained data on permissible Kz value, technical balancing means should be applied for AM when K_{2U} = 1.3% is reached; if K_{2U} exceeds this value, it is necessary to reduce the AM load. So, for example for 4A series AM with Pn = 7.5 kW at K_{2U} = 3%, it is necessary to reduce the load by 50%; for 4A series AM with Pn = 200 kW at K_{2U} = 3% - by 35%. When the motor load is reduced, range of normal operating mode is doubled. If the technological process of the enterprise and drive mechanisms connected to the AM care able reduce the load, this method can be more cost-effective in the short production term than installation of technical balancing means or purchasing of more powerful AM,

3) in the AM operating conditions, when power supply network containing electric motor equipment is designing, it is necessary to take into account influence of permissible K_{2U}, Kz values. This will increase reliability and efficiency of asynchronous motors and power supply systems in general.

Conclusion

By means of computer simulation experiments, using Matlab software package and Simulink package, permissible operating parameters of asynchronous motors were defined. Quantitative evaluation of main factors influencing AM operation mode is shown with examples. This study results can be used to solve problems of power quality management optimization at enterprises.

The work was executed within the framework of FSBEI HPE Trans-Baikal State University Council for Research and Innovative Activities Scientific Grant No. 276-GR of 01/02/2019 implementation.

References

1. N. Kuznetsov, Reliability of electric machines. – M.: Publishing house MEI, **432** (2006).
2. N. Ermolin, I. Zherikhin, Reliability of electric machines. – L.: Energy, **248** (1976).
3. I. Zhezhelenko, Yu. Sayenko, A. Gorpinich, I. Shvetsova, Priazovsk National Technical University Reporter: Collection of scientific articles / PNTU. – Mariupol Issue 15, P.2., 25 – 29 (2005).
4. M. Sorkind, Electric Engineering News. Information and Reference Publishing House No. 2 (32), (2005).
5. S. German-Galkin, G. Kardonov, Electrical machines: Laboratory works with PC. – SBR: Korona-print, **256** (2003).
6. I. Chernykh, Simulation of electrical devices using MATLAB, SimPowerSystems and Simulink. – SBR: Piter M.: DMK Press, **288** (2008).
7. http:// principact.ru/content/view/166/72/
8. http://www.elektrikii.ru/publ/6-1-0-55
9. www.bmrzzakharov.narod.ru/raschet/overload.html
10. V. Kogorodsky, S. Kuzhekov, L. Paperno, Relay protection of motors with voltage above 1 kV. – M.: Energoatomizdat, **248** (1987).
11. GOST 32144 – 2013. Electrical power. Electromagnetic compatibility of technical equipment. Electrical power quality standards in the general-purpose power supply systems. – M.: Standartinform, **19** (2014).

Self-oscillations at the frequency of subharmonics in nonlinear electric chains and systems

M.I. Ibadullaev[1], *A.N.* Tovbaev[2], *A.Zh.* Esenbekov[3]

[1]Tashkent State Technical University named after Islam Karimov, 100095, Uzbekistan, Tashkent, University St. 2A
[2]Navoi state mining institute
[3]Karakalpak State University, 230100, Uzbekistan, Nukus, Ch. Abdirov St.1

Abstract. It is known that the occurrence and existence of autoparametric oscillations (AIC) at the subharmonic frequency (GHC) in power lines (power lines) and in power supply systems is extremely undesirable, since they cause ferroresonant overvoltages at different frequencies. At the same time, there is an extensive class of nonlinear electric circuits in which the excitation of the AIC at the frequency of the SGC forms the basis of frequency-converting devices serving as secondary power sources. It is shown that single-phase-three-phase nonlinear systems are, to one degree or another, equivalent circuits of power lines, the main elements of which are: longitudinal compensation capacitors, transverse compensation reactors, and transformers with non-linear characteristics. The regularities of the excitation of the GCC at the frequency (ω / 3) of the power lines were studied, theoretical and experimental studies of the equivalent model of single-phase-three-phase circuits with nonlinear inductance were carried out. For a theoretical analysis of the steady-state mode of SGK at a frequency (ω / 3) with inductive coupling, the frequency-energy approach is used. The conditions of existence and critical parameters of the circuit are determined, and the mechanism of the appearance of the SGC at the frequency (ω / 3) is also studied.

1 Introduction

Electric systems in large quantities contain elements with significant inductance (generators, transformers, reactors, etc.). On the other hand, power lines have capacities relative to the ground and between phases. Often, to control the voltage and increase the stability of parallel operation, additional capacities relative to the ground are included in the line cut (Fig. 1).

Combinations of such inductances and capacitances create a number of complex oscillatory circuits in the circuit of the electrical system. In normal operation, the capacitance and inductance systems of these circuits are shunted by the load or connected directly to the terminals of a powerful source so that free oscillations cannot develop in them.

With various commutations in the system, part of the oscillatory circuits can be determined and energetic oscillations develop in them, leading to significant overvoltages [1-7,10-12]. These overvoltages are resonant in nature and, in principle, can have a very significant duration. Resonant overvoltage is one of the most difficult to analyze types of overvoltages, since in most cases there are steel cores, the magnetization characteristics of which are nonlinear.

Recently, the study of nonlinear oscillations in electrical systems has been developed, since the presence of longitudinal compensation capacitors and chokes with steel cores in the power lines open up wide possibilities for the formation of various oscillatory circuits and highlights resonant overvoltages [1-7,10,11].

2 Subharmonic resonance in power lines (power lines)

In practice, longitudinal capacitive compensation is used to increase line throughput. Capacities are switched on longitudinally at one or several points of the line. Usually capacitors are installed on one of the switching points of the line.

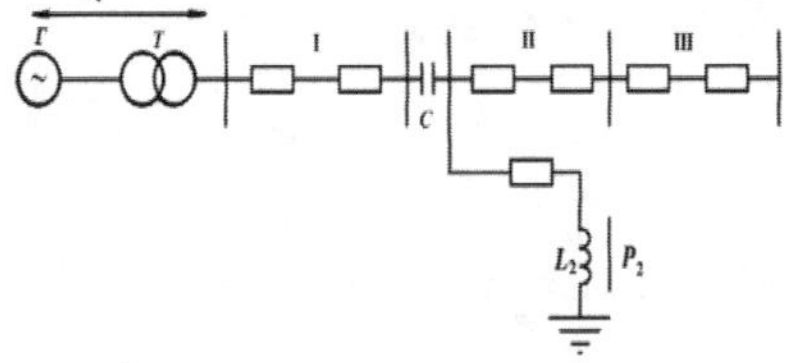

Fig. 1. Scheme of a power line with longitudinal compensation.

Suppose a transmission break occurred in section II so that only section I, compensation capacitance C, and reactor P remained in operation. The equivalent circuit is shown in Fig.2a.

a)

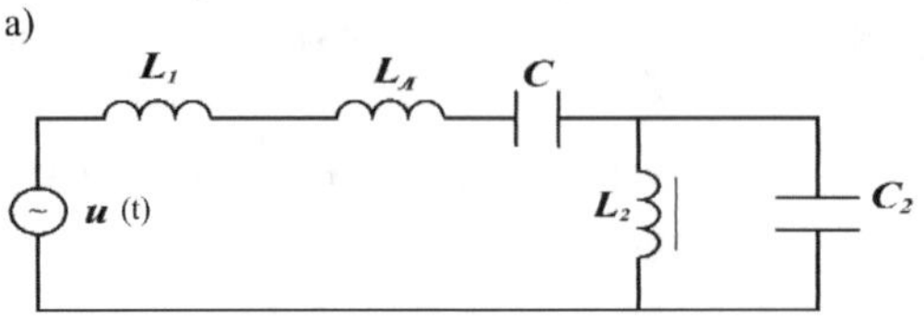

b)

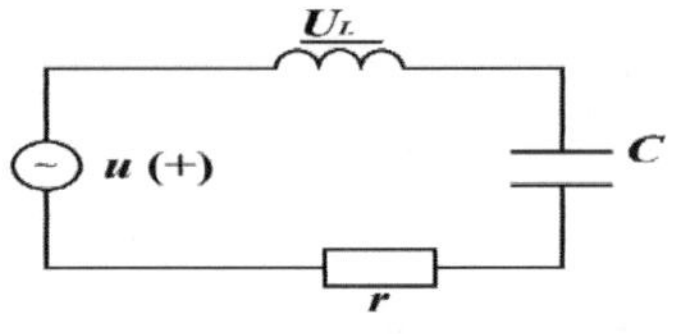

Fig. 2. Equivalent equivalent circuit.

The natural frequency of the circuit, consisting of C and L_2, is: $\omega_0 = \frac{\omega}{\sqrt{X_L/X_C}}$;
where ω_0 is the network frequency, $X_L = \omega L_2$ is the inductive reactance of the reactor, $X_C = \frac{1}{\omega C}$ is the capacitive resistance of the longitudinal compensation.
If the reactor power is $Q_P = 0{,}2$, then the lines $\left(\frac{X_1}{X_L} \approx 5\right)$, and the capacitor resistance is $X_C \approx \frac{Z_L}{3}$,, then $\frac{\omega}{\omega_0} = \frac{1}{\sqrt{3}} < \frac{1}{3}$. Therefore, in the oscillatory circuit L, C, subharmonic oscillations (SHO) with a frequency of $\left(\frac{1}{3}\right)$ are possible. The line capacitance is much smaller than the longitudinal compensation capacitance, therefore, for a subharmonic current, it represents a very large resistance and should not have a significant effect on the development of SHO.

The analysis of overvoltages in the case under consideration can well be carried out using a single-frequency equivalent circuit, converting it to the form in Fig. 2b, where the line and source inductance are sequentially combined with the nonlinear reactor inductance in the form of the equivalent inductance L(i), C is the longitudinal compensation capacity and R - active resistance equivalent inductance.

3 Analysis of the steady state SHO mode at a frequency $\left(\frac{\omega}{3}\right)$.

To study the patterns of excitation of subharmonic oscillations of power lines, theoretical and experimental studies of the equivalent model in single-phase-three-phase circuits with non-linear inductance were carried out (Fig. 2b).

The frequency-energy approach was used to analyze the steady-state regime of the SGC at a frequency $\left(\frac{\omega}{3}\right)$ with inductive coupling [8, 9, 10].

Subsequently, the energy equilibrium equations are compiled for the frequency divider and the condition is imposed that the ferromagnetic element converts current energy with a frequency of 3ω into current energy with a frequency of ω.

Consider the stea(t) ate excitation of third-order subharmonic oscilla $\left(\frac{\omega}{3}\right)$ in a single-phase ferroresonant circuit using energy ratios with the aim of quantitative and qualitative assessment of electromagnetic processes.

The core magnetization characteristic can be approximated by a cubic polynomial:

$$\psi = ai - bi^3 \quad (1)$$

We take a certain point as a base point on the dependence $\psi = f(i)$ of inductance. The corresponding flux linkage value will be called basic $\psi_Б$, current $i_Б$. The dependence curve $\psi = f(i)$ is depicted in relative units. Denoting: $\psi_Б = \sqrt{\frac{a^3}{b}}$; $i_Б = \frac{\psi_Б}{a}$; $i_Б^3 = \frac{\psi_Б}{b}$; or $i_Б = \sqrt[3]{\frac{\psi_Б}{b}}$ the magnetization curve in relative units from (1), we obtain the current-voltage characteristic of the ferromagnetic element in relative units, we present in the form: $\psi = i - i^3$.

Given the law of current change:

$$i = I_1\cos(\omega t + \varphi_1) + I_3\cos(3\omega t + \varphi_3) \quad (2)$$

Then:

$$\Psi = i - i^3 = A_1\cos(\omega t + \varphi_1) + B_1\cos(3\omega t + \varphi_3) -$$
$$-A_2 sin(\omega t + \varphi_1) + B_2\, sin(3\omega t + \varphi_3), \quad (3)$$

Where:

$A_1 = I_1 - \frac{3}{4}\cdot I_1^3 - \frac{3}{2}\cdot I_1\cdot I_3^2 - \frac{3}{4}\cdot I_1^2\cdot I_3\cdot cos\cdot(3\varphi_1 - \varphi_3)$

$B_1 = I_3 - \frac{3}{4}\cdot I_3^3 - \frac{3}{2}\cdot I_1^2\cdot I_3 - \frac{1}{4}\cdot I_1^3\cdot cos(3\varphi_1 - \varphi_3)$

$A_2 = \frac{3}{4}\cdot I_1^2\cdot I_3\cdot sin(3\varphi_1 - \varphi_3)$;

$B_2 = \frac{1}{4}\cdot I_1^3\cdot sin(3\varphi_1 - \varphi_3)$

Inductance Voltage:

$$U_L = \frac{d\Psi}{dt} = -\omega A_1 sin(\omega t + \varphi_1) - \omega A_2 cos(\omega t + \varphi_1) -$$
$$-3\omega_1 B_1 sin(3\omega t + +\varphi_3) - 3\omega B_2(3\omega t + \varphi_3) \quad (4)$$

Replace currents and voltage with their complex images:

$\dot{I}_1 = j\dot{I}_1\cdot e^{j\varphi_1}$; $\dot{I}_3 = j\dot{I}_3\cdot e^{j\varphi_3}$

$$\dot{U}_1 = -\omega\cdot A_1\cdot e^{j\varphi_1} - j\omega\cdot A_2\cdot e^{j\varphi_1} = j\omega(-A_2 + jA_1)e^{j\varphi_1}$$
$$(5)$$
$$\dot{U}_3 = -3\omega\cdot B_1\cdot e^{j\varphi_3} - j3\omega\cdot B_2\cdot e^{j\varphi_3} = -j3\omega(B_2 + jB_1)e^{j\varphi_3}$$

Integrated capacities entering the inductance:

$$\dot{S}_1 = \frac{1}{2}\dot{U}_1 = \frac{1}{2}\omega I_1^*(-A_2 + jA_1);$$
$$\dot{S}_3 = \frac{1}{2}\dot{U}_3 I_3^* = \frac{1}{2}\omega \dot{I}_3(B_2 - jB_1) \quad (6)$$

or: $P_1 = -\frac{1}{2}\omega I_1 A_2 = -\frac{3}{8}\omega I_1^3 I_3 sin(3\varphi_1 - \varphi_3)$ (7a)

$Q_1 = \frac{1}{2}\omega I_1 A_1 = \frac{1}{2}\omega I_1^2 \left[1 - \frac{3}{4}I_1^2 - \frac{3}{2}I_3^2 - \frac{3}{4}I_1 I_3 cos(3\varphi_1 - \varphi_3) I_6\right]$ (7b)

$P_3 = \frac{3}{2}\omega I_3 B_2 = \frac{3}{8}\omega I_1^3 I_3 sin(3\psi_1 - \varphi_3) = -P_1$ (7c)

$Q_3 = \frac{3}{2} I_3 B_1 = \frac{3}{2}\omega I_3 \left[I_3 - \frac{3}{4}I_3^2 - \frac{3}{2}I_1^2 I_3 - \frac{1}{4}I_1^3 cos(3\psi_1 - \varphi_3)\right]$ (7d)

Introducing the notation: $3\varphi_1 - \varphi_3 = \chi$ we rewrite equation (7):

$P_1 = -\frac{3}{8}\omega I_1^3 I_3 \sin\chi = -a\sin\chi = -P_3$ (8a)

$Q_1 = \frac{1}{2}\omega I_1^2\left(1 - \frac{3}{4}I_1^2 - \frac{3}{2}I_3^2\right) - \frac{3}{8}\omega I_1^3 I_3 \cos\chi = b_1 - a\cos\chi$ (8b)

$Q_3 = \frac{3}{2}\omega I_3^2\left(1 - \frac{3}{4}I_3^2 - \frac{3}{2}I_1^2\right) - \frac{3}{8}\omega I_1^3 I_3 \cos\chi = b_3 - a\cos\chi$ (8c)

It follows from (8a) that $P_1 = -P_3$, this confirms the possibility of using a nonlinear inductance as a frequency divider $\left(\frac{\omega}{3}\right)$. Moreover, the division of the frequency ω by $\left(\frac{\omega}{3}\right)$ will be at $P_1 < 0$; $P_3 > 0$ ($\chi = 0 \div 180^0$), and the multiplication of the frequency ω by 3ω, respectively, for:

$$P_1 > 0;\ P_3 < 0\ \ (\chi = 0 \div -180^0).$$

By eliminating χ from (8a) and (8b) we have:

$P_1^2 + (Q_1 - b_1)^2 - a^2 = 0$ (9)

By squaring equation (10) and adding, we get

$K_p^2 + (K_q - b_1')^2 - a'^2 = 0$ (11)

Transforming equation (11) taking into account (9) and (10), we obtain:

$36\frac{I_3^4}{(1-K_q)^2} + 27\frac{I_1^2 I_3^2}{(1-K_q)^2} + 9\frac{I_1^4}{(1-K_q)^2} - 48\frac{I_3^2}{(1-K_q)} - 24\frac{I_1^2}{(1-K_q)} + 16\left[\frac{K_p^2}{(1-K_q)^2} + 1\right] = 0$ (12)

Expression (12) is an equation of second-order curves (ellipse), which describes the relationship between the squares of the amplitude of the current of the fundamental harmonic i_3^2 and subharmonic i_1^2, i.e. it expresses a curve whose invariants are equal to:

$$\delta = \frac{567}{y} > 0\ \ S=45;\ \Delta = 324\,[-8 + 7(k^2+1)],$$

where: $k = \frac{k_p}{1-k_q}$ (13)

It is known that $\Delta S = 324 \cdot 45[-8 + 7(k^2+1)] \le 0.$ (14)

We have equations of real ellipses with coordinates $k=0.378$; $x_0 = \frac{8}{21} = 0{,}38$; $y_0 = \frac{16}{21} = 0{,}762$ with an ellipse slope $\alpha = 22^0 30'$.

From Fig.3b, the upper parts of the ellipses physically correspond to the region of existence of subharmonic oscillations at a frequency $\left(\frac{\omega}{3}\right)$.

Inequality (14) allows us to determine the parameters of the frequency divider at which stable frequency-dependent SHO can exist in the frequency converting devices at the frequency $\left(\frac{\omega}{3}\right)$. The direct AS separates the region of stable and unstable (dashed lines) the existence of SHO.

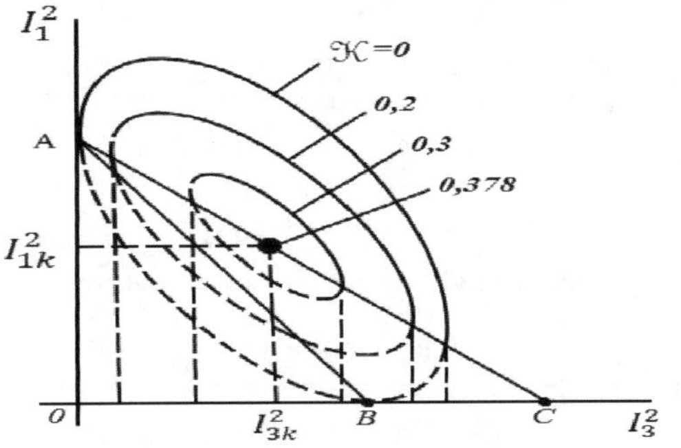

Fig. 3. Graphic representation of the input-output characteristics of the SHO $\left(\frac{\omega}{3}\right)$.

It should be noted that the determinant - Δ is a function of the parameters of the frequency divider circuit, within which the ferromagnetic element can convert the energy of a current of frequency 3ω coming to it from the network into current energy with a frequency of ω. $\Delta \le 0$ corresponds to the steady-state mode of excitation of the SHO, and $\Delta > 0$ corresponds to an imaginary ellipse: physically in this case they cannot exist in the GHS scheme, i.e. Because of the imbalance in the power of conversion and scattering in the system, the SHO breaks down.

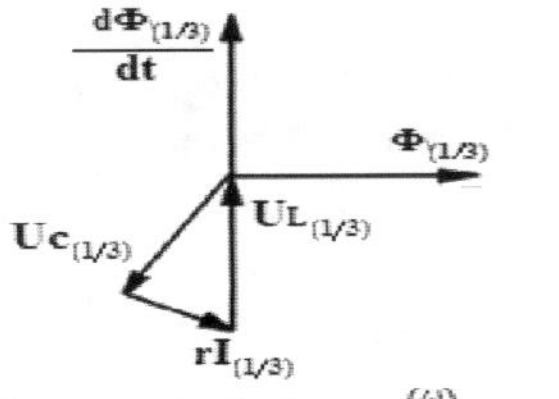

Fig. 4. Vector diagram in the SGK mode $\left(\frac{\omega}{3}\right)$.

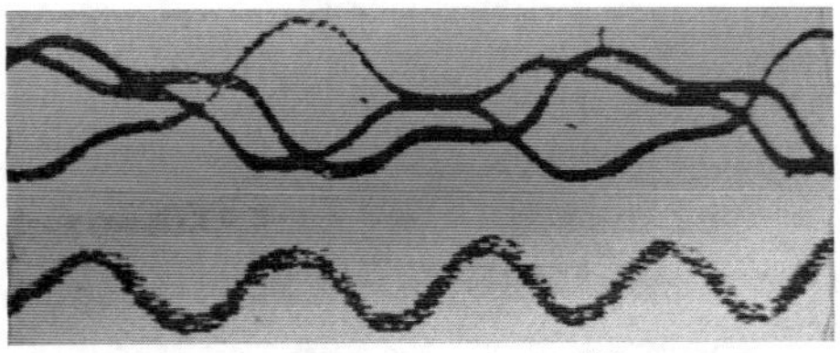

Fig. 5. Oscillograms of the SGK curves $\left(\frac{\omega}{3}\right)$ in three-phase circuits.

For the physical justification of the excitation of subharmonic oscillations according to the results of experimental and theoretical studies [8, 10, 12], the following conclusion can be made:

1. If a purely sinusoidal voltage with a frequency ω is applied to the nonlinear oscillatory circuit, then the SHO can arise in the circuit with a period equal to $\left(\frac{n}{m}\right)\omega$, where n,m are integers and n<m. Most often, oscillations occur with a frequency $\left(\frac{\omega}{m}\right)$, where m is an odd number.

The fact is that the energy for the development of the SHO comes from the network to the beat with fluctuations. It follows that the frequency of the SHO should be in a simple ratio with the frequency of the voltage source.

2. SHO is generated by the circuit itself, since they are not contained in the applied voltage. For the stable existence of the SHO with a certain frequency$\left(\frac{\omega}{m}\right)$, the natural frequency of the circuit should be equal to $\left(\frac{\omega}{m}\right)$. Therefore, in the presence of SHO, the resulting flow acquires such a value that provides the desired natural frequency of oscillations.

3. Since the voltage source does not contain subharmonics, the sum of the subharmonic components of the voltage drop in the coil $\left(U_{L\left(\frac{1}{3}\right)}\right)$, resistance $\left(I_{\left(\frac{1}{3}\right)} \cdot R\right)$ and capacitance $\left(U_{C\left(\frac{1}{3}\right)}\right)$ should be equal to zero, the vector diagram of which is shown in Fig. 4. From the vector diagram it can be seen that this is possible only if the subharmonic component of the current lags behind the subharmonic component of the flow in the coil by an acute angle. Consequently, SHO is excited at relatively lower resistances. Depending on the magnetization curve, subharmonics of odd orders may appear in the circuit under consideration, and subharmonics of even order may appear in systems with magnetization.

4. In contrast to single-phase electro-ferromagnetic circuits in three-phase systems, the higher and lower harmonics create direct, reverse and zero sequence vectors. In general, the phase shifts of the third-order SHO $\frac{\omega}{3}$ between phases in three-phase systems can be represented in three versions: 0^0, 40^0, 80^0; 0^0, 80^0, 160^0; 0^0, 160^0, 320^0, 0^0 (Fig. 5).

5. If phase shifts or phase ratios are measured with respect to the GHS (ω / 3), then the first and third variants of the GHS phase shifts in a three-phase system correspond to a direct phase sequence, and the second variant is the reverse phase sequence. In the general case, phase shifts of n-order subharmonics in an m-phase system can be determined from the expression; $\varphi_k = \frac{2\pi}{mn}(k-1)$, where k is the phase order. Therefore, with symmetric systems, SGCs can create various combinations of asymmetric systems. Depending on the structure and connection diagram of three-phase systems, only higher and lower harmonics can appear in phase, linear and neutral wires.

References

1. Evdokunin T.A., Titenkov S.S. Ferroresonant overvoltages with asymmetrical switching on of power transformers 6-10 kV. // Izv. Academy of Sciences, Energy, 2001, No. 6. p. 67–75.

2. Selivanov V.N. Equivalent network circuit with isolated neutral with protection against ferroresonance. // Izv. Academy of Sciences, Energy, 2001, No. 1. p. 94–99.

3. Vasin V.P. Low-frequency oscillations in extended power systems with variable dynamic parameters. // Izv. RAS, Energy, 1998, No. 3. p. 87–96.

4. Maslennikov V.A. Damping of low-frequency oscillations in power systems. // Izv. Universities, Energy, 1995, No. 5-6, p. 9-14.

5. Kadomskaya K.P., Levkovsky A.I. Ferroresonant overvoltage and protection against them during switching of the autotransformer block 750/500 kV - idle 500 kV, 2003, No. 2, Power plants, p. 61–65.

6. High voltage technique. /Under. Ed. Razevich D.V., M .: Energy, 1976, p. 325.

7. Bazutkin V.V., Kadomskaya K.P., Kostenko M.V., Mikhailov Yu.A. Overvoltage in electrical systems and protection against them. SPb. Energoatomizdat, 2010, p. 275.

8. Tovboyev A.N., Nazarov A.I., Esenbekov A.Zh., Ibodullaev M.I. Frequency-energy relations in the analysis of autoparametric oscillations. Scientific and technical and production journal "Mountain Herald", Uzbekistan, 2017, No. 2, p. 165-170.

9. Bessonov L.A. Nonlinear electrical circuits. M.: Higher school, 1977, 264 p.

10. Ibodullaev M.I. Subharmonic oscillations in three-phase electro-ferromagnetic circuits without a neutral wire. Tashkent: Vestnik TashGTU, 1993, No. 1. p. 77–83.

11. Wright Jan A. Subharmonic oscillations in power systems. Discuss JEEE Trans Power Apper and System, 1970, No. 8. p. 61–65.

New approach to modeling the nonlinear loads

L.I. Kovernikova[1,2,*], *Van Chung* Luong[2]

[1]Melentiev Energy Systems Institute SB RAS, Irkutsk, Russian Federation
[2]Irkutsk National Research Technical University, Irkutsk, Vietnam

Abstract. Parameters of operating conditions of electric power systems shall ensure efficient and reliable operation of both power generating companies and of consumers, which to a great extent depends on power quality. Most urgent problem in this area is currently non-sinusoidal voltage. Power quality control, development of technical measures to sustain parameters of harmonic modes in conformity with provisions of GOST 32144-2013 require calculations that, in turn, need models adequately representing the electric networks and nonlinear loads in the computing program. Results of studies on parameters of harmonic modes that were obtained by measurements have shown that harmonic modes are random, and their parameters have probabilistic nature. Due to unpredictability of harmonic modes, nonlinear loads can be modeled using measured parameters. The paper presents methodological approach, algorithms and a computing program for modeling the nonlinear loads using parameters measured in the nodes of their connection to the power supply, and gives examples of modeling the nonlinear loads.

1 Introduction

Parameters of electric power systems conditions shall ensure efficient and reliable operation of both power generating companies that supply power to consumers, and of consumers, which to a great extent depends on power quality. Low power quality raises power losses in the electric networks, reduces service life of electric equipment, and efficiency of process equipment of industrial enterprises. Non-sinusoidal voltage is currently a challenge related to power quality [1]. This problem is urgent not only for Russia, but for other countries as well. The amount of nonlinear loads (sources of current harmonics) is rapidly and continuously growing. Levels of current and voltage harmonics in the electric networks grow as well.

For controlling the non-sinusoidal modes in the electric networks, for their analysis, for developing the technical measures to sustain the harmonic mode parameters in conformity with requirements [2], the harmonic modes need to be calculated. Calculations can be performed provided the models adequately representing electric networks and nonlinear loads are used for modeling the harmonic modes. Nonlinear load models have been developed for a long time. Both Russian and foreign scientists have made considerable contribution into the theory of modeling the nonlinear loads [3-5].

Results of studies on parameters of harmonic modes have shown that harmonic modes are random, and their parameters have probabilistic nature [6-8]. They depend on the network configuration, structure of its elements, on the amount of nonlinear loads, wave and frequency properties of the network, on values of phase angles of harmonic currents of nonlinear loads, on voltage values on the clamps of nonlinear equipment, on changes of operating modes and loads, and many other factors. Due to their unpredictability, the nonlinear loads can be modeled based on measurements of harmonics only.

Calculation of non-sinusoidal mode of an electric network is an aggregation of mode calculations for harmonics from 2 to 40 in conformity with [2]. In practice, the number of harmonics can be lower, which is determined in the result of analysis of measured values of current and voltage harmonics. Calculations at each harmonic are performed using a system of equations for nodal voltages

$$\boldsymbol{U}_n = \boldsymbol{Z}_n \boldsymbol{I}_n , \qquad (1)$$

where n – harmonic number, $\boldsymbol{U}_n$ – a column-matrix of values of nodal voltages to be determined, $\boldsymbol{Z}_n$ – a square matrix of self- and mutual impedances of the network nodes, $\boldsymbol{I}_n$ – a column-matrix of currents in the nodes that must be determined based on the results of measurements.

Each element of matrix $\boldsymbol{I}_n$ is a complex number $\dot{I}_{ni} = I_{ani} + jI_{rni}$, where i – row number of the matrix corresponding to the node number of the network; I_{an} – active current of n-th harmonic; I_{rn} – reactive current of n-th harmonic. Active and reactive currents are computed based on the measured parameters as

$$I_{an} = I_n \cos\varphi_{UI(n)} \qquad (2)$$

$$I_{rn} = I_n \sin\varphi_{UI(n)} \qquad (3)$$

where I_n – measured rms value of the n-th harmonic current; $\varphi_{UI(\mathrm{n})}$ – measured value of the phase angle between voltage and current of the n-th harmonic. Currents I_{an} and I_{rn} similarly to measured ones I_n and $\varphi_{UI(n)}$, are series of random values X: $x_1, x_2, \ldots, x_i, \ldots, x_m$, where m is the number of elements in the series. A model of nonlinear load is a set of distribution functions for currents I_{an} and I_{rn} that allow computation of values of active and reactive current harmonics of nonlinear loads with desired probability. For solving the system (1) in conformity with [2], probability of currents values shall be 95%.

Subject of the present paper is methodological approach to modeling the nonlinear loads that takes into account features of parameters of harmonic modes in the electric networks, an algorithm for modeling the distribution functions of active and reactive current harmonics, an algorithm of a computer program for identification of distribution functions of random values of current harmonics and computation of their values with the desired probability, it also gives examples of using the developed methodological approach and computer program.

2 Methodological approach to modeling the nonlinear loads

Methodological approach takes into account features of parameters of harmonic modes [6-9] and includes: 1) check of measured series of random values of harmonic mode parameters for availability of abnormal elements (rough outliers), whose values considerably exceed the values of the remaining elements; 2) analysis of measured parameters of harmonic modes for determination of time intervals characterized by different values of parameter; 3) analysis of correlation relations between current and voltage harmonics; 4) analysis of coefficients of harmonic components of voltages ($K_{U(\mathrm{n})}$) for selecting the harmonics for modeling; 5) analysis of directions of harmonic flows of active and reactive currents through a node connecting to the network of nonlinear load; 6) modeling the distribution functions of harmonics of active and reactive currents; 7) check of correctness of computed values of active and reactive currents with desired probability. The methodological approach is described further.

1) The series of measured parameters of the harmonic mode may contain elements whose values are many times different from the values of other elements (rough outliers). They distort the picture of the mode, do not allow constructing a real histogram and correctly determine the law of the distribution function. Outliers can be visually determined using scatter plot. When processing the results of measurements they are replaced with neighboring elements, a mean value of neighboring elements or by other ways proposed in [10-11].

2) Analysis of measured data demonstrated that the measured values of harmonic mode parameters notably differ from one another on some time intervals. Those differences are caused by changes in the network structure, in the parameters of its elements, in loads, etc., that occur during operation. In case of considerable differences in parameters on different time intervals the period of measurements shall be divided into intervals for which models of nonlinear loads should be developed.

3) Identification of availability or lack of correlation relationships between current and voltage harmonics allows use of appropriate mathematical methods during modeling. Lack of strong correlation relationship proves independence or weak dependence of values of voltage harmonics on current harmonics, i.e., evidences probabilistic nature of both parameters. Analysis of measured parameters of harmonic modes [7-8] confirmed that there is either no correlation between current and voltage harmonics or it is negligible. Lack of correlation or weak dependence is accounted for by the fact that measured parameters of harmonic modes are averaged values on some time interval. In [2] the averaging interval is 10 minutes.

4) Not all the harmonics out of the interval from 2 to 40 can be modeled, but only harmonics the value $K_{U(\mathrm{n})}$ of which exceeds limit values [2]. When selecting the harmonics for modeling, the measured values $K_{U(\mathrm{n})}$ should be compared to limit values for 95% and 100% of measurement time.

5) The values of current harmonics with numbers $n > 1$ are much smaller than the values of current harmonics for $n = 1$. Direction of their flow may change due to a lot of insignificant causes for the first harmonic. Therefore, for each time interval defined in Item 2) the directions of flow of active and reactive current harmonics through a node for connection of a nonlinear load to electric network shall be analyzed. Analysis can be performed using a phase angle between harmonics of voltage and current $\varphi_{UI(\mathrm{n})}$ [12]. It is assumed that the active current has a positive sign if it flows from the network to the load. For the reactive current, the same direction is taken as for the reactive power of the load, which is inductive. Active and reactive currents flow from the network to the load if angle $\varphi_{UI(\mathrm{n})}$ lies in the first quadrant of the complex plane. Active current flows from the load to the network, whereas reactive current flows from the network to the load if angle $\varphi_{UI(\mathrm{n})}$ lies in the second quadrant. Both currents flow from the load to the network if angle $\varphi_{UI(\mathrm{n})}$ lies in the third quadrant. Active current flows from the network to the load, whereas reactive current flows in the reverse direction if angle $\varphi_{UI(\mathrm{n})}$ lies in the fourth quadrant. Thus, intervals of modeling identified in Item 2) can be further divided into four time intervals (variants) corresponding to positions of angle $\varphi_{UI(\mathrm{n})}$ on the complex plane. Variants with prevailing number of measurements (out

of their total number of measurements) are preferable for modeling.

6) Identification of distribution functions of active and reactive current harmonics for computation with 95% probability or any other. Analysis of numerous measurements of current harmonics allows us to conclude that two different approaches (represented as two algorithms in Fig. 1) must be used for identification of distribution functions.

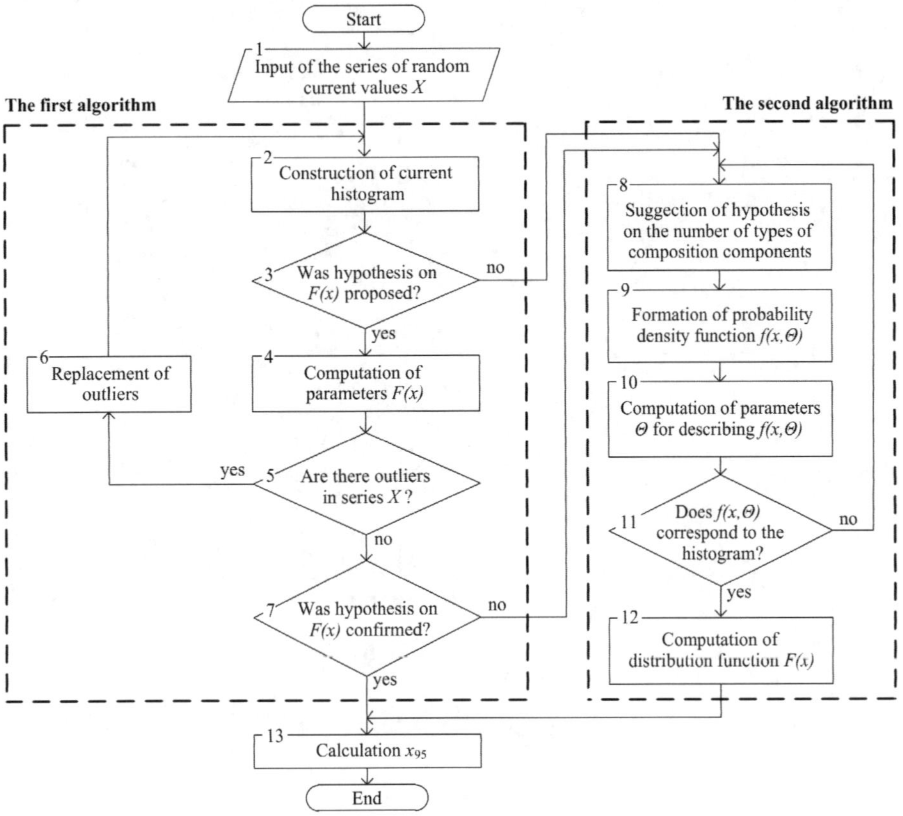

Fig. 1. Algorithms for modeling a current harmonics.

Notation: X – a series of measured random values of current harmonic, $F(x)$ – a distribution function, $f(x,\Theta)$ – probability density function, x_{95} – value of current harmonic with 95% probability.

The first algorithm (Blocks 2-7) is intended for identification of distribution functions of series X corresponding to well-known distribution laws. The second algorithm (Blocks 8-12) is intended for identification of distribution functions of series X corresponding to compositions of known distributions. Computer program has been developed on the base of algorithms presented.

7) After determination of I_{an} and I_{rn} the correctness of values obtained shall be checked. To this end, the current value I_n, is computed using I_{an} and I_{rn}, and it shall not go beyond the measured currents.

3 An algorithm of computer program for modeling the current harmonics

A diagram of the computational algorithm is given in Fig. 2. It was developed on the base of algorithms for modeling the current harmonics given in Fig. 1.

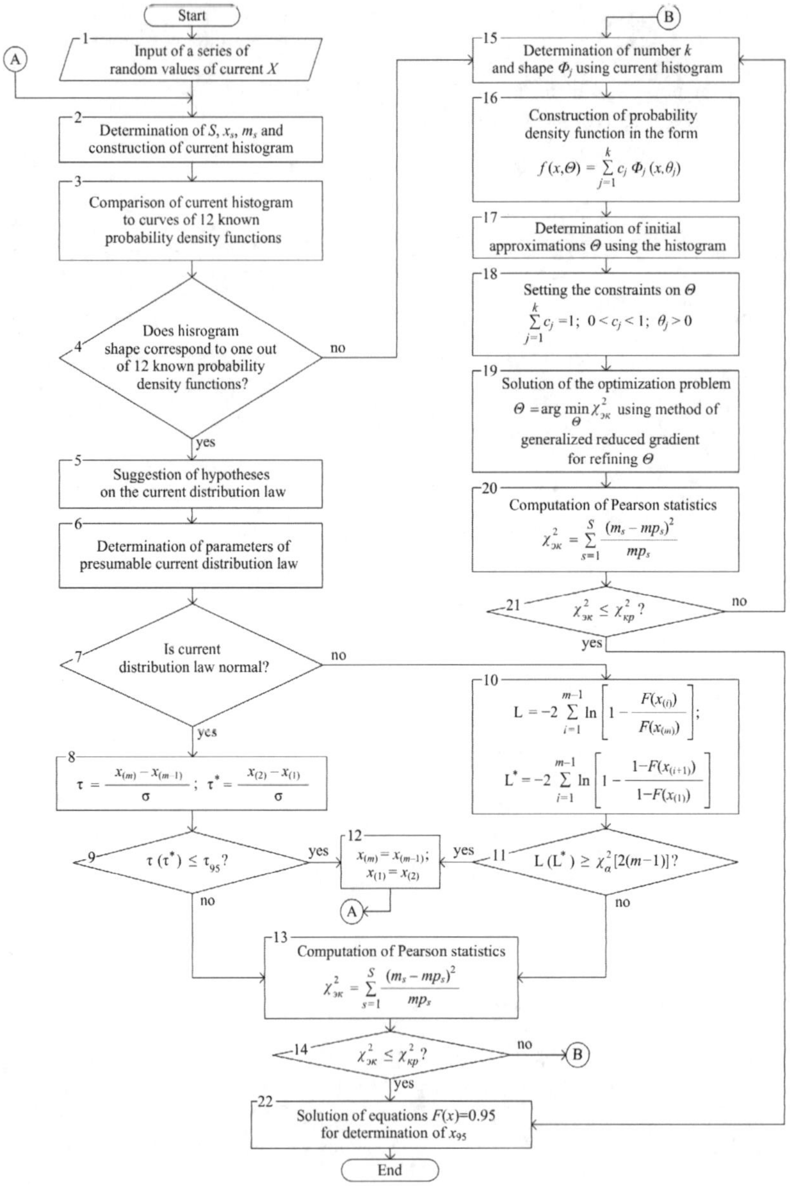

Fig. 2. Block-diagram of computer program.

Computer program contains the following blocks.

Block 1. Input to the program of a series of active and reactive current harmonics X: $x_1, x_2, \ldots, x_i, \ldots, x_m$, where m – the number of elements in the series.

Block 2. Determination of the number of intervals of the histogram (S); their boundary values (x_s); the number of values X in the series that occurred in the s-th interval (m_s); histogram construction.

Blocks 3-5. Visual comparison of the shape of the current histogram to the shape of curves of twelve probability density functions used in the computer program to hypothesize the law of current distribution. Twelve laws are proposed to be used in the procedure: Gaussian law, Weibull law, exponential law, gamma-distribution, Rayleigh law, logarithmically normal law, beta-distribution, minimum value distribution law, maximum value distribution law, logistic law, and Maxwell and Cauchy distribution law. If the shape of the histogram does not allow hypothesizing the distribution law then go to Block 15. If current distribution law has been hypothesized, the hypotheses shall be checked following the algorithm by going over to Block 6.

Block 6. Computation of parameters describing the proposed distribution law of the series X. For example, Gaussian distribution is described by two parameters:

mathematical expectation (μ), and standard deviation (σ).

Block 7. Selection of a criterion for checking the series *X* for availability of outliers depending on the presumable distribution law.

Methodological approach already contains Item 1) for detection and replacement of rough outliers. But in the series *X* can be elements that are not included into a compact group of the main mass of series elements and, at the same time, they are slightly removed from it. Such elements can be referred to as presumable outliers. They cannot be seen on the scatter plot. For detecting such outliers special statistical criteria shall be applied [10]. If presumable distribution law of the current is normal, then the availability of outliers can be checked using Irwin criterion, Von Neumann test, and others. If presumable law differs from normal one, then modification of Darling criterion is used [10].

Blocks 8-11. Check if there are the large x_m and very small x_1 values of the series *X* ordered in ascending order using the Irwin criterion (for that purpose statistics τ, τ^* are computed and compared to critical value τ_{95}) or using the Darling criterion modification (for that purpose statistics L, L^* are computed and compared to the value of the α-quantile of χ^2-distribution.

Block 12. Replacement of outliers by neighboring elements of the series *X* and then to Block 2, as replacement of outliers may change the shape of a histogram.

Blocks 13-14. Check of concordance of presumable current distribution with the histogram using Pearson goodness-of-fit criterion.

Blocks 15-21. Identification of probability density function of the series *X* using method of separating mixtures of probability distributions [13] that is used when distribution of series *X* cannot be described by known functions due to its complexity. The amount of components in the distribution composition and types of their probability density functions used in the computational procedure are hypothesized based on the histogram shape. Then the probability density function is formed as a weight sum of components of distribution composition

$$f(x,\Theta) = \sum_{j=1}^{k} c_j \Phi_j(x, \theta_j), \tag{4}$$

where $f(x,\Theta)$ – a desired function of the current probability density; $k \geq 1$ – a positive integer; Φ_j – a known function of probability density of the *j*-th component of the composition; $\Theta = (c_1,...,c_k,\theta_1,...,\theta_k)$ – a vector of parameters of composition components that are to be determined; c_j – weighting factor of the *j*-th component, $c_j > 0$, $j = 1, ..., k$, $c_1 + ... + c_k = 1$; θ_j – a vector of parameters of the *j*-th component of the composition.

The task of distribution description consists in determination of the vector Θ. Initial approximations of parameters of the vector Θ, determined in the course of visual analysis of the histogram are defined using the series *X*.

A histogram of current with two peaks is given in Fig. 3 as an example. A hypothesis is suggested that a histogram has two components whose probability density functions follow Gaussian distribution.

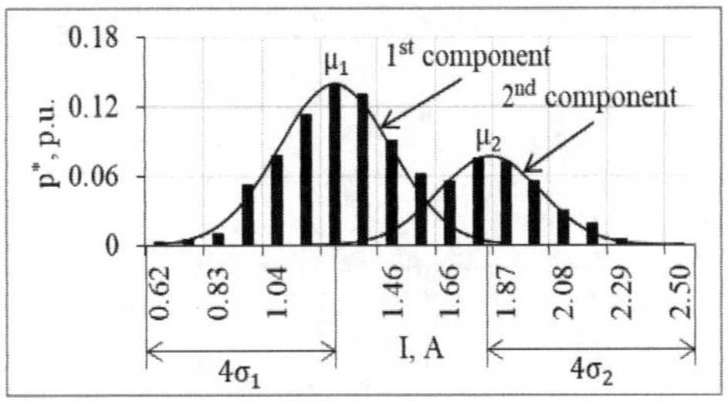

Fig. 3. A histogram of current with two peaks.

Vector θ_j consists of two parameters: $\theta_1 = (\mu_1, \sigma_1)$, $\theta_2 = (\mu_2, \sigma_2)$, and vector Θ consists of six parameters, i.e., $\Theta = (c_1, c_2, \mu_1, \mu_2, \sigma_1, \sigma_2)$. Initial values μ_1, σ_1, μ_2, σ_2 are determined from the histogram. For determination of c_1, c_2 the total area of the histogram under the curves of composition components is assumed to be equal to 1. Values c_1, c_2 are set equal to fractions of the areas under the curves of components.

At normal distribution the position of histogram peaks relative to the abscissa axis allows determination of mathematical expectation for each component. A half of the figure width under the curve of the component allows determination of standard deviation [14].

Further refinement of Θ is performed by solving the optimization problem

$$\Theta = \arg\min_{\Theta} \sum_{s=1}^{S} \frac{(m_s - mp_s)^2}{mp_s} \tag{5}$$

subject to constraints

$$g_1(\Theta) = \sum_{j=1}^{k} c_j - 1 = 0, \tag{6}$$

$$h_1(\Theta) = -c_j < 0, \tag{7}$$

$$h_2(\Theta) = c_j - 1 < 0, \tag{8}$$

$$h_3(\Theta) = -\theta_j < 0, \tag{9}$$

where p_s – theoretical probability of occurrence of random values of series *X* into *s*-th interval; mp_s – theoretical frequency of occurrence of random values of series *X* into the *s*-th interval.

Block 22. Computation with the given probability, e.g., 95%, of value of current harmonic by solving the equation $F(x)$=0.95. The modeling of one harmonic of active or reactive current of nonlinear load terminates with computation of value of the current harmonic. Computer program has been implemented in MS Excel and Visual Basic programming environment for Windows-applications [15].

4 An example of modeling the current harmonics at the nodes of traction transformers connection to the network

Railway substations with traction transformers get power from 110-220 kV public distribution network. Each substation has two 40 MWA transformers. One transformer is backup. 25kV winding of a transformer powers the electric-traction network that powers electric locomotives. Locomotives are operated by DC engines powered from single-phase two-pulse rectification circuit that originate current harmonics penetrating into the power network through transformers.

Modeling the 3-rd and 5-th harmonics of active and reactive currents at the nodes of traction transformers connection to the network at four substations of East-Siberia Railway (Mysovaya, Tataurovo, Zaigraevo, Novo-Il'insk) has shown that laws of harmonics distribution of active and reactive currents obey the known distribution laws: Gaussian law, exponential law, gamma-distribution, Weibull law, and have more complicated forms that are compositions of known distribution laws.

5 Conclusions

Methodological approach, algorithms of modeling, computer program algorithm, and computer program for modeling nonlinear loads using the measured parameters of harmonic modes in nodes connecting nonlinear loads to the electric network, which allow representing nonlinear loads as a set of distribution functions of current harmonics, has been developed.

Third and fifth current harmonics at the nodes of traction transformers connection to the network at substations of East-Siberia Railway were studied and modeled. The results have shown that the harmonics of the currents are distributed according to various laws: Gaussian law, exponential law, gamma distribution, Weibull law, and are also described by compositions of known distribution laws.

The research is conducted in the framework of the research projects № AAAA-A17-117030310432-9 and № AAAA-A17-117030310438-1 of the program of fundamental research of SB RAS III.17.4.

References

1. L.I. Kovernikova, V.V. Sudnova, R.G. Shamonov, et al., *Electric power quality: present-day state, problems and proposals to their solutions* (Nauka, 2017) (in Russian)
2. State standard 32144-2013. Electric energy. Electromagnetic compatibility of technical equipment. Power quality limits in public power supply systems (2014) (in Russian)
3. I.V. Zhezhelenko, *High harmonics in the power supply systems of industrial plants* (Energoatomizdat, 2000) (in Russian)
4. J.D. Greene, C.A. Gross, *Nonlinear Modeling of transformers*, IEEE Transactions on Industry Applications, Vol. **24**, No. **3**, 434-438 (1988)
5. Probabilistic aspects task force of the harmonics working group subcommittee of the transmission and distribution committee, *"Time-varying harmonics: Part I, characterizing measured data"*, IEEE Transactions on Power Delivery, Vol. **13**, No. **3**, 938-944 (1998)
6. E.W. Kazibwe, T.H. Ortmeyer, M.S.A.A. Hammam, *Summation of probabilistic harmonic vectors*, IEEE Transactions on Power Delivery, Vol. **4**, No. **1**, 621-628 (1989)
7. L.I. Kovernikova, *Some results of research into harmonics in the high voltage networks with distributed nonlinear loads*, Przeglad Elektrotexchniczny **11**, 239-243 (2013)
8. L.I. Kovernikova, *Some properties of harmonics mode parameters in electrical network with distributed nonlinear loads, Proceeding of International Conference "Power Quality Management"*, 101-108 (Moscow, 2014) (in Russian)
9. L.I. Kovernikova, Van Chung Luong, *Algorithm for high-voltage network nonlinear loads modeling on measured parameters*, Electrichestvo **10**, 16-25 (2017) (in Russian)
10. A.I. Kobzar, *Applied mathematical statistics. For engineers and researchers* (M.: FIZMATLIT, 2012) (in Russian)
11. B.Yu. Lemeshko, S.B. Lemeshko, S.N. Postovalov, E.V. Chimitova, *Statistical data analysis, simulation and study of probability regularities. Computer approach* (Novosibirsk: NSTU Publisher, 2011) (in Russian)
12. A.E. Emanuel, *Power definitions and physical mechanism of power flow* (John Wiley & Sons, 2010)
13. O.K. Isaenko, V.Yu. Urbakh, *Division of probability distribution mixtures into their components*, *Itogi nauki i techn. Ser. Probability Theory. Math. Stat. Theor Cybern.*, V. **13**, 37-58 (VINITI, M., 1976) (in Russian)
14. L. Zacks, *Statistical estimation* (Moscow, "Statistics", 1976) (in Russian)
15. State Registration Certificate of Computer Program No. 2019619614. Software for modeling the current harmonics of nonlinear loads using measured parameters. (in Russian)

E3S Web of Conferences **139**, 01056 (2019) https://doi.org/10.1051/e3sconf/201913901056
RSES 2019

Increase the Reliability of Operation of Protection DC traction substation

Leonid L. Alekseev[1], *Anton A.* Kolesnikov[2,*], *Boris V.* Papkov[1], and *Sergey* Osokin[3]

[1]Nizhny Novgorod State University of Engineering and Economics, 606340, Knyaginino, Russian Federation
[2]Altair Ltd, 603163, Nizhny Novgorod, Russian Federation
[3] NSTU n.a. R.E. Alekseev, Nizhniy Novgorod, Russian Federation

Abstract. For protection of direct current (DC) traction substation is proposed to use centralized differential protection. Estimation of reliability indicators is carried out on the basis of the method of Markov chains. The obtained results allow to make a conclusion about the prospects of implementing this protection.

1 Introduction

To protect special electrical installations of DC traction substation, current protections of transformers and rectifiers [1], special DC protections [2], and ground protections are used. These technical solutions are characterized by low selectivity and outdated element base. The purpose of research is the development of highly reliable centralized differential protection (CDP) for DC traction substation.

2 Realization of proposed protection

The CDP proposed by the authors is based on the differential principle in combination with the double entry method [3, 4]. To describe the principle of protection, consider the fragment of the DC traction substation shown in fig. 1.

For the scheme (fig. 1) construct a unidirectional graph (fig. 2a) with vertices and arcs, where vertices V1 to V13 of the graph are protected items, and arcs e1 – e14, respectively, the branches of current transformers (CT), current sensors (CS) and circuit breakers. The arcs that characterize the branch of the CT, CS and switches, display the facts switching and have a weight that represents information about the magnitude flowing in the branch current obtained by measurement using the appropriate CT and CS.

The next step is the definition of range of protection based on the topology of scheme with the position of the isolators. As a result, possible transition to a new form of the graph (fig. 2b) by deleting the arcs e3-e6, e9, e10 and merge the vertices V6-8 (V6*) and V9-12 (V7*).

For determining the location of fault, protection compares the currents on differential principle for particular regions and detects potentially damaged item. For all graph vertices with degree >1 compile equations.

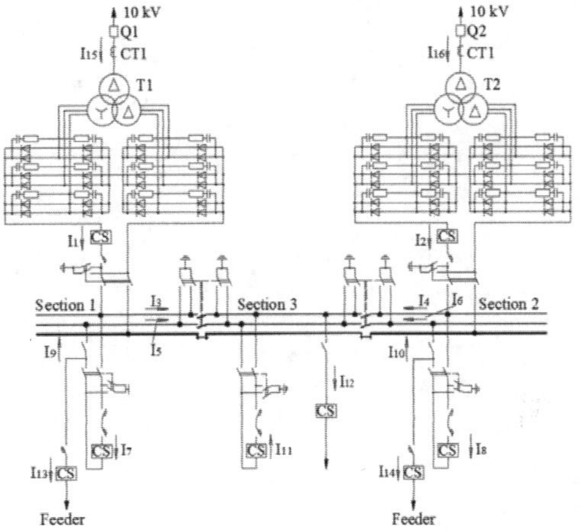

Fig. 1. Scheme of fragment of the DC traction substation.

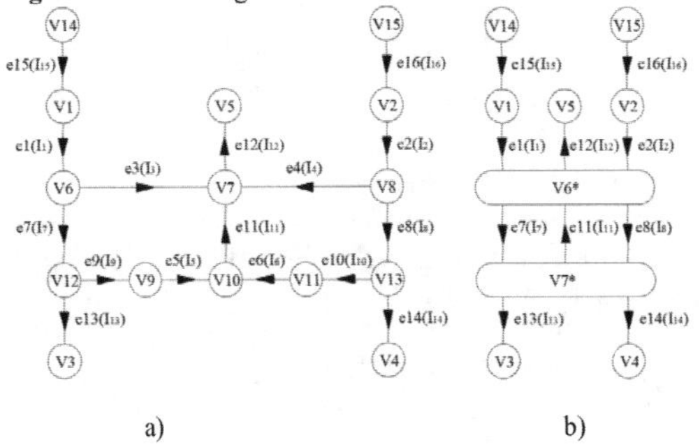

Fig. 2. Unidirectional graph for the fragment of the DC traction substation. a) initial graph; b) final graph

Then it is determined whether to trigger starting on fault in the region, or there is a CT, CS or communication lines (CL) fault. The presence of a fault is determined using the method of double entry. Since each arc of the graph is reflected with the same weight (current value) in the matrices twice: as an arc, associated with the vertex directed to one matrix, and as the arc associated with the vertex directed from another matrix, in the case getting wrong values is the sum of incoming and outgoing currents in the two matrices becomes incorrect, but the total differential current of the whole network remains equal to zero. In case of observance of rule of detection of the fault and maintaining a sum of currents for the entire network is equal to zero, a CT, CS or CL fault is detected.

Signs of the functioning of the differential protection for different ratios of currents and the results of performing matrix operations for this scheme is shown in Table 1. Variables SMV1...SMV7, obtained as a result of matrix operations, determine the conditions for the presence of short circuit in the protection zone. The SUM variable determines the total differential current.

Table 1. Results matrix operations and signs of the functioning of the protection.

The resulting value matrix operations					Signs of fault
SMV 1	*SMV* 2	*SMV* 6*	*SMV* 7*	*SUM*	
0	0	0	0	0	Fault is missing, CTs is functioning.
0	0	≠0	0	≠0	Fault on *V*6*.
0	0	0	≠0	≠0	Fault on *V*7*
0	0	≠0	≠0	≠0	Fault on e_7 (e_8, e_{11})
0	0	≠0	≠0	0	CT Fault on e_7 (e_8, e_{11}).
0	0	≠0	0	0	Fault on CS common for *V*6*,*V*1, (*V*2, *V*5) e_1 (e_2, e_{12}).
0	0	0	≠0	0	Fault on CS common for *V*7*, *V*3, or *V*4 e_{13} (e_{14}).
≠0	0	0	0	≠0	Fault on *V*1
0	≠0	0	0	≠0	Fault on *V*2
≠0	0	0	0	0	CT Fault on e_{15}.
0	≠0	0	0	0	CT Fault on e_{16}.
0	0	0	0	≠0	Fault on one of CS, (CT) e_{13}, e_{14}, e_{15}, e_{16}

As a result, depending on the ratio of the currents in the scheme as well as results of operations on matrices it can implement the reliable operation of the differential protection of DC traction substation. This provides not only the action of protection in case of damage on each site, but excluded its excessive action when damaged current transformers and current sensors.

3 Quantitative assessment of the reliability of the proposed protection

To quantify the advantages of the proposed technical solutions from the point of view of reliability, using the method of Markov chains. This method is often used to describe the processes of failure and repair with the elementary streams, and is most suitable for calculating reliability of system of relay protection (SRP) [5, 6]. The distribution laws of failure and repair will accept exponential.

Taking into account specificity of the analyzed differential protection, define two types of protection failures [7]: undesired-tripping protection failures (in the absence of fault on the protected object), and fail-to-operate protection failures (in case of fault on the protected object).This approach is used in several papers, e.g. [8].

In fig. 3 presents two variants of the relay protection system organization is presented. In first case (fig. 3a) the SRP, consisting of current protections (A1, A2), earth protections (B1, D), DC current protection – (C1); in the second case (fig. 3b) - a system with application CDP (F).

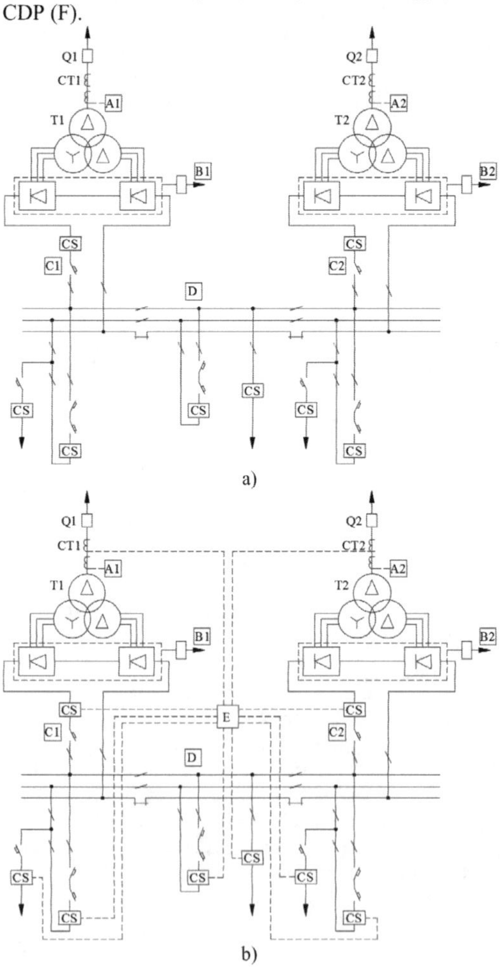

Fig. 3. Protection scheme of DC traction substation with connection of protections to CT and CS, a) standard protection scheme; b) system with the usage of CDP.

We will calculate reliability indexes for both cases. The initial data for calculations are obtained on the basis of [9, 10] and from manufacturers of protection devices. The data is presented in Table 2.

Table 2. Initial data for reliable indexes calculation.

Parameter	Value
Mean time between failures of protection (hour)	100000
Mean restoration time of protection (hour)	2
Queueing time (hour)	2
Rate of false alarms of CDP (1/year)	0,0307
Rate of failures to trip of CDP (1/year)	0,0876
Rate of false alarms of current protection (1/year)	0,0307
Rate of failures to trip of current protection (1/year)	0,0876
Rate of false alarms of earth protection (1/year)	0,00338
Rate of failures to trip of earth protection (1/year)	0,00349
Rate of false alarms of DC protection (1/year)	0,03
Rate of failures to trip of DC protection (1/year)	0,03
Mean time between failures of CT (hour)	400000
Rate of CT failures including secondary circuits (1/year)	0,0429
Rate of busbars 3,3 kV failures (1/year)	0,03
Rate of rectifier unit failures (1/year)	0,25

3.1 Mode 1: absence of fault on the protected object

In fig. 4 shows graphs of states and transitions in an absence of fault on the protected object mode. Here: Ew – state SRP without failures; $A1_1$, $A2_1$,.. E_1 - the state of the SRP in the presence of defects that could lead to undesired-tripping protection failures of devices of relay protection A, B…F, and, as a result, failure of the SRP in general; μ_1 – repair rate of the protection, CS and CT.

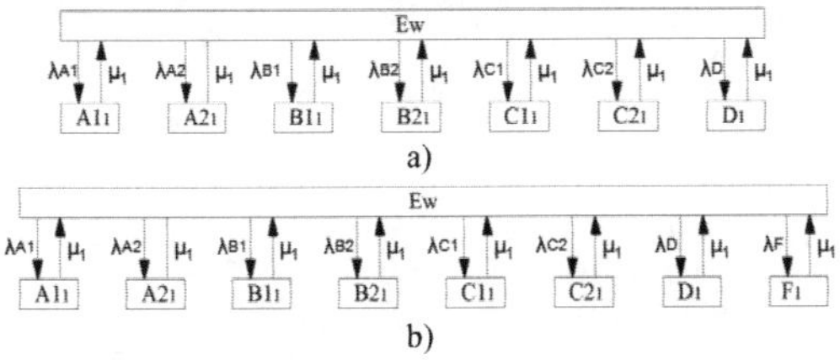

Fig. 4. Graphs of states and transitions for an absence of fault on the protected object mode a) system without CDP, b) system with CDP.

Convert the graphs (fig. 4) to the form shown in fig. 5, where E_1 - failure state of SRP, λ_1 – the resulting failure rate of protection.

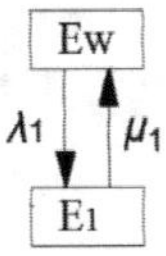

Fig. 5. Converted graph of states and transitions for an absence of fault on the protected.

Define the probability $P_1(t)$ that the system is in state E_1, which is an emergency, for random time t. Form a system of differential equations describing a graph, where $P_w(t)$ - the probability of location the system in state without failures E_w.

$$\left.\begin{aligned} \frac{dP_w(t)}{dt} &= -\lambda_1 \cdot P_w(t) + P_1(t) \cdot \mu_1 \\ \frac{dP_1(t)}{dt} &= \lambda_1 \cdot P_w(t) - P_1(t) \cdot \mu_1 \end{aligned}\right\} \qquad (1)$$

Normalizing expression, the meaning of which is that the researched system located in state E_1 or E_w as constituting a complete group of events, has the form

$$P_w(t) + P_1(t) = 1$$

As at the initial moment of operation of the system at $t = 0$ the system is in state without failures:

$$P_w(0) = 1,\ P_1(0) = 0$$

As a result of solve the system of differential equations a function of the unreadiness probability

$$q(t) = P_1(t) = \frac{\lambda_1}{\lambda_1 + \mu_1} \cdot (1 - \exp[-(\lambda_1 + \mu_1)t]) \qquad (2)$$

Make an assessment of the probability of failure-free operation of the SRP. The graph for the calculation will look similar as the graph in fig. 5 with the difference that it will not be possible to transition from state E_1 to Ew, that is excludes repair rate μ_1. This circumstance is due to the fact that when system fails (transition to absorbing state E_1), the experiment is finished – the system cannot leave this state.

Due to absence of repair instead of a system of differential equations (1), the result will be an expression of the probability of failure-free operation (PFFO), with exponential distribution law:

$$R(t) = P_w(t) = \exp[-(\lambda_1 \cdot t)] \qquad (3)$$

In fig. 6 shown the dependences of the function of the unreadiness probability of SRP, and functions of the PFFO on the average time between checks.

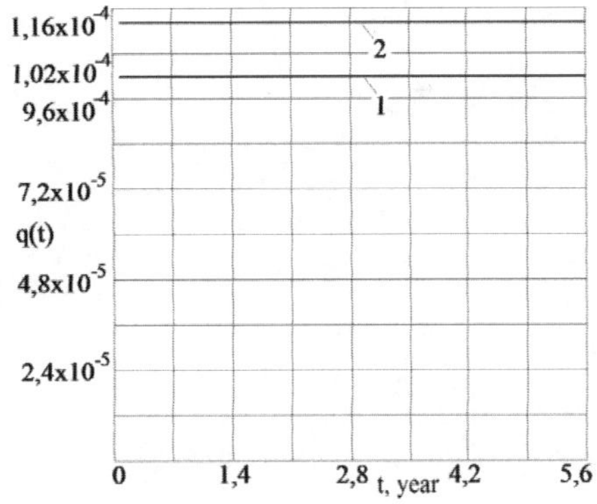

a)

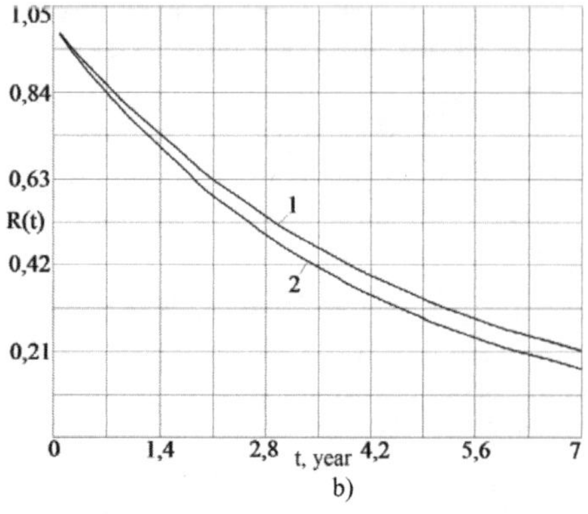

Fig. 6. Function of the unreadiness of SRP (a) and function PFFO of SRP (b). 1 - system without CDP; 2 system with CDP.

3.2 Mode 2: in case of fault on the protected object

The calculation of reliability indicators for the damage mode on the protected object is per-formed separately for the following damages using the approach described above: short circuit in the transformer, short circuit in the rectifier, short circuit on 3.3 kV buses.

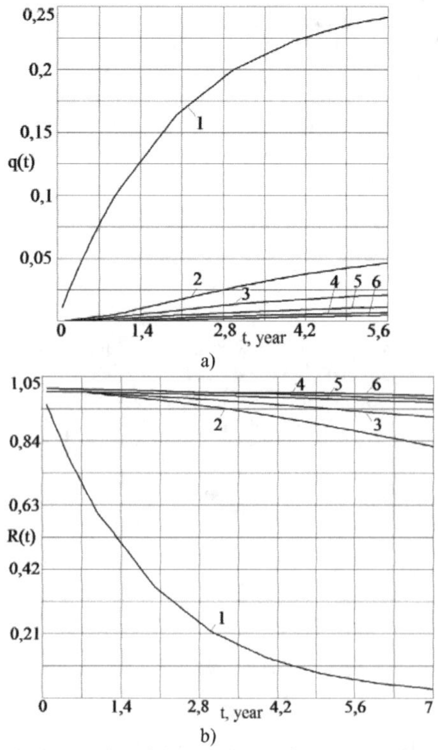

Fig. 7. Function of the unreadiness of SRP (a) and function PFFO of SRP (b).1 – standard scheme, fault in transformer; 2 – CDP, fault in transformer; 3 – standard scheme, fault in rectifier; 4 – CDP, fault in rectifier; 5 – standard scheme, fault on bus 3,3 kV; 6 – CDP, fault on bus 3,3 kV.

Analysis of numerical calculations of indicators of reliability of the relay protection (fig. 6, 7) shows:

- developed protection in the mode 2 has higher indicators of reliability than standard technical solution on average up to 5 times;
- the obtained calculation ratio can be used in the methods of practical reliability analysis of centralized relay protection of DC traction substation.

4 Conclusions

1. Proposed a new principle of organization of the centralized differential protection of DC traction substation.
2. Developed technical solutions allow to detect the current transformers, current sensors and communication channels faults, and having high reliability.
3. Protection can be adapted to changes in the configuration of the DC traction substation.
4. The proposed technical solution can increase the reliability of the differential relay protection, what follows from by the results of practical calculations.

References

1. K.G. Markvardt, Jelektrosnabzhenie jelektrificirovannyh zheleznyh dorog (Moscow,Transport, 1982)
2. V.F. Harisov, Zashhita kontaktnoj seti postojannogo toka ot korotkih zamykanij (Moscow,Transport, 1987)
3. A.L. Kulikov, V.Ju. Vukolov, A.A. Kolesnikov, Vestnik NGIEI. **2 (69)**, 71 (2017)
4. A.L. Kulikov, V.Ju. Vukolov, A.A. Kolesnikov, M.D Obalin, Jelektricheskie stancii. **1 (1038)**, 40–51 (2018)
5. B.H. Smeets, M.H.J. Bollen, *Stochastic modelling of protection systems: comparison of four mathematical techniques* (Eindhoven, TU/e, 1995)
6. B.V. Papkov, A.L. Kulikov, *Teorija sistemy i sistemnyj analiz dlja jelektrojenergetikov* (Jurajt, 2016)
7. A.I. Shalin, *Nadezhnost' i diagnostika relejnoj zashhity jenergosistem* (Novosibirsk, NGTU, 2002)
8. Je.P. Smirnov, Elektrichestvo, **9,** 44 (1965)
9. A. Khurram, H. Ali, A. Tariq, O. Hasan, Proc. of int. workshop *Formal Methods for Industrial Critical Systems*, 169–183 (2013)
10. A.V. Efimov, A.G. Galkin Nadezhnost' i diagnostika sistem jelektrosnabzhenija zhe-leznyh dorog: Ucheb. dlja studentov vuzov zh. - d. transp. (Moscow, UMK MPS Rossii, 2000)

Direct flow drawing equipment performance reliability and increase of its effectiveness

Veronica Ivanova[1] and *Daniyar* Madrakhimov[2,*]

[1]Tashkent State Technical University named after I. Karimov, Tashkent, Uzbekistan
[2] Tashkent State Technical University named after I. Karimov, Tashkent, Uzbekistan

Abstract. It is an analytic investigation of issues related to improving the efficiency and reliability of the drawing technological process, which are to be considered. After analytical efforts, a technical solution is proposed and justified. This technical solution is meant to improve the operational parameters of the cable products current-carrying conductor – reducing electrical resistance by reducing the fractional deformation and recalculating of the drawing sequence.

Keywords: Drawing, Cable And Wire Products, Drawing Machine, Drawing Mill, Technological Tool, Wire-Drawing Die, Drawing Block, and Electric Drive.

1 Introduction

Wire-drawing operations are the beginning of the manufacturing technology for all types of cable-conductor products (CAT). This technical process provides a production of high-quality cable conductors [1]. An analysis of application of modern drawing technologies among manufacturers of cable-conductor equipment made it possible to conclude that jet drawing is the most common way among existing technologies.

All basic research in the field of increasing the reliability and efficiency of drawing technologies [2, 3, 4] led to the conclusion that the solution of the problems existing for the current production is provided mainly by technical re-equipment of enterprises, by equipping the production process with advanced technological equipment and modernizing the installed plant and equipment.

Actually efficiency improving of the drawing technological process may be achieved by using modern production lines in the production process, combining several technological operations; using so called multiwire drawing, when 8 until 56 wires are simultaneously being drawn; reduction of power consumption by technological equipment; the choice of the optimal parameters of the emulsion; service life extension of the technological tool - wire-drawing die; using high-quality die billets (wire rods); improvement processes of technologies and modernization of equipment with the introduction of resource-saving and energy-saving electromechanical systems (EMC) [4, 8].

An ultimate goal of the manufacturing process is to provide high operational characteristics of the technical checkpoint, including the low value of the electrical (true ohmic) resistance of the current-carrying conductor (CCC). At the same time, a well-established drawing technology, as an initial technological operation, has the greatest influence on the process of formation of the structure of the current-carrying conductor.

2 Proposed technical solution description

An analytic investigation of the methods which have been used to increase the efficiency of modern drawing machines (DMs) made it possible to conclude that the joint influence of factors to improve the quality of the drawn wire and simultaneously increase the energy and resource saving of the drawing technology cannot be solved at the same time.

In-depth study aiming to improve the quality of conductor resistance CR (undertaken by Sh. Berin and N.Z. Dniestrovsky) [5] allowed to reduce the electrical resistance of the wire during cold plastic deformation for copper to 3%, and for aluminum to 1.5%. This fact can be explained by the Drawing Theory (Fig. 1) for conductor resistance CR, which depends on the magnitude of the drawn wire fractional deformation when passage through drawing sequences.

In reliance on technological equipment, the distribution of the applied drawing force, which is the total for the entire process (Fig. 2), is determined by the technical parameters of the main rod breakdown and wire-drawing machines. This process ensures the maintenance of tension in the pulling system of the wire-drawing machines. The calculation of the partial drawing forces of the intermediate pulling blocks and their proportional distribution depends on the calculated drawing ratio, the magnitude of the backward tension acting between the pulling and drawing blocks and is carried out through the drawing sequences for the entire process.

In compliance with the above suggested solution to the problem of reducing the electrical resistance of the current-carrying conductor (CCC), it is possible to achieve by reducing the fractional deformation at the price of reducing the number of passages through the drawing sequences, i.e. a decrease in the number of drawing passes and working zones of plastic deformation (technological tool – wire dies). Metal structural changes occur just at that spot.

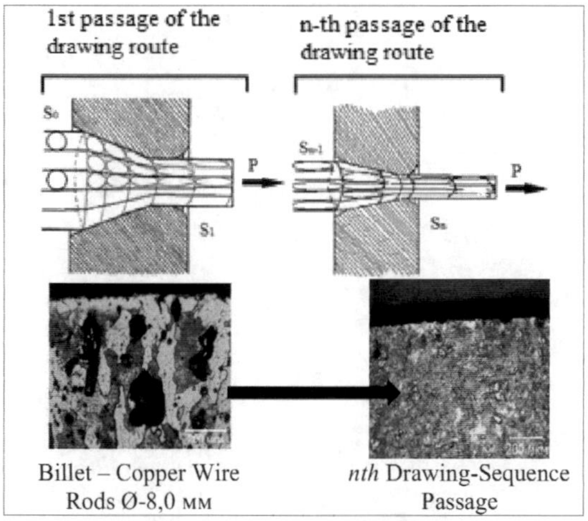

Fig. 1. Drawn Wire Structure Changes Diagram When Passage Through Drawing Sequence and Fractional Deformation Effect on Copper Wire Samples Micro Specimen.

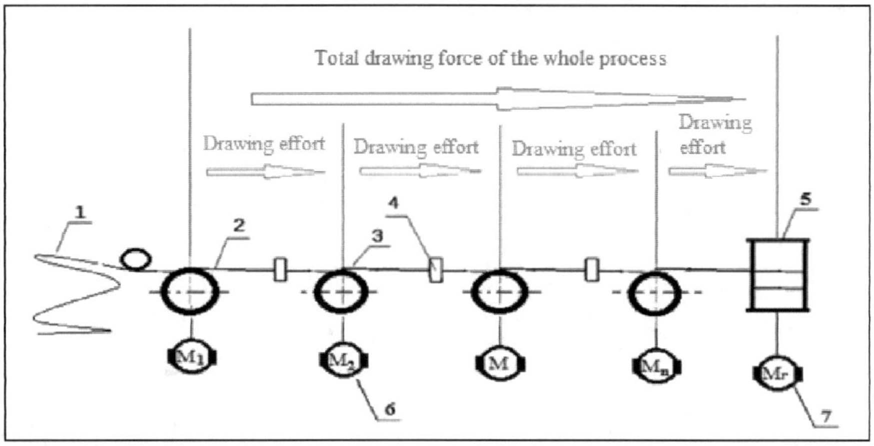

Fig. 2. Straight-Through Type Drawing Machine: 1- pay-off stand; 2- wire; 3- intermediate block; 4- process tool (wire drawing die); 5-reciever; 6- drawing block electro drive; 7- main drawing machine electro drive.

3 Investigations' experimental part

Exploratory tests were carried out on copper wire drawn samples in the testing laboratory of NAVOICABLECONNECTOR LLC (City of Navoi, Uzbekistan). During conducting a structural analysis of the microslices (Fig. 3.4), it was established that the microstructure of the sample undergoes compaction, which is explained by the presence of technological drawing action on the line of application of the drawing force. The measurements of the ohmic resistance of the samples confirmed [5] the deterioration of the electrical conductivity of the current-carrying conductor (CCC) due to the influence of the fractional strain that causes the compaction of the wire microstructure.

The experimental part has been carried out on a direct-flow VM (Fig. 2). It includes: pay-off stand (1), receiver (5) intermediate blocks system (3), as well as the main drawing electro drive machines (7), which produce the necessary drawing force and maintain the required tension in the electromechanical system "Pay-Off Stand – Receiver". There are also systems of individual electronic drives (6) mounted on the pulling blocks (3), and carrying out the wire pulling with predetermined intermediate wire diameters along the drawing passages.

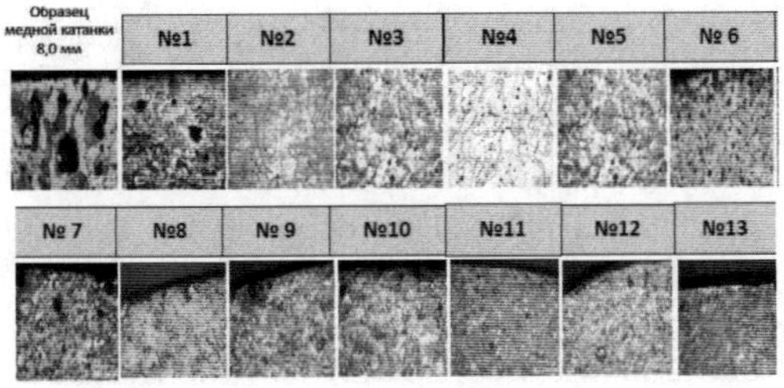

Fig. 3. Microstructure of Drawn Copper Current-Carrying Conductor Sample at VM VSK-13 (The example 8 mm copper rod).

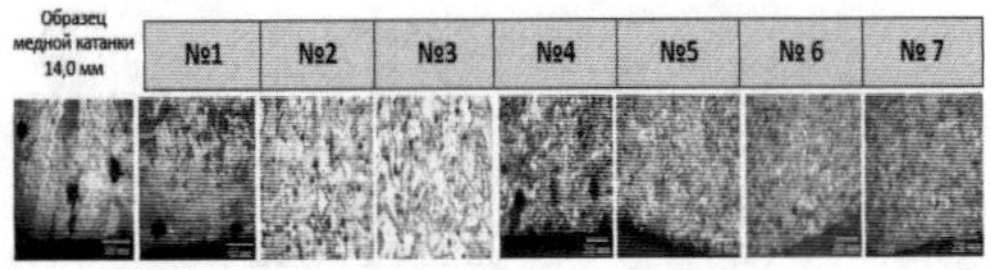

Fig. 4. Microstructure of Drawn Copper Shaped Wire Sample (The example 14 mm copper rod).

A new drawing passages readjustment to a smaller number of passages allowed to reduce the number of drawing passages from 10 to 6 and to exclude idle drawing blocks (DB) with individual electromechanical systems (EMC) from the technology. The obtained clipped samples studies results showed an improvement of the copper wire samples quality indicators (electrical conductivity) by an average of 15% (Fig. 5). The reduction in the number of unused EMCs led to a 27% decrease in energy consumption for straight-through type drawing machine (STTDM). It is a ground to consider the proposed technical solution as energy-saving.

The survey results obtained during researches, in terms of electrical conductivity of current-carrying conductor (CCC), will create good conditions for the efficient and low-cost transportation of electric energy in the power supply system and, accordingly, they can be considered as energy efficiency in relation to power transmission systems.

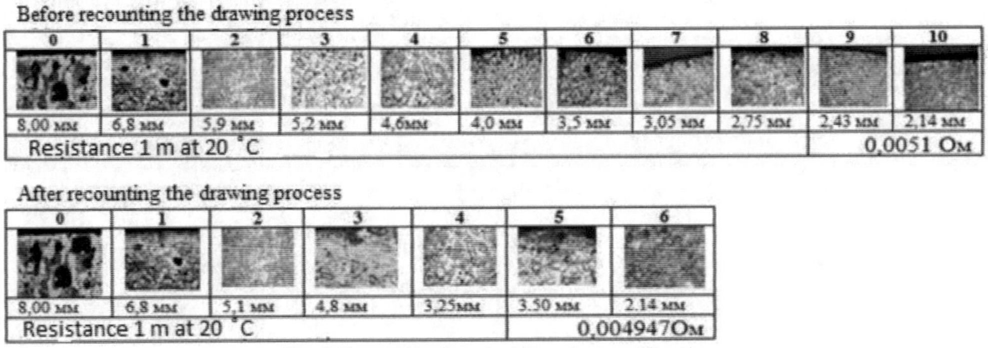

Fig. 5. Round Wire Samples Microstructure Alterations

4 Mathematical modeling results

In virtue of a generalized mathematical model [2, 9, 10] of a straight-through type drawing machine (STTDM) (Fig. 6) and generalized requirements for electric drives (ED) [7], at the Electrical Machines Department of the Tashkent State Technical University, simulation results were obtained (1), taking into account the main drawing process parameters [6, 7, 8] , which made it possible to consider various options for the operation of a straight-flow drawing bench and to determine the optimal multiplicity of the drawing process while ensuring the energy efficiency of the technology as a whole.

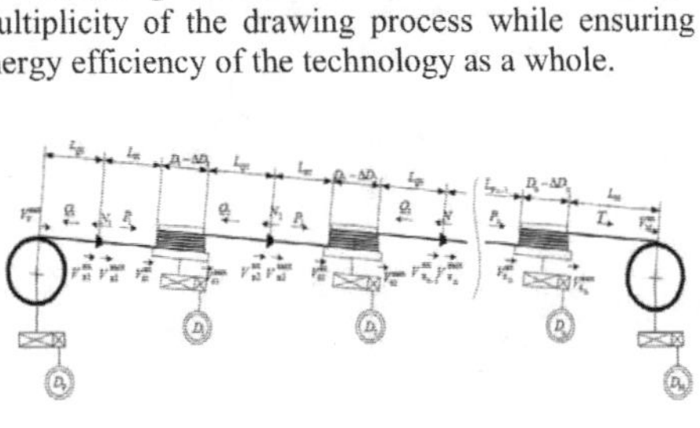

Fig. 6. The N-fold Straight-Flow Drawing Bench Design Scheme for Copper Wire Drawing [2, 7].

We investigated the EMC of the nth inter-drum span in a straight-through type drawing machine (STTDM) in relation to the present conditions under consideration of using the system of differential equations in the operator form [2].

$$\begin{aligned}
&\mathrm{M}_{\text{дин}_n}(p) = M_{\text{дв}\ n}(p) - P_n(p)\frac{R^{\text{вх}}_{\text{б}\ n}}{j_{\text{ред}\ n}} + \\
&+Q_{(n-1)}(p)\frac{R^{\text{вых}}_{\text{б}\ n}}{j_{\text{ред}\ n}} - \mathrm{M}_{\text{xx}\ n}(p); \\
&\omega_{\text{дв}_n}(p) = M_{\text{дин}\ n}(p)\frac{1}{J_{\Sigma n}\, p}; \\
&v^{\text{вх}}_{\text{б}_n}(p) = \omega_{\text{дв}\ n}(p)\frac{R^{\text{вх}}_{\text{б}\ n}}{j_{\text{ред}\ n}}; \\
&v^{\text{вых}}_{\text{б}_n}(p) = \omega_{\text{дв}\ n}(p)\frac{R^{\text{вых}}_{\text{б}\ n}}{j_{\text{ред}\ n}}; \\
&P_i(p) = (v^{\text{вх}}_{\text{б}\ n} - v^{\text{вых}}_{\text{в}_n})\frac{EF_n}{L_{p\ n}p} \\
&v^{\text{вых}}_{\text{в}_n} = v^{\text{вых}}_{\text{в}_n}\mu
\end{aligned} \qquad (1)$$

$$v_{\text{в}n}^{\text{вх}} = Q_i(p)\frac{L_{Q_n}p}{EF_{(n-1)}} + v_{\text{б}(n-1)}^{\text{вых}}$$

$$Q_n(p) = P_n(p) - N_n(p)$$

All experiments with numeric values to simulate the process of copper wire drawing, in order to determine the multiplicity of the drawing passages, were carried out using a generalized block diagram (Fig. 7) compiled on the basis of the system of equations (1).

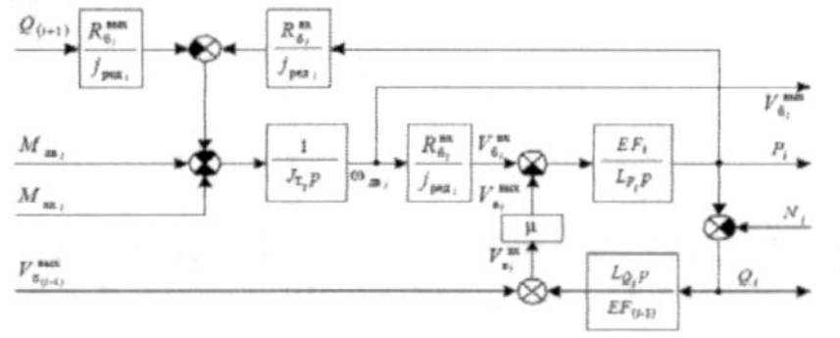

Fig.7. Mathematical Model Structure Diagram of the Inter-Drum Space [2].

There has been simulated the full cycle of the straight-flow drawing bench. Fig. 8 shows fragments of simulation results (Fig. 8, a) and industrial experiments (Fig. 8b, c) illustrating the processes of starting (acceleration) and drawing at operating speed. The start-up and acceleration time to a given operating speed corresponds to 25-30 seconds, respectively.

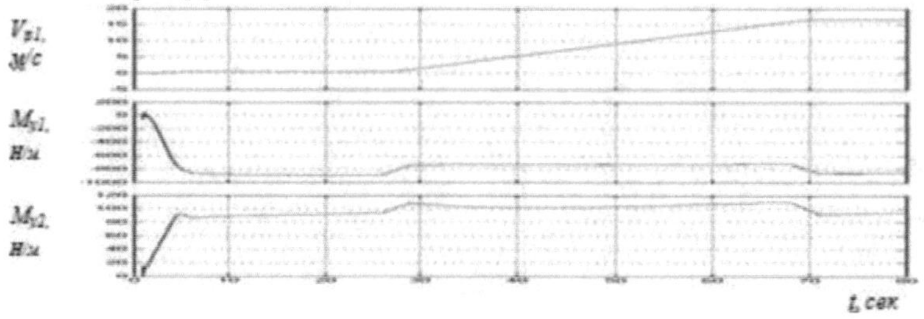

a)

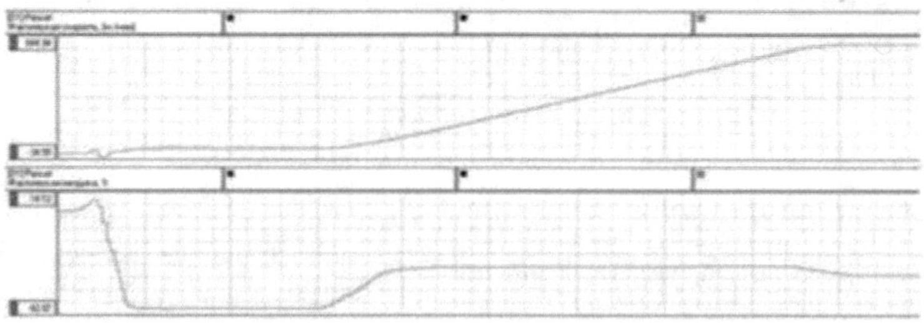

b)

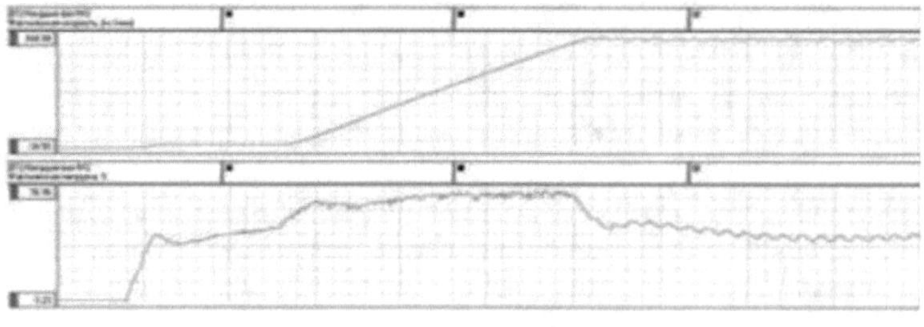

c)

Fig. 8. Simulation Results Fragments (a) and Industrial Experiments (b, c).

5 Conclusions

The calculated data which we obtained and the results of an industrial experiment confirmed the correctness of the decision to turn off the electric motors of the wire-drawing machines pulling blocks (table-1). An industrial experiment was also carried out for drawing machines with a main electric drive (type VSK-13), however, positive results were obtained only by improving the quality parameters of stripped copper wire samples (reducing ohmic resistance). And, in relation to increasing the energy efficiency of the process for this particular case, other parameters remained unchanged. The reason – in the wire-drawing machine there is one main drive, which is, in kinematic sense of word, connected with the pulling drums, and therefore changing the passages to a given output diameter will be reduced to the number of dressed pulling washers.

Table 1. Comparative Analysis of Drawing Process Parameters

Drawing Process Parameters	Total ED for VC		Individual ED for every TB	
	Current	Calculated	Current	Calculated
Passages q-ty, number.	10	6	7	5
Electric Motor Quantity	1		7	5
Power Input, kWt	252		45	33,2
Material	Copper			
Drawn Product	Round Wire		Shaped Conductor	

The fact that the quality of process has been improved is not the only advantage of the proposed method. Significant savings can also be considered. Savings are due to:

• Reducing of drawing bench metal consumption;

• Reduction in the number of loaded gearboxes (gears), which led to a clear reduction in losses in the gear system of the machine as a whole;

• Reduction of expensive drawing tools details - drawing blocks, and, consequently, reduction of contact friction losses of tool surfaces with the object to be processed.

6 References

1. A.K. Bulkhin, V.F. Kidyaev, S.A. Kizhaev, *Drawing Equipment Electric Drive and Automation*, **Part 1**, 11, 381-394 (Samara, 2002)
2. S.A. Linkov, Mult. DF Draw. Bench Math. Mod. **volume 159. Part 2** (St. Petersburg, 2004)
3. V.I. Zyuzin, V.A. Kharitonov, A.A. Radionov et al. *Resource Saving in Wire Products: Collective Monograph* 160 (Magnitogorsk, 2001)
4. V.V. Tsypkina, A.Sh. Akhmedov, D.B. Riksitullaeva, V.P. Tsypkina (Ivanova) Dev. of a Meth. for Impr. the Param. of Cop. Wire DDP // Energy and Res. Sav. Prob. (Special Issue) **No. 3-4**, 205 (Tashkent, 2013)
5. I.Sh. Berin, N.Z. Dniester, *Copper and Aluminum Wire Production, Metallurgy Publishing House.* 16-17 (Moscow, 1974)
6. V.P. Ivanova, B.E. Yakubov Dev. of an Inf. Mod. for a Res.-Saving DM, **No. 3,** 36, (2017) URL: http://7universum.com/ru/tech/archive/item/4550 (accessed: 02.25.2019).
7. V.V. Tsypkina, A.A. Ivanov Dev. of Gen. Req. for Modern. of the DM Elec. Dr. for Cable Prod. **No. 4,** 37 (Moscow, 2017)
8. O.A. Troitsky, V.I. Stashenko, V.G. Ryzhkov, V.P. Lyashenko, E.B. Kobylskaya El.-plastic Dr. Proc. and New Tech. for CLW, **No 4**, 111-117 (2011)
9. Ivanova V.P., Kakhramonov Sh.K. Exp. Res. of a direct-flow DM. NAU, **No. 43, Part 1**, (2019)
10. Pirmatov N.B, Ivanova V. P., Madrakhimov D.B En. Ef. of Dr. Eq. thr. E&R Sav. Int. Jour. of Adv. Res. in Sc., Eng. and Tech. **Vol. 6, Issue 3,** (March, 2019)

Method of resource saving of cable production recycables

Victoria Tsypkina[1] and *Daniyar* Madrakhimov[2,*]

[1]Tashkent State Technical University named after I. Karimov, Tashkent, Uzbekistan
[2] Tashkent State Technical University named after I. Karimov, Tashkent, Uzbekistan

Abstract. The article discusses development of recycling technologies of cabling and wiring products with justification for the relevance of the cable waste use in the process as part of the closed technological cycle of cabling and wiring products manufacture. Sources of waste generation are analysed. Issues of uninterrupted operation and reliability of technological equipment, efficiency of the recycling process and problems of the development of processing technologies in cable production are highlighted.

Keywords: Material resources, cable production, cable waste, recycling, waste use, copper rod, polymer waste, closed process chain.

Introduction

Production of cabling and wiring products is a highly material-intensive process in which copper content in some groups reaches 90% (enameled wires). Therefore, reduction in the price of copper rod is a fundamental economic parameter, ensuring reduction in cost and increase of cable production competitiveness. Technological cycle, which includes production of copper rod from cable waste in comparison with production of copper rod from copper ore, determines high cost- efficiency of copper recycling, used in the process chain of cable production.

Organization of closed process chain of cabling and wiring products (CWP) is a priority for the development of any cable production, as it determines great economic benefit and high efficiency of all process and therefore development of cable waste processing (recycling) technologies is effective.

Closed process chain of production of cabling and wiring products includes all technological operations along the production route, from the production of wire rod (blanks, for electrical conductor) and ending with recycling of cable waste with its mandatory use in CWP production cycle.

Cable waste is a mandatory part of any technological operation and consists of entering ends (leaders) and rejected cable products, derived from the process cycle (table-1). Part of cable waste is recyclable products that can be used in technology with established percentage of the total raw material weight and materials, included in cable products design.

Table 1. Scrap of SWP cable, net mass 299.2 kg.

№	Components	Content, kg	Content, %
1	Copper waste	122,07	40,8
2	Waste of insulating polyethylene	36,8	12,3
3	Waste of hose polyethylene	84,67	28,3
4	Waste of plastic dye	1,5	0,5
5	Aluminium foil waste	22,74	7,6
6	Waste of tinned wire	1,14	0,38
7	Waste of solder POS-40	0,04	0,01
8	PVC film waste	23,33	7,8
9	PETF film waste	2,39	0,8
10	Caprone thread waste	0,48	0,16
11	Other waste (garbage)	4,04	1,36

Cable waste management process has evolved in the following stages [6]:

- waste burial, together with other wastes of industrial products and human life;
- burning to extract metal elements (First of all, copper);
- dismantling of cable structure elements;
- chopping and splitting into metal and non-metal parts;
- new ways to recycle cable waste [1].

Technological process of processing cable waste, used in practice, can be carried out by three methods [6]:

- Cable annealing on fire–this is the most primitive way, using which, polymer insulation is burnt simply to produce metal coated wires. This technology is the most harmful for both the environment and the processors themselves. In addition, a significant part of the surface layers that conduct wires, "goes" into scrap.
- Separating the cable from insulation manually, which is time-consuming and labour-intensive and recycles small volumes of cable.

https://doi.org/10.1051/e3sconf/201913901058

- Mechanical recycling of the cable on high-performance, specialized installation - such recycling consists of crushing the cable, after which the polymers and metals are separated in various ways, including electromagnetic way.

At that, the latter method is subdivided into methods that allow dividing and separating fragmented segments of cables and wiring into polymer and metal fractions, by electromagnetic separation, air vibro-separation or combined electromagnetic-air vibro-separation.

To date, in the production structure of cable companies in Uzbekistan only JV "Uzkabel" JSC (Tashkent) has a site for processing of cable waste (production of Switzerland) (Figure 1), which includes:

- sorting of cable waste;
- crushing into factions;
- separation;
- Recycling of waste in the production cycle;
- Waste disposal.

and refers to the combined method of electro-magnetic air vibro-separation.

Average performance of the applied technology at the inlet depends on the type of recycled cable and level of automation of auxiliary operations (table1).

Cable waste recycling line, designed for integrated cable waste recycling with the removal of magnetized fractions and obtaining copper and aluminum granulate, with simultaneous sorting and separation into fractions of electrical conductive metal (copper, aluminum), insulation material and ferrous scrap and waste [6].

Table 2. Average performance of the combined method of electro-magnetic air vibro-separation.

№	Type of cabling and wiring products	Average performance, kg/h
1	for mixed copper cable	700
2	for telephone wire	650
3	for power cable	1100

Table 3. At the same time, purity degree of copper particles, obtained at the output, for different types of cables is on average.

№	Type of cabling and wiring products	Purity degree of copper particles, obtained at the output, %
1	for mixed copper cable	99,5
2	for telephone wire	99,8
3	for power cable	99, 9

The principle of the line operation is to extract copper (aluminum) from the waste of cabling and wiring products (diameter from 0.1 mm or more) by crushing the raw materials into granules, and then extracting copper (aluminum) fractions from it. The line is a set of separate machines, connected by process chain through a system of air transport and belt conveyors, as well as a single system of air supply and purification (fig.1).

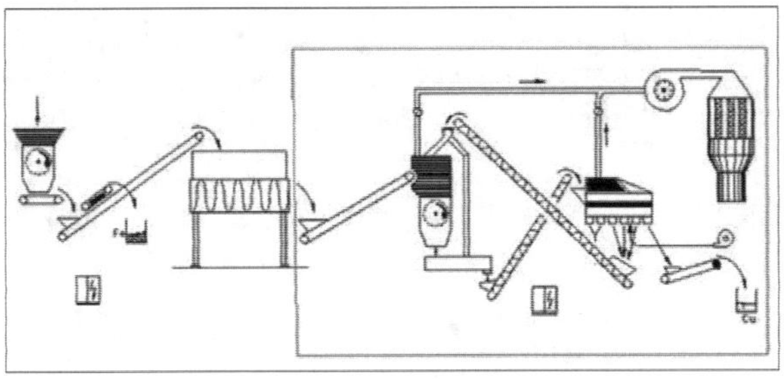

Fig. 1. Basic diagram of process line for recycling of cable waste

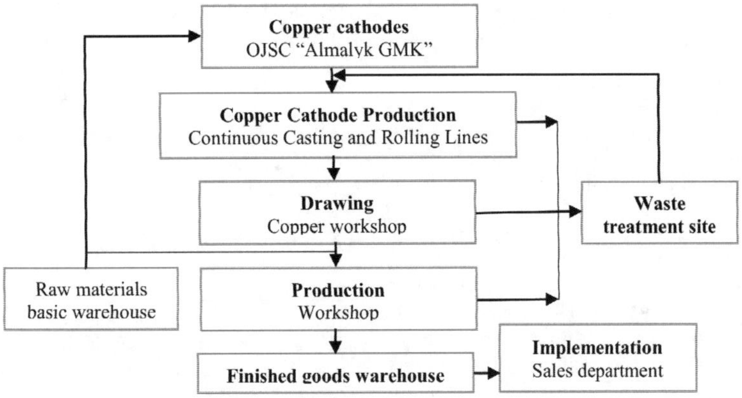

Fig. 2. Cabling and wiring products at "UZKABEL" JV JSC taking into account waste processing site

Efficiency of the closed process cycle in the production of cabling and wiring products is in the amount of raw materials (copper and aluminum), used in the technology of obtaining copper and aluminum wire rod. Technological reworks of used raw materials look like this (pic. 2).

Technological cycles of producing copper wire rod from cathodes and waste of raw materials contain the same number of reworks, but energy consumption, labor and capital intensity of technologies are not comparable. Only consumption of electrical energy in the processing of copper waste into anode copper is 6 times less than in the processing of aluminum waste, which when compared with the cost of obtaining primary anode copper, starting from the mine, 20 times more [2].

Copper cable waste, received as a result of disposal, is fairly homogeneous, which is provided by modern purification technology, which is based on modern methods of recycling of recyclable materials waste.

Effectiveness of recycling is determined by economic indicators, and its assessment is obtained from formula [3, 7]:

$$I_{er} = F(E, E_c\, M) \rightarrow 1{,}0 \quad (1)$$

where I_{er} – Recycling Effectiveness Index; E – Recycling Effectiveness Index, showing cost ratio of products manufacturing from waste and from fresh raw materials (The cost of recycling waste should aim to minimum); E_c – indicator of environmental importance of recycling, which reflects environmental harm, done by technology of producing product out of waste; ideally, recycling should not harm the environment; M – indicator of relative volume of alleged recycling; ideally, 100% of waste should be recycled.

If these conditions are observed as much as possible, I_{er} tends to one. *Indicator* I_{er} speaks of the effectiveness of recycling for society as a whole, which in turn speaks on need for deep government involvement in the formulation of regulatory stimulating documents, related to the stages of waste formation, storage, recycling. This is necessary, in particular, due to the fact that for individual performers of works issues of ecology and depth of waste processing remain out of the sphere of attention [7].

In recycling of cable waste, environmental issues, completeness and depth of recycling are also relevant. However, for many cable manufacturers, financial efficiency comes to the fore, and the solution to the problem of obtaining recycled copper and aluminum comes down to a lighter way - burning off the waste of production.

The amount of cable waste is formed from production waste and time interval that corresponds to the amount of cables and wires consumption. Time interval is determined by the lifespan of the cable product (usually 25-30 years). However, lifespan depends on the operating conditions of cable products and can be up to one year, so for example, for oil-submersible cables it is associated with a significant change in working conditions in geophysical wells.

Currently, Uzbekistan is experiencing a sharp increase in the consumption of imported electrical equipment and cable products. According to experts, in general, due to own production and copper receipts as part of imported cables and electrical products, copper consumption increased by about 20%. This indicator should include cable products with expired service life - 25 years, which need to be replaced. All this gives reason to expect an increase in copper waste inflow into recycling, which can reach 80%, which will amount to 280.0 thousand tons of pure copper [8, 9, 10].

However, in the process of cable waste recycling, there are problematic issues that need to be addressed by application of new technologies:

- recycling of thermoplastic polymer waste and its further use in production;
- Use of thermoreactive waste, as secondary raw material;
- Thermal treatment of polymers when burning rejected lengths and entering ends (leads) of cable products in compliance with sanitary standards and subsequent use of heat energy, released at the same time;
- deep separation of the polymer fraction, with the aim of increasing the percentage of secondary copper and aluminum;
- Increase in reliability of technological equipment operation, taking into account climatic features of the region (high air temperature in the summer).
- Process of using recycled copper and aluminum in the production of wire rod is conducting successive technological operations, starting with stripping cable waste, crushing with the separation of granular metal (armor and shell) and not metal (insulation and hose) fractions (chip scrap). Constraining factor in the use of heat plastic polymers waste for the production of copper and aluminum wire rod is the high dirtiness of waste with finely dispersed metal and non-metallic impurities and oil-containing fillers, technical complexity of removing them during recycling. For example, when recycling oil-submersible cables, copper is clogged with metal and non-metal materials, reservoir fluids.

Improvement of waste recycling technology should cover the process of cable waste generation. In this case, it is possible to use pre-selection of waste, prepare them for further processing, which will significantly reduce the clogging of waste by related materials. Such an organization will reduce labor intensity of waste recycling technology, eliminate clogging of copper and aluminum waste with excess impurities. Pre-selection technology is widely used in world practice in collection of plastic waste, in particular.

Various types of recycling technologies have been developed now that provide a comprehensive approach to recycling in cable industry and will remove all technical and organizational constraints.

Modern recycling technologies are presented in the market of technological equipment in the form of industrial plants, operation of which does not have a negative impact on the environment. However, effective process organization is possible only in large enterprises that specialize in collecting and recycling cable waste.

These enterprises have sufficient financial, production and human resources to comprehensively address the existing problems in centralized waste collection, recycling and sale of produced commodity products with stable maintaining of required production volume and environmental frequency of the process, ensuring high quality of secondary product and low cost [4, 7, 8].

Based on the analysis of existing recycling technologies, it can be noted that the development of recycling in cable production has a number of features:

– high social significance;

–Economically motivated for recycling plants and consumers of materials;

–modern recycling technologies are possible only with significant capital costs and are primarily available for large cable companies [5, 9, 10].

With regard to the issue of material resources, natural resources are the determining factors. In recent decades, we can see exhaustion of some deposits, emergence of new deposits that is accompanied by a significant change in the geography of their location and, as a result, significant geopolitical upheavals. A negative global trend is the steady depletion of natural resources. Exploration of new deposits, conservation and efficient use of natural reserves is a vital task of the state.

References

1. G.I. Meschanov, Env.-ment. Asp. of CP, **№ 6** (2000)
2. O.P. Petrukov, The conc. of opt. int. man. of dom. Garb. in Moscow reg. / O.P. Petrukov, L.Ya. Shubov, F.F. Gaev // Dom. Garb., **№ 9** (2007)
3. G.I. Tsutskareva, Refl. on rec. waste, sci. and hi-tech., **№ 2**, 56 (2015)
4. A.V. Sechina, Ov. of meth. of ind. Rec. of cab. waste // Writings of SSA, **№ 7**, 80–89 (2009)
5. O.S. Kusrayeva, Form. man. mech. of wast. rec. of ind. Ent.: Auth. Abs. of Ph.D. thesis, 20 (St. Petersburg, 2012)
6. A.V. Sechina, Ov. of meth. of ind. Rec. of CW. https://docplayer.ru/33672417-Menedzhment-obzor-metodov-promyshlennoy-pererabotki-othodov-kabeley.html
7. A.V. Abramov, Asses. of rec. eff., MTUH, 36-38 (St. Petersburg, 2012)
8. B.B. Bobovich, V.V. Detkin Rec. of prod. and cons. waste. Int. Eng. (2000)
9. Yu.S. Yusfin, L.I. Leontiev, P.I. Chernousov. Ind. and the env. (2002)
10. Rec. of waste, **№14 – 16** (2008)

Using smart meters for checking the topology and power flow calculation of a secondary distribution network

Irina Golub[1,3,*], *Evgeny* Boloev[1,2], and *Yana* Kuzkina[1,3]

[1] Energy Systems Institute of Siberian Branch of the Russian Academy of Sciences, Department of Electric Power System, 664033 Lermontov str., 130, Irkutsk, Russia
[2] Irkutsk National Research Technical University, Institute of Power Engineering, Department of Electric Power System, 664074 Lermontov str., 83, Irkutsk, Russia
[3] Engineering Center «Irkutskenergo», Department of relay protection and automation, 664043 Ryabikova str., 67, Irkutsk, Russia

Abstract. The paper analyzes options of using smart meters for power flow calculation and for assessing the state of a real three-phase four-wire secondary distribution network based on measurements of average values of active and reactive power and of voltages. The work is based on the authors' research on allocation of measurements to ensure secondary distribution network observability and on selection of the most efficient method for linear and non-linear state estimation. The paper illustrates solution of a problem on identification composition of load nodes in the phases and reveals challenges related to voltage account in the neutral wire and in its grounding.

Introduction

The first experience in studying the operation conditions of primary and secondary distribution networks (DN) of medium and low voltage consisted in calculation of energy losses using the deterministic method [1] whose efficiency was confirmed by numerous practical calculations.

Power consumption data were obtained using power meters records at a primary substation, permitted power of consumers in the secondary network, and the number of consumption phases. Total produced power was determined using power supplied to the primary network that was distributed in proportion to the rated power of secondary transformers and adjusted with account of power losses in transformers and in the primary network. Total load of each feeder was evenly distributed between load nodes or according to the requested consumers' demand.

A primary network was modeled using a single-line equivalent circuit, and a three-phase four-wire equivalent circuit proposed in [2] was used for the secondary network that allowed power flow calculation in the phase coordinates. Unlike Carson method [3] where three equations are constructed for the feeder section that connect phase voltages to the nodal currents, method [2] in the general case requires constructing five equations corresponding to three phases, a neural wire and grounding. Use of five equations instead of three is caused by the fact that parameters of phase wires in method [2] do not include corrective additives that take into account the impact of ground (ground is assumed to be an ideal conductor) and of a neutral wire.

Modeling of a secondary distribution network revealed significant differences between its characteristics and characteristics of a high voltage network.

1. A secondary distribution network operates as an open network, power flows in the feeders being directed from a distribution substation to the load nodes. This allows power flow calculation using a backward-forward method [4] following the network graph with detailed representation of models of DN elements, which is similar to direct and reverse run in the universal Gaussian method.

2. Loads in the secondary network can be single-phase, two-phase and three-phase ones that results in asymmetry loads under symmetry of parameters of the network equivalent circuit.

3. Imaginary components of voltage drops in the feeder sections are practically equal to zero, which is due to high impedance ration R/X and small lengths of individual sections.

4. Strong dependence of voltages both on reactive and active power is an important feature of the network.

5. Peculiarities of a three-phase four-wire network that are due to the impact of current running in the neutral wire and voltage drop in the phase wires caused by this current are also noteworthy.

Ref. [5] mentions that currents and voltages in the non-grounded neutral wire are the higher the greater the asymmetry of phase loads; the current curve profile in this case notably differs from the profile of voltage curve. Current in the neutral wire in the general case can exceed current in phases. Current drop in the neutral wire is due to its numerous groundings that allow distribution of this current between a neutral wire and ground in inverse proportion to their resistances. So far as current in the neutral wire effects phase voltages, then groundings and asymmetry loads have an impact on the operation condition of an electric network.

Ref. [6] shows that current in the neutral wire and voltage drop in it cause voltage change in the nodes of phase wires which is due to connection of loads between phase and zero wires, rather than between a phase wire and ground. It should be noted that voltages in the phase wires are measured in reference to the neutral wire rather than in reference to the ground. Numerous groundings of a neutral wire approximate voltage between a phase and a neutral wire to the voltage between a phase and a ground, still, they do not ensure equality to the voltage zero in the neutral wire and do not allow us to neglect it when calculating power flows in the network with asymmetry loads.

Conversion of a passive network into an active one is primarily due to introduction of uncontrollable sources of renewable generation that redirect power flows during the day. Accurate and reliable state estimation is the main method that ensures flexibility of such a transition. As compared to the state estimation using synchronized vector measurements (PMU) in the HV transmission networks, state estimation of a distribution network is a new more complicated task that requires detailed study. Moreover, synchronized PMU measurements are not used in the distribution networks due to their high cost.

Data on the state of a LV distribution network can be obtained by smart meters. In the distribution networks abroad an Advanced Metering Infrastructure (AMI) [7] connecting a distribution network with a communication network and with a modern metering structure is responsible for data transfer from a Smart Meter to the data acquisition system.

Automated System for Commercial Accounting of Electric Energy (ASCAEE) [8] is a Russian analogue of AMI. Data from consumers' smart meters and from the balance meter installed on the LV side of a secondary substation are transmitted to the upper level of ASCAEE. Along with consumed energy, these meters allow measurements of average values of active and reactive power, magnitudes of currents and voltages. These measurements are used for state estimation of a secondary distribution network and ensure its gradual transition from a passive network to an active one.

The lesser the time interval between measurements in case of coincidence of time interval for load measurements, the higher the accuracy of power flow calculation based on ASCAEE data. Nevertheless, smart meter measurements are, as a rule, non-synchronized due to errors in adjustment of a clock recording the time of measurements, and due to time delay of data transmission by communication channels.

Errors in the data (obtained from specialized geographic systems) on connection of low-voltage feeders to transformers as well as errors in the data on network topology that may vary in emergencies or during scheduled maintenances, shall also be avoided. Data on the length of conductors and their parameters need checking and, probably, clarification prior to the state estimation procedure.

Data on connection of loads to feeder phases is of primary importance and, as a rule, needs special study. This information is needed for modeling each phase of a distribution network, for determining the composition of load nodes in it and for specifying the measured values of power and voltage in the nodes for further calculation of power flows, power losses, and for state estimation.

Phase identification

There are several approaches to phase identification, the most interesting, promising and simple one being the method proposed in [9] that consists in simultaneous measurement during a long period of time, $t = 1,..,n$, where n – the number of voltage magnitude observations $U_{1a}(t)$, $U_{1b}(t)$, $U_{1c}(t)$ in the power node with known phases a, b, c, and voltage magnitude $Ui_X(t)$ in the phase x of a load node i, that needs identification.

Phase x is identified by maximum value of correlation coefficients between two vectors for measuring the voltage magnitudes in nodes 1 and i – $U_{1a}(t)$ and $Ui_X(t)$, $U_{1b}(t)$ and $Ui_X(t)$, $U_{1c}(t)$ and $Ui_X(t)$. If the maximum correlation coefficient corresponds to the firs pair $U_{1a}(t)$ and $Ui_X(t)$, then x corresponds to phase a.

It is noteworthy that not only measurements of a specially organized experimental study can be used for this method of identification, but hourly daily measurements of average values of voltage magnitudes, and daily instant measurements of voltage within one or several months as well. On the assumption that voltage measurements in nodes 1 and i are normally distributed random values, the coefficient of mutual correlation for them is determined by:

$$R_{U_{1a},Ui_X} = \frac{\sum_{t=1}^{n}\left(U_{1at} - \overline{U_{1a}}\right)\left(Ui_{Xt} - \overline{Ui_X}\right)}{\sqrt{\sum_{t=1}^{n}\left(U_{1at} - \overline{U_{1a}}\right)^2 \sum_{t=1}^{n}\left(Ui_{Xt} - \overline{Ui_X}\right)^2}} \,. \quad (1)$$

State estimation of a three-phase four-wire distribution network

Assuming that topology of a secondary distribution network that includes information on the composition of load nodes in each phase and parameters of elements of a distribution network scheme are known, let us consider the main stages of its state estimation using records of meters that include active $\bar{z}_P^{a,b,c}$ and reactive $\bar{z}_Q^{a,b,c}$ nodal powers and voltage magnitudes $\bar{z}_U^{a,b,c}$ in phases a, b, and c.

Let the problem of state estimation of a secondary distribution network with a zero wire be written similarly to the problem considered in [2] for calculation of power flow in a three-phase four-wire distribution network, and let a non-linear system of equations for measurements be solved using the simple iteration method [10]. State variables would include real $u_i'^{a,b,c}, u_i'^{n}, u_i'^{g}$ and imaginary components $u_i''^{a,b,c}$, $u_i''^{n}$, $u_i''^{g}$ of the vectors of phase voltages, neutral wire voltages, and ground voltages for the grounded neutral wire, where i is the node number.

State estimation procedure in the simple iteration method consists of two steps repeated iteratively. On the first step of the first iteration we determine pseudo measurements $\bar{z}_{Jai}^{a,b,c,n,g} + j\bar{z}_{Jpi}^{a,b,c,n,g}$ of nodal currents in each phase and current in the neutral wire using measured values of active and reactive nodal power and initial approximations of voltage magnitudes, and currents in the ground in case of a grounded neutral wire

$$\begin{pmatrix} \bar{z}_{Jai}^{a} + j\bar{z}_{Jpi}^{a} \\ \bar{z}_{Jai}^{b} + j\bar{z}_{Jpi}^{b} \\ \bar{z}_{Jai}^{c} + j\bar{z}_{Jpi}^{c} \\ \bar{z}_{Jai}^{n} + j\bar{z}_{Jpi}^{n} \\ \bar{z}_{Jai}^{g} + j\bar{z}_{Jpi}^{g} \end{pmatrix}^{k} = \begin{pmatrix} \left(\bar{z}_{iP}^{a} - j\bar{z}_{iQ}^{a}\right)/U_i'^{a} \\ \left(\bar{z}_{iP}^{b} - j\bar{z}_{iQ}^{b}\right)/U_i'^{b} \\ \left(\bar{z}_{iP}^{c} - j\bar{z}_{iQ}^{c}\right)/U_i'^{c} \\ -\left(J_{ai}^{k} + J_{bi}^{k}\cdot a2 + J_{ci}^{k}\cdot a1\right)Z_{gi}/\left(Z_{nni} + Z_{gi}\right) \\ -\left(J_{ai}^{k} + J_{bi}^{k}\cdot a2 + J_{ci}^{k}\cdot a1\right)Z_{nni}/\left(Z_{nni} + Z_{gi}\right) \end{pmatrix}, \quad (2)$$

where, according to [2], $Z_{gi} = Z_{ggi} + Z_{gri}$ equals to the sum of ground impedance Z_{ggi} and grounding impedance Z_{gri}, Z_{nni} is impedance of neutral wire at node i, $a1 = -1/2 + j\sqrt{3}/2$, $a2 = -1/2 - j\sqrt{3}/2$.

In equation (2) we assumed that admittances of shunt elements of phase wires and that of a neutral wire are equal to zero.

At subsequent iterations, instead of initial approximations of voltages $U_i'^{a,b,c}$, the estimates of phase voltage vectors $\hat{U}^{a,b,c} = \left(\hat{u}'^{\,a,b,c}_i - j\hat{u}''^{\,a,b,c}_i\right)$ obtained at a previous iteration and refined taking into account the voltages in the neutral wire U^n are inserted into denominators of the first three equations

$$\hat{U}^{a,b,c} = U^{a,b,c} - U^{n}. \quad (3)$$

Linear equations of measurements (4) corresponding to pseudo measurements of active and reactive components of currents and real components of voltages are specified as equal to measured voltage magnitudes [10]:

$$\begin{pmatrix} g_i^{a,b,c,n,g} & b_i^{a,b,c,n,g} \\ -b_i^{a,b,c,n,g} & g_i^{a,b,c,n,g} \\ \mathrm{I}^{a,b,c,n,g} & 0^{a,b,c,n,g} \end{pmatrix} \begin{pmatrix} u_i'^{a,b,c,n,g} \\ u_i''^{a,b,c,n,g} \end{pmatrix} = \begin{pmatrix} \bar{z}_{Jai}^{a,b,c,n,g} \\ \bar{z}_{Jpi}^{a,b,c,n,g} \\ \bar{z}_{Ui}^{a,b,c} \end{pmatrix}, \quad (4)$$

where $g_i^{a,b,c,n,g}$ and $b_i^{a,b,c,n,g}$ are nodal admittance matrices; $I^{a,b,c}$ and $0^{a,b,c}$ are identity and zero matrices.

If all the mutual resistances between phase wires, neutral wire and ground are assumed to be equal to zero, equations (4) can be solved at each iteration either simultaneously or for each phase, for neutral wire and for ground separately.

In the general form the system (4) can be put down as:

$$R^{-1/2}H \cdot u = R^{-1/2}\bar{z}, \quad (5)$$

where R is diagonal matrix whose elements are equal to the variances of measurement errors.

Under availability of redundant measurements the $R^{-1/2}H$ is over determined matrix, therefore, for solving (5) the weighted least-squares method is used

$$J(u) = (\bar{z} - Hu)^{T} R^{-1} (\bar{z} - Hu).$$

Results of calculations on the example of an actual distribution network

Let us first illustrate the preparation of data on referring the loads to phases of a real main feeder of a distribution network to nine nodes of which 24 private houses are connected (numbered from 12 to 35 in Fig. 1).

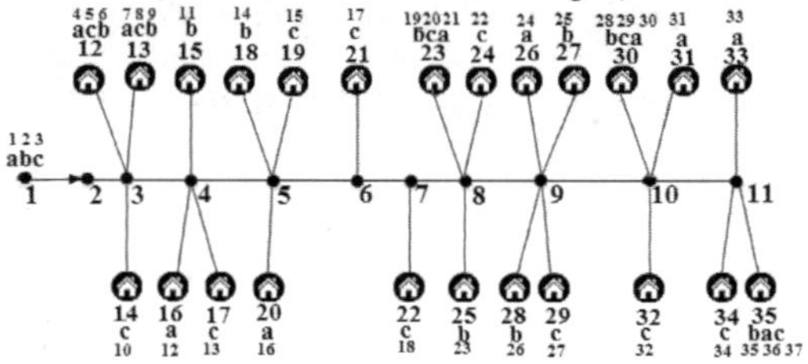

Fig. 1. Scheme of the main feeder of a distribution network

Electric power consumed by five three-phase loads 12, 13, 23, 30, and 35 of houses is measured by three-phase meters MIR C-04, and electric power consumed by remaining single-phase consumers is measured by MIR C-05 meters. A balance three-phase meter MIR C-07 whose connection phases are known is installed in the power node of the feeder. All the meters are installed on the supports of the main feeder, e.g., in node 3 there are three meters that measure consumed power of 12-14 houses and power losses in branches to those houses.

Phases for meters connection were identified using data of ASCAEE protocols on daily measurements of 37 instant values of voltage magnitudes within two winter months (once at approximately the same time of a day) and measurements of average hourly values of voltage magnitudes in a spring and summer day.

Correlation coefficients were computed using MATLAB procedure *corrcoef* that along with determination of mutual correlation coefficients R gives information on probability P of their occurrence in the 95% default confidence range whose lower RLO and upper RUP boundaries are defined. Correlation is assumed to be valid if probability that the correlation coefficient exceeds the boundaries of a confidence range is less than 0.05.

Fig. 2 presents graphs of mutual correlation coefficients between each out of three measurements of voltage magnitudes for the sampling of instant measurements of voltage magnitudes for two winter months: 1 – in phase a, 2 – in phase b, 3 – in phase c of a power supply node with every (from 4 to 37) measurement of voltage magnitudes in the unknown phases of load nodes. Maximum positive values of mutual correlation coefficients allowed determination of phases of each metering (from 4 to 37) shown in Fig. 1.

Thus, maximum correlation coefficient for a three-phase meter related to node 12 corresponds to its first phase and phase a of the power source; subsequent maximum correlation coefficients relate the second phase of the meter to phase c of the power source, and the third phase of a meter to phase b.

Analysis of correlation coefficients for other samplings showed that all of them identify the same phases of load connections. But more certain assessment of load connection phases was obtained for daily hourly measurements of voltage magnitudes the major share of coefficient for which was in the strong correlation zone, and all the coefficients were relevant.

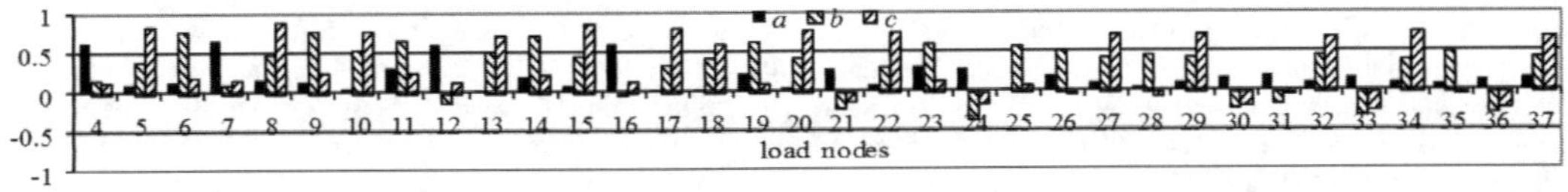

Fig. 2. Mutual correlation coefficients between measurements of voltages in phases a, b, c in the power supply node and in the phases of each of the load nodes

Assuming that impedances of phase and neutral wires are set correctly, let us illustrate the process of state estimation in the distribution network (Fig. 1) for one of the operation conditions, for which 37 sets of phase measurements of average values of active and reactive power and of voltage magnitudes are known. Average values of power measurements (Fig. 3) and voltage magnitudes in phases with reference to the voltage in the neutral wire (Fig. 4) were determined for meters installed in the same node. Average value of measurements in phase a were determined in nodes 10 and 11; in phase b – in nodes 3, 8, 9, and in phase c – in nodes 3, 8, 10, and 11.

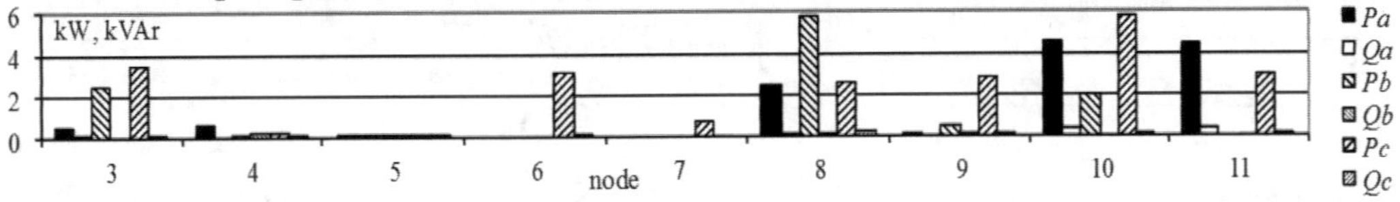

Fig. 3. Measurements of active and reactive loads in the feeder phases

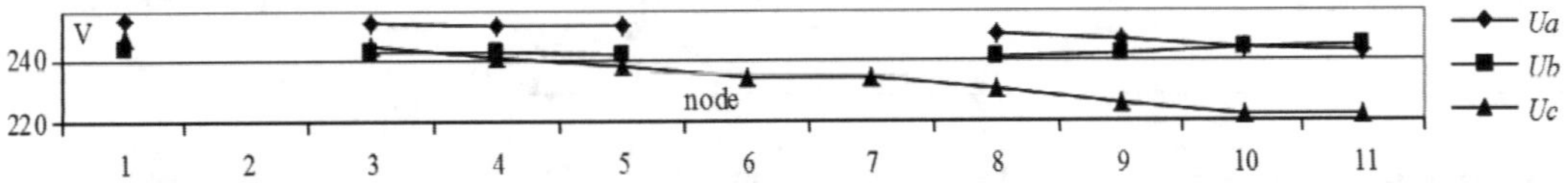

Fig. 4. Measurements of average values of voltage magnitudes in the feeder phases

State estimation was done for the basic set of measurements that included voltage measurement in power supply node 1 and measurements of active and reactive power in the load nodes. Fig. 5 shows current magnitudes in the phases and currents in the neutral wire caused by asymmetry loads that were obtained in the course of calculations. Estimates of voltages in phases with reference to ground that were obtained without account of voltages in the neutral wire, and voltages in the neutral wire with reference to ground are shown in Figs. 6 and 7.

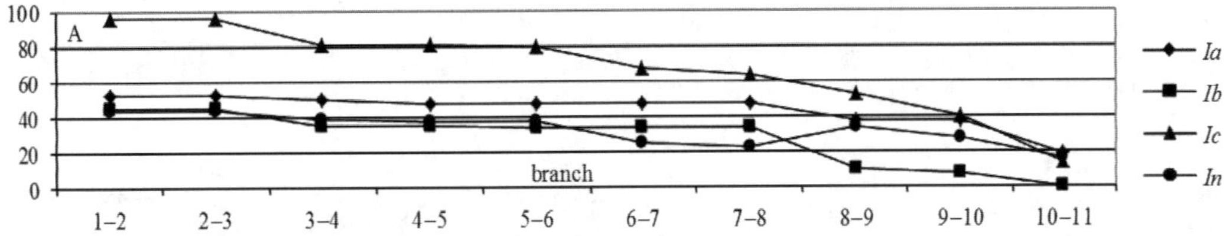

Fig. 5. Currents in phases Ia , Ib , Ic of the feeder and in the neutral wire - In

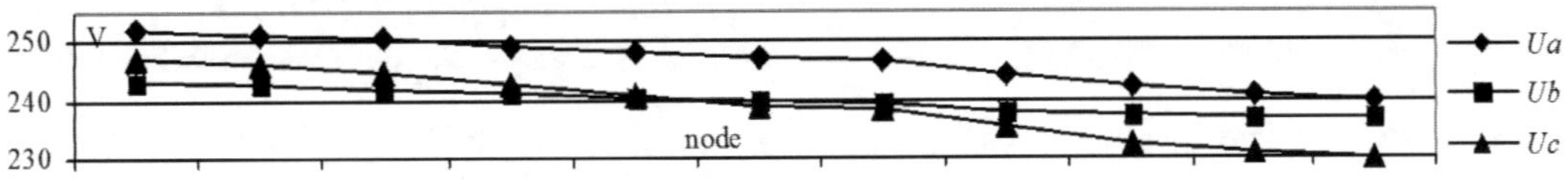

Fig. 6. Voltages in phases Ua, Ub, Uc of the feeder to ground

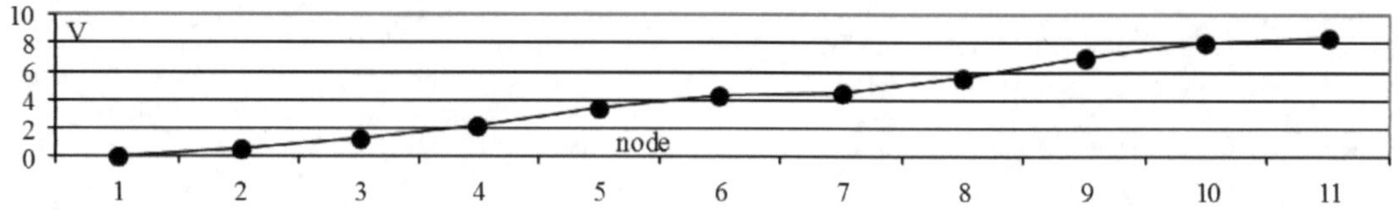

Fig. 7. Voltages in the neutral wire to ground

If voltage profiles in phases with reference to ground are similar, then voltage estimates in phases with reference to the neutral wire that were adjusted with account of voltages in the neutral wire (Fig. 8) differ from one another, especially at the end of the feeder. Estimates of phase voltages with reference to the neutral wire (Fig. 8) can be compared to measured voltage magnitudes and show that the maximum estimate error against the measurement is 1.5V.

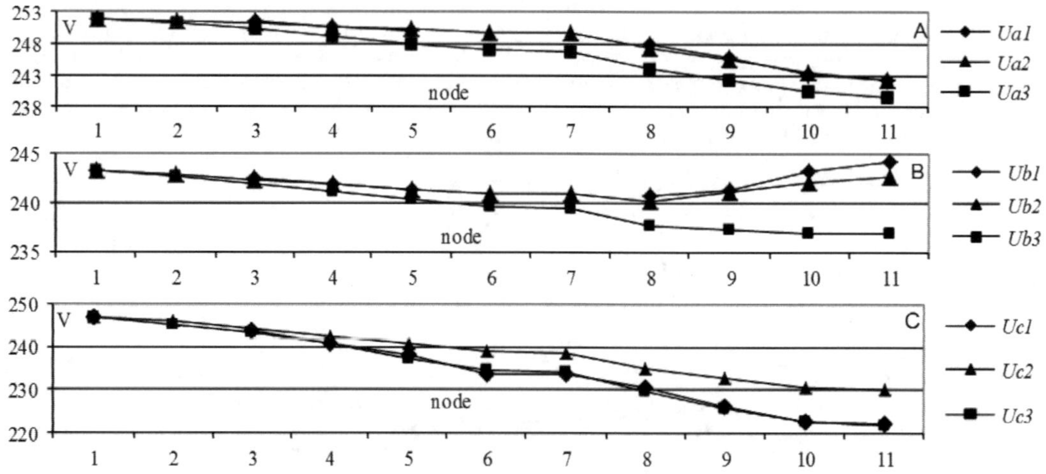

Fig. 8. Voltage curves for phases a, b, c: measured ones (phase to neutral) – $Ua1$, $Ub1$, $Uc1$; calculated ones (phase to ground) – $Ua2$, $Ub2$, $Uc2$; refined ones with account of voltage in the neutral wire (phase to neutral) – $Ua3$, $Ub3$, $Uc3$

Conclusion

Comparison of the results of voltage assessment to their measured values demonstrates a small error of estimates that, hopefully, can be lowered by increasing the number of measurement samplings; there is no need to compute average voltage values in the nodes using mismatched measurements from several meters, and to clarify data on resistances of wires and groundings. Another important problem for the analyzed feeder is reduction of voltage asymmetry as the maximum coefficient of phase voltage asymmetry for measured voltage magnitudes made up 8%, and that for linear voltages made up 9%.

The research was done under support of Scientific Project III.17.4.2 of Fundamental Research Program of SO RAN, reg. No. № AAAA-A17-117030310438-1

References

1. O.N. Voitov, I.I. Golub, L.V. Semenova, Electricity **9**, 38 (2010).
2. R.M. Ciric, A.P. Feltrin, L.F. Ochoa, IEEE Trans. Power Syst. **18**, 1283 (2003)
3. J.R. Carson, Bell Labs Technical Journal. **5**, 539 (1926)
4. C.S. Cheng, D.A. Shirmohammadi, IEEE Trans. Power Syst. **10**, 671 (1995)
5. L.N. Baptidanov, Electric networks and stations (Gosenergoizdat, Moskow-Leningrad, 1963)
6. R.R. Mohassel, A.S. Fung, F. Mohammadi, K.A. Raahemifar, CCECE **27**, 8 (2014)
7. A.M. Maksimova, D. K. Yemelianova, Molodoy Uchenyi **21**, 177 (2016)
8. .R. Mohassel, A.S. Fung, F. Mohammadi, K.A. Raahemifar, CCECE **27**, 8 (2014)
9. H. Pezeshki, P. Wolfs, ISGT Europe **3**, 7 (2012)
10. E.V. Boloev, I.I. Golub, V.V. Fedchishin, Vestnik IRGTU **22**, 95 (2018)

Development of the logical part of the intellectual multi-parameter relay protection

Anton Loskutov [1,*], *Pavel* Pelevin [1] and *Mile* Mitrovoc [1]

[1]Department of Electric Power Engineering, Power Supply and Power Electronics Nizhny Novgorod State Technical University n.a. R.E. Alekseev 603950, Nizhny Novgorod, Minin St., 24, Russia

Abstract. The issues of increasing the sensitivity and reliability of multi-parameter relay protection by sharing more than one information feature (current module, voltage module and phase, active and reactive power) are considered. In this case, the response parameters of individual one-dimensional measuring fault detectors based on the accumulation of statistical data during simulation in the Matlab / Similink software package are determined. A method for combining the signals of one-dimensional measuring fault detectors to increase the sensitivity of protection is proposed. The reliability of the organization of the logical part of multi-parameter relay protection was estimated using the theory of Markov processes, the principles of "2 out of 3" and "1 out of 2".

1 Introduction

The introduction of modern electrical equipment, the development of renewable energy sources and the integration of active elements for regulating the parameters of the electric network affect the complexity of its configuration and operating modes, which leads to increased requirements for the sensitivity and reliability of relay protection devices (RP) [1]. Currently, there is a massive transition of RP devices to a microprocessor elemental base, which allows a radical change in the methods and approaches of organizing the logical part of RP, improving technical excellence.

Until now, the logic of modern relay protection has remained unchanged. It practically corresponds to traditional electromechanical RP devices. This leads to the underutilization of the potential and computational power of digital RP devices.

In contrast to electromechanical RP, in microprocessor RP, in addition to the hardware, there is a software part that implements the logic of the protection device. The hardware and software of a modern RP device have different failure and recovery processes, which affects the adequacy of the overall operational reliability assessment. There are many methods that estimate the hardware reliability of devices [2,3]. However, today there are no stable methods for calculating reliability that estimate the operation of the logical part of a RP device.

In western countries, there are now modern and trendy approaches for organizing the logical part of "Special Protection Schemes (SPS)" [4,19] and "Protection Redundancy" [5], which use redundancy to ensure high indicators of technical excellence of RP devices.

2 The relevance and purpose of the research

The generally accepted method of increasing the sensitivity of RP and the recognition of emergency conditions is to expand the number of observed parameters and is associated with the construction of multi-parameter RP [6-9,13,18]. However, it is difficult to organize decision-making procedures with a large number of observed parameters and there are no unambiguous implementation methods for that.

The closest approaches to the practical implementation of decision-making procedures for multi-parameter RP are the information approach [6–9] and statistical methods for recognizing modes [10–14]. The disadvantages of these approaches is the complicated principle of constructing multi-parameter RP through multi-dimensional set-point spaces of measuring fault detectors. By the increase of the dimension of more than two observed parameters, we can see a sharp and unjustified increase in the computational cost of implementing such protections, their significant increase in cost, the decrease in reliability and a possible limitation of practical use.

The aim of this work is to build multi-parameter RP with the union of individual measuring fault detectors (FD) through the logical part of RP to increase sensitivity and ensure the maximum possible reliability of RP devices. This implies the development of special algorithms that are different from those that are currently intuitively used and that provide the processing of information to a fuller extent.

To achieve this aim it is necessary to know the probabilistic characteristics of each FD that affect the system of random disturbances (signals). The

* Corresponding author: loskutov.nnov@gmail.com

 https://doi.org/10.1051/e3sconf/201913901060

organization of the logical part is based on the voting scheme, including the principles of "1 out of 2" and "2 out of 3", and the reliability of the proposed structure is calculated using the theory of Markov processes.

3 Statement of the task

To train relay protection and obtain statistics on normal and emergency modes, a simulation of a 220 kV network with adjacent overhead lines was carried out (Figure 1). Network simulation is implemented in the Matlab/Simulink software package [15].

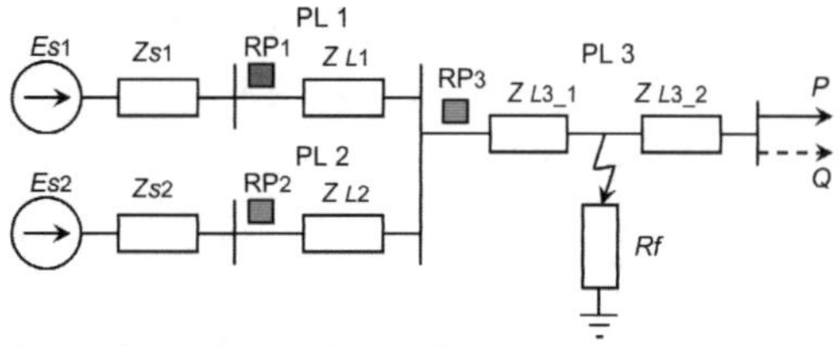

Fig. 1. The equivalent circuit of the observed section of the electric network 220 kV

The statistics on the modes were accumulated at the site of the observed RP_1, where 10,000 iterations of single-phase faults along overhead line 3 (VL3) were examined. One of the factors affecting the reduction in the probability of recognition of emergency conditions and the sensitivity of RP is the contact resistance at the site of fault (R_f). Therefore, the simulation of short circuits (SC) was carried out taking into account R_f. Thus, in the set-point space of RP, there is a region of intersection of normal and emergency modes, which leads to unrecognizability and complication of making the right decision on shutdown. Figure 2 shows an example implementation of a measurement FD based on measurements of current and voltage modules.

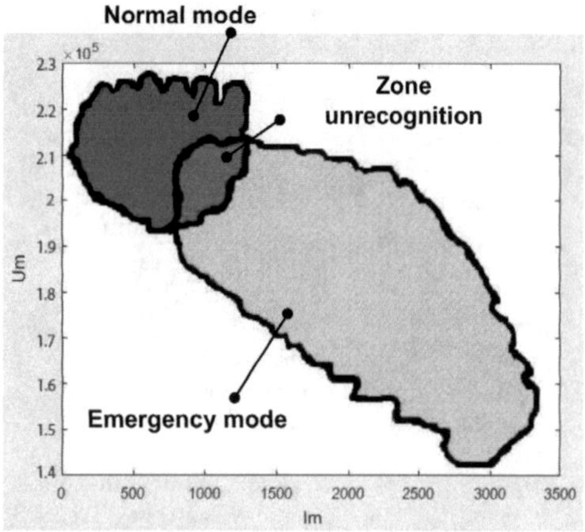

Fig. 2. Areas of normal and emergency modes of the current module and voltage module for simulated network

Presentation of the obtained statistics on the modes is possible in the form of graphs of the density of conditional probability, where the horizontal axis is the measured current in the normal and emergency modes in the place of the observed RP_1 (Figure 3). To simplify the calculation, the approximation of the probability by the normal distribution was carried out. The Bayesian criterion [14, 16] was used to make decisions.

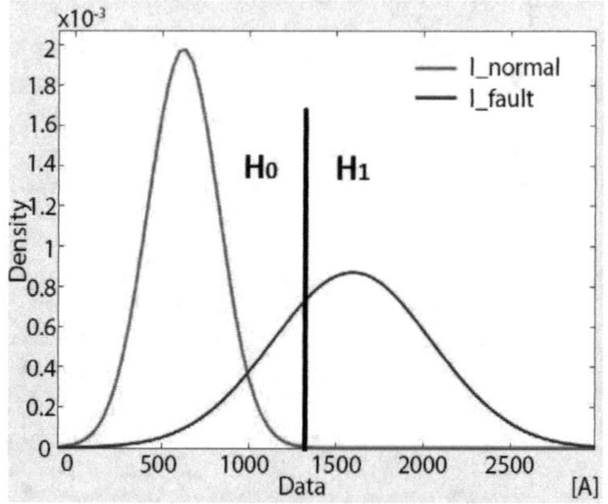

Fig. 3. Conditional Density Chart probabilities of normal permissible and current module emergency modes

To determine the response parameters, a solution of the two-hypothesis problem is proposed. It is accepted that the recognition problem has only two possible solutions and involves testing hypotheses H0 and H1: H0 - means that a measured signal of normal mode (zero) is formed at the output of the solving device (FD), H1 - at the output of the solving device (FD) a measured alarm signal is generated (unit) [11].

To increase the sensitivity of RP there was used the multidimensional approach. There were considered for example five separate one-dimensional FD: module current (Im); module voltage (Um); phase voltage (φ); active power (P); reactive power (Q). The Bayes criterion was used to calculate the settings and form two hypotheses task for each FD.

The total signal received from all FD and fed to the input of the logical part of the protection is a fixed sample of elementary discrete binary signals (0 or 1) observed at the same time: vector A ($A = A_{1i}, A_{2i}, \ldots A_{pi}$, where $i=1 \ldots n$, n - the number of observed modes (iterations), p - the number of FD). To develop a decision rule for the logical part of RP, it is necessary to process the incoming logical signals in accordance with the results of preliminary simulation [11,16].

The formation of an algorithm for the recognition of RP of increased sensitivity is possible when the logical operation "OR" combines discrete signals - elements of vector A. The essence of the combination in this way is triggering the protection if it is recognized by at least one FD. Simulation showed that when using five FDs, the probability of recognition of emergency conditions increases by 67.5% in relation to the FD with the lowest probability of triggering. It also affects the overall increase in the sensitivity of the RP.

To estimate the sensitivity, an object characteristic was used - the dependence of the contact resistance on the distance of damage on the line at which RP provides short-circuit recognition (Figure 4) [17]. In accordance with the object characteristics, it is possible to establish at what contact resistances and remoteness from the substation there is a short circuit unrecognition by individual FD and the protection as a whole.

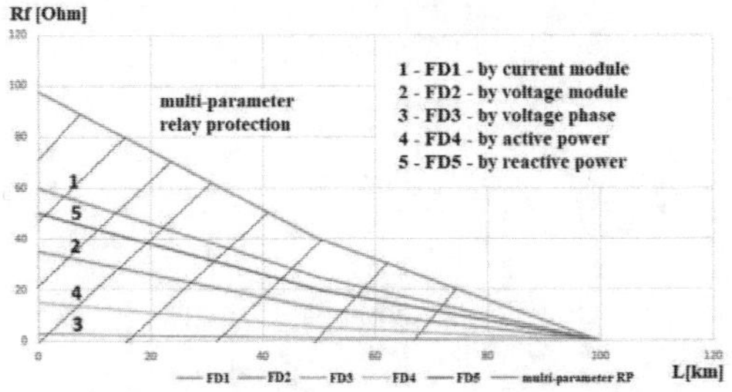

Fig. 4. Object characteristic of recognition of multi-parameter relay protection

For the circuit (Figure 1), simulation experiments were carried out at the beginning of the VL3 line (0 km), in the middle (50 km) and at the end (100 km). Figure 4 shows that multi-parameter RP provides an increase in sensitivity compared to the most sensitive FD, i.e. it recognizes emergency conditions for a given contact resistance.

4 The proposed method of organizing the logical part of multi-parameter relay protection

The proposed approach that increases the sensitivity of relay protection is focused on the functioning of RP in the conditions of absolute reliability of individual FDs. In practice there can occur failures, leading to failure of the relay protection in emergency mode or excessive operation in normal conditions. In practice, failures are possible that lead to failure of the RP in emergency mode or excessive triggering in normal conditions.

4.1 Principle "2 out of 3"

To organize the logical part (LP) of RP, it is proposed to use voting schemes with combining the logical signals of vector A on the principle of "2 out of 3" [18,19]. For this, three identical "OR" blocks are used, providing redundancy in information processing. The unstable operation of one of the logical elements does not affect the functioning of the RP, which ensures an increase in the necessary reliability. The structural diagram of the organization of the LP of the proposed algorithm is shown in Figure 5.

To estimate the reliability of the proposed structure, there was used Markov model of transitions from one state to another. It should be emphasized that the reliability estimation was carried out indirectly due to the fact that the source of the failure and recovery rate was the data from the published literature on the reliability of special protection schemes [19].

Table 1 shows the values of the failure rate and recovery, where: λ_{fd} - the failure rate for detecting the failure of a logical element; λ_{fnd} - failure rate for not detecting a logic element failure; μ_{rd} - recovery rate for detecting the failure of a logical element per year; μ_{rnd} - recovery rate for not detecting a logic element failure per year.

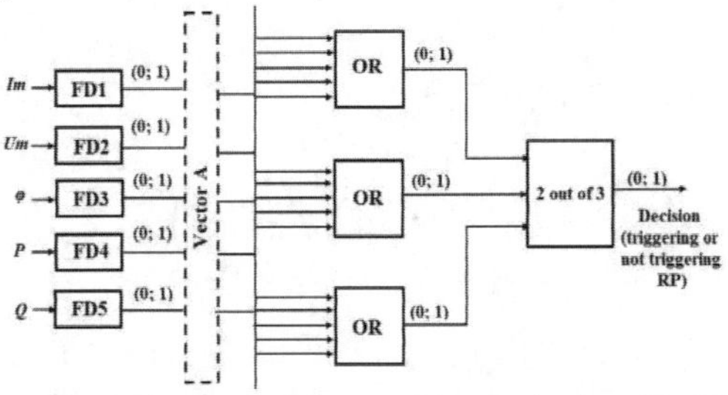

Fig. 5. The structural diagram of the organization of the LP when combining the totality of the signals of individual FD with the principle of "2 out of 3"

Table 1. Failure and recovery rates

Parameter	Values
λ_{fd}	0.018 (once every 25 years)
λ_{fnd}	0.026 (once every 35 years)
μ_{rd}	1095 (8 hours a year)
μ_{rnd}	4 (test 6 months)

4.2 Reliability estimation of the proposed structure using the theory of Markov processes

The Markov model of states of the proposed organization scheme of the logical part is shown in Figure 6. For this model, five states are identified: 1 - all three logical elements work without fail; 2 - one of the logical elements has failed and a failure has been detected; 3 - one of the logical elements has failed and no failure has been detected; 4 - two of the logic elements failed and a failure was detected; 5 - two of the logic elements failed and no failure was detected.

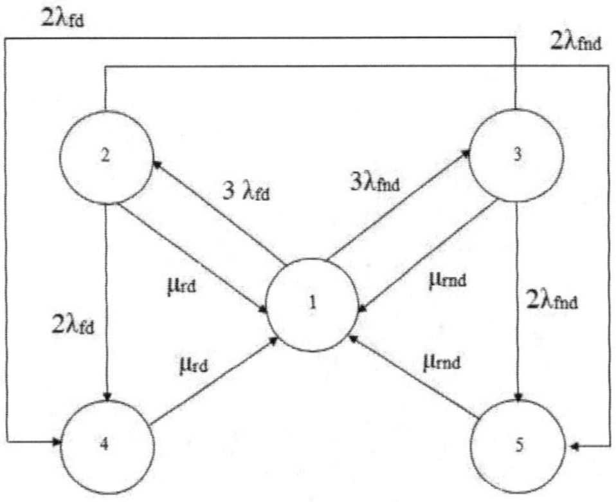

Fig. 6. Markov model of the state of the proposed organization of the logical part

For the Markov model, the determination of the probabilities $P_i(t)$ ($i = 1,2, \ldots n$) is performed by solving the Kolmogorov-Champen system of differential equations. Due to the fact that in the theory of random processes it was proved that homogeneous Markov processes without absorbing states (states from which there is no way out) have a stationary mode that occurs at sufficiently large times (t → ∞), we obtain a system of algebraic equations for determining the reliability exponents [20]. Thus, the following system of equations is formed:

$$\begin{gathered} -\ 3\lambda_{fd} + 3\lambda_{fnd}\ P_1 + \mu_{rd}P_2 + \mu_{rnd}P_3 + \\ \mu_{rd}P_4 + +\mu_{rnd}P_5 = 0; \\ -(2\lambda_{fd} + 2\lambda_{fnd} + \mu_{rd})P_2 + 3\lambda_{fd}P_1 = 0; \\ -(2\lambda_{fd} + 2\lambda_{fnd} + \mu_{rnd})P_3 + 3\lambda_{fnd}P_1 = 0; \\ -\mu_{rd}P_4 + 2\lambda_{fd}P_2 + 2\lambda_{fd}P_3 = 0; \\ -\mu_{rnd}P_5 + 2\lambda_{fnd}P_2 + 2\lambda_{fnd}P_3 = 0; \\ P_1 + P_2 + P_3 + P_4 + P_5 = 1. \end{gathered} \quad (1)$$

So that the resulting system is not degenerate, a rationing condition is added ${}_{i=1}^{n} P_i$.

Further, after summing the first and last equations, we obtain the following system:

$$\begin{gathered} 1 -\ 3\lambda_{fd} + 3\lambda_{fnd}\ \ P_1 +\ 1 + \mu_{rd}\ P_2 + \\ +\ 1 + \mu_{rnd}\ P_3 +\ 1 + \mu_{rd}\ P_4 + (1 + \mu_{rnd})P_5 = \\ 1; \\ -(2\lambda_{fd} + 2\lambda_{fnd} + \mu_{rd})P_2 + 3\lambda_{fd}P_1 = 0; \\ -(2\lambda_{fd} + 2\lambda_{fnd} + \mu_{rnd})P_3 + 3\lambda_{fnd}P_1 = 0; \\ -\mu_{rd}P_4 + 2\lambda_{fd}P_2 + 2\lambda_{fd}P_3 = 0; \\ -\mu_{rnd}P_5 + 2\lambda_{fnd}P_2 + 2\lambda_{fnd}P_3 = 0. \end{gathered} \quad (2)$$

By the substitution of numerical values of the probability of each state, we can calculate the following:

$$\begin{matrix} P_1 \\ P_2 \\ P_3 \\ P_4 \\ P_5 \end{matrix} = \begin{matrix} 0.868 & 1096 & 5 & 1096 & 5 \\ 0.054 & -1095.088 & 0 & 0 & 0 \\ 0.078 & 0 & -4.06 & 0 & 0 \\ 0 & 0.036 & 0,02 & -1095 & 0 \\ 0 & 0.052 & 0.04 & 0 & -4 \end{matrix}^{-1} \begin{matrix} 1 \\ 0 \\ 0 \\ 0 \\ 0 \end{matrix} = \begin{matrix} 0,981 \\ 4,8374 \cdot 10^{-5} \\ 0,0187 \\ 6,1696 \cdot 10^{-6} \\ 2,4396 \cdot 10^{-4} \end{matrix} \quad (3)$$

Calculations show that when using the "2 out of 3" principle, a sufficiently high reliability is obtained and the protection will work in 98.1% of cases without fail.

4.3 An additional increase in the reliability of the logic part of the relay protection

In order to further increase reliability, the "1 out of 2" principle was used. It does not affect the achieved maximum sensitivity and does not violate the essence of combining information from FDs [5,18]. This approach implies the redundancy of the system of logical elements, united by the principle of "2 out of 3" and supplemented by the logic "1 out of 2" (Figure 7).

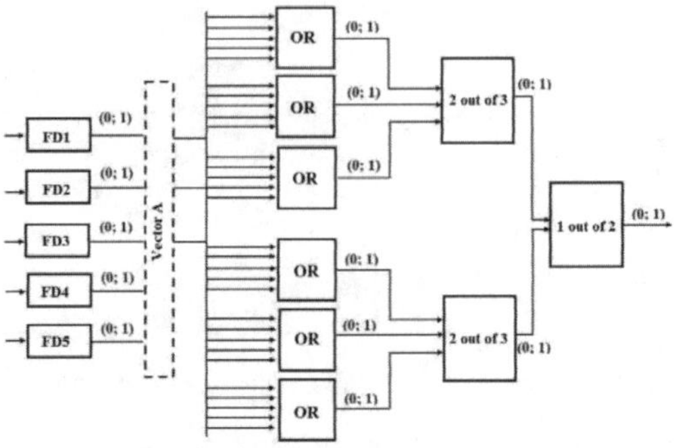

Fig. 7. The structural diagram of the organization of the logical part of the principle of "1 of 2"

To evaluate such a system, the failure model of non-renewable systems, parallel connection of elements was used. Suppose that both systems have the same characteristics and that their probability of failure-free operation *p(t)* is equal to the probability obtained by the calculation of the Markov model - $p_1(t) = p_2(t)$ = 98.1% = 0.981 p.u. According to this, the failure probability is equal to *q(t)* = 1-*p(t)*, where $q_1(t) = q_2(t)$ = 1-0.981 = 0.019p.u. With a parallel connection of elements, the probability of failure of the system is:

$$q(t) = {}_{i=1}^{n} q_i\ t\ . \quad (4)$$

Based on the expression (4), we find that the probability of failure of the system of the logical part based on the principle of "1 out of 2" is *q(t)* = 0.000361. According to this, the probability of uptime is *p(t)* = 1-*q(t)* = 1-0.000361 = 0.999639. The organization of a logical organ supplemented by the principle of "1 out of 2" leads to a probability of correct operation of up to 99.96%. Thus, an additional increase in reliability is provided, without violating the achieved level of sensitivity and recognition of emergency conditions.

Conclusion

1. An approach to organizing the logical part of multi-parameter relay protection in order to improve sensitivity and reliability has been proposed.

2. To determine the response parameters of various FD, a statistical approach was used with the formation of distribution laws, based on the application of the Bayes criterion and detuning from all admissible modes.

3. The method of organizing the logical part of the relay protection is based on the operation "OR" in order to increase the sensitivity and recognition of emergency modes by relay protection.

4. The organization of the relay protection logic by the principles of "2 out of 3" and "1 out of 2" ensured

high reliability of the device, which was evaluated using the theory of Markov processes.

The presented research results were obtained with the support of a grant from the President of the Russian Federation for state support of young Russian scientists (MK-3210.2019.8). Agreement No. 075-15-2019-337 of 11.06.2019.

References

1. Loskutov A B 2018 Solving Problems in Transition of Electrical Industry to Digital Technologies *Smart Electrical Engineering* **1**(**1**) 9–27
2. STO 34.01- 4.1-008-2018. Mikroprotsessornyye ustroystva re-leynoy zashchity i avtomatiki. Metodicheskiye ukazaniya po raschetu nadezhnosti. Standart organizatsii. - Vved. 28.04.2018. – PAO «Rosseti, 2018. – p. 41 (in Russian)
3. Shalin A.I. Nadezhnost' i diagnostika releynoy zashchity energosistem, Novosibirsk, NGTU, 2002. - p. 384 (in Russian)
4. Ahmed Sh. K., Karthikeyan S.P., Sahoo S.K. Special Protection Schemes: A Survey and Vision for the future / *Applied Mechanics and Materials* **839**, 2016, pp 49-53.
5. IEEE PSRC 2008 WG I-19 Redundancy Considerations for Protective Relaying Systems
6. Lyamets Yu Ya, Nudel'man G S Zinov'yev D V, Kerzhayev D V, Romanov YU V 2009 Mnogomernaya releynaya zashchita. CH.1. Teoreticheskiye predposylki *Elektrichestvo* **10** pp 17-25 (in Russian)
7. Lyamets Yu Ya, Nudel'man G S, Podshivalina I S , Romanov Yu V 2011 Effekty Mnogomernosti v Releynoy Zashchite *Elektrichestvo* **9**, pp 48-54 (in Russian)
8. Ivanov S V, Lyamets Yu Ya. 2016 Metod Informatsionnogo Analiza. Raspoznavaniye Zamykaniy v Zadannoy Zone Dvukhtsepnoy Elektroperedachi *Izvestiya akademii nauk energetika* **1** pp 47-57 (in Russian)
9. Nagay I V, Nagay V I 2013 Postroyeniye Mnogoparametricheskikh Rezervnykh Zashchit Elektricheskikh Raspredelitel'nykh Setey 6–10 kV *Energetik* **2** pp 18-21 (in Russian)
10. Sharygin M.V., Kulikov A.L. Zashchita i avtomatika sistem elektrosnabzheniya s aktivnymi promyshlennymi potrebitelyami. Nizhniy Novgorod, 2017, pp. 40-68. (in Russian)
11. Sharygin M.V., Kulikov A.L. Ob"yedineniye signalov sovokupnosti otdel'nykh puskovykh organov releynoy zashity // *Elektricheskiye stantsii*, **9** 2018, pp. 42-48. (in Russian)
12. Sharygin M.V. Primeneniye statisticheskikh kriteriyev raspo-znavaniya rezhima releynoy zashchity setey elektrosnabzheniya / M.V. Sharygin, A.L. Kulikov // *Elektrotekhnika* **2** 2019 pp. 58-64. (in Russian)
13. Sharygin M V, Kulikov A L 2018 Statistical Methods of Mode Recognition in Relay Protection and Automation of Power Supply Networks Power Tech. and Eng **52** (**2**) pp 235-41
14. Kulikov A.L., Sharygin M.V. Opredeleniye ustavok releynoy zashchity i avtomatiki, osnovannoye na statisticheskom bayyesovskom metode proverki gipotez // *Elektrichestvo*, **7** 2017 pp. 20 – 29. (in Russian)
15. Svidetel'stvo o gos. registratsii programmy dlya EVM № 2019619654. Programma formirovaniya statisticheskikh dannykh po normal'nym i avariynym rezhimam raboty v elektricheskoy seti dlya izmeritel'nykh organov mnogoparametricheskoy releynoy zashchity / Kulikov A.L., Loskutov A.A., Mitrovich M. – zayavl. 08.07.2019; opubl. 22.07. 2019. (in Russian)
16. Kulikov A L, Loskutov A A, Mitrovic M 2019 Method of automated synthesis of the logic part of relay protection device which increases its sensitivity *Int. Sc-Tech Conf. ISEPC-2019* (23-24 May 2019, St. Petersburg, Russia)
17. Lyamets YU. YA. Metod ob"yektnykh kharakteristik dlya analiza i sinteza distantsionnoy zashchity / YU. YA. Lyamets, V. I. Antonov, G. S. Nu-del'man // Izvestiya vysshikh uchebnykh zavedeniy. *Elektromekhanika* **1** 1999. p. 95. (in Russian)
18. Kulikov A L, Loskutov A A, Mitrovic M 2019 Improvement of the technical excellence of multiparameter relay protection by combining the signals of the measuring fault detectors using artificial intelligence methods *Int. Sc-Tech Conf. SES-2019* (18-20 September 2019, Kazan, Russia)
19. McCalley J, Oluwaseyi O, Krishnan V, Dai R, Singh C, Jiang K 2010 System Protection Schemes: Limitations, Risks, and Management *PSERC* pp 10-9
20. Viktorova V.S., Stepanyants A.S. Modeli i metody rascheta nadezhnosti tekhnicheskikh sistem // Moskva, 2013, pp. 126-136. (in Russian)

E3S Web of Conferences **139**, 01061 (2019)
RSES 2019
https://doi.org/10.1051/e3sconf/201913901061

Improvement of algorithms for voltage circuits fault detection in relay protection terminal of 6-35 kV electrical networks

Vladimir Vukolov[1], *Mikhail* Obalin[1], *Anton* Petrov [2,*]

[1]R. E. Alekseev Nizhnii Novgorod State Technical University, 603950, Minin Str. 24, Nizhnii Novgorod, Russia
[2]JSC NIPOM, advanced development department, 603140, Lenin Ave., 20, Nizhnii Novgorod, Russia

Abstract. Issues of increase of consumer power supply reliability by prevention of protection device false operation and unreasonable tripping of the protected network element at malfunction of measuring voltage transformer secondary circuits are considered. Options for implementation in relay protection microprocessor terminals of advanced blocking algorithms in case of voltage circuits fault are proposed.

1 Introduction

Reliability, along with speed of response, sensitivity and selectivity, is the main characteristic of relay protection devices. Reliability of modern microprocessing terminals of relay protection devices is understood so that the protection device should operate at all types of damages provided by design and should not-operate in all other cases for which protection with this device is not provided. Unreliable relay protection device itself becomes a source of accidents, and false triggering of protective relays lead to unreasonable network tripping which leads to economic losses due to production or technological failures.

Damages to the voltage transformer circuits leads to a false triggering of relay protection devices, which use the voltage signal. Fault detection in the measuring transformer circuits is carried out automatically using the voltage circuit lock function.

Voltage transformer secondary circuit damages are short circuit and a circuit interruption. To protect a measuring transformer against short-circuit overcurrent, circuit breakers or fuses which disconnect the damaged circuit can be added into its secondary circuits. The very moment of damage and subsequent triggering of the circuit breakers or fuses lead to distortion of voltage signal and improper operation by relay protection devices. Primary circuits' interruption without short circuit is also a kind of voltage transformer damage. A fact of voltage circuits' damage should be detected by voltage circuit lock function which sends a fault signal and blocks operation of relay protection device which use information from the voltage measuring elements.

2 Connection of relay protection terminals to voltage transformer secondary circuits

In practice, relay protection devices can be connected to secondary windings of voltage transformers worked in the network with isolated neutral in several ways (Fig. 1) depending on the scheme of secondary voltage circuits at the substation and necessity of information about the signal parameters (Table 1).

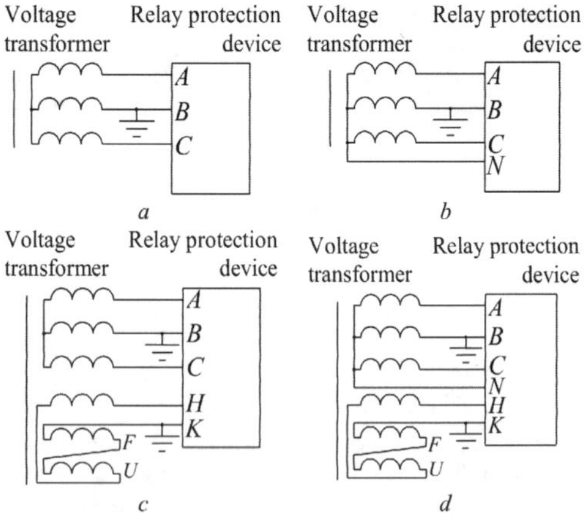

Fig. 1. Connection of relay protection terminal to secondary voltage circuits: *a* – A, B, C phases; *b* –A, B, C phases with N neutral conductor; *c* – A, B, C phases and circuits of open delta circuit H, K; *d* –A, B, C phases with N neutral conductor and circuits of open delta circuit H, K

Positive and negative sequence of phase-to-ground voltages, and also phase-to-phase voltage values, can be reached for any type of terminal connecting to the Y-winding circuit of transformer secondary voltage (Fig. 1). Zero sequence measurement is carried out through outputs H and K of additional open delta winding (Fig. 1, *c*, *d*). Calculation of zero sequence value and measurement of phase-to-ground voltages are possible when the terminal is connected to output N of Y-circuits (Fig. 1, *b*, *d*).

Table 1. Connection of relay protection terminals to voltage transformer secondary circuits

Voltage signal characteristic value	Connection to voltage transformer circuits			
	A,B,C	A,B,C, N	A,B,C, H, K	A,B,C, N, H, K
Phase-to-ground value	–	+	+	+
Phase-to-phase value	+	+	+	+
Positive sequence	+	+	+	+
Negative sequence	+	+	+	+
Calculated zero sequence	–	+	–	+
Measured zero sequence	–	–	+	+

Note. To measure zero sequence voltage, **H** and **K** circuits are used as open delta winding terminals.

3 Existing algorithms for detecting damage to voltage transformer circuits

Currently, two algorithms to detect damages in voltage transformer circuits are widely used in practice [1, 2]:

1. Comparison of zero sequence voltages, one of which is measured at open delta winding of voltage transformer, and another is calculated by vector adding of Y-winding phase-to-ground voltages. It is obvious that, at this approach implementation, relay protection terminal requires connection to all secondary circuits of the voltage transformer (Fig. 1, *d*). Its serious drawback is the fundamental inability to detect damage in the primary circuits of measuring voltage transformer.

2. Change monitoring in voltage circuits in the change absence in current circuits. One of conditions for protecting device triggering is asymmetry in voltage circuits (negative or zero sequence voltages) while current circuits are symmetric. The second condition is loss of voltage in all three phases in the absence of short-term asymmetry in current circuits that appears even at symmetrical three-phase metal short circuit [3]. This algorithm can be implemented for any variant of terminal connection to secondary voltage circuits of voltage transformer, but impossibility to calculate some parameters of the signal (for example, zero sequence voltage at absence of Y-winding N conductor) limits its application. The undoubted advantage of the approach is possibility to detect damages, including those in primary circuits of voltage transformer. Its implementation additionally requires a signal from current transformers.

4 Identifying voltage transformer circuit damage for various ways of terminal connecting

Capabilities of microprocessor-based relay protection devices allow implementing both approaches for identifying damage, and this is why automation equipment can fully control status of phase conductors of voltage transformer circuits. Depending on operating mode, requirements to protection of a particular connection, structure of secondary voltage networks, etc. Voltage circuit lock function is able to work in accordance with any of mentioned approaches both individually and simultaneously.

Up-to-date relay protection and automation devices usually have symmetrical input resistance. This preconditions certain features at detection of some types of voltage circuit damages. Reaction of the algorithms on typical faults in the voltage transformers circuits is given in tables 2, 3.

Table 2. Action of the algorithm which compares calculated and measured zero sequence voltage at various faults in voltage transformer's circuits

Damage	Required action	Characterization of the action
		A, B, C, N, H, K
Phase interruption of secondary Y-winding	Operation	True
Two phases interruption of secondary Y-winding	Operation	True
Phase and zero conductor interruption of secondary Y-winding	Operation	True
Opening of Y-winding circuit breaker	Operation	True
Short-circuit of phase and zero conductors of secondary Y-winding	Operation	True
Two-phase short circuit of secondary Y-winding with circuit breaker opening	Operation	True
Three-phase short circuit of secondary Y-winding with circuit breaker opening	Operation	True
Zero conductor interruption of secondary Y-winding	Operation	False
Phase interruption of voltage transformer primary circuit	Operation	False
Single-phase earthing in the primary network	Disregard	True
Short circuit of two voltage transformer phases in the primary circuit	Disregard	True

Note. "True" value corresponds to voltage circuit lock function behaviour complying with the requirements. "False" value corresponds to incorrect voltage circuit lock function action.

Due to symmetry of input impedance of relay protection device voltage circuits, in normal electrical network operation, zero conductor interruption will not cause changes in symmetrical voltage components and this will not allow working of voltage circuit lock function at such damage (Tables 2, 3). If it is necessary to control the neutral conductor integrity, asymmetry of input resistance should be created artificially, for example, by connecting, in general, of three different resistors between neutral and phase conductors of the device terminals (Fig. 2). In this case, breakage of zero wire will be accompanied by a neutral shift in Y-winding circuit to the point defined by the resistances relations and appearance of zero sequence design voltage, which is calculated as geometric sum of its phase voltages. This phenomenon will cause voltage circuit lock function triggering when using either of two approaches: in the case of comparison of zero sequence calculated and

https://doi.org/10.1051/e3sconf/201913901061

measured values a difference between these values will appear which is a condition for the algorithm operation; appearance of zero sequence voltage in Y-winding in the absence of disturbances in the current circuits is a prerequisite for operation of the second mentioned approach.

Table 3. Action of change monitoring in voltage circuits in the change absence in current circuits

Damage	Required action	Characterization of the action			
		A,B,C	A,B,C, N	A,B,C, H, K	A,B,C, N,H,K
Phase interruption of secondary Y-winding	Operation	True	True	True	True
Two phases interruption of secondary Y-winding	Operation	True	True	True	True
Phase and zero conductor interruption of secondary Y-winding	Operation	–	True	–	True
Opening of Y-winding circuit breaker	Operation	True	True	True	True
Short-circuit of phase and zero conductors of secondary Y-winding	Operation	–	True	–	True
Two-phase short circuit of secondary Y-winding with circuit breaker opening	Operation	True	True	True	True
Three-phase short circuit of secondary Y-winding with circuit breaker opening	Operation	True	True	True	True
Zero conductor interruption of secondary Y-winding	Operation	–	False	–	False
Phase interruption of voltage transformer primary circuit	Operation	True	True	True	True
Single-phase earthing in the primary network	Disregard	True	True	True	True
Short circuit of two voltage transformer phases in the primary circuit	Disregard	True	True	True	True

Note. "True" value corresponds to voltage circuit lock function behaviour complying with the requirements. "False" value corresponds to incorrect voltage circuit lock function action. Dash means impossibility of this type of damage to exist in the considered system of secondary voltage circuits.

Use of open delta scheme to provide measured zero sequence voltage and its further use in relay protection algorithms imposes the need to monitor state of the conductors and this winding. As a solution to this problem for medium voltage networks, connection to U conductor of additional voltage transformer winding (Fig. 1, *d*) with monitoring and comparison of voltages between voltage transformer K and U phases, and between H and U, can be use, similarly to110-220 kV networks [4]. However, design of three-phase voltage transformers which are most common used in networks with isolated neutral does not allow installation at substation of additional U voltage busbar due to lack of output of this winding [5].

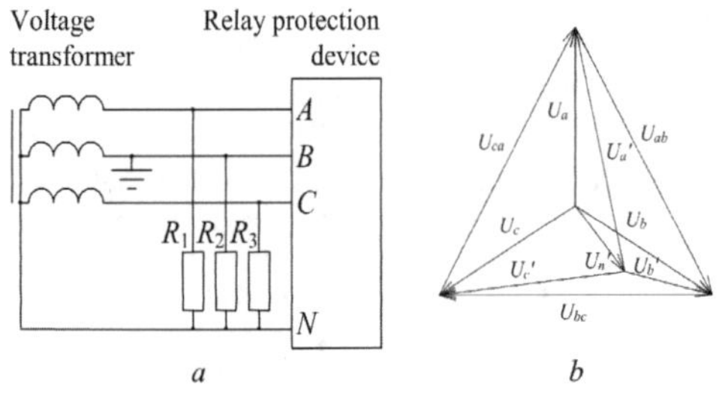

Fig. 2. Connection of additional resistors: *a* - wiring diagram; *b* - neutral shift in Y-winding at zero conductor disconnection; U_i - phase voltages at normal network operation; U_i' - phase voltages at zero conductor discontinuity; U_{ij} - interphase voltages; R_i - additional resistance connected to the phase voltage; U_n' - position of the shifted neutral

Thus, if it is necessary to use zero sequence voltage in logic of relay protection device, it is advisable to define it by calculating phase voltages of Y-winding voltage transformer. In this case, measured zero sequence signal of open delta winding will be used only to monitor the state of Y-winding N conductor.

5 Feature of operation of isolated neutral networks

One of the features of 6-35 kV electrical networks working with isolated neutral is that the neutral point position is defined by capacitance conductivities of each of the three phases and is virtually the same for the entire network [6]. If the power source of this network operates with unloaded buses or short network, the neutral conductor shift can occur due to asymmetric position of conductors, both relative to the ground and relative to each other. Another reason for the neutral point shift may be non-simultaneous switching of the switch phases that leads to saturation of voltage transformer magnetic circuit and ferroresonance processes with asymmetric change in phase conductivities relative to the ground of the measuring transformer [7, 8].

At such operation modes, zero sequence voltage appears even in the absence of any damage to the network, to what the device protecting against single-phase earth faults reacts as on grounding fault while actually there is not an accident. This phenomenon is called "false earthing".

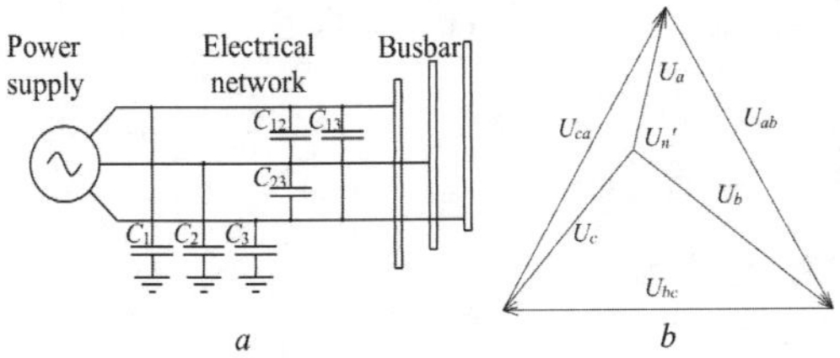

Fig. 3. Operation of the power supply with unloaded buses: a - electrical circuit; b - vector voltage diagram; Ci - capacitance of the phase relative to the ground; Cij - capacitance between two phases; Ui - phase voltage; Uij - interphase voltage; Un' - shifted neutral position

Algorithms used in different approaches of voltage circuit lock function implementation behave differently when a "false earthing" appears in the electrical network. Erroneous work of the algorithm based on comparison of zero sequence measured and a calculated value when a "false earthing" appears is excluded. This is due to the fact that zero sequence voltage appears in both secondary windings of the voltage transformer and difference between them is minimal. On the contrary, without additional modifications, algorithm, which controls disturbances in the current and voltage circuits, will work erroneously and will block part of the protection and automation devices.

6 Improvements of voltage circuit lock function

As mentioned above, the neutral shift and appearance of zero sequence voltage in 6-35 kV electrical networks is possible even in the absence of damage. Strictly speaking, this mode is not normal, but voltage circuit lock function should not react to it due to lack of damage in the voltage transformer circuits. The simplest way to improve the voltage circuit lock function algorithms is to cancel zero sequence check in secondary voltage circuits for the approach based on change monitoring in voltage circuits in the change absence in current circuits. In this case, simultaneous action in relay protection device of both approaches will allow identifying all damages in the voltage transformer circuits, provided that terminal connection scheme shown in Fig. 1, *d* is used. It is worth noting that application of the voltage circuit lock function approach, which compares measured and calculated zero sequence values of, is possible only for this secondary circuit's scheme.

Application of the secondary voltage circuit diagram shown in Fig. 1, in which terminals of open delta winding are not used, requires revision. Connecting asymmetric resistor system, which causes a neutral shift at zero conductor discontinuity (Fig. 2), will result in zero sequence voltage. Moreover, due to symmetry of the voltage inputs, the neutral point shift value at zero conductor discontinuity will be defined by the resistors parameters and will be positioned in a well-defined point on voltage vector diagram. This fact can be used to detect damage to zero conductor in the Y-windings. The logic diagram of the block, which is created using this proposal, is shown in Fig. 4. Detection of zero conductor discontinuity will be carried out at zero sequence voltage jump (operation of zero sequence pulse starting element) with shift of neutral to the complex plane defined by setpoints (operation of U_n' pulse starting element), at that changes in the symmetrical current components should not occur (non-operation of dI_1, dI_2, dI_0 pulse starting elements). The algorithm reset is carried out when the neutral point leaves the specified area (return of U_n' starting element). The algorithm work is shown in Fig. 5, 6.

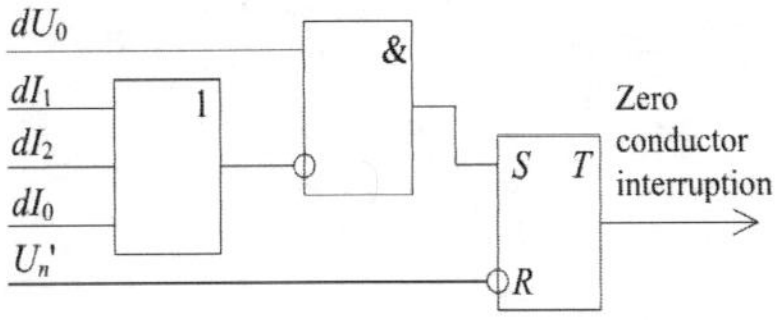

Fig. 4. Detection of zero conductor discontinuity in Y-windings voltage transformer circuits: dU0 – element for starting change in zero sequence voltage; dI1, dI2, dI0 – element for starting change of respective sequences current; Un' – element for starting calculation of neutral shift to a point defined by a set of asymmetrical resistances

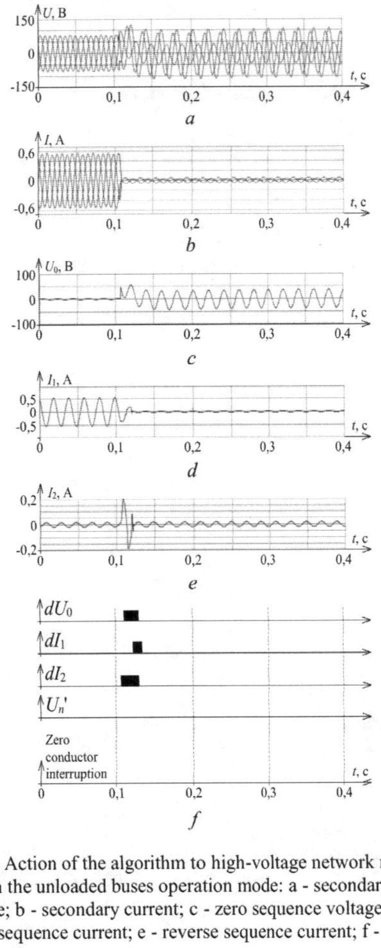

Fig. 5. Action of the algorithm to high-voltage network neutral shift in the unloaded buses operation mode: a - secondary voltage; b - secondary current; c - zero sequence voltage; d - direct sequence current; e - reverse sequence current; f - logic of operation

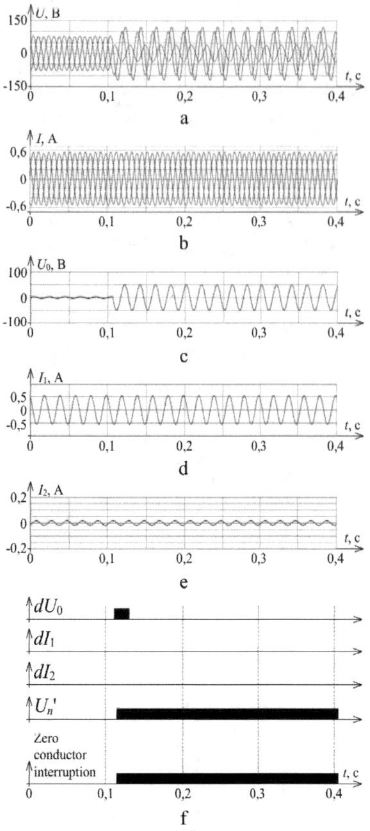

Fig. 6. Action of the algorithm to interruption of zero conductor in secondary Y-windings a - secondary voltage; b – secondary current; c - zero sequence voltage; d –direct sequence current; e – reverse sequence current; f – logic of operation

Application of secondary circuit voltage diagram shown in Fig. 1, *c*, which uses open delta winding to define zero sequence values, is impractical due to the need for additional control of its circuit's state.

Use of the scheme shown in Fig. 1, *a* requires no modification, but has a limitation on application because it does not allow defining a zero sequence voltage.

Conclusion

Analysis of secondary voltage schemes and fault detection algorithms for such circuits allows drawing the following conclusions:

1. Application areas of fault detection algorithms are defined for different variants of secondary voltage circuits schemes used at substations.
2. Use of the open delta winding is advisable only when implementing the algorithm comparing zero sequence voltages in star and open delta circuits.
3. To detect disconnection of star circuit neutral conductor, it is proposed to create an artificial asymmetry in the input circuits of relay protection device and to develop group of starting elements, which will analyses shift of the neutral position in the complex plane of voltages.

This work was supported by the Ministry of Science and Higher Education of the Russian Federation (agreement No. 075-15-2019-1209 of May 31, 2019. Unique project identifier RFMEFI57717X0244).

References

1. S. Ya. Petrov, Releyshik J., **1(6)**, 49 (2010).
2. D. Gallyamov, A. Mikhailov, Elektroenergetika glazami molodezhi J., **IV**, 129 (2015).
3. D. N. Kokoulin, V. A. Shuin, Vestnik ISPU J., **3**, 18 (2012).
4. N. A. Doni, A. V. Sdobin, Releyshik J., **4(5)**, 73 (2009).
5. V. S. Fishman, News Elteh J., **6**, 32-35 (2010).
6. V. A. Shuin, A, V. Gusenkov, Bibliotechka elektrotekhnika J., **11**, 104 (2001).
7. N. A. Kolechitskaya, N. S. Lazarev, R. N. Shulga, K. A. Zmieva, Russian Electrical Engineering J., **4**, 2 (2013).
8. A. Yu. Emel'yantsev, News Elteh J., **4**, 34 (2009).

Improving the reliability of combined overhead and cable power lines by organizing smart autoreclosing

Pavel Pelevin [1,*], *Anton* Loskutov [1] and *Timur* Sharafeev [1]

[1]Department of Electric Power Engineering, Power Supply and Power Electronics Nizhny Novgorod State Technical University n.a. R.E. Alekseev 603950, Nizhny Novgorod, Minin St., 24, Russia

Abstract. The application of combined overhead and cable power lines (OCPLs) of 110 kV voltage rate and higher call for new features for power system protection and automation devices. One of the most urgent is the problem of autoreclosing (AR) of OCPL. One or another solution to this problem significantly affects the reliability and efficiency of the OCPL operating. In many countries, both non-selective AR and selective AR are used. A method of selective AR of OCPL based on traveling waves is proposed, which increases the reliability of OCPL protection and automation. The method uses a special procedure of traveling wave pattern classification for faulted section identification.

1 Introduction

Nowadays, OCPLs are becoming more common in large cities. This is due to the need to replace some overhead lines (OL) with underground cable lines (UCL). Another reason is constraints to the construction of new high-voltage OLs in dense urban areas. Capital costs for UCL construction is higher than for OL construction. This causes the practice of hybrid lines construction comprising OL sections and UCL sections. In contrast, many challenges are appearing connected with OCPL operation in normal and emergency conditions. One of the most crucial challenges is AR of OCPL.

It is well known that AR of power lines is an effective way to increase the reliability and the sustainability of power transmission. However, there is additional damage may be caused by re-energization OCPL without faulted section identification (non-selective AR of OCPL) in case of faults on UCL. Therefore, power line dead-time reduction achievement due to AR is counterbalanced by faulted UCL re-energization probability. Accordingly, the non-selective AR of OCPL is not the right way for power supply reliability increasing.

Thereby it is relevant to employ selective AR of OCPL, which performs specified faulted section identification procedure before reclosing attempts and block AR if a fault on UCL. In this meaning, the paper's goal is to analyze and outlook existing methods of AR of OCPL and to evaluate prospects for their application.

2 Classification of methods of selective AR of OCPL

There are several faulted section identification methods for selective AR of OCPL plenty large enough to classify them by operating principle. (Fig. 1):

1) The first class of methods is differential in which faulted section identified with signals measurements by the ends of controlled zones (for example, by ends of cable section) [1]–[3]. There are differential methods:

a) Based on measuring of phase currents [1],[2];

b) Based on measuring currents in shields of cables [3];

2) Another class of selective AR of OCPL is distant methods. Distant methods based on remote measurements of controlled signals by one or two ends of the line. The faulted section can be estimated using fault locating methods, or it can be identified with the use of specified methods and algorithms of faulted section identification [4]–[7]. Distant methods classified on:

a) Passive methods, which also can be classified on:

– Single-ended [4], [5];

– Double-ended (or multi-ended in case of many junctions on OL) [6];

b) Active methods [7].

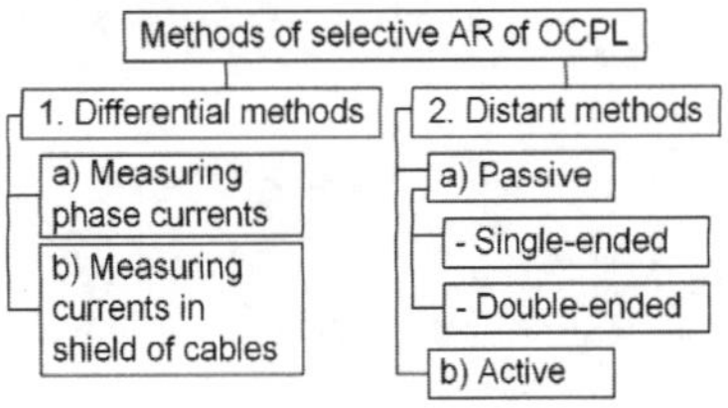

Fig. 1. Classification of methods of selective AR of OCPL

Let us consider the features of the various methods of AR of OCPL and make a comparative analysis.

3 Differential methods of selective AR of OCPL

Differential methods are based on the comparison of signals (usually current signals) by the ends of UCL or OL sections (Figs. 3-5, a, b). This class includes a method for determining the faulted section of OCPL by comparing currents by the ends of UCL sections of OCPL [1]. Reference [1] proposes installing current transformers (CTs) at each OL to UCL transition. There are current sensors (CSs) connected to secondary windings of these CTs. CSs measure signals and transmit information through a special communication channel to the main device installed at the power substation (PS).

In reference [2], a technical solution based on special optical current sensors (using the Faraday effect) was applied. These sensors are sequentially connected by optical fiber with an electronic-optical device (main device) installed at one of the line terminals. The main device thus obtains the current measurements at the UCL section ends via an optical fiber connection. It calculates the differential current of the UCL section and determines the external or internal fault. If there is an internal fault for the UCL section, the AR is blocked.

Another differential method is to supervise the currents in the cable shields. In this case, CTs are used installed on the cable shield grounding conductors [3]. CT supply current to a measuring device, which transmits the current signal parameters via an optical fiber link to the main device with selective AR logic.

The advantage of differential methods is the high security of determining the faulted section (ensuring the almost absolute selectivity of the AR of OCPL) and the applicability for various configurations of the OCPL. The main drawback is the need to install additional equipment both at the line terminal and at the OL to UCL transition (current transformers, current sensors, data collection and data transmission devices) and to organize a special communicational channel. All this makes the application of differential methods an expensive technical solution. Also, it will be shown below that the number of equipment required is directly proportional to the number of UCL sections. Therefore, the reliability of these technical solutions decreases with the growth of UCL sections, and the engineering-and-economical costs of installation, commissioning and maintenance increase. Also, it may be required for power supply and heating the devices at the OL to UCL transition, which leads to even more complications and costs of equipment used.

4 Distant methods of AR of OCPL

Distant methods of selective AR of OCPL can be described in terms of their operational principle as methods based on measurements of signals at the line terminals, not at ends of UCL or OL sections (i.e. the measuring point is remote from the controlled sections of the OCPL). Distant methods can be classified according to the type of controlled signals into passive methods (Figs. 3–5, c–e) and active methods (Figs. 3 and 5, f). Passive methods based on monitoring current and voltage transients naturally occurring due to faults on the line (for example [4]–[6]). The faulted section can be estimated via fault locating methods or by the use of specified faulted section identification algorithms [5]. Active methods based on the injection of specified signals (high-frequency signals as a rule) into the line or its section (for example into the shield of UCL). The faulted section is monitored through artificial signal parameters (presence or absence of signal, signal energy, etc.).

For the simplest implementation of the single-ended distant passive method of selective AR, one can use zones of nonpilot stepped distance protection or impedance-based fault locating information. The main drawback of this method is the AR blocking in a large part of the OL sections due to the need to apply margin when setting the blocking region. The reason for this is the insufficient accuracy of impedance-based techniques.

The way for eliminating drawbacks mentioned before is to use methods based on electromagnetic transients monitoring and traveling wave fault locating having enough accuracy for selective AR. There is a technique of adaptive (selective) AR based on double-ended traveling wave fault locating adapted for OCPL [6] (Figs. 3-5, e). Calculated fault location used for decision making about faulted section and, therefore, AR blocking if fault detected in block region.

4.1 Selective AR of OCPL based on transients monitoring

It is known that fault cause rapid discharge of the line that generates a steplike change of voltage in the fault location. Thus, the first stage of the transient process begins, which connected with traveling waves propagation along the line length from the fault location. Inhomogeneity of OCPL, which is generally due to OL to UCL transitions, leads to many reflections of traveling waves during propagation along the line. Therefore, there are additional high-frequency transients in the voltage and current signals. These transients carry information about the faulted section of the OCPL and fault location. Studies conducted on simulation models [5] are showing that behavior of electromagnetic transient processes and particularly traveling wave transients depend on the faulted section of OCPL. Thus, it is possible to identify the faulted section by classification of transient current and voltage signals.

Therefore, distant passive selective AR methods based on transient process estimation are looking promising. It should be noted that it is possible to use both single-ended measurements (Figs. 3-5, d) and double-ended (or multi-ended in case of junctions on OL section) synchronizeless measurements (Figs. 3-5, e). At the same time, conventional traveling wave fault locating methods aren't suitable for OCPL. Hence it is advantageous to develop special signal processing algorithms and faulted section classification methods.

In addition to the above, single-ended measurements based methods require more complex signal processing and classification procedure in contrast to double-ended based methods having much fewer requirements for signal processing. Nevertheless, double-ended methods require communication channels between substations. Thus, the benefit of single-ended methods application is the number of equipment reduction down to one device (except measuring transformers), and the drawback is signal processing algorithms computational costs increasing. It is worth noting that the highlighted drawback is not crucial at present due to the ongoing growth of the computational power of microprocessor-based units.

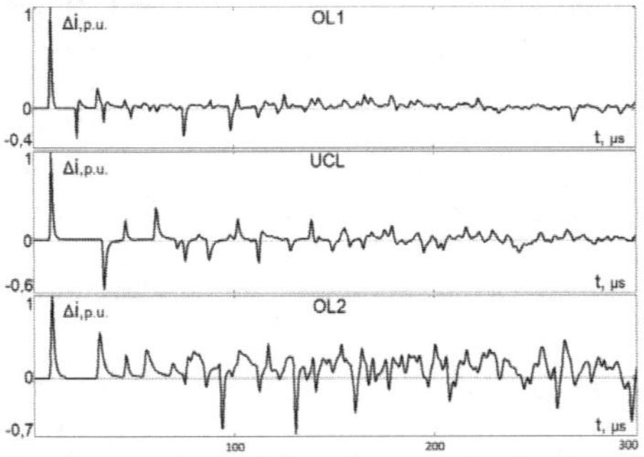

Fig. 2. Dependence of filtered current signal measured at one of the line terminals from the faulted section for OCPL configuration OL-UCL-OL.

Several faulted section identification algorithms were developed based on the above principle. These methods are classified as distant passive methods of selective AR. One of the methods is to apply single-ended measurements and pattern classification techniques [5]. The method essence is special signal processing of current or voltage transients and traveling wave pattern forming (like shown in fig. 2). This traveling wave pattern then compared with reference patterns preliminary computed via simulation modeling of faults in OCPL and stored. The comparison is performed with correlation signal processing [5].

4.2 Distant active methods of selective AR

Active distant methods include the selective AR method with the induction of a superimposed signal into the cable shield [7] (pict. 3,5, f). According to this method, a high-frequency signal generator is connected to the cable sheath through an inductive coupler. The signal with a selected frequency of higher harmonics is induced into the cable sheath. Thereby there are high-frequency electrical oscillations in the cable sheath. During normal operating mode of OCPL, and also in the case of fault on OL, there is no path for the current and superimposed signal is not induced to the cable core. In opposite, if there is insulation breakdown and cable core having been shorted to the sheath, the imposed electrical signal is propagating along the line to the power substation and can be measured by a protection device. The analog-to-digital conversion with a high sampling rate is required, 80-256 samples per cycle as an instance.

It should be noted that the impedance of an external electrical system increases with the rise of imposed signal frequency, therefore, it is rather to measure voltage signals than current signals. At the same time, the frequency of the imposed signal is limited by a voltage transformer bandpass and a protection device sampling rate.

The drawbacks of the method are the need to use special high-frequency signal source and the constraints on OCPL configuration (only UCL-OL and UCL-OL-UCL configurations).

5 Comparison of selective AR schematics

For comparison methods of selective AR in terms of reliability, it is appropriate to consider corresponding technical solutions. Evaluation of the reliability of the faulted section identification algorithms is difficult and depends on the consideration of particular cases and the probabilities of events. Therefore, the methods of selective AR will be evaluated by comparison of a set of the elements constituting the schemes of selective AR. To carry out such an assessment, Figures 3–5 show the schematic solutions of selective AR for three OCPL configurations: UCL-OL, OL-UCL-OL, UCL-OL-UCL. Figures 3-5 show the differential methods of selective AR (a, b) and the distant methods: passive (c, d, e) and active (f). These selective AR methods are shown in the figures in the order of their description in the text above. It is worth noting that in the case of the OCPL configuration OL-UCL-OL, the distant active method [7] is inappropriate to use due to the constraints of the organization of the imposed signal source at the UCL-OL transition that is remoted from the power substation.

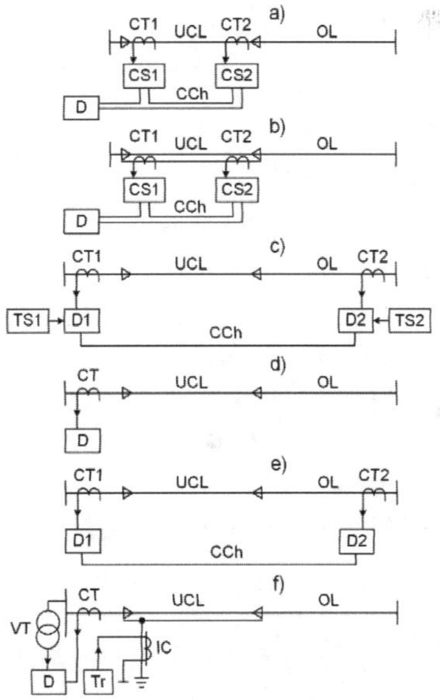

Fig. 3. Schemes of selective AR for OCPL configuration: UCL-OL.

In figs. 3-5, the following abbreviations are introduced: UCL - underground cable line, OL - overhead line, CT - current transformer, VT - voltage transformer, CS - current sensor (signals measuring, processing and transmitting device), D - main device of selective AR, CCh - communication channel, Tr - transmitter (imposed signal generator), IC - special signal transformer (inductive coupler), CTR - common time reference system.

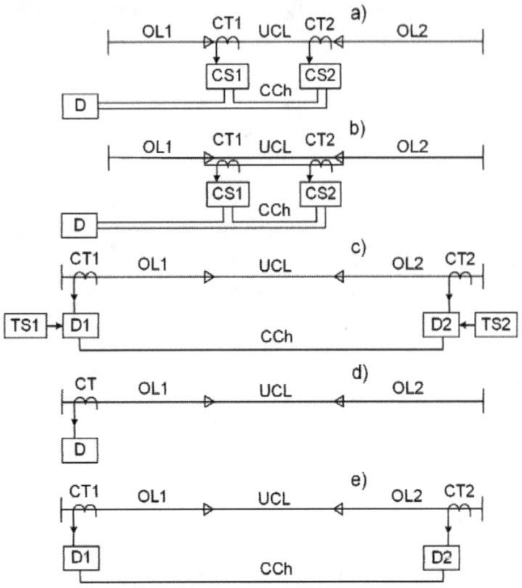

Fig. 4. Schemes of selective AR for OCPL configuration: OL-UCL-OL.

We also assume that the elements of the circuit for the reliability evaluation are connected in series. Let us write down the expressions of the probability of failure-free operation for the schemes of selective AR (Figs. 3-5, a – f). We also assume that the failures of elements are independent in occurrence. Then for the schemes on figs. 3-5 (a – f), we can write expressions (1) - (3).

For the «UCL-OL» configuration:

$$p_{sch.a,b} = p_D (p_{CT} \cdot p_{CS})^2 \cdot p_{CCh} \quad (1.1)$$

$$p_{sch.c} = (p_D \cdot p_{CT} \cdot p_{TS})^2 \cdot p_{CCh} \quad (1.2)$$

$$p_{sch.d} = p_D \cdot p_{CT} \quad (1.3)$$

$$p_{sch.e} = (p_D \cdot p_{CT})^2 \cdot p_{CCh} \quad (1.4)$$

$$p_{sch.f} = p_D \cdot p_{CT} \cdot p_{VT}) \cdot p_{Tr} \cdot p_{IC} \quad (1.5)$$

For the «OL-UCL-OL» configuration:

$$p_{sch.a,b} = p_D (p_{CT} \cdot p_{CS})^2 \cdot p_{CCh} \quad (2.1)$$

$$p_{sch.c} = (p_D \cdot p_{CT} \cdot p_{TS})^2 \cdot p_{CCh} \quad (2.2)$$

$$p_{sch.d} = p_D \cdot p_{CT} \quad (2.3)$$

$$p_{sch.e} = (p_D \cdot p_{CT})^2 \cdot p_{CCh} \quad (2.4)$$

For the «UCL-OL-UCL» configuration:

$$p_{sch.a,b} = p_D (p_{CT} \cdot p_{CS})^4 \cdot p_{CCh1} \cdot p_{CCh2} \quad (3.1)$$

$$p_{sch.c} = (p_D \cdot p_{CT} \cdot p_{TS})^2 \cdot p_{CCh} \quad (3.2)$$

$$p_{sch.d} = p_D \cdot p_{CT} \quad (3.3)$$

$$p_{sch.e} = (p_D \cdot p_{CT})^2 \cdot p_{CCh} \quad (3.4)$$

$$p_{sch.f} = (p_D \cdot p_{CT} \cdot p_{VT}) \cdot p_{Tr} \cdot p_{IC})^2 \cdot p_{CCh} \quad (3.5)$$

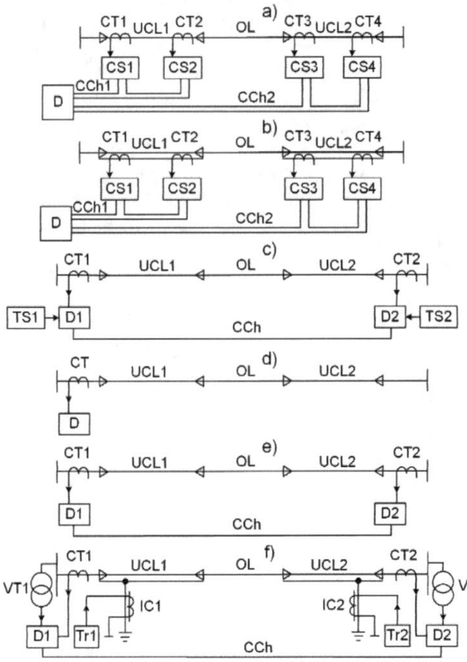

Fig. 5. Schemes of selective AR for OCPL configuration: UCL-OL-UCL.

For analyzing figures 3-5 and expressions (1) - (3), we assume the probability of failure-free operation of the similar elements is the same. Thuswise, it can be seen that for the differential methods the amount of equipment required, as well as the probability of scheme failure, greater than for distant methods. It can also be noted that in the case of distant methods of selective AR (Figs. 3-5, c-e), with the exception of the active method with signal induction into the cable shield (Figs. 3-5, f), the number of equipment required does not change when increasing the number of cable sections. In contrast, for the differential methods (Figs. 3-5, a, b), the number of required equipment increases with the growth of cable sections number, and the probability of failure-free operation decreases.

The conducted comparison and estimation of the selective AR methods seems to be sufficient for a preliminary conclusion that distant methods of selective AR are the most promising for their further development. This is due to the distant methods have less cost and seem more reliable for technique implementation. It is also worth noting that the development of traveling-wave based methods of selective AR makes it possible to further use signal estimation algorithms for implementation in advanced traveling-wave based protection devices.

6 The logic of distant methods of selective AR

It is possible to implement selective AR methods with the use of the device of selective AR together with existing microprocessor devices of power system protection with breaker automatic control features. Thereby general logic of selective AR is as shown in fig. 6. The logic of selective AR provides capability for using both the fault locating algorithms and the faulted section identification algorithms. The setting of operating mode provides the choice. The calculated fault location (FL) is applied to inputs of comparison elements (fig. 6) where the fault location entrance in the blocking region is checked [6]. The blocking region is established by two settings: the first blocking region initial point (BRI1) and the first blocking region end (BRE1). Comparison elements are joint by the AND element, which output connected with one of the inputs of the OR element. The other inputs of the OR element is for different blocking regions joining up in the main logic of selective AR. If the faulted section identification algorithm will provide the "fault on UCL" signal or calculated fault location will hit into one of the blocking regions, then AR will be blocked.

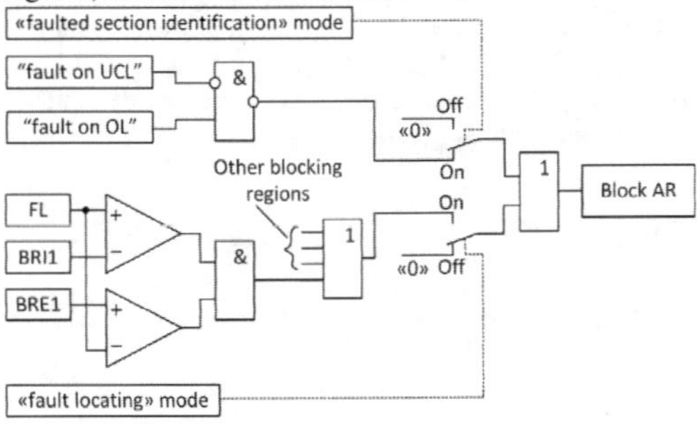

Fig. 6. The block diagram of the logic of selective AR.

Developed logic scheme for AR blocking is flexible and allows one to use any fault locating techniques or faulted section identification techniques.

Conclusion

The methods of selective AR were classified by operating principles and considered in the paper with particular technique examples. Based on conducted comparison analysis there a conclusion can be made that differential methods of selective AR (fig. 3-5, a-b) can be used for the "UCL-OL" line configuration (fig. 3) and the "UCL-OL-UCL" line configuration (fig. 4). On the contrary, in the case of other line configurations (ex. fig.5), the differential methods application is unbeneficial, except particular cases when distant methods inefficient.

It is important to note that distant methods of selective AR are more viable and can be applied in most cases of OCPL configuration. However, it is advisable to use the distance active method (Fig. 3-5, f) only with the "UCL-OL" and "UCL-OL-UCL" line configurations. In the case of relatively short OCPL sections, some single-ended distant passive methods of selective AR can be ineffective, so the use of double-ended methods is promising.

The presented research results were obtained with the support of a grant from the President of the Russian Federation for state support of young Russian scientists (MK-3210.2019.8). Agreement No. 075-15-2019-337 of 11.06.2019.

References

1. Dogadkin D., Marin R., Shirshova Ye., Ismukov G., KulikovA., Lint M., Podshivalin A. Elektroenergiya. Peredacha i raspredeleniye – in Russ. (Electric Power Transmission and distribution), 2016, № 5 (38), p. 114–119
2. Patent of invention WO2015033001 (A1). Optical system for identifying faults in mixed power transmission lines. Date of publication: 12.03.2015
3. Patent of invention RU2669542 (C1). System of selective blocking of automatic reclosing on combined aerial cable transmission lines. Date of publication: 11.10.2018
4. Patent of invention RU 2658673 (C1). Method of automatic reclosure of overhead transmission line. Date of publication: 22.06.2018
5. Kulikov A.L., Loskutov A.A., Pelevin. P.S. Algorithm for Identifying the Faulty Section in Combined Cable and Overhead Power Lines Based on Surge Portrait Recognition. Elektrichestvo (Electricity), 2018, No. 3, pp. 11–17
6. Kasztenny B. Locating faults before the breaker opens – adaptive autoreclosing based on the location of the fault / Kasztenny B., Guzman A., Mangapathirao V. Mynam, and Titiksha Joshi // 44th Annual Western Protective Relay Conference. October 2017
7. Patent of invention RU 2663413 (C1). Overhead power transmission line automatic re-closure method. Date of publication: 06.08.2018
8. Schweitzer E.O. Locating Faults by the Traveling Waves They Launch / Schweitzer E.O., Guzman A., Mynam M.V., Skendzic V. // 67th Annual Conference for Protective Relay Engineers, College Station, TX, USA. March 2014

Power electrical network topology detection using changes of phasor measurements in transient states

Natalja Gotman[1]*, *Galina* Shumilova[1]

[1]Federal Research Center "Komi Scientific Center of the Ural Branch Russian Academy of Sciences", ISE and EPN, 167000, Russia

Abstract. The solution of the problem of a topology detection of an electrical network on changing voltage and current phasors obtained from the phasor measurement units (PMUs) in a transient state using artificial neural networks (ANNs) is considered. Experimental results for the 14-bus test system to detect the failed line after short circuit and the line was turned on by an auto-reclosing device are presented.

1 Introduction

Detecting the electrical network connection diagram in real time is important for continuous monitoring of power lines to ensure uninterrupted operation of the power system. Topology changes caused, for example, by line fault, can influence other lines (leading to certain lines overload) and destabilize a whole network. The fast identification of such changes allows taking up necessary measures and for reducing the risks of cascading failures that lead to large-scale blackouts.

Various tools for monitoring electrical networks, such as SCADA (Supervisory Control and Data Acquisition), smart meters and traditional devices for measuring voltage and current have been developed and implemented. However, ensuring network observability with such unsynchronized data flows, received from these devices, remains the problem. In addition, operators can not have sufficient information about the state of the power system in real time, that is especially important when large violations occur.

The successful introduction PMUs [1] allows implementing entirely new algorithms for the power network topology detection. PMUs, unlike the traditional SCADA system, have an ability to record such fast transients with high accuracy providing synchronized measurements of current and voltage phasors from GPS/GLONASS with a high sampling rate at various geographically remote points of a power system.

Many approaches using PMU were proposed to detect topological network changes caused by line outages [2–6]. The starting point in these approaches is that the line outage causes changes of the voltage phasor (voltage magnitude and phase angle) at all buses of a power network, and these changes are larger than the variation of voltage phasor during normal operation, for example, during swings.

The key feature that makes the line outage identification possible is that voltage phasor measurements reported by PMUs are different for different line-outage scenarios, i.e., these are a kind of "signatures" that is used by authors of considered works to detect changes in the power network topology. Each method has its own specific features but in all methods detecting topology is carried out at quasi-steady-state conditions, after the end of the transient state, caused by some unforeseen circumstances, that increases the time of the line outage identification.

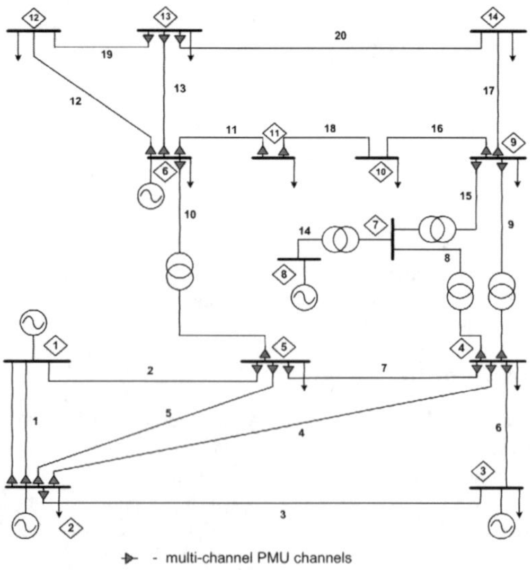

Fig. 1. 14-bus test scheme "IEEE 14 Bus Test Case" with the indicating PMU placement.

In the authors' papers [7-9], the network topology was detected using ANNs where input variables were represented by a magnitude and phase of the voltage phasor at buses at which PMUs are attached and active and reactive power flows at the end of lines near the location bus at the steady state.

 https://doi.org/10.1051/e3sconf/201913901063

In this paper, to reduce the time necessary for detecting the line outage, it is proposed to detect the topology of a power network at the transient state and use the values of phasor measurement change reported by PMUs during the transient process. Experimental studies were carried out at the 14-bus test system. Fig. 1 shows the PMU placement at the buses 2, 4, 5, 6, 9, 11 and 13 obtained as a result of the optimal PMU deployment according to the criterion of the minimum number of devices [10] that ensure a topology observability of the power system. An observability of the bus 8 is provided due to zero injection of the transit bus 7.

2 ANN configuration and input parameters

ANN configuration for the problem under consideration is shown in Fig. 2. It represents a three-layer perceptron. In the input layer of the network there are 27 neurons *NI* (or 54, depending on the calculation option), in the output layer – *NK* = 15 (the number of failed lines in the test scheme). The number of neurons in the hidden layer is defined for each case by the minimum error of the identification result.

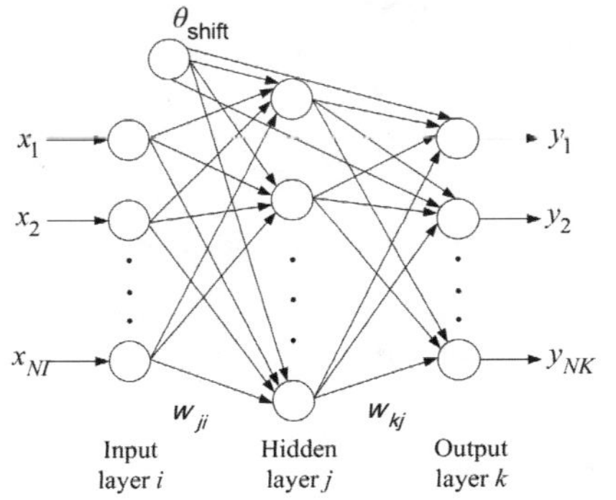

Fig. 2. Structure of the artificial neural network used to identify power network topology.

The input variables are represented by the change of a voltage phase at buses endowed with PMU and the change of a current magnitude in lines incident to buses at which PMU are attached using seven voltage phasor measurements at buses are instrumented with PMU and twenty current phasor measurements in network lines. With the above placing PMUs we get 27 phasor measurement values. When there are measurements available at both line ends, one of them could be ignored. Therefore, seven current magnitudes from 27 values were not used. Also, voltage magnitude measurements were not taken into account since most of them did not change much when there was an outage of one of the lines. The changes in the voltage phase and current magnitudes were calculated as a difference between measurement values in 0.1 s.

3 Experimental studies

To obtain the data base, the calculations were carried out using program-computed complex (PCC) "RastrWin 3" (v 1.80.0.1485) which includes the program module "RUSTab" intended for the calculation of electromechanical transients. PCC "RastrWin 3" was used to simulate the 14-bus test scheme of the electrical network "IEEE 14 Bus Test Case" using data on network parameters and generators from the PCC "DigSILENT PowerFactory".

500 modes were calculated by means of the load changes in all load buses in the range of 70 - 150 percent of the base level and adding to the obtained values a random value equal to 0 - 20 percent of the base load of the bus. For each mode the calculations of transients when there was an outage of one of the lines and turning on the line by an auto-reclosing device after three seconds were performed.

Given the fact that transformer outages were not considered, 15 emergencies were calculated for each of 500 modes (if transformer outages were taken into account there would be 20 emergency).

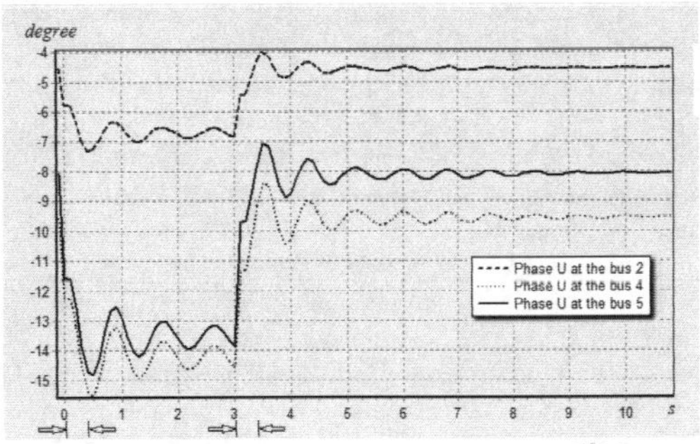

Fig. 3. Changes of the voltage phases at buses 2, 4, 5 during emergency situations for the line number 2.

4 Research results

To detect the network topology with the permitted error, it is necessary to know the time period of the transient phenomena during which the neural network provides for such error. To solve this problem, calculations were carried out with two data sets. The first of them used the changes in the measurement values for one time slice with 27 input parameters of ANN. The second – the changes in measurement values for two consecutive time slices with 54 parameters.

The number of samples for ANN training and testing was varied depending on the calculation option and the number of time slices. The calculations were carried out according to the program developed in the software environment Embacadero C++ Builder. The results are given in Table 1. For the time instants corresponding to the line outage or turning on the line by an auto-reclosing device both ANNs correctly detect the network topology.

Table 1. The comparison of the results between two calculation options of the topology detection.

Options	Data (the number of time slices when there was an outage of the line and line was turned on by an auto-reclosing device)	The number of samples		The number of misclassified samples at testing	Relative error (%)
		for training	for testing		
First option	1	6000	9000	0	0
	2	12000	18000	432	2.4
	3	18000	27000	1078	3.99
	4	24000	36000	5616	15.6
Second option	by 2 from 2	6000	9000	0	0
	by 2 from 3	12000	18000	358	1.99
	by 2 from 4	18000	27000	1185	4.39
	by 2 from 5	24000	36000	5914	16.43

The results for both options are quite close to each other. The admissible relative error 4.39% for the second option corresponds to the data usage for four time slices when there was an outage of the line and four time slices when the line was turned on by an auto-reclosing device, while for the first case the relative error 15.6% corresponds to these data. On this basis one can prefer using the second version of the calculations in which the values of the changes of voltage phases and current magnitudes for two consecutive time slices are used.

Table 2 presents average values of voltage phase changes at buses are instrumented with PMUs for all modes and all emergency conditions for a time interval from 0.1 s to 0.5 s since the moment of the line outage and the moment when it is turned on by an auto-reclosing device.

Table 2. Average values of change in voltage phases at the buses are instrumented with PMU when there was an outage of the line and the line was turned on by an auto-reclosing device.

Average value of voltage phase change at the buses (degrees)	Time, s				
	0.1	0.2	0.3	0.4	0.5
2	0.542	0.167	0.102	0.071	0.116
4	1.165	0.354	0.247	0.083	0.229
5	1.027	0.296	0.208	0.076	0.186
6	0.925	0.540	0.391	0.109	0.337
9	1.091	0.491	0.378	0.111	0.281
11	1.007	0.504	0.387	0.109	0.302
13	1.043	0.532	0.393	0.121	0.334

For time intervals 0.3 s and 0.4 s (the time for which the maximum admissible error for ANN with 54 parameters was obtained) the average values are in a range [0.071; 0.393] (degrees). If one considers the value of the average change in the voltage phase at buses as a threshold of ANN application for detecting network topology, the value of 0.4 degrees can be regarded as a threshold for the studied system. That is, if the maximum change in the voltage phase among the buses with PMUs is less than the specified one, then it results in a large error of the network topology detection. The least average values correspond to an extreme points of curves of a voltage phase change. For these particular points one can observe a sharp increase in error of this solution. Fig. 3 shows the intervals of the admissible ANN calculation error when the network topology is detected (with the arrows).

5 Conclusions

The method of detecting electrical network topology based on ANN using phasor measurements is proposed. It is in a transient. The method allows detecting the topology at the beginning of emergency conditions associated with both line outage and turning it on by an auto-reclosing device.

Regardless of why the topology changes (line outage/turning on line) the same ANN can be used responding to changes in voltage phases and current magnitudes received from PMU.

The area for future work on this topic is the consideration of other methods for solving the problem of detecting topology in the transient phenomena as well as an inclusion of a wider range of possible events, such as generator trips, two or more lines outages.

References

1. Мокеев, А.В. Особенности разработки, испытаний и внедрения устройств синхронизированных векторных измерений // Современные подходы к обеспечению надежности электроэнергетических систем / под ред. М.В. Хохлова. Сыктывкар: Коми научный центр УрО РАН, 2014, с. 56–62.
2. Tate, J. E. Line outage detection using phasor angle measurements / J. E. Tate, T. J. Overbye // IEEE Trans. Power Syst., 2008, Vol. 23, No. 4, pp. 1644-1652.
3. Kim, T. PMU Placement for Line Outage Identification via Multiclass Logistic Regression / T. Kim, S.J. Wright // arXiv:1409.3832v1 [math.OC] 12 Sep 2014. pp. 1-8.
4. Cavraro, G. Power Distribution Network Topology Detection with Time-Series Signature Verification Method / G. Cavraro, R. Arghandeh // IEEE Trans. on PS, 2018, Vol. 33, No. 4, pp. 3500-3509.
5. Ponce, C. FLiER: Practical Topology Update Detection Using Sparse PMUs / C. Ponce, D. S. Bindel // IEEE Trans. Power Syst., 2017, Vol. 32, No. 6, pp. 4222-4232.
6. *Srikumar*, M. S. Line Outage Detection Using Phasor Measurement Units / M. S. Srikumar, Dr. T. Ananthapadmanbha, F. Z. Khan, V. Girish // Procedia Technology 21 (2015), SMART GRID Technologies, 2015, August 6-8, pp. 88 – 95.
7. Готман, Н.Э. Идентификация топологии электрической сети на основе искусственных

нейронных сетей с использованием векторных измерений / Готман Н.Э., Шумилова Г.П., Старцева Т.Б. // Методические вопросы исследования надежности больших систем энергетики. Вып. 66. Актуальные проблемы надежности систем энергетики / под ред. Н.И. Воропая, М.А. Короткевича, А.А. Михалевича. Минск: БНТУ, 2015, с. 251-257.

8. Шумилова, Г.П. Определение топологии электрической сети при делении ее на участки с привязкой к устройствам PMU / Шумилова Г.П., Готман Н.Э., Старцева Т.Б. // Методические вопросы исследования надежности больших систем энергетики. Вып. 67. Проблемы надежности систем энергетики / под ред. Н.И. Воропая, Ю.Я. Чукреева. Сыктывкар: Издательство ООО «Коми республиканская типография», 2016, с. 250-255.
9. Готман, Н.Э. Определение топологии электрической сети с использованием синхронизированных векторных измерений / Готман Н.Э., Шумилова Г.П., Старцева Т.Б. // Актуальные проблемы, направления и механизмы развития производительных сил Севера – 2016: Материалы Пятого Всероссийского научного семинара (21-23 сентября 2016 г., Сыктывкар): в 2 ч. – Сыктывкар: ООО «Коми республиканская типография», 2016, Ч. II, с. 115-123.
10. Хохлов М.В., Голуб И.И. Унифицированный подход к оптимизации размещения PMU в сети для обеспечения надежности наблюдаемости ЭЭС // Методические вопросы исследования надежности больших систем энергетики: Вып. 65. Надежность либерализованных систем энергетики / под ред. Н.И. Воропая, А.Н. Назарычева. Иркутск: ИСЭМ СО РАН, 2015, с. 591-601.

On application of semi-Markov processes superposition to energy systems modeling

Yuriy E. Obzherin[1,*], *Stanislav M.* Sidorov[1], and *Mikhail M.* Nikitin[1]

[1]Sevastopol State University, Higher Mathematics Chair, 299053 Sevastopol, Russian Federation

Abstract. One of the important tasks of the theory of reliability and efficiency of energy systems is the task of creating information systems for managing energy systems and the transition to intelligent management and engineering. The solution to this problem is possible based on the construction of mathematical models relating to various aspects of the structure and functioning of these systems. The paper, using the example of a superposition of independent semi-Markov processes constructed in the works of V.S. Korolyuk, A.F. Turbin, examines the possibilities of using semi-Markov processes with a common phase state space for modeling energy systems; an illustrative example of the application of this approach is given.

1 Introduction

Semi-Markov processes (SMP) are widely used to build models and analyze systems for various purposes: technical, industrial, energy, information, economic, biological, etc.

The leading world publishing houses Elsevier, Springer, World Scientific have published a number of monographs on the subject in recent years. [1-5].

Semi-Markov processes can be effectively used to build models relating to various aspects of the structure and functioning of energy systems: reliability, efficiency, monitoring, diagnostics, maintenance and forecasting.

In this paper, by the example of a superposition of independent semi-Markov processes constructed in the works of V.S. Korolyuk, A.F. Turbin and their apprentices [6, 7], the possibility of using semi-Markov processes with a common phase state space for energy systems modeling is considered.

2 Description of independent semi-Markov processes superposition

Let us consider the semi-Markov model of independent SMPs superposition, following to [6,7].

The system consists of N independent elements (subsystems), functioning of each of which is described by SMP $\xi^{(i)}(t)$ with arbitrary phase space of states $Z^{(i)}$, $i=\overline{1,N}$.

For each $Z^{(i)}$ the decomposition $Z^{(i)}=Z_1^{(i)}\cup Z_0^{(i)}$, $Z_1^{(i)}\cap Z_0^{(i)}\neq\varnothing$, $i=\overline{1,N}$, takes place where $Z^{(i)}$ is interpreting as the set of i-th element's(subsystem's) up states, $Z_0^{(i)}$ is the set of i-th element's(subsystem's) down states.

Let

$$\varsigma_{(i)}(t)=\chi_{Z_1^{(i)}}(\xi^{(i)}(t))=\begin{cases}1, \text{ if } \xi^{(i)}(t)\in Z_1^{(i)},\\ 0, \text{ if } \xi^{(i)}(t)\in Z_0^{(i)},\end{cases}$$

$$\varsigma(t)=\{\varsigma_1(t),\varsigma_2(t),\dots,\varsigma_N(t)\}.$$

Vector process $\zeta(t)$ characterize the system's states at the instant t by its element's (subsystem's) up or down states. The set of possible $\zeta(t)$ process values is the set of binary vectors $\overline{d}=(d_1,d_2,\dots,d_N)$ from (00...0) to (11...1), define it as $D=\{\overline{d}\}$.

Let us consider the concept of system failure: on the set of values of binary vectors D, a function $g(\overline{d})$ is defined such that:

$$g(\overline{d})=\begin{cases}1, \text{ if the system is operational with this}\\ \quad\text{combination of its elements state,}\\ 0, \text{ otherwise.}\end{cases}$$

Let us define the set of values of the vector $\overline{d}$, at which the system is operational, by D_1, and the set of values at which the system is inoperable, by D_0, i.e. $D_1=\{\overline{d}:\ g(\overline{d})=1\}$, $D_0=\{\overline{d}:\ g(\overline{d})=0\}$, by assumption, $D=D_1\cup D_0$, $D_1\cap D_0\neq\varnothing$.

In [6, 7] a superposition of independent processes ПМП $\xi^{(i)}(t)$, $i=\overline{1,N}$ was constructed, and it was shown that the average stationary time to failure T_1, the average stationary renewal time T_0 and the stationary availability factor K_a can be determined by the following formulas:

https://doi.org/10.1051/e3sconf/201913901064

$$T_1 = \sum_{\bar{d}\in D_1}\prod_{i=1}^{N} T_{d_i}^{(i)}\left\{\sum_{\bar{d}\in D_1'}\prod_{i=1}^{N} T_{d_i}^{(i)}\sum_{j\in G(\bar{d})}\frac{1}{T_1^{(j)}}\right\}^{-1} =$$

$$= \sum_{\bar{d}\in D_1}\prod_{i=1}^{N} T_{d_i}^{(i)}\left\{\sum_{\bar{d}\in D_0'}\prod_{i=1}^{N} T_{d_i}^{(i)}\sum_{j\in I(\bar{d})}\frac{1}{T_0^{(j)}}\right\}^{-1}, \quad (1)$$

$$T_0 = \sum_{\bar{d}\in D_0}\prod_{i=1}^{N} T_{d_i}^{(i)}\left\{\sum_{\bar{d}\in D_1'}\prod_{i=1}^{N} T_{d_i}^{(i)}\sum_{j\in G(\bar{d})}\frac{1}{T_1^{(j)}}\right\}^{-1} =$$

$$= \sum_{\bar{d}\in D_0}\prod_{i=1}^{N} T_{d_i}^{(i)}\left\{\sum_{\bar{d}\in D_0'}\prod_{i=1}^{N} T_{d_i}^{(i)}\sum_{j\in I(\bar{d})}\frac{1}{T_0^{(j)}}\right\}^{-1}, \quad (2)$$

$$K_a = \sum_{\bar{d}\in D_1}\prod_{i=1}^{N} T_{d_i}^{(i)}\left\{\prod_{i=1}^{N}\left(T_1^{(i)}+T_0^{(i)}\right)\right\}^{-1}, \quad (3)$$

where $T_1^{(i)}$ is the average stationary time to failure of the i-th element (subsystem); $T_0^{(i)}$ is the average stationary renewal time of the i-th element (subsystem);

$$T_{d_i}^{(i)} = \begin{cases} T_1^{(i)}, & \text{if } d_i = 1, \\ T_0^{(i)}, & \text{if } d_i = 0; \end{cases}$$

D_1' is the set of vectors $\bar{d}\in D_1$ such that changing the value of some single component from one to zero brings the vector $\bar{d}$ into the set D_0 (D_1' is the set of boundary operational states of the system);

D_0' is the set of vectors $\bar{d}\in D_0$ such that changing the value of some single component from zero to one brings the vector $\bar{d}$ into the set D_1 (D_1' is the set of boundary system failure states);

$G(\bar{d})$ is the set of vector $\bar{d}\in D_1'$ component numbers, changing the value of each of which from unit to zero brings the vector $\bar{d}$ into the set D_0;

$I(\bar{d})$ is the set of vector $\bar{d}\in D_0'$ component numbers, changing the value of each of which from zero to unit brings the vector $\bar{d}$ into the set D_1.

The paper [7] shows that the formulae (1) – (3) can be used for immediate calculation of reliability indicators of various typical structures and also, using these formulas and step-by step structure merging algorithm [7], for reliability characteristics estimating for the systems with high number of elements.

Using the formulae (1) – (3) one can also take into account the presence of unloaded reserve, inspection, maintenance, availability of time reserve, etc.

To do this one should find the appropriate expressions for T_1, T_0 of subsystems elements, e.g. from [8 – 11] and use them in the formulae (1) – (3).

The paper [7] represents approximate formulae for T_1, T_0, K_a, valid under the condition of system's elements (subsystems) rapid recovery. Using the semi-Markov superposition model built in [6, 7] and the embedded Markov chain stationary distribution found there one can also find other characteristics of the system including that determine the effectiveness of system's operation.

3 Example of electric energy system modeling

As an example to illustrate the application of the formulae (1) – (3) to model the electric energy systems let us determine the reliability characteristics of the electric energy system with the structural diagram represented in the Figure 1 and the elements reliability characteristics represented in the Table 1.

Table 1. Reliability characteristics of the electric energy system's elements

№	Element	T_0, yr.	T_1, yr.	K_a
1	Electric power line	3.0	0.04	0.9868
2	Substation A input bus	1.5	0.031	0.9797
3	Substation A transformer 1	1.7	0.027	0.9843
4	Substation A transformer 2	0.9	0.033	0.9646
5	Substation B input bus	2.1	0.02	0.9906
6	Substation B transformer 1	1.1	0.018	0.9839
7	Substation B transformer 2	1.5	0.01	0.9934
8	Substation A output bus	1.4	0.014	0.9901
9	Substation B output bus	1.3	0.017	0.9871
10	Switch	0.5	0.037	0.9311
11	Switch	0.36	0.016	0.9574
12	Switch	0.6	0.025	0.9600
13	Switch	0.8	0.015	0.9816
14	Electric power line	1.2	0.0014	0.9988
15	Electric power line	2.1	0.0015	0.9993

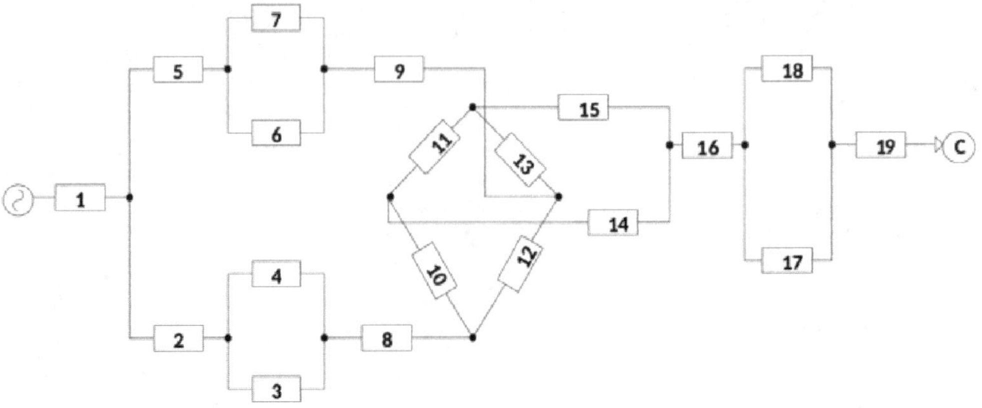

Fig. 1. Structural diagram of the electric energy system.

Table. 2 Reliability characteristics of the electric energy system's elements (continuation)

16	Substation C input bus	0.5	0.028	0.9470
17	Substation C transformer 1	1.0	0.021	0.9794
18	Substation C transformer 2	1.1	0.024	0.9786
19	Substation C output bus	2.0	0.009	0.9955

Finding the characteristics of the system considered with the aid help of the formulae (1) - (3) is carried out according to the following scheme:

1. Parallel elements 3, 4, and also 6, 7 are merged to one element.

2. Sequential elements 2, 8 and the result of the merging of 3 and 4, as well as 5, 9 and the result of 6 and 7 merging are merged to one element.

As a result, we obtain a merged subsystem, the block diagram of which is presented in Fig. 2.

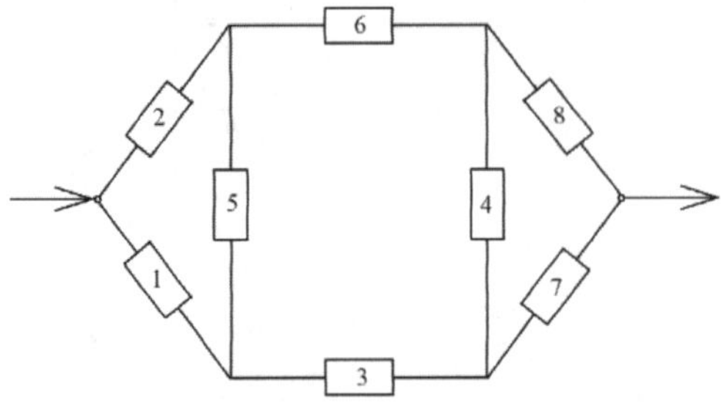

Fig. 2. Structural diagram of the aggregated subsystem.

In the figure 2 element 1 corresponds to the result of the aggregation of initial diagram elements 2, 3, 4, 8 while element 2 – to the results of the aggregation of elements 5, 6, 7, 9. Elements 3 – 8 correspond to elements 10 - 15 of the initial scheme. The reliability characteristics of the aggregated subsystem obtained are represented in Table 2.

Table 3. Reliability characteristics of the merged subsystem's elements.

Element	T_1, yr.	T_0,yr.	K_a
1	0.7051	0.02217	0.9695
2	0.7924	0.01809	0.9777
3	0.5000	0.03700	0.9311
4	0.3600	0.01600	0.9574
5	0.6000	0.02500	0.9600
6	0.8000	0.01500	0.9816
7	1.2000	0.001400	0.9988
8	2.1000	0.001500	0.9993

3. Applying the formulae (1) – (3) to the subsystem where the set D_1 contains 84 vectors, D_0 – 172 vectors one can obtain the next reliability characteristics of the subsystem T_1= 4,978 yr. ,T_0= 0,01013 yr. , K_a = 0,9980.

4. Subsystem of parallel elements 17 – 18 is merged to one element with T_1= 25,4911 yr. ,T_0= 0,01120 yr. , K_a = 0,9996.

Representing the initial system as consequential connection of two previously calculated subsystems with each other and with elements 1,16 and 19 of the initial system let us merge them to one element with reliability characteristics T_1= 0,3254 yr.,T_0=0,02523 yr., K_a = 0,9280.

Thus, for the system considered

$$T_1= 0{,}3254 \text{ yr.}, T_0=0{,}02523 \text{ yr}, \; K_r = 0{,}9280.$$

The same result for of the system considered is obtained with the help of the structural functions method [10, 11].

4 Conclusion

The paper shows the possibilities of semi-Markov processes with common phase space of states application to model construction and operation analysis of energy systems using an example of independent SMP superposition. This approach lets to avoid of assumption on exponential distribution law and also take into account inspection, maintenance, etc. in the model of system. Unlike some other methods for finding the reliability and efficiency characteristics of energy systems, it allows to find not only probabilistic characteristics of the system, but also temporary ones. To solve the dimension problem one can use the algorithms of asymptotic and stationary phase merging of the systems [4 – 7].

The research was carried out within the state assignment of the Minobrnauki of Russia (№ 1.10513.2018/11.12), with financial support by RFBR (project № 18-01-00392a).

References

1. Grabski F. *Semi-Markov Processes Applications in System Reliability and Maintenance* (Elsevier, 2015)
2. Obzherin Yu.E., Boyko E.G., *Semi-Markov Models: Control of Restorable Systems with Latent Failures* (Elsevier, Academic Press, 2015)
3. Janssen J., Manca R. *Applied Semi-Markov Processes* (Springer Science + Business Media, 2006)
4. Korolyuk V.S., Limnios N. *Stochastic Systems in Merging Phase Space* (World Scientific, Imperial College Press, 2005)
5. Silvestrov D., Silvestrov S. *Nonlinearly Perturbed Semi-Markov Processes* (Springer, 2017)
6. Koroluk V.S., Turbin A.F. *Markovian Restoration Processes in the Problems of System Reliability* (Naukova Dumka, Kiev, 1982)
7. Korlat A.N., Kuznetsov V.N., Novikov M.M., Turbin A.F., *Semi-Markov Models of Recoverable Systems and Queuing Systems* (Shtiinta, Chisinau, 1991)
8. Kashtanov V.A., Medvedev V.I. *Theory of Complex Systems Reliability, 2nd ed.* (Fizmatlit, Moscow, 2010)
9. Rudenko Yu, N., Ushakov I.A. *Reliability of Energy Systems, 2nd ed* (Nauka, Novosibirsk, 1989)
10. Beichelt F., Franken P. *Reliability and Maintenance - Mathematical Methods. Radio i Svjas* (Moscow, 1988)
11. Beichelt F., Tittmann P. *Reliability and Maintenance – Networks and Systems* (Mathematical Models. Chapman & Hall, CRC Press, Boca Raton, London, New York, 2012)

Semi-Markov model of a multi-component energy system with component-wise storages

Yuriy E. Obzherin[1], *Stanislav M.* Sidorov[1,*], and *Mikhail M.* Nikitin[1]

[1]Sevastopol State University, Higher mathematics Department, Sevastopol, Russia

Abstract. To increase the reliability and efficiency of the functioning of various purposes systems (in particular, energy systems), time redundancy is used. In this paper, a semi-Markov model of a multi-component system with component-wise storage devices, which are the sources of the time reserve, is constructed. The stationary characteristics of reliability and efficiency of the system under consideration are found. The analysis of the influence of storage capacities on the reliability and efficiency of the system is been.

Introduction

To ensure the reliability and efficiency of systems, time redundancy is used.

Time redundancy [1-6] provides the system with the ability to use up some additional time (time reserve) to restore performance during operation. For systems with time redundancy, a malfunction of the system is not necessarily accompanied by a system failure, since it is possible to restore the system's operability for a reserve time.

Time redundancy is used in gas transmission systems, in which underground gas storages are the source of time reserve, in electric power systems, time reserve is realized due to energy storage [5, 6].

There are problems of modelling systems, and solving optimization problems, taking into account the availability of a reserve of time.

In [6], a semi-Markov model of a two-component system with component-wise storage devices was constructed; stationary characteristics of reliability were found.

In this paper, on the basis of the theory of semi-Markov processes with a common phase state space [7–10, 12], a semi-Markov model of a multi-component system with a component-wise instantly replenished time reserve is constructed. The stationary characteristics of the reliability and efficiency of the system are found, the influence of the time reserve value on the obtained characteristics is analysed.

1 Description of the system functioning

Consider the system S, consisting of N components, time to failure of which are random variables (RVs) α_k with the distribution functions (DFs) $F_k(t)$, a restoration times are RVs β_k with DFs $G_k(t)$, $k=\overline{1,N}$. Each component of the system has a random instantly replenished time reserve [1-3] τ_k with DF $R_k(t)$. The time reserve starts to be used at the moment the component recovery starts. A component failure occurs when the component is restored and the time reserve ($\beta_k > \tau_k$) has been completely used up and continues until the component is restored. By the time the component is restored, the time reserve is replenished to its original value. RVs α_k, β_k, τ_k are assumed to be independent in aggregate, having finite mathematical expectations; DFs $F_k(t)$, $G_k(t)$, $R_k(t)$ have distribution densities $f_k(t)$, $g_k(t)$, $r_k(t)$. The failure of the system S is determined based on the analysis of the structure of the system. Recovery of system components is assumed unlimited.

2 Semi-Markov model building

To describe the functioning of the system S, we use the semi-Markov process (SMP) [8, 9] $\xi(t)$. We introduce a discrete-continuous phase state space of the form:

$$E=\{i\bar{d}\bar{x}:\bar{d}=(d_1,d_2,...,d_k,...,d_N),\\ \bar{x}=(x_1,x_2,...,x_k,...,x_N),\ x_k\geq 0,\ k=\overline{1,N}\},\qquad (1)$$

where $i=\overline{1,N}$ is the number of the component in which there was a change in physical condition. Component d_k of the vector $\bar{d}$ describes the physical state of the element with the number k:

$$d_k=\begin{cases}1, \text{ if k}-th \text{ component is operational},\\ \bar{1}, \text{ if k}-th \text{ component is restored and functions due to the time reserve},\\ 0, \text{ if k}-th \text{ component is in failure}.\end{cases}$$

The continuous component x_k of vector $\bar{x}$ indicates the elapsed time since the last change in the physical state in the component with the number k; note that $x_i = 0$.

Find the residence times in the states of the system. To do this, we introduce the RV:

$$\delta_z^{(k)} = \begin{cases} \alpha_k, \text{ if } z = 1, \\ \beta_k \wedge \tau_k, \text{ if } z = \bar{1}, \\ \left[\beta_k - \tau_k\right]^+, \text{if } z = 0, \end{cases} \quad (2)$$

where $\wedge$ is the minimum sign, $\left[\beta_k - \tau_k\right]^+$ is RV, which DF is determined by the equality

$$P\left(\left[\beta_k - \tau_k\right]^+ \le t\right) = 1 - \frac{\int_0^\infty \bar{G}_k(y+t) r_k(y)dy}{P(\beta_k > \tau_k)}.$$

We denote by $V_z^{(k)}(t) = P\left(\delta_z^{(k)} < t\right)$ the DFs of RVs $\delta_z^{(k)}$, $\bar{V}_z^{(k)}(t) = 1 - V_z^{(k)}(t)$, $v_z^{(k)}(t)$ is probability densities of RVs $\delta_z^{(k)}$:

$$v_1^{(k)}(t) = f_k(t), \quad v_{\bar{1}}^{(k)}(t) = \bar{R}_k(t) g_k(t) + \bar{G}_k(t) r_k(t),$$

$$v_0^{(k)}(t) = \frac{\int_0^\infty g_k(x+t) r_k(x)dx}{P(\beta_k > \tau_k)}.$$

Then the mean residence time in the state $i\bar{d}\bar{x}$ is determined by the equality:

$$\theta_{i\bar{d}\bar{x}} = \bigwedge_{k=1}^{N} \left[\delta_{d_k}^{(k)} - x_k\right]^+, \; x_i = 0. \quad (3)$$

Define the transition probabilities of the embedded Markov chain (EMC) $\{\xi_n; n \ge 0\}$.

We introduce additional functions:

$$w_{1,z}^{(k)}(t) = \begin{cases} v_1^{(k)}(t), \; z = \bar{1}, \\ 0, \quad z = 0, 1, \end{cases}$$

$$w_{\bar{1},z}^{(k)}(t) = \begin{cases} \bar{R}_k(t) g_k(t), \; z = 1, \\ \bar{G}_k(t) r_k(t), \; z = 0, \\ 0, \quad z = \bar{1}, \end{cases} \quad w_{0,z}^{(k)}(t) = \begin{cases} v_0^{(k)}(t), \; z = 1, \\ 0, \quad z = 0, \bar{1}. \end{cases}$$

Note that the functions $f_{\bar{1},1}^{(k)}(t) = \dfrac{w_{\bar{1},1}^{(k)}(t)}{P(\tau_k > \beta_k)}$, $f_{\bar{1},0}^{(k)}(t) = \dfrac{w_{\bar{1},0}^{(k)}(t)}{P(\tau_k < \beta_k)}$, $k = \overline{1, N}$ are the densities of the distributions.

The probability densities of the transitions of the EMC $\{\xi_n; n \ge 0\}$ are as follows:

$$p_{i\bar{d}\bar{x}}^{j\bar{d}\bar{y}} = \begin{cases} \dfrac{w_{d_j,d_j'}^{(j)}(x_j + y_i) \prod_{k=1, k \ne j}^{N} \bar{V}_{d_k}^{(k)}(x_k + y_i)}{\prod_{s=1, s \ne i}^{N} \bar{V}_{d_s}^{(s)}(x_s)}, & j \ne i, y_k = x_k + y_i, \; k \ne j, y_j = 0, \\ \dfrac{w_{d_i,d_i'}^{(i)}(y_k - x_k) \prod_{k=1, k \ne i}^{N} \bar{V}_{d_k}^{(k)}(y_k)}{\prod_{s=1, s \ne i}^{N} \bar{V}_{d_s}^{(s)}(x_s)}, & j = i, y_k = x_k + t, \; k \ne i, y_i = x_i = 0, \end{cases} \quad (4)$$

vectors $\bar{d}$ and $\bar{d}'$ differ only in the j-th component.

3 Finding the stationary distribution of an embedded Markov chain

To find the stationary characteristics of the reliability and efficiency of the system S, we find the stationary distribution of the EMC $\{\xi_n; n \ge 0\}$.

We introduce the following notation: let $\bar{x} = (x_1, x_2, ..., x_N)$, then vectors $(\bar{x} - t)^{(i)}$, $\left[(\bar{x} - x_j)^{(i)}, t\right]$, where $x_j = \bigwedge_{k=1, k \ne i}^{N} x_k$, are defined as follows:

$$(\bar{x} - t)_k^{(i)} = \begin{cases} x_k - t, \; k \ne i, \\ 0, \quad k = i, \end{cases} \quad \left[(\bar{x} - x_j)^{(i)}, t\right]_k = \begin{cases} x_k - x_j, \; k \ne i, \\ t, \quad k = i. \end{cases}$$

Suppose that the stationary distribution of the EMC $\{\xi_n; n \ge 0\}$ has densities $\rho(i\bar{d}\bar{x})$, using (4), we compose a system of integral equations that they satisfy.

1. In the case of states $i\bar{d}\bar{x}$, $\bar{d}_i = \bar{1}$ next transitions are possible:
   a) $i\bar{d}^{(i)}(\bar{x} - t)^{(i)} \to i\bar{d}\bar{x}, 0 < t < x_j, \; \bar{d}_i^{(i)} = 1,$
   b) $j\bar{d}^{(i)}\left[(\bar{x} - x_j)^{(i)}, t\right] \to i\bar{d}\bar{x}, \; t > 0, \; \bar{d}_i^{(i)} = 1.$

Consequently

$$\rho\left(i\bar{d}\bar{x}\right) = \int_0^{x_j} \frac{v_1^{(i)}(t) \prod_{k=1, k \ne i}^{N} \bar{V}_{d_k}^{(k)}(x_k)}{\prod_{s=1, s \ne i}^{N} \bar{V}_{d_s}^{(s)}(x_s - t)} \rho\left(i\bar{d}^{(i)}(\bar{x} - t)^{(i)}\right) dt + \int_0^\infty \frac{v_1^{(i)}(x_j + t) \prod_{k=1, k \ne i}^{N} \bar{V}_{d_k}^{(k)}(x_k)}{\prod_{s=1, s \ne j}^{N} \bar{V}_{d_s}^{(s)}(x_s - x_j)} \rho\left(j\bar{d}^{(i)}\left[(\bar{x} - x_j)^{(i)}, t\right]\right) dt. \quad (5)$$

2. In the case of states $i\bar{d}\bar{x}$, $\bar{d}_i = 0$, transitions will be:
   a) $i\bar{d}^{(i)}(\bar{x} - t)^{(i)} \to i\bar{d}\bar{x}, 0 < t < x_j, \; \bar{d}_i^{(i)} = \bar{1},$

b) $j\bar{d}^{(i)}\left[(\bar{x}-x_j)^{(i)},t\right]\to i\bar{d}\bar{x},\ t>0,\ \bar{d}_i^{(i)}=\bar{1}.$

Hence,

$$\rho\left(i\bar{d}\bar{x}\right)=\int_0^{x_j}\frac{w_{\bar{1},0}^{(i)}(t)\prod_{k=1,\,k\neq i}^{N}\bar{V}_{d_k}^{(k)}(x_k)}{\prod_{s=1,\,s\neq i}^{N}\bar{V}_{d_s}^{(s)}(x_s-t)}\rho\left(i\bar{d}^{(i)}(\bar{x}-t)^{(i)}\right)dt+$$
$$+\int_0^{\infty}\frac{w_{\bar{1},0}^{(i)}(x_j+t)\prod_{k=1,\,k\neq i}^{N}\bar{V}_{d_k}^{(k)}(x_k)}{\prod_{s=1,\,s\neq j}^{N}\bar{V}_{d_s}^{(s)}(x_s-x_j)}\rho\left(j\bar{d}^{(i)}\left[(\bar{x}-x_j)^{(i)},t\right]\right)dt. \tag{6}$$

3. For $i\bar{d}\bar{x},\ \bar{d}_i=1$ transitions will be:

a) $i\bar{d}^{(i)}(\bar{x}-t)^{(i)}\to i\bar{d}\bar{x},0<t<x_j,\ \bar{d}_i^{(i)}=\bar{1},$

b) $i\bar{d}^{(i)}(\bar{x}-t)^{(i)}\to i\bar{d}\bar{x},0<t<x_j,\ \bar{d}_i^{(i)}=0,$

c) $j\bar{d}^{(i)}\left[(\bar{x}-x_j)^{(i)},t\right]\to i\bar{d}\bar{x},\ t>0,\ \bar{d}_i^{(i)}=\bar{1},$

d) $j\bar{d}^{(i)}\left[(\bar{x}-x_j)^{(i)},t\right]\to i\bar{d}\bar{x},\ t>0,\ \bar{d}_i^{(i)}=0.$

Therefore,

$$\rho\left(i\bar{d}\bar{x}\right)=\int_0^{x_j}\frac{w_{\bar{1},1}^{(i)}(t)\prod_{k=1,\,k\neq i}^{N}\bar{V}_{d_k}^{(k)}(x_k)}{\prod_{s=1,\,s\neq i}^{N}\bar{V}_{d_s}^{(s)}(x_s-t)}\rho\left(i\bar{d}^{(i)}(\bar{x}-t)^{(i)}\right)dt+$$
$$+\int_0^{x_j}\frac{w_{0,1}^{(i)}(t)\prod_{k=1,\,k\neq i}^{N}\bar{V}_{d_k}^{(k)}(x_k)}{\prod_{s=1,\,s\neq i}^{N}\bar{V}_{d_s}^{(s)}(x_s-t)}\rho\left(i\bar{d}^{(i)}(\bar{x}-t)^{(i)}\right)dt+$$
$$+\int_0^{\infty}\frac{w_{\bar{1},1}^{(i)}(x_j+t)\prod_{k=1,\,k\neq i}^{N}\bar{V}_{d_k}^{(k)}(x_k)}{\prod_{s=1,\,s\neq j}^{N}\bar{V}_{d_s}^{(s)}(x_s-x_j)}\rho\left(j\bar{d}^{(i)}\left[(\bar{x}-x_j)^{(i)},t\right]\right)dt+$$
$$+\int_0^{\infty}\frac{w_{0,1}^{(i)}(x_j+t)\prod_{k=1,\,k\neq i}^{N}\bar{V}_{d_k}^{(k)}(x_k)}{\prod_{s=1,\,s\neq j}^{N}\bar{V}_{d_s}^{(s)}(x_s-x_j)}\rho\left(j\bar{d}^{(i)}\left[(\bar{x}-x_j)^{(i)},t\right]\right)dt. \tag{7}$$

4. $$\sum_{d\in D}\sum_{i=1}^{N}\int_{R_+^{(N,i)}}\rho\left(i\bar{d}\bar{x}\right)d\bar{x}^{(i)}=1 \tag{8}$$

(normalization condition).

Denote by $\tilde{\rho}\left(i\bar{d}\bar{x}\right)=\dfrac{\rho\left(i\bar{d}\bar{x}\right)}{\prod_{k=1}^{N}\bar{V}_{d_k}^{(k)}(x_k)}$. Then the system of equations (5) – (8) will have the following form:

1. In the case of conditions $i\bar{d}\bar{x},\ \bar{d}_i=\bar{1}$

$$\tilde{\rho}\left(i\bar{d}\bar{x}\right)=\int_0^{x_j}v_{\bar{1}}^{(i)}(t)\tilde{\rho}\left(i\bar{d}^{(i)}(\bar{x}-t)^{(i)}\right)dt+$$
$$+\int_0^{\infty}v_{\bar{1}}^{(i)}(x_j+t)\tilde{\rho}\left(j\bar{d}^{(i)}\left[(\bar{x}-x_j)^{(i)},t\right]\right)dt. \tag{9}$$

2. If $i\bar{d}\bar{x},\ \bar{d}_i=0$, then

$$\tilde{\rho}\left(i\bar{d}\bar{x}\right)=\int_0^{x_j}w_{\bar{1},0}^{(i)}(t)\tilde{\rho}\left(i\bar{d}^{(i)}(\bar{x}-t)^{(i)}\right)dt+$$
$$+\int_0^{\infty}w_{\bar{1},0}^{(i)}(x_j+t)\tilde{\rho}\left(j\bar{d}^{(i)}\left[(\bar{x}-x_j)^{(i)},t\right]\right)dt. \tag{10}$$

3. In the case of states $i\bar{d}\bar{x},\ \bar{d}_i=1$

$$\tilde{\rho}\left(i\bar{d}\bar{x}\right)=\int_0^{x_j}w_{\bar{1},1}^{(i)}(t)\tilde{\rho}\left(i\bar{d}^{(i)}(\bar{x}-t)^{(i)}\right)dt+$$
$$+\int_0^{x_j}w_{0,1}^{(i)}(t)\tilde{\rho}\left(i\bar{d}^{(i)}(\bar{x}-t)^{(i)}\right)dt+$$
$$+\int_0^{\infty}w_{\bar{1},1}^{(i)}(x_j+t)\tilde{\rho}\left(j\bar{d}^{(i)}\left[(\bar{x}-x_j)^{(i)},t\right]\right)dt+$$
$$+\int_0^{\infty}w_{0,1}^{(i)}(x_j+t)\tilde{\rho}\left(j\bar{d}^{(i)}\left[(\bar{x}-x_j)^{(i)},t\right]\right)dt. \tag{11}$$

4. $$\sum_{d\in D}\sum_{i=1}^{N}\int_{R_+^{(N,i)}}\tilde{\rho}\left(i\bar{d}\bar{x}\right)\prod_{k=1}^{N}\bar{V}_{d_k}^{(k)}(x_k)d\bar{x}^{(i)}=1 \tag{12}$$

(normalization condition).

Let`s us introduce the following notation:

$$\rho_z^{(k)}=\begin{cases}\rho^{(k)},\ z=1,\bar{1},\\ \bar{\rho}^{(k)},\ z=0,\end{cases}\quad \rho^{(k)}=\frac{1}{2+P(\beta_k>\tau_k)},$$
$$\bar{\rho}^{(k)}=\frac{P(\beta_k>\tau_k)}{2+P(\beta_k>\tau_k)},\ \bar{\rho}^{(k)}=\rho^{(k)}\cdot P(\beta_k>\tau_k).$$

Lemma. The solution to the system of equations (9) – (12) has the following form:

$$\tilde{\rho}(i\bar{d}\bar{x})=\rho_0\prod_{k=1}^{N}\rho_{d_k}^{(k)}, \tag{13}$$

where the constant ρ_0 is found from the normalization condition (12).

Proof. By substitution, we verify that the functions $\tilde{\rho}(i\bar{d}\bar{x})=\rho_0\prod_{k=1}^{N}\rho_{d_k}^{(k)}$ satisfy the system of equations (9) – (12) for any value ρ_0.

1. In the case $i\overline{dx},\ \overline{d}_i=\overline{1}$

$$\tilde{\rho}(i\overline{dx})=\rho_0\prod_{k=1}^{N}\rho_{d_k}^{(k)}=\int_0^{x_j}v_1^{(i)}(t)\rho_0\prod_{k=1}^{N}\rho_{d_k}^{(k)}dt+\int_0^{\infty}v_1^{(i)}(x_j+t)\rho_0\prod_{k=1}^{N}\rho_{d_k}^{(k)}dt=\rho_0\prod_{k=1}^{N}\rho_{d_k}^{(k)}\left(\int_0^{x_j}v_1^{(i)}(t)dt+\int_0^{\infty}v_1^{(i)}(x_j+t)dt\right)=$$

$$=\rho_0\prod_{k=1}^{N}\rho_{d_k}^{(k)}\left(\int_0^{x_j}v_1^{(i)}(t)dt+\int_{x_j}^{\infty}v_1^{(i)}(t')dt'\right)=\rho_0\prod_{k=1}^{N}\rho_{d_k}^{(k)}\int_0^{\infty}v_1^{(i)}(t)dt=\rho_0\prod_{k=1}^{N}\rho_{d_k}^{(k)}\int_0^{\infty}f_i(t)dt=\rho_0\prod_{k=1}^{N}\rho_{d_k}^{(k)}.$$

2. If $i\overline{dx},\ \overline{d}_i=0$ then

$$\tilde{\rho}\left(i\overline{dx}\right)=\rho_0\prod_{k=1}^{N}\rho_{d_k}^{(k)}=\int_0^{x_j}w_{\overline{1},0}^{(i)}(t)\rho_0\prod_{k=1}^{N}\rho_{d_k}^{(k)}dt+\int_0^{\infty}w_{\overline{1},0}^{(i)}(x_j+t)\rho_0\prod_{k=1}^{N}\rho_{d_k}^{(k)}dt=\rho_0\prod_{k=1}^{N}\rho_{d_k}^{(k)}\left(\int_0^{x_j}w_{\overline{1},0}^{(i)}(t)dt+\int_0^{\infty}w_{\overline{1},0}^{(i)}(x_j+t)dt\right)=$$

$$=\rho_0\prod_{k=1}^{N}\rho_{d_k}^{(k)}\left(\int_0^{x_j}w_{\overline{1},0}^{(i)}(t)dt+\int_{x_j}^{\infty}w_{\overline{1},0}^{(i)}(t')dt'\right)=\rho_0\prod_{k=1}^{N}\rho_{d_k}^{(k)}\int_0^{\infty}w_{\overline{1},0}^{(i)}(t)dt=\rho_0\prod_{k=1}^{N}\rho_{d_k}^{(k)}\int_0^{\infty}f_{\overline{1},0}^{(i)}(t)dt=\rho_0\prod_{k=1}^{N}\rho_{d_k}^{(k)}.$$

3. For $i\overline{dx},\ \overline{d}_i=1$

$$\tilde{\rho}\left(i\overline{dx}\right)=\rho_0\prod_{k=1}^{N}\rho_{d_k}^{(k)}=\int_0^{x_j}w_{\overline{1},1}^{(i)}(t)\rho_0\prod_{k=1}^{N}\rho_{d_k}^{(k)}dt+\int_0^{x_j}w_{0,1}^{(i)}(t)\rho_0\prod_{k=1}^{N}\rho_{d_k}^{(k)}dt+\int_0^{\infty}w_{\overline{1},1}^{(i)}(x_j+t)\rho_0\prod_{k=1}^{N}\rho_{d_k}^{(k)}dt+$$

$$+\int_0^{\infty}w_{0,1}^{(i)}(x_j+t)\rho_0\prod_{k=1}^{N}\rho_{d_k}^{(k)}dt=\rho_0\prod_{k=1}^{N}\rho_{d_k}^{(k)}\left(\int_0^{x_j}w_{\overline{1},1}^{(i)}(t)dt+\int_0^{x_j}w_{0,1}^{(i)}(t)dt+\int_0^{\infty}w_{\overline{1},1}^{(i)}(x_j+t)dt+\int_0^{\infty}w_{0,1}^{(i)}(x_j+t)dt\right)=$$

$$=\rho_0\prod_{k=1}^{N}\rho_{d_k}^{(k)}\left(\int_0^{x_j}w_{\overline{1},1}^{(i)}(t)dt+\int_{x_j}^{\infty}w_{\overline{1},1}^{(i)}(t')dt'+\int_0^{x_j}w_{0,1}^{(i)}(t)dt+\int_{x_j}^{\infty}w_{0,1}^{(i)}(t')dt'\right)==\rho_0\prod_{k=1}^{N}\rho_{d_k}^{(k)}\left(\int_0^{\infty}w_{\overline{1},1}^{(i)}(t)dt+\int_0^{\infty}w_{0,1}^{(i)}(t)dt\right)=$$

$$=\rho_0\prod_{k=1}^{N}\rho_{d_k}^{(k)}\left(\int_0^{\infty}\overline{R}_i(t)g_i(t)dt+\int_0^{\infty}v_0^{(i)}(t)dt\right)=\rho_0\prod_{k=1}^{N}\rho_{d_k}^{(k)}\left(P(\tau_i>\beta_i)+\int_0^{\infty}dt\int_0^{\infty}g_i(y+t)r_i(y)dy\right)=$$

$$=\rho_0\prod_{k=1}^{N}\rho_{d_k}^{(k)}\left(P(\tau_i>\beta_i)+\int_0^{\infty}r_i(t)\overline{G}_i(t)dt\right)=\rho_0\prod_{k=1}^{N}\rho_{d_k}^{(k)}\left(P(\tau_i>\beta_i)+P(\tau_i\le\beta_i)\right)=\rho_0\prod_{k=1}^{N}\rho_{d_k}^{(k)}.$$

The lemma is proved. □

Consequently, the stationary distribution of EMC $\{\xi_n;n\ge 0\}$ of the system in question has the following form:

$$\rho(i\overline{dx})=\rho_0\prod_{k=1}^{N}\rho_{d_k}^{(k)}\overline{V}_{d_k}^{(k)}(x_k),\qquad(14)$$

where the constant ρ_0 is found from the normalization condition.

4 Finding the stationary characteristics of the system

To find the stationary availability factor K_a, mean stationary operating time of the system to failure T_+, mean stationary restoration time T_-, we use the following formulas presented in [8, 12]:

$$K_a=\frac{\int_{E_+}m(e)\rho(de)}{\int_E m(e)\rho(de)}=\frac{T_+}{T_++T_-},$$

$$T_+=\frac{\int_{E_+}m(e)\rho(de)}{\int_{E_+}P(e,E_-)\rho(de)},\quad T_-=\frac{\int_{E_-}m(e)\rho(de)}{\int_{E_+}P(e,E_-)\rho(de)},\qquad(15)$$

where $\rho(de)$ is stationary distribution of EMC $\{\xi_n;n\ge 0\}$, $P(e,E_-)$ – transition probabilities of the EMC $\{\xi_n;n\ge 0\}$ to a subset of fault states E_-, $m(e)$ is mean residence time of the semi-Markov process in the state $e\in E$.

Consider the concept of system failure: based on an analysis of the structure of the system, on the set of all the vectors $D=\{\overline{d}\}$ a structural function [11] $g(\overline{d})$ is defined as:

$$g(\overline{d})=\begin{cases}1, & \text{if the system is operational for a given combination of } \overline{d} \text{ components;}\\ 0, & \text{if the system is in failure.}\end{cases}$$

The set of vector $\bar{d}$ values at which the system is operational, we denote D_+, and the set of values at which the system is in failure, by D_-, i.e. $D_+ = \{\bar{d} : g(\bar{d}) = 1\}$, $D_- = \{\bar{d} : g(\bar{d}) = 0\}$. By assumption, $D = D_+ \cup D_-$, $D_+ \cap D_- = \varnothing$.

In accordance with the choice of D_+ and D_-, the phase state space E is divided into sets E_+ and E_- of operable and fault states. $E = E_+ \cup E_-$, $E_+ \cap E_- = \varnothing$.

We introduce the following notations for finding T_+, T_-:

D_+^0 is a set of vectors $\bar{d} \in D_+$ such that a change in the state of one component puts the system in a failure state (in D_-),

$G(\bar{d})$ is the set of numbers of the components of the vector $\bar{d} \in D_+^0$, changing the value of each of which translates the vector $\bar{d}$ into D_-.

We find the numerator of the first of formulas (15). Using (2), we determine the mean residence times in the states:

$$E\theta_{i\bar{d}\bar{x}} = \frac{\int_0^\infty \prod_{k=1}^{N} \bar{V}_{d_k}^{(k)}(x_k + t)dt}{\prod_{\substack{s=1,\\ s\neq i}}^{N} \bar{V}_{d_s}^{(s)}(x_s)}, x_i = 0. \tag{16}$$

Using (16) and stationary distribution (14), we obtain:

$$\int_{E_+} m(e)\rho(de) = \sum_{\bar{d}\in D_+} \sum_{i=1}^{N} \underbrace{\int_0^\infty \cdots \int_0^\infty}_{N-1} \rho_0 \prod_{k=1}^{N} \rho_{d_k}^{(k)} \prod_{\substack{j=1,\\ j\neq i}}^{N} \bar{V}_{d_j}^{(j)}(x_j) d\bar{x}^{(i)} \int_0^\infty \frac{\prod_{k=1}^{N} \bar{V}_{d_k}^{(k)}(x_k+t)}{\prod_{\substack{s=1,\\ s\neq i}}^{N} \bar{V}_{d_s}^{(s)}(x_s)} dt = \sum_{\bar{d}\in D_+} \rho_0 \prod_{k=1}^{N} \rho_{d_k}^{(k)} \sum_{i=1}^{N} \underbrace{\int_0^\infty \cdots \int_0^\infty}_{N-1} d\bar{x}^{(i)} \int_0^\infty \prod_{k=1}^{N} \bar{V}_{d_k}^{(k)}(x_k+t)dt =$$

$$= \sum_{\bar{d}\in D_+} \rho_0 \prod_{k=1}^{N} \rho_{d_k}^{(k)} \prod_{k=1}^{N} \int_0^\infty \bar{V}_{d_k}^{(k)}(x_k)dx_k = \rho_0 \sum_{\bar{d}\in D_+} \prod_{k=1}^{N} \rho_{d_k}^{(k)} E\delta_{d_k}^{(k)} = \rho_0 \prod_{k=1}^{N} \rho^{(k)} \cdot \sum_{\bar{d}\in D_+} \prod_{k:d_k=1} E\alpha_k \cdot \prod_{k:d_k=\bar{1}} E(\beta_k \wedge \tau_k) \cdot \prod_{k:d_k=0} P(\beta_k > \tau_k) E\left([\beta_k - \tau_k]^+\right),$$

$$\int_{E_-} m(e)\rho(de) = \rho_0 \sum_{\bar{d}\in D_-} \prod_{k=1}^{N} \rho_{d_k}^{(k)} E\delta_{d_k}^{(k)} = \rho_0 \prod_{k=1}^{N} \rho^{(k)} \cdot \sum_{\bar{d}\in D_-} \prod_{k:d_k=1} E\alpha_k \cdot \prod_{k:d_k=\bar{1}} E(\beta_k \wedge \tau_k) \cdot \prod_{k:d_k=0} P(\beta_k > \tau_k) E\left([\beta_k - \tau_k]^+\right).$$

$$\int_{E_+} P(e, E_-)\rho(de) = \rho_0 \prod_{k=1}^{N} \rho^{(k)} \cdot \sum_{\bar{d}\in D_+^0} \sum_{j\in G(\bar{d})} I(d_j) \cdot \prod_{\substack{k:d_k=1,\\ k\neq j}} E\alpha_k \cdot \prod_{\substack{k:d_k=\bar{1},\\ k\neq j}} E(\beta_k \wedge \tau_k) \cdot \prod_{\substack{k:d_k=0,\\ k\neq j}} P(\beta_k > \tau_k) E\left([\beta_k - \tau_k]^+\right),$$

where $I(d_j) = \begin{cases} P(\beta_j > \tau_j), & \text{if } d_j = \bar{1}, \\ 1, & \text{if } d_j = 1, 0. \end{cases}$

$$K_a = \frac{\sum_{\bar{d}\in D_+} \prod_{k=1}^{N} \rho_{d_k}^{(k)} \prod_{k=1}^{N} E\delta_{d_k}^{(k)}}{\sum_{\bar{d}\in D} \prod_{k=1}^{N} \rho_{d_k}^{(k)} \prod_{k=1}^{N} E\delta_{d_k}^{(k)}} = \frac{\sum_{\bar{d}\in D_+} \prod_{k=1}^{N} \rho_{d_k}^{(k)} E\delta_{d_k}^{(k)}}{\prod_{k=1}^{N} \rho_{d_k}^{(k)} \left(E\delta_1^{(k)} + E\delta_{\bar{1}}^{(k)} + P(\beta_k > \tau_k) E\delta_0^{(k)}\right)} = \frac{\sum_{\bar{d}\in D_+} \prod_{k=1}^{N} \rho_{d_k}^{(k)} E\delta_{d_k}^{(k)}}{\prod_{k=1}^{N} \rho_{d_k}^{(k)} \left(E\alpha_k + E(\beta_k \wedge \tau_k) + P(\beta_k > \tau_k) E[\beta_k - \tau_k]^+\right)} =$$

$$= \frac{\sum_{\bar{d}\in D_+} \prod_{k:d_k=1} E\alpha_k \cdot \prod_{k:d_k=\bar{1}} E(\beta_k \wedge \tau_k) \cdot \prod_{k:d_k=0} P(\beta_k > \tau_k) E\left([\beta_k - \tau_k]^+\right)}{\prod_{k=1}^{N} (E\alpha_k + E\beta_k)}, \tag{17}$$

$$T_+ = \frac{\sum_{\bar{d}\in D_+} \prod_{k:d_k=1} E\alpha_k \cdot \prod_{k:d_k=\bar{1}} E(\beta_k \wedge \tau_k) \cdot \prod_{k:d_k=0} P(\beta_k > \tau_k) E\left([\beta_k - \tau_k]^+\right)}{\sum_{\bar{d}\in D_+^0} \sum_{j\in G(\bar{d})} I(d_j) \cdot \prod_{\substack{k:d_k=1,\\ k\neq j}} E\alpha_k \cdot \prod_{\substack{k:d_k=\bar{1},\\ k\neq j}} E(\beta_k \wedge \tau_k) \cdot \prod_{\substack{k:d_k=0,\\ k\neq j}} P(\beta_k > \tau_k) E\left([\beta_k - \tau_k]^+\right)}, \tag{18}$$

$$T_- = \frac{\sum_{\bar{d}\in D_-} \prod_{k:d_k=1} E\alpha_k \cdot \prod_{k:d_k=\bar{1}} E(\beta_k \wedge \tau_k) \cdot \prod_{k:d_k=0} P(\beta_k > \tau_k) E\left([\beta_k - \tau_k]^+\right)}{\sum_{\bar{d}\in D_+^0} \sum_{j\in G(\bar{d})} I(d_j) \cdot \prod_{\substack{k:d_k=1,\\ k\neq j}} E\alpha_k \cdot \prod_{\substack{k:d_k=\bar{1},\\ k\neq j}} E(\beta_k \wedge \tau_k) \cdot \prod_{\substack{k:d_k=0,\\ k\neq j}} P(\beta_k > \tau_k) E\left([\beta_k - \tau_k]^+\right)}, \tag{19}$$

where

$$E(\beta_k \wedge \tau_k) = \int_0^\infty \bar{G}_k(t)\bar{R}_k(t)dt, \; E\left([\beta_k - \tau_k]^+\right) = \frac{\int_0^\infty \bar{G}_k(t)R_k(t)dt}{P(\beta_k > \tau_k)}.$$

In the transformations, we use the following formulas [8, 12]:

$$\sum_{j=1}^{N}\int_0^\infty \ldots \int_0^\infty \bar{F}_j(t)\left(\prod_{\substack{k=1,\\ k\neq j}}^{N}\bar{F}_k(y_k+t)dy_k\right)dt = \prod_{j=1}^{N}E\alpha_j,$$

$$\sum_{k=1}^{N}\int_{R_+^{N,k}}\int_0^\infty f_i(x_i+t)\prod_{\substack{r=1,\\ r\neq i}}^{N}\bar{F}_r(x_r+t)d\bar{x}^{(k)}dt = \prod_{\substack{k=1,\\ k\neq i}}^{N}E\alpha_k,$$

where $F_k(t)$ are distribution functions of independent non négative random variables α_k with mathematical expectations $E\alpha_k$.

We find the efficiency characteristics of this system, such as the mean specific income per calendar time unit (S) and the mean specific expenses per time unit of up-state (C).To determine them, we use the formulas [10]:

$$S = \frac{\int_E m(e)f_s(e)\rho(de)}{\int_E m(e)\rho(de)}, \quad C = \frac{\int_E m(e)f_c(e)\rho(de)}{\int_{E_+} m(e)\rho(de)},$$

where $f_s(e)$, $f_c(e)$ are functions denoting income and expenses in each state.

Let $f_s(e) = \begin{cases} c_1, & e \in E_+, \\ -c_2, & e \in E_-, \end{cases} \quad f_c(e) = \begin{cases} 0, & e \in E_+, \\ c_2, & e \in E_-. \end{cases}$

where c_1 is the income per time unit of the system up-state, c_2 denotes expenses per time system failure.

Then

$$S = \frac{c_1\sum\limits_{\bar{d}\in D_+}\prod\limits_{k:d_k=1}E\alpha_k \cdot \prod\limits_{k:d_k=\bar{1}}E(\beta_k\wedge\tau_k)\cdot\prod\limits_{k:d_k=0}P(\beta_k>\tau_k)E\left([\beta_k-\tau_k]^+\right)}{\prod\limits_{k=1}^{N}(E\alpha_k+E\beta_k)} - \tag{20}$$
$$-\frac{-c_2\sum\limits_{\bar{d}\in D_-}\prod\limits_{k:d_k=1}E\alpha_k \cdot \prod\limits_{k:d_k=\bar{1}}E(\beta_k\wedge\tau_k)\cdot\prod\limits_{k:d_k=0}P(\beta_k>\tau_k)E\left([\beta_k-\tau_k]^+\right)}{\prod\limits_{k=1}^{N}(E\alpha_k+E\beta_k)},$$

$$C = \frac{c_2\sum\limits_{\bar{d}\in D_-}\prod\limits_{k:d_k=1}E\alpha_k \cdot \prod\limits_{k:d_k=\bar{1}}E(\beta_k\wedge\tau_k)\cdot\prod\limits_{k:d_k=0}P(\beta_k>\tau_k)E\left([\beta_k-\tau_k]^+\right)}{\sum\limits_{\bar{d}\in D_+}\prod\limits_{k:d_k=1}E\alpha_k \cdot \prod\limits_{k:d_k=\bar{1}}E(\beta_k\wedge\tau_k)\cdot\prod\limits_{k:d_k=0}P(\beta_k>\tau_k)E\left([\beta_k-\tau_k]^+\right)}. \tag{21}$$

Consider special cases of serial and parallel connection of system components.

1. Parallel connection.

In this case, a system failure occurs only if all components of the system are in failure. Formulas (17) – (19) take the form:

$$K_a = \frac{\sum\limits_{i=1}^{N}P(\beta_i>\tau_i)E\left([\beta_i-\tau_i]^+\right)\prod\limits_{\substack{k=1,\\ k\neq i}}^{N}\left(E\alpha_k+E(\beta_k\wedge\tau_k)\right)}{\prod\limits_{k=1}^{N}(E\alpha_k+E\beta_k)} + \frac{\prod\limits_{i=1}^{N}\left(E\alpha_i+E(\beta_i\wedge\tau_i)\right)}{\prod\limits_{k=1}^{N}(E\alpha_k+E\beta_k)},$$

$$T_- = \frac{\prod\limits_{i=1}^{N}P(\beta_i>\tau_i)E\left([\beta_i-\tau_i]^+\right)}{\sum\limits_{i=1}^{N}\left(P(\beta_i>\tau_i)\prod\limits_{\substack{k=1,\\ k\neq i}}^{N}P(\beta_k>\tau_k)E\left([\beta_k-\tau_k]^+\right)\right)},$$

$$T_+ = \frac{\sum\limits_{i=1}^{N}P(\beta_i>\tau_i)E\left([\beta_i-\tau_i]^+\right)\prod\limits_{\substack{k=1,\\ k\neq i}}^{N}\left(E\alpha_k+E(\beta_k\wedge\tau_k)\right)}{\sum\limits_{i=1}^{N}\left(P(\beta_i>\tau_i)\prod\limits_{\substack{k=1,\\ k\neq i}}^{N}P(\beta_k>\tau_k)E\left([\beta_k-\tau_k]^+\right)\right)} + \frac{\prod\limits_{i=1}^{N}\left(E\alpha_i+E(\beta_i\wedge\tau_i)\right)}{\sum\limits_{i=1}^{N}\left(P(\beta_i>\tau_i)\prod\limits_{\substack{k=1,\\ k\neq i}}^{N}P(\beta_k>\tau_k)E\left([\beta_k-\tau_k]^+\right)\right)}.$$

2. Serial connection.

In this case, a system failure occurs if at least one of the components is in a failure. Formulas (17) – (19) take the form:

$$K_a = \frac{\prod\limits_{i=1}^{N}\left(E\alpha_i+E(\beta_i\wedge\tau_i)\right)}{\prod\limits_{k=1}^{N}(E\alpha_k+E\beta_k)},$$

$$T_+ = \frac{\prod\limits_{i=1}^{N}\left(E\alpha_i+E(\beta_i\wedge\tau_i)\right)}{\sum\limits_{i=1}^{N}\left(P(\beta_i>\tau_i)\prod\limits_{\substack{k=1,\\ k\neq i}}^{N}\left(E\alpha_k+E(\beta_k\wedge\tau_k)\right)\right)},$$

$$T_- = \frac{\sum\limits_{i=1}^{N}P(\beta_i>\tau_i)E\left([\beta_i-\tau_i]^+\right)\prod\limits_{\substack{k=1,\\ k\neq i}}^{N}\left(E\alpha_k+E(\beta_k\wedge\tau_k)\right)}{\sum\limits_{i=1}^{N}\left(P(\beta_i>\tau_i)\prod\limits_{\substack{k=1,\\ k\neq i}}^{N}\left(E\alpha_k+E(\beta_k\wedge\tau_k)\right)\right)} + \frac{\prod\limits_{i=1}^{N}P(\beta_i>\tau_i)E\left([\beta_i-\tau_i]^+\right)}{\sum\limits_{i=1}^{N}\left(P(\beta_i>\tau_i)\prod\limits_{\substack{k=1,\\ k\neq i}}^{N}\left(E\alpha_k+E(\beta_k\wedge\tau_k)\right)\right)}.$$

In the case $\tau_k = h_k = const$,

$$E(\beta_k \wedge \tau_k) = E(\beta_k \wedge h_k) = \int_0^{h_k} \bar{G}_k(t)dt,$$

$$E\left([\beta_k - \tau_k]^+\right) = E\left([\beta_k - h_k]^+\right) = \int_{h_k}^{\infty} \bar{G}_k(t)dt \Big/ P(\beta_k > h_k).$$

As an illustrative example of the use of formulas (17) – (21), consider a system consisting of 3 components with parallel connection in which operating time of components K_1, K_2, K_3 are $E\alpha_1 = 8$ h, $E\alpha_2$ = $E\alpha_3 = 6$ h; recovery times of K_1, K_2, K_3 are $E\beta_1 = 0.71$ h, $E\beta_2$ = $E\beta_3 = 0.83$ h, RVs α_i, β_i have 5th order Erlang distribution; c_1=10 c.u., c_2=15 c.u. Each component has a non-random time reserve ($R_i(t) = 1(t - h_i)$, $h_1 = h_2 = h_3 = h$), which varies from 0 to 0.7 hours in 0.1 increments. The corresponding values of $T_+(h_1, h_2, h_3)$, $T_-(h_1, h_2, h_3)$, $K_a(h_1, h_2, h_3)$, $C(h_1, h_2, h_3)$, $S(h_1, h_2, h_3)$ of the system for the specified distribution were calculated. The results are presented in Table 1.

Table 1. The influence of the time reserve value on the system reliability and efficiency characteristics.

h, h	T_-, h	T_+, h	K_a	C, c.u./h *10^{-3}	S, c.u./h
0	0.2632	316.51	0.99917	12.47	9.9792
0.1	0.2297	388.01	0.99941	8.882	9.9852
0.2	0.1982	499.30	0.9996	5.953	9.9901
0.3	0.1712	693.22	0.99975	3.705	9.9938
0.4	0.1499	1.06*10^3	0.99986	2.122	9.9965
0.5	0.1336	1.798*10^3	0.99993	1.114	9.9981
0.6	0.1211	3.384*10^3	0.99996	0.537	9.9991
0.7	0.1115	7.023*10^3	0.99998	0.238	9.9996

Analysis of the data in the table shows the significant effect of the time reserve on reliability and efficiency characteristics.

These formulas can be used to analyse the reliability of multi-component systems "gas pipeline - underground gas storage - consumer"; oil pipeline with tank farms (for storing intermediate oil reserves); electric power systems, where the time reserve for the component is ensured by the presence of an energy storage device.

Conclusion

In this paper, a semi-Markov model of a multi-component system with a component-wise random instantly replenished time reserve is constructed. The analysis of the influence of the amount of time reserve on the obtained reliability and efficiency characteristics is carried out.

In the future, it is planned to build hidden semi-Markov models of multi-component systems with an element-wise time reserve for analysing the functioning of energy systems.

The results of this work can be used to build semi-Markov models of systems with various types and strategies for using the time reserve, engineering calculations and solving optimization problems associated with the use of the time reserve.

The research was carried out within the state assignment of the Minobrnauki of Russia (No. 1.10513.2018/11.12), with financial support by RFBR (project No. 19-01-00704a).

References

1. G.N. Cherkesov, *Reliability of Technical Systems with Time Redundancy* (Sovietskoe Radio, Moscow, 1974)
2. B.P. Kredentser, *Prediction of Reliability of Systems with Time Redundancy* (Naukova Dumka, Kiev, 1978)
3. V.Y. Kopp, Yu.E. Obzherin, A.I. Peschanskiy, *Stochastic models of automized system with time reservation* (Publishing of Sevastopol State Technical University, Sevastopol, 2000)
4. I.A. Ushakov, *Probabilistic Reliability Models* (Wiley, San Diego, Calif., 2012)
5. Yu. N. Rudenko, I.A. Ushakov, *Reliability of Energy Systems*, 2nd ed. (Nauka, Novosibirsk, 1989)
6. Yu.E. Obzherin, S.M. Sidorov, M.M. Nikitin, Analysis of reliability of systems with component-wise storages, E3S Web of Conférences **58**, 02024. (2018).
7. N. Limnios, G. Oprisan, *Semi-Markov Processes and Reliability* (Springer Science+Business Media, New York, 2001)
8. V.S. Korolyuk, A.F. Turbin, *Markovian restoration processes in the problems of system reliability* (Naukova Dumka, Kiev, 1982)
9. V.S. Korolyuk, V.V. Korolyuk, *Stochastic Models of Systems* (Springer Science+Business Media, Dordrecht, 1999)
10. V.M. Shurenkov, *Ergodic Markovian processes* (Nauka, Moscow, 1989)
11. F. Beichelt, P. Tittmann, *Reliability and Maintenance—Networks and Systems. Mathematical Models* (Chapman & Hall, CRC Press, Boca Raton, London, New York, 2012)
12. A.N. Korlat, V.N. Kuznetsov, M.M. Novikov, A.F. Turbin, *Semi-Markov Models of Recoverable Systems and Queuing Systems* (Stiinta, Chisinau, 1991)

Topological algorithm for forming nodal stresses of complex networks energy systems

D.S. Akhmetbaev[1*] , *A.R.* Dzhandigulov[2] , *A.D.* Akhmetbaev[3]

[1]Seifullin Kazakh Agro Technical University, Nur-Sultan, Republic of Kazakhstan
[2]L.N. Gumilyov Eurasian National University, Nur-Sultan, Republic of Kazakhstan
[3]Information Systems Directorate, «Kazakhtelecom» JSC, Almaty, Republic of Kazakhsta

Abstract. The paper presents a new topological algorithm for the formation of nodal stresses of complex networks of power systems. The first step of the described algorithm is the search and determination of the values of all possible and specific trees of the graph corresponding to a given network of the power system. The well-known advantage of the topological approach compared to matrix methods, which allows one to obtain the final solution of non-linear equations of the steady state, has led to the development of many methods and corresponding software implementations. The paper provides a comparative analysis of the advantages and disadvantages of these methods. The main computational complexity of these methods is the search for 2 trees for each tree, which are obtained from it by dividing into two parts by removing any branch. A completely new topological approach that does not require finding 2 trees was proposed by one of the authors. The process of this method consists of the following steps: finding all possible graph trees, selecting specific graph trees, calculating the network node voltage. The work offers a unique algorithm for the implementation of these stages, which are implemented and tested at the software level. The result of the execution of the software package is the calculation of the steady state of a complex electric network using the distribution coefficients of the driving currents.

1 Introduction

The main task of the topological analysis of the electric network is to determine all of its possible trees [1-3]. From the theory of graphs [4] it is known that as the number of nodes increases, the number of trees in the graph of a real electric network increases quite quickly. Therefore, it is necessary to develop more efficient algorithms search all possible trees.

In [5, 6], three methods are realized for finding possible trees of complex electric networks. The essence of all these methods is to iterate over and study the subgraphs for whether they are trees or not. In this case, in one method, iteration is carried out along branches, the second - through nodes, and in the third - along contours. Another approach is implemented in the theory of structural numbers [7]. The disadvantages of these methods is the need to store a large data array in memory.

One of the reasons for the limited practical distribution of these methods is the need to find the so-called 2-trees for each tree found, which significantly increase the number of necessary operations [8-10].

In this paper, a new topological approach is implemented that does not require finding 2-trees. The first stage, as in existing methods, calculates all possible trees of the graph. The search algorithm is as follows: it is performed on a directed search of various samples of n-1 edges from all edges of the graph with simultaneous verification of the resulting subgraph for connectivity. The advantage of the proposed algorithm is the elimination of repetition due to the orientation of the sample.

Further, in the second stage, instead of finding 2 trees, specific graphs are distinguished from the number of all possible trees of the graph found in the first stage.

In fact, the trees themselves are not needed, but so-called tree sizes are needed, which are a product of the weights of the branches that make up the trees. Therefore, there is no need to remember these trees, it is enough to remember only their weights.

Knowing these weights is enough to calculate the current distribution coefficients, without solving any systems of equations, as is done in matrix methods. Current distribution coefficients are the initial data for the third stage - the calculation of the node-voltage of the network.

2 The first stage of the algorithm: determining the weights of all possible trees of the graph

https://doi.org/10.1051/e3sconf/201913901066

The search algorithm for all possible trees of the graph with the calculation of their values.

Let m be the number of all branches of the original graph. n is the number of nodes (vertices) of the graph.
Step 1. We carry out directed enumeration of samples of $n-1$ elements from m elements.

Step 2. Checking the end of the search options: yes - complete the algorithm, no - continue the calculation.

Step 3. For the next sampling, we carry out a connection test.

Verification is carried out as follows. Starting from the first node (vertex), we begin the movement along the branches from the selection. Three options are possible: 1) either there are no new paths from the next vertex, 2) either the already selected vertex is returned (that is, a cycle has formed), 3) or all vertices are selected. In the first two cases, the check stops - the selection of branches does not form a tree.

Step 4. Calculation of tree sizes. Let G_k be the set of edge numbers of the kth tree. Y_i is the conductivity of the i-th branch, which is expressed from the given complex resistance of the i -th branch of Z_i according to the formula $Y_i = 1 / Z_i$.

Then the weight of the (kth) tree is calculated by the formula:

$$Q_k = \prod_{i \in G_k} Y_i \qquad (1)$$

The resulting value of all possible trees is formed as the sum of the weights of the individual trees of the graph. According to this algorithm, a program is compiled that reads the source data from an external file and performs a preliminary check of the correctness of the source data.

As an illustration, we present the result of the program for determining all possible trees of a real 220 kV network, shown in Fig. 1 (Scheme with the parameters of the selected part of the network voltage of 220 kV of the current energy system of the Republic of Kazakhstan).

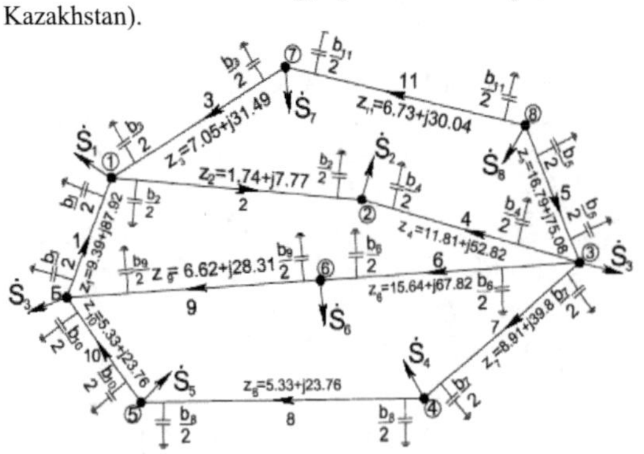

Fig. 1. The equivalent circuit

In the network under consideration, the number of all branches $m=11$, the number of nodes $n=9$.

The number of all possible trees is 85, and they are presented in table 1.

Table 1. All possible graph trees

tree №	List of branches	tree №	List of branches	tree №	List of branches
1	1 2 3 4 5 6 7 8	29	1 2 4 5 6 7 8 11	57	1 4 5 6 7 8 10 11
2	1 2 3 4 5 6 7 10	30	1 2 4 5 6 7 10 11	58	1 4 5 6 7 9 10 11
3	1 2 3 4 5 6 8 10	31	1 2 4 5 6 8 10 11	59	1 4 5 6 8 9 10 11
4	1 2 3 4 5 7 8 9	32	1 2 4 5 7 8 9 11	60	1 4 5 7 8 9 10 11
5	1 2 3 4 5 7 9 10	33	1 2 4 5 7 9 10 11	61	2 3 4 5 6 7 8 9
6	1 2 3 4 5 8 9 10	34	1 2 4 5 8 9 10 11	62	2 3 4 5 6 7 8 10
7	1 2 3 4 6 7 8 11	35	1 2 5 6 7 8 9 11	63	2 3 4 5 6 7 9 10
8	1 2 3 4 6 7 10 11	36	1 2 5 6 7 8 10 11	64	2 3 4 5 6 8 9 10
9	1 2 3 4 6 8 10 11	37	1 2 5 6 7 9 10 11	65	2 3 4 5 7 8 9 10
10	1 2 3 4 7 8 9 11	38	1 2 5 6 8 9 10 11	66	2 3 4 6 7 8 9 11
11	1 2 3 4 7 9 10 11	39	1 2 5 7 8 9 10 11	67	2 3 4 6 7 8 10 11
12	1 2 3 4 8 9 10 11	40	1 3 4 5 6 7 8 9	68	2 3 4 6 7 9 10 11
13	1 2 3 5 6 7 8 9	41	1 3 4 5 6 7 8 10	69	2 3 4 6 8 9 10 11
14	1 2 3 5 6 7 8 10	42	1 3 4 5 6 7 8 11	70	2 3 4 7 8 9 10 11
15	1 2 3 5 6 7 8 11	43	1 3 4 5 6 7 9 10	71	2 3 5 6 7 8 9 11
16	1 2 3 5 6 7 9 10	44	1 3 4 5 6 7 10 11	72	2 3 5 6 7 8 10 11
17	1 2 3 5 6 7 10 11	45	1 3 4 5 6 8 9 10	73	2 3 5 6 7 9 10 11
18	1 2 3 5 6 8 9 10	46	1 3 4 5 6 8 10 11	74	2 3 5 6 8 9 10 11
19	1 2 3 5 6 8 10 11	47	1 3 4 5 7 8 9 10	75	2 3 5 7 8 9 10 11
20	1 2 3 5 7 8 9 10	48	1 3 4 5 7 8 9 11	76	2 4 5 6 7 8 9 11
21	1 2 3 5 7 8 9 11	49	1 3 4 5 7 9 10 11	77	2 4 5 6 7 8 10 11
22	1 2 3 5 7 9 10 11	50	1 3 4 5 8 9 10 11	78	2 4 5 6 7 9 10 11
23	1 2 3 5 8 9 10 11	51	1 3 4 6 7 8 9 11	79	2 4 5 6 8 9 10 11
24	1 2 3 6 7 8 9 11	52	1 3 4 6 7 8 10 11	80	2 4 5 7 8 9 10 11
25	1 2 3 6 7 8 10 11	53	1 3 4 6 7 9 10 11	81	3 4 5 6 7 8 9 11
26	1 2 3 6 7 9 10 11	54	1 3 4 6 8 9 10 11	82	3 4 5 6 7 8 10 11
27	1 2 3 6 8 9 10 11	55	1 3 4 7 8 9 10 11	83	3 4 5 6 7 9 10 11
28	1 2 3 7 8 9 10 11	56	1 4 5 6 7 8 9 11	84	3 4 5 6 8 9 10 11
				85	3 4 5 7 8 9 10 11

3 Calculation of distribution coefficients of driving currents

The distribution coefficients of the driving currents are calculated on the basis of expressions obtained using matrix methods or network circuit topology methods. In [12], to determine the coefficients of current distribution by the matrix method, the following expression was obtained:

$$C = Z_B^{-1} M^T (M Z_B^{-1} M^T)^{-1} \quad (2)$$

where M is a rectangular matrix of the connection of nodes of dimension $m \times n$; Z_B^{-1} – diagonal branch conductivity matrix; T – sign transpose matrix. The elements of the current distribution matrix C are complex numbers.

The fundamental complexity of calculations using (1) lies in the need to perform the matrix inversion operation of a sufficiently large dimension. In addition, in some cases the problem is complicated by the poor conditionality of the matrix $M Z_B^{-1} M^T$. To increase the efficiency of computing this kind of matrix, the methods of factorization and matrix triangulation are used in practice [13].

If the matrix C is sought in the form of a solution to the matrix equation, then, in the general case, it can be found in the form

$$C = A/B \quad (3)$$

where B is the determinant of the matrix of the equation, A is the matrix of algebraic complements. The essence of the topological method is to present the expressions for A and B through the parameters of the network considered as a graph [5, 6].

In [8], the numerator of formula (3) is defined as the difference in the sums of weights of specific 2-trees, determined with respect to the index of the element of the desired matrix C, that is, the corresponding branch of the graph and with respect to the basis node. The main computational complexity of the method under consideration is finding 2 trees for each tree, which are obtained from it by dividing into two parts by removing any branch. Due to the need for cumbersome calculations, this method has not found wide practical application [9].

A completely new topological approach was proposed in the work of the first author [14–16]. Moreover, to calculate matrix A in formula (3), explicit expressions are obtained that are computed without the need to find 2-trees. Namely, its element A_{ij} is represented as the sum of the products of the conductivities of the branches of specific trees of the graph that contain the i-th branch in the path from the j-th node to the base node, taken taking into account the orientation of the i-th branch. The advantage of this approach is the absence of the need to calculate all 2-trees, which reduces the number of operations by n^2 times [17]. Due to the fact that we already calculated the module of specific trees at the first stage and it remains only to determine its sign, these two stages can be combined into one cycle, which further reduces the number of operations and the required memory size of the computing equipment.

To illustrate the operation of the program, we present in table 3 the results of calculating the current distribution coefficients in a real 220 kV network, shown in Fig. 1. Initial data - the matrix of complex resistances is presented in table 2.

Table 2. Complex resistance of branches

branch № j	1	2	3	4	5	6
ReZ_j	9.39	1.74	7.05	11.81	16.79	15.64
ImZ_j	87.9 2	7.77	31.49	52.82	75.08	67.82
branch № j	7	8	9	10	11	
ReZ_j	8.91	5.33	6.62	5.33	6.73	
ImZ_j	39.8	23.76	28.31	23.75	30.04	

The current distribution matrix has the dimension n x n. In our case, 9x9.

The accuracy of the calculations is estimated by the accuracy of the values of possible and specific trees, as well as the determination of modules and phase-node network voltages.

Table 3. Matrix of current distribution coefficients

-0.5043+ 0.0289*i	-0.4733+ 0.0272*i	-0.2631+ 0.0156*i	-0.1432+ 0.0085*i	-0.0716+ 0.0043*i	-0.0775+ 0.0048*i	-0.4487+ 0.0259*i	-0.3956+ 0.0229*i
0.3434+ 0.0200*i	-0.5957+ 0.0188*i	-0.1822+ 0.0108*i	-0.0992+ 0.0059*i	-0.0496+ 0.0030*i	-0.0537+ 0.0033*i	0.2222+ 0.0179*i	0.1066+ 0.0159*i
-0.1523- 0.0089*i	-0.1224- 0.0084*i	0.0808- 0.0048*i	0.0440- 0.0026*i	0.0220- 0.0013*i	0.0238- 0.0015*i	0.6709- 0.0079*i	0.5023- 0.0070*i
-0.3434- 0.0200*i	-0.4043- 0.0188*i	0.1822- 0.0108*i	0.0992- 0.0059*i	0.0496- 0.0030*i	0.0537- 0.0033*i	-0.2222- 0.0179*i	-0.1066- 0.0159*i
0.1523+ 0.0089*i	0.1224+ 0.0084*i	-0.0808+ 0.0048*i	-0.0440+ 0.0026*i	-0.0220+ 0.0013*i	-0.0238+ 0.0015*i	0.3291+ 0.0079*i	0.4977+ 0.0070*i
0.2357+ 0.0146*i	0.2504+ 0.0139*i	0.3504+ 0.0087*i	0.1907+ 0.0047*i	0.0953+ 0.0023*i	-0.1914+ 0.0030*i	0.2621+ 0.0133*i	0.2874+ 0.0120*i
0.2600+ 0.0143*i	0.2763+ 0.0133*i	0.3865+ 0.0069*i	-0.3339+ 0.0038*i	-0.1669+ 0.0019*i	0.1139+ 0.0018*i	0.2892+ 0.0126*i	0.3170+ 0.0109*i
0.2600+ 0.0143*i	0.2763+ 0.0133*i	0.3865+ 0.0069*i	0.6661+ 0.0038*i	-0.1669+ 0.0019*i	0.1139+ 0.0018*i	0.2892+ 0.0126*i	0.3170+ 0.0109*i
0.2357+ 0.0146*i	0.2504+ 0.0139*i	0.3504+ 0.0087*i	0.1907+ 0.0047*i	0.0953+ 0.0023*i	0.8086+ 0.0030*i	0.2621+ 0.0133*i	0.2874+ 0.0120*i
0.2600+ 0.0143*i	0.2763+ 0.0133*i	0.3865+ 0.0069*i	0.6661+ 0.0038*i	0.8331+ 0.0019*i	0.1139+ 0.0018*i	0.2892+ 0.0126*i	0.3170+ 0.0109*i
-0.1523- 0.0089*i	-0.1224- 0.0084*i	0.0808- 0.0048*i	0.0440- 0.0026*i	0.0220- 0.0013*i	0.0238- 0.0015*i	-0.3291- 0.0079*i	0.5023- 0.0070*i

4 Calculation of voltage nodes

Existing software for calculating and analyzing electric modes is based on classical mathematical models of the steady state of electric power systems; an analysis of their advantages and disadvantages can be found in [18]. In the work, the matrix equation developed in [19,20] is taken as the basis for calculating the steady states of complex networks

$$\dot{U} = U_0 + C^T Z_B C \widehat{U}^{-1} \hat{S} \quad (4)$$

where C is the matrix of current distribution coefficients; Z_B is the diagonal matrix of branch resistances; $\widehat{U}$– is the diagonal matrix of nodal conjugate voltages; $\hat{S}$ – is the column matrix of the conjugate powers of the nodal loads and generators.

The algorithms for implementing the calculation of node voltages depends on the form of representation of the nodal parameters, which are represented by constant active and reactive powers or active powers and voltages [21-23]. If the nodal parameters are given by powers, then the following expressions are valid for determining the node stresses:

$$U_k' = U_0 + \sum_{j=1}^{n} Z_{kj} U_j^{-1} \left(P_j \cos(\delta_k + \psi_{kj}) + Q_j \sin(\delta_k + \psi_{kj}) \right)$$

$$U''_k = \sum_{j=1}^{n} Z_{kj} U_j^{-1} \left(P_j \sin(\delta_k + \psi_{kj}) - Q_j \cos(\delta_k + \psi_{kj}) \right)$$

where $U_k = \sqrt{(U_k')^2 + (U_k'')^2}$ - voltage module of the k–th node;

$\delta_k = \operatorname{arctg} \dfrac{U_k''}{U_k'}$ - voltage phase of the k–th node;

$\underline{Z}_{kj} = \sqrt{(\mathrm{Re}\sum_{j=1}^{m} \underline{C}_{kj}{}^t \underline{Z}_j \underline{C}_{jk})^2 + (\mathrm{Im}\sum_{j=1}^{m} \underline{C}_{kj}{}^t \underline{Z}_j \underline{C}_{jk})^2}$ – mutual nodal resistance module;

$\psi_{kj} = \operatorname{arctg} \dfrac{\mathrm{Im}\sum_{j=1}^{m} \underline{C}_{kj}{}^t \underline{Z}_j \underline{C}_{jk}}{\mathrm{Re}\sum_{j=1}^{m} \underline{C}_{kj}{}^t \underline{Z}_j \underline{C}_{jk}}$ – phase of complex mutual nodal resistance.

If the nodal parameters are set by active powers and voltages, then the following expressions are valid for determining the node voltages:

$$Q_n^k = \frac{U_n^2 \cos\delta_n^{k-1} - U_0 U_n - Z_{nn} P_n \cos(\delta_n^{k-1} + \psi_{nn})}{Z_{nn} \sin(\delta_n^{k-1} + \psi_{nn})} - \frac{\sum_{j=1}^{n-1} Z_{nj} U_n U_j^{-1(k-1)} \left(P_j \cos(\delta_j^{k-1} + \psi_{nj}) + Q_j \sin(\delta_j^{k-1} + \psi_{nj}) \right)}{Z_{nn} \sin(\delta_n^{k-1} + \psi_{nn})}$$

The imaginary part of the voltage of the node under consideration, respectively, is equal to:

$$U_n'' = U_n \sin\delta_n^k = \sum_{j=1}^{n-1} Z_{nj} U_j^{-1(k-1)} \left(P_j \sin(\delta_j^{k-1} + \psi_{nj}) - Q_j \cos(\delta_j^{k-1} + \psi_{nj}) \right) + Z_{nn} U_n^{-1} \left(P_n \sin(\delta_n^{k-1} + \psi_{nn}) - Q_n^k \cos(\delta_n^{k-1} + \psi_{nn}) \right)$$

Hence, the value of the phase of the nodal stress for the kth iteration has the form:

As can be seen, from the above expressions, for calculating the voltage of the nodes it is enough to know the current distribution coefficients in the electric network. The algorithms developed above for the topological functions of a complex electric network are brought up to a calculation program for its steady state.

As an illustration, a comparative calculation of the circuit with the parameters of the selected part of the network with a voltage of 220 kV of the real power system of the Republic of Kazakhstan is made.

In order to verify the correctness of the results obtained, voltage calculations were performed at the nodes of the network at steady state using the proposed method and the industrial program RastrWin3.

The comparison results are presented in table 4.

Table 4. Results of comparative calculations of node voltages

	Calculation with applying a matrix C		Program verification calculation RASTRWin3		Deviations		
Nodal number	U, кV	δ, degree	U, кV	δ, degree	ΔU, кV	ΔU, %	Δδ, degree
1	241.43	-4.8315	240.65	-4.79	0,78	0,32	-0,04
2	241.53	-4.8433	240.78	-4.81	0,75	0,31	-0,03
3	240.89	-4.3086	240.36	-4.28	0,53	0,22	-0,03
4	237.38	-3.0902	237.1	-3.08	0,28	0,12	-0,01
5	235.60	-1.8466	235.45	-1.84	0,15	0,06	-0,01
6	236.27	-1.9230	236.12	-1.92	0,15	0,06	0,00
7	239.60	-6.2297	238.33	-6.17	1,27	0,53	-0,06
8	239.12	-6.5359	238.03	-6.49	1,09	0,46	-0,05

Deviations of the calculated values by the proposed method and using the industrial program RastrWin3 are less than 1%.

The execution time of the proposed program was 10-1 seconds. This means that it becomes possible to build a program complex for calculating the steady state of a complex electrical network of a power system in real time.

5 Conclusion

1. A program for calculating the possible trees of a complex directed graph has been developed.

2. A program for calculating current distribution coefficients based on the topology of a complex electric network has been developed.

3. A comprehensive program for calculating the steady-state regime of a complex electric network of a power system has been developed.

4. The software implementation of the algorithm for the formation of a steady-state regime significantly increases the efficiency of research as the number of nodes and branches of the electric network of the power system grows.

References

1. Reza F., Seeley S. Modern analysis of electrical circuits. Energiya. Moscow (1964).
2. Byerly R., Long R., King C. Logic for applying topological methods to electric networks. AIEE Transactions, Nov.1958, pt. I. v. 77.
3. Sato N. Digital calculation of network and mesh transformation matrices. AIEE Transactions, Oct. 1960, pt.III. v.79.
4. Berzh K. Graph theory and its applications. Publishing house of foreign literature, 1962. - 320 p..
5. Geraskin O. Finding a tree graph is an electric network using digital. Izvestiya AN USSR, Energetika i transport, №4, 53-60 (1965).
6. Geraskin O. Topological content of nodes and circuits of the electrical network and the calculation of their size with the help of a digital computer // Izvestia ANSSR. Energy and transport, 1966. №2. - pp. 59-70.
7. Bellert S., Wozniacki G. Analysis and synthesis of electrical circuits by the method of structural numbers. Mir. Moscow (1972).
8. Geraskin O. Methods for the determination of 2 - trees of the graph in the topology of electrical networks. Izvestiya AN USSR, Energetika i transport, №5, 106-112 (1967).
9. Geraskin O. Topological analysis of current distribution coefficients in electric networks. Izvestiya vuzov. Energetika, No.5, p. 20–28 (1967)
10. Christofides N. Graph Theory. Algorithmic approach. Mir. Moscow (1978).
11. Dzhandigulov A., Akhmetbayev D. Finding all spanning graphs of a given graph. The program for the computer. Certificate Republic of Kazakhstan on entering information into the state register of rights to objects protected by copyright (Republic of Kazakhstan). No. 1551 dated January 31, 2019
12. Melnikov N. Matrix method of analysis of electric circuits. Energiya. Moscow (1972).
13. Brameller A., Allan R., Hammam Y. Weakly filled matrices. Energiya. Moscow (1979).
14. Akhmetbayev D., Aubakir D., Sarsikeyev Y., Bainiyazov B., Surkov M., Rozhkov V., Ansabekova G., Yerbolova A., Suleimenov A., Tokasheva M. Development of Topological Method for Calculating Current Distribution Coefficients in Complex Power// Networks, Results in Physics, 2017, 7, pp. 1644–1649.
15. Akhmetbayev D., Akhmetbayev A., Zhumazhanov S., Zhakishev B. New Modeling of Steady-State Modes of Complex Electrical Grids of Power Systems MATEC Web of Conferences 155, 01043 (2018).
16. Akhmetbayev D., Akhmetbayev A., Aidarova A. Determination of rational transformation coefficients transformers distribution networks//E3S Web of conferences 25.04003 (2018)
17. Dzhandigulov A., Akhmetbaev D. Topological implementation of a new algorithm to calculate the coefficients of the current distribution in complex electrical networks. computer program. Certificate of entry into the state register of rights to objects protected by copyright. (Republic of Kazakhstan). No. 1552 dated January 31, 2019
18. Nepsha F., Otdelnova G., Savinkina O. Comparison of the functionality of existing software tools for calculating and analyzing electrical modes. // Vestnik of KuzGTU. Electrical complexes and systems (series), 2013 (2) p. 116–118.
19. Akhmetbaev D. The method of calculating the steady-state modes of electrical networks based on current distribution coefficients. Elektrichestvo. №11. 23-27(2010).
20. Akhmetbayev D., Akhmetbayev A., Suleymenov A., Kolcun M. Modeling the set mode of complex power grid, based on infeed coefficients// Proceedings of the 9th International Scientific Symposium Elektroenergetika – 2017, 12–14. 9. 2017, Stará Lesná, Slovak Republic.
21. Akhmetbayev D., Akhmetbayev A., Zhantlessova A. Formation of the Z-Form of Equations of Steady-State Modes of Energy Systems' Complex Electric Networks. Web of Conferences58, 02021 (2018)
22. Akhmetbaev D., Akhmetbaev A., Berdygozhin A. A topological method for the formation of nodal equations in the inverse form for electric power systems // Elektrichestvo. Moscow. No. 5 (2018) pp. 18–27.
23. Akhmetbayev D., Dzhandigulov A. Development of algorithms for a new topological method for calculating current distribution coefficients in complex electrical networks//Eurasian Journal of Mathematical and Computer Applications ISSN 2306–6172 Volume 7, Issue 3 (2019) 4–12.

Electromagnetic safety of high voltage traction networks

Natalya Buyakova[1*], *Vasilij* Zakaryukin[2], *Andrey* Kryukov[2, 3]

[1] Angarsk State Technical University, 665835, 60 Tchaikovsky St., Angarsk, Russia
[2] Irkutsk State Transport University, 664074, 15 Chernyshevsky St., Irkutsk, Russia
[3] Irkutsk National Research Technical University, 664074, 83 Lermontov St., Irkutsk, Russia

Abstract. The article provides procedure for simulation modeling of traction power supply systems which allows analyzing electromagnetic safety conditions in perspective high voltage traction networks. Modeling of electromagnetic fields in the Fazonord software application proved that 50 kV traction system is characterized by the highest electromagnetic strength at a height of 1.8 m above the ground surface, while 2x25 kV system is characterized by the lowest one. Traditional traction system with 27.5 kV voltage is the worst in terms of magnetic field strength, while 2x50 kV is the best; autotransformer systems of traction power supply create magnetic fields due to demagnetizing effect of power leads.

1 Introduction

Low-frequency electromagnetic fields (EMF) created by high voltage transmission lines and railroads traction networks can generate interference causing disturbances of electrical and electronic devices' normal functioning [1, 2] and result in serious personal injuries when operations are conducted on disconnected power supply lines or communication lines when personnel is subject to induced voltage.

The main effects are caused by 50 Hz electromagnetic fields on power facilities. Traction networks (TN) of alternating current electrified railroads are one of the major sources of EMF due to electromagnetic imbalance [3 – 11]. High voltage values can occur on accessory devices due to TN electromagnetic influence. Such voltages can result in, in the first turn, serious damage of the equipment and deadly electrocution cases. When the railroad track goes through a residential area, the levels of electromagnetic fields strengths (EMF) created by a electrical traction network, can, in a number of cases, exceed the permissible norms.

Operation modes of electrified railroad differ in complicated and non-linear dynamics which significantly hinders obtaining experimental data complying with EMF extreme voltage levels [1, 11]. That is why, computer-based modeling methods are used to analyze electromagnetic safety conditions in traction power supply systems. Methods for modeling multi-wire power lines of electromagnetic fields, developed by the authors, allowed obtaining strengths quantitative assessments of electrical and magnetic fields created by traction networks of alternating current electrified railroads [11 – 17]. Results of prospective high voltage traction systems modeling are given below.

2 94 kV autotransformer traction networks

The work [18] describes autotransformer traction power supplying system (TPSS) 94 kV which can be used in electrification projects designed for new long-distance railroads and for reconstruction of the existing traction networks. In this TPSS main traction substations (TS) are equipped with balancing transformers, and intermediate substations are provided with one-phase transformers (fig. 1). Distance between main TS can reach 300...350 km.

On intermediate TS are located at a distance of 30…45 km between each other, one-phase transformers with windings voltage 93.9/27.5 kV are installed. Balancing transformers can be supplied in two versions. In the first option, seven-winding transformer is used that has secondary windings connected as per triangles and incomplete stars circuits. In the second option five-winding transformer is used; the secondary windings are designed for 66.4 kV and 27.5 kV. The advantage of this transformer is that it can be connected with 'two wires – rail' transmission line, however, the power of traction windings is one third higher in the second option, than in the first one. That is why seven-winding transformer should be preferred, which is further analyzed.

TPSS 94 kV modeling was carried out in Fazonord software application [12], whose current version shall not contain more than five windings. That is consequently seven-winding transformer model is implemented as two four-winding transformers: 40000-230/66.4/66.4 and 40000-230/27.5/27.5. In addition, program software does not provides for asymmetrical windings modeling, that is why, for implementation of 'incomplete star' circuit, windings models were used that were connected as incomplete star, one lead of which was not used.

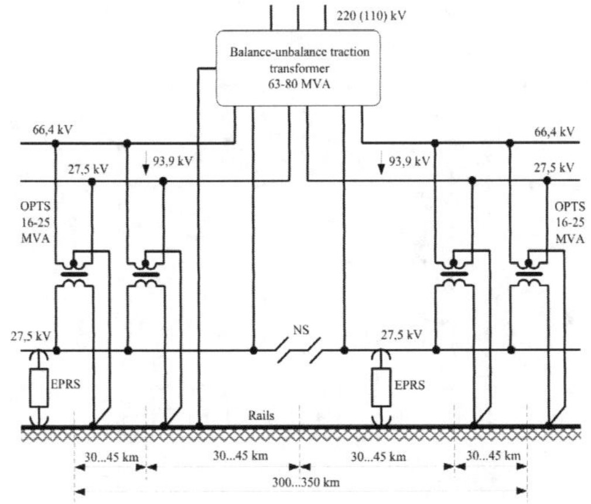

Fig. 1. High voltage TPSS diagram with balance-unbalance transformers: OPTS – one-phase traction substation: EPRS – electrically-powered rolling stock; NS – neutral section

The modeled traction network included the following elements:

- wires PBSM-95+MF-100 of double-track road section overhead catenary;
- wire A-185 of 27.5 kV line;
- wire A-185 of 66.4kV power line;
- four rail lines R-65.

In order to reduce traction network electrical influence, 66.4 kV power wires were located above overhead catenaries.

Fazonord software application calculating circuit included models of balancing and one-phase transformers, multi-wire traction network section having 2 km length, and impedance bond with secondary winding which are needed to correctly determine current distribution in rail lines. Loads of $8 + j8$ MVA of each track were connected between the contact net and center tap of impedance bond.

Results of electrical and magnetic fields strengths at the beginning of traction network at a height of 1.8 from the ground surface for traction system options 27.5 kV and 94 kV are provided in fig. 2 and 3. XOY coordinate plane is perpendicular to railroad axis, origin of coordinates is located on the track axis, OY axis is perpendicular to the ground surface.

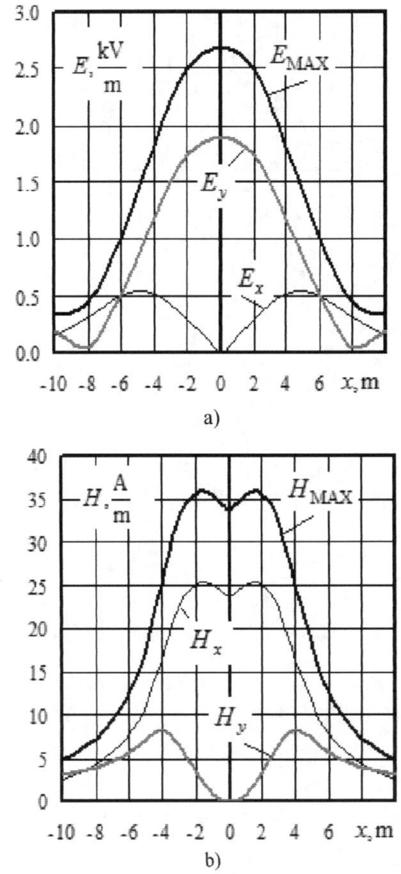

Fig. 2. Dependences of EMF strength components from x coordinate for 94 kV TPSS: a – electrical field; б – magnetic field

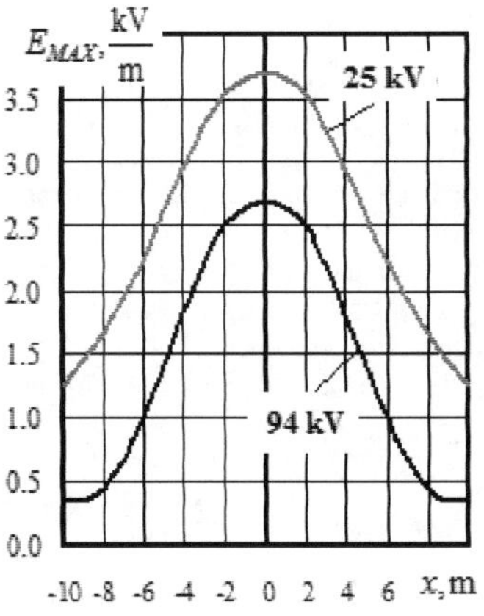

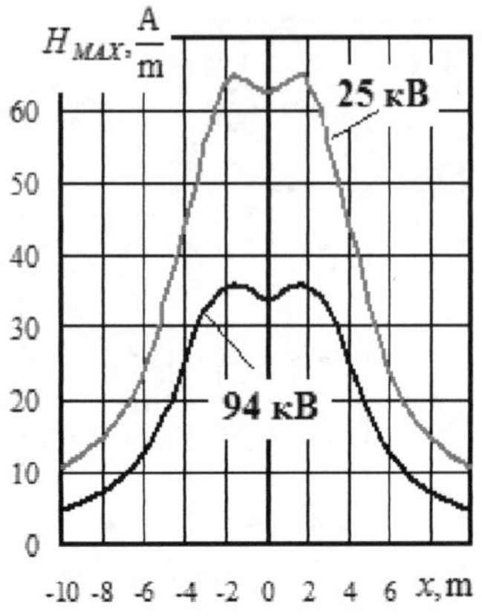

b)

Fig. 3. Dependences of EMF strength amplitude from the *x* coordinate.

The modeling results allow formulation of the following conclusions.

1. Due to reducing effect of power wires, 94 kV TPSS creates lesser EMF than the traditional 27.5 kV TPSS.
2. Perspective system of 94 kV traction power supply ensures acceptable conditions of electromagnetic safety levels for railroad workers and population.
3. In points with coordinates $x = -10$ m, $y = 1.8$ m and $x = 10$ m $y = 1.8$ m, electrical field 94 kV TPSS have linear polarization, whereas magnetic field – elliptic polarization. Such nature of magnetic field polarization is associated with return of a part of traction current through the ground and various inductances of direct and reverse current circuits.

3 Traction networks with 50 kV contact net

Traction power supply systems with high voltage overhead catenaries are applied abroad: 50 kV, 2×50 kV and 50+110 kV [19]. Higher supply voltage of electrically-powered rolling stock enhances loading capability of traction power supply system, but can result in deterioration of the railroad electromagnetic conditions. Increase in voltage would lead to increase of electrical field strength, though, compared with 27.5 kV system and comparable loads, magnetic field strength would decrease.

In order to perform quantitative analysis of the above factors, design study of high voltage traction networks 50 kV, 2×50 kV and 50+110 kV electromagnetic field structure was carried out.

3.1 50 kV Traction network

Traction transformer 40000/230/50, a 2 km section of double-track traction network, impedance bonds to calculate currents in rail lines and two dividing transformers with low losses to allow connection of loads between nodes and the earth were represented in the design model circuit.

Calculations of EMF voltages are given for double-track road section conditions when power of 8+ *j*8 MVA is transmitted in each track road section traction network. TN parameters distribution was taken into account here and further.

The results are represented in fig.4 where electrical and magnetic fields strengths profiles are depicted at a height of 1.8 m from the ground surface.

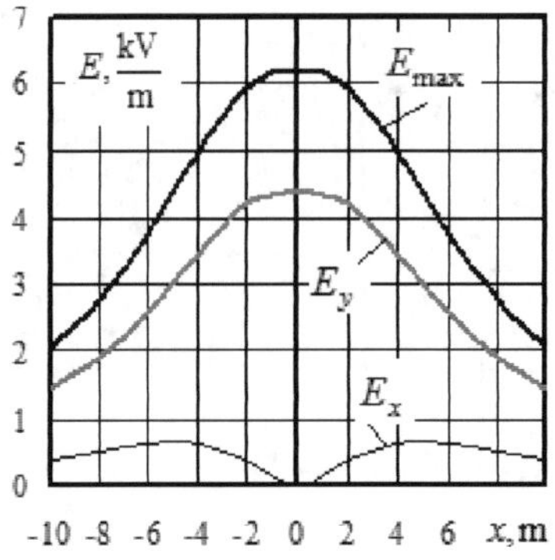

a)

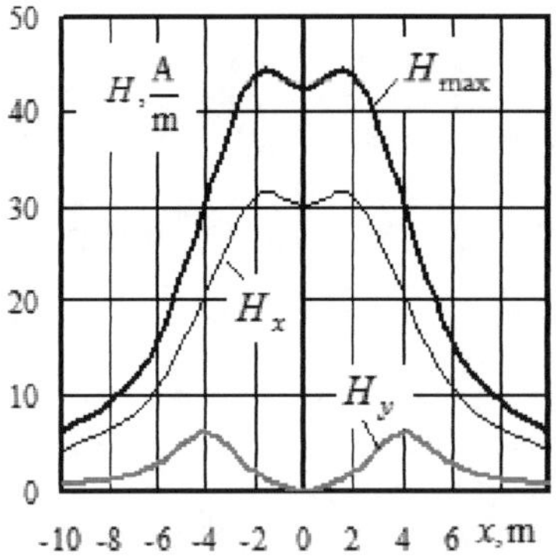

b)

Fig. 4. Dependences of EMF strength components of 50 kV traction network from *x* coordinate: a – electrical field; b – magnetic field

We can infer the following based on 50 kV TPSS modeling results review:

1. Despite a high voltage at a height of 1.8 m, electromagnetic safety conditions with regard to electrical field strength are not violated, strength effective value makes 4.4 kV/m.
2. Magnetic field strength for 240 A of each track's overhead catenary does not exceed the permissible limits.
3. Electrical field has linear polarization, whereas magnetic field beyond the limits of the central zone, limited by X coordinate with location of overhead catenarys, is elliptically polarized which is associated with return of a significant current portion through the ground

 https://doi.org/10.1051/e3sconf/201913901067

(390 A current flows through return feeder in the model, while 112 A – through the substation earth).

3.2 2x50 kV traction network

Modeling circuit accepted for analysis included models of 25000/230/50/50 one-phase transformer and 16000/110 autotransformers. TPSS 2x50 kV traction network differs from 50 kV network in that it has power wires.

Results of electromagnetic field modeling at a height of 1.8 m for double-track road section loads 8 + *j*8 MVA each, are given in fig. 5.

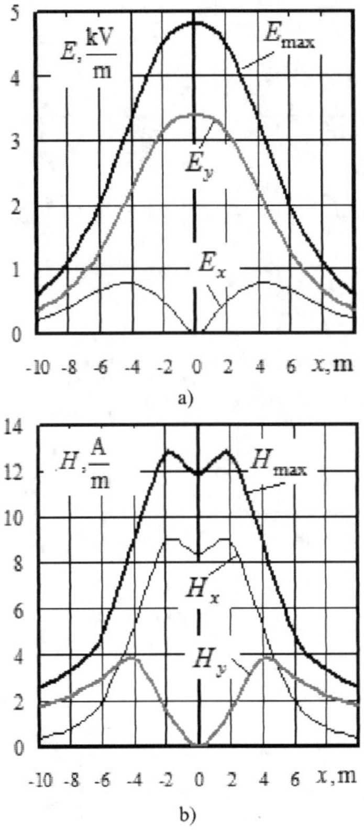

Fig. 5. Dependences of EMF strength components in 2x50 kV TN on *x* coordinate: a – electrical field; b – magnetic field

The results provided are, in the first turn, subject to the fact that the main flow of electromagnetic energy is concentrated within an interval between the power wires and its tracks overhead catenaries. It is due to this fact, that electrical and magnetic fields strengths do not exceed electromagnetic safety regulatory values at a height of 1.8 m, and fields polarization is virtually linear due to low ground currents.

3.3 50+110 kV traction network

Traction network circuit for 50+110 kV voltages differs from 2x50 kV network only in transformer parameters. The design model contained one-phase traction transformer having power 25000 kV·A with voltages 230/115/55 kV and autotransformers with power 16000 kV·A 110/50 kV.

Results of electromagnetic field modeling at a height of 1.8 m for each track loads 8 + *j*8 MVA each, are given in fig. 6.

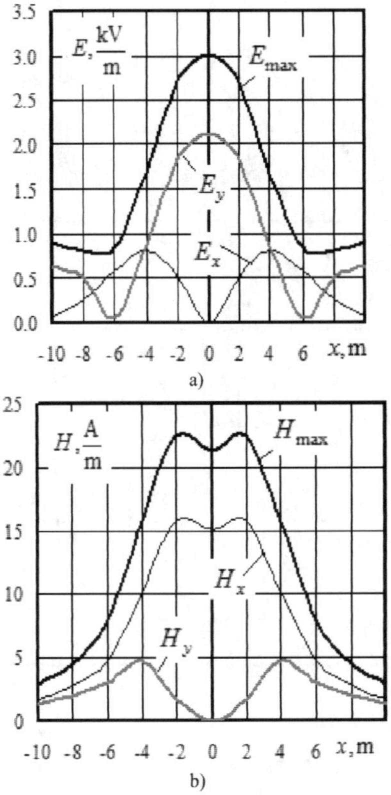

Fig. 6. Dependences of EMF strength components in 50+100 kV TN on *x* coordinate: a – electrical field; б – magnetic field

Modeling-based results are the following.

1. Electrical and magnetic fields strengths at a height of 1.8 m in the given operating mode of traction network do not exceed electromagnetic safety regulations.
2. Traction network electrical field at a height of 1.8 m has linear polarization, whereas magnetic field beyond the leads location is characterized by elliptical polarization. It is associated with irregular inductances of 110 kV and 50 kV circuits which result in different angular shift of currents in relevant wires.

4 Comparison of high voltage traction networks

Results of high voltage TPSS comparison by electrical safety criterion for similar loads at a height of 1.8 m from the ground surface are provided in fig. 7 – 10. Dependences for 25 kV traction power supply system are

https://doi.org/10.1051/e3sconf/201913901067

provided in fig. 7. Fig. 9, 10 are shown as comparison diagrams for all AC traction systems.

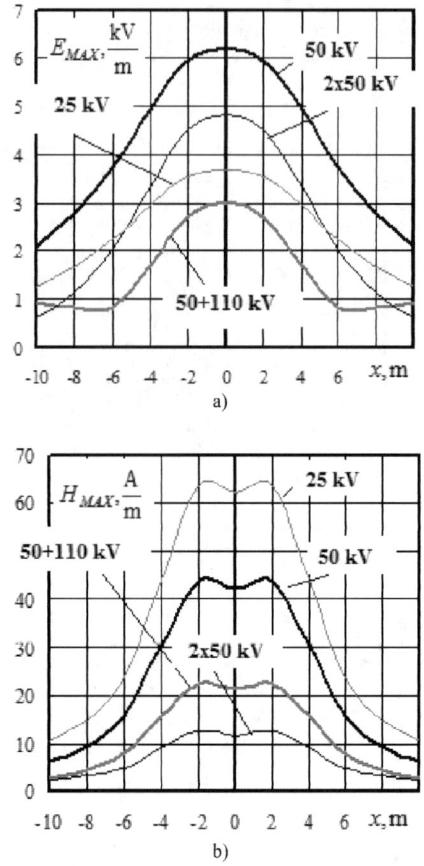

Fig. 7. Dependences of EMF strengths amplitude from *x* coordinate for systems with overhead catenary 50 kV voltage. a – electrical field; б – magnetic field

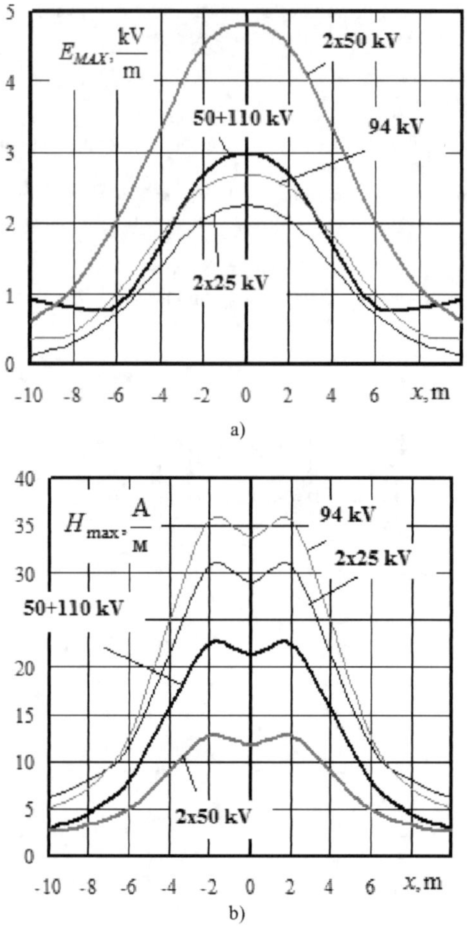

Fig. 8. Dependences of EMF strength amplitudes from *x* coordinate for autotransformer systems: a – electrical field; б – magnetic field

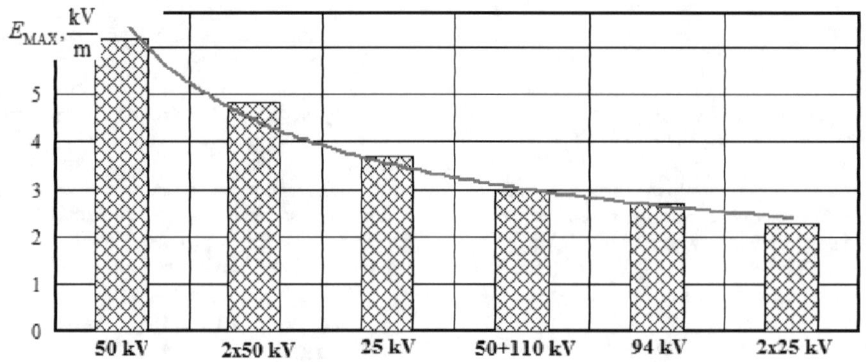

Fig. 9. Comparison of traction power supply by electric field strength

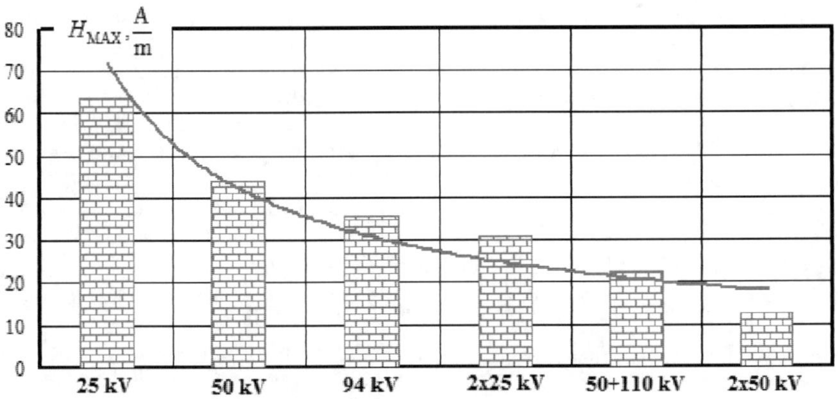

Fig. 10. Comparison of traction power supply systems by magnetic field strength

Comparison of different traction systems electromagnetic fields reveals the following:

- the highest electrical field strength at a height of 1.8 m above the ground surface is created by 50 kV traction system, 2x25 kV system is characterized by the lowest strength;
- traditional traction system with 27.5 kV voltage is the worst in terms of magnetic field strength, while 2x50 kV is the best; autotransformer systems of traction power supply create magnetic fields due to demagnetizing effect of power leads.

5 Conclusions

1. Due to reducing effect of power цшкуы to EMF, prospective traction power supply system 94 kV creates lesser electromagnetic fields than the traditional 27.5 kV TPSS. 94 kV system ensures conditions of electromagnetic safety for railroad workers and population.

At 10 m distance from the track axis, at a height of 1.8 m, 94 kV TPSS electrical field has linear polarization, while the magnetic one – elliptical polarization. Such magnetic field polarization is associated with return of a portion of traction current through the ground and with various inductances of direct and reverse current circuits.

2. Despite the stepped up voltage of 50 kV traction network at a height of 1.8 m, electromagnetic safety conditions with regard to electrical field strength are not violated. Magnetic field strength for 240 A of each track's overhead catenary does not exceed the permissible limits.

3. In 2x50 kV traction system at a height of 1.8 m electrical safety conditions with regard to electrical and magnetic fields strengths criteria, are complied with. In the first turn, this is subject to the fact that the main flow of electromagnetic energy is concentrated within an interval between the power wires and its tracks overhead catenaries.

4. In 50+110 kV traction power supply system the strengths of electrical and magnetic fields at a height of 1.8 m do not exceed electrical safety regulations. Traction network electrical field at a height of 1.8 m has linear polarization, whereas magnetic field beyond the wire locations is characterized by elliptical polarization. It is associated with different inductances of 110 kV and 50 kV circuits which result in different angular shift of currents in relevant wires.

5. 50 kV traction system is characterized by the highest electrical field strength at a height of 1.8 m above the ground surface, while 2x25 kV system is characterized by the lowest one. Traditional traction system with 27.5 kV voltage is the worst in terms of magnetic field strength, while 2x50 kV is the best; autotransformer systems of traction power supply create magnetic fields due to demagnetizing effect of power wires.

References

1. S. Apollonsky, *Issues of electromagnetic safety on electrified railroad. V. II. Electromagnetic safety on the railroad with alternating current in traction network* (RUSAINS, Moskow, 2017)
2. A.I. Sidorov, I.S. Okrainskaya, *Electromagnetic fields near ultrahigh voltage electrical installations*. (Chelyabinsk, 2008)
3. A. Ogunsola, A. Mariscotti *Electromagnetic Compatibility in Railways* (London: Springer, 2013)
4. A. Ogunsola, U. Reggiani, L. Sandrolini. International Symposium on Electromagnetic Compatibility, EMC'09. (2009)
5. Sheilah Frey. *Railway Electrification Systems & Engineering* (White Word Publications, Delhi, 2012)
6. H. Biesenack, E. Braun, G. George, etc. *Energieversorgung elektrischer bannen* (Wiesbaden: B.G. Teubner Verlag, 2006)
7. A. Steimel *Electric traction motive power and energy supply. Basics and practical experience.* (Munchen: Oldenbourg Industrieverlag, 2008)
8. R. Kircher, J. Klühspies, R. Palka, et al. Transportation Systems and Technology. 4(2) (2018)

https://doi.org/10.1051/e3sconf/201913901067

9. Luan Xiaotian, Zhu Haijing, Qiu Bo, Han Bochong. CUE2016-Applied Energy Symposium and Forum (2016)
10. M.P. Bader Electromagnetic compatibility (Moskow, 2002)
11. N. Buyakova, V. Zakaryukin, A. Kryukov *Electromagnetic safety in railroads power supply systems: modelling and control* (Angarsk, 2018)
12. V.P. Zakaryukin, A.V. Kryukov *Complex asymmetrical modes of electrical systems* (Irkutsk, 2005)
13. V.P. Zakaryukin, A.V. Kryukov. Power Technology and Engineering Vol. 49, no. 4 (2015)
14. V.P. Zakaryukin, A.V. Kryukov. Innovation & Sustainability of Modern Railway – Proceedings of ISMR'2008 (2008)
15. N. Buyakova, V. Zaharukin, A. Kryukov. Advances in Intelligent Systems Research. Vol. 158. IWCI 2018 (2018)
16. Natal'ya Buyakova, Vasiliy Zakaryukin, Andrey Kryukov, Tu Nguyen. E3S, Web of Conferences 58, 01006(2018) RSES 2018 (2018)
17. N.V. Buyakova, V.P. Zakaryukin, A.V. Kryukov Advances in Engineering Research. Vol. 158. International Conference on Aviamechanical Engineering and Transport. (2018)
18. A.M. Vasilyansky, P.P. Mamoshin, G.B. Yakimov World's railroads. No. 8. (2002)
19. Kotelnikov A.V. *Railroads electrification. World trends and prospects* (Moskow, 2002)

E3S Web of Conferences **139**, 01068 (2019) https://doi.org/10.1051/e3sconf/201913901068
RSES 2019

The assessment of integration effects accounting the stochasticity of wind and solar power plants generation in the Asian Power Grid

Leonid Trofimov[1], *Ivan* Trofimov [1,*]

[1]Melentiev Energy Systems Institute of Siberian Branch of the Russian Academy of Sciences, 664033 Irkutsk, Russia

Abstract. Electricity generation by wind and solar power plants (WPP and SPP) is characterized by the uncertainty of the time and generation volumes, and does not correlate with the demand determined by the graph of daily electric load. By using of the ORIRES optimization model, the integration effects associated with various scenarios of interstate power grid expansion are calculated. The stochastic generation of WPP and SPP creates a certain variation in the model solutions, but does not affect to the total integration effects. The article presents the calculations carried out on a large number of variations for wind and solar power generation. The stability (constancy) of total integration effects is shown, despite the stochastic generation of WPP and SPP in the considered energy power systems.

1 Introduction

To study different scenarios the Interstate Power Grid expansion (IPG) of the Northeast Asia & Central Asia countries in the Melentiev Energy Systems Institute, the ORIRES optimization model is used (hereinafter – the Model). In this Model, a linear optimization method is used to find a solution, which determines the optimal installed capacity and generation type mix of the energy power systems (EPS) and their interconnections, the intersystem and interstate electric ties transfer capabilities, and the operating modes of these capacities and ties. Each EPS is represented in the Model by power demand characterized by a graph of daily electrical load, which determines how much electricity needs to generate for each hour. The problem of determining the actual power for each type of power plants is considered in the modeled EPSs, taking into account the electricity transfer between these systems. The main result is a solution, which takes into account the minimum annualized costs, computing the optimal structure of various capacity types in each EPS and their optimal operating modes at each hour of the day and in different seasons of the year (winter / spring / summer / fall, weekdays / weekends) [1-3].

The model takes into account the most common types of electricity generation by types of power plants – Co-generation, hydro-power plants, nuclear, solar (SPP) and wind power plants (WPP). The WPP and SPP capacities give generation volumes unpredictable in time and power to the design scheme, which can lead to variations in the Model solution. In this paper, a quantitative assessment of these variations is carried out.

2 Data representation features of wind and solar power plants capacities

As an example, in Fig. 1 presents graphs and diagrams for the **Northeast & Central China** energy power system (hereinafter – **"NE&C China EPS"**), for the estimated 2035 target year, constructed by the Computing and Geo-Information System, developed by the authors [4-8].

On the left side of Fig. 1*(a)* the middle column shows the total optimized values obtained by the Model of new installed capacities by generation type mix. Some of constraints (max and min installed capacities) are shown at the columns to the left and to the right sides of the optimized values column.

On the right side of Fig. 1*(b)* shows the graph of daily electrical load (by hours) in the "NE&C China" node. The diagrams below the daily load graph show the optimized contribution to the generation type mix (by different type of power plants).

* Corresponding author: t_john88@mail.ru

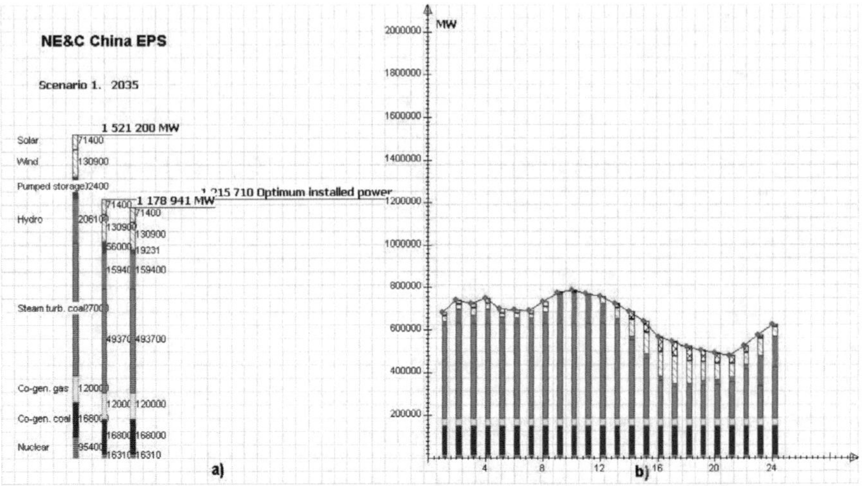

Fig. 1. a) Max, min and optimized installed capacities; b) The graph of daily electrical load "NE&C China EPS", winter, weekday, 2035.

Daily load graphs vary depending on the season of the year (winter / spring / summer / autumn) and the type of day (weekday / holyday). For all graphs the optimal diagrams with the optimizing costs are constructed.

The generation of wind and solar power plants cannot be controlled by a human (system operator). Optimization of its contribution to covering the daily load graph does not make sense. In the Model, it is specified by some fixed set of values for each hour of daily generation, which are constants when solving the optimization problem. Only contributions to the generation of other types of power plants are optimized. Obviously, changes in the volumes of the hourly generation of WPP and SPP in the daily load graph can change (vary) the Model's solutions and, in particular, the value of the objective function.

The aim of this work is to assess the impact of variations in the contribution of WPP and SPP generation on the total solution of the optimization model. Installed capacities of WPP and SPP are not optimized, and are set (vary) to form various solutions of the model. Variations of the WPP and SPP contributions for "NE&C China EPS", as the largest node in the design scheme, are considered below. The share of installed capacity of WPP and SPP in it is only 16-17% of the total optimized capacity.

3 Comparison of scenarios of the IPG expansion

Consider the results of the calculations of three scenarios of the Northeast and Central Asia countries in the 11-node model for 2035. In Fig. 2, the main indicators of the solution of optimization problem for scenario 1 are plotted on the map (isolated operation of each EPS).

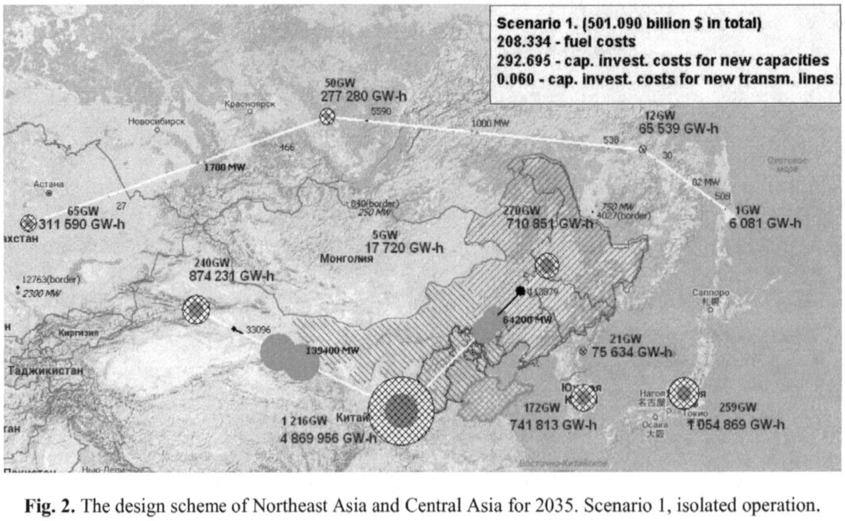

Fig. 2. The design scheme of Northeast Asia and Central Asia for 2035. Scenario 1, isolated operation.

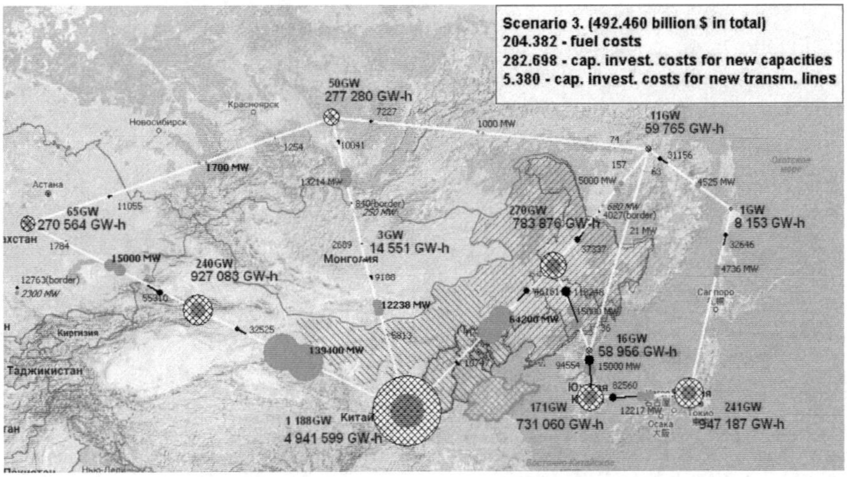

Fig. 3. The design scheme of Northeast Asia and Central Asia for 2035. Scenario 3, interconnected operation.

Some of the nodes on Fig. 2 are connected by electric ties, but only inside the countries. Near each node, there are values of indicators obtained by the Model solution. The total optimized installed capacities for each node (*in GW*), the total annual power generation in each node (*in GWh*) are indicated. The connections indicate their transfer capabilities, and annual electricity flows. In the upper right corner of the map shows the total financial costs in billions of dollars, divided into three components: fuel costs, capital investments & operating costs for new capacities and costs for new electric ties.

Fig. 3 shows the map with design scheme for scenario 3, which differs from scenario 1 by adding new interstate electric ties between all nodes. Comparing these two scenarios (isolated and interconnected operation) you can notice the less value of new installed capacities for all nodes in scenario 3.

In addition, the total investment costs (the value of the objective function) are decreased, primarily due to investments in new installed capacities by generation type mix. Table 1 shows the total costs for three scenarios.

Table 1. The total investment costs, original variant.

Scenarios	Total costs, billion $
Scenario 1, isolated operation	501.089
Scenario 2, interconnected operation without Cenrtal Asia	493.535
Scenario 3, Northeast Asia and Central Asia interconnected operation	492.46

In this table, there is solution for scenario 2, which differs from scenario 3 in "Central Asia EPS - Northwest China EPS" connection has not been included. The total costs in it, are slightly larger than in scenario 3. Thus, we can note the one of the integration effects [9] – the total costs decreasing in interconnected operation scenarios. It can be assumed that, due to the lack of installed capacities is compensated by the import of electricity from other connected nodes.

3 Series of computational experiments with variations on the WPP contribution to total generation

All calculations for the three scenarios from Table 1 were carried out with fixed WPP generation for "NE&C China EPS". For each season: winter, spring, summer and autumn has its own graph. These four graphs are shown on the Fig. 4.

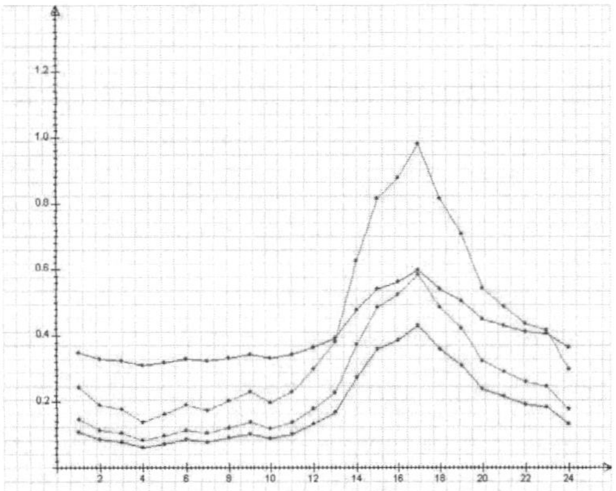

Fig. 4. "NE&C China EPS" daily load graphs with WPP generation by 4 seasons, original variant.

In the next series of calculations, we assume that by 2035 in the NE&C of China will be windier and the WPP generation will be more. So the WPP generation for all 4 seasons was increased up to 40%, without changing the shape of these graphs, Fig. 5.

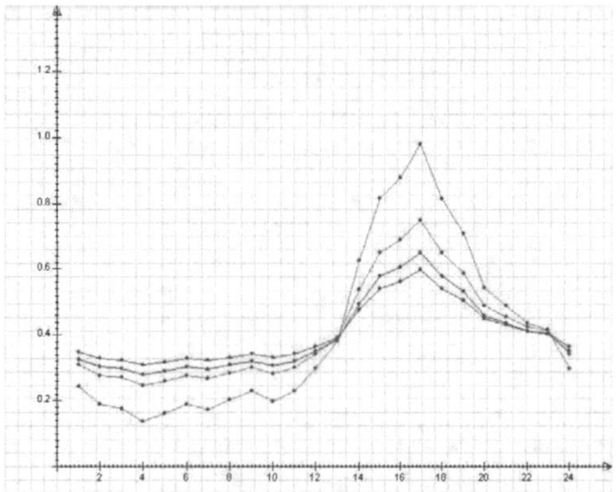

Fig. 5. "NE&C China EPS" daily load graphs. WPP generation up to 40%.

Since "NE&C China EPS" is the largest node in the design scheme and has the largest WPP capacity, significant changes in the total costs are noticeable. However, the overall integration effect has changed insignificantly, see the Table 2. Analysis of changes in the cost structure for this variant showed that the overall reduction in costs was mainly due to decreasing the fuel component. Which is quite predictable, since the generation from non-renewable fuel power plants was supplanted by WPP.

In the next series of calculations we performed to test the impact of shape of WPP generation graphs. The WPP generation peak was changed from evening to morning, Fig. 6. Note that with these changes, the WPP generation for all 4 seasons remained the same 40%. In the next variant, the WPP generation graphs were set in a stochastic manner, Fig. 7.

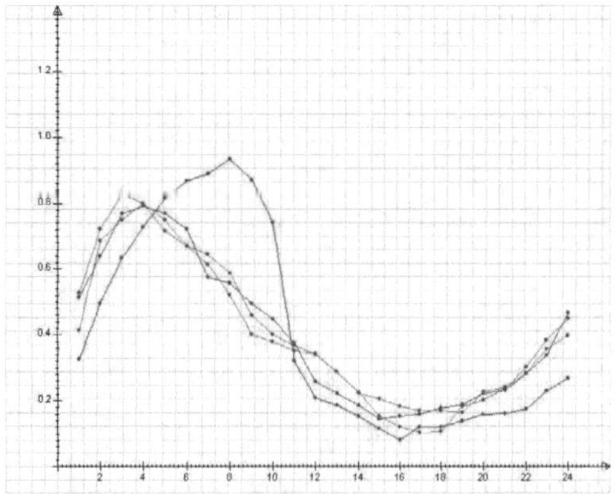

Fig. 6. "NE&C China EPS" daily load graphs. WPP generation peak changed to the morning.

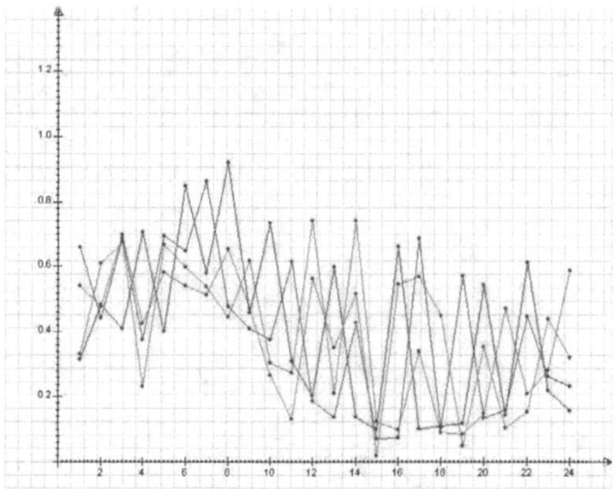

Fig. 7. "NE&C China EPS" daily load graphs. WPP generation is set stochastically.

The calculation results for all variants are shown in table 2.

Table 2. The total investment costs for all variants.

Scenarios	**Total costs, billion $**			
	Original	**WPP gen. up to 40%**	**WPP gen. peak changed**	**WPP gen. set stochastic.**
Scenario 1, isolated operation	501.089	498.812	498,777	498,795
Scenario 2, interconnecte d operation without Cenrtal Asia	493.535	491.244	491,145	491,191
Scenario 3, Northeast and Central Asia interconnecte d operation	492.46	490.175	490,052	490,096

Comparing many variants, we can notice that with changing the WPP generation volumes, the values of objective function varies insignificantly – 0.1-0.2%. Herewith, the integration effects obtained between scenarios of isolated and interconnected operation of nodes also changes insignificantly.

It can be concluded that the stochasticity of the WPP generation can only slightly affect to the total integration effects. The stochasticity of wind and solar power generation can only affects to the changing in fuel costs, due to the replacement capacities obtained by non-renewable fuel power plants.

Conclusion

With the scenario of interconnected operation of all nodes the total investment costs are reduced. The ability to compensate for the maximum daily load due to the import from connected nodes allows reducing the added capacities and, as a result, the investment costs. This is one of the most important integration effects.

With changing weather conditions, for example, with an increase in windy days in a year, the WPP and SPP generation may increase (at the same time, the fuel costs from non-renewable power will decrease). But this only slightly affects the overall integration effects.

It is obvious that the share of WPP and SPP generation in the daily electrical load cannot be too large, since the risk of possible emergency situations with a shortage or overproduction of electricity in the daily balance increases. The integration of energy power systems can reduce these risks and, as a result, increase the share of WPP and SPP capacities.

This study was carried out at the Melentiev Energy Systems Institute in the framework of SB RAS Program No III. 17.6.2 (AAAA-A17-117030310447-3).

References

1. L. Belyaev, S. Podkovalnikov, V. Savelyev, L. Chudinova, *The effectiveness of interstate poewr grids* (Novosibirsk: Nauka, 2008) [in Russian]
2. L. Belyaev, S. Podkovalnikov, V. Savelyev, *Scientific report. The program for the development of hydropower in Russia to 2030 until 2050,* Irkutsk (2015) [in Russian]
3. L. Belyaev, L. Chudinova, O. Khamisov, G. Kovalev, L. Lebedeva, S. Podkovalnikov and V. Savelev, Studies of interstate electric ties in Northeast Asia, International Journal of Global Energy Issues, **17**, 3 (2002) pp. 228-249
4. S. Podkovalnikov, I. Trofimov, L. Trofimov, Technological aspects of the geographic information computing system to study the interstate power grids expansion, Journal of Software Engineering, **1** (2018) pp. 349-360 [in Russian]
5. I. Trofimov, L. Trofimov, S. Podkovalnikov, *Data Representation from Energy Balances by Using Geo-information System*, in Vth International workshop "Critical Infrastructures: Contingency Management, Intelligent, Agent-based, Cloud Computing and Cyber Security, IWCI 2018", Advances in Intelligent Systems Research. Atlantis Press, **158** (2018) pp. 177-182
6. I. Trofimov, L. Trofimov, L. Chudinova, S. Podkovalnikov, *Output Forms for Calculation Results in the Computing & Geo-Information System*, in VI-th International workshop "Critical Infrastructures: Contingency Management, Intelligent, Agent-based, Cloud Computing and Cyber Security, IWCI 2019", Advances in Intelligent Systems Research. Atlantis Press, **169** (2019) pp. 25-29
7. S. Podkovalnikov, I. Trofimov, L. Trofimov. *Data processing and optimization system to study prospective interstate power interconnections,* in 10th International Conference on Asian Energy Cooperation "AEC 2017". E3S Web Conf., **27** (2018) pp. 32-40
8. S. Podkovalnikov, I. Trofimov, L. Trofimov, L. Chudinova, The Specialized System for Processing of Interstate Power Grids Data, Energy Systems Research Journal, **1,** 1 (2018) pp. 61-67
9. S.V. Podkovalnikov, V.A. Savelev, O.V. Khamisov, L.Yu. Chudinova, "Justification of Effectiveness of International Power Interconnections with Separation of Effects between Participants, Automation and remote control", vol. 10, pp. 1756-1765, 2018.

Synthesis of fast and superfast solvers of large systems of linear algebraic equations in Power Flow Calculation

Magomed Gadzhiev[1], *Ksenia* Zhgun[1], *Nikolai* Zubov[2], and *Vladimir* Ryabchenko*[1,2]

[1] Moscow Power Engineering Institute, Department of Electric Power Systems, Krasnokazarmennaya 14, Moscow, Russia
[2] Bauman State Technical University, 2nd Baumanskaya str., 5, Moscow, Russia

Abstract. Algorithms for fast and superfast solvers of large systems of linear algebraic equations, used in solving mathematical problems of electric power industry, are proposed. These algorithms are constructed based on a novel method of multistep decomposition of a multidimensional linear dynamic system. Examples of the analytical synthesis of iterative solvers for matrices of the general form and for large numerical systems of linear algebraic equations are presented.

1 Problem description

One of the central objects in linear algebra is a large system of linear algebraic equations (SLAEs)

$$Ax = b, \tag{1}$$

where $A \in F^{n\times n}$ – Jacobi matrix of size n×n, $b \in F^n$ – previously given vector of dimension n. Here and below, F denotes either the field of real numbers R or/and the field of complex numbers C.

There is a large number of iterative methods and software packages of iterative solvers designed for system of linear equations Eq (1). However, these methods are effective under fairly strong conditions on the properties of the matrix A, among which there are cyclicity, symmetry, and positive definiteness [1–3]. Furthermore, no iterative solvers that could solve the equations with the general matrix A. are known. For example, the methods based on Krylov's subspaces, such as the conjugate gradient method or the generalized minimal residual method of order r GMRES(r), demonstrate good global convergence for positive definite and normal matrices [1]. In the absence of the normality property of the convergence condition become fairly complex, maybe even chaotic, and they are poorly understood. For example, in [4] one can find examples of simple equations (1), for which the iterations of GMRES(1) converge rapidly reaching an accuracy 10–15, while the error of the process GMRES(2) demonstrates chaotic behavior within the range 0,1…1.

The aim of this work is to develop fast and superfast iterative solvers based on a novel method of multistep decomposition and synthesis of control of a multidimensional dynamic system [5] for SLAEs. These solvers must ensure rapidly converging iterative processes in numerical computations. In other words, we want these iterative solvers to be stable and converging within several first iterations independently of the properties of the matrices in the SLAEs. The minimization of algorithmic complexity is not considered here.

2 Analysis of iterative solvers based on control theory

Parameters of iterative solvers, such as shift, relaxation, etc., can be considered as controls, and the solvers themselves can be considered as controlled dynamic systems. One of the first approaches in this direction can be found in the early book by Bellman [6]: «…We have already referred to the fact that computing can be considered to be a control process in which we want to blend complexity, time, and accuracy in some appropriate way. It should, in addition, be treated in many cases as an adaptive control process in which the results of the previous calculation are used to guide the subsequent calculations, producing not only a choice of parameters but a choice of algorithms. »

In recent times a number of numerical algorithms have been studied from the viewpoint of stabilization of dynamic systems using feedback control synthesis. Within this direction of research, [1] cites the papers [7, 8], where the choice of the step size in ODE-solvers was improved by finding an appropriate control. In [9–11] , the concept of controllability was applied to finding eigenvalues matrix pencils and solvers of linear matrix equations (1).

In [12,13] methods of the robust stability theory were used to optimize an SLAE solver.

Let us consider some popular iterative solvers for the linear matrix equation (1), based on the simple iteration

* Corresponding author ryabchenko.vn@yandex.ru

(Richardson's) method and the methods of polynomial and linear iteration [1].
For invertible matrix A Eq.(1) has the well-known solution

$$x = A^{-1}b \tag{2}$$

The iterative solver based on Richardson's method can be classified as a discrete bilinear control system with a single input and multiple outputs (SIMO), defined over Fn:

$$x_{k+1} = x_k + u_k(b - Ax_k), x_0 \in F^n \tag{3}$$

Here u_k – is the control parameter (scalar control) and x_0 – is the initial condition. The fixed (steady) point in n-dimensional space of this iterative process is solution (2) to the linear equation (1).
Various strategies for u_k and certain families of matrices were proposed [14, 15]. In particular, for $u_k = u = \text{const}$ the discrete system (3) has a steady-state value (is asymptotically stable), if the set of eigenvalues (denoted by) of the matrix $E_n - uA$, where E_n – is the identity matrix of order n, is within the unit circle on the complex plane, i.e.$\forall\lambda \in \text{eig}(E_n - uA), |\lambda| < 1$.
If u_k is a feedback control, then $u_k = \frac{e_k^* A e_k}{\|Ae_k\|^2}$, $e_k = b - Ax_k$, where e_k – is the current residual, then This approach is used in the method GMRES(1). It is known that the dynamic system(3) with this control – is asymptotically stable and the iterative process converges if the matrix$A + A^* \succ 0$ is positive definite. Clearly, this condition is not satisfied for arbitrary matrices.
In [13], an alternative to the discrete SIMO- system (3) was proposed in the form of a multi input multi output (MIMO) linear system:

$$x_{k+1} = (E_n - A)x_k + Gu_k + b, x_0 \in F^n, \tag{4}$$

where is a given matrix that generally is not directly related to The asymptotic stability and, therefore, the convergence of the iterative process (4) can be ensured using the feedback control

$$u_k = Kx_k, \tag{5}$$

that can be designed by linear-quadratic optimization. This yields the known LQRES [1] solver.

3 Controllability of iterative solvers

It is known that a necessary and sufficient condition for the efficiency of feedback control (5) is the complete controllability condition of the dynamic system [6]. In [1], the controllability of the iterative solver based on Richardson's method (3) was analyzed. It was shown that, for every input sequence (u_k), the trajectory of the state (x_k) converges to the solution of Eq. (2), if and only if the sequence of residuals

$$e_{k+1} = (E_n - u_k A)e_k, \tag{6}$$

converges to zero. Thus, the dynamics of the discrete system (3) is equivalent to the dynamics of (6).
The iterative solver (3) based on Richardson's method is completely controllable, when A has n different real eigenvalues [1]. Otherwise, controllability is not guaranteed.
The controllability of the iterative solver based on the discrete MIMO- system (4) can be determined using the well-known tests [5]. It is clear that (4) is completely controllable if and only if the Kalman rank condition is satisfied

$$\text{rank}[G \quad (E_n - A)G \quad \cdots \quad (E_n - A)^{n-1}G] = n \tag{7}$$

4. Linear control of iterative solvers based on the multistep decomposition method
Consider a method of linear control of linear multidimensional dynamic systems that can be used for designing fast and superfast iterative solvers of SLAEs (1) [17].
Under the condition that the matrix A, is invertible, solution (2) can be written in the form of the power series [3]

$$x = A^{-1}b = \sum_{j=0}^{n-1} \alpha_j (E_n - A)^j b$$

for certain $\alpha_j \in$, $j = 0,1,\dots,n-1$. Thus, the inverse matrix A^{-1} can be determined using the MIMO system (4). The behavior of residual e_k is described by the linear system [1]

$$e_{k+1} = (E_n - A)e_k - AGu_k, e_0 \in F^n \tag{8}$$

An analysis of (8) proves the following propositions.
4) The discrete MIMO system (8) is completely controllable if and only if the pair $(E_n - A, -AG)$ is completely controllable in the sense of criterion (7).
5) If the MIMO system (8) is completely controllable, then there exists a feedback control (5) (which is not unique in the case m>1), such that the asymptotic stability condition holds:

$$\forall\lambda \in \text{eig}(E_n - A + AGK), |\lambda| < 1. \tag{9}$$

Let, as before, $\tilde{A} = E_n - A$, $\tilde{G} = -AG$. Denote by $\tilde{G}_\perp$ the matrix that is a left divisor of the matrix G and, therefore, satisfies the conditions [5]:

$$\tilde{G}_\perp \tilde{G} = 0, \tilde{G}_\perp \tilde{G}_\perp^+ = E_m,$$

$$\tilde{G}_\perp \tilde{G}_\perp^+ = \left(\tilde{G}_\perp \tilde{G}_\perp^+\right)^T,$$

$$\tilde{G}_\perp^+ \tilde{G}_\perp = \left(\tilde{G}_\perp^+ \tilde{G}_\perp\right)^T,$$

where $\tilde{G}_\perp^+$ is a pseudoinverse matrix for $\tilde{G}_\perp$.
Determine the controller matrix K in the formula for the feedback control (5) by

$$K = \tilde{G}^+ \tilde{A} - \Lambda \tilde{G}^+ \tag{10}$$

Here, Λ is a matrix with the prescribed (desired) eigenvalues.
It can be shown that

$$\mathrm{eig}(\widetilde{A} + \widetilde{G}K) = \mathrm{eig}(\widetilde{G}_{\perp}\widetilde{A}\widetilde{G}_{\perp}^{+}) \cup \mathrm{eig}(\Lambda)$$

A drawback of this feedback control (10) is that the stability of the matrix $\widetilde{G}_{\perp}\widetilde{A}\widetilde{G}_{\perp}^{+}$ cannot be guaranteed in the general case. For this reason, we consider the multistep decomposition of system determined as follows:
initial state

$$\widetilde{A}_0 = \widetilde{A}, \widetilde{G}_0 = \widetilde{G}, \quad (11)$$

the first step

$$\widetilde{A}_1 = \widetilde{G}_{\perp}\widetilde{A}\widetilde{G}_{\perp}^{+}, \widetilde{G}_1 = \widetilde{G}_{\perp}\widetilde{A}\widetilde{G}, \ldots, \quad (12)$$

the kth (intermediate) step

$$\widetilde{A}_k = \widetilde{G}_{\perp k-1}\widetilde{A}_{k-1}\widetilde{G}_{\perp k-1}^{+}, \widetilde{G}_k = \widetilde{G}_{\perp k-1}\widetilde{A}_{k-1}\widetilde{G}_{k-1}, \quad (13)$$

the Sth (final) step

$$\widetilde{A}_S = \widetilde{G}_{\perp S-1}\widetilde{A}_{S-1}\widetilde{G}_{\perp S-1}^{+}, \widetilde{G}_S = \widetilde{G}_{\perp S-1}\widetilde{A}_{S-1}\widetilde{G}_{S-1} \quad (14)$$

Here $S=\mathrm{floor}\left(\frac{n}{m}\right) - 1$; floor – the operation of rounding a number n/m to the nearest integer in the smaller side, e.g., $\mathrm{floor}(0{,}4) = 0$, $\mathrm{floor}(2{,}9) = 2$, etc.
If we also determine in the step-by-step manner the matrices

$$K_0 = \Lambda_0\widetilde{G}_0^{-} - \widetilde{G}_0^{-}\widetilde{A}_0, \widetilde{G}_0^{-} = \widetilde{G}_0^{+} - K_1\widetilde{G}_{\perp 0}, \quad (15)$$

$$K_1 = \Lambda_1\widetilde{G}_1^{-} - \widetilde{G}_1^{-}\widetilde{A}_1, \widetilde{G}_1^{-} = \widetilde{G}_1^{+} - K_2\widetilde{G}_{\perp 1}, \ldots \quad (16)$$

$$K_k = \Lambda_k\widetilde{G}_k^{-} - \widetilde{G}_k^{-}\widetilde{A}_k, \widetilde{G}_k^{-} = \widetilde{G}_k^{+} - K_{k+1}\widetilde{G}_{\perp k}, \quad (17)$$

$$K_S = \Lambda_S\widetilde{G}_S^{-} - \widetilde{G}_S^{+}\widetilde{A}_S, \quad (18)$$

Then, after closing the system by the feedback control (5), with the matrix ***K*** given by

$$K = \Lambda(\widetilde{G}^{+} - K_1\widetilde{G}_{\perp}) - (\widetilde{G}^{+} - K_1\widetilde{G}_{\perp})\widetilde{A}, \quad (19)$$

we obtain the equality of the sets of eigenvalues:

$$\mathrm{eig}(\widetilde{A} - \widetilde{G}K) = \cup_{i=1}^{S+1}\mathrm{eig}(\Lambda_{i-1}) \quad (20)$$

Here $\Lambda_{i-1}(i-1 = 0,1,\ldots,S)$ – are the matrices with the prescribed eigenvalues at each level of decomposition and $\cup_{i=1}^{S+1}\mathrm{eig}(\Lambda_{i-1}) = \mathrm{eig}(\Lambda)$.
Thus, the multistep decomposition of system (11) – (14) and the procedure of determining the matrix K (19) based on (15) – (18) ensure the prescribed location of the eigenvalues $\Lambda = \{\lambda_1, \lambda_2, \ldots \lambda_n\}$, as specified in (20).
Consider the problem of designing control (5), that guarantees that the transient process of the discrete MIMO system takes a finite time. In this case, the matrix $\widetilde{A} - \widetilde{G}K$ has only zeros as its eigenvalues [18], i.e.,

$$\mathrm{eig}(\widetilde{A} - \widetilde{G}K) = \{0,0,\ldots,0\} \quad (21)$$

This requirement implies that we can take any nilpotent matrices with the index of nilpotency not exceeding r as the matrices Λ_i [2, 5]. In this case, the convergence of the designed solvers considered as discrete systems will take place not more than n steps. We call such solvers fast.
If we use the zero matrices as Λ_i, then this case can be assigned the discrete system $x_{k+1} = b$, in which the convergence is obviously already at the second step, i.e., $x_0 = 0$, $x_1 = b$. Such solvers can be called superfast.
For the zero matrices Λ_i of the corresponding size, the final stage of algorithm (15) – (18) is

$$K_0 = -(\widetilde{G}_0^{+} - K_1\widetilde{G}_{\perp 0})\widetilde{A}_0, K_1 = -(\widetilde{G}_1^{+} - K_2\widetilde{G}_{\perp 1})\widetilde{A}_1, \ldots, \quad (22)$$

$$K_k = -(\widetilde{G}_k^{+} - K_{k+1}\widetilde{G}_{\perp k})\widetilde{A}_k, \ldots K_S = -\widetilde{G}_S^{+}\widetilde{A}_S. \quad (23)$$

It follows from (22) – (23), that, if the number of decomposition steps is $S = 1$, then the formula for the matrix K in control (5), that guarantees not only the finite duration of the transient process but also the superfast convergence (during the first two iterations independently of the size of the discrete MIMO system) takes the form:

$$K = -(\widetilde{G}^{+} + \widetilde{G}_1^{+}\widetilde{A}_1\widetilde{G}_{\perp})\widetilde{A}. \quad (24)$$

For $S = 2$ и $S = 3$, the formulas for K are somewhat more complicated:

$$K = -(\widetilde{G}^{+} + (\widetilde{G}_1^{+} + \widetilde{G}_2^{+}\widetilde{A}_2\widetilde{G}_{\perp 1})\widetilde{A}_1\widetilde{G}_{\perp})\widetilde{A}, \quad (25)$$

$$K = -(\widetilde{G}^{+} + (\widetilde{G}_1^{+} + (\widetilde{G}_2^{+} + \widetilde{G}_3^{+}\widetilde{A}_3\widetilde{G}_{\perp 2})\widetilde{A}_2\widetilde{G}_{\perp 1})\widetilde{A}_1\widetilde{G}_{\perp})\widetilde{A}. \quad (26)$$

The analysis of the final formulas (24) – (26), shows that the matrices K obtained using the multistep decomposition contain enclosures whose form is determined both by the matrix A of system (1), and by an arbitrarily chosen matrix G. and by an arbitrarily chosen matrix G , we have a wide choice of the matrices K. For example, if we require in advance that G has not less than n/2, columns, then control with the matrix K (24) guarantees that the iterative process has fast convergence rate:

$$x_{k+1} = \widetilde{A}x_k - G(\widetilde{G}^{+} + \widetilde{G}_1^{+}\widetilde{A}_1\widetilde{G}_{\perp})\widetilde{A}(b - Ax_k) + b \quad (27)$$

4 Examples of the synthesis of superfast iterative solvers

First, consider SLAEs (1) in the general form with a noncyclic matrix not satisfying the normality conditions

$$A = \begin{bmatrix} r & w & 0 & 0 \\ -w & r & 0 & 0 \\ 0 & 0 & \lambda & 1 \\ 0 & 0 & 0 & \lambda \end{bmatrix}, \quad (28)$$

https://doi.org/10.1051/e3sconf/201913901069

where r, w, λ are given numbers. The nonciclicity of matrix (28) is directly indicated by the diagonal 2×2 blocks the first of which is a Jordan block and the second one is an upper triangular matrix. The fact that this matrix is not normal is verified straightforwardly: $A^T A \neq AA^T$. Let the vector b be also given in the general form

$$\mathbf{b} = [b_1 \quad b_2 \quad b_3 \quad b_4]^T \tag{29}$$

Specify the matrix G by:

$$G = \begin{bmatrix} 1 & 0 & 0 & 0 \\ 0 & 0 & 0 & 1 \end{bmatrix}^T \tag{30}$$

We construct the iterative solver using algorithm (11) – (18), (22) – (23). It is seen from the analysis of matrices (28) и (30), that in this case S=floor$(n/m) - 1 =$ floor$(4/2) - 1 = 2 - 1 = 1$ therefore, after performing the two-step decomposition consisting of the zero and the first step, we may use formula for the matrix K (24).

Thus, according to (24), we obtain

$$K = \begin{bmatrix} -\frac{r^2+w^2-1}{r^2+w^2} & \frac{r^3-2r^2+rw^2+r-2w^2}{r^2+w^2} & 0 & 0 \\ 0 & 0 & -\frac{(\lambda-1)^2}{\lambda} & -\frac{\lambda^2-1}{\lambda^2} \end{bmatrix}.$$

In this case $\text{eig}(\tilde{A} - \tilde{G}K) = \{0,0,0,0\}$, which was to be proved.

By running the iterative solver (27) with the parameters for the zero initial conditions indicated above, we obtain the following sequence of residuals

$$\mathbf{e}_0 = [b_1 \quad b_2 \quad b_3 \quad b_4]^T,$$

$$e_1 = \begin{bmatrix} -\frac{(r^2-r+w^2)(-b_2r^2+b_2r-b_2w^2+b_1w)}{w(r^2+w^2)} \\ -\frac{-b_2r^2+b_2r-b_2w^2+b_1w}{r^2+w^2} \\ \frac{b_3\lambda^2-b_3\lambda+b_4}{\lambda^2} \\ -\frac{(\lambda-1)(b_3\lambda^2-b_3\lambda+b_4)}{\lambda} \end{bmatrix},$$

$$e_2 = [0 \quad 0 \quad 0 \quad 0]^T, \ldots, e_k = [0 \quad 0 \quad 0 \quad 0]^T,\ k > 2.$$

We see that the iterative process converged to the exact solution already at the second iteration step.

An analysis of the proposed superfast solvers of linear equations with the matrices and vectors the elements of which are distributed normally and the eigenvalues of the matrices A re distributed by Girko's circular law [19]:

$$A = \text{randn}(1000 \times 1000), G = \text{randn}(1000 \times 500), b = \text{randn}(1000 \times 1),$$

$$A = \text{randn}(5000 \times 5000), G = \text{randn}(5000 \times 2500), b = \text{randn}(5000 \times 1),$$

showed that in all the cases the iterative process converged at the third or fourth iteration step independently of the size of the problem. For each value of the problem size, at least n+1 examples were, n – dimension of the vector in the right part (1).

Superfast iterative solver for power flow calculation of large electric power system

Let's consider the example the calculation of superfast iterative solver for a model of a real power system [20]. It's scheme includes several concentrated subsystems connected by relatively weak connections. Electric power system considers 286 buses, 531 branches, 129 synchronous generators. There are several variants of the Jacobi matrix A for this system, including size 258×258. «Portrait» of the specified matrix for one of the regimes of electric power system is shown in Fig. 1.

Computing the determinant of the Jacobi matrix yields a number that close to infinite $det\,A = a_0 = \lambda_1\lambda_2 \cdots \lambda_{258}$= 6,3108·10258, i.e. it's very difficult to solving SLA (1) by the standard method by calculating the attached matrix. In addition, the Jacobi matrix demonstrates weak scalability, which means large differences in the values of elements (in this case items vary from –500 to 1500).

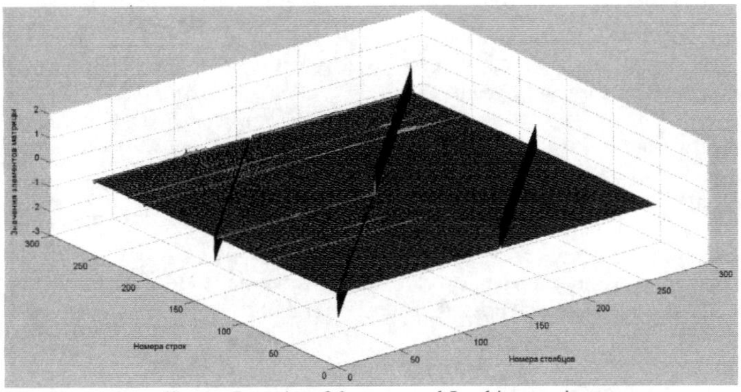

Fig. 1. «Portrait» of the corrected Jacobian matrix.

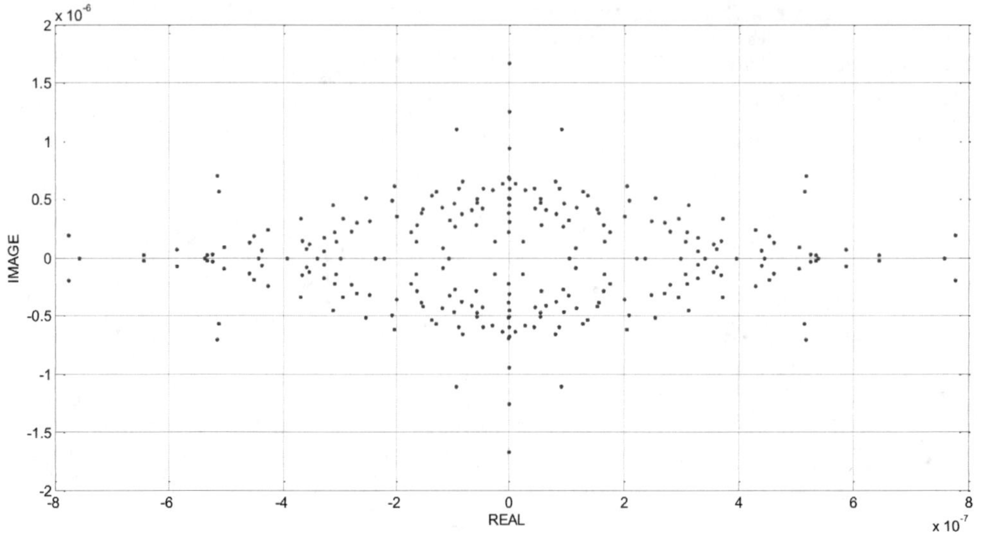

Fig. 2. Eigenvalues of the corrected Jacobian.

It is necessary to adjust the matrix A in such a way that fast convergence of iterative processes in multivariate calculations of electric power systems regimes is carried out. The «Portrait» of the corrected Jacobian matrix of electric power system and the distribution of its eigenvalues are shown in Fig. 1 and 2, respectively. As we see at the Fig. 1 the scaling of the matrix is quite satisfactory (the values of the elements vary from -3 to 2). From Fig. 2 it follows that all eigenvalues of the corrected matrix with high accuracy (10–7…10–6) are located at the beginning of the coordinates of the com-Plex plane. This provides the highest rate of convergence of the iterative process with the error rate of 10-13...10-12. In this case, the convergence of the iterative process in the analyzed case occurs on the second iteration independently of the right part of the SLA (1).

5 Conclusion

Algorithms for fast and superfast solvers of large SLAEs based on a novel method of the multistep decomposition of a linear multidimensional dynamic system are proposed.

Examples of the analytical design of iterative solvers for matrices of the general form demonstrated that the iterative processes converge already at the second iteration step. The statistical analysis of the constructed superfast solvers of linear equations with matrices and vectors consisting of normally distributed elements showed that in all the cases the iterative process converged at the third or fourth iteration independently of the size of the problem. The difficulty of the implementation of the proposed iterative algorithm could be caused by the absence of efficient procedures for computing pseudoinverse matrices of rank deficient matrices. However, this difficulty can be overcome as described in [21].

Application of the fast solver in the practical problem of calculation of the real EPS mode it is shown that the convergence of the iterative process with the residual norm 10–13…10–12 starts on the second iteration independently of the right side of the SLA (1).

References

1. U. Helmke, J. Jordan, *Control and Stabilization of Linear Equation Solvers* (Lecture Notes in Control and Information Sciences. Springer-Verlag Berlin. 2010)
2. D.S. Bernstein, *Matrix Mathematics* (Princeton Univ. Press, 2009)
3. G.H. Golub and Ch.F. van Loan, *Matrix Computations* (Johns Hopkins Univ. Press, Baltimore, MD, 1996)
4. M. Embree, SIAM Review, **45** (2003)
5. N. E. Zubov, E. A. Mikrin, and V. N. Ryabchenko, *Matrix Methods in the Theory and Practice of Aircraft Automatic Control Systems* (MGTU im. N. E. Baumana, Moscow, 2016) [in Russian].
6. R. E. Bellman, Nonlinear processes, **2** (1971)
7. K. Gustafsson, M. Lundh, G.S. Soderlind, BIT, **2** (1988)
8. K. Gustafsson, ACM Trans. Math. Software, **17(4)** (1991)
9. U. Helmke, P.A. Fuhrmann, Systems & Control Letters, **41** (2000)
10. U. Helmke, F. Wirth, Systems & Control Letters, **41** (2001)

11. J. Jordan, *Reachable sets of numerical iteration schemes.* (Ph.D. thesis. University of Wurzburg, 2008)
12. A. Bhagya, E. Kaszkurewicz, Advances in Design and Control, (2006)
13. Y. Wakasa, Y. Yamamoto, Proc. of SICE first Annual Conference on Control Systems, (2001)
14. D. Calvetti, L. Reichel, Numerical Algorithms, **12** (1996)
15. D.C. Smolarski, P.E. Saylor, BIT, **28** (1988)
16. A. A. Samarskii and A. V. Gulin, *Numerical Methods* (Nauka, Moscow, 1989) [in Russian].
17. E. A. Mikrin, N. E. Zubov, D. E. Ivanov, V. N. Ryabchenko, Reports of the Academy of Sciences, **3** (2018)
18. B. T. Polyak and P. S. Shcherbakov, *Robust Stability and Control* (Nauka, Moscow, 2002) [in Russian].
19. V.L. Girko, Theory Probab. Appl., **29** (1984)
20. M. Sh. Misrikhanov and V. N. Ryabchenko, Autom. Remote Control, **72:10** (2011)
21. N. E. Zubov, E. A. Mikrin, and V. N. Ryabchenko, Vestn. MGTU Baumana, Estestv. Nauki, **3** (2018)

https://doi.org/10.1051/e3sconf/201913901070

Assessment of the Impact of the Rigidity of Nodes on the Operating Reliability of EPSs

Nailya Chemborisova[1*]

[1]National Research University Moscow Power Engineering Institute, Moscow, Russia

Abstract. The operating reliability of electrical power systems is mostly based on the topology and parameters of the schemes considered in this article. The preliminary (before the calculation of operating conditions) analysis of the rigidity or sensitivity of nodes, based on the matrix of node conductivity, allows for evaluating and increasing the controllability of conditions by voltages and limiting short-circuit currents, making the reduction of active power losses in the grid more efficient. A conclusion was made regarding the consistent effects of changes in the node rigidity on increasing the EPS operating reliability.

1 Introduction

In the operation and development of complex electrical power systems (EPSs), one of the illustrative examples being metropolis power supply systems (MPSSs), it is becoming more and more difficult to enhance reliability, improve control conditions and ensure resilience. This is partially related to the following factors: an increased number of interrelated energy facilities; the growing maximum power of electrical systems; complicated grid layouts; the improved coherence of individual elements of the system; complications in ensuring the required controllability levels of facilities; and increased levels of short-circuit (SC) currents.

It should be noted that several factors must be taken into account when designing MPSSs. Furthermore, there are a plethora of grid layout configurations that have various capabilities, properties, and indicators defining various levels of reliability. The scheme presented in [1] is deemed to be a near-optimal representation of a metropolis's power supply. However, in due to less than ideal urban planning and building construction, in practice it is impossible to use all recommended principles of optimal electrical grid organization, which entails reduced operating reliability of power systems and of power supply to consumers. Therefore, given a pre-determined set of common properties, a structural, grid-specific preliminary analysis is required. This analysis will allow for the design of future similar grids, and will correct the errant properties of existing grids.

Based on the specifics of metropolis development, some accompanying unresolved challenges can be considered [2]:

1. Improving the reliability of power supply to consumers;
2. Re-distribution of power exchanges and reduction of loading on over-loaded grid elements (OHLs, transformers);
3. Improving the capacity of system-forming and distribution grids of metropolises;
4. Improving the breaking capacity of 110 kV+ switches in order to eliminate the need for sectioning in grids.

To keep up with these challenges, control must be optimized and grid capabilities must be used more efficiently. A one-linear metropolis model can serve as an appropriate basis for the preliminary assessment of MPSSs functioning in various conditions, to arrange a fault-current limiting device (FCLD) or other control devices. A typical example of an MPSS is the Moscow Power System, a part of the Central Integrated Power System providing power to Moscow and the Moscow Region.

2 Substantiation of the Method for Analyzing Modes and Their Reliability

One of the EPS properties that must be accounted for is the rigidity of its nodes. For a specific point, it is characterized by loading growth for which the voltage level and phase change by a unit: the higher the growth, the more the EPS rigidity in this point [3] and the less the response behavior (sensitivity).

For the quantitative assessment of rigidity and response behavior, a node sensitivity factor can be used [4, 5]. It shows the extent of voltage change in the node relative to the nominal voltage when the loading increased, in the considered case, by 1 MW and 1 MVAr [8].

$$k_s = \frac{U_0 - U_i}{U_{nom} \cdot S_{in}} \cdot 100\% \qquad (1)$$

Here, U_0 is the initial voltage in the node before loading increased, U_i is the initial voltage in the node after loading increased, with the difference between them in the numerator ΔU, U_{nom} is the nominal voltage in the node, S_{in} is the loading increase increment in terms of power.

In rigid nodes, the sensitivity factor will be less than in the sensitive ones. Using this factor, one can evaluate the sensitivity of the considered node to changes in the loading parameters. There are sensitive groups of elements, the loading parameters of which significantly vary due to random changes in the layout or conditions, and rigid groups of elements, the loading parameters of which almost do not vary due to external action [3]. It should be noted that when using a more precise method for searching sensitive nodes, as suggested in [6], one needs the results of singular or spectral analysis of the Jacobian matrix (written in power imbalances in the polar coordinate system) used in calculations of steady-state conditions (that do not always give a solution, even if it exists). The analysis of the properties of electrical power systems is related to the issue of marginally sensitive and rigid nodes caused by the irregularity of the EPS structure.

In designing and operating EPSs with a larger number of grid elements, it is rather hard to calculate and analyze electrical conditions without using a systemic approach. Furthermore, the cost-effectiveness of the current conditions can be decreased; the total losses of active power in the grid ΔP_t can rise due to the irregular areas of schemes that usually impair power system conditions. The control solution can be simplified by preliminarily analyzing the properties of the EPS substitution scheme and its parameters.

EPS rigidity primarily defines a high-voltage grid. It must be sufficiently developed, redundant and connect the primary parts of the power systems and nodes with high loading. Closed schemes of electrical grids having several power supplies are the most reliable ones. However, short-circuit currents (SCCs) grow in closed grids, which reduce reliability and become dangerous for switches and equipment installed during previous periods of EPS development [7]. To limit SC currents, current-limiting reactors can be installed and the grid can be sectioned; when sectioning, SC currents are reduced, but the electrical bond between the sectioned parts of the grid is impaired, which may result in the reduced reliability of power supply to consumers. Sectioning disadvantages also include reduced rigidity of the scheme; worsened normal and post-emergency conditions; complicated voltage control; increased power and energy losses in the grid, so sectioning is considered as a forced temporary measure.

In this manner, the goal of increasing EPS controllability has contradictory requirements. On the one hand, to enhance reliability, it is necessary to increase the scheme rigidity, while on the other hand, it is required to reduce it to limit SC currents that also affect reliability. One of the options is using compensators in sensor nodes and SCC limiters (reactors, sectioning) in the most rigid nodes that can be considered as control nodes. While this results in somewhat reduced sensitivity and rigidity of marginal nodes, increased regularity of the system, it also leads to the limitation of short-circuit currents.

However, another issue arises when reducing the rigidity of nodes. The voltage levels and capacity of system-forming and distribution grids start going down [8], which influences the performance reliability.

Since real EPSs require a system to control their steady-state and emergency conditions, it is required to preliminarily (before calculating the steady-state conditions or short-circuit currents) analyze the grid structure, its layout and operating parameters in order to have a benchmarking of the extent of changes in operating parameters in various nodes.

3 Method of Balance Conductivities

When forming the matrix of nodal conductivities $[Y]$ necessary to solve the equations of the steady-state condition [7], one can acquire additional information on nodes based on the differences ΔY (balance conductivities) between the inherent conductivity in each node Y_{ii} and the sum of its mutual conductivities Y_{ij}. The elements comprising the matrix $[\Delta Y]$ must be considered on a complex plane when ranking the acquired results in the descending (ascending) order. The nodes where the maximum values of the reactive components $[\Delta Y]$ $+j\Delta B$, $-j\Delta B$ are obtained require increased attention when arranging control, during calculations and the analysis of conditions. For example, when controlling the conditions, one of the tasks is to find the place of inputting the most efficient control action (CA) to change voltages and consumption of reactive power in the electrical system, and to reduce losses of active power in the grid. If the resulting difference $j\Delta B$ has a capacitive nature, the voltage in the node is maintained rather efficiently. The nodes with the highest capacitive difference will be the most rigid as compared with other nodes. If ΔB has the highest values in modulus and an inductive nature, voltage will be maintained less efficiently than in a rigid node, since this node will consume additional reactive power, and the node is considered to be sensitive [8]. When comparing two nodes with an inductive difference of conductivities, the voltage in the node is better maintained with a less difference in modulus. Then one can rank the nodes by the extent of changes in the difference of conductivities and acquire a list of nodes ranked by the level of rigidity. This preliminary analysis of the grid parameters identifies rigid and sensitive nodes and allows for the rational arrangement of reactive power compensators in the grid or SCC limiters.

Using a rather simple analysis, one can find problematic nodes in a formalized manner (with the highest SCC values, with the highest changes of voltage in the node, etc.). This analysis requires no time-consuming calculations and information which is rather conventional for calculations of steady-state conditions. It allows developing a systemic approach to the condition control, making it the most efficient, which is especially urgent in existing grids of metropolises.

It is known that, in a regular grid, the natural distribution of powers is also economical [1, 9], and close electrical grids are usually irregular (with various X/R ratios in various sites). The highest irregularity is found in outlines containing various nominal voltages, transformers, which results in changed power exchanges by branches and voltage in nodes, with power losses increased.

In these conditions, to compare schemes by the assessment results of some quality indicators of supplied power energy, voltages in nodes and losses in the grid, a sufficiently simple approach will be required, which allows taking into account the specifics of operation of complex irregular systems.

A comparison of several schemes by voltage levels in nodes may seem necessary in designing EPSs. A scheme with a large number of sensitive nodes will have reduced voltages as compared with the voltages in the scheme with more rigid nodes, but also decreased SCCs [10]. The power system can be considered as several regions with prevailing nodes having specific properties (sensitive or rigid).

4 Calculation Results for Increasing Reliability

The analysis of changes in operating parameters is interesting first of all in the most sensitive nodes, since they respond better during changes. Rigid nodes are almost of no interest in voltage normalization, since volt-age levels in them are only slightly changed in the case of similar impacts.

To demonstrate the capabilities of such an approach, a famous testing scheme was considered as given in Fig. 1, as well as a modified annular scheme based on it (Fig. 2).

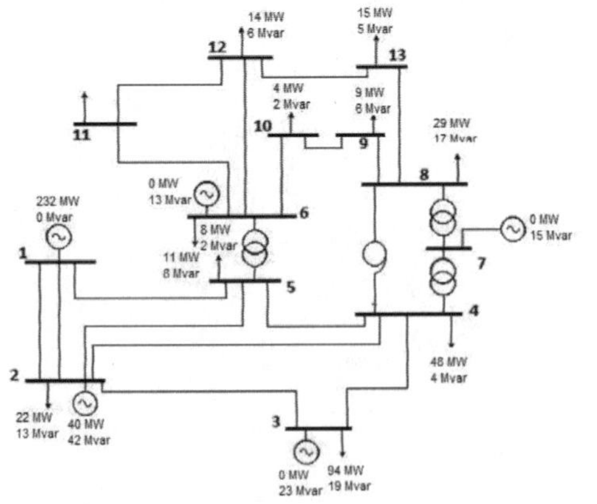

Fig. 1. Option 1 – Test scheme (complex closed).

Table 1 gives the calculation results for the most rigid (6) and sensitive (11) nodes 110 kV for sensitivity factors and SCC (three-phase SC).

To conveniently compare the options, currents are given in relative units with regard to the maximum current in the most rigid node (6) for option 1 (complex closed scheme). ks are calculated using the above formula.

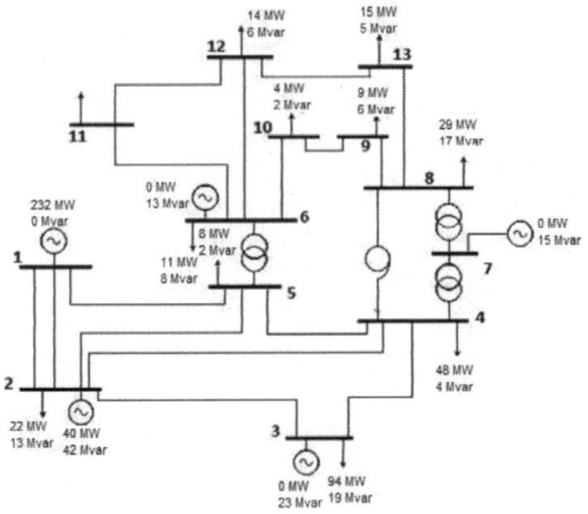

Fig. 2. Option 2 – Modified scheme (annular).

Table 1. Results of comparable calculations of k_s and SCC.

Scheme No.	Node no./type	k_s	I_{3ph} / r.u.
1	6/rigid	0.118	1.000
1	11/sensor	0.132	0.853
2	6/rigid	0.188	0.905
2	11/sensor	0.214	0.743

It should be noted that the numbers of the most rigid and sensitive nodes acquired using the matrix of nodal conductivities have fully coincided with the calculation results of worsening, e.g., the capabilities of various schemes in maintaining voltage can be compared using the minimal values of balance conductivities: the higher the balance conductivity by modulus in the most sensitive node, the higher the predicted voltage change (the sensitivity factor), but the lower SCC. If it is necessary to maintain voltage in the sensitive node for two schemes with various values of such balance conductivities, the preferable scheme is the one with a lower value in modulus, since the voltage will be better maintained. For example, in sensitive node 11, the sensitivity factor is higher in option 2, so the voltage in the most sensitive node will be maintained less efficiently. This approach allows for the choosing of nodes, the voltage control of which should receive special attention, and comparing schemes by their capability to maintain voltage in the most sensitive nodes.

If a scheme with the minimal SCC is required, the comparison of the most rigid nodes will be the most informative. Balance conductivities lower in modulus in the most rigid node will characterize lower SCC. Since the nodes were ranked using the balance conductivity method before calculations of the steady-state conditions and SCC, followed by the calculation of worsening for the selected nodes and of the sensitivity factor that also allowed finding rigid and sensitivity nodes in the options (k_s in rigid nodes is lower than in the sensitive ones, as per the definition of rigid nodes), the using of this method can be deemed correct in the preliminary evaluation of the rigidity of nodes. When selecting a scheme with the lowest SCC, the best option in rigidity evaluation is option 2, since for rigid node 6, the sensitivity factor is higher, e.g., the node is less rigid than in option 1. When comparing after SCC calculation using the option having lower SCC values, option 2 is also the choice. It should be noted that SCC values for the considered options are higher in rigid nodes than in the sensitive ones. This proves the opportunity to use the method of balance conductivities for preliminary benchmarking of nodes in SCC under a single option and to compare schemes by relative values of SCC in various options.

To confirm the conclusions made, a retrospective scheme of the Moscow Energy System was considered, with a similar analysis performed. The comparison revealed that the most rigid node maintains the voltage very well, since ΔU is below 1 kV, and ΔU in the most sensitive node during voltage changes is within (+8.5)-(-5.7) kV relative to the nominal level. It should be noted that these nodes have approximately equal basic load and function at the same level of voltage (110 kV) in annular grids. Then, in the grids typical of annular schemes, this approach will operate very efficiently. When comparing SCCs and places of the node in the ranked list of 100 kV nodes, the following current values have been obtained: Butyrki (most rigid) – 114.3 kA, Ochakovo (least rigid) – 77.1 kA, which is significantly higher than the breaking capacity of switches at this voltage level. For sensitive nodes 100 kV located on the opposite end of the ranked list of nodes, these values are significantly lower, for example, Beskudnikovo – 31.2 kA, Vladimir (most sensitive) – 28.1 kA.

The method of balance conductivities resulting in the ranked list of nodes as suggested herein allows for significant substantiation of promising schemes in designing or repair schemes in operation, finding possible options of the grid scheme that ensure grid controllability and power supply reliability of its individual nodes.

5 Conclusion

The goal of increasing EPS controllability has contradictory requirements. On the one hand, to enhance reliability, it is necessary to increase the scheme rigidity, while on the other hand, it is required to reduce it to limit SC currents, which also affect reliability. To assess the potential for increasing the controllability of the electrical grid (by maintaining voltage in the most sensitive nodes and reducing SCCs in the most rigid nodes) and improving the reliability of individual nodes, a preliminary analysis is required to form a ranked list of nodes based on the calculation of balance conductivities. This analysis should precede the calculation of steady-state conditions and SCCs and should use elements of the node admittance matrix. The same approach allows benchmarking various configurations of grid schemes to find options having enhanced grid controllability and power supply reliability of individual nodes, and it can also substantially reduce the number of considered scheme options during design and operation.

The opportunities for using the suggested approach were checked by calculations for the known test scheme and the Moscow Energy System, which proved it to be correct.

References

1. D.P. Faibisovich (Ed.), *Guidebook for designing electrical grids*, 2nd ed. (ENAS, Moscow, 2012)
2. V.V. Sergeev, V.N. Savinov, V.S. Pavlikov, *Electrical Stations*, **11**, 59-67 (2007)
3. N.E. Savoskin, *Reliability of electrical systems* (Publishing House of the Penza State University, Penza, 2004)
4. S. Conti, S. Raiti, G. Vagliasindi, *Industrial Electronics (ISIE), 2010 IEEE International Symposium on*, 2548-2554 (IEEE, 2010)
5. Q. Zhou, J. Bialek, *Proc. of the 16th Power Systems Computation Conference (PSCC2008), Glasgow, Scotland* (2008).
6. O.N. Voitov, N.I. Voropai, A.Z. Gamm, I.I. Golub, D.N. Efimov, *Irregularities analysis of electrical power systems* (Nauka, Sibirskaya izdatelskaya firma RAN, Novosibirsk, 1999)
7. V.I. Idelchik, *Electrical systems and grids: Study book for high schools* (Energoatomizdat, Moscow, 1989)
8. O.V. Frolov, N.Sh. Chemborisova, *Energy industry: Control, quality, and efficiency of using energy resources: Collection of works of the Sixth All-Russian Scientific Conference with International Participants*, **1**, 100-104 (AmGU Publishing House, Blagoveshensk, 2011)
9. O.A. Lopatin, N.Sh. Chemborisova, Electricity, **3**, 10-12 (2011)
10. O.V. Frolov, N.Sh. Chemborisova, New in the Russian Electrical Power-Engineering, **4**, 33-40 (2011)

Risks in the energy sector: management practices analysis in the electricity market

Yulia Zhilkina [1,*]

[1] Federal Grid Company of Unified Energy System, Russia, 117630, Moscow, st. Akademika Chelomeya, 5A

Abstract. Energy security is today one of the most discussed event in international economic relations. It is being actively discussed by all governments and leading international organizations. There doesn't exist any universal definition of energy sector. In addition to the lack of such a universal definition, analysis of decisions in this area, one notices the lack of attention to such phenomena as risks in the energy sector, especially in the management of investment projects in the industry. For most industrial countries, energy insecurity means import dependency and aging infrastructure, while many emerging economies have additional vulnerabilities such as insufficient capacity, high energy intensity, and rapid demand growth. In many low-income countries, multiple vulnerabilities overlap, making them especially insecure.

1 Introduction

In the Energy Sector, critical infrastructure security and resilience goals and priorities are developed and achieved through a collaborative effort between the industry and the government. These goals include assessing security risks and threats, securing critical infrastructure from all hazards, enhancing critical infrastructure resilience, sharing information, and promoting learning and adaptation. There are numerous activities and programs that support these goals, developed and maintained by a wide variety of public and private organizations. The vast majority of energy infrastructure by the private sector is owned and operated, the Energy Sector security and resilience efforts are a shared responsibility between the public and private sectors. This section highlights some of the approaches and efforts that are underway in the Energy Sector to help achieve the national critical infrastructure security and resilience goals [1].

Throughout history, energy security has been viewed as protection from disruptions of essential energy systems. The notion of "essential energy systems" evolved from supplies of oil for military purposes to encompass various energy sources, infrastructure, and end-use sectors. The idea of "protection from disruptions" has also evolved from securing military or political control over energy resources to setting up complex policies and measures of strategically managing risks that affect all elements of energy systems. Though it is possible to discuss energy security at household, community, and other levels, most political concerns and scholarly research are about energy security of individual countries. This is because nation-states have a historic responsibility for security, national energy systems provide appropriate units of analysis of key risks and vulnerabilities, and the majority of policy interventions to maintain energy security occur at the national level. Consequently, the analysis presented in this chapter primarily concentrates on the national level of energy security. [2-4]

Life safety is the state of security of the material world and human society from the negative impacts of various kinds. Its objects can be classified as follows are nature and society (see Figure 1).

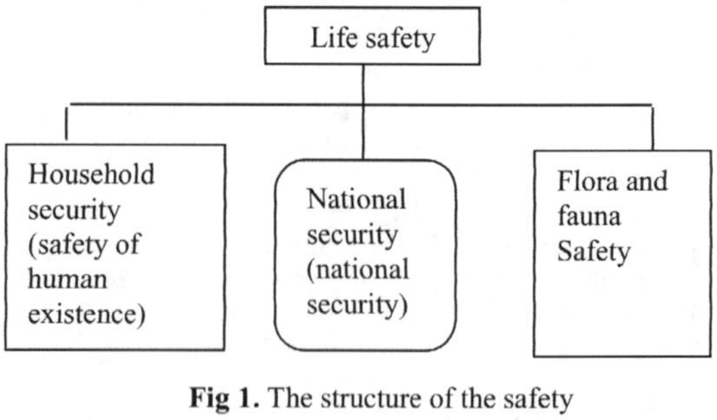

Fig 1. The structure of the safety

Uninterrupted provision of vital energy services – energy security – is a high priority of every nation. Energy security concerns are a key driving force of energy policy. These concerns relate to the robustness (sufficiency of resources, reliability of infrastructure, and stable and affordable prices); sovereignty (protection from potential threats from external agents); and resilience (the ability to withstand diverse disruptions) of energy systems.

Our analysis of energy security issues in over 130 countries shows that the absolute majority of them are vulnerable from at least one of these three perspectives [5, 6].

2 The measures ensuring energy security of Russia

Thus, energy security – is, first, energy complex security, the lack of significant threats to its existence, and secondly, the lack of energy threats to the life of individuals and their various communities and organizations at the macro level. Energy security includes both a low level of security threats to the functioning and development of the energy complex of the country, and a low level of probability of threats to the development of society from the energy complex itself. These threats are closely associated with the problems of prevention and liquidation of emergency situations at the facilities of the energy complex of the country. The measures ensuring energy security of Russia includes measures on prevention of emergency situations through accidents or acts of sabotage at power facilities in peacetime and wartime and in case of their occurrence – measures for liquidation of emergency situations, security and protection of population, environment and decreasing material losses.[7-11]

Energy security meets the needs of the individual, society, state in the field of energy. The material and technical basis for its provision is the fuel and energy complex (FEC) of the Russian Federation. Thus, energy aspects of national security are a broader concept than energy security and include the latter.

3 Analysis of the main threats

In the nature of risk and his omnipresence across every field of effort and mitigation, management and supervising risk, involved with matters of financial nature requires a discipline dedicated towards this end. Identifies 3 types of risk:

1. Risks that an organisation can face with a result to environmental changes. These changes target factors such as changes to interest rates, market and change rates.
2. Risks born outside an organisation's engagement and interaction with other organisations. These engagements involve things like horizontally and vertically transactions with sellers, customers and other parties.
3. Risks arising from an organisation and target specifically people processes and internal systems of the organisation. [11].

As objects are:

- individual, society and state;
- certain types of resources: natural, material (objects of labor), means of production, labor, financial, organizational and managerial resources of production activities, spheres and sectors of the economy, such as production, distribution, exchange and consumption, industries and sub-sectors of the national economy, markets for various goods and services [12].

Processes as subjects of threats can be: macro - and microeconomic processes; processes of technical and economic, organizational, managerial and socio-economic changes; processes of production, distribution, exchange and consumption, saving, accumulation and investment, innovation processes and General processes of formation, development, degradation, etc. [14].

Financial risk management is looking to give an answer and understand as how the risks cited above can affect organisations. If financial risk management as a discipline, it is concerned to risk cited above in general with managing exposure, and also helps with market's sensibility risks and credit risks organisations.

The authors simply define and state that risk management is "the continual process of risk reduction" and of which financial risk management is "really about how firms actively select the type and level of risk that it appropriate for them to assure". As we can see, the author would suggest that the attempt to define the topic appear with different factors, is superficial and simply differ as to the level of detail at which the definition is provided.

Financial risk management does not exclude the need to engage in risky activities and therefore risk management and risks that have been taking are not mutually exclusive. The accentuation is about deciding which risks to assure, by which is meant against which types of risk assurance will be given.

In this context, the term assurance means the capability of providing an assessment of these risks and also providing information that organisation can count to make decisions that preoccupy them. Therefore, the purpose of the activity is about providing information, based upon analysis and estimates, as to the level of reward in relation to risks that are endemic to any business operation. [10].

The national economy states, imperfection system organization of state power and civil society, the socio-political polarization of Russian society and social relations criminalization, the growth crime organized and increase in the scale of terrorism; international relations aggravation and complication of international relations create a wide range of internal and external threats to the national security of the country.

In the economic sphere, the threats are complex and are primarily due to a significant reduction in the gross domestic product, a decrease in investment, innovation and scientific and technical potential, stagnation of the agricultural sector, an imbalance in the banking system, an increase in external and domestic public debt, a tendency to dominate in the export of fuel and energy components, and in imports - food and consumer goods, including basic necessities.

The weakening of the country's scientific, technical and technological potential, the reduction of research in strategically important areas of scientific and technological development, the outflow of specialists and intellectual property threaten Russia with the loss of leading positions in the world, the degradation of science-intensive industries, the strengthening of external technological dependence and the undermining of Russia's defence capability .

The current threats to the energy security of the Russian Federation was analysing, the author considers it legitimate to highlight the main ones:

• reduction of natural fuel and energy resources;
• inefficient state policy in the field of energy;

• lack of investment resources;
• equipment aging;
• man-made and natural threats;
• terrorism;
• increased pressure on the environment;
• critical state of the scientific and design industry complex;
• foreign economic threats.

Thus, based on the real situation in the country, taking into account global energy trends, Russia needs to attract investment for the integrated development of mineral and non-traditional resources of the regions. Investment is a vital need for the energy security of the Russian Federation. In this regard, you should pay attention to the risks.

4 Economic risks

First of all, some clarification of the concept of «risk» is required. Risk - an event that can occur under uncertainty with some probability, with three possible economic results, estimated in economic and financial indicators:
• negative – damage, loss, loss;
• positive – a gain, profit, win;
• zero – no harm, no gain.

The existence of risk is associated with the inability to predict the future with 100% accuracy. It is necessary to highlight the main property of risk based on this: risk occurs only in relation to the future and is inextricably linked to forecasting and planning, and hence to decision-making in General (the word «risk» literally means «decision-making», the result of which is unknown). It is also worth noting that the categories «risk» and «uncertainty» are closely related and are often used as synonyms.

However, it is useful to distinguish between «risk» and «uncertainty». First, risk occurs only when a decision is necessary (if it is not, there is no point in taking the risk). In other words, it is the need to make decisions in the face of uncertainty that generates risk, in the absence of such a need, and there is no risk.

Second, risk is subjective and uncertainty is objective. For example, an objective lack of reliable information on the potential volume of electricity demand leads to a range of risks for project participants. Risk assessment should take into account individual risk tolerance, which is described by indifference or utility curves. Therefore, it is recommended to describe the risk with the three above parameters:

$$Risk = \{P * L * Y\} \quad (1)$$

5 Risks in PJSC «ROSSETI»

Industrial risk. Power industry is an infrastructural branch of the economy. The majority of end users of the industry products are located in the Russian Federation. Thus, the projected dynamics of the industry is determined by the overall dynamics of socio-economic development of all other sectors of the economy of the Russian Federation, as well as to some extent - climatic (water content levels) and weather (average annual temperature) conditions in Russia.

Country risk. The economy of the Russian Federation is still characterized by some features of the emerging market, in particular: the non-convertibility of the Russian ruble in most countries, currency control, as well as the level of inflation. The existing Russian tax, currency and customs legislation is subject to varying interpretations and frequent changes.

Economic risk. Economic factors are overwhelmingly beyond the Issuer's control. Although the economic situation has shown signs of improvement, the economic Outlook of the Russian Federation depends largely on the effectiveness of economic measures, financial mechanisms and monetary policies undertaken by the Government, as well as the development of fiscal, legal and political systems.

Financial risk. Servicing of previously attracted loans and borrowings by PJSC «Rosseti» is carried out strictly in accordance with the established service schedules or ahead of them.

Environmental risk. Taking into account the current activities of PJSC «Rosseti», the risks are associated with the implementation of investment programs approved by the Board of Directors, in terms of compliance with the norms in construction.

Cybersecurity risk. The Federal Grid Company implements measures for automatic delivery of all necessary information to consumers (including relay maintenance crews). A step-by-step implementation of specific measures combined with continuous threat analysis should decrease the losses from possible cyber-attacks as well as the costs of the security measures. In the course of the roundtable on cyber security at the Relay Protection and Automation exhibition in 2016, the need was realized to implement technical measures for cyber security under the assumption that cyber-attacks would take place. The necessity of analysing the impact of those measures on the core functions of the power facility was not even considered during the talks. Like the following statement in order 31: "Cyber security solutions must not impair the functionality of the substation automation systems." We must not forget this while considering cyber threats. The utilities have been placed in conditions where cash expenses on the hypothetical threat elimination are demanded from them, while it is unknown what impact the offered cyber security solutions will have on the core relay protection and control functions. Moreover, the vendors will bear no responsibility if normal operation is undermined.

6 Conclusion

World experience shows that there is no single, optimal organizational structure for all. It is necessary to choose the management structure that is adequate to the current economic conditions of the company and allows it to achieve its goals. Within any structure, emphasis can be placed on the decentralization of authority, allowing decision-makers at lower levels to make their own decisions. Decentralized structures are recommended when a company has access to dynamic markets, diversified production, competitors, and rapidly changing technologies. Project risk management methods in the power industry can and should become a means of effective implementation of projects at all levels of management – Federal, regional, local. It is hoped that the problem of minimizing risks in the energy sector will be reflected.

References

1. Decree of the President of the Russian Federation of May 13, 2019 № 216 "On Approving the Doctrine of the Energy Security of the Russian Federation".

2. Zhilkina Yu. V. Economic mechanism of innovative management of energy security // III Manyakin readings. Materials of the international scientific-practical conference. 2014. Publishing house: Russian state University of trade and Economics, Omsk Institute (branch)

3. Vasenkov A. V., A. A. Prokhozhev Security of Russia: modern understanding of security. – M.: Rosecolored "Creation", 1998. – P. 22.

4. Marco Kerste, Matthijs Gerritsen, Jarst Weda, Bert Tieben Systemic Risk in the Energy Sector – Is There Need for Financial Regulation? // Energy Policy № 78, March 2015 (https://doi.org/10.1016/j.jebo.2017.09.013)

5. M. Beccali, M. Cellura, V. Lo Brano, A. Marvuglia, "Forecasting daily urban electric load profiles using artificial neural networks", in Journal on Energy Conversion and Management, Vol. 45, No. 18, 2004, pp. 2879 -2900.

6. Nikitenko, S.M., Goosen, E.V. The problems and prospects of the public-private partnership in the Russian fuel and energy sector IOP Conference Series: Earth and Environmental Science Volume 53, Issue 1, 9 February 2017

7. Peter Burgherr, Jennifer Giroux and Matteo Spada. Accidents in the Energy Sector and Energy Infrastructure Attacks in the Context of Energy Security // European Journal of Risk Regulation. Volume 6, Issue 2, 2015, pp. 271-283 (DOI: https://doi.org/10.1017/S1867299X00004578)

8. Tatiana Grynko, Tatiana Gviniashvili Organisational and economic mechanism of business entities' innovative development management // Economic Annals-XXI, October 2017.

9. T.T. Tatietse, P. Villeneuve, J. Ngundam, F. Kenfack, "Contribution to the analysis of urban residential electrical energy demand in developing countries", in Energy, Vol. 27, No. 6, 2002, pp. 591-606.

10. Dudin, M.N., Lyasnikov, N.V., Sekerin, V.D., Gorokhova, A.E., Danko, T.P., Bank, O.A. Technological changes as the development factor of the global and Russian energy sector International Journal of Energy Economics and Policy Volume 7, Issue 1, 2017, pp. 209-215

11. Muscas C. "Power quality monitoring in modern electric distribution systems," IEEE Instrum. Meas. Mag., vol. 13, no. 5, 2010. pp. 19–27.

12. Zhilkina Yu.V. Economic security of the Russian Federation: threats, ways to overcome them // Bulletin of Academy of economic security of the Ministry of Internal Affairs of the Russian Federation. 2010. № 7. pp. 44-50.

Intellectualization of determination of electrical loads in city electric networks

A. G. Saidkhodjaev[1], *B. Kh.* Ametova[2], *M. M.* Mamutov[2]

[1]Tashkent State Technical University named after Islam Karimov, 100095, Uzbekistan, Tashkent, University St. 2A.
[2]Karakalpak State University, 230100, Uzbekistan, Nukus, Ch. Abdirov St.1.

Abstract. This article illustrates new methods for automatically fixing and determining the calculation loads of electrical consumers, in particular the maximum load. The accuracy in the calculations is increased taking into account several factors affecting the maximum load values. It also offers a method and algorithm determination of the maximum electrical loads in urban electrical loads.

The increase of electric stations and electric transferring channel capacities, also the advancement of culture of power system operation, the application of new computing technologies and automatic system controlling the technological processes, all the above leads to the necessity of achieving precise calculation, control and increased accuracy of power measurements.

There are large number of "levels" of power control, that could include simple and detailed researches, that are often accompanied by trial tests and planned examinations. The preliminary tests of electric power represent important step in accumulating the prior data.

In order to carry out the examinations there are eight main steps have to be taken:

1. Work out the projected plan, determine important zones and centers of the company and household, also planning the effective timing.
2. Compile major data on company's (household's) power consumption, production (household) information, using standard format.
3. Carry out necessary changes and take measures to obtain the data concerning the consumption and its profitability.
4. Calculate the power balances and efficiency per consuming element and also per whole consuming unit.
5. Establish the technique of technical service using the up-to-day recommendations that would lead to power savings and raise the profitability.
6. Determine the possibilities and ways of power savings, without requiring considerable investments.
7. Work out an exact plan of actions and events in order to execute recommendations on power savings, economy and reduction of power losses.
8. Prepare the report with the given analysis of power control and necessary recommendations on actives concerning the power economy. [4]

For control over the electric power usage, there are over 540 thousands of users in Tashkent city, who has meters of following types: active and reactive power source, inductive, single and triple phase, three and four transmitters, with the current and voltage transformers.

The technical data of the meters according to the certification proved that they have accuracy level 2, in other words the maximum inaccuracy of meter measurement allowed **is** up to 2 percent's, which in Republic scale totals in 5,16 billion per year. The improvement in accuracy of measuring the output and power requires the optimal exploiting of power systems. Also it needs control over the peculiarities of power processes in power saving» systems during the heaving loads that deviate the shape of the current line and cause the current vibrations and change in directions. The accuracy measurement in the power system, depends not only on the class of the meter, but also on measuring structure of the device, i.e. how precise the device can indicate the distorted load characteristics. Therefore, one of the main problems of control and registration is the issue of choosing the registration and control systems with the accuracy levels of 0.2, 0.5, 1. [5]

Technology wise this requirement are met in the following registration systems: "ALFA"; automatic control and registration system of electric power (ACRSE); "EMKOS"- automatic system of commercial tele registration of gas, steam, electricity, coal, coke, water, petroleum products, and other various types of devices presented by "SIEMENS" and «NEVA» systems. The multifunctional micro processing electric power meter of "ALFA" type is accuracy class 0.2 and 0.5, designed for measuring the active and reactive power, current, cosine cp, current and full power.

The main functions of the "ALFA" MET program are:

* Corresponding author: anvarsaidkhodjaev@yandex.com

- Reading of commercial data from meter in the automatic mode;
- Establishing and supporting the database, storing the full information on the company's ACRSE system;
- Displaying the calculated for the definite time period power data;
- Creating various types of reports, using the read of the meter data.

The "ALFA" system is used on large industrial enterprises with 5, 10, 50, 100 consumers. Quite safe and convenient in operating, the system supplies accurate measurement of all major electric data of electric power consumption for any time period, during respectively the last 3 years of operating. The system could get connected to the control computer of Uzbekistan State Power Control through the modem, which allows charging the users of electricity on time. Currently such kinds of device are installed on large enterprises.

The "EMKOS» is the more multi-purpose system of commercial tele registration of power, designed to set automatic distant registration for consumption of various types of power and has the following operating features:

- Carries out the mathematical analysis of registration results, consequently identifying the electric power consumption tendency and organizing the recommendations on more rational usage of electric power resources;
- Manages the presented tasks with the same efficiency for any type of power and power sources, including gas, steam, electricity, coal, coke, water and petroleum products.

This system is mainly installed on large industrial enterprises, large public and institutions, that consume considerable amount of various types of electric power. This system is widely used by the electric power producing and distributing enterprises. It allows to the 5-25 percent of electric power.

The information once downloaded into the "EMKOS" database, can be then repeatedly used and processed by various programs sub-module algorithms, as for instance, for creating reports, graphs, carrying out graphic a mathematic analysis.

Thus, the "EMKOS» system carries out constant multi-channel accumulation, registration and safe-keeping of the measured data, obtained from registering devices of electric resources (meters, usage meters), processes the data to present it in the form convenient for commercial inter-accounting and electric power consumption analysis.

Since the "ALFA" and "EMKOS" systems are expensive registration systems, they are currently installed only at large electric power sources and at industrial enterprises. [3]

The most interesting system for the public as well as for household sector electric power service registration and control can be the "Israelis" model of commercial registration based on the usage of plastic cards, similar to VISA cards, that after preliminary payment to the back, can be inserted into the meter. The meter indicators the electric power consumption, also about the sum left on the card. The disadvantage of the system is its high price and uncertified crediting system of meters.

Combined, multi-functional | Dialog" meters from | SIEMENS" have flexible and module meter structure, mechanism of reading off the optical interface, digital measurement system, communication channels, memory, clock and remote controlled information systems. These devices have a digital measurement system and allow measuring and controlling of the voltage, current, frequency, power, with the accuracy level of 0.2, -0.5, -1. The disadvantage is the high price again. [3]

More than ten years are used the newest, technological and multipurpose information systems "NEVA" of SPF "Energosoyuz". The basis of the system "NEVA" is the block of registration and the control of normal and emergency operation and waste-account of energy carriers (BRCW). The elementary case of application of system "NEVA" is the modernization of old systems and creation of new systems of registration of emergency cases. Up to 64 analogs and up to 288 discrete signals can be connected to one oscillography. Signals will be automatically written down, transferred to the personal computer and submitted on the screen of the monitor in a comfortable way to the user. By the signals of currents and voltage from gauges, their working values in any point before-emergency, emergency and after- emergency process is calculated inside an oscillography by the programming way. Having connected to BRCW signals from switches and disconnection s, it is possible to display a condition of the circuit of object on the screen of the personal computer. On any discrete entrance of BRKC it is possible to connect pulse inputs from counters of the electric power, water, gas, etc. which work under the program "ASKUE". In the system "NEVA" automatic formation of the daily sheet of a production schedule is made. The device for oscillography, measurements and the analysis of electric modes and quality of the electric power is included into the complete set. Power inspections ("Energy audit") of an electric part of industrial and energy supplying enterprises, inclusion of system "NEVA" in a local network of the enterprise, definition of a place of damage, registration of operations of the switching equipment and devises of relay protection and automatics, are opened for system "NEVA". Any events fixed by the system "NEVA", inclusion or switching - off of switches, operation of protection and automatics, start of an oscillography and many other functions are sounded with the help of the program "INFORMATOR". [3]

For the public and household sector of Uzbekistan the most acceptable is the Moscow "automatic integral measuring system of electric power economy" (AISE), which is cheaper and can be used for bot h the technical and commercial registrations of electric power resources and water expenditure. The AISE is installed in Moscow since 1998. All new apartment buildings built for the past 3 years have been supplied by the AISE. In Moscow this system is installed on 1600 objects and serving over

160000 apartments, as well as in the industry. Simultaneously the large-scale measures are taken on transferring the city electric circuit to double tariff payment system. The nighttime period is established from 23:00 p.m. to 7:00 a.m... The discount night time tariff is 4 times less than the ordinary tariff. [2]

For the complex solution of task on electric power and water usage, from the actual consumption to the registration, the current data from the meters can be transmitted from AISE to the system of consumer payment reception at any time.

AISE represents the computerized radio-engineering complex with the adaptable structure for the particular application, by locating in the served territory.

AISE operates in the automatic mode (without the assisting personnel) on the level of apartments and enterprises, displaying the information on the computer screens in the controller's office. It uses the Complex of technical means (CTM), mass produced by the "ENEPEKO" for the technical and commercial registration and control over the electric power resource and water consumption. [3]

In this system old inductive meters can be replaced, without replacing the electric cable. In addition, the electronic system block is installed together with the main meter, thus increasing the accuracy of measurement. That's why this electronic single-phase, single tariff meter of type "SEB-1 M-l" is mostly used in the apartments. The electric power supply is impossible without accurate measurement of amounts produced, transferred and consumed electric power. Up to present days all the enterprises of Ministry of Electric power were structured subdivisions with the incomplete balance. The general balance was organized in general on the Ministry of Electric Power level and all the imbalances were covered up by each other. Today, after the recon structuring of the branch, when the balance of production, transportation and distribution is organized on each independent enterprise, the non-coverable imbalance will appear only due to the accuracy of measurement "level" of accuracy of the devices.

The important problem of registration and control is the issue of the measurement device (meter) ownership. At present time the meters are the properties of consumers. Thus the individual is not seriously interested in improving the existing situation of the system of registration, production, transportation and output of electric power. To restructure the registration it is necessary to put the registration means of electric power, heat , gas and water on the balance of power sellers. This would allow them replace the equipment with the technologically new system of registrations and control over the expenditures and payments.

The other, of no small importance peculiarity in electric power consumption is the change in the consumption structure.

According to the Hea State Electric Power control data, consumption structure of electric power for the years of 1970-1980 was composed of:

Population -	12-18%
Public (communal) -	22-28%
Industrial -	60%

The main electric power system loads were regulated mainly by the large power consuming industrial customers. In the past 9-10 years the consumption structure has rapidly changed and looks like this:

Population -	25-30%
Public (communal) -	30-35%
Industrial -	40%

In the general maximum the main burden is from the population. That is not limited according to the civil rights code, public and industrial burden, such as state financed institutions, water supply, transportation, heating plants, city electric, which have that same maximums with the general one. The real decrease of maximums is impossible; regulation is possible with the various tariffs application. [3]

The issue of identifying the necessary number of measurements for electric in household and public institutions should be based on the probability theory of a mathematical analysis, because the urban circuit load has a random nature and is supported by these laws. It should be mentioned that the change in the schedule of electric power loads in the household buildings depends on the normal law of distribution and requires a large number of measurements. As for the common consumers of public sector, where the nark schedule and the nature of the electric power loading have little change in time because it has insignificant difference configuration and schedule, the student distribution law could apply with taking into consideration the small portion out of large volume of data. Where the distribution S(t) is supported by the law of:

$$S(t) = C\left(1 + \frac{t^2}{n-1}\right)^{-\frac{n}{2}}, \quad (1)$$

The number "n" in the student's distribution may lake any large or small number and the S(t) approaches the normal distribution law P(t) with the increasing "n":

$$P(t) = \frac{1}{\sqrt{2\pi}} \cdot e^{-\frac{t^2}{2}}, \quad (2)$$

During the number of years there have been carried out a lot of test researches on electric power consumption for all the categories of household and public institutions considering various factors. Based on the measurements there was worked out an established whole set of common daily schedules of for household and public institutions.

The required number of measurements to obtain reliable initial data for calculating the load of the same type of consumers, according to the Student's distribution, is more than 60. Therefore, measurements were carried out continuously at the same point of the electrical network for 8 days and at 8 objects of the same type, i.e. 64 measurements at the same point of the consumer.

The researcher has showed that the load on any element of city current is the functional of large number of random factors, their numerous combinations and time changes. Therefore, all the values, calculated of the base of electric power loads, will be values of probability functions for the random variable electric loads.

The structure and the nature of the electric load graphs, obtained by the experimental way for all the kindred consumer have the same tendency of changes. Thus the probable values of electric loads could be considered as the numeral characteristics of statistic distribution.

Conclusion

1. On the basis of researches, by the author, using the experimental way of typical daily diagrams and calculation method of the total mixed loading of non-uniform consumer's probable-statistical way, a technique and the program of calculation on computers of city networks was developed under a pressure voltage of 0, 38 kV and 6-10 kV and at probable task of loadings. [1]

2. Thus, the full cycle of calculation of the systems of electro supply of cities on lower and average hierarchy is created and is fully automated, starting from the initial data, calculation of city electric networks. The offered technique allows us to reveal a real mode regime of operation of a network for each hour and to make necessary calculations, including those that are related to optimization the mode selection of operations and elements of an electro supply system, and relying on the most authentic loadings considering changes in dynamics in time. The perspective power consumption and electric loadings can be received relying on the forecast methods, stated in [2]

3. We have worked out the algorithm programs that would allow completely automate the whole design process and system projects of city electric power saving, with highly accurate data based on the probability statistical methods. There algorithms programs are based on the new technologies of control and registration the consumption of electric power.

4. The process of projecting electric networks is automated & application of computers for calculating and projecting of electro supply systems considerably allows us to reduce terms of development of the project and to exempt the qualified no production personnel from mechanical non- creative work. Besides, quality of the project documentation and labor productivity of workers of the project organizations impartment settlement expenses are increases, in comparison with the expenses at projecting systems by a traditional technique.

References

1. A.Saidkhodjaev, B.Teshabaev, Leu. Patent number IAP 04216 (IAP 2006 0404), (2010).
2. A.Saidkhodjaev, Power supply of cities. Textbook. – T.: "Fan va texnologiya", **328**, (2015)
3. A.Saidkhodjaev, Fundamentals of energy saving. Textbook. – T.: "Lesson Press", **320**, (2015).
4. A.Saidkhodjaev, Methods and tools of the energy audit. Textbook. – T.: "Noshirlik yog'dusi", **182**, (2015).
5. A.Saidkhodjaev, Issues and prospects of urban power supply systems development. Monograph. – T.: "Fan", **256**, (2012).

E3S Web of Conferences **139**, 01073 (2019)
RSES 2019

https://doi.org/10.1051/e3sconf/201913901073

Modeling of a resource-saving method of drawing

Victoria Tsypkina[1*], *Veronica* Ivanova[1]

[1]Tashkent State Technical University named after I. Karimov, Tashkent, Uzbekistan

Abstract. In this article examines the possibility of modernization of the drawing of the machine by means of preliminary modeling of resource-saving method of drawing (building information and mathematical models) and the definition of expediency of the further implementation of the physical model of the process equipment.

Improving the efficiency of cable production is an urgent problem, the solution of which is based on the resource-saving technology of drawing process.

The wire-drawing machine is a sophisticated electromechanical complex that consists of various objects, combined into a single control system. One of the main characteristics of these machines is productivity [1]. The main source of productivity gain is increase in drawing speed and reduction of time and shutdown rate, which is achieved by mechanization and automation of production process [1, 2].

The problem of increasing the economic efficiency of the technical process of drawing can be solved by combining of die and dieless drawing [7]. This method has a number of advantages, as it produces economic effect of about 25% due to increase in the lifetime of expensive imported diamond tools. Change in the design of a drawing machine (modernization) is done by adding the auxiliary unit, which creates an extra force in the area of wire output from the finishing die to obtain additional stretching to ensure compliance with the requirements imposed on finished products, i.e. match of the output diameter of a finished product. However, the proposed modernization of the drawing machine leads to a change in the design, and the solution of this problem resolves itself into analysis of the calculations performed and determination of feasibility of the structure change by modelling [9], namely:

- Building of an information (virtual) model of drawing according to both the existing technology and the proposed one;
- Building of a mathematical model of the object - drawing machine, both working (actual) and modernized.

The information model will provide a real opportunity to improve the management of wire-drawing process taking into account the geometric, mechanical and physical similarity with their observance. The actual physical model is to take into account the basic parameters of the physical effects of drawing, including secondary ones, the effect of which is taken into account in the development of effective technical solutions [5].

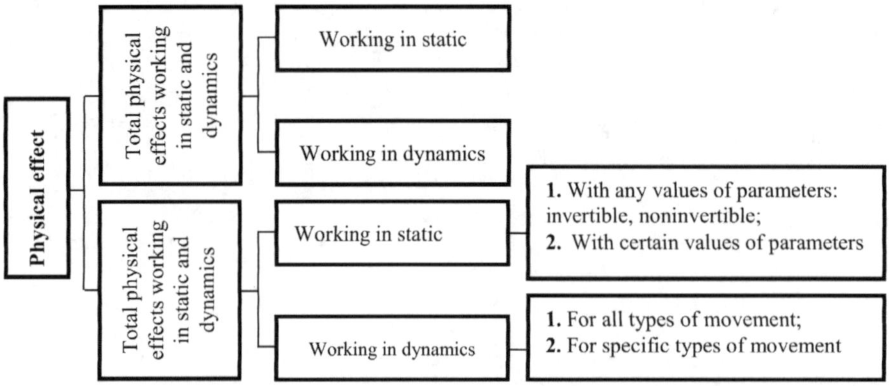

Fig.1. The scheme of creating an information model built on the basis of a phased selection of data on relevant physical effects

Development of technical solutions and analysis of the impact of physical effect is are based on a hierarchical classification of physical effects – a static information model (Fig.1), through a phased selection of data on relevant physical effects that occur during drawing: diameter of the object (product), the chemical composition of the material, and physical effect parameters are: the diameter of the rod; the diameter the wire drawn; maintenance of the constant volume of material. Deviation of a single value from established norms will result in defective products [8,10].

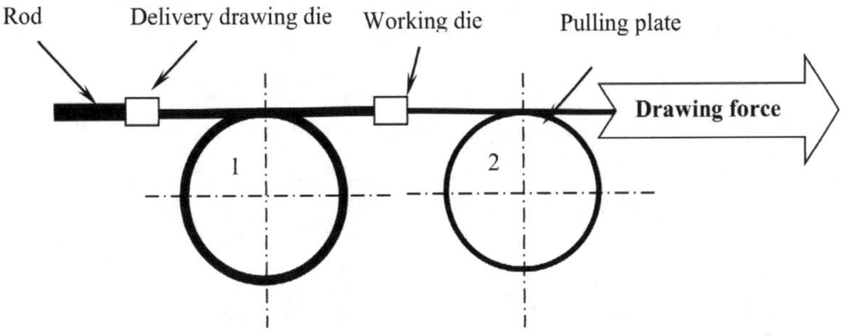

Fig. 2. Virtual model of the technological process of drawing

Research of the technology of drawing process is considered a series circuit of stages - links, each of which, acting on the metal passing through them, transmits to or extracts energy (heat) from it. In practice, the progress of the process is determined by three factors: a pulling plate, a drawing tool, and a rod.

The pulling plate is used to create a pulling force, defined as the power of drawing; the technological tool forms required dimensions and ensures clean surface [4, 5, 7]. The object affected by the mechanism is metal that is also a link of the entire model of drawing as a whole. Analyzing physical phenomena, the created information model allows evaluating a method of combining of die and dieless drawing as metal forming, taking into account increased load upon the equipment, since testing of these processes may cause accidents and damage to equipment. Reliability assurance is determined by the necessary parameters of forming by modelling with the use of similarity theory, which is represented in the general procedure of analyzing physical phenomena on the basis of similarity theory [6, 9, 10].

Thus, the virtual model of the process takes the form (Fig.2), and the listed working modes are realized by using a generalized mathematical model (1).

$$\begin{cases} W(z,t,T) = f_{11}(z)f_2(T) \\ W(z,t,T) = f_{12}(z)f_2(T) \\ f(z) = \dfrac{dp}{dl} qS(l-z) \\ f = ES\dfrac{du}{dz} \\ N = \sum_{v=0}^{m} \frac{dp}{dl}\frac{dv}{dt} \\ f(t) = \sum\limits_{n=1}^{m} d_m dt \\ \Omega = (0 < z < l, 0 < r < r_0, 0 < t < t_0); T(r,z,0) \\ \qquad = T_0(z,r) \in \Omega \end{cases} \quad (1)$$

References

1. I.M. Razumov, I.L. Merlin. *Tech. rate set. in non-ferrous metal-working industry* (1951)
2. I.L. Perlin. Non-ferrous metals, 1959, #3, p.66
3. Rep. on a gr. of exp. business trip to France on the sub. "Prod. of col., sec. and troll. alloys of AC, Tashkentkabel (Deutsche Kabel AG Tashkent).
4. L.A. Krasilnikov, A.G. Lysenko Wire Drawer, (1987)
5. L.V. Aleksandrov, N.P. Shepelev Mod. as a st. of creat. eff. Tech.sol. SPA "Poisk", (Moscow, 1992)
6. L.S. Eigenson. MPH "Sov. Nauka" (1952)
7. Util. Model Patent No. FAR 00659 dated July 22, 2010г "WDM", the Republic of Uzbekistan.
8. N.B. Pirmatov, V.P.Ivanova, D.B. Madrakhimov Energy Eff. of Draw. Eq. through En. and Res. Saving Int. Jour.of Adv.Res. in Sc., Eng. and Tech. **Vol. 6, Issue 3** (March 2019)
9. V.P. Ivanova, B.E. Yakubov Dev. of an Inf. Mod. for a Res.-Saving DM, **No. 3,** 36, (2017) URL: http://7universum.com/ru/tech/archive/item/4550 (accessed: 02.25.2019).
10. V.V. Tsypkina, A.Sh. Akhmedov, D.B. Riksitullaeva, V.P. Tsypkina (Ivanova) Dev. of a Meth. for Impr. the Param. of Cop. Wire DDP // Energy and Res. Sav. Prob. (Special Issue) No. 3-4, 205 (Tashkent, 2013)

Development of method for normalizing electricity consumption

F. A. Hoshimov[1*], *I. I.* Bakhadirov[1], *M. T.* Erejepov[2], *B. A.* Djumamuratov[2]

[1]Tashkent State Technical University named after Islam Karimov, 100095, Uzbekistan, Tashkent, University St. 2A.
[2]Karakalpak State University, 230100, Uzbekistan, Nukus, Ch. Abdirov St.1.

Abstract. The article deals with the calculation of specific indicators of electricity consumption for the production of auxiliary components the technological process. Based on the results of the authors, a method for calculating the normalization of energy in the absence of source materials is proposed. In addition, the method makes it possible to determine the norms of energy consumption given the deviation of the actual values from theoretical.

One of the determining conditions of cost reductions in industrial enterprises and improve economic efficiency of production in general are the rational use of energy resources. However, energy-efficient way of development of the domestic economy is only possible when the formation and implementation of energy efficiency programs in individual enterprises, for which it is necessary to establish an appropriate methodological and methodical base. Delaying the implementation of energy efficiency measures causes significant economic damage to enterprises and negatively affects the overall environmental and socio-economic situation [1].

Energy efficiency is one of the most important indicators of the efficiency of the enterprise as a whole, and for the steel industry, with the characteristic of great energy, and one of the reasons for its survival.

In today's day industries become more popping up new production facilities, equipped with the latest standards. Many companies have built a production process on the basis of compressor equipment. One of the main indicators of the efficiency of the compressor is the specific consumption of energy expended to produce compressed air. To assess the feasibility of using electricity compressor machine requires that its consumption is not coming out of the limits of the rules.

The refore, all companies with compressors, required to establish technically sound consumption rate of electricity to produce compressed air for each compressor machine [2-3].

The specific consumption rate of the electric power on production of compressed air consists of the following elements:

$$E = E_1 + E_2 + E_3 \tag{1}$$

where E_1 — power consumption on the compressor drive;

E_2 — power consumption on the drive of pumps of the cooling system;

E_3 — power consumption for auxiliary needs of compressor installation (lighting, ventilation, the electro crane, machines, etc.)

The electric power expense can be determined by pumps of the cooling water system:

- With counters of the electric power consumed electrical engines of pumps or by amount of the cooling water, its pressure and to the coefficient of performance of pumps on a formula

$$E = \frac{NQ_w}{3600 \cdot 102\eta_p \eta_e \eta_{trans}} \tag{2}$$

Where N — a water pressure, including also height of absorption, m of water column;

Q_w — an hour consumption of water;

η_p — Coefficient of performance of the pump;

η_{trans} - coefficient of performance of transfer is accepted on the following.

To data belt 0, 94 — 0, 98, to V-belt 0, 88 — 0, 98, to reducer 0, 88 — 0, 96.

The norm of a specific expense on production of compressed air to be established with a single compressor and the station in general needs for two periods of year, winter and summer [4-6].

For establishment of a consumption rate of the electric power on production of compressed air the materials of tests of compressor machines given to operation, based on the correct indications of instrumentations and theoretical calculations have to serve. The initial value of a specific expense of the electric power is determined by these materials and normal service conditions are established according to which for receiving norm necessary amendments to it are entered [7].

* Corresponding author: ilider1987@yandex.com

In case of lack of materials of test and data of operation the initial value of a specific expense of the electric power is determined by a formula

$$S_{ini} = \frac{1000\, L_{is} a_{cor}}{3600 \cdot 102 \eta_{is} \eta_e \eta_{trans}} \quad (3)$$

here L_{is}- isothermal operation of the compressor, m kg,

$$L_{is} = 23000\, \rho_1 V_1 \lg \frac{\rho_2}{\rho_1} \quad (4)$$

where P_1 - pressure of absorption, ama;

P_2 — the final pressure of compression, ama;

$\lg \frac{\rho_2}{\rho_1}$ — relation logarithm value;

V_1 — the initial soaked-up air volume, equal to 1 m.

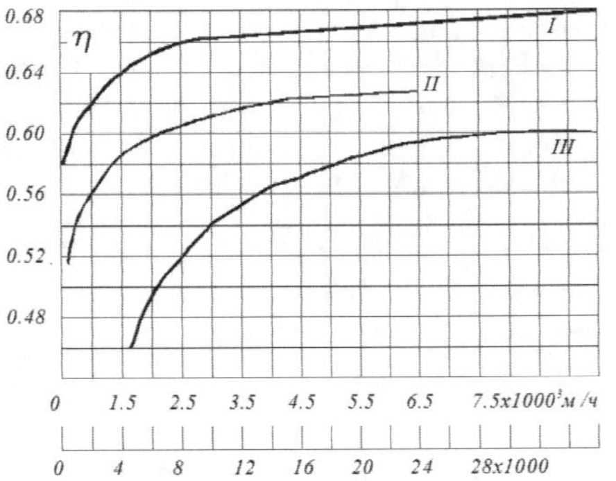

Fig. 1 Average value of isothermal coefficient of performance for various types of compressors depending on productivity at pressure of compression of 8-9 ama of loading

I - piston compressors; II - rotational compressors; III - turbo compressors.

η_{is} — isothermal coefficient of performance of the compressor, we define; with test of the compressor or fig. 1;

η_e— coefficient of performance of the electric motor;

η_{trans}— coefficient of performance of transfers from the electric motor to the compressor;

a_{cor}— the correction coefficient for transfer in normal cubic meters (nm^3), is on a formula;

$$a_{cor} = \frac{1{,}205}{\gamma_v} \quad (5)$$

where γ_v the specific weight of the soaked-up air in the valid conditions,

$$\gamma_v = 0{,}465 \frac{B_{av}}{273 + t} \quad (6)$$

where B_{av} — average barometric pressure, mm of mercury

t - temperature of the soaked-up air for the period of rationing ,°C.

The initial value of a specific expense of the electric power won't be final as the found value belongs to certain conditions: to pressure, loading and to a condition of the compressor. In practical conditions it is necessary to make a number of the amendments considering a deviation of the valid conditions from the theoretical to the found initial value of a specific expense [8].

These amendments have to be considered by the following coefficients:

a) In the coefficient considering wear of the compressor. For new compressors it is equal to unit, and for old machines of piston and rotational types not lower than 0, 90, for turbo compressors not lower than 0, 95. Decline in production below the given norms isn't allowed, and the compressor has to be exposed to capital repairs.

b) In the coefficient considering the final pressure of compression.

c) In correction coefficient, on absorption temperature.

Thus, the economy of the electric power on compressor installations has to be based on the correct organization of rationing and creation of progressive consumption rates of the electric power. Norms have to be under construction on the basis of the technical calculations corresponding to a current state of equipment, features of production in the production technology.

References

1. F.Hoshimov, K.Allayev. Energy saving at the industrial enterprises, **210** (2011)
2. G. Hoffman. Rationing consumption of energy and power balances of the industrial enterprises. **126** (1966).
3. F.Hoshimov. Methodical bases of energy saving in the industry. Tashkent, 82 (2010)
4. F.Hoshimov, I.Rahmonov. Rationing of electricity productation in the rolling of ferrous metallurgy. European Science review Scientific journal. **11-12** Austria, Vienna. (2014)
5.F.Hoshimov, I.Rahmonov. Analysis of the optimal energy indicators of electric arc furnace. «Austrian Journal of Technical and Natural Sciences», **3-4** Austria, Vienna. (2015)
6. I.Rahmonov Reducing electricity losses in elements of ferrous metallurgy power plants. Electronic scientific journal "Researches of technical sciences" **4(18)** 18-21 (2015).
7. A.Taslimov, I.Rakhmonov. The calculation of energy savings when replacing underloaded motors departments of ferrous metallurgy. Electronic scientific journal "Branch aspects of technical sciences". **2(50)** (2016).
8. F.Hoshimov, A.Rasulov, A.Taslimov, I.Rakhmonov. The current state of electrometallurgy in Uzbekistan: Monograph, "East West" Association Studies and Higher Education GmbH.– Vienna, Austria, **84**, (2017).

E3S Web of Conferences **139**, 01075 (2019) https://doi.org/10.1051/e3sconf/201913901075
RSES 2019

Features stabilization of the consumer current using electro-ferromagnetic circuits

G.R. Rafikova[1], *M.R.* Ruzinazarov[1], *S.K.* Makhmutkhonov[1]

[1]Tashkent State Technical University named after Islam Karimov, 100095, Uzbekistan, Tashkent, University St. 2A

Abstract. Electromagnetic ferromagnetic oscillatory circuits having falling sections on the amplitude characteristics formed the basis of stabilizers based on the principle of summing the currents of individual branches. A distinctive feature of such stabilizers is that the current remains unchanged both when the load resistance changes, so when the input voltage varies over a wide range. This is required by devices such as electromagnets of physical instruments, electromagnetic lenses of electron microscopes, instruments for emitting nuclear paramagnetic resonance, instruments and devices of various automation schemes, measuring and computing equipment, as well as many electrotechnological processes. Therefore, current stabilization has long attracted the attention of many researchers, and this is still one of the important problems of electrical engineering. The curve shape of the proposed current stabilizer is close to a sinusoid, the power factor is very high, since the device operates in capacitive mode.

In the modern system, in the power supply circuit of various devices, electro-ferromagnetic circuits are widely used as current and voltage stabilizers, phase and frequency converters, contactless switching devices, since devices created on the basis of electro-ferromagnetic oscillatory circuits have high operational reliability and resistance to mechanical overloads, fairly high efficiency, cosφ, manufacturing limit from several watts to tens of kW, low cost.

Interest in electromagnetic circuits has also increased with the rapid development of semiconductor technology, because in these circuits peculiar phenomena arise that are not fundamentally observed in linear systems, in particular, autoparametric oscillations are possible at the source frequency and at other frequencies (higher, subharmonic and Raman oscillations).

In this case, currents and voltages on individual elements of the circuit abruptly change, or stable negative sections can be observed on the amplitude and phase characteristics of the circuit [1, 2, 5].

The practical use of phenomena in electromagnetic circuits, where single-circuit, double-circuit circuits are considered, opens up new possibilities for their application in the creation of electrical devices.

In electromagnetic vibrational circuits (EFVCC), a coil with a ferrimagnetic core introduces nonlinearity, which leads to the occurrence of phenomena not observed in linear circuits. The simplest EFVCC, consisting of a nonlinear inductance and a linear capacitance connected in parallel, has an "S" -shaped current-voltage characteristic. By sequentially switching on a parallel line of linear inductance, it is possible to expand the zone of the volt-ampere characteristic of the negative section [3,7]. Based on the compensation of the incident section, a new principle of current stabilization is proposed by summing the current-voltage characteristics of individual circuit elements, ensuring a minimum specific gravity of circuit elements and an almost sinusoidal shape of the stabilized current curve.

The use of the incident section on the amplitude characteristics as sensitive organs in the control circuits of amplifying and thyristor devices can significantly improve the overall dimensions of stabilizing devices, while retaining the advantages of controlled semiconductor, ferrimagnetic elements and ferroresonant circuits [7-10].

To analyze the steady state of a three-element oscillatory circuit, we use the circuit equivalent circuit shown in Fig. 1. where the following assumptions are made:

Losses in EFVCC are taken into account by the active conductivity g connected in parallel with the nonlinear element. The g value is considered constant over the entire range of input voltage variation.

1. Active resistances of linear elements are taken into account.

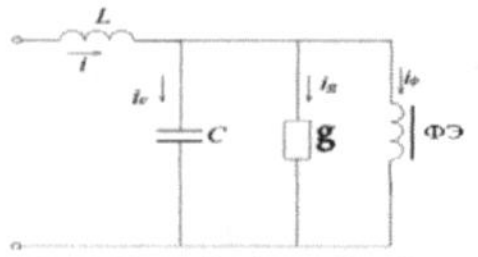

Fig. 1. EFVCC equivalent circuit

$$LCW\frac{d^3\Phi}{dt^3}+LgW\frac{d^2\Phi}{dt^2}+W\frac{d\Phi}{dt}+\frac{LK}{W}\frac{d\Phi^n}{dt}=u \quad (1)$$

here is the u-applied voltage;
W- the number of turns of the winding of the ferromagnetic element;
Φ- main magnetic flux in the core of the ferromagnetic element;

$\phi=\Phi_m \sin\omega t$;

n- degree of approximating function. For electrical steel, it is advisable to take n = 7 or n = 9.
We solve the problem for the general case:
After reducing to relative units and introducing some notation from (1.1) we get:

$$y=\frac{d^3x}{d\tau^3}+\delta\frac{d^2x}{d\tau^2}+\frac{1}{A}\frac{dx^n}{d\tau}+\beta\frac{dx}{d\tau}, \quad (2)$$

Here

$$y=\frac{u}{u_б};\quad x=\frac{\Phi}{\Phi_б};\quad б=\frac{g}{\omega c};\quad \beta=\frac{1}{\omega^2 LC};\quad \tau=\omega t;\quad U_б=\omega^3 WLC\Phi_б$$

$$\Phi_б=\sqrt[n-1]{\frac{W^2\omega^2 C}{AK}}$$

The chain equation (2), taking into account dimensionless coefficients, has the following form:

$$Y^2{}_m=\delta^2 X^2{}_m+X^2{}_m(X^{n-1}{}_m-1+\beta)^2 \quad (3)$$

$$tg\psi=\frac{\delta}{X_m^{n-1}-1+\beta} \quad (4)$$

We find the current-voltage characteristic using the expression of current in a dimensionless form $Z_m=\dfrac{I_m}{I_\delta}$:

$$Z_m=X_m\sqrt{(X^{n-1}{}_m-1)^2+\delta^2},\quad tg\psi=\frac{\delta}{X_m^{n-1}-1} \quad (5)$$

In fig. 2 - 5 show a series of characteristics for various circuit parameters and the influence of the source frequency.

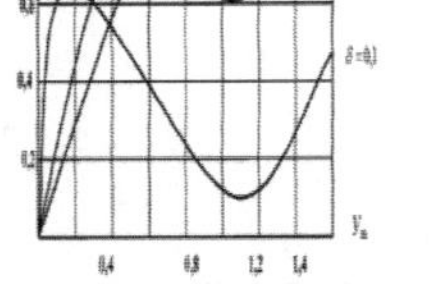

Fig. 2. Dependencies $Z_m=f(Y_m)$

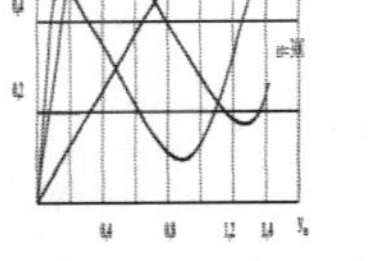

Fig. 3. Dependences $Z_m=f(Y_m)$ for various ω

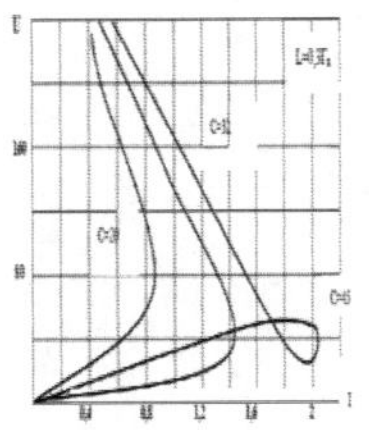

Fig. 4. Dependencies *U=f(I)* for various C

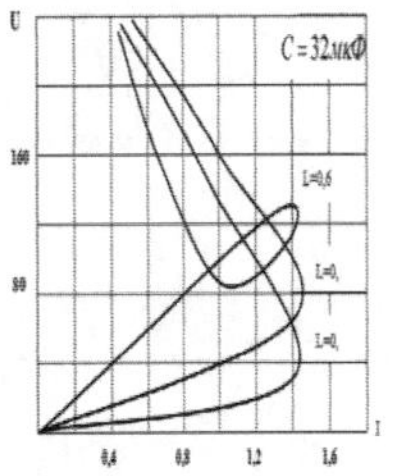

Fig. 5. Dependences *U=f(I)* for various L

The following conclusions can be drawn from an analysis of the results:

1 All parameters of the circuit and the frequency of the power source significantly affect the width of the zone of the falling part of the current-voltage characteristics of the EFVCC;

2. The width of the falling part of the characteristic substantially depends on the value of the capacitance, with an increase in this parameter the resonance point shifts to the region of high stresses and this leads to the expansion of the zone of the negative part of the characteristic;

3. By varying the value of the parameter L, it is possible to change the slope of the falling part of the amplitude characteristic.

Here is an analysis of the current stabilizer (Fig. 6.) based on the principle of summing the currents of two mutually parallel branches, the stabilization principle makes it possible to increase the output power, improve the stabilization coefficient, expand the allowable range of input voltage variation and reduce the proportion of higher harmonic components in the load current.

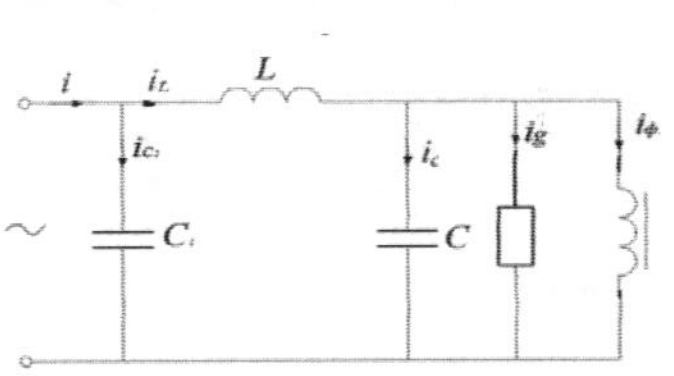

Fig. 6. Equivalent circuit CS

$$i=i_{c1}+i_c+i_\phi+i_g \quad (6)$$

Where

$$i_g=gW\frac{d\Phi}{dt};\quad i_C=CW\frac{d^2\Phi}{dt^2};\quad i_{C1}C_1\frac{du}{dt} \quad (7)$$

We have

$$i=C_1\frac{du}{dt}+C\,W\frac{d^2\Phi}{dt^2}+gW\frac{d\Phi}{dt}+\frac{K}{W}\Phi^7 \quad (8)$$

After the introduction of relative values (8) takes the following form:

$$z = \frac{1}{\beta_1}\frac{dy}{d\tau} + \frac{d^2x}{d\tau^2} + \delta\frac{dx}{dt} + \frac{x^n}{A} \qquad (9)$$

Here

$$i_б = \omega^2 W C_2 \Phi_б; \; (\beta_1 = 1/\omega^2 L C_1)$$

Taking $x = X_m \cdot \sin\tau$, when $y = Y_m \cdot \cos(\tau + \psi_n)$, from (9) we obtain

$$z_m^2 = (-\frac{y_m}{\beta_1}\sin\psi_n + \delta X_m)^2 cos\tau + (-\frac{y_m}{\beta_1}cos\psi_n + X''_m - X_m)^2 \qquad (10)$$

$$tg\psi_i = \frac{-\frac{y_m}{\beta_1}\sin\psi_n + \delta X_m}{-\frac{y_m}{\beta_1}cos\psi_n + X_m^n - X_m} \qquad (11)$$

Assuming δ=0 from (10), we have $\psi_n = 0$

$$y_m = X_m(X_m^{n-1} - 1 + \beta), \qquad (12)$$

Where $\beta = \frac{1}{\omega^2 LC}$

Calculations and experimental studies have shown that it is convenient to take the value of this coefficient equal to unity.

From (12) we have $y_m = X''_m$ (13), for this case from (10)

$$z = \frac{y_m}{\beta_1} + X_m - X^7{}_m \qquad (13)$$

Based on dependencies (13) and (14), we construct the functions

$X_m = f(y_m)$ и $Z_m = f(y_m)$ (fig. 7) for various values β_1 for n=7.

The effect of current stabilization is clearly noticeable when $\beta_1 = 1.3$. Therefore, when choosing the parameters of the FST scheme, it is necessary to ensure the value of the coefficients for $\beta = 1$ and $\beta_1 = 1.3$. Under these conditions, there is no trigger effect and there is a stable falling section on the current – voltage characteristic of a three-element ferroresonant circuit in a wide range of input voltage variations [1,2,7].

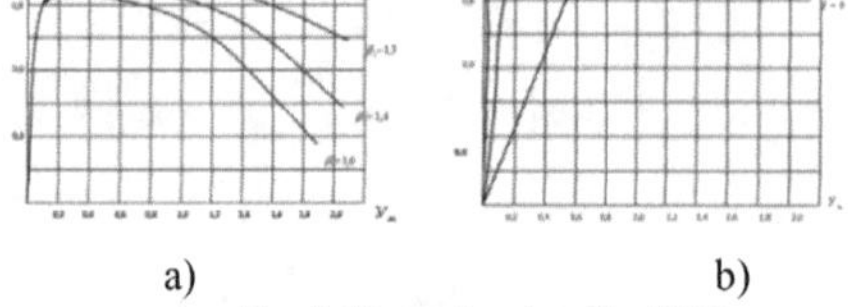

a) b)

Fig. 7. Dependencies $Z_m = f(Y_m)$

Thus, the capacitance C1 with its linear characteristic provides compensation for the falling part of the characteristic of the three-element oscillatory circuit and significantly affects the nature of the change in the amount of current consumed from the network. The obtained curves (Fig. 7) show that current stabilization, which begins with the appearance of a falling section on the current – voltage characteristic of the EFVCC, can continue even after the resonance point, if the circuit parameters provide parallel extension of the "S" -shaped characteristic with a linear characteristic of capacitance C1. Therefore, with the right choice of circuit parameters, current stabilization is observed in a very wide range of input voltage changes, which leads to an increase in the output power of the current stabilizer [6].

The proposed FST scheme has a filtering method, which is an important advantage compared to other similar devices. The shape of the stabilized current curve is slightly different from the sine wave, the proportion of higher harmonics is negligible.

References

1. Filipov E. Nonlinear electrical engineering. –M. Energy 1978-496 s
2. Bessonov L.A. Nonlinear electrical circuits. –M. High school, 1977. -343 p.
3. Kadyrov T.M., Rasulov A.N. Single-phase ferroresonant current stabilizer with a three-phase input. University News. Energy 1976, No. 11, pp. 122-124.
4. Kadyrov T.M., Rasulov A.N. Ferroresonant current stabilizer with a sinusoidal current curve. Automation and Telemechanics 1977, No. 11, from 197-200.
5. Milyakh A.N., Volkov I.V. Constant current systems based on inductive capacitive preamplifiers. - Kiev. Naukova Dumka, 1974 - 216 p.
6. M.I. Ibbadullaev, Rasulov A.N. Ruzinazarov M.R. Stabilization of current in two-phase-single-phase circuits. Problems of energy and resource saving No. 3-4. Toshkent 2018.135 - 137 p.
7. Rasulov A.N., Karimov R.CH. Operating mode stabilizer of current on active and inductive loading. European Science Review №9-10. Vina 2015. 140 – 143 l.
8. I.Rakhmonov *Technical science and innovation*: Vol.2018 Iss. 1 , Article 10. (2018)
9. F.Hoshimov, .Rahmonov, AJTNS, 3-4, 52-55 (2015).
10. F.Khoshimov, I.Rakhmonov , *Technical science and innovation*: Vol. 2018 : Iss. 1 , Article 9. (2018)

 https://doi.org/10.1051/e3sconf/201913901076

Development of optimal modes and mathematical models of energy performance of electric steelmaking production

H.T. Khakimov[1*], *Z.M.* Shayumova[1,3], *Z.Kh.* Kurbanbaeva[2], *B.M.* Khusanov[1]

[1]Tashkent State Technical University named after Islam Karimov, 100095, Uzbekistan, Tashkent, University St. 2A
[2]Karakalpak State University, 230100, Uzbekistan, Nukus, Ch. Abdirov St.1
[3]Institute of Energy of the Academy of Sciences of the Republic of Uzbekistan, 700100, Tashkent city, Republic of Uzbekistan

Abstract. The article deals with the issues of determining the optimal modes of electric steelmaking production on the basis of analysis of the graph of changes in the average power and voltage of the arc furnace in one meltdown cycle. In addition, empirical equations for determining the flow rate and specific energy consumption during the operation of process equipment under normal conditions were obtained for the main equipment of the electric arc-furnace shop using the mathematical method.

In recent years, the consumption of electric energy of electric arc furnaces (EAF) increased sharply. With the continuous growth of the specific capacity and performance of electric furnaces the efficient uses of electric energy in smelting become great importance. In the article, the experience of operating EAF capacity of 100 tons of "Uzmetkombinat" is discussed [1-3].

Melt shop JSC "Uzmetkombinat" is a sophisticated technological complex of a number of interconnected units. The workshop is composed of the 100 - tons electric arc furnaces each with transformer capacity of 95 MV · A. The shop is IME-unit complex processing of steel (ICCO), which allows you to heat the metal and implement a large number of fine-tuning operations to the required quality steel in the ladle. In addition, the shop is equipped with varietal continuous casting machines (UNRS), each of which has four streams. The cross section blanks 250x320 mm.

Furnaces operate continuously. Prolonged shutdown occurs only when the furnace repair. Load of EAF is continuous-cyclic. Cyclical nature of the work is characterized by alternating heats with stops furnace drain metal filling the furnace and the filling mixture. There are three successive passing periods for melting steel: melting, oxidation, reduction [4-6].

Figure 1 shows a plot of the average power of the EAF-100 with transformer capacity of 95 MV · A for one smelting process cycle lasts about 60 minutes. The graph indicates the average load. In fact, the load has erratic dramatically alternating character, uneven in phases.

EAF load fluctuations, particularly reactive power fluctuations cause significant voltage fluctuations in the mains, which is the larger, the greater the power of the furnace transformer and a lower short-circuit power at the point of attachment of the arc furnace. Furthermore, EAF is the source of the higher harmonic components in the network and generates 3,5,7,11 etc. harmonics [7].

Selection of the main electrical parameters of arc furnaces is complex and ambiguous. Furnaces operate in various conditions, and they conducted processes can also be different. Key performance indicators are production and specific energy consumption depend on many factors give an analytical expression for optional values of these parameters, suitable for practical calculations and it is't represent possibility. Therefore, the definition of the main parameters of electric arc furnaces based on data from these operating systems and virtually experimental formulas derived on the basis of statistical data on existing furnaces [8-9].

In this workshop achieved an increase in productivity and reduction in specific energy consumption as well as reducing the time of melting by increasing the capacity of the furnace transformer with 45 MV · A · up to 95 MV A.

Furnace , EAF -100 "Uzmetkombinat" has the following parameters.

* Corresponding author: ilider1987@yandex.com

https://doi.org/10.1051/e3sconf/201913901076

Table 1. Parameters of EAF -100 "Uzmetkombinat"

EAF-100 "Uzmetkombinat"	Project characteristics	actual characteristics
Capacity, tons	100	120
The average weight of melting, tons	100	114
The total duration of melting, min	90	60
Performance of work under the current, min	60	46
Power transformer, MVA	80	95
Limits secondary voltage, V	250-600	250-750
Diameter, mm	610	610
Nozzles of natural gas, MV	2x20	4x3,5

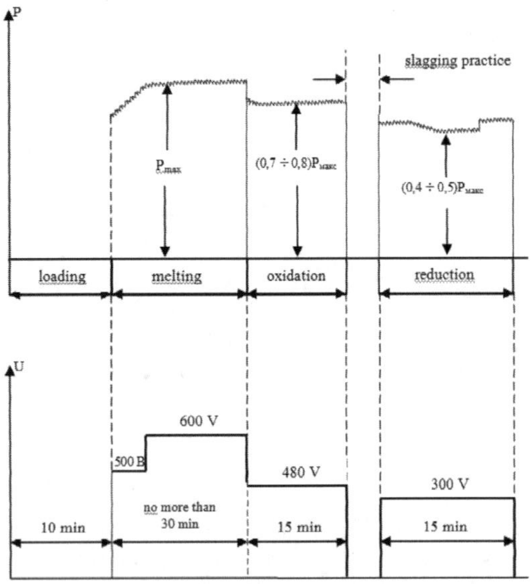

Fig.1. Schedule changes in the average power and voltage of arc furnace EAF for one technological cycle of melting

Conversion of electrical energy into heat is carried out in these arc furnaces in electric arcs, burning in the furnace between the electrodes and metal covered with slag. In EAF electric arc burn on metal and represent the load, circuit is connected on Y to the location of the zero point or a strong charge, either liquid metal.

Efficiency of power consumption EAF is provided with optimal electric mode, ie, when the electric arc furnace operates with the highest performance at the lowest power consumption. Both of these conditions do not coincide with the same operating current. However, there is a relatively small range of operating currents, within which one can get close to the optimum values for both the main indicators [10].

For determining the energy saving mode, the optimal active power is set in furnace installation.

Changing the active power is achieved by changing the operating voltage and the alternation of continuous operation at different voltages.

The following Table 2 shows the actual value of specific energy consumption by type of equipment shop EAFP for 2018.

Table 2. The actual value of specific energy consumption by type of equipment shop EAFP for 2018

№	N 2018	The specific energy consumption, kW · h / t			
		EAF	(ICCO)	UNRS	Total EAFP
1	January	363,4	35,8	27,2	426,4
2	February	354,9	28,5	26,6	410
3	March	358	31,8	26	415,8
4	April	361,3	28,6	25,7	415,6
5	May	360,1	28,3	23	411,4
6	June	357,4	27,5	24,6	409,5
7	July	355,8	30,4	23,6	409,8
8	August	360	29,5	24,9	414,4
9	September	358,5	28,6	25,1	412,2
10	October	351,6	28,5	26,9	407
11	November	352,2	25,6	25,8	403,6
12	December	369,2	22,3	27,8	419,3
Average Months		**359**	**28,8**	**25,6**	**412,9**

The table shows that 87% of electricity consumption is used in the furnace EAF, 7% in the ICCO, and 6% in the continuous caster.

Based on the above data on the energy indicators of main equipment of ESMS by using empirical equations are mathematical methods to determine the flow rate and specific energy consumption during operation of technological equipment under normal conditions.

Equipment - EAF

$$W_{EAF} = 0{,}335 \cdot \Pi_{EAF} + 1844 \text{ , kW} \cdot \text{h / month}$$

$$e_{EAF} = 0{,}335 + \frac{1844}{\Pi_{EAF}} \text{ , kW} \cdot \text{h / t}$$

Equipment – ICCO

$$W_{ICCO} = 0{,}019 \cdot \Pi_{ICCO} + 619 \text{ , kW} \cdot \text{h / month}$$

$$e_{ICCO} = 0{,}019 + \frac{619}{\Pi_{ICCO}}, \quad \text{kW} \cdot \text{h / t}$$

Equipment – UNRS

$$W_{UNRS} = 0{,}019 \cdot \Pi_{UNRS} + 371{,}45 \text{ , kW} \cdot \text{h / month}$$

$$e_{UNRS} = 0{,}019 + \frac{371{,}45}{\Pi_{UNRS}}, \text{kW} \cdot \text{h / t}$$

For lime-burning site - YOW

$$W_{YOW} = 0{,}0208 \cdot \Pi_{YOW} + 55{,}457, \quad \text{kW} \cdot \text{h / month}$$

$$e_{YOW} = 0{,}0208 + \frac{55{,}457}{\Pi_{YOW}}, \text{kW} \cdot \text{h / t}$$

The resulting mathematical models of flow and the specific energy consumption for production equipment EAFP shop allow a sufficient accuracy for practical purpose to analyze and predict the energy performance and evaluate the efficiency of their work.

Thus, analysis, optimum energy performance of electric arc furnaces leads to the following conclusions:

1.Maintain at each stage of steelmaking optimal modes EAF-100 minimizes the duration of melting and power consumption.

2. Standards of electricity consumption in the smelting of various steel grades obtained during the examination EAF -100.

3. Statistics by sampling the energy performance EAF -100 include information about melting, corresponding to the normal conditions of their conduct.

References

1. B.Kudrin, Electrician, **9**, 35-45 (2003).
2. B.Kudrin, Ind. Ener., **6**, 22-25 (2006).
3. V.Klyunya, FET, **238** (2006).
4. L.Koptsev, A.Koptsev, Ind. Ener., **1**, 18-23 (2011).
5. A.Grinev, Ind. Ener., **3**, 19-22 (2012).
6. L.Kazarinova, T.Barbasova, O.Kolesnikova, A.Zakharova, CTME, **14**, 5–11 (2014).
7. M.Bazaar, K.Shetty, NPTA, **583** (1982).
8. I.Rakhmonov *Technical science and innovation*: Vol. 2018 Iss. 1 , Article 10. (2018)
9. F.Hoshimov, .Rahmonov, AJTNS, **3-4**, 52-55 (2015).
10. F. Khoshimov, I.Rakhmonov , *Technical science and innovation*: Vol. 2018 : Iss. 1 , Article 9. (2018)

Optimization setting of steel-smelting industry in the issue of alloy steels

I.U. Rakhmonov[1,2*], *N.N.* Niyozov[1]

[1]Tashkent State Technical University named after Islam Karimov, 100095, Uzbekistan, Tashkent, University St. 2A.
[2]Karakalpak State University, 230100, Uzbekistan, Nukus, Ch. Abdirov St.1.

Abstract. It is very difficult to find any area of the state economy where metals and, in particular, steel are not required. The range of varieties has become enormous, and accordingly, the task of saving energy, fuel and resources for its manufacture becomes extremely urgent. Below in this context, we consider the procedures for the production of alloy steels, when each alloying additive has its own individual price, and the steel scheme is a narrowly specialized purpose.

The metallurgical industry consumes up to 7% of the total industrial electricity consumption of the Republic of Uzbekistan. At the same time, about 20% of the total energy consumption is spent on the production of steel and rolled products, and about 50% of the total energy consumption of resources in the industry is spent on the production of pig iron. The costs of providing energy resources account for a significant share of the cost of production, while improving the quality indicators and consumer properties of products, reducing its cost are the most important priorities for the activities of metallurgical enterprises [1].

At the same time, an increase in the quality indicators of products can be ensured only due to the additional consumption of energy resources by the corresponding new units without increasing the volume of production, which leads to an increase in its energy intensity. And at the same time, the specific energy intensity of products remains indicators of the efficiency of energy use at enterprises and refers to the main indicators of their performance. Reducing the energy intensity of products and financial costs, providing energy resources (which often does not coincide) is one of the foundations for the survival of enterprises, since it largely determines their competitiveness [2].

Alloy steels in the infrastructure of the state economy have an extremely wide range of applications. Special steel is required for the manufacture of firearms barrels, only slightly less "stringent" requirements are imposed on steel grades used for the manufacture of working elements of metal-cutting machines and other mechanisms of high accuracy and strength.

Parts and elements manufactured at a non-"precision" level for machine tools were used on "consumer goods": cars, tractors, vehicles, etc. On a statewide scale, the shares of "demand" for various types of steel can be obtained from an analysis of the reporting of their use in previous years, as well as taking into account trends in the accelerated development of certain areas [3-4].

Of course, the "quota" of state power for the manufacture of firearms is legally defined and the composition of the steel production related to this part is strictly regulated, in no way connected with the production that allows optimization. That is what we will deal with below.

Imagine the income of steelmaking in the form of the Lagrange function [2];

$$L = \sum_{i=1}^{n} C_i(b_i)T_i + \lambda W \quad (1)$$

where the T_i tariff per ton of steel with the i-th alloying additive, and the relationship of income with the corresponding production resources b_i is expressed by the dependence $C_i\ (b_i)$. λ-indefinite Lagrange multiplier; W-auxiliary Lagrange function.

Since the payment for the consumed electricity at the proposed tariffs is determined by the financial condition of the state and therefore is quite stable (close to constant), the consumer will strive to provide a predetermined volume of production of finished goods with minimal own costs ($L \rightarrow min$) and, accordingly, with maximum profit. Achieving a minimum of costs is possible with the help of resource b_i manipulation, known in economic theory [2].

By sequentially manipulating the tariff, the manufacturer provides a predetermined (demanded) volume of production of finished products with minimal resource costs. In [1], it was proved that in the optimum region of function (1), denoted by $L\ [C(b)]$, the derivative of (1) with respect to the resource corresponds to an indefinite Lagrange multiplier

* Corresponding author: ilider1987@yandex.com

$$\lambda_i = \partial L[C(b_i)] / \partial b_i \quad (2)$$

and is a marginal change in the optimal value of the objective function when changing the resource b_i. In [1], it is proposed to call λ_i an estimate of the *i*-th resource.

Consider the auxiliary Lagrange function of the W-expression (1). As applied to the problem of finding the minimum of functions, W can be a limiting balance between the resources used (fully) and their limiting values, i.e.

$$W = B_{1add} + B_{2add} + ... + B_{nadd} - B_1 - B_2 - ... - B_n = 0 \quad (3)$$

Taking into account (3), we find the conditions for achieving the minimum of expression (1). We find the partial derivatives with respect to resources and equate them to zero:

$$\left.\begin{aligned} \partial L / \partial b_1 &= [\partial C_1(b_1) / \partial b_1](T_1 - \lambda) = 0; \\ \partial L / \partial b_2 &= [\partial C_2(b_2) / \partial b_2](T_2 - \lambda) = 0; \\ &\dots\dots\dots\dots\dots\dots \\ \partial L / \partial b_t &= [\partial C_t(b_t) / \partial b_t](T_t - \lambda) = 0. \end{aligned}\right\} \quad (4)$$

From (4) we determine the conditions for obtaining a minimum:

$$\lambda = [\partial C_1(b_1)\partial b_1]T_1 / = [\partial C_2(b_2) / \partial b_2]T_2 = ...$$
$$... = [\partial C_t(b_t) / \partial b_t]T_t$$

or

$$\lambda = [\partial C_i(b_i) / \partial b_i]T_i = const \quad (5)$$

those. the minimum costs for the production of finished products by the enterprise of a given quantity will take place when the growth rates of income are equal for all resources used (or for all types of tariffs associated with them). It also means the equality of all "marginal workshops".

Consider the optimization problem that is solved by all electric consumers, while realizing that they should be as busy as possible, and electric consumers of steelmaking tend to produce the maximum amount of finished products necessary for the proposed tariffs [8].

We compose maximize the Lagrange function for different layers of consumers:

$$L' = \sum_{i=1}^{n} C_i(T_i)T_i^* + \lambda' W' \Rightarrow \max \quad (6)$$

where $\sum C_i(T_i)T_i^*$ - payments for consumed electricity, steelmaking as a result of market balancing between supply and demand (optimized part).

The relationship between demand and current tariffs T_i is characterized by the dependence $C_i(T_i)$; T_i - tariff as a constant proposed to electric consumers (each in its own time interval), to which the electric consumer answers with demand depending on its capabilities; λ' - an indefinite factor; W' – is the auxiliary function of Lagrange, for which it is advisable to use the maximum balanced correspondence between the tariffs Ti offered by energy supply companies and the tariffs T_i^* "demanded" by consumers [5-6].

In this way,

$$W = T_1^* + T_2^* + ...T_n^* - T_1 - T_2 - ... - T_n = 0 \quad (7)$$

It should be noted that the tariffs "demanded" by T_i^* electric consumers are constants for a given period of society development, since they are characterized by the general level of production and consumption of the main and auxiliary production of steelmaking.

In view of (7), we determine the conditions for attaining the maximum of the Lagrange function (6). We take the partial derivatives of this function at the regulated tariffs T_i and equate them to zero:

$$\left.\begin{aligned} \partial L' / \partial T_1 &= [\partial C_1(T_1) / \partial T_1](T_1^* - \lambda') = 0; \\ \partial L' / T_2 &= [\partial C_2(T_2) / \partial T_2](T_2^* - \lambda') = 0; \\ &\dots\dots\dots\dots\dots\dots \\ \partial L' / \partial T_t &= [\partial C_t(T_t) / \partial T_t](T_t^* - \lambda') = 0. \end{aligned}\right\} \quad (8)$$

Find the undetermined Lagrange multiplier λ' from system (8):

$$\lambda = [\partial C_1(T_1)\partial T_1]T_1^* = [\partial C_2(T_2) / \partial T_2]T_2^* = ...$$
$$... = [\partial C_t(T_t) / \partial T_t]T_t^*$$

or

$$\lambda = [\partial C_i(T_i) / \partial T_i]T_i^* = const \quad (9)$$

Condition (9) allows us to conclude that from the position of electric consumers in the steelmaking shop, the maximization of electricity consumption takes place with a set of tariffs that ensures equality of the increment rates of income of the main and auxiliary production for all types of tariffs. This indicates the "optimality" of the set of tariffs for electricity consumption and the "perception" of this set by all categories of consumers [7-8].

Consider the indefinite Lagrange multipliers λ and λ' in functions optimized by (1) and (6). If the coefficient λ is a factor balancing the distribution of resources, then the coefficient λ' is a factor balancing the distribution of tariffs for services. These coefficients are called dual variables (1). We accept the assumption:

$$\lambda' = k\lambda \quad (10)$$

where *k* is a linear coefficient that establishes for a balanced case the proportionality between electricity tariffs and the cost of energy used to produce finished products.

Using expressions (5), (9) and (10), we obtain

$$[\partial C_i(b_i) / \partial b_i]T_i = k[\partial C_i(T_i) / \partial C_i(T_i) / \partial T_i]T_i^* \quad (11)$$

If in the process of market balancing of interests all proposed electricity tariffs turn out to be in demand (T = T), then expression (11) is simplified:

$$\partial C_i(b_i) / \partial b_i = k[\partial C_i(T_i) / \partial T_i] \quad (12)$$

What can be interpreted as a proportional change in the speed of demand for services depending on the energy consumed [9-10].

Returning to the issue of equality of income of enterprises of primary and secondary production, i.e. L = L', we summarize the findings:

$$\min_{b_i \geq 0} L[\partial C_i(b_i), T_i = \max_{C} L'[C_i(T_i), T_i^*] \quad (13)$$

The presence of double equality indicates an economically competitive equilibrium at some "Saddle" point. Finding a set of tariffs that meet condition (13) is considered a mathematical solution to the dual problem with Kuhn-Tucker conditions (1).

Conclusion

The set of tariffs defined in the article for electric smelters of steelmaking is optimal. Satisfaction of consumer demand, in turn, becomes the best incentive for social development in a competitive market in society as a whole.

For steelmaking, the task of increasing revenues and profitability of the main and auxiliary production translates into the strategic task of minimizing costs, which implies a consistent transition from the cheapest to the more expensive resources.

References

1. B.Kudrin, Electrician, **9**, 35-45 (2003).
2. B.Kudrin, Ind. Ener., **6**, 22-25 (2006).
3. V.Klyunya, FET, **238** (2006).
4. L.Koptsev, A.Koptsev, Ind. Ener., **1**, 18-23 (2011).
5. A.Grinev, Ind. Ener., **3**, 19-22 (2012).
6. L.Kazarinova, T.Barbasova, O.Kolesnikova, A.Zakharova, CTME, **14**, 5–11 (2014).
7. M.Bazaar, K.Shetty, NPTA, **583** (1982).
8. F. Hoshimov, I.Rahmonov, JESR, **11-12**, 56-59 (2014).
9. F.Hoshimov, .Rahmonov, AJTNS, **3-4**, 52-55 (2015).
10. I.Rakhmonov, IJARSET, **6**, 8183-8185 (2019).

Method for determining the maximum load of consumers in city power supply systems

A.G. Saidkhodjaev[1*], *A.M.* Najimova[2], *A.K.* Bijanov[2]

[1]Tashkent State Technical University named after Islam Karimov, 100095, Uzbekistan, Tashkent, University St. 2A.
[2]Karakalpak State University, 230100, Uzbekistan, Nukus, Ch. Abdirov St.1.

Abstract. In this article, we propose a new method for determining the maximum load of electric consumers in urban electric networks, which differs from existing methods in more accurate and reliable determination of the maximum loads. Based on the determination of the maximum loads of the objects of urban electrical networks, it is concluded that the proposed methods are determined by high accuracy and minor errors.

The rapid growth of social and technical development, the widespread computerization of all aspects of life in the Republic of Uzbekistan places at the forefront not only the reliability of electricity supply, but also the creation, re-equipment and reconstruction of the entire power supply system of cities. In recent years, with the growth of the people's welfare, the electrical load in urban networks has increased.

The main problem is the accurate and reliable determination of electrical loads of urban consumers at all levels of the power supply system, since electrical loads are the basis of all electrical engineering calculations. Existing methods for calculating electrical loads: demand factor, ordered diagrams, specific load, etc. they have a large calculation error, however the size largely determines the network configuration and the choice of its parameters.

The concept of design electrical load is taken as the maximum load in the time interval of half an hour. Professors A. Glazunov and B. Kudrin offer an hour maximum load, and the author justifies in urban networks the total electrical load on the outgoing feeders of transformer substations (TS), distribution points (RP) and so on. Have a daily schedule when overlapping more flat, have cross sections of wires and cables of large quantities and heating time $3T_0$ more than 30 minutes and meet the hourly maximum load. [2]

The patented two new ways of determining the maximum (calculated) electrical loads based on the formula of the daily schedule curve for each type of consumer are proposed. To obtain characteristic (typical) daily schedules of electrical loads, numerous on-site experimental studies of electrical loads in residential and public buildings were carried out, with the right to select the objects to be examined and their number, as well as the required number of measurements to obtain initial data with a sufficient degree of accuracy. The measurement time is selected during periods of maximum load maximum, depending on the nature of electricity consumption, climatic factors and the type of consumers, are surveyed in winter and summer maximums. The required number of measurements to obtain reliable initial data for calculating the load of the same type of consumers, according to the Student's distribution, is within [1]:

$$60 < n < \infty$$

Therefore, measurements were carried out continuously at the same point of the electrical network for 8 days and at 8 objects of the same type, i.e. 64 measurements at the same point of the consumer. The essence of the proposed integrated method for calculating electrical loads is as follows:

- first, on the basis of numerous experimental studies of urban power consumers and consumers, using the mathematical apparatus of probability theory, characteristic (typical) daily graphs of electrical loads (almost a complete set) of city power consumers and power consumers were constructed.

- further on the basis of these characteristic daily schedules are determined

the equations of the curve of the function of changing the characteristic (typical) daily power consumption with 30 and 60-minute maximum (interval) of each type of electrical consumer individually, using the methods of Lagrange, Newton, Stirling, Bessel, Chebyshev, with line functions, interval method and other.

Electrical load in any element of the urban network is a probabilistic value - a function of random variable electrical loads. Therefore, the calculation of urban electrical networks should be carried out by a probabilistic-statistical method. In this case, the probability values of electrical loads can be considered as numerical characteristics of statistical distributions.

* Corresponding author: anvarsaidkhodjaev@yandex.com

Thus, the formula is defined that describes the curve of the daily load curve, which closely matches the characteristic (typical) schedule. For each type of consumer, its own curve equation and its own calculation formula. This allows to determine the maximum (rated) electrical load at the low voltage levels of the electrical networks of the city's power supply system using a set of characteristic (typical) daily load patterns and the installed (nominal) power with probability 0.7 - 0.9 to determine the 30 and 60 minute maximum electrical load, that is, the maximum (calculated) value of the electrical load and individual electrical consumers , using probabilistic-statistical calculation methods using the formula:

$$P_p = \frac{\sum\limits_{\beta=1}^{m} \cdot \sum\limits_{\gamma=1}^{l} \cdot \sum\limits_{j=1}^{n} \cdot \overline{P}_{cp}}{m \cdot l \cdot n} + \alpha \cdot \sqrt{\frac{\sum\limits_{\beta=1}^{m} \cdot \sum\limits_{\gamma=1}^{l} \cdot \sum\limits_{j=1}^{n} \cdot p^2}{m \cdot l \cdot n} - \left(\frac{\sum\limits_{\beta=1}^{m} \cdot \sum\limits_{\gamma=1}^{l} \cdot \sum\limits_{j=1}^{n} \cdot p}{m \cdot l \cdot n} \right)^2} \quad (1)$$

Where $P_{p.}$ - average maximum of average maximum loads;

σ - standard deviation;

α - normalized deviation

n - is the number of points of the measured value of the daily electrical load graph for the base time interval,

l - is the number of days of load measurement at each point of the electrical network,

m - the number of similar objects of consumers.

For each type of consumer, its own curve equation, its own maximum (calculated) electrical load.

The total electrical load transformer substations (TS) distribution point (DP) we determine by summing up daily schedules of all consumers included in this network, for each half hour or hour (48 or 24 options), regardless of the number of electrical consumers included in this electric network, defined as the largest load of all summed periods, characteristic daily schedules with a maximum of electrical loads in a half-hour or hourly time interval (48 or 24 variants). [2]

The addition of active power must be carried out according to the probabilistic-statistical method for calculating electrical loads using the formula:

$$\overline{P}_{\Sigma i} = C_{i1}\overline{P}_1 + C_{i2}\overline{P}_2 + C_{i3}\overline{P}_3 + \cdots + C_{in}\overline{P}_n = \sum_{i=1}^{n} C_i \overline{P}_i$$

$$\sigma_{\overline{P}_i} = \sqrt{C_{i1}^2\sigma_{\overline{P}1}^2 + C_{i2}^2\sigma_{\overline{P}2}^2 + \cdots + C_{in}^2\sigma_{\overline{P}n}^2} \quad (2)$$

where: C_i - to the distribution coefficients of the schemes;

Pi - mathematical expectation (standard) of the load of the i-th consumer;

σ_p - the standard deviation of the load of the i- th consumer.

The numerical characteristics of the full load will be equal to:

a) mathematical expectation

$$\overline{S} = \sqrt{\overline{P}^2 + \overline{Q}^2} \quad (3)$$

b) standard deviation

$$\sigma_{\overline{S}} = \sqrt{\cos^2 \overline{\varphi} \sigma_{\overline{P}}^2 + \sin^2 \overline{\varphi} \sigma_{\overline{Q}}^2}$$

When determining the total load of non-uniform consumers, the following expression is used for each calculated hour:

a) mathematical expectation

$$\overline{S_\Sigma} = \sqrt{\left(\sum_{i=1}^{n} \overline{P_i}\right)^2 + \left(\sum_{i=1}^{n} \overline{Q_i}\right)^2} \quad (4)$$

b) standard deviation

$$\sigma_{\overline{S_\Sigma}} = \sqrt{\cos_\Sigma^2 \overline{\varphi} \sum_{i=1}^{n} \sigma_{\overline{P_i}}^2 + \sin_\Sigma^2 \overline{\varphi} \sum_{i=1}^{n} \sigma_{\overline{Q_i}}^2}$$

Conclusion

1. New methods of calculation of electrical loads proposed by A. Saidkhodjaev. On the basis of numerous experiments, a complete set of typical daily schedules of electrical loads was developed. The method is developed and the formulas of the function of the curve of the daily schedule of electrical loads of the same type of consumers are obtained.

2. A method and technique for determining the total maximum (calculated) electrical load in the nodes of the power supply system of cities.

3. The program of calculation of schedules of daily loadings of consumers providing calculation of numerical characteristics and mathematical expectation and a standard deviation, daily loadings with half-hour intervals of time is created.

References

1. A.Saidkhodjaev, B.Teshabaev, Leu. Patent number IAP 04216 (IAP 2006 0404), (2010).
2. A.Saidkhodjaev, Power supply of cities. Textbook. – T.: "Fan va texnologiya", **328**, (2015)
3. A.Saidkhodjaev, Fundamentals of energy saving. Textbook. – T.: "Lesson Press", **320**, (2015).
4. A.Saidkhodjaev, Methods and tools of the energy audit. Textbook. – T.: "Noshirlik yog'dusi", **182**, (2015).
5. A.Saidkhodjaev, Issues and prospects of urban power supply systems development. Monograph. – T.: "Fan", **256**, (2012).

E3S Web of Conferences **139**, 01079 (2019) https://doi.org/10.1051/e3sconf/201913901079
RSES 2019

Non-contact voltage relay for switching windings of a boost transformer

E.G. Usmanov [1], *A.N.* Rasulov[1], *M.K.* Bobojanov [1], *R.Ch.* Karimov [1*]

[1]Tashkent State Technical University named after Islam Karimov, 100095, Uzbekistan, Tashkent, University St. 2A

Abstract. In power supply systems the change in the magnitude of the voltage degrades the operation of electrical appliances right up to their failure. The proposed scheme of a contactless voltage relay based on a contactless voltage relay with switching the windings of a booster step-up transformer with a time delay allows switching the windings of a magnetizing booster transformer with a voltage change at the consumer terminals. Analytical studies of a non-autonomous non-linear dynamic circuit, which is the basis of the proposed device based on the classical method of solving a differential equation, which determines the nature of the voltage change for controlling thruster switches. The change of energy parameters of the proposed stabilizer depending on the voltage is given. The article considers the principle of operation of the proposed scheme of a contactless switching device based on a contactless voltage relay that switches the windings of a booster step-up transformer. An analytical study of a non-autonomous nonlinear dynamic circuit, which is the basis of the proposed device and the characteristics $u_c = f\ t$ required for switching the thruster is presented.

One of the important indicators of the quality of electricity is the stability of the current voltage value. Using special technical means of regulation namely transformers with voltage regulation under load or step-up transformers it is possible to improve the quality of the voltage. In this case, it is necessary to change the number of turns the primary windings of the transformer or switch the magnetizing windings of the amplifier amplifier transformers. For this purpose we propose to use contactless voltage relays based on resistive nonlinear circuits [1-3].

Consider the circuit (Fig.1) of the proposed contactless switching device based on an optoelectronic voltage relay. The device consists of two proximity relays. The first (I) relay is used to turn on the voltage boost transformer (BT) winding. The second (II) relay serves to disconnect the winding of the boost transformer.

The proposed circuit works as follows: when the input voltage is reduced the first relay is activated and a control signal is applied to open the power thruster, which turns on the winding of the boost transformer in the network. With a further increase in the input voltage, the second voltage relay is triggered by shunting the optocoupler diode circuit of the first relay with its thruster optocoupler circuit terminating the access of the control signal to the power thruster; such tripping of the power thruster is still achieved as soon as the load current passes through zero, thereby discon-necting the boost coil transformer [4-5].

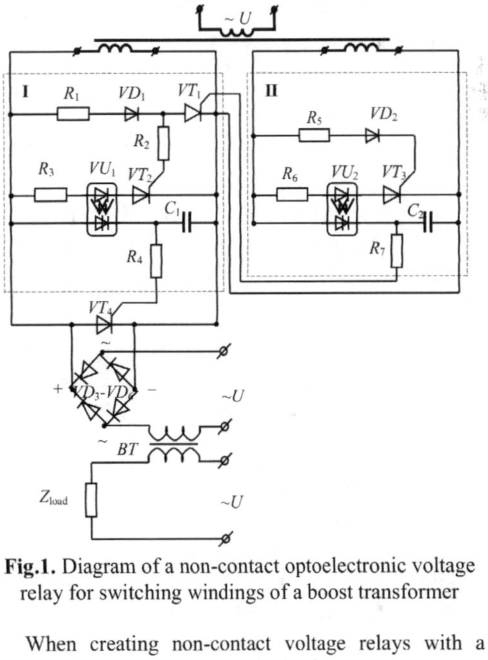

Fig.1. Diagram of a non-contact optoelectronic voltage relay for switching windings of a boost transformer

When creating non-contact voltage relays with a sinusoidal form of voltage at the load a non-autonomous non-linear dynamic circuit consisting of a diode VD_1,

* Corresponding author: raxmatillo82@mail.ru

active resistance R_1, capacitor C and resistance R_2 is used. Also on the output side there are optocouplers VU, diode bridge VD_2-VD_5 and, load Z_{load} (Fig.2) [6-7, 9].

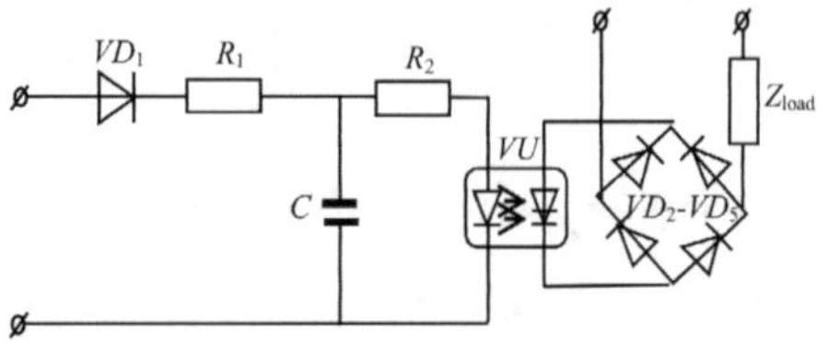

Fig.2. Non-autonomous non-linear dynamic circuit

Currently, various methods of analysis of such chains are widely used. For a theoretical analysis the equation of state of the circuit shown in Fig.2, it is proposed to use the numerical Euler method. In this case it is necessary to determine an approximate solution of the equation on a certain interval [6, 12].

$$\frac{dy}{dt} = f(t, y) \quad (1)$$

Denote by:

$$t_1 - t_0 = t_2 - t_1 = \cdots = t_n - t_{n-1} = \Delta t = h \quad (2)$$

From expression (2), each point $t_0\, t_1\, t_2 \ldots t_n$ is replaced by the differences in the relations of the derivative, then:

$$\Delta y = f(t, y)\Delta t \quad (3)$$

In the first section, $y_1 - y_0 = \Delta y$ and $\Delta t = h$, hence

$$y_1 - y_0 = f\ t_0, y_0\ h \quad (4)$$

From (4) it is known $y_0,\ t_0\ h$, then:

$$y_1 = y_0 + f\ t_0, y_0\ h \quad (5)$$

At t=t, from equation (3) we get:

$$\Delta y = f\ t, y\ h \quad (5^*)$$

From here,

$$y_2 - y_1 = f\ t_0, y_0\ h \quad (6)$$

Then,

$$y_2 = y_1 + f\ t_1, y_1\ h \quad (7)$$

For the remaining sections of the equation is written similarly,

$$y_3 = y_2 + f\ t_2, y_2\ h \quad (8)$$

$$y_n = y_{n-1} + f\ t_{n-1}, y_{n-1}\ h \quad (8^*)$$

Thus, we determine the approximate values of the points $t_0, t_1, t_2, \ldots t_n$ of the function "y" connecting these points, we obtain linear Euler curves.

We analyze this chain using the classical method [1, 12]. We take the characteristic of the diode as ideal and assume that $u = U_m \sin \omega t$. Then from the moment t_0 to t_1 the diode is open and the circuit equation has the following form:

$$U_m \sin \omega t = R_1 i + U_c, \quad (9)$$

Given that $i_c = C\frac{du_c}{dc}$ we have:

$$U_m \sin \omega t = R_1\ c\frac{du_c}{dt} + \frac{U_c}{R_2}\ + U_c \quad (10)$$

From here,

$$\frac{du_c}{dt} = \frac{1}{R_1 C}\ U_m \sin t - U_c(1 + \frac{R_1}{R_2} \quad (11)$$

Equation (11) is solved by the numerical Euler method

$$U_{c(k+1)} = U_{ck} + f\ U_{ck} * t_k\ * h \quad (12)$$

From here,

$$\mathrm{f}(U_{ck}, t_k) = \frac{1}{R_1 C}\ U_m \sin \omega t - U_c\ \ 1 + \frac{R_1}{R_2} \quad (13)$$

Here $k = 0, 1, 2, \ldots$; h - is the integration step.

Until t= t_1, the voltage across the capacitance is determined by (12), taking into account the initial conditions. From the moment t= t_1 the diode opens and until the moment t_2 the voltage on the capacitor remains at the voltage level for the moment t_2, the voltage on the capacitor remains at the voltage level for the moment t_1, from the moment t_2 the diode opens again and the voltage on the capacitance is again described by dependences (12) with other initial condition.

Figure 3-5 shows the curves of the voltage across the capacitance for various values of the circuit parameters at $U_m = Const$ and the integration step h=0,001 [11-12].

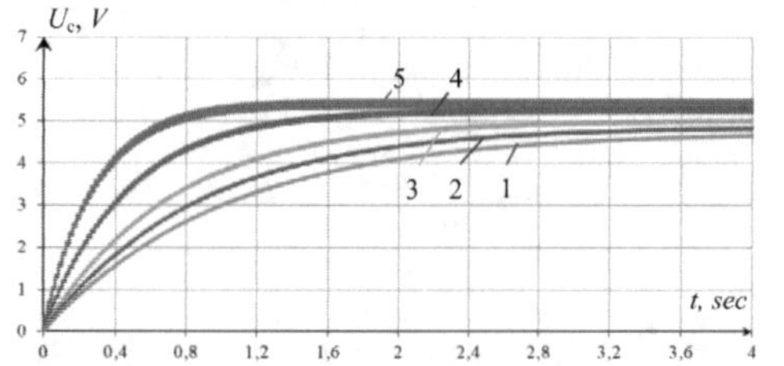

Fig.3. A family of curves of variation of voltage across a capacitance as a function of time at C=1 μF; R_2 = 3,0 kOhm, for resistance values R_1: 1 - 0,3 kOhm; 2 - 0,5 kOhm; 3 - 0,8 kOhm; 4 - 1,0 kOhm; 5 - 1,2 kOhm

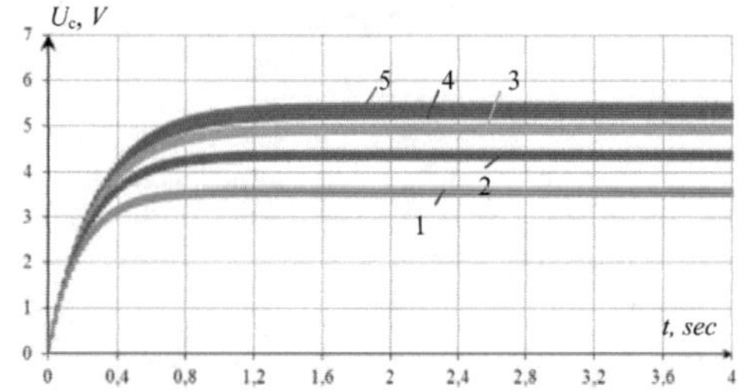

Fig.4. A family of curves of variation of voltage across a capacitance as a function of time at C=1 μF; R_1 = 3,0 kOhm for resistance values R_2: 1 - 3,0 kOhm; 2 - 2,0 kOhm; 3 - 1,0 kOhm; 4 - 0,5 kOhm; 5 - 0,25 kOhm

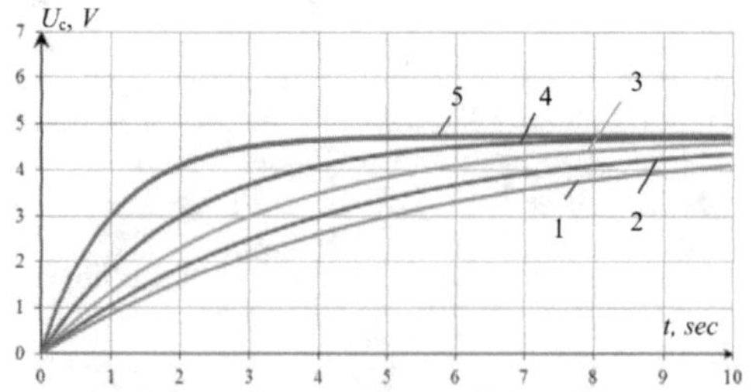

Fig.5. A family of voltage variation curves for capacitance as a function of time at R_1 = 2,0 kOhm; R_2 = 3,0 kOhm, for capacitance C: 1 - 1,0 μF; 2 - 2,0 μF; 3 - 3,0 μF; 4 - 4,0 μF; 5 - 5,0 μF

It can be seen from the curves that by changing the parameters R_1 and C of the circuit in question, it is possible to adjust within relatively large limits, the time the capacitor is fully charged, which is necessary for the LEDs of the optocoupler shown in Fig.2, to reduce the false response of the relay during switching changes in the supply voltage.

https://doi.org/10.1051/e3sconf/201913901079

Figure 6 shows the voltage change across the capacitance of a dynamic circuit during simulation in the Matlab Simulink package.

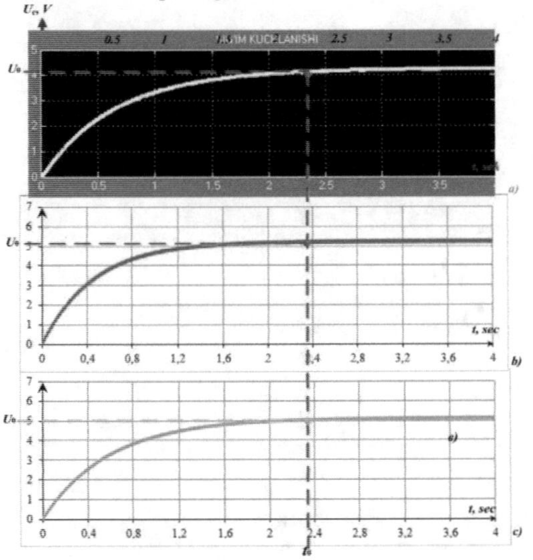

Fig.6. Comparative analysis of a nonlinear electric circuit based on modeling: *a* - in the Matlab Simulink package; *b* - Electronics Workbench; *c* - a curve built on the basis of theoretical research

A comparative analysis of the dependences $U_c = f\ t$ (Fig.6) shows that the difference in characteristics is 1.5–2.0%. This indicates that the used method of theoretical research is acceptable for this chain [7-8, 10].

Fig.7 shows the waveforms of the voltage input-output voltage stabilizer. With U_{input}=175 V (voltage deviation from the nominal -20,4%), U_{output}=214 V (-2,72%). With U_{input}=241 V (+8,71%), U_{output}=224 V (+1,78%). Thus, when the input voltage varies from -20,4% to +8,71% of the nominal value using the proposed device, we obtain the output voltage in the range from -2,7% to +1,7% of the nominal voltage, that meets the required standards.

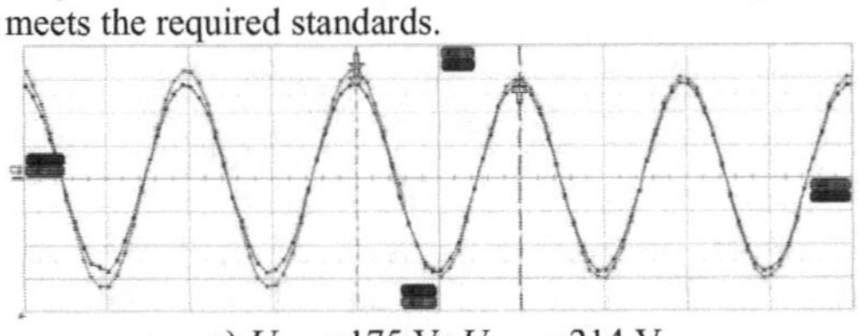

a) U_{input}=175 V; U_{output}=214 V

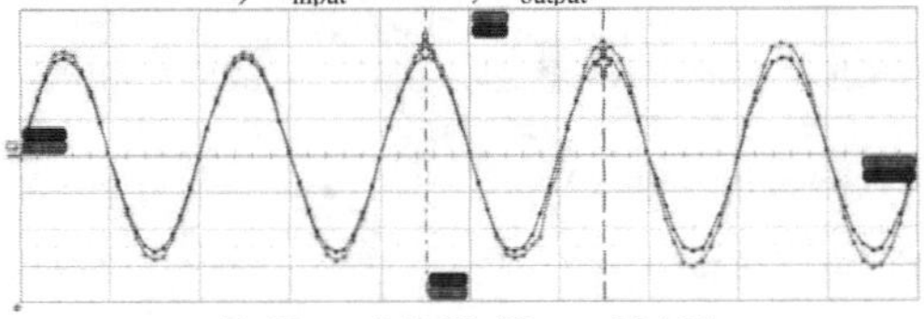

b) U_{input}=241 V; U_{output}=224 V

Fig.7. Waveform of voltage input-output voltage stabilizer

Figure 8 shows the main energy indicators of the boost booster transformer. These characteristics show that when the load changes over a wide range the value of energy indicators changes within $cos\varphi$=0,96; η=0,8÷0,95.

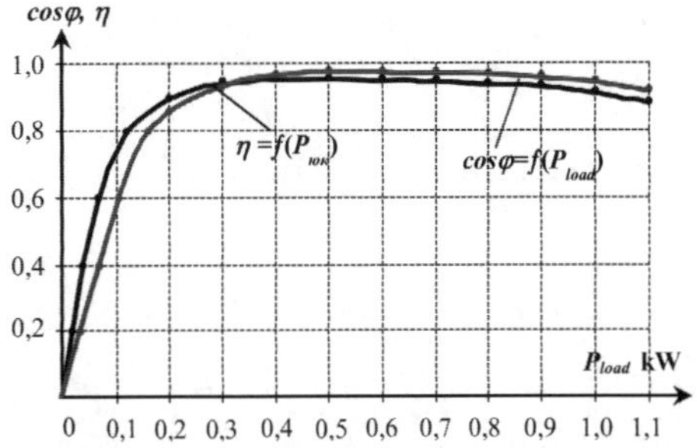

Fig.8. Characteristics of the dependence of $cos\varphi$ and the coefficient of performance of the boost voltage transformer

Conclusion

The investigated voltage stabilizer circuit provides voltage stability of consumers within acceptable limits of ± 5% of the nominal value with a change in the input voltage within 175 ÷ 241 V.

The control of the windings of the boost booster transformer using non-contact switching show high energy characteristics.

The proposed voltage relay circuit provides non-contact switching of the boost transformer windings when the input voltage changes over a wide range the output voltage has a sinusoidal shape and varies within ± 2% of the rated voltage.

References

1. EG Usmanov, EH Abduraimov, RCh Karimov, Bulletin of TSTU, **3-4**, 48-51 (2012).
2. RCh Karimov, Collection of materials of the IV international scientific conference "Modern Trends in Technical Sciences", 30-33 (2015).
3. EG Usmanov, EH Abduraimov, RCh Karimov, Problems of computer science and energy, **2-3**, 37-41 (2006).
4. MK Samoxvalov, Textbook, UlSTU,126 (2003).
5. MK Bobojanov, EG Usmanov, EH Abduraimov, RCh Karimov, JESR, **1-2**, 210-212 (2018).
6. RCh Karimov, ICh Karimov, JYS USA, **5**, 106-109 (2016).
7. UD Karix, 107 (2002). http://www/fineprint/com/
8. RCh Karimov, JESR, **9-10**, 144-146 (2015).
9. RCh Karimov, Collection of materials of the IV international scientific conference "Modern Trends in Technical Sciences", 33-35 (2015).
10. AN Rasulov, RCh Karimov, JEE, **4**, 179-183 (2015).
11. MQ Bobojanov, RCh Karimov, Bulletin of TSTU, **4(101)**, 53-57 (2017).
12. RCh Karimov, EG Usmonov, JA Xolbekov, Bulletin of TSTU, **2(107)**, 75-82 (2019).

The role of information in power management tasks

I.U. Rakhmonov[1,2*], *K.M.* Reymov[2], *Z.M.* Shayumova[1,3]

[1]Tashkent State Technical University named after Islam Karimov, 100095, Uzbekistan, Tashkent, University St. 2A
[2]Karakalpak State University, 230100, Uzbekistan, Nukus, Ch. Abdirov St.1
[3]Institute of Energy of the Academy of Sciences of the Republic of Uzbekistan,700100, Tashkent city, Republic of Uzbekistan

Abstract. The more complex the system designed to serve the person, the higher the requirements for its high-quality functioning and its reliability itself. Reliability monitoring should reach the level of the highest "survivability" if we consider systems such as "power systems", i.e. systems whose services mankind does not intend to refuse in the foreseeable future. The presence of naturally wearing elements and seed details actualizes the issue of developing a certain "reliability index" and, accordingly, the procedure for its calculation and current display. Below we will consider possible solutions to the problems posed by the example of the electric power system. Since we are talking about "information", first of all, on the basis of rigorous mathematical calculations, we will try to establish a connection between the measurement information and the so-called Shannon information. Summing up the above, we can outline the goals of research in the near future.

The advantage of the canonical, Shannon theory of information is its universality, independence from the specific semantic content of information. The amount of information, determined by Hartley and Shannon and associated with a change in the probability of the state of the object, usually ignores the coordination with the control tasks, with the achievement of certain goals [1].

In technical systems, the control task is to maintain some adjustable parameters (for example, the frequency and voltage of nodes in power systems) at the level of the required values, and the regulation itself is based on the so-called measurement information. The fundamental question is whether there is a connection between the canonical form of information and its measuring, parametric form. In our opinion, such a connection should be sought by analyzing the value characteristics of information.

The value of information entirely depends on its reception, i.e. from the consequences of its perception by the receptor (in particular, an automated controlled power system). The value of information was the subject of research by a number of scientists - A.A. Kharkevich, M.M.Bongard, R.L. Stratonovich, B.N. Petrov and others. The generality of their concept lies in a statistical approach to information, and the wording of M. M. Bongard should apparently be recognized as the most specific interpretation of the value for the conditions for the further use of information for management [2-4].

$$V = \log_2(P^{/} / P) \qquad (1)$$

where V is the value of information,
P and $P^{/}$- the probability of achieving the goal before and after receiving information.

If the entropy measure of information, determined by Hartley or Shannon (through probabilities P and P '), is an unambiguous and universal quantity, then this universality does not extend to the concept of the value of information (1), if only because the effect of goal achievement in various control systems can be different (incommensurable), and the analysis of the value of information without regard to its receptor becomes abstract [5-7].

In complex technical systems, to which energy systems (ES) should be attributed, the goal of control can be considered to be the achievement of a certain state that is optimal according to some previously justified criteria (in particular, economic). Note that for an ES at each instant of time, such an optimal state is unique. Any «removal» from the optimal state can be estimated by the corresponding damage (E), and «approaching» to the optimum (from «under the optimal» state) by the wins (*E*), which are similar in meaning and significance. In particular, damage and gain may well be purely economical. The distance from the optimum itself can be expressed in terms of deviations of the controlled parameters from their «optimal» (corresponding to the system optimum) values, i.e. through parametric (measuring) information [8-9].

Having determined the optimum of the system, i.e. outlining the purpose of control, we specified the use of parametric information (regulation parameters), the very meaning of the information and the method of its measurement, which is fully consistent with the approach [10]. Using now the concept of the system control effect, we find the relationship between the parametric information defined above and the Shannon

* Corresponding author: ilider1987@yandex.com

information, i.e. between "management" information and canonical information.

We will describe the state of the ES near the optimum by a certain linear economic characteristic, which is a function of a set of adjustable parameters π:

$$E = f(\pi_1^{'}, \pi_2^{'}, ... \pi_j^{'}, ..., \pi_m^{'}) = \alpha_1\pi_1^{'} + \alpha_2\pi_2^{'} + ... + \alpha_j\pi_j^{'} + ... \alpha_m\pi_m^{'} \quad , \quad (2)$$

Where: E - is the current (actual) value of the economic indicator of the ES;

$\pi_j^{'}$ - current (actual) values of the regulation parameters;

α_j - weighting factors (statistical, regression - in particular).

We will consider the considered state sufficiently «close» to the optimal one, in which the ES functions most economically (with the economic indicator E_o):

$$E_o = f(\pi_1^o, \pi_2^o, ... \pi_j^o, ..., \pi_m^o) = \alpha_1\pi_1^o + \alpha_2\pi_2^o + ... + \alpha_j\pi_j^o + ... \alpha_m\pi_m^o \quad (3)$$

Here π_j^o - are the system parameters that determine its optimal state, we will call optimal.

The assumption of the proximity of the current state of the ES to the optimal E_o is necessary to comply with the linearity of the problem, i.e. invariance of coefficients α_j.

Consider a situation where all but one of the control parameters has their optimal value, and the j-th parameter has its current value that is different from the optimal one α_j. Denote the deviation of this parameter from the optimal by

$$\pi_j = \left|\pi_j^{'} - \pi_j^o\right| \quad (4)$$

Obviously, this will also lead to a deviation of the economic performance indicator of ES E_j from its optimal value:

$$E^{/}_j = E_j - E_o = \alpha_j = \left|\pi_j^{'} - \pi_j^o\right| = \alpha_j\pi_j \quad (5)$$

For brevity, we will call the $E_j^{/}$ effect of regulating the ES with respect to the parameter π_j. If the parameter π_j can be in n specific discrete states $P(\pi_{ji})$, then the resulting efficiency $E_{j\Sigma}$, as is known, can be calculated as the sum:

$$E_{j\Sigma} = \alpha_j \sum_{i=1}^{n} P(\pi_{ji})\pi_{ji} \quad (6).$$

Analyzing the structure of expression (2), we come to the conclusion that the system-wide *(E_{ES})* regulation effect can be obtained by summing the regulation effects on individual components, i.e. it has the property of additivity:

$$E_{ES} = \sum_{j=1}^{m} \alpha_j \sum_{i=1}^{n} P(\pi_{ji})\pi_{ji} \quad (7).$$

Since the sum (7) completely repeats the properties of the (6), for simplicity we return to the analysis of the j-th term. Note that in the transition from the case of a discrete change of a parameter π_j to its continuous change, expression (6) will change

$$E_{j\Sigma} = \alpha_j \int_o^{\pi_{j\max}} P(\pi_j)\pi_j d\pi_j \quad (8)$$

As follows from (5), $E_j = 0$ for the optimal value of the parameter π_j^o, i.e. with its very specific meaning, the probability of reaching which with targeted regulation $P(\pi_j^o) = 1$. Recall E_j that it can be interpreted as an optimization effect if an ES transfers from a suboptimal state to an optimal one, or as damage from nonoptimality if there is a reverse transition. The optimal state should be considered as completely ordered, whose entropy is equal to zero. The entropy of a disordered state, as is known, is associated with the probability of this state and for our example can be represented as follows:

$$H_j = \sum_{i=1}^{n} P(\pi_{ji}) \log_2 P(\pi_{ji}) \quad (9)$$

and for the continuous case

$$H_j = - \int_o^{\pi_{j\max}} P(\pi_j) \log_2 P(\pi_j) d\pi_{ji} \quad (10)$$

Recall the property of additivity of entropy, which allows the system to calculate the entropy as the sum of the enropies of its individual elements, which for our example gives

- in the discrete case

$$H_{ES} = -\sum_{j=1}^{m}\sum_{i=1}^{n} P(\pi_{ji}) \log_2 P(\pi_{ji}) \quad (11)$$

continuously

$$H_{ES} = \sum_{j=1}^{m} \int_o^{\pi_{j\max}} P(\pi_j) \log_2 P(\pi_j) d\pi_j \quad (12)$$

The last formulas (9-12) are well-known expressions for Shannon's informational entropy, which is defined for a set of random variables. When the probability of the state changes from $P(\pi_j)$ to 1 (the system transitions from under the optimal state to the optimal one), the Shannon information change

$$I = O - H_{ES} = \sum_{j=1}^{m} \int_o^{\pi_{j\max}} P(\pi_j) \log_2 P(\pi_j) d\pi_j \quad (13)$$

It is this information by Shannon that we use to transfer ES from under the optimal state to the optimal one and at the same time, the effect of the E_{ES} control is achieved. Naturally, practical regulation is carried out not according to Shannon information, but according to parametric (measuring), by minimizing the parameters π_j - see (5). Obviously, there is some correspondence between these two forms of information, which can be revealed through the effect of regulation (control).

Considering the property of the additive effect of system regulation noted above and the well-known property of additive entropy, we consider the correspondence (== -is the sign of correspondence) of the effect and information using the example of one (*j*-th) parameter:

$E_{j\Sigma} == I_j$ or

$$\alpha_j \int_o^{\pi_{j\max}} P(\pi_j)\pi_j d\pi_j == \int_o^{\pi_{j\max}} P(\pi_j)\log_2 P(\pi_j) d\pi_j \quad (14)$$

To simplify the analysis, we differentiate the right and left sides of the expression (14):

$$\alpha_j P(\pi_j)\pi_j == P(\pi_j)\log_2 P(\pi_j) \quad (15)$$

Comparing the right and left sides of expressions (14,15) with respect to probability factors, we can conclude that they are affinely similar, i.e.

$$\frac{P(\pi_j)}{P(\pi_j)\log_2 P(\pi_j)} = idem \quad (16)$$

The factor αj of the left-hand side of expression (14) can be considered as a scale factor providing a geometric similarity. The only element in (14) that does not have a similar one is the regulation parameter π_j. This means that, up to a certain constant (*C*), the parametric (measuring) information, on the basis of which targeted optimization control in the ES is carried out, corresponds to the information according to Shannon:

$$C\pi_j = I_j \quad (17)$$

We can note the validity of relations (14) and (17) for the boundary conditions. Let the ES with respect to parameter *j* be brought to the optimum state. Moreover $\pi_j = 0,\quad P(\pi_j) = 1,\quad \log_2 = P(\pi_j) = 0,\quad I_j = 0,$ the right and left sides of expressions (14) and (17) are identically equal to zero. This means that in the optimum state there is no (and not needed!) information for further control. Consider another example - a certain parameter π_v does not affect the formation of the economic effect of management, i.e. its coefficient $\alpha_v = 0$ or we are dealing with useless information. In this case, the definition of the goal has not been formulated with respect to this parameter, therefore, from the point of view of the ES control, the parameter π_v value can be any. It is known that the probability of any parameter value is always equal $I(P(\pi_j) = 1$, $\log_2 1 = 0$ and there is no regulation information for control $I_v = 0,\quad C\pi_j = 0,\quad t.i.\quad C = 0$ for this parameter should be performed.

The presence on the left side of expressions (14) and (17) of a physically measurable parameter is logical in the sense that a transition from abstract, conical information to concrete, physically measurable is necessary. An important conclusion that follows from the above is that such a «bridge» between the conical form of information (according to Shannon) and its measurable, parametric form exists for pre-agreed conditions and lies in some constant *C*.

Thus, when fixing the method of measuring and using information (for control purposes) and when determining the control goal as achieving a certain (optimal) state of a controlled object, Shannon information corresponds to measuring, parametric information (or vice versa) up to a constant.

Fixing the method of using information logically leads to the question of the effect of its use, of its value from a management perspective. If information on Shannon is a factor for determining the bandwidth of information transmission channels in the automated control system, then the determining factor in the organization of the management itself is its value. In this regard, it is legitimate to talk about the development of the informational aspects of management, about the construction of management systems adequate to the processed information. In theoretical terms, the justification of «informational» management methods and the development of an informational management theory are required.

References

1. V.Sokolov, E&T, **2**, 46-52 (1982).
2. V.Sokolov, E&T, **2**, 41-49 (1985).
3. V.Venikov, Electricity, **1**, 1-5 (1986).
4. L.Koptsev, A.Koptsev, Ind. Ener., **1**, 18-23 (2011).
5. A.Grinev, Ind. Ener., **3**, 19-22 (2012).
6. L.Kazarinova, T.Barbasova, O.Kolesnikova, A.Zakharova, CTME, **14**, 5–11 (2014).
7. M.Bazaar, K.Shetty, NPTA, **583** (1982).
8. I.Rakhmonov *Technical science and innovation*: Vol. 2018 Iss. 1 , Article 10. (2018)
9. F.Hoshimov, .Rahmonov, AJTNS, **3-4**, 52-55 (2015).
10. F.Khoshimov, I.Rakhmonov , *Technical science and innovation*: Vol. 2018 : Iss. 1 , Article 9. (2018)

Method of choosing the unification of cable sections of electric network cables under conditions of load development uncertainty

A.D. Taslimov[1], *A.S.* Berdishev[2], *M.V.* Melikuziev[1], *F.M.* Rakhimov[2]

[1]Tashkent State Technical University named after Islam Karimov, 100095, Uzbekistan, Tashkent, University St. 2A
[2]Tashkent institute of irrigation and agricultural mechanization engineers

Abstract. The article analyzes the optimization criteria used in the conditions of uncertainty of the initial information, selects the number and values of standard cable sections under uncertainty, the development of the load over time using the minimax criteria of Wald and Savage, and also makes recommendations on the number and values of standard sections of distribution cables electrical networks

When predicting the optimal development and construction of distribution electric networks (DEN), one should take into account the technical and economic possibilities of choosing the optimal parameters with an increase in electrical loads over a fairly wide range. In this case, on the basis of mathematical models of DEN, the optimal network parameters are identified that correspond to the series of changes in the initial information and the optimal construction of the DEN for a number of parameters is revealed, which on a certain scale reflects the development of this network in time. The resulting sets of optimal parameters determine the rational construction of the network and the development strategy of the parameters [1].

Under these conditions, the problem of choosing the optimal parameters of the DEN is solved taking into account the uncertainty of a complex of factors characterizing the development of the load in the considered time period. Such factors are: the normative coefficient of capital investment efficiency, the initial load ratio in fractions of the project, the laws of load growth over time; annual uniform and uneven load growth; the year of reaching the design load from the moment the transformer substation (TS) was commissioned; the considered period of time, for example, the estimated life of the designed network before its reconstruction [2].

According to [3], the choice of the optimal values of the parameters of the DEN should be based on the optimization criterion in the conditions of uncertainty of the initial information. Depending on a varying degree of uncertainty, information is classified as “information situations”. The information situation is the separation of the uncertainty of the states of nature from a given set S = {$S_1,\ldots,S_n$} of which decision-maker (DM) has at the time of obtaining the solution.

The categories of information situations are developed and seven categories of information situations are proposed. If the information situation is characterized by a fuzzy set of states of nature, then the choice problem is solved by methods of the theory of fuzzy sets [4]. Depending on whether the information belongs to a specific information situation, it is necessary to use one or another criterion for choosing a solution that is widely used in solving practical problems [5; 6].

When the probabilities of all values of the indefinite factor are known (the first information situation), the methods of the theory of statistical decisions are used, namely the Bayes criterion. Moreover, the optimal solution corresponds to the maximum (minimum) of the mathematical expectation of the efficiency vector, weighted by the probabilities of the state of the medium. For discrete given probabilities of states of nature, the Bayes criterion is written as follows [5]:

$$\sum_{j=1}^{n}(p_j \cdot Y_{jk}) \to \min ,$$

where p_j - is the probability of the state of nature, and $\sum_{j=1}^{n} p_j = 1$, where n - is the number of all considered states of nature; Y_{jk} - is the efficiency vector (functional) for the κ-th strategy under the j-th state of nature.

Usually, under conditions of uncertainty in the development of the load, the distribution density $F(\beta)$ of the annual growth rate of the load β, which is regarded as a continuously distributed random variable, can be known. In the analytical representation of the dependence of the efficiency vector $Y(\beta)$ on the random variable β, the Bayes criterion is written as follows [5]:

$$\int_{q_{\min}}^{q_{\max}} F(\beta)\cdot Y(\beta)\cdot d\beta \to \min$$

In the first and second information situations, the optimal solution is selected based on the Bayes criterion.

In the third informational situation, the solution is chosen based on the theory of relations [3].

The Bernoulli – Laplace criterion is used to select a solution in the fourth information situation. Moreover, if there is no data on the state of nature, then it is believed that one of the states of nature is more likely than any other. This criterion is recommended as the simplest one for determining the probabilities of the state of nature. The Bernoulli – Laplace criterion is mathematically described as follows [3]:

$$\frac{1}{n}\cdot\sum_{j=1}^{n} Y_{jk} \rightarrow \min$$

In the fifth informational situation, with the antagonistic interests of nature, the Wald or Savage criterion is used for the solution.

According to Wald's criterion, the optimal solution is selected by the condition:

$$\max_j Y_{jk} \rightarrow \min$$

When applying the Savage criterion, the concept of risk R_{jk} is used and the optimal solution is selected by the condition [3,6]:

$$R_{jk} = Y_{jk} - \min_j Y_{jk}$$

According to the Savage criterion, the optimal strategy is the one in which the risk is minimal in the worst conditions [6]:

$$\max_j R_{jk} \rightarrow \min$$

A common drawback of choosing a solution for both criteria is that the solutions obtained are not very sensitive to additional information about uncertain factors.

In the sixth informational situation is the Hurwitz criterion. The condition for choosing a strategy for this criterion is [3]:

$$(\xi \cdot \max_j Y_{jk} + (1-\xi)\cdot \min_j Y_{jk}) \rightarrow \min$$

where ξ is the pessimism coefficient chosen by the DM between zero and one.

As a result of the analysis and processing of additional information, the probabilities of a particular growth rate of the load (state of nature) were obtained, therefore, the Bayes criterion is the criterion for choosing a solution. But the question arises about the reliability of the results: there are a lot of urban DEN, they are built in a variety of conditions (the area of the city district, surface load density, consumers, urban planning requirements) and with various parameters (voltage, power, cross-section). Under these conditions, it is impossible to reliably determine the probabilities of the pace of development.

Thus, to make a more informed decision, one can propose using the following method: in any information situation from the first to the fourth (it is these information situations that characterize the state of the initial information of the technical system and, in particular, the power supply system), the decision is determined by three criteria: Walda, Savage, and Bernoulli-Laplace, which do not take into account the probability of uncertain factors, despite the fact that the criteria of Wald and Savage are the selection criteria in the fifth information situation (antagonistic nature). If the decisions obtained by these three criteria coincide, then this decision is taken as rational; if the obtained solutions do not coincide, then the rational solution is determined by the zoning method.

When applying minimax criteria, the task of choosing the optimal number of cable cross-sections is solved during the "game with nature", in which the number of cable cross-sections are strategies, and the laws of load growth, estimated period, time to reach design load, etc. - vague states of nature. Many states of nature are limited by the range of possible values of the minimum and maximum limits of change of the resulting uncertain factor.

When choosing the number of cable cross-sections, the optimization criterion is [7.8]:

$$Z(N_F,\Phi)=\delta_1+\delta_2+\delta_3 N_{F,i}^{-1}+\delta_4\Phi N_F^{0,3}$$
$$N_F \in \{1, N\}, \qquad \Phi \in S,$$
$$S: \Phi_{\min} \subseteq \Phi \subseteq \Phi_{\max} \qquad (1)$$

where δ_1, δ_2, δ_3, δ_4 - are the constant components of the optimization criterion.

Φ - resulting uncertain factor; S - is the set of realizations of the indefinite factor corresponding to the set of combinations of "states of nature"; N_F - is a set of strategies corresponding to the number of cable cross-sections.

Wald's criterion: In this method, the optimal strategy is that for which a minimum of maximum values of the optimization criterion is achieved:

$$\min_{N_F \in N} \ \max_{\Phi \in S} \ Z(N_F,\Phi)$$

The above criterion focuses on the most unfavorable circumstances by an uncertain factor. In this case, it is guaranteed that the optimization criterion - for any state of nature - will not be greater than some guaranteed value.

Since, with an increase in the value of the uncertain factor, the optimization criterion increases, we have:

$$\max_{\Phi\in S} Z(N_F,\Phi)=\delta_1+\delta_2+\delta_3 N_{F,i}^{-1}+\delta_4\Phi_{\max} N_F^{0,3}$$

Then the Wald criterion takes the form:

$$\min_{N_F \in N}\left[\delta_1+\delta_2+\delta_3 N_{F,i}^{-1}+\delta_4\Phi_{\max} N_F^{0,3}\right] \qquad (2)$$

Thus, according to the Wald criterion, the problem of choosing the optimal number of cable cross sections is reduced to determining the minimum total costs at the maximum value of the uncertain factor.

Savage criterion: The optimal strategy is the one for which the minimum risk value (R) from not being informed about the values of the resulting uncertain factor:

$$\min_{N_F \in N} \ \max_{\Phi \in S} \ R(N_F,\Phi) \qquad (3)$$

where $\max\limits_{\Phi\in S} R(N_F,\Phi)=\max\limits_{\Phi\in S}\left[Z(N_F,\Phi)-\min\limits_{N_F\in N} Z(N_F,\Phi)\right]$,

- the maximum possible loss relative to the minimum total cost in the field of possible values Φ.

Expression (3) can be reduced to the form:

$$\min_{N_F \in N} \ \max_{\bar{N}_F \in N} \ \max_{\Phi \in S}\left[Z(N_F,\Phi)-Z(\bar{N}_F \in N)\right], \qquad (4)$$

where $\bar{N}_F$ are the values of the number of sections corresponding to the minimum of total costs at the limit values of Φ.

The maximum difference in the optimization criterion when using the N_F strategy compared to the $\bar{N}_F(\max_{\Phi} R_{N_F})$ strategy depends on both the maximum and minimum values of Φ.

Then, taking into account (1), expression (4) takes the form:

$$\min_{N_F \in N} \max_{\bar{N}_F \in N} \max_{\Phi \in S}\left[\delta_3(N_F^{-1} - \bar{N}_F^{-1}) + \delta_4(N_F^{0.3} - \bar{N}_F^{0.3})\Phi\right], \quad (5)$$

Thus, in the problem of choosing the optimal number of cable cross-sections according to the Savage criterion, the method for pairwise comparison of strategies is used [7].

The numerical solution of equations (2) and (5) gives the optimal values of the number of cable cross sections DEN 0.38 and 10 kV, determined by the criteria of Wald and Savage, respectively. With the accepted initial data [2], the optimal values of the number of cable cross sections DEN determined by the Wald and Savage criteria are given in the table.

As can be seen from the table, the influence of various laws of load growth on the choice of the optimal number of cable cross-sections for DEN is not the same when evaluated according to the Wald and Savage criteria. So, according to the Wald criterion, the optimal values of the number of cable cross sections are one or two - with a uniform law of load growth (in the entire range of electrical load density). And according to the Savage criterion, the optimal values of the number of cable cross sections practically do not depend on the law of load growth.

This character of the influence of the laws of load growth is explained by the fact that in this problem the values of the cable cross-sections that are optimal according to the Wald criterion are determined by Φ^{max} taking into account the large role of energy losses; the latter are different for different laws of load growth. According to the Savage criterion, the optimal number of cable cross-sections is determined by Φ^{min} (a small role of energy losses), the values of which are approximately the same for the considered laws of load growth (table).

Table 1. Optimal according to the criteria of Wald and Savage, the number of cross-sections of DEN cables

Load growth law	Criteria	Loop network diagrams	Dual beam networks
		σ<10 мВА/км²	σ>10 мВА/км²
Even	Wald	2	1
	Savage	2	1
Uneven	Wald	1	1
	Savage	2	1

Based on the foregoing and analysis of the results by minimax criteria given in the table, the following conclusions can be drawn.

In DEN - at low electrical load densities (up to 10-12 mVA/km²), it is optimal to use two cable sections (240 and 95 mm² or 185 and 70 mm²); - high load densities (more than 10-12 mVA/km²), the use of one cable section (150 or 120 mm²) is optimal.

Calculation of electricity losses in cable lines for real residential areas of the city due to the use of a limited number of standard cable sections. Calculations show that the use of two standard sections instead of 6-8, selected in accordance with current standards, leads to a reduction in electricity losses in cable lines by 12-15% per year. At the same time, the decrease in specific losses of electricity on average is 20 kWh/year per 1 kW of load on the tires of transformer substations.

References

1. A.D. Taslimov. Mathematical models in the optimization of parameters and structures of urban electricity supply systems // Scientific and Technical Journal "Problems of Energy and Resource Saving". Tashkent, 2017. - No. 3-4. S. 62-65.
2. A.D. Taslimov. Unification of parameters of urban distribution electric networks in conditions of uncertainty: Monograph. - T .: TashGTU, 2019 .-- 148 p.
3. Trukhaev R.I. Decision making models in the face of uncertainty. - M .: Nauka, 1981. - 258 p.
4. Leshchinskaya T.B. Methods of multicriteria optimization of rural power supply systems in the conditions of uncertainty of initial information. - M .: Agroconsult, 1998 .-- 120 p.
5. Levin M.S., Leshchinskaya T.B. Methods of solution theory in problems of optimization of power supply systems. - M.: Publishing House VIESH, part 1. 2006. - 130p.
6. Ventzel E.S. Operations research. - M.: Higher School, 2001.- 208 p.
7. Taslimov A.D., Melikuziev M.V., Murodov B.K. Development of a technical and economic model for optimizing the distribution network parameters // Journal "Vestnik TashGTU". Tashkent, 2018 .-- No. 3. P. 53-57.

Bereznev Yu. I. The choice of wire cross-sections of lines 0.38 and 10 kV in conditions of uncertainty and the problem of unification of their scale // News of higher educational institutions. Energy issues: scientific, technical and industrial journal. - 2009. - No. 9/10. S. 38-40.

Method of selecting parameters of cable lines distributive networks 10 kv in uncertainty conditions

A.D. Taslimov[1], *A.S.* Berdishev[2], *M.V.* Melikuziev[1], *F.M.* Rakhimov[2]

[1]Tashkent State Technical University named after Islam Karimov, 100095, Uzbekistan, Tashkent, University St. 2A
[2]Tashkent institute of irrigation and agricultural mechanization engineers

Abstract. The article discusses the task of selecting the optimal number of types and values of standard cable sections of urban distribution electric power supply systems under conditions of uncertainty of electric load development using zoning method and proves the expediency of using a very limited number of standard cable sections (application of one or two standard cable sections at wide range of load DNsity variation).

The construction and development of distributive networks(DN) in time is determined by the influence of a number of factors and the choice of parameters of the elements of DN is carried out taking into account the uncertainty of a complex of factors characterizing the development of the load in the considered time period.

Construction and development of distributive networks(DNs) in time are caused by the influence of a number of factors and selection of parameters of DNs elements is carried out taking into account uncertainty of the set of factors characterizing the development of load in the considered period of time. At the same time, such factors are usually considered: normative coefficient of efficiency of capital investments, coefficient of initial load in shaDN of the project, laws of increase of load over time; Annual uniform and uneven load growth; Year of achieving design load from the moment of transformer substation (TS) commissioning; Considered period of time, for example, estimated operation time of the designed network before its reconstruction [1].

Selection of parameters values of avionics cables shall be performed on the basis of optimization criterion under conditions of uncertainty of initial information [2]. At the same time, depending on whether the information belongs to a certain information situation, it is necessary to use one or another criterion of decision selection [3,4]: Bayes criterion, Bernoulli - Laplace criterion, Wald and Savage criterion, Gurwitz criterion.

The general disadvantage of choosing a solution according to these criteria is that the obtained solutions are little sensitive to additional information about uncertain factors [4]. And if the DNults of selecting the optimal number of cable sections according to the Wald, Savage, and zoning criteria are the same, the final decision is applied. Otherwise, different criteria [5,6] are used.

The main disadvantage of choosing a solution according to Wald and Savage 's criteria is the subjective evaluation of the optimal solutions obtained, which leads to different DNults. Thus, at a given value of electrical load DNsity, the quantity sections of SES cables found on the Wald criterion are less than those obtained on the Savage criterion or equal to them. This is due to the fact that the use of the Wald principle is oriented to the maximum value of the DNulting undefined factor, and on the Savage criterion the DNult depends on both the maximum and minimum values of its values. In this case, the values of the undefined factors given by the intervals must be supplemented by reasonable assumptions about the reliable values of the undefined factors. An illustration of such an addition is the method of zoning a plurality of vectors of the state of nature [1,9].

For the task of selecting the number of cable sections, the zoning method [1,9] consists in dividing the set S of possible values of the DNulting undefine d factor F into areas, in each of which a certain strategy is optimal (in our case, some number of cable sections from the set N under consideration).

The zoning method is based on the operation of dividing the set Sinto subclasses of two NFiNF 1 strategies.

The pairwise boundary equation is defined by the equality of optimization criteria for the NFiNF 1 strategy, i.e.:

$$Z(N_F, \Phi) = Z(N_F + 1, \Phi). \quad (1)$$

If the paired boundary is outside the state vector field [4], i.e.

$$\max_{\Phi} \quad \Delta Z(\Phi) \prec 0, \qquad \min_{\Phi} \quad \Delta Z(\Phi) \geq 0,$$

Where $\Delta Z(\Phi) = Z(N_F, \Phi) - Z(N_F + 1, \Phi)$,

Then one of the strategies cannot be optimal under any circumstances and is excluded from consideration. If the paired boundary intersects the field of nature state vectors, i.e.

$$\max_{\Phi} \quad \Delta Z(\Phi) \geq 0, \qquad \min_{\Phi} \quad \Delta Z(\Phi) \prec 0$$

then the pair boundary equation (1) is compiled, the solution of which with respect to Φ determines the optimality zone of a certain strategy, i.e. a certain number of used cable sections.

In the problem under consideration, due to the linear dependence of the optimization criterion on the resulting uncertain factor Φ outside the pair boundary, one strategy is strictly optimized over another. To select the number of cable sections, in accordance with [1], the equation of the pair boundary (1) takes the form:

$$\delta_1+\delta_2+\delta_3 N_F^{-1}+\delta_4 \Phi_{nap} N_F^{-0,3}=\delta_1+\delta_2+\delta_3 (N_F+1)^{-1}+\delta_4 \Phi_{nap} (N_F+1)^{0,3}, \quad (2)$$

Where Fpar - characterizes the boundaries of the area by an uncertain factor in which a certain strategy is optimal.

Decision (2) concerning Fpar determines the area of optimum application of a certain number of cable sections:

$$\Phi_{nap} = \frac{\delta_3}{\delta_4} \frac{(N_F+1)^{-1} - N_F^{-1}}{N_F^{0.3} - (N_F+1)^{0.3}}.$$

An algorithm for selecting the optimum number of cable sections under conditions of uncertainty by zoning method has been developed and a program for selecting the optimum number of cable sections under conditions of uncertainty by zoning method has been created based on the algorithm [7].

At the received initial data [1], Figure 1 shows the zoning of the DNulting undefined factor F by the number of sections of electrical distribution cables 10kV, at that for each power value of transformer substation (TS) the uncertainty zone corDNponding to the maximum and minimum value is defined.

With the accepted initial data [1], in Fig. Figure 1 shows the zoning of the resulting indeterminate factor Φ by the number of cable cross-sections of 10 kV distribution electric networks, and for each power value of the transformer substation (TS), an uncertainty zone corresponding to the maximum and minimum values of the uncertain factor is determined. The zones between the curves determine the optimal number of used cross-sections of 10 kV DN cables.

If, in the uncertainty zone Φmin ÷ Φmax, the optimality zones of several numbers of cable sections fall, then it's impossible to get a definite solution. As can be seen from fig. 1a, such a case arises of the capacities of transformer substations (400-2000kVA) when loop circuits of 10kV networks are used. For example, at 630 kVA, two optimality zones of different numbers of cable cross-sections NF = 2 and 3 immediately fall into the uncertainty zone Φmin = 1.7 Φmax = 2.8.

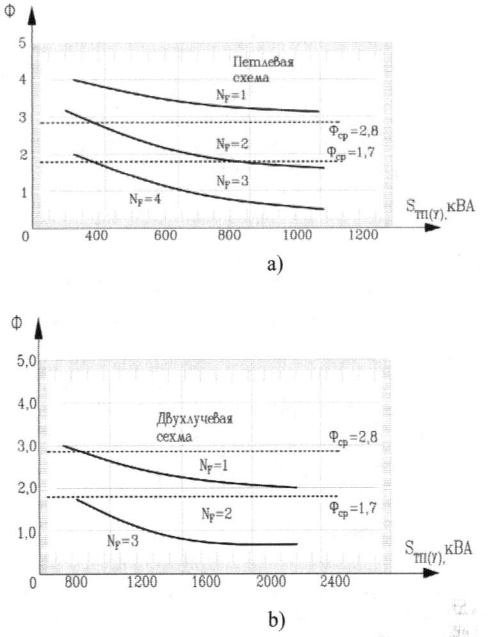

Fig. 1. Zoning of an uncertain factor in the number of used cable cross-sections of DN 10kV.

If the uncertainty interval Φmin ÷ Φmax found for specific conditions falls into the optimality zone of only one value of the number of cable cross-sections, then we can obtain an unambiguous solution to the problem. As can be seen from fig. 1b, such an unambiguous choice is possible with a power of TS 20000kVA or more, when two-beam schemes of 10kV networks are used. In this case, the uncertainty zone Φmin ÷ Φmax falls only in the optimality zone for the use of one cable section for any TS power.

In the above cases, it is recommended to partially or by density remove the uncertainty based on additional information on the values of the uncertainty based on additional information on the values of the uncertain factors [1]. Expert information is used as additional information about the values of uncertain factors.

To determine the most probable (real) values of the considered uncertain factors, a survey of leading specialists in the design and operation of urban DN.

The ratios of weighting coefficients of additional criteria obtained from experts were processed by the direct assessment method [8]. The total ranks for all ratios are determined and the ratios are adjusted. In this case, the ratio corresponding to the lowest rank, i.e.

$$X_{мин} = \min(x_1, x_2, ..., x_n),$$

where x1, x2, ..., xn are different ratios of weighting coefficients of additional criteria.

In this case, the minimum and maximum values of the resulting uncertain factor were determined according to [1]; moreover, Φmin corresponded to the values β2min (υ) = 1.01 (0.6), tpr (min) = 2 years, Tp (min) =

10 years, and at Фmax-- β2min (υ) = 1.1 (0.6), tpr (max) = 10 years, Tr (max) = 25 years

Using expert assessments, the minimum and maximum values of the normative coefficient of efficiency of capital investments, the initial load coefficient in fractions of the design factor, the factor of uniform (uneven) increase in load β (υ), the values of the estimated period Tr and the value of the time to reach the project load tpr are determined, and also determined their most likely values. As a result, the most likely (average) values of the factors are:

κ_{np}= 0,8; $p_{н}$ =0,2; $\beta_{cp}=(\upsilon_{cp})=1,02\ (0,4)$, $t_{np.cp.}$=5years, $T_{p.cp}$=25 years, and then the value of the resulting indeterminate factor Фcp = 1.7 (2.8).

As can be seen from fig. 1a, for small capacities of TS for loop network circuits (σ <10mVA / km2), at Fsr = 1.7, which corresponds to the uniform law of load growth, it is optimal to use one or two cable sections in 10 kV DN, and at Fsr = 2, 8, which corresponds to the uneven law of load growth, one cable section is optimal. Thus, the results confirm the advisability of using a very limited number of cable cross-sections of urban 10 kV DN and lead to the same quantitative results: it is optimal to use one and two standard cable cross-sections in the networks under consideration. At the same time, it is recommended that at high load densities and double-beam network circuits - use a section of 120 mm^2; at low load densities, when loop circuits are used, one section is 120 mm^2 or two sections 185 and 70 ^{mm2} In addition, the results obtained in terms of the number of cable cross-sections used in RES correspond to similar foreign technical solutions (France, Germany, Poland, Russia, etc.), which is a certain confirmation of the reliability of the results.

References

1. A.D. Taslimov. Unification of parameters of urban distribution electric networks in conditions of uncertainty: Monograph. - T.: TashGTU, 2019.- 148 p.
2.Trukhaev R.I. Decision making models in the face of uncertainty. - M.: Nauka, 1981. - 258 p.
3.Levin M.S., Leshchinskaya T.B. Methods of solution theory in problems of optimization of power supply systems. - M.: Publishing House VIESH, part 1. 2006. - 130p.
4. Ventzel E.S. Operations research. - M.: Higher School, 2001.- 208 p.
5. Bereznev Yu. I. The choice of wire cross-sections of lines 0.38 and 10 kV in conditions of uncertainty and the problem of unification of their scale // News of higher educational institutions. Energy issues: scientific, technical and industrial journal. - 2009. - No. 9/10. S. 38-40.
6. Bereznev Yu. I. Economic criterion parameter of an element of a rural electric network in conditions of uncertainty / Yu. I. Bereznev // News of higher educational institutions. Energy issues: scientific, technical and industrial journal. - 2009. - No. 5/6. - S. 85-91.
7.Taslimov A.D., Rakhimov F.M., Melikuziev M.V. The program for selecting the optimal number of cable cross-sections of distribution networks in the face of uncertainty in the development of the load // Intellectual Property Agency of the Republic of Uzbekistan. DGU certificate No. 05889 16.1.2018.
8. Beshelev S.D., Gurvich S.G. Mathematical and statistical methods of expert estimates. - M.: Statistics, 2010.- 264s.
9. Diner I.Ya. Zoning of sets of state vectors of nature and the problem of choosing a solution. // In the book. : Operations Research, -M .: Nauka, 1972. S. 183-194.

The composition of natural waters of some source rivers of the republic of Uzbekistan, used in the thermal power engineering and the results of the experimental researches at preliminary and ion exchange treatment of water

R. Yusupaliev[1], *B.Kh.* Yunusov[1], *M.M.* Azimova[1]

[1]Tashkent State Technical University named after Islam Karimov, 100095, Uzbekistan, Tashkent, University St. 2A

Abstract. The article describes the possibility of use of the natural waters for generation of steam at thermal electric power stations. The chemical composition of some source rivers of the Republic of Uzbekistan is provided as well as its influence to the processing of water and contamination of the steam in the thermal energy engineering. The conditions of the preliminary chemical treatment of the natural water at TashTPP with the use of aluminum sulphate are described, as well as the experimental unit for production of the coagulant in the preliminary treatment and clarification of water; the experimental data on treatment of water with ions of chloride using the new ionites of type ASH-64 and ASH-64M, which were synthetized by the scientists of the Tashkent national university.

Introduction

The removal of the colloids and clarification of water at Tashkent thermal electric power station is performed by introduction of the technical coagulants of the aluminum sulphate $Al_2(SO_4)_3$ in to the being treated water. In connection with the impossibility of regular provision of the water treatment unit (WPU) of the TashTPP with the required agent $Al_2(SO_4)_3$, the researches are performed on production of the solution of this agent at industrial conditions.

For the deep desalting of water at TPP in the Republic of Uzbekistan, different ionites are used: KU-1, KU-2, AN-17, AV-18 and others. These ionites are imported to Uzbekistan from other countries and Russia. In order to fulfill the set tasks, the study of the existing approaches on water clarification for the TPP needs was performed.

Chemical composition of the natural waters of the Republic of Uzbekistan and their use at TPP

A complex of the processes on treatment of the deeply desalted water is among the important processes of the modern TPP that provides the normed content of the mixes in the vapor and the feed water of the boilers. For this reason each TPP has a system of WPU units. The WPU system includes the units on chemical treatment of the initial water, as well as preparation of the make up water and treatment of the turbine condensate using the ionites at TPP.

In the thermal energy engineering at TPP of the Republic of Uzbekistan, in order to produce the vapor, the waters from the natural sources, like rivers, water reservoirs are mainly used. The quality of these waters is instable and depends on the components that contact the soil. The bigger content of salts in the soil, the bigger content of the mixtures in the natural water and it has good dissolving properties. There are easily and hardly dissolving mixtures in the natural water of organic and non-organic origin. The mixtures, contaminating the natural waters, are of different forms, and classification of them is provided on the pic.1.

Hence, the economic expenditure increases considerably during the treatment of water at TPP.

The content of ions of chloride Cl, sulphate $SO^{2-}{}_4$, nitrate $NO^-{}_3$ is observed increasing in the river water and characterize high content of salts in the water. It is required to underline that the compounds of carbon in the river water are of different forms, in the form of free carbonic acid CO_2, molecular H_2CO_3 and in the form of anions HCO_3^- and CO_3^{2-}. The formation of the carbon compounds in the natural waters depends on the pH value of the service and influences the quality of water, alkali content, acidity and the carbonate hardness of water. [1]

The table 1 contains the chemical composition of the natural waters of some of the rivers of the Republic of Uzbekistan used at TPP.

Table 1.

Chemical content	Unit of msnt	Amu-Dariya	Sir-Dariya	Narin	Chir-chik	Kashka-dariya	Surkhan-dariya	Kara-dariya	Akhan-garan
Nonal hardness	mg-eq./l	4.3-10.1	3.8-8.8	3.0-5.1	1.8-2.6	2.8-3.4	2.0-5.1	4.1-7.6	1.5-2.9
Ca2+	mg/l	60.3-109.3	53.1-87.5	42.9-62.0	30.1-39.1	43.1-53.7	38.9-74.4	39.7-80.1	20.9-48.4
Mg^{2+}	mg/l	12.4-36.3	11.9-54.5	7.1-23.4	3.3-9.7	4.6-11.3	5.4-19.1	17,543,1	1.7-4.5
Na^{++} K^{+}	mg/l	49.2-164.2	16.8-81.5	9.5-37.1	0,8-38	0.2-12.0	0.8-12.5	12.2-5.05	1.5-7.2
Fe^{2++} Fe^{3+}	mg/l	0.01-0.04	0.01-0.04	0.01-0.04	0.01-0.05	0.01-0.07	0.01-0.07	0.01-0.05	0.01-0.04
HCO_3^-	mg/l	103.7-166.1	120-180.5	114-172	97-140	163.3-195.2	90.2-192.2	134,-220	62.8-136.6
SO_4^{2-}	mg/l	112.9-237.2	102.5-333	47.7-19.5	10-15,7	4.4-7.0	30.0-93.9	101,-210	7.6-20.0
Cl-	mg/l	58.5-242.2	15.6-51.1	9.2-322.6	3.1-4.5	1.8-3.9	2.2-10.0	12.8-28	1.4-3.6
NO_3^-	mg/l	0.09-1.98	0.15-10	0.04-0.99	0.1-099	0.9-3.0	0.03-8.1	0.9-10	0.5-2.1
pH	mg/l	7.9-8.9	7.6-8	7.6-88.3	7.6-8.1	8.1-8.4	8.0-8.3	7.4-8.0	7.5-8.8
Oxy-da-tion	mg/l	0.7-1.9	0.7-1.1	0.5-1.9	0.4-15	0.7-1.6	0.4-1.5	0.7-1.5	0.3-1.3

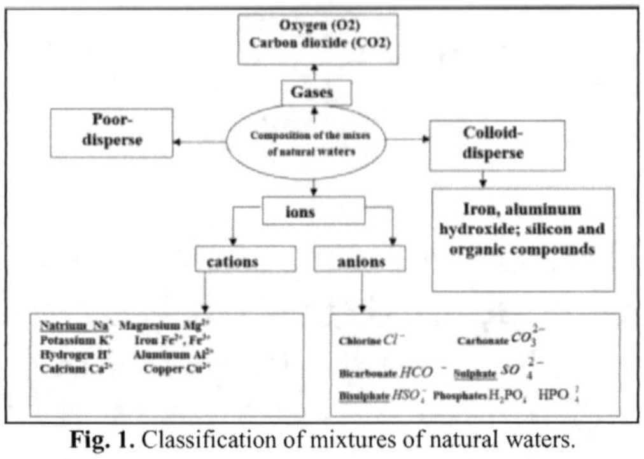

Fig. 1. Classification of mixtures of natural waters.

As seen from the table 1 the widely spread compounds of the natural waters of the Republic of Uzbekistan are mainly compounds of calcium Ca and magnesium Mg, the percentage composition of which sharply changes during the year.

Dependence of composition of different forms of carbon compounds from pH index is provided on the picture 2.

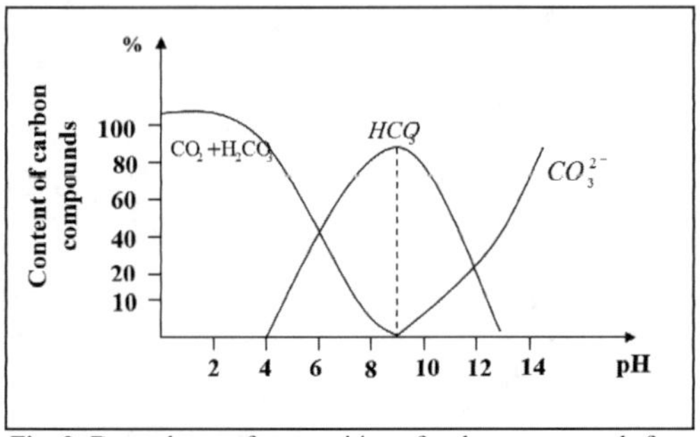

Fig. 2. Dependence of composition of carbon compounds from pH level.

As seen from the diagram, in the case of increase of pH of water from 3 to 12 the degree of molecular dissociation of the carbonic acid: (H_2CO_3), and at pH from 1.0 to 4.5-95% of the carbonic acid practically is in the type of molecules in non-dissociation form.

At pH from 8.5 to 14 the dissociation of the bicarbonate anions increases: $HCO_3^- \rightarrow H^+ + CO_3^{2-}$ with formation of carbonate ion CO_3^{2-}. It is required to underline that in order to produce the vapor in the steam boilers, the pH of the feed water shall be within 9-10.5 (1). In this conditions the carbon acid compounds in the feed water of the steam boilers can be in the form of carbon ions CO_3^{2-}.

In the operating service, when the temperature of the heat medium > is 200°C, the formation of the hardly dissolvable compounds on the inner surfaces of the steam boilers occur in the form of calcium of carbonate $CaCO_3$. These sediments have low heat transfer values (0.6-0.7 W/mK), impact the heat exchange between the metal and the operating medium. This leads to the increase of the temperature of the inner surface of the boiler, consumption of fuel and speed of the corrosion process on the surface of the heater. These processes lead to the emergencies, capacity decrease of the boiler and impairment of the modes of normative operation in the TPP.

The silicone compounds, similar to the carbon acid compounds are also present in the natural waters in different forms: molecular $nSiO_2 \cdot mH_2O$, H_2SiO_3 and in the form of anions $HSiO_3^-$, SiO_3^{2-}. The form of formation of the silocone compounds in the natural water also depends on the pH value.

At pH from 6 to 8 - 9% of the molecules of H_2SiO_3 dissociate with formation of $HSiO_3^-$, and the rest in the type of molecules of H_2SiO_3 or in the colloid condition $nSiO_2 \cdot mH_2O$. When pH is bigger than 8, the silicone compounds, present in the natural water, mainly are in the type of $HSiO_3^-$ and partially as SiO_3^{-2}. In this way, the bigger the content of the silicone in the steam water, the bigger the contamination of the vapor with the silicone compounds. It is caused by the silicone compounds positive temperature rate of dissolution both in water and in the vapor[2].

Hence the bigger content of the silicone compounds in the boiler water, the bigger possibility of contamination of the vapor in the result of steam generation of the boiler water.

The contamination of vapor with the mixes of the silicone compounds form the insolvable sediments in the wheel space of the turbine that push the corrosion and erosion processes. The insolvable silicone sediments mainly consists of the iron oxides, zinc, copper, silicone acid, calcium compounds, magnesium and aluminum. In the composition of the sediment there are sulfates, chlorides, silicates, carbonates, sodium phosphates and even sodium hydroxide. The increase of such sediments pushes the corrosion of the metal, especially in the side of LPC. All types of sediments in the steam turbines decrease their capacity, limit the nominal power of the turbine.

Based on the above mentioned it is obvious that to increase the reliability of the operation and economy of the thermal equipment of the thermal power plants, it is required to provide the feed of the steam boilers with highly desalted and silicone free water.

In relation to this, the treatment of the natural waters to produce the vapor at all TPP of the Republic of Uzbekistan is implemented in two stages: at the first stage the reagent treatment such as the coagulation and liming and the second stage the deep desalting by thermal and ionite treatment of the water.

The economic expenditure at deep desalting of the water at TPP depends on many factors: quality of the initial water, selection of the methods and used reagents. At ionite treatment - the type of the used ionites and technologies of the ionite treatment of water.

Electrocoagulation technology for production of coagulant at TashTPP

In the WTU of the Tashkent TPP the component is the preliminary treatment of water from silt and different organic contaminants. The preliminary treatment is in clarification of the initial water (Boz-su canal) by removing the coarsely dispersed contaminants.

At the present time the clarification of water at TashTPP is done by method of coagulation which is the introduction of the special coagulant of the aluminum sulphate ($Al_2(SO_4)_3$) into the being treated water.

Due to the impossibility of regular provision of the WTU of TashTPP with the required reagent - the aluminum sulphate. To solve the problem, the electrocoagulation technology is proposed for produce the aluminum hydroxide during the preliminary treatment and clarification of the water.

This technology provide high efficiency of the treatment of water and possesses the advantages compared with the reagent methods such as compactness, absence of huge reagent station, simple maintenance, possibility of complete automation and low capital expenditures.

The researches showed that the coagulant, produced by electrochemical method, possesses with the bigger sorption capacity than the aluminum hydroxide formed as a result of hydrolysis of its salts. The later can be explained with the special conditions of formation of the coagulating structures in the electrolysis, as well as lower influence of the anions of the strong acids the concentration of which, in this case, is lower than the one during the introduction of the chemical reagents into the water [3].

The essence of the electrocoagulation method is in the anodic dissolution of the metals under the influence of the electric current with further formation of the colloid mixtures.

In the being developed technology, it is proposed to use the aluminum and duraluminum wastes.

This technology of production of the coagulant was introduced for operation. It diagram is provided on the fig. 3.

The consumption of the electric energy can be minimized by selecting the inter-electrode intervals and the time of commutation of the electrodes. The calculations, based on the experience at pilot unit on production of the coagulant by electrochemical method, showed that the capacity of the unit at the density of the current at electrodes is 50A/m2 and the temperature of the electrolyte is 20^0C and equates to 10l/hour. The capacity of the unit can be regulated by the density of the current, as well as the temperature of the electrolyte.

The produced coagulant with the required composition is accumulated in the coagulant tank (available at TashTPP). From this tank the coagulant through the dosing unit is mixed with the initial water (water from Boz-su) and is fed to the inlet of the clarification column.

To better the process of treatment and clarification, the being treated water is added with the acid to receive the required pH. The pH detectors, installed at the inlet and outlet of the clarification column, and the control diagram of the dosing unit, regulate the supply of the required volume of the acid through the dosing unit.

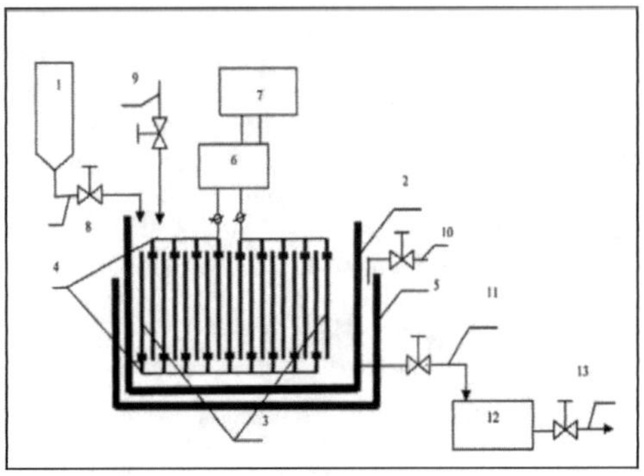

Fig. 3. The diagram of the pilot unit of production of the coagulant at TashTPP.

1-metering tank for the acid; 2-coagulant tank; 3-aluminum plates; 4-connecting buses; 5-cooling tank; 6-ballast resistance; 7- power source; 8-acid supply piping; 10-cooling water supply piping; 11-ready coagulant drain piping; 12-coagulant tank; 13-supply of coagulant to the water treatment line.

The implementation of the proposed project is justified by the pilot sample of the electrocoagulator, that showed its operability. The produced coagulant on this unit, completely satisfies the requirements of the treatment and clarification of the initial water. The calculations, based on the experience of the operation on the pilot unit on production of the coagulant by electrochemical method, showed that the productivity of the proposed unit at the density of the current at electrodes of 50A/m2 and temperature of the electrolyte of 20^{0}C equates to 20l/hour of coagulant with the chemical composition: density 1.3 gr/cm3; oxidation of aluminum Al_2O_3 -7,6%; in conversion to the aluminum sulphate $Al_2(SO_4)_3$- 25,5 %. The capacity of the unit can be regulated by the density of the current, as well as the temperature of the electrolyte.

The mass consumption of the substances used for coagulation is decreased 2.5-3 folds compared to the reagent coagulation.

Based on the results of the researches on this unit, we proposed and approved with TashTPP the sketch design. "Development of the highly efficient automated electrocoagulation technology of preliminary treatment and clarification of the water at TashTPP".

The research of the ion-exchange capabilities of the new ionites

One of the reliable methods of preparation of the deep desalting of water for the feed of the steam generators of TPP is the ionite treatment of water. Hence during the preparation of the make up water and cleaning of the turbine condensate at TPP, the big amount of synthetic ionites is required. [4].

At the present time, the National University of Uzbekistan proposed a technological regulation of production of the new ion-exchange materials on the base of the polyacrylonitrile, polyacrylamide, that are produced by the Navoiy checmial combinate. The successful synthesis of the pilot lot of the samples of ion-exchange materials of the type ASH-64 and ASH-64M was performed.

These ionites were received by modification of the polyacrylonitrile, produced at the Navoiy chemical combinate through the polymerization.

The synthesized ionites of the type ASH-64 and ASH-64M are resistant to the influence of the 4% solution of the salt acid and fully preserve the absorption ability within the range of the temperatures of 6-60degr.C and above.

The preliminary experimental researches showed, that the ionites ASH-64 and ASH-64M successfully clean the water from the ionites of the chlorine as a result of the following reaction:

RIOH+Cl$^-$ =RICI+OH-

The tables show the results of the experiment researches of the ion-exchange ability of the ionites ASH-64 and ASH-64M against the ions of the chlorine.

As can be seen from the results of the research, the Cl ions are well absorbed on ionites ASH-64 and ASH-64M, however, with the increase of the content of the ions in the being treated water, the volume of the carried water before the breakthrough of the chlorine decreases. It is related to the size of the exchange capacity of the ionites. The values of the nominal exchange capacit of the ionites mainly differs from the quantity of the introduced functional groups such as the primary (-NH2), secondary (=NH) and tertiary (=N) amine group. The more there are such tied amine groups in the received ionites, the bigger value of the ion-exchange ability (E). [5].

Table 2. The results of the research of the ion-exchange ability of the ionite ASH-64 in the water with the content of the chlorine ions of 10mg/l is shown in the table 2.

Volume of the carried water, ml.	Volume of NaOH consumed for the titrating, ml.	The residual content of the chlorine ions in the filter cake where the initial content of the chlorine is 10mg/l.	Residual equivalent content of chlorine ions mgeq/l.
200	0	0	0
400	0	0	0
1500	0	0	0
2500	0	0	0
5000	0	0	0
6300	0	0	0
6350	0.1	0.8	0.02
6400	0.15	0.18	0.06
6450	0.20	2.7	0.07
6500	0.30	4.7	0.12
6550	0.35	5.4	0.15
6000	0.40	6.3	0.17

https://doi.org/10.1051/e3sconf/201913901083

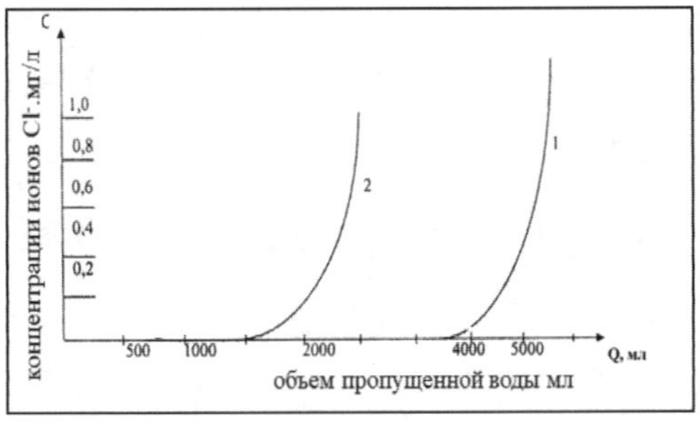

Fig. 4. Absorption of the chlorine ions on ionites ASH-64 and ASH-64M with content of the chlorine in the being studied water of 1-10mg/l, 2-36mg/l.

Table 3. The results of the research of the ion-exchange ability of the ionite ASH-64M in the water with the content of the chlorine ions is 36mg/l.

Volume of the carried water, ml.	Volume of NaOH consumed for the titrating, ml.	The residual content of the chlorine ions in the filter cake where the initial content of the chlorine is 36mg/l.	Residual equivalent content of chlorine ions mgeq/l.
200	0	0	0
400	0	0	0
600	0	0	0
750	0	0	0
800	0.3	4.5	0.12
850	0.4	6.3	0.17
900	0.5	8.1	0.22
950	0.6	9.9	0.27
1000	0.8	13.5	0.36
1050	1.1	18.5	0.51
1100	1.7	29.7	0.81
1150	2.1	36.9	1.01

The experimental results of adsorption of the chlorine ions on ionites of type ASH-64 with content of the chlorine 10.20 and 36 mg/l is shown on the picture.

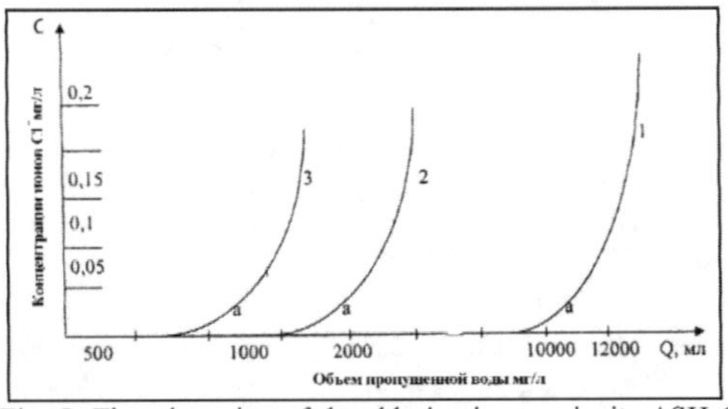

Fig. 5. The adsorption of the chlorine ions on ionite ASH-64 with content of 1-10mg/l, 2-20mg/l, 3-36mg/l of chlorine in the being studied water.

The value E_p against chlorine ions is calculated on the formula mgr eq/m:

$E_p=(C_{исх.}-C_Ф)$ x Q/m

where $C_{исд.}$- concentration of chlorine ions in the being studied water, mg/l.

$C_Ф$- concentration of chlorine ions in the filter cake, mg/l.

Q- volume of the being filtered water before the breakthrough of chlorine ions, ml.

m- consumption of NaOH solution, ml.

As a result, the experimental researches on study of the adsorption of ionites ASH-64 and ASH-64M showed, that these ionites can be used during the treatment of the anions of strong acids, such as chlorine, sulphate and nitrate, hence these ionites can be used at water processing units of TPP.

Conclusion

At TPP to produce vapor, the deep cleaned desalted water is required. Hence the economical consumption during preparation of water at TPP depends on the following factors: chemical composition of the initial water, capacity of the units, used reagents and types of ionite materials during the processing of the water.

At TashTPP the technical aluminium sulphate with the clearance of 65-70% is used in the preliminary treatment of water for the elimination of the colloido-disperse substances. By using it, the formation of the sludge compounds and consumption of the reagent increases dramatically.

Hence, we have developed the economical electrocoagulating technology on production of the required reagent using the aluminum wastes as the electrodes. Also the experimental results of the ion-exchange ability of the new ionites ASH-64, ASH-64M against the chlorine ions are provided. The results of the research showed that the indicated ionites well absorb the chlorine ions.

References

1. Lifchits O.F. “Reference book on water treatment of the boiler units” M.Pub. Energy, 1986.
2. Guseva O.V. Physico-chemical bases of calculation of the water-chemical modes of the heat networks and evaporators.- M, MEI. 2001. -48 p.
3. Yusupaliev R.M. , Usmonov N.O. “Pilot electrocoagulating unit of the preliminary treatment of the natural water” Energy preservation energy engineering energy audit No.10 (152), Moscow, 2016.
4. Kopilov A.S., Ochkov V.F. Water treatment in the energy engineering. M. MEI. 2006.
5. Yusupaliev R.M Study book “Control of the water treatment techniques and use of the chemical agents in the Heat energy engineering”. Tashkent, 2013.

E3S Web of Conferences **139**, 01084 (2019) https://doi.org/10.1051/e3sconf/201913901084
RSES 2019

Research of the basic characteristics of the new magnetic elastic force sensors

Sulton Amirov[1], *Kamila* Jurayeva[1]

[1]Tashkent institute of railway engineering, Department of Power Supply of Railways, Adilkhodjayev – 1, Tashkent, 100167, Uzbekistan

Abstract. The article investigated the static, dynamic and metrological characteristics of new magnetoelastic force sensors taking into account the distribution of parameters of electrical and magnetic circuits. It is shown that taking into account the distribution of circuit parameters during calculations significantly reduces the error in calculating the characteristics of magnetoelastic sensors.

Introduction

All over the world, research work is underway aimed at developing new power measuring elements and devices for monitoring and control systems with high sensitivity, accuracy and linearity of conversion characteristics, high reliability and stability of characteristics under extreme operating conditions. At the same time, magnetoelastic force sensors (MFS) are the most fully meeting the requirements of monitoring and control systems.

1.1 New magnetoelastic force sensors

New designs of MFS have been developed [1,2]. It is shown that in a MFS containing coaxially located annular magnetic circuits with sections of measuring windings and through slots made along the circumference, dividing each annular magnetic circuit into two rods, elastic conical bushings located coaxially with the annular magnetic circuits and interacting with the bases with the surfaces of the annular magnetic circuits, performing ring magnetic cores are identical and the introduction of additional, concentrically and mutually mirrored elastic conical bushings, lips Installed with the possibility of simultaneous interaction with the inner and outer surfaces of the annular magnetic cores, as well as the inclusion of the measuring windings of the annular magnetic cores with each other in series, increases the sensitivity, accuracy and linearity of the sensor conversion characteristics. In addition, it has been shown that the implementation of through slots on the forming surfaces of the annular magnetic cores and the reinforcement of the horseshoe-shaped ring casings of inelastic material on the inner and outer forming surfaces of the annular magnetic cores, on which the bases of the conical elastic bushings act, increases the sensitivity of the MFS.

1.2 MFS magnetic circuits

The expressions of magnetic fluxes in coaxially arranged annular magnetic cores and magnetic voltages between them are determined depending on the length of the magnetic circuit, taking into account the distributed nature of the magnetic resistance of the magnetic circuit, the magnetic capacity of the annular gap and sections of the magnetizing windings (Fig. 1).

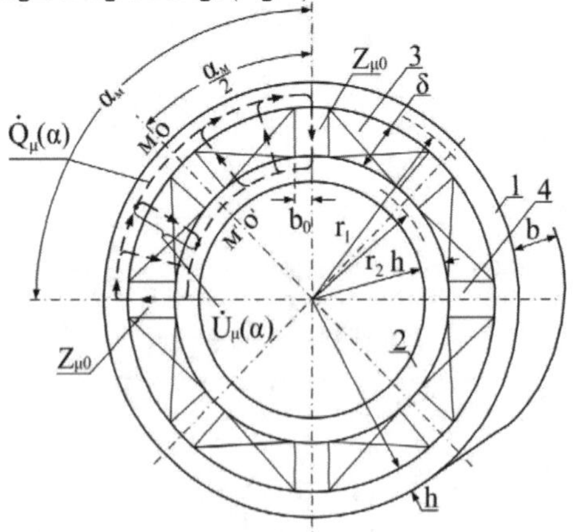

Fig. 1. The design of the magnetic circuit developed by the MFS

1, 2 - coaxially located annular magnetic cores; 3 - section magnetizing windings; 4 - ferromagnetic jumpers; r_1, r_2, h, b - respectively, the average radius, thickness and height of the annular magnetic circuits 1 and 2; δ - air gap between the ring magnetic circuits; b_0 - width of the ferromagnetic jumper.

In order to simplify the analysis of the magnetic circuit of the MFS under study, we neglect the lateral scattering fluxes and assume that the ferromagnetic jumpers with the corresponding sections of the magnetizing winding are identical. In addition, we assume that the magnetic circuit of the investigated MFS operates on a linear section of the main magnetization curve of the magnetic circuit material. These assumptions, as shown by experimental studies, introduce minor inaccuracies in the calculations, but significantly simplify the analysis of the magnetic circuit under study [3, 4].

In the study of the magnetic circuits of the new MFS, the method of constructing an equivalent circuit (Fig. 2) and the classical method of compiling and solving differential equations were used [5].

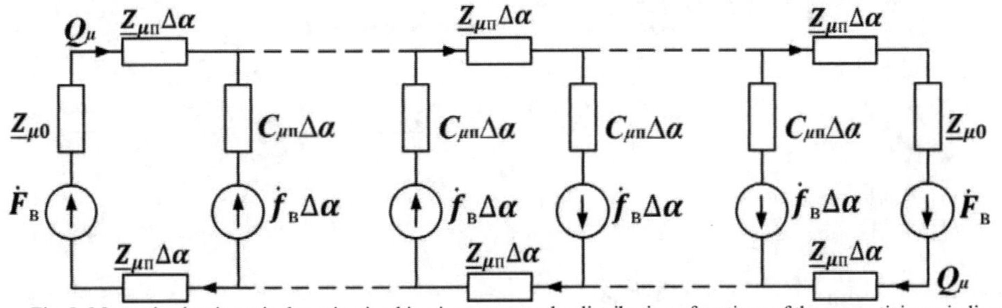

Fig. 2. Magnetic circuit equivalent circuit taking into account the distribution of sections of the magnetizing winding

Expressions for magnetic voltage and magnetic flux of magnetic circuits developed by MFS with concentrated sections of magnetizing windings:

$$\dot{U}_\mu(\alpha) = \frac{\dot{F}_в}{\underline{\Delta}}\Big\{\Big\{sh\left[\underline{\gamma}(\alpha_м - \alpha)\right] - sh\left(\underline{\gamma}\alpha\right)\Big\} + $$
$$+\frac{\underline{\gamma}\underline{Z}_{\mu 0}}{2\underline{Z}_{\mu п}}\Big\{ch\left[\underline{\gamma}(\alpha_м - \alpha)\right] - ch\left(\underline{\gamma}\alpha\right)\Big\}\Big\}, \quad (1)$$

$$\dot{Q}_\mu(\alpha) = \frac{\underline{\gamma}\dot{F}_в}{2\underline{Z}_{\mu п}\underline{\Delta}}\Big\{\Big\{ch\left[\underline{\gamma}(\alpha_м - \alpha)\right] + +ch\left(\underline{\gamma}\alpha\right)\Big\} + $$
$$+\frac{\underline{\gamma}\underline{Z}_{\mu 0}}{2\underline{Z}_{\mu п}}\Big\{sh\left[\underline{\gamma}(\alpha_м - \alpha)\right] + sh\left(\underline{\gamma}\alpha\right)\Big\}\Big\}. \quad (2)$$

where is $\dot{F}_в$ - the magnetomotive force (MMF) of the magnetizing winding section; $\underline{\gamma} = \sqrt{2\underline{Z}_{\mu п}C_{\mu п}}$ - the complex value of the coefficient of propagation of the magnetic flux along the magnetic circuit, $\frac{1}{grad}$; $\underline{Z}_{\mu п} = \frac{1}{\dot{\mu}_{б.п.}\mu_0 hb}$, $C_{\mu п} = \mu_0\frac{b}{\delta}$ - linear values of the complex magnetic resistance of the ring magnetic circuits and the magnetic capacity of the annular gap per unit angle of the magnetic circuit; $\underline{\Delta} = \left(1 + \frac{\underline{\gamma}^2\underline{Z}_{\mu 0}^2}{4\underline{Z}_{\mu п}^2}\right)sh\left(\underline{\gamma}\alpha_м\right) + \frac{\underline{\gamma}\underline{Z}_{\mu 0}}{\underline{Z}_{\mu п}}ch(\underline{\gamma}\alpha_м)$; $\underline{Z}_{\mu 0}$ - complex magnetic resistance of ferromagnetic jumpers, 1/Gn; $\dot{\mu}_{б.п.}$ - the relative magnetic permeability of the ferromagnetic material without taking into account energy losses; $\alpha_м$ - maximum value of the angular coordinate α.

Expressions for magnetic voltage and magnetic flux of magnetic circuits developed by MFS with distributed sections of magnetizing windings:

$$\dot{U}_\mu(\alpha) = \frac{\dot{F}_в}{\underline{\Delta}}\Big\{sh\left[\underline{\gamma}(\alpha_м - \alpha)\right] - sh\left(\underline{\gamma}\alpha\right)\Big\} + \dot{F}_в\frac{\underline{\gamma}\underline{Z}_{\mu 0}}{2\underline{Z}_{\mu п}\underline{\Delta}}\cdot$$
$$\cdot\Big\{ch\left[\underline{\gamma}(\alpha_м - \alpha)\right] - ch\left(\underline{\gamma}\alpha\right)\Big\} + \dot{f}_в\frac{\underline{Z}_{\mu 0}}{2\underline{Z}_{\mu п}\underline{\Delta}}\cdot$$
$$\cdot\Big\{sh\left[\underline{\gamma}(\alpha_м - \alpha)\right] - sh\left(\underline{\gamma}\alpha\right)\Big\} + \dot{f}_в\frac{\underline{\gamma}\underline{Z}_{\mu 0}^2}{4\underline{Z}_{\mu п}^2\underline{\Delta}}\cdot$$
$$\cdot\Big\{ch\left[\underline{\gamma}(\alpha_м - \alpha)\right] - ch\left(\underline{\gamma}\alpha\right)\Big\} \quad (3)$$

$$\dot{Q}_\mu(\alpha) = \frac{\underline{\gamma}\dot{F}_в}{2\underline{Z}_{\mu п}\underline{\Delta}}\Big\{ch\left[\underline{\gamma}(\alpha_м - \alpha)\right] + ch\left(\underline{\gamma}\alpha\right)\Big\} + $$
$$+\dot{F}_в\frac{\underline{\gamma}^2\underline{Z}_{\mu 0}}{4\underline{Z}_{\mu п}^2\underline{\Delta}}\Big\{sh\left[\underline{\gamma}(\alpha_м - \alpha)\right] + sh\left(\underline{\gamma}\alpha\right)\Big\} + $$
$$+\dot{f}_в\frac{\underline{\gamma}\underline{Z}_{\mu 0}}{4\underline{Z}_{\mu п}^2\underline{\Delta}}\Big\{ch\left[\underline{\gamma}(\alpha_м - \alpha)\right] + ch\left(\underline{\gamma}\alpha\right)\Big\} + $$
$$+\dot{f}_в\frac{\underline{\gamma}^2\underline{Z}_{\mu 0}^2}{8\underline{\Delta}\underline{Z}_{\mu п}\underline{\Delta}}\Big\{sh\left[\underline{\gamma}(\alpha_м - \alpha)\right] + sh\left(\underline{\gamma}\alpha\right)\Big\} - \frac{\dot{f}_в}{2\underline{Z}_{\mu п}} \quad (4)$$

here $\dot{f}_в = const$ - is the linear value of the magnetizing force of a uniformly distributed section of the field winding per unit angle α.

An analysis of equations (1) - (4) and their curves (Fig. 3) shows that magnetic the stresses along the MFS magnetic circuit of the efforts, respectively, with concentrated and distributed magnetizing forces are distributed nonlinearly and change their signs at the magnetic neutral point, and the corresponding magnetic fluxes are unstable and have a minimum value at the magnetic neutral point, and with increasing magnetic flux β attenuation coefficient the degree of nonlinearity of the magnetic voltage distribution and the inconstancy of the magnetic flux along the length of the magnetic circuit increases.

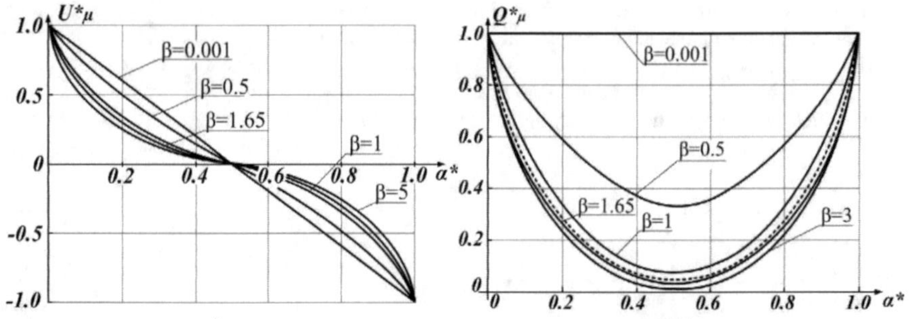

Fig. 3. Dependence curves $U_\mu^* = f(\alpha^*)$ (a) and $Q_\mu^* = f(\alpha^*)$ (b) for different values of the magnetic flux attenuation coefficient β: solid curves are -calculated and dashed curves are - experimental data; $U_\mu^*(\alpha^*) = \frac{U_\mu(\alpha^*)}{U_\mu(0)}$; $Q_\mu^*(\alpha^*) = \frac{Q_\mu(\alpha^*)}{Q_\mu(0)}$; $\alpha^* = \frac{\alpha}{\alpha_м}$.

A comparative analysis of the calculated and experimental dependence curves $U_\mu^* = f(\alpha^*)$ and $Q_\mu^* = f(\alpha^*)$ magnetic circuits, respectively, without taking into account and taking into account the distribution of magnetizing forces, showed that taking into account the distribution of the magnetizing forces of the excitation coil sections significantly (up to 20%) reduces the error in calculating the magnetic circuits of the MFS forces. Note that expressions (1) - (4) are mathematical models of magnetic circuits developed by MFS.

1.3 Static characteristics of new MFS

The analytical equation of the static characteristic of the magnetoelastic transformation stage, i.e. the dependence of the relative magnetic permeability of the magnetic circuit developed by MFS on external mechanical stresses and magnetic field strength created by the excitation winding with alternating current, obtained in the form of the following approximating function [6]:

$$\mu_{6.п.} = a_1 \pm a_2\sigma, \quad (5)$$

where, a_1, a_2 are the approximation coefficients; σ - the absolute value of compressive mechanical stresses. Dependency approximation $\mu_{6.п.} = f(\sigma)$ for electrical steel E3 in the initial section of the main magnetization curve gave the following values of the approximation coefficients: $a_1 = 1085$ and $a_2 = 9{,}77$ $мм^2 \cdot N^{-1}$.

In the fig. 4 shown the dependency graphs $\mu = f(U_м)$ and $\Delta\mu = f(U_м)$, $U_м$ - is the mechanical force. An analysis of the obtained analytical and experimental dependences of the relative magnetic permeability of the MFS ferromagnetic core on the applied mechanical stresses shows that for MFS whose magnetic circuit operates on a linear portion of the main magnetization curve, the above dependence can be approximated as a linear function with a negative (positive) coefficient, sign which depends on the brand used of magnetoelastic material.

The expression of the output signal at the output of one concentrated section of the measuring winding with the concentrated section of the magnetizing winding for the developed MFS has the following form:

$$E_{out} = 2j\omega W_{out} \frac{\underline{\beta}\,(U_м)\dot{F}_в}{\underline{Z}_{\mu n}(U_м)\alpha_м\underline{\Delta}^*}\left[ch\left(\frac{1}{2}\underline{\beta}\,(U_м)\right) + \right.$$
$$\left. + \underline{\beta}\,(U_м)\underline{k}_0 sh\left(\tfrac{1}{2}\underline{\beta}\right)\right] ch\left[\left(\tfrac{1}{2}\underline{\beta}\,(U_м)\right)\right],$$
(6)

where W_{out} - is the number of turns of a uniformly distributed measuring winding.

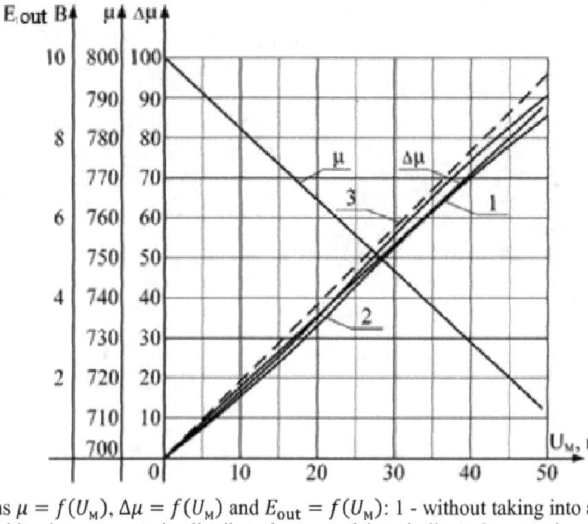

Fig. 4. Dependency graphs $\mu = f(U_м)$, $\Delta\mu = f(U_м)$ and $E_{out} = f(U_м)$: 1 - without taking into account the distributed nature of the windings; 2 - taking into account the distributed nature of the windings; 3 - experimental characteristic

The expression of the output signal at the output of one distributed section of the measuring winding with the distributed section of the magnetizing winding has the following form:

$$E_{\text{out}} = 2j\omega W_{\text{out sp.}} \left\{ \frac{\dot{F}_{\text{в}}}{2\underline{Z}_{\mu\text{п}}(U_{\text{м}})\alpha_{\text{м}}} - \frac{\underline{\beta}(U_{\text{м}})}{2\underline{Z}_{\mu\text{п}}\alpha_{\text{м}}} \left(2\frac{\dot{F}_{\text{в}}}{\underline{\Delta}^*} + \right. \right.$$
$$\left. + \dot{f}_{\text{в}} \frac{\alpha_{\text{м}}\underline{k}_0}{\underline{\Delta}^*} \right) \cdot \left[ch\left(\frac{1}{2}\underline{\beta}(U_{\text{м}})\right) + \underline{\beta}(U_{\text{м}})\underline{k}_0 sh\left(\frac{1}{2}\underline{\beta}(U_{\text{м}})\right) \right] \cdot$$
$$\left. \cdot sh\left(\frac{1}{2}\underline{\beta}(U_{\text{м}})\right) \right\}$$
(7)

where $W_{\mathbf{out\ sp.}}$ - is the specific value of the number of turns of a uniformly distributed measuring winding.

1.4 Dynamic characteristics of the developed MFS

The dynamic characteristics of the developed MFS were studied under the following six typical modes (influences): 1) a constant force acts on the MFS and it is connected to a sinusoidal voltage source; 2) MFS is connected to a constant voltage source, and a single force is applied to the sensor input; 3) MFS is connected to a constant voltage source, and a linearly increasing force is applied to the sensor input; 4) MFS is connected to a sinusoidal voltage source, and a single force is applied to the sensor input; 5) with a sinusoidal voltage of the power source and supplying a linearly increasing force to the sensor input; 6) with a sinusoidal voltage of the power source and supply to the input of the sensor a sinusoidal force [7]. Expressions of the transient voltage are obtained for these modes, and for the first mode and expressions of the amplitude and phase-frequency characteristics. In particular, for the fourth mode, the expression has the form (Fig. 5):

$$U_{\text{э.out}}(t) = U_{\text{эп}}(t)\frac{k_2 U_{\text{м}}}{T_{\text{э}\mu}} e^{-\frac{t}{T_{\text{э}\mu}}} = = \frac{k_2 U_{\text{м}} U_{\text{эпm}}}{T_{\text{э}\mu}} e^{-\frac{t}{T_{\text{э}\mu}}} sin\omega_{\text{э}} t. \quad (8)$$

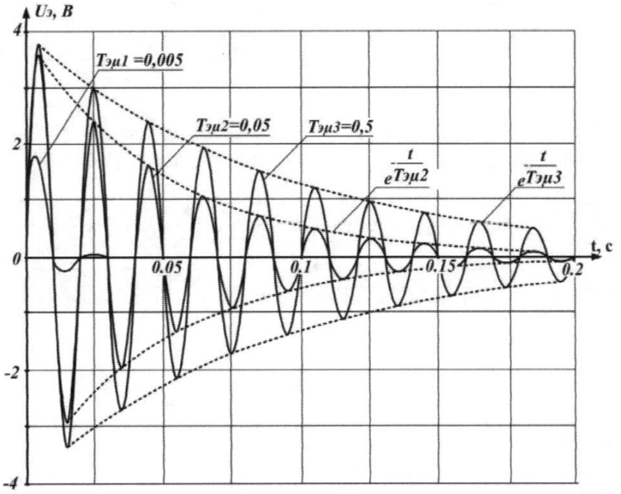

Fig.5. Transient voltage curves for the mode when the developed MFS is connected to a sinusoidal voltage source, and a single force is applied to the sensor input

where $k_2 = G_{\text{э}} K_{I_{\text{э}}U_{\mu}} K_{I_{\mu}U_{\text{э}}} K_{U_{\text{м}}C_{\mu}}$; $G_{\text{э}}$ - electrical conductivity of the magnetization circuit; $K_{I_{\text{э}}U_{\mu}} = w_{inp}$ - the number of turns of the input winding is the conversion coefficient of the interchain physical and technical effect (PTE) between the electric current $I_{\text{э}}$ and the magnetic voltage U_{μ}; $K_{I_{\mu}U_{\text{э}}} = w_{out}$ - the number of turns of the input (measuring) winding - the conversion coefficient of the inter-circuit PTE of electromagnetic mutual induction between the magnetic current I_{μ} and the output electrical voltage U_{out}; $K_{U_{\text{м}}C_{\mu}}$ - the coefficient of interchain PTE (magnetoelastic effect) between the mechanical force $U_{\text{м}}$ and magnetic capacity (magnetic conductivity by the classical analogy of chains); $T_{\text{э}\mu} = T_{\text{э}} + T_{\mu}$ - time constant of the electromagnetic circuit developed by MFS; $T_{\mu} = R_{\mu} C_{\mu\Sigma}$ and $T_{\text{э}} = G_{\text{э}} K_{I_{\text{э}}U_{\mu}} K_{I_{\mu}U_{\text{э}}} C_{\mu\Sigma} = w_{inp}^2 C_{\mu\Sigma} G_{\text{э}} = \frac{L_{\text{э}.inp}}{R_{\text{э}}}$ - respectively, the time constant of the magnetic and electrical circuits, $L_{\text{э}.inp}$ - electrical inductance of the excitation circuit; $C_{\mu\Sigma}$ - the total magnetic capacity of the magnetic circuit in the path of the working magnetic flux Q_{μ}; $U_{\text{эпm}}$ and $\omega_{\text{э}}$ - respectively, the amplitude and frequency of the voltage of the power source.

An analysis of the obtained expressions and their curves for dynamic modes showed that the developed MFS in the monitoring and control system correspond to real differentiating links with statism, and at high frequencies of the voltage of the power source, i.e. when $\omega_{\text{э}} T_{\text{э}\mu} \gg 1$, MFS from a differentiating link turns into a proportional one.

1.5 The error of the new MUD

The analysis of the compiled classification of possible sources of errors developed by the MFS showed that methodological, technological, operational errors are systematic sources, and internal, external and operational errors are sources of random errors. To identify the sources of errors and their analysis, the apparatus of parametric structural schemes was used, as well as the well-known concepts of additive and multiplicative errors [8, 9]. A quantitative assessment of errors showed that the non-linearity of the operating characteristic (the error is up to ±7.0%), magnetoelastic hysteresis (up to ±2.0%), changes in the ambient temperature (up to ±2.5%) and the instability of the parameters of the power supply (up to ±15.0%). It has been found that the most profitable ways to reduce the error: namely, from the nonlinearity of the operating characteristic of the MUD is the conversion of the analog signal to a proportional frequency (the error decreases to ±0.1%); from glue-elastic hysteresis - production of a magnetic circuit monolithic (up to ±0.2%); from changes in ambient temperature - the use of a second, stress-free MFS (up to ±0,1% per 10^0C of its additive component) and the use of differential circuits of two MFS, loaded with forces of different signs (compression and tension) (up to ±0.015 per 10^0C of its multiplicative component); from instability of power supply parameters - conversion of highly stable DC voltage into rectangular pulses (up to ±0.01%). Calculations and experiments have shown that the relative reduced error of the developed MFS does not exceed ±0.5%. The main characteristics of the developed MFS shown in Table 1.

Table 1. Main characteristics of the developed MFS

Range of the measured efforts, kN	$5 \cdot 10^2 \div 5 \cdot 10^4$
Sensitivity, mV/kN	0.2
Conversion error,%	0.5
Nonlinearity,%, no more than	0.5
Range of output voltage variation, V	0÷10
Operating temperature range, °C	-30…+60
Speed, s	10^{-4}
Overall dimensions, mm	120x100x100
Weight, kg, no more than	0.5

Conclusion

Thus, the main technical characteristics of the new magnetoelastic force sensors taking into account the distribution of the parameters of their electric and magnetic circuits are investigated in the article.

References

1. Patent RUz (UZ) №IAP 04866. Magnitouprugiy datchik usiliy/ Amirov S.F., Nazirova Z.G., Jurayeva K.K., Boltaev O.T., Sharapov Sh.A., Fayzullaev J.S.// Offitsialniy vestnik, 2014. - №4.
2. Patent RUz (UZ) №IAP 05432. Magnitouprugiy datchik usiliy/ Amirov S.F., Turdibekov K.Kh., Juraeva K.K., Boltaev O.T., Fayzullaev J.S.// Offitsialniy vestnik, 2017. - №7.
3. Zaripov M.F. Preobrazovateli s raspredelyonnimi parametrami dlya avtomatiki i informatsionno-izmeritelnoy texniki. - Moskva: Energiya, 1969. - 176 s.
4. Konyukhov N.E., Mednikov F.M., Nechaevsky M.L. Elektromagnitnie datchiki mexanicheskix velichin. - Moskva: Mashinostoenie, 1987. - 256 s.
5. Bessonov L.A. Teoreticheskie osnovi elektrotexniki. Elektricheskie tsepi. - Moskva: Gardariki, 2006.- 701 s.
6. Sharapov V. Piezoceramic sensors. - Springer, 2011 .-- 498 p.
7. Shandrov B.V. Texnicheskie sredstva avtomatizatsii: uchebnik dlya studentov visshix uchebnix zavedeniy/ B.V. Shandrov, A.D. Chudakov. - Moskva: Izdatelskiy tsentr "Academiya", 2007. - 368 s.
8. Zaripov M.F., Zaynullin N.R., Petrova I.Yu. Energoinformatsionniy metod nauchno-texnicheskogo tvorchechstvo. - Moskva: VNIIIPI GKNT, 1988.- 124 s.
9. Golovanova A.M., Kravtsov A.V. Teoreticheskie osnovi elektrotexniki. Elektricheskie izmereniya: - Uchebnoe posobie dlya studentov elektrotexnicheskix spetsiolnastey. - Moskva: FGOU VPO MGAU, 2006 .- 96 s.

Two approaches for automating analysis of electromagnetic processes in non-linear circuits with valves

Khushnud Sapaev[1] and *Shukhrat* Umarov[2]*

[1] Technical Sciences Doctor, Professor, Tashkent State Technical University named after I. Karimov, Tashkent, Uzbekistan
[2] Technical Sciences Candidate, Assistant Professor, Tashkent State Technical University named after I. Karimov, Tashkent, Uzbekistan

Annotation: This article analyses and compares two approaches related to automating modeling of valve electric circuits with piecewise linear approximation of valve characteristics. The first approach is based on operator equivalent circuits' analytical formulas and on analytical expressions programming which describe equivalent circuits on each conduction interval of valve elements. The second approach provides implementation of a system for modeling valve converters based on implicit numerical integration formulas.

Keywords: converter, power supply, mathematical model, operator method, transient process, rectifier, inverter, loading.

As far as is known, the confident solutions of various problems concerning analysis and parametric synthesis of valve converters are associated with the implementation of a large volume of calculations. In this connection, the following should be noted that a significant part of failures related to electronic equipment occurs because of circuit and design flaws that arise mainly due to the use of manual methods for calculating circuits in engineering practice. This is due to the fact that manual calculation methods are based, as a rule, on a number of simplified assumptions that significantly reduce the quality of design and suggest the need for experimental prototyping - a very time-consuming and expensive process. Therefore, a widespread use of digital computers (DC) in the design process is a prerequisite for improving the quality, efficiency and shortening the development time of valve converters.

The modern technological development horizon of computer technology allows us to raise the question of not only performing separate calculations, but it also has made it possible to determine the optimal parameters of circuit components, to simulate various dynamic modes, to explore any combination of circuit elements, and thereby provides automation of design and development work related to the implementation into production of various complex schemes of valve converters.

The greater part of programs for analyzing electrical circuits with valves use valve element models idealized on the basis of a number of assumptions, which are practicality confirmed by practice. In this case, the current-voltage characteristics (CVC) of essentially nonlinear elements – valves correspond to the characteristics of an ideal key or have a different form, reflecting a piecewise linear approximation (PLA). The PLA of nonlinear characteristics and the use of digital computers for calculations allow widespread use in the analysis of circuits the fitting method for the intervals of valve conductivity [1], in which the electric circuit is considered as a linear on the segments between the break points of the approximated characteristics, and the structure or parameters of the circuit change when moving from one segment to another.

Possible ways to fitting method implementation on a digital computer (DC) may be:

- The program development based on the classical methods of transient analysis;
- The creation and use of universal programs, including models of electric circuits power elements and their control systems.

A correlative study of the two approaches which has been described above is carried out on the example of calculating a parallel autonomous current inverter [1, 2, 3] with a compensating device. Schematic diagram of a stand-alone current inverter (SACI) with a compensating device (CD) is shown in the Fig. 1.

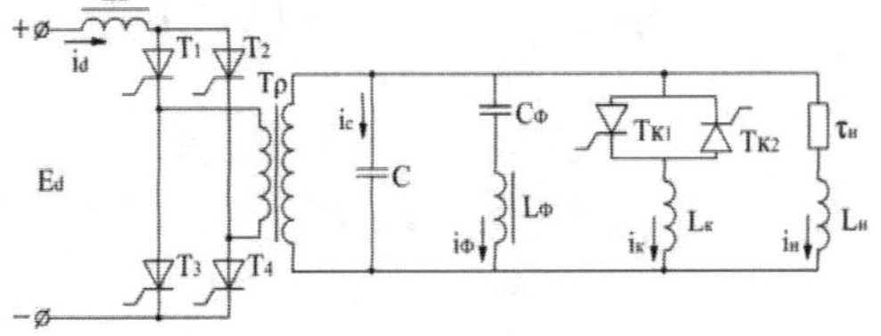

Fig. 1. Parallel SACI with CD Schematic Diagram.

The circuit operate main valves are in operation in pairs (T1, T4 - T2, T3). They do switch at a frequency of 50 Hz. The pair T1 – T4 is at the initial moment in a conducting state. Phase shift between the output voltage $u_н$ and the output current of the inverter $i_н$ is $_β$ = 45 °. The compensating device, consisting of the elements *Lк, Rк, T5, T6,* is turned on for the sixth period with an angle α = *29°,* counted to the left along the time axis from the transition of the voltage curve through *"Zero"*.

Initial approximation to the circuit analysis machine implementation has been based on analytical formulas of operator equivalent circuits and programming of analytical expressions describing equivalent circuits on each conduction interval of valve elements whose characteristics correspond to ideal keys [4, 5, 6]. Also programmable is the key state management algorithm. The mathematical models development is to be carried out in the following sequence:

a) Many possible types of equivalent circuits are to be determined. That will occur with the selected method of excitation and stabilization;

b) The equivalent universal circuits are to be compiled. Current images and originals and, also, required voltages in the calculation are to be displayed;

c) Different possible ways of the process development are to be analyzed and the sequence of equivalent circuits changing types at the clock intervals will be revealed. The boundary conditions for the process development stages change are to be formed this way;

d) An algorithm block diagram for calculating the transient will be compiled this way. The algorithm block diagram is based on the search for the fulfillment of boundary conditions taking into account the selected method of excitation and regulation.

The analysis of possible operator equivalent circuits (OECs) shows that a complete calculation of the transient process of parallel SACI with CD may be carried out on the basis of two universal OECs that differ only in operating state of CD. One of them corresponds to the switched-off state; the other corresponds to switched-on state. Their using for the analysis of various operational situations is determined by the corresponding initial conditions. The calculation formulas of instantaneous values of the sought currents and voltages are to be derived for each of these universal OECs [3].

The currents and voltages originals are to be expressed as:

$$i_d(t)=\tau_1 E_d/a_{12}+(B_2/D_1)exp(-б_2 t)+$$
$$+(B_3/D_2)\,exp(-б_3 t)+A_1K_1+B_1K_2;$$
$$u_c(t)=\tau_H\tau_K E_d/a_{12}+(B_5/D_1)exp(-б_2 t)+$$
$$+(B_6/D_2)\,exp(-б_3 t)+A_4K_1+B_4K_2;$$
$$i_к(t)=\tau_H E_d/a_{12}+(B_5/D_3)exp(-б_2 t)+$$
$$+(B_6/D_4)\,exp(-б_3 t)+A_4K_3+B_4K_4;$$
$$i_H(t)=\tau_H E_d/a_{12}+(B_5/D_5)exp(-б_2 t)+$$
$$+(B_6/D_6)\,exp(-б_3 t)+A_4K_5+B_4K_6;$$

The obtained analytical expressions have been programmed and a process development algorithm has been determined. In result we obtained a mathematical model, with the help of which transient processes were calculated, and sought currents and voltages time diagrams were constructed, shown in Fig. 2, 3 and 4.

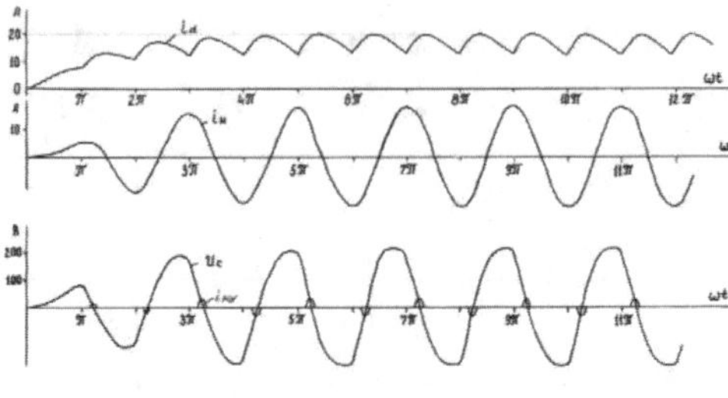

Fig. 2. Currents and Voltages Timing Diagrams at Start-up

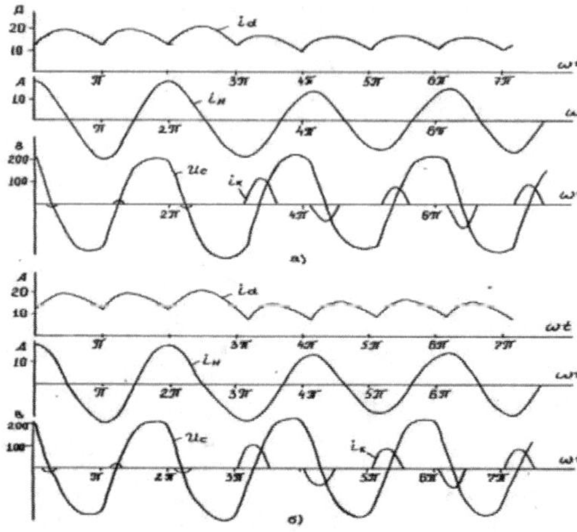

Fig. 3. Currents and Voltages Timing Diagrams during Load Shedding: a) Independent Excitation; b) Combined Excitation.

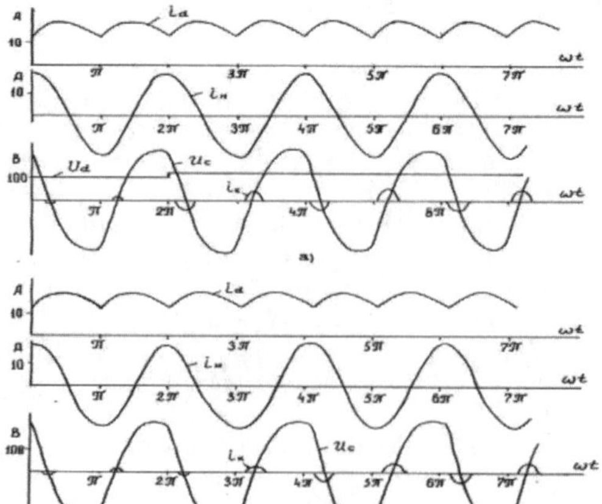

Fig. 4. Currents and Voltages Timing Diagrams during Input Voltage Increase: a) Independent Excitation; b) Combined Excitation.

We have undertaken a series of computer studies using the developed mathematical model (algorithm and program). The recommended parameters and part types of the SIP circuit elements based on a parallel inverter operating in a continuous mode of the input current of the inverter were obtained.

The second approach is implemented using a simulation system of valve converters based on implicit numerical integration formulas (PSIs) [7], while the valve characteristics VC of the valves take into account on-state losses (differential resistance r_d = 0.001 Ohm, threshold voltage u_0 = 0.1 B). The initial data prepared during the statement of the problem carry information about the circuit corresponding to one initial state of the valve elements. In the process of machine solution, after each switching of valves, a new equivalent circuit of the simulated circuit is automatically formed. The state of the valves is controlled using the developed algorithm. The minimum integration step is assumed to be 2 μs. The maximum value of the step automatically obtained during the operation of self-accelerating PSIs is 2150 μs.

A comparison of two approaches to automating analysis showed that the discrepancy in the results does not exceed tenths of a percent. The error was estimated by instantaneous values (the values were obtained by the analytical method and taken as a standard). It is calculated by the formula:

$$\varepsilon = \left[\frac{a-x}{a}\right],$$

Here ***a*** is the value obtained by the analytical method, and ***x*** is the value obtained in the process of numerical integration.

Table 1. Calculation Selected Results

Time max.	$u_н$ (standard) B	$u_н$ (PSIs) B	ε %	$i_н$ (standard) A	$i_н$ (PSIs) A	ε %
1667	4,287	4,279	0,19	2,085	2,082	0,19
3333	15,90	15,88	0,13	3,993	3,985	0,20
…	…	…	…	…	…	…
61667	-70,38	-70,24	0,19	17,9	17,88	0,11
63333	58,73	58,62	0,19	20,06	20,03	0,15
…	…	…	…	…	…	…
121667	71,17	71,04	0,18	17,98	17,96	0,11
123333	-58,75	-58,65	0,17	20,15	20,12	0,15

During the calculation process, the error at the integration step has been taken equal to 0.001. It should be noted that a decrease in this step had a weak effect on the calculation results.

We can observe from the table data, Table 1: coincidence of the results obtained by the first method (standard) and those obtained by implicit integration formulas is high – the relative error of instantaneous values over significant time periods (there were six periods) lies within 1%.

Summarizing the above we can say that, a design engineer may be in a puzzle: what analysis method should be selected in terms of efficiency and reliability? In this connection, we may say that most methods have a certain relationship and complement to each other [8, 9, 10]. In order to obtain reliable results that ensure the receipt of valve converters technical parameters, ensuring the fulfillment of the requirements of the technical specifications for the design of an object (taking into account the significant non-linearity of valve elements with a variety of control systems), it is advisable to use a combined modeling technique.

References

1. B.D. Bedford, R.G. Hoft. Princ. of Inv. Circ. (Wiley, New York, 1964)
2. Bose, Bimal K. Modern power elec. and AC dr. (Prentice Hall PTR, 2002).
3. Rik De Doncker, Duco W.J. Pulle, Andre Veltman. Adv. Elect. Drives: Analysis, Mod., Contr., 474 (Springer, 2011)
4. Sh.B. Umarov Math. mod. for calc. of trans. in stab. PS on the basis of single-phase seq. inv. cur. Jour. “Probl. of energy. and res. saving”, **Vol 4**, 112-115, (2015)
5. Sh.B. Umarov Math. Mod. of Stab. Power Sup. Based On Cur. Inv. //IJARSET: **Vol 6, Issue 8**, (August 2019)
6. Sh.B. Umarov, I.A. Abdullabekov Alg. of calc. of trans. of the stab. power supp. on the bas. of a single-ph. seq. autonom. cur. inv. with freq. reg. Int. sci. journal. "Young scientist", **No. 21**, 228-232, (2016)
7. Van Bakkoven W. Linear Imp. Diff. For. of Var. Step and Order// JEEE Frans. On CAS. **Vol 2**, 109-115, (1975)
8. Yu. Zinin, V.Tereshkin, A.Karamov Trans. of thyr. bridge inv. start-up with freq. doub. Power electronics, **No. 5**, (2011)
9. S.G.German-Galkin Comp. sim. of semicon. sys. in MATLAB env. Textbook, 320, (2001).
10. D.Tarnapowicz, S.German-Galkin “Ener, optim. of mech. sys. with PMSG”, “3rd Int. Conf. on Energy and Env. Prot.”, **Vol.46**, 1-8, (2018)

 https://doi.org/10.1051/e3sconf/201913901086

Power System Mode optimization by piecewise-linear approximation of energy characteristics of Power Plants

Tulkin Gayibov[1,*], *Sherxon* Latipov[1], and *Bakhadir* Uzakov[2]

[1]Tashkent State Technical University named after Islam Karimov, Uzbekistan
[2]Karakalpak State University, Uzbekistan

Abstract. The issues of improving of methods and algorithms for power system short-term mode optimal planning are considered. The main disadvantages and calculation difficulties typical for traditional algorithms are described. An effective algorithm for power system mode optimization based on a piecewise-linear approximation of nonlinear dependencies is proposed. Application of the proposed algorithm can effectively increase the accuracy and overcome many calculation difficulties associated with discontinuity of energy characteristics of power plants and various constraints in the form of equality and inequality.

1 Introduction

Solving the problem of short-term mode optimization of the power system provides definition for each interval of the considered planning period the active loads of the power plants, which ensure reliable supply of high-quality electricity to all consumers with minimal costs for their generation and transmission. Plan development is usually carried out on the basis of solving the corresponding optimization problem, where costs are taken into account by the objective function, and the requirements for the quality of electricity and reliability of power supply are taken into account by simple and functional constraints in the form of equality and inequality. Such problem for the modern power systems presents the difficult nonlinear mathematical programming problems, characterized by the presence of discontinuities in the energy characteristics of the power plants involved in optimization, as well as the presence of a large number of various constraints in the form of equality and inequality.

The most widespread and widely practically used algorithms for such problem solution in basis of gradient methods [1–5] and Newton's method [6-8] provide for the approximation of the energy characteristics of power plants by polynomials of various degrees. Application of a quadratic polynomial facilitates convenient solution of the problem. However, due to the low approximation accuracy a result can be obtained with a significant error, reducing optimization effect. With increase of the approximating polynomial degree, the approximation accuracy and, correspondingly, optimization increase as well. But the problem can become multi-extreme, which solution is associated with additional difficulties.

Another problem in the problem solution is related to taking into account the constraints in the form of inequality. Currently, in the power system mode optimization algorithms based on the gradient methods [1-5], simple constraints imposed on adjustable parameters are taken into account quite effectively, fixing in each iteration corresponding adjustable parameters within the violated limits. Taking into account of complex functional constraints in them and in the optimization algorithms by the Newton's method [6-8] is carried out by the penalty function method, characterized by a number of advantages such as universality, ease of implementation, and the calculation possibility from any initial approximation. At the same time, in some cases, when the penalty factors increase in these algorithms, computational difficulties arise due to the deterioration and thereby loss of reliability of optimization process convergence. The same difficulty occurs in the problem solution by the method of incremental rates equality [8, 9]. In [10], by the example of the electric network mode optimization, it was shown that in cases of application of the straight-linear approximation of the objective function, the iterative process converges is slow and unreliable.

In this regard, the problem of effective algorithm development and implementation for power system mode optimization with effective consideration of the real energy characteristics of power plants and constrains remains actual.

2 The optimization algorithm

In the problems with a large number of simple and complex constraints and with discontinuous energy characteristics of plants, the power system mode optimization can be effectively solved by the method of successive piecewise-linear approximation of nonlinear dependencies. In this case, the nonlinear programming problem is reduced to the linear programming one and is solved by the simplex method.

In this paper, we present an power system mode optimization algorithm based on a piecewise-linear approximation of the nonlinear energy characteristics of power plants and constraint functions free from the indicated calculation difficulties.

To reveal the essence of the algorithm, we consider the problem of the power system active load optimal distribution between thermal power plants (TPPs), taking into account the constraints, which is formed as follows:

minimize the function of the total fuel equivalent consumption in TPPs

$$B = \sum_{i=0}^{n} B_i(P_i) \quad (1)$$

subject to constrains

$$W(P) = \sum_{i=1}^{n} P_i - P_H - \pi = 0, \quad (2)$$

$$h_l(P) \le h_{ls.}, \qquad l = 1,2,..L \quad (3)$$

$$P_i^{min} \le P_i \le P_i^{max}, \qquad i = 1,2,..n, \quad (4)$$

where n is the number of rated power plants; P_i is the active load of the i-th TPP; $B_i(P_i)$ is equivalent fuel consumption in i-th TPP at its load P_i; P_L is the total load of the power system; π is the total active power loss in electric networks of the power system; $W(P)$ is the unbalance active power function in power system; h_l, h_{ls} are the calculated and specified limit values of the l-th function of the load of power plants (for example, the active power flow through a controlled transmission line).

The main condition under which the proposed algorithm can be applied is the separability of functions, i.e. the possibility of presenting them as a sum of several functions, each of which depends on only one variable [11]. The objective function in the adopted optimization model for the power system mode is always separable. Functional constraints for which these conditions are not satisfied are preliminarily reduced to a separable form using special methods, as in [11].

Normally, in the iterative calculation optimization cycle, the value of the total active power loss in electric networks is taken from the calculation of its steady state mode, and therefore the constraint (2) is linear. (3) represents a lot of constraints in the form of inequality on active power flow in some monitored transmission lines, determined from the conditions of ensuring the static stability of the power system, and on the permissible currents of individual transmission lines. The flows of active power in monitored lines in the proposed algorithm is expressed through the distribution coefficients, which determine their values with sufficient accuracy for practical purposes. As a result, these constraints turn into a linear form.

During optimization by this algorithm, a piecewise-linear approximation of the objective function and nonlinear functional constraints is made by choosing in a given interval [$P_{i,min}$; $P_{i,max}$] m_i points with coordinates [B_{ik}; P_{ik}] and replacing it in each segment [$P_{i,k}$; $P_{i,k+1}$] by the linear functions:

$$\bar{B}_{ik}(P_i) = B_i(P_{ik}) + \frac{B_i(P_{ik+1}) - B_i(P_{ik})}{P_{ik+1} - P_{ik}}(P_i - P_{ik})$$

$$\bar{h}_{ik}(P) = h_i(P_{ik}) + \frac{h_i(P_{ik+1}) - h_i(P_{ik})}{P_{ik+1} - P_{ik}}(P_i - P_{ik}).$$

representing the values P_i in the interval $[P_{ik}; P_{ik+1}]$ as

$$P_i = \lambda_k P_{ik} + \lambda_{k+1} P_{ik+1} \quad (5)$$

and choosing λ_k and λ_{k+1} under the condition $\lambda_k + \lambda_{k+1} = 1$, we obtain the following approximate linear programming problem for (1)–(4):

minimize function

$$B = \sum_{i=1}^{n} \sum_{k=1}^{mi} \lambda_{ik} \bar{B}_{ik}(P_i) \quad (6)$$

subject to constraints

$$\left.\begin{aligned} &\bar{W} = \sum_{i=1}^{n} \sum_{k=1}^{mi} \lambda_{ik} W_{ik}(P) = 0, \\ &\bar{h}_i = \sum_{i=1}^{n} \sum_{k=1}^{mi} \lambda_{ik} \bar{h}_{ik}(P) \le 0, \qquad i = 1,2,.L, \\ &\sum_{k=1}^{mi} \lambda_{ik} = 1; \quad i = 1,2,.n, \\ &\lambda_{ik} \ge 0; \quad i = 1,2,.n \quad k = 1,2,.m_i. \end{aligned}\right\} \quad (7)$$

where λ_{ik} are the unknown parameters of the obtained linear programing problem.

Problem (6)–(7) is solved by the simplex method, taking into account constraints on the basis choice, which consists of inadmissibility of including two vectors in the basis that correspond to two non-neighboring ones.

After solving the problem (6)–(7), approximate optimal values of TPP loads P_i are determined by (5). If it is necessary, in order to find the exact solution, we can perform the next step with a more accurate piecewise-linear approximation of the nonlinear dependences of the separable problem in the vicinity of the point obtained in the previous step.

3 The results of calculation experiments

We illustrate the computational qualities of the algorithm described above by optimizing of mode of the power system with testing circuit [8] shown in Fig. 1, where 0 (balancing), 1, 6, 7 are nodes with TPPs.

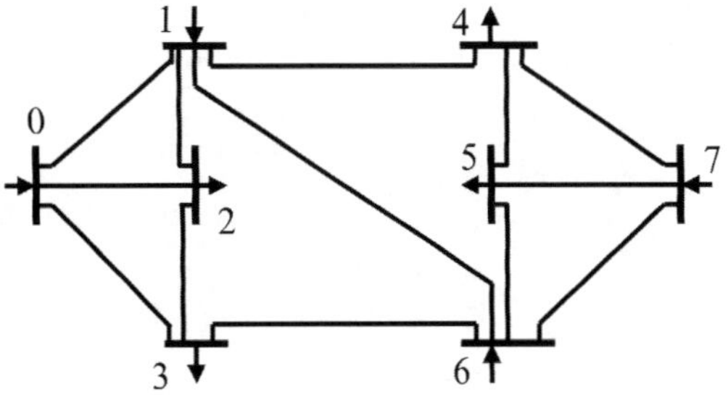

Fig. 1. The scheme of the power system

The characteristics of the incremental rates of the fuel equivalent consumption of these TPPs are given in table 1. Where B_{00}, B_{10}, B_{60}, B_{70} are fuel equivalent consumptions per hour at minimum loads of TPPs. The power of the load nodes are, MW: $P_2 = 400$, $P_3 = 600$, $P_4 = 200$, $P_5 = 500$. In three power transmission lines (PTLs) the power flows are limited, MW: $P_{6\text{-}3} \leq 450$, $P_{6\text{-}5} \leq 150$, $P_{0\text{-}3} \leq 100$. Matrices of power distribution coefficients of the nodes through these lines are given in table 2.

Table 1. Characteristics of the incremental rate of fuel equivalent consumption in TPPs.

TPP–0 $B_{00} = 114\, t.f.e./h.$		TPP – 1 $B_{10} = 32\, t.f.e./h.$		TPP–6 $B_{60} = 114{,}8\, t.f.e./h.$		TPP– 7 $B_{70} = 97\, t.f.e./h.$	
b_0 *t.f.e./MW.h*	P_0 *MW*	b_1 *t.f.e./MW.h*	P_1 *MW*	b_6 *t.f.e./MW.h*	P_6 *MW*	b_7 *t.f.e./MW.h*	P_7 *MW*
0,316	320	0,34	70	0,2705	400	0,22	320
0,317	384	0,359	77	0,2836	565	0,27	398
0,319	427	0,366	92	0,2967	583	0,295	438
0,334	428	0,524	92	0,3098	595	0,32	450
0,3341	640	0,533	100	0,3229	647	0,345	495
				0,336	715	0,395	502
				0,3491	759		
				0,3622	790		

Table 2. Power distribution coefficients of the nodes through controlled PTLs.

PTL	Nodes						
	1	2	3	4	5	6	7
6-3	0,2536	-0,0713	-0,172	0,4106	0,439	0,6343	0,4986
6-5	-0,0701	0,02025	0,0484	-0,281	-0,4223	0,1029	-0,2986
0 3	0,14	0,1735	0,267	-0,1744	-0,1917	-0,224	-0,1939

For comparison, presented the reference result of the optimal distribution of the power system total load between TPPs taking into account the given simple and functional constraints:
$P_0 = 545{,}023$ MW; $P_1 = 70$ MW; $P_6 = 634{,}977$ MW; $P_7 = 450$ MW; $P_{6\text{-}3} = 450$ MW; $P_{6\text{-}5} = 150$ MW; $P_{0\text{-}3} = 68{,}5$ MW; $B = 532{,}57$ t.e.f./h.,
where *t.f.e./h.* is tones of fuel equivalent per hour.

By numerical integration of incremental rate characteristics of TPPs at given points (table 1), we perform a piecewise-linear approximation of fuel equivalent consumption characteristics of TPPs. As a result, the function of total fuel equivalent consumption in power system is expressed as follows:

$$\begin{aligned}\overline{B} &= 114\lambda_{11} + 134{,}56\lambda_{12} + 147{,}93\lambda_{13} + 148{,}256\lambda_{14} + 219{,}75\lambda_{15} + \\ &+ 32\lambda_{21} + 34{,}446\lambda_{22} + 39{,}883\lambda_{23} + 39{,}883\lambda_{24} + 44{,}112\lambda_{25} + \\ &+ 97\lambda_{31} + 116{,}11\lambda_{32} + 127{,}41\lambda_{33} + 131{,}1\lambda_{34} + 146{,}062\lambda_{35} + \\ &+ 148{,}652\lambda_{36} + 44{,}8\lambda_{41} + 90{,}513\lambda_{42} + 95{,}736\lambda_{43} + 111{,}505\lambda_{44} + \\ &+ 133{,}017\lambda_{45} + 148{,}069\lambda_{46} + 158{,}403\lambda_{47}\end{aligned} \tag{8}$$

We express the loads of individual TPPs as

$$\left.\begin{aligned} P_0 &= 320\lambda_{01} + 384\lambda_{02} + 427\lambda_{03} + 428\lambda_{04} + 640\lambda_{05}, \\ P_1 &= 70\lambda_{11} + 77\lambda_{12} + 92\lambda_{13} + 92\lambda_{14} + 100\lambda_{15}, \\ P_6 &= 320\lambda_{61} + 398\lambda_{62} + 438\lambda_{63} + 450\lambda_{64} + 495\lambda_{65} + 502\lambda_{66}, \\ P_7 &= 400\lambda_{71} + 565\lambda_{72} + 583\lambda_{73} + 595\lambda_{74} + 647\lambda_{75} + \\ &+ 715\lambda_{74} + 759\lambda_{75} \end{aligned}\right\} \tag{9}$$

subject to conditions

$$\begin{aligned} &\lambda_{01} + \lambda_{02} + \lambda_{03} + \lambda_{04} + \lambda_{05} = 1, \\ &\lambda_{11} + \lambda_{12} + \lambda_{13} + \lambda_{14} + \lambda_{15} = 1, \\ &\lambda_{61} + \lambda_{62} + \lambda_{63} + \lambda_{64} + \lambda_{65} + \lambda_{66} = 1, \\ &\lambda_{71} + \lambda_{72} + \lambda_{73} + \lambda_{74} + \lambda_{75} + \lambda_{76} + \lambda_{77} = 1. \end{aligned}$$

Substituting these expressions of the power plant loads into the power balance equation in power system and the formulas for the power flows along the controlled PTLs expressed through the given distribution coefficients, we obtain a linear programming problem that is effectively solved by the simplex method. Then, substituting the obtained values of the variables λ_{ik} into expressions (9), we find the optimal loads of TPPs:
$P_0 = 541{,}487$ MW; $P_1 = 70$ MW; $P_6 = 621{,}985$ MW; $P_7 = 466{,}528$ MW.
At these loads of TPPs we have:
$P_{6\text{-}3} = 450$ MW; $P_{6\text{-}5} = 150$ MW; $P_{0\text{-}3} = 68{,}4$ MW; $B = 532{,}68$ *t.f.e./h.*
Having compared the obtained solution with the reference result, it can be argued that there is no need to perform the following optimization step for its refinement. At the same time, the TPPs loads turned out to be quite close to the reference ones, the fuel equivalent consumption overrun is negligible.

Improving the piecewise-linear approximation accuracy of the problem can be achieved by increasing the number of given points in incremental rate characteristics of fuel equivalent consumption. In this case, the dimension of the resulting linear programming problem increases accordingly.

On the basis of numerous experimental calculations, it has been established that in real-life problems of power system mode optimization, it is expedient to carry out a calculation with piecewise-linear approximation of non-linear dependencies using all the given points in incremental rate characteristics of fuel equivalent consumption of power plants. This enables to ensure the absolute convergence of calculation process and obtain the optimal result in one step with sufficient accuracy.

Another positive feature of the described algorithm is that it successfully overcomes the difficulties associated with the discontinuity of the energy characteristics of power plants.

4 Conclusion

Optimization of power system modes by piecewise-linear approximation of nonlinear dependencies allows us to consider effectively the simple and functional constraints in the form of equality and inequality.

In optimization of a power system mode by piecewise-linear approximation of nonlinear dependences, difficulties associated with the discontinuities in energy characteristics of power plants are automatically overcome.

References

1. *Methods for power system mode optimizaion.* V.M. Gornstein, B.P. Miroshnichenko, A.V. Ponomarev and others; Edited by V.M.Gornstein (Energy, Moscow, 1981).

2. Krumm L.A. *Methods of optimization and control of electric power systems* (Nauka, Novosibirsk, 1980).

3. *Automation of dispatch control in the electric power systems* / Edited by Yu.N. Rudenko and V.A. Semenov (Publishing MPEI, Moscow, 2000).

4. Murty, P. S. R., *Optimization and Control in Power Systems* (BS Publications, Headarabad, 2008).

5. M.R. Irving and Y.H. Song. *Optimization Methods for Electric Power Systems*, Part 1, Mathematical Optimization Methods. *IEE Power Engineering Journal*, 14 (5): 245-254, (2000).

6. D.I. Sun, B. Ashley, B. Brewer, A. Hughes and W.F. Tinney. Optimal Power Flow by Newton Approach. *IEEE Trans. Power Apparatus and Systems*, PAS-103(2): 2864 – 2878, (1984).

7. Arzamastsev D.A., Bartolomei P.I., Kholyan A.M. *ACS and optimization of power system mode: Textbook for university students* (Higher school, Moscow, 1983).

8. Fazylov H.F., Nasyrov T.Kh. *Steady state mode of power systems and its optimization* (Moliya, Tashkent, 1999).

9. Nasirov T.Kh., Gayibov T.Sh. *Theoretical foundations of power system mode optimization* (Fan va texnologiya, Tashkent, 2014).

10. Gayibov T.Sh., Gafurov T.F. *Problems of Energy and Resource Saving*, No 2-3, 16-23 (2005).

11. G.V. Reklaitis, A. Ravindran, K. Ragsdel. *Engineering optimization. Methods and Applications* (John Wiley and Sons, NY, 1983).

Portable autonomous solar power plant for individual use

Javoxir Toshov [1], *Elyor* Saitov [2]

[1]DSc, Associate Professor, Dean of Energy Engineering faculty, Tashkent State Technical University, Uzbekistan
[2]PhD, Associate Professor, Deputy Dean of Energy Engineering faculty, Tashkent State Technical University, Uzbekistan

Abstract. Development and design of low power mobile and inexpensive solar stations are important for farmers, individuals, geologists and trotters. The most important aspect is ensuring reliability of units in times of transportation, convenience and operational ease as well as minimum dimensions, low price. All the above will ensure large-scale implementation of solar stations in various applications and utilizations.

1. Introduction

Our modern society is constantly moving somewhere, and this movement, otherwise called "constant physical inactivity" - makes some sections of the population retire from society and its modern achievements. But, probably, nobody succeeds in completely retreating from all its features and achievements in the modern world, and therefore, although some minimal connection with it is necessary for everyone, even the most avid modern "aborigines" [1].

In this article we will consider mobile solar power plants based on solar batteries - through which, any of us, while away from home, is able to establish normal communication with family and friends, as well as provide our other, minimal energy needs.

The sun is a source of inexhaustible energy reserves, and the widespread use of this energy is one of the most urgent tasks for the Republic of Uzbekistan, which has significant solar energy resources. Other alternative sources of energy (ASE) are also significant.

The main supplier of solar electric energy are solar cells, the principle of which is based on the direct conversion of solar radiation energy directly into electrical energy. Solar cells generate energy at low operating costs and do not pollute the environment. Due to the increase in the cost of petroleum products, a sharp increase in demand for alternative energy sources is forecasted. The environmental and social effects of the use of solar power in remote and inaccessible areas should also be considered. The widespread use of alternative sources of energy will save natural resources, improve the ecology of the environment, reduce carbon dioxide emissions, improve living conditions of the population and stimulate the training of specialists in the field of alternative sources of energy [5-6].

2. Main Body

2.1 Theoretical analysis

Calculations show that the use of alternative sources of energy in remote areas of the republic gives not only a social and environmental effect, but also a direct economic effect.

The aim of this work was the development of portable solar stations, providing lighting for at least 8 hours, charging mobile phones. At the same time, these stations should be convenient for transportation. In this case, it was necessary to accurately calculate the power of the solar flux, which would not only have to provide the necessary energy for the system for illumination at night with a duration of 8-10 hours, but also recharge the battery. However, the device should have been optimally sized for ease of transportation and placement [3].

In accordance with the requirements of [4], we developed a universal portable autonomous solar power plant for individual use, intended for the electrification of farms and summer cottages, as well as for use in remote and difficult to reach places where there is no traditional power supply.

The power of the solar panel is 20 watts. Inverter power maximum 300 watts. The effective output voltage of the inverter is 220 V × with a frequency of 50 hertz. Battery capacity 14 A × hour. The maximum duration of one TV and room lighting for 4 hours. The maximum duration of lighting a room for 20 hours. The equipment is placed in a case with a size of 15×20×30 cm^3 in a protected version.

2.2. Experimental part

Currently, the world is actively developing the production of efficient solar cells for the development and commissioning of similar mini-solar stations [7-8].

However, it should be noted that the direct use of these sources is not possible due to a number of factors, since the energy generated by the photoelectric generator has a number of specific features:

• The non-stationary nature of energy production, which means that a system of energy storage (electrochemical battery) is needed.

• Typically, electrical energy is generated as a constant voltage in the range of 12 to 48 volts.

https://doi.org/10.1051/e3sconf/201913901087

At the same time, most household consumers (refrigerators, televisions, pumps, and others) require an alternating voltage of 50 hertz and a voltage of 220 volts.

To convert the DC voltage of solar stations to AC with the necessary characteristics, inverters are required [9-10].

Inverters for solar stations in their characteristics should differ significantly from the simplest inverters and uninterruptible power supplies, which are widely represented on the market. They should be distinguished by a high level of reliability, high overload capacity, built-in electronic protection systems [11-12]. Currently, "solar" inverters in the Republic of Uzbekistan are not produced by industry, and they have to be purchased for currency funds. There is also no production of some other nodes of solar and wind stations - battery charge controllers, supporting structures, there is a significant lag in this area. The development of the in-house production of such inverters is necessary for the successful implementation of small (up to 1 kW) solar power plants for farmers and the population remote from industrial electrical networks. We developed at the Department of Digital Electronics and Microelectronics comprehensive solutions for creating a whole line of "solar" inverters with a capacity of up to 2 kW for various applications and installations of various capacities (charging cell phones, mobile solar power plants, etc.). The operational characteristics of the developed inverter with power up to 2 kW are given in the table .1.

2.3. The results and discussion

Mini solar stations with such inverters have been successfully tested in real conditions and have established themselves as reliable devices with a sufficiently long service life.

With the introduction of the developed inverter and the transition to its industrial production, it will be possible to refuse to purchase inverters for solar power plants abroad. This will significantly save foreign currency. Perhaps the widespread use of the developed inverter in all regions of Uzbekistan.

Table 1. Performance characteristics of the developed inverter

Parameter Name	Parameter Value
Rated Power	2000 W
Maximum power (within 3 minutes)	2500 W
Short-term (up to 5 seconds) overload capacity from the nominal value	50%
DC input voltage	21 to 30 V
AC output voltage (50 Hz)	220V
Protection against short circuit of loading 220B	Automatic switch 15A
The output voltage form	sinusoidal, with the peaks limited by a protection device
Harmonic coefficient (at rated power)	Not more than 5%
Stability of the output voltage (when the battery is fully charged) and maximum load	From +5 to -10%
Weight	not more than 20 kg
Service life	Not less than 10 years
Continuous Operation Time	Unlimited
Maintenance Need	Check Cooling Fan and Air Filter

A simplified diagram of the power part of the inverter is shown in Fig. 1.

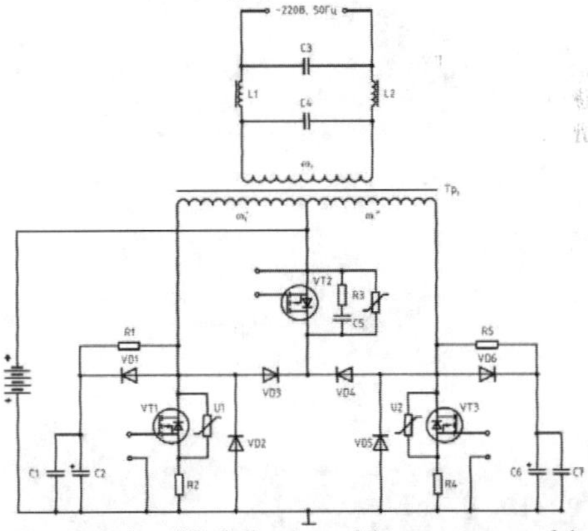

Fig. 1. A simplified diagram of the power part of the inverter.

The inverter power unit (Fig. 2) was made on the basis of 4 mm thick aluminum, on which insulated heat sink radiators of power transistors and diodes, as well as other elements of the power unit, were fixed.

The whole control unit circuit is divided into four parts according to the functional-node principle.

The following nodes were allocated:

- the microcontroller part, combining the

microcontroller, the elements of the display of operating modes and controls,
- node high-speed protection and the formation of pulses of control power transistors,
- galvanic isolation node with three channels,
- power node, forming the necessary parameters of the supply voltage.

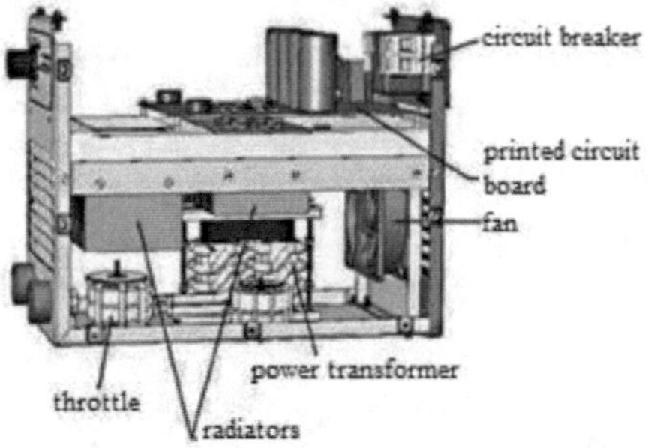

a)

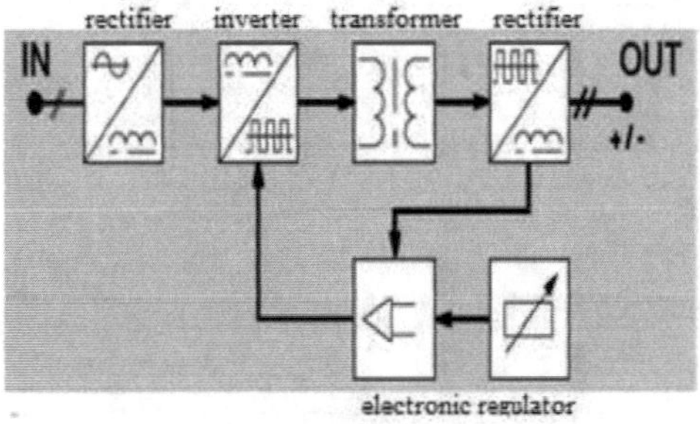

b)

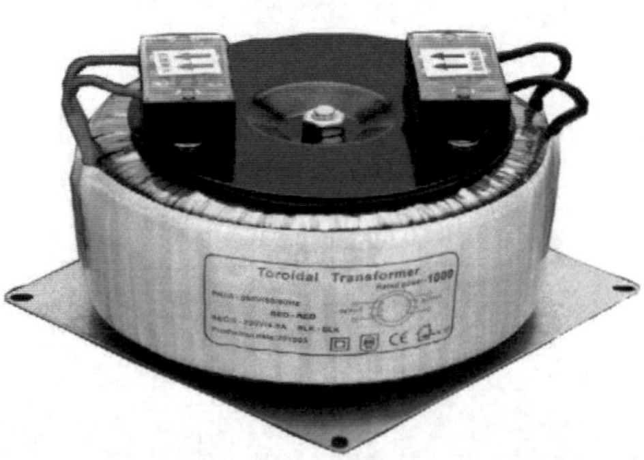

c)

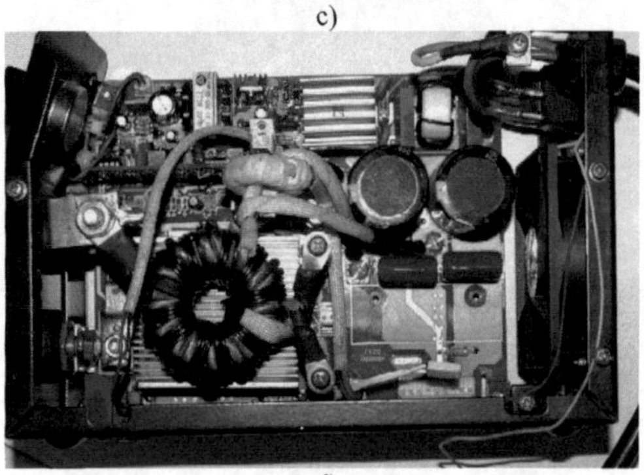

d)

Fig. 2. Power block inverter.

a) the power part design of inverter;
b) control and management scheme;
c) toroidal output transformer;
d) the appearance of the power part of inverter.

3. Equations and mathematics

As for the batteries, the rechargeable battery (assembled from finished batteries in parallel-serial connection to the desired voltage and capacity).

It should be noted here that the solar battery is a source of electric power, the properties of which are highly dependent on illumination. For the case of an ideal p-n junction $I = I_s\left(e^{\frac{U}{kT}} - 1\right) - I_{\Phi}$. The main dependencies follow from this formula: $U_{o.c.c} \sim \ln\Phi$, and $I_{sh.c.c} \sim \Phi$, where Φ is the illumination of the solar battery (SB).

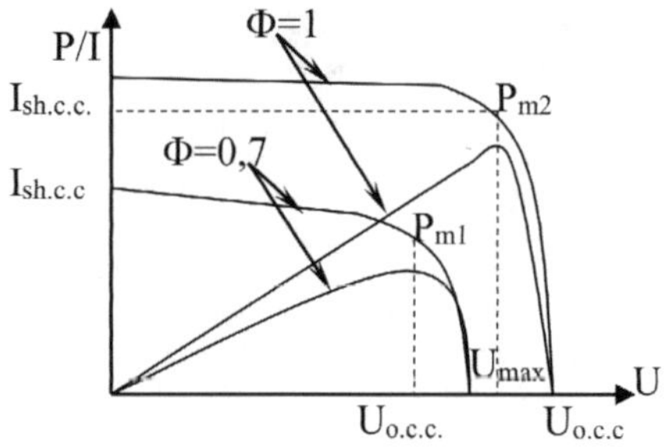

Fig. 3. Volt-ampere characteristic of the solar battery.

On (Fig. 3) the Volt-Ampere characteristic of the SB there is always a point at which its output power is maximum, but the position of this point changes (see the figure) from the illumination Φ, which, in the case of a fixed SB installation, changes greatly during the day.

To increase the total efficiency. You can use a special charging regulator that optimizes the operating point of the SB.

4. Conclusions

Such a solar station provides continuous operation of one TV and lighting the room for 4 hours, and without a TV it illuminates the room and provides charging of cell phones for 20 hours. To ensure the operation of such stations, it is necessary to place a solar panel in the sun from 6 to 10 hours, depending on weather conditions. Our experiments have shown that the service life of such solar stations is at least 20 years, from time to time there is a need only to replace the batteries. Fig. 4. Below is the appearance and arrangement of such solar stations.

Fig. 4. General and appearance of the solar panel device.

This solar station is intended for individual use, as well as for use in remote and far apart places where there is no traditional power supply. Possible consumers of the products may be "Uzgydromet" Center JSC, Ministry of Agriculture, "Uzbekgeofizika" JSC, "Uzbekenergo" JSC, as well as organizations or individuals whose specific activity is to work in remote and hard-to-reach places where there is no traditional power supply.

Thanks to Academician M.K.Bahadirkhanov for discussing the results and valuable advice.

References

[1]. I.N. Sidorov, M.F. Biinatov, L.G. Shvedova. Reference: Inductive elements of electronic equipment. M .: Radio and Communication, 1992.

[2]. S.A. Eranosyan. Network power supplies with high-frequency converters. L .: Energoatomizdat, 1991.

[3]. Razevig V. D .. The use of P-CAD and PSpice programs for circuit simulation on a PC. Issue 3. Moscow: Radio and Communications, 1992.

[4]. N.F. Zikrillaev., S.A. Tachilin. E.B.Saitov. Efficient inverters for alternative energy sources. International scientific-practical conference "Problems of modern engineering" Andijan May 16-17, 2013.

[5]. E.B. Saitov., O. O. Ponomareva, H. Turaev. Solar radiation and wind in various regions. International scientific conference. Karshi April 15 - 17, 2016.

[6]. Zikrillaev N.F., Abdurakhmonov B.A., E.B. Saitov. The ability to create integrated photocells. International Scientific Conference "Innovations in Materials Science" (Tashkent-2014).

[7]. Kovalev F.I., Usachev V.A. The basics of power electronics. Power semiconductor devices. M.: Radio and Communication, 2012.

[8]. Saitov E. Study of Quantitative and Qualitative Characteristics of Nickel Clusters and Semiconductor Structures. International Journal of Advanced Research in Science, Engineering and Technology, Vol. 3, Issue 5, May 2016. pp. 1952-1956.

[9]. B. A. Abdurakhmanov, M. K. Bakhadirkhanov, K. S. Ayupov, H. M. Iliyev, E. B. Saitov, A. Mavlyanov, H. U. Kamalov. Formation of Clusters of Impurity Atoms of Nickel in Silicon and Controlling Their Parameters. Nanoscience and Nanotechnology, Vol. 4 No. 2, 2014. pp. 23-26.

[10]. Pavel Khavanov and Anatoly Chulenyov. Energy-efficient solar power plants for autonomous heat supply in Russia. MATEC Web of Conferences 251, 03012 IPICSE-2018.

[11]. Paul Breeze. Solar Power Generation. ISBN 978-0-12-804004-1. Copyright © 2016 Elsevier Ltd. All rights reserved.

[12]. Tadashi Tanuma. Advances in Steam Turbines for Modern Power Plants. ISBN 978-0-08-100314-5. Published 2017. Copyright © 2017 Elsevier Ltd. All rights reserved. Imprint Woodhead Publishing No. of pages 568.

Self-starting of pump units when extinguishing voltage in the energy saving mode

Kudrat Abidov [1], *Odiljon* Zaripov [2]

[1]Head of Department, Tashkent State Technical University, Tashkent, Universitetskaya str. 2, Uzbekistan
[2]Assistant, Tashkent State Technical University, Tashkent, Universitetskaya str.2, Uzbekistan

Abstract. The article discusses the main processes that occur during the self-start of pumping units of pumping stations of systems of machine lifting for irrigation, ensuring the rational use of electricity, pumping equipment and irrigation water.

1 Introduction

One of the main consumers of electricity in Uzbekistan is the pumping stations of the agricultural industry. Of the generated 47 billion kWh / year of electricity, an average of 8 billion kWh / year or about 16% is spent on covering the loads of 1,588 state reclamation pumping stations, which are currently on the balance sheet of the Ministry of Agriculture and Water Resources of the Republic of Uzbekistan. In recent years, the ever-increasing number of small pumping stations owned by farms and tenants has led to the fact that the energy costs of pumping stations are becoming even more significant.

The use of pumping units, taking into account the specific features of irrigated agriculture, made it possible to solve complex engineering and reclamation problems. The construction and commissioning of large machine channels with unique cascades of pumping stations and complex hydraulic structures made it possible to transfer the flow of high-water rivers to large irrigated massifs located in the lower reaches of the low-water rivers Zarafshan, Kashkadarya, Surkhandarya, Isfayramsay, etc., where there is a lack of water and systematically repeated water impeded the development of agriculture [1, 2].

With the help of pumping stations, 2.418 million hectares of 4.3 hectares of irrigated land are currently irrigated. About 59 billion m3 of irrigation water is pumped annually with all pumps, with 27 billion m3 being pumped by the head pumping stations, and 32 billion m3 by the second and subsequent rises.

The largest pumping stations were built and commissioned in the complex of Amu-Bukhara, Sherabad, Karshi, Dzhizak, Amu-Zang machine channels, which made it possible to sharply increase the water supply of irrigated lands and master hundreds of thousands of hectares of new lands. About 2,000 m3 / s of water is pumped by pumps only in these five channels, and the power of the main electric motors is 1.2 million kW.

As you know, transients are more dangerous for electrical equipment, including a pump installation, than established ones. The main damages and accidents of pumping units occur during transients, starts, stops, which, in accordance with the work schedule, can be assigned several times a day, as well as during emergency disconnection of electric motors from the power system. Significant dynamic loads arise on the elements of structures and hydro-power equipment: water hammer, a sharp change in the power effects on the impeller and the blade drive systems, accompanied by flow pulsations, vibration. Therefore, when operating pumping stations, it is necessary to comprehensively take into account the work of the pressure path and technological equipment and assign schemes and the composition of structures based on calculations and analysis of transients taking into account the static and dynamic characteristics of pumps and electric motors.

2 Experimental research

Pumping units of an open irrigation system operate on pipelines of relatively short lengths ending in siphon outlets. With a planned shutdown of the pump unit, the control valve closes at the beginning, then the drive motor is disconnected from the network. Due to this, the pump unit stops and does not rotate in the opposite direction.

During [3,4] emergency shutdowns due to the fact that the control valve is not closed, when there is no check valve or it does not work, the water in the pipeline will go back after shutting down the unit. At the beginning, the speed of the unit decreases and reaches zero, then under the influence of water in the pipeline it changes direction, it rotates in the opposite direction and the value of the reverse stealing speed becomes more than the nominal value. When the pump rotates in the opposite direction, the pump seals fail. This leads to a violation of the pump sealing system. Therefore, after

 https://doi.org/10.1051/e3sconf/201913901088

this mode, before starting the pump, the plugs of the pump are removed, the burnt out oil seals are removed, and new oil seals are clogged. The vacuum pump is started to prepare the pump for start-up, since the control valve does not close tightly, filling the collector with water is required.

The number of installed pumping units at reclamation pumping stations varies widely (2-16 pieces). The pump drive motors are powered by a step-down substation of the pumping station. Two-winding and three-winding transformers are used at these substations. The number of simultaneously self-starting motors of the pumping station is determined based on the permissible value of the voltage drop at the time of self-starting. Knowing the mains voltage and the resistance of the power source, we determine the value of the recovery voltage on the motors.

When the pumps are of vertical design, axial, where there are no gates, any voltage deviation is associated with rotation in the opposite direction, which negatively affects the operation of the thrust bearing, the fastenings of the pump and motor windings. All this is associated with the consumption of scarce materials, it requires a lot of time and qualified repairmen. If the pump works with suction, then during a power outage it may lose water, turn on empty. Experience shows that the loss of a column of water occurs, as a rule, only when the speed drops below 50%. After each cancellation or landing of the mains voltage, the pumps are switched off. It takes a long time to restore the station to its original working condition. A simple pumping station during an intensive irrigation period is unacceptable, therefore, it is of great technical and economic importance to reduce the downtime of pumping units of the pumping station using self-starting. Calculations of self-starting currents of induction motors must be performed both during design and during operation of electric power facilities.

When power is applied, self-starting motors of auxiliary motors are operated when the rotation frequency increases. Self-starting will be successful if the pumping units participating in this mode are deployed to the operating speed in an acceptable time. The success of self-start depends on the time of a power outage, the parameters of the supply circuit, the total power of the disconnected motors and their load, the mechanical characteristics of the mechanisms and other factors.

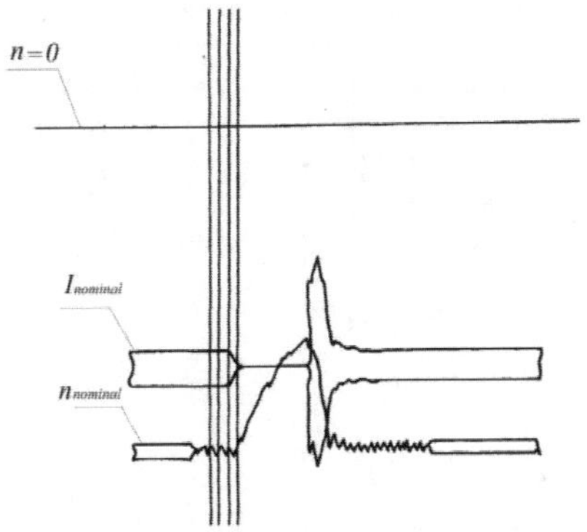

Fig.1. Self-starting at $t_{shutdown}$ = 1.5 s

In all cases, self-start must be justified. The main task of self-start-up is to maintain the efficiency of the pumping station with short-term repayment. If the capacity of the power supply system is sufficient, all pumping units for which it is necessary can participate in the simultaneous self-start. If the limited capacity of the power supply system does not allow this, then several stages of self-start are provided, that is, alternate self-start of several groups of pumping units. In this case, the question also arises of the mutual influence of the power supply system and the engines of pumping units involved in self-starting. A more significant limitation in the use of self-starting pumping installations is caused by the fear of the possibility of water hammer in the pressure pipe when the loaded pump is suddenly turned on.

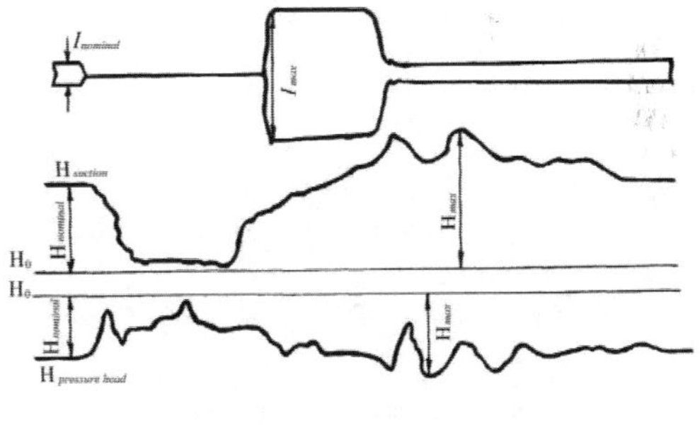

Fig.2. Self-start when $t_{shutdown}$ = 4.5 s

In the proposed work, the time of the on state of the engine during repayments is determined using a sensor recording a pressure gauge installed at the pump outlet (Fig. 1) or by calculation based on four square characteristics of the pump and parameters of the pump installation. The self-starting zone is in the range of the pressure change from the nominal value to the minimum value. In this work, after the repayment, the disappearance of the voltage in the network, it is possible to determine the allowable area of the engine self-starting based on

a) design data of the pumping station and the four square characteristics of the pump (by calculation);
b) pressure measurement using sensors or a recording pressure gauge installed at the pump outlet (to increase accuracy) or a reference pressure gauge (Fig. 2);
d) the characteristics of the run-out of the pump unit, which determine the zone for successful self-start (Fig. 3).

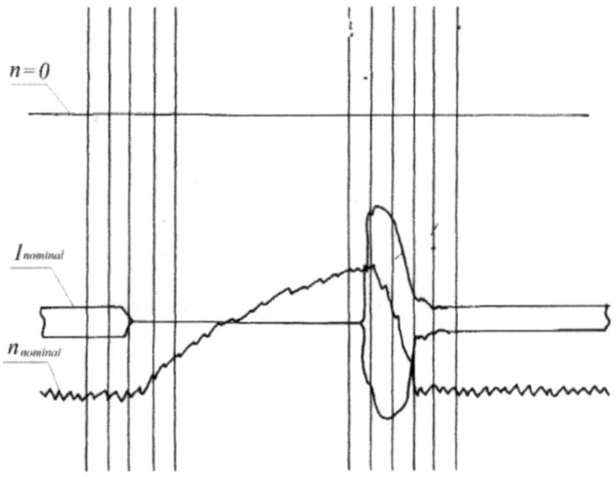

Fig. 3. Self-starting at $t_{shutdown} = 2.2$ s

The reliability of the pump unit as a whole is increased, downtime is reduced, and the consumption of scarce materials is reduced. Using the method allows you to free the labor of skilled workers or specialists, because there is no chance of an emergency at pumping stations [7.8].

High-voltage power lines of Amu-Zang pumping stations and electric motors are operated in difficult climatic conditions. During the irrigation season, there is a decrease in the mains voltage and a short-term decrease in the mains voltage. This leads to the shutdown of all pumping units of pumping stations [9,10].

The pumping station of the first lift is equipped with horizontal centrifugal pumps of the 24НДС type with a turned impeller diameter D_{RC} = 375 mm. As a drive motor, three-phase asynchronous motors of the ДАЗО type were used.

The first stage pumping station consists of 16 24НДС pumps with an impeller diameter of 875 mm and ДАЗО – 15-59-10 asynchronous motors in the same quantity.

The pumping station of the second stage consists of similar pumps (24НДС with D_{RC} = 375 mm) in the same amount of 16 pieces, but ДАЗО-15-69-10 electric motors were used as an electric drive. All of these pumping station motors are powered by a single substation. These motors are powered by transformer No. 1 type ТДТН-2500/110/6/6.

Experimental oscillograms of the self-start process in field conditions were taken at the Amu-Zang pump station of the 1st lift of the 2nd stage. Self-starting processes of one pumping unit No. 11 were oscillated at various values of the time delay, which varied from 1.5-3 seconds. The pump rotation speed, stator current I, run-out time and self-start time were recorded. The results of the oscillogram of a field study are shown in Fig. 1. and fig. 3. The oscillogram shows that the multiplicity of the starting current and the duration of the self-start increase with increasing repayment time.

3.Design Studies

A number of programs for calculating the specified mode for industrial enterprises have been developed, that is, development is taking the path of programmatically implementing the mathematical model [5,6]. Organizations involved in this area use different methods of software implementation. The calculation of the processes occurring in such systems can be represented in the form of nonlinear equations describing the operation of pumping units for an individual and common pipeline, and the loss of electrical energy to overcome hydraulic resistance in pipelines under various conditions.

To calculate the mode, a mathematical model of the pumping unit was developed to study the operating modes of the pumping station. The specifics of transients in pumping stations is the loss of drive. As a typical example in Fig. 4. The calculated process curve after the loss of the drive of the Amu-Zang pumping station in the Surkhandarya region of the Republic of Uzbekistan is given.

At time "a", the engine was disconnected from the mains and its M_{eng} moment instantly drops to zero, while the hydraulic moment of resistance of the M_{pum} pump remains. This leads to a decrease in the rotational speed n and the developed head H, which results in a decrease in the supplied flow rate Q_H.

At time "b", the rotation frequency decreases so much that the flow rate becomes equal to zero, and then the movement of water through the pipelines and the impeller from the upstream to the downstream, i.e. in the "turbine" direction. The direction of rotation is still kept "pumping". This is a counterflow mode. Due to the fact that in this mode the pump capacity drops sharply, there is an increase in pressure in the spiral chamber, pressure, as well as hydraulic torque.

 https://doi.org/10.1051/e3sconf/201913901088

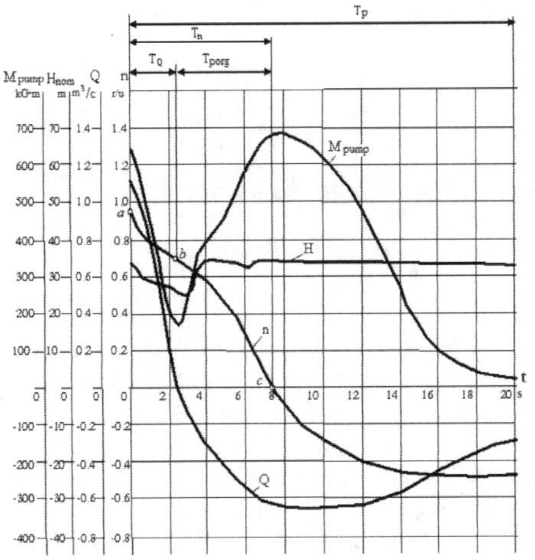

Fig. 4. The nature of the transition process when turning off the electric motor.

At time "c", the rotation frequency becomes equal to zero, and then, under the influence of the flow, the impeller begins to rotate in the opposite direction. The hydraulic torque reaches its maximum value and begins to decline. When reaching M_{pum}= 0, a steady acceleration mode occurs. The speed and flow rate are determined by the impeller, pressure and position of the guide vanes.

The above shows that the loss of the drive of the unit is accompanied by intense dynamic effects on pressure pipelines and equipment elements, which can be limited by choosing the optimal control mode. This must be taken into account when designing pumping stations.

4. Conclusion

1. The increase in the exposure time leads to an increase in the frequency of the starting current and the duration of the start. The results of field studies show that the minimum starting current will be with a decrease in the repayment time to 1.5-2 seconds. The shorter the power interruption, the less the motors have time to brake, the lower their starting currents and the greater the initial voltage on the tires after switching on the backup power and, therefore, the faster the motors start self-starting.

2. The materials of the article and the recommendations expressed in it can be the basis for innovative projects dedicated to improving the operational reliability and energy efficiency of electric motors for drives of pumping units.

References

1. V.KH. Georgiadi, *Behavior of TPP power units during interruptions in power supply for their own needs*, in Russian (NTF Energoprogress **2**. 88, 2003)
2. V.I. Vissarionov, *Energy equipment for the use of alternative and renewable energy sources,* in Russian (RES Publishing House 448,2004)
3. O.Y. Glovatsky, R.R.Ergashev, B.KH.Norov, *Failure Analysis of Elements of Machine Water-Raising Systems,* in Uzbekistan (Materials of the TIIM International Conference November, 27,2008)
4. O.Y.Glovatsky, A.A. Gufranov, B.R. Uralov, Ergashev R.R. *Improving the reliability of operation of water supply facilities of large pumping stations*, in Uzbekistan (Materials of the international conference TIIM November, 29, 2008)
5. T.S.Kamalov, K.Muminov, O.KH.Ishnazarov, *Transmission coefficient of a functional converter of an asynchronous electric drive system with frequency control,* in Uzbekistan (Uzbek magazine "Problems of Informatics and Energy". №**4**,2003, pp. 32-37)
6. T.S. Kamalov, *Current state and development trend of electric drives of industrial and agricultural installations,* in Uzbekistan (Uzbek magazine "Problems of Informatics and Energy", №**1**,2004, pp.50-58)
7. E.I. Sokolov, A.Y.Baru, Y.L. Shindneset, *Experience in the development and implementation of frequency converters for a controlled electric drive of pumping units of oil trunk pipelines,* in Russian (Electrical Engineering,№**7**, 2004,pp.52-57)
8. Λ.D. Ernst, *Self starting of asynchronous electric motors,Tutorial.* in Russian (Omsk, Publishing House of Omsk State University,46, 2006)
9. K.G. Abidov, *Determination of the critical time for a power outage for self-starting of pumping units,* in Uzbekistan (Journal "Problems of Energy and Resource Saving",**3**,160,2017)

Generation of electrical energy at gas pipelines using a transported natural gas technological pressure drop

M.D. Buranov [1], *A.A.* Mukolyants [1,*], *I.V.* Sotnikova [1,**]

[1]Tashkent State Technical University, Tashkent, Uzbekistan

Abstract. The article discusses the possibilities of generating electricity without burning fuel by expanding high-pressure natural gas at gas distribution stations with lower specific capital costs. It is proposed to reduce the pressure of the transported natural gas using expander-generator units instead of traditional throttle devices.

Currently, renewable forms of energy are one of the promising areas of scientific and technological progress. Renewable types of energy include solar, wind, heat, tidal energy, as well as the development of principles for the return of part of the energy spent on transporting natural and associated petroleum gases through main pipelines from source to consumer.

The problem of energy conservation, being one of the most important problems in all developed countries, is becoming especially acute in Uzbekistan.

With the growth of the economy and the standard of living of the population, the demand for energy has also increased. In this regard, in recent times, the generating capacities of the country's power system have noticeably increased. A 300 MW power unit was commissioned at New Angren Thermal Power Plant (TPP), as well as an 800 MW unit at Talimarjan TPP; a project was completed to expand Navoi TPP with the construction of a combined cycle gas turbine unit (CCGT) with a capacity of 478 MW, and a cogeneration gas turbine unit was introduced at the Tashkent Thermal Power Plant. with a capacity of 27 MW. In the framework of projects on the use of alternative and renewable energy sources, expansion units were introduced at Syrdarya and Talimarjan TPPs. At the beginning of 2019, Epsilon Development Company commenced re-commissioning of Khujum well No. 2 (Kultak-Kamash investment block) and the overhaul was initiated, the work on perforation in the range of 3660-3718 meters was performed. After acid treatment of the well, an industrial influx of natural gas in the amount of about 145-180 thousand cubic meters per day was obtained.

From the discovered Khujum field, along the newly built 5.1 km gas pipeline connecting the Khujum field with the gas collection point of Nazarkuduk gas pipeline, gas production was started at the headquarters of Shurtanneftegaz LLC. The company also received an industrial flow of gas from two wells in Talimarjan field. Thus, it is expected that the total volume of hydrocarbons produced by the company will be 1-1.25 million cubic meters per day.

Determining the prospects for improving the gas transmission system of the republic, Uztransgaz joint stock company, in terms of the technical properties of transit transportation, can operate in stages using various options for modernizing production and improving technological processes [1].

Maintaining a high level of energy intensity of the national economy can lead to the fact that unmet demand will make up a significant part of the current energy consumption in the republic. It will be impossible to cover this demand, given the deterioration of the fixed assets of the existing energy system, its high capital intensity and inertia, without intensifying work in the field of energy conservation.

One of the areas of energy conservation is the use of expander-generator units (EGU) for the production of environmentally friendly electricity (without burning fuel) through the use of a technological pressure drop of gas in the systems of transportation via main pipelines and distribution of natural gas. With the gas supply system existing in the republic, the pressure of the transported natural gas is usually reduced in two stages - at gas distribution stations (GDS) and gas control points (GCP) and is performed by throttling.

Currently, projects to use excess energy of gas pressure during its reduction in gas distribution and consumption systems in a number of EU countries are aimed at the production of electric energy. However, to date, no practical measures have been taken for the large-scale and effective practical use of this technology in the Central Asian republics, including Uzbekistan.

In this regard, for uninterrupted power supply on the linear part of gas pipelines, gas metering devices at gas distribution stations and gas storage facilities, and other gas supply facilities, the authors consider it appropriate to use EGU to generate clean electricity by utilizing compressed natural gas energy.

Studies conducted at Urga, Northern Shurtan and Northern Berdak fields (tab. 1,2,3) showed that the use of transported reservoir gas in turboexpander units requires additional processing and bringing it to the relevant technical standards.

Table 1 Initial composition of the reservoir gas of Urga field (% mol.)

Gas components	Horizons					
	J_3^2		J_3^3	J_3^{5+6}	J_3^7	
	Well 7	Well 8	Well 4	Well 11	Well 6	Well 8
CH_4	86,75	87,40	88,30	92,54	89,18	88,41
C_2H_6	7,05	6,60	5,57	4,09	5,98	6,52
C_3H_8	1,85	1,85	1,98	1,01	1,54	1,81
iC_4H_{10}	0,24	0,25	0,34	0,16	0.26	0,31
nC_4H_{10}	0,38	0,35	0,48	0,19	0.26	0.38
$C_5H_{12+higher}$	0,47	0,45	0,57	0,42	0,37	0,63
N_{2+rare}	3,12	2,30	2,49	0,70	0,74	1.28
CO_2	0,14	0,80	0,27	0,89	0,67	0,66
H_2S	-	-	-	-	-	-
Content of $C_{5+higher}$: - % mol. - g/m³	 0.47 22,3	 0,45 22,3	 0,57 22,3	 0,42 22,3	 0,37 22,3	 0,63 22,3
Molecular mass of $C_{5+higher}$	114,062	119,132	94,051	94,051	94,051	85,094
Gas density (760 mm Hg 20⁰C), kg/m³	0,777	0,778	0,771	0,742	0,772	0,771
Relative gas density (760 mm Hg 20⁰C)	0,645	0,646	0,640	0,613	0,641	0,640

Analysis of the obtained data suggests the need to install an additional filter before entering the expander unit for removing mechanical impurities from the gas during transportation and for reducing the percentage of condensate.

A characteristic feature of the turbo-expander design developed by the authors is location of the turbine and compressor wheels on one shaft, which determines the predominant use of radial or radial-axial runners in order to simplify the design of the supply and removal of the working medium.

In this regard, to use the physical energy of the gas, obtained by reducing the pressure on GDS and GCP, instead of traditional throttle devices, it is advisable to use EGU, which allow to generate electricity due to the pressure drop of the gas [2-5].

Table 2. Initial composition of the reservoir gas of ShimoliyBerdak gas condensate field

Composition of the gas (molar %)	Res. J_3^{11}	Res. J_3^9				Res. J_3^8	Res. J_2^7		
	well 9 Perforation range 1577-1566, (m)	well 3 Perforation range 1988-1980, (m)	well 7 Perforation range 2024-2017, (m)	well 14 Perforation range 1991-1988, (m)	Average	well 6 Perforation range 2226-2217, (m)	well 7 Perforation range 2222-2218, (m)	well 10 Perforation range 2139-2136, (m)	Average
CH_4	87,12	92,21	92,31	89,95	91,49	92,95	92,90	94,08	93,49
C_2H_6	6,36	3,81	3,05	4,0	3,62	3,78	3,10	2,41	2,76
C_3H_8	3,05	1,53	1,50	1,72	1,58	1,31	1,48	1,23	1,36
iso-C_4H_{10}	0,78	0,24	0,20	0,50	0,31	0,22	0,16	0,32	0,24
n-C_4H_{10}	0,82	0,23	0,17	0,42	0,27	0,22	0,15	0,30	0,23
$C_5H_{12+higher}$	1,33	0,49	0,47	1,19	0,72	0,42	0,42	0,63	0,53
N_2	0,43	1,33	1,53	1,49	1,45	0,75	1,07	0,93	1,0
H_2S	-	-	-	-	-	-	-	-	-
CO_2	0,11	0,16	0,77	0,73	0,55	0,35	0,72	0,10	0,41
Total	100	100	100	100	100	100	100	100	100
Potential condensate content, g/m³									
- in 1 m³ of reservoir gas	57,48	23,7	22,4	51,83	32,31	20	20,3	29,32	24,81
- in 1 m³ of dry gas	58,25	23,8	22,5	51,44	32,58	20,1	20,4	29,51	24,96
Potential ethane content, g/m³	79,55	47,65	38,15	50,03	45,28	47,28	38,77	30,14	34,52
Critical parameters:									
- $P_{cr.}$, kgf/cm²	46,57	46,73	46,88	46,73	46,79	46,89	46,94	46,71	46,82
- $T_{cr.}$, K	211,53	199,9	199,26	199,95	199,7	199,7	199,20	198,75	198,97
Gas density at 20⁰C, 760 mm Hg, kg/m³	0,824	0,746	0,746	0,746	0,746	0,739	0,741	0,739	0,740
Relative gas density	0,684	0,62	0,620	0,619	0,619	0,613	0,615	0,613	0,614

Continuation of Table 2

Composition of the gas (molar %)	Res. J_2^{3a}			Res. J_2^{2a}			Res. J_2^{26}
	well 14	well 16	Average	well 9	well 15	Average	well 16
	Perforation range 2468-2463 (m)	Perforation range 2452-2444 (m)		Perforation range 2552-2545 (m)	Perforation range 2516 -2504 (m)		Perforation range 2615-2605 (m)
CH_4	92,76	89,14	90,95	92,39	90,99	91,69	92,60
C_2H_6	2,56	4,50	3,53	2,46	4,57	3,52	3,44
C_3H_8	1,89	2,21	2,05	1,92	0,65	1,29	0,72
iso-C_4H_{10}	0,45	0,51	0,48	0,32	0,37	0,35	0,20
n-C_4H_{10}	0,44	0,54	0,49	0,27	0,41	0,34	0,25
$C_5H_{12+higher}$	0,90	1,08	0,99	1,20	0,80	1,0	0,66
N_2	0,48	1,24	0,86	0,71	1,58	1,15	1,34
H_2S	-	-	-	-	-	-	-
CO_2	0,52	0,78	0,65	0,72	0,63	0,63	0,79
Total	100	100	100	100	100	100	100
Potential condensate content, g/m^3							
- in 1 m^3 of reservoir gas	39,24	48,73	35,68	57,81	32,32	45,07	32,11
- in 1 m^3 of dry gas	39,60	49,23	35,96	58,51	32,58	45,55	32,32
Potential ethane content, g/m^3	32,02	56,28	44,15	30,77	57,16	44,03	43,03
Critical parameters:							
- $P_{cr.}$, kgf/cm^2	46,80	46,84	46,81	46,74	46,81	46,77	46,90
- $T_{cr.}$, K	202,32	205,09	203,8	203,08	201,14	202,40	199,34
Gas density at 20^{0}C, 760 mm Hg, kg/m^3	0,763	0,782	0,773	0,777	0,760	0,764	0,748
Relative gas density	0,634	0,649	0,642	0,645	0,631	0,634	0,621

Table 3. Physical and chemical characteristics of the gases of North Shurtan field

	Free gas	Oil dissolved gas
Density (by air)	0,652	0,706
Content, in volume %:		
methane	87,8	81,62
ethane	4,98	6,34
propane	2,48	3,84
isobutane	0,56	1,25
normal butane	0,64	1,79
isopentane	0,18	0,7
normalpentane	0,09	0,45
hydrogensulfide	0,15	0,05
carbondioxide	1,68	1,58
nitrogen	1,44	2,68
Calorificvalue, kcal/nm^3	8712	9670

The pressure reduction of the transported natural gas is carried out in two stages. At the first stage - at gas distribution stations - gas pressure decreases from the pressure in the main gas pipeline from 5.5 MPa to 1.2 MPa, at the second stage - at gas control points - from 1.2 to 0.15 MPa. In the process of expansion of high-pressure natural gas in EGU, its pressure decreases to 1.2 MPa and temperature – to 10 °C. [6-9].

Based on the above, a statement of the problem of optimization was formed and presented in the following form: annual electricity generation

$$\max E = f(P, G), \qquad (1)$$

where $P \in [P_1, P_2]$; $G \in [G_1, G_2]$.

Limitations

$T_2 = 0;\ T_1 \leq T_{Max};\ Q_{TO} \leq Q_{Max};\ N_{ЭГ} \leq N_{Max}.(2)$

Due to the fact that expansion with the help of an expander leads to a more significant decrease in the gas temperature, for economic comparison the fuel gas consumption for a fire heater with a characteristic thermal h = 40% was taken into account in the calculations.

The calculations were performed using the GasCondOil thermodynamic calculation program [10], and the results are presented in Table 4.

Table 4. Comparison of natural gas expansion parameters on throttle and expander

Type	Parameter	Outlet gas pressure, MPa		
		0,3	0,6	1,2
Throttle	Transition gas temperature after throttle, °C	-11,8	-9,6	-6,2
	Heat quantity to heat transit gas after throttle (t_k = 10 °C), kJ/hour (kW)	693000 (190,5)	639700 (170,7)	532500 (140,0)
	Consumption of fuel gas for heating transit gas after throttle (h = 40%), nm^3	50	46	40
Expander (heating after)	Transit gas temperature after expander, °C	-102,4	-81,2	-50,3
	Expander capacity, kW	830	650	450
	Heat quantity to heat transit gas after expander (t_k=10°C), kJ/hour (kW)	700790 (1025)	3027500 (835)	2186600 (600)
	Consumption of fuel gas for heating transit gas after expander (h = 40 %), nm^3	270	220	160
Expander (heating before)	Heat quantity to heat transit gas before expander (t_k=10°C), kJ/hour (kW)	5948900 (1650)	4341900 (1200)	2795800 (770)
	Expander capacity, kW	1400	1000	600
	Consumption of fuel gas for heating transit gas before expander (h = 40 %), nm^3	440	320	200
	Transit gas temperature before expander,°C	160,5	120,2	83,5

Thus, the calculations and operating experience of expander-generator sets confirm the value of relative power generation in the amount of 40 ... 60 kW/thousand nm3 and make it possible not only to introduce secondary energy resources into the economy and provide the generation of electricity, but also to reduce the level of harmful emissions in comparison with traditional technologies.

References

1.URL: http: //old.uztransgaz.uz/ru/content/osnovnye-pokazateli
2. V.Gurov, EES, **4**, 22-24 (2006).
3. O.Malkhanov, RSE, **1,** 39-41 (2008).
4. E.Marchenko, O.Malkhanov, RSE, **6**, 38-42 (2009).
5. H.Fasold, GEG, **6,** 261-269 (1995).
6. A.Gafurov, BKSEU, **1**, 28-36 (2014)
7. I.Kharisov, STSSPbSPU, **1**, 119-123 (2012).
8. B.Agababov, ESWT, **1,** 71-73 (2011).
9. O.Sorokin, IPE, **6**, 36-41 (2005).
10. // URL: http://gascondoil.com/GazCondOil

Drying and disaggregation of cotton powder cellulose in the flash drying unit

Bakhtiar Yunusov [1]

[1]The Tashkent state technical university, after I. Karimov, Tashkent city, Uzbekistan

Abstract. The issues related with the drying and disaggregation of the cotton powder cellulose (CPC) in the flash drying unit are being discussed, as well as the properties of the cotton powder cellulose are being analyzed as the subject of drying and processing treatment.

Introduction

The produced mechanically and chemically the cotton powder cellulose differs with the degree of order, change in the morphological structure and the macroscopic properties [1].

During drying of the powder cellulose with the initial 400% wetness, the parts of material stick with each other and form big clogs of size 0.0024-0.02 m. The equivalent diameter of its clogs after the drying, before grinding, makes up d_{eq}=0.008 m. That is why the dried cotton powder cellulose is then milled at the ball grinder for 2 hours, sieved through the sieve No.12 and then packed. The purpose of the work: Inspection of the CPC properties, selection of the type of the equipment for drying, disaggregation and separation of CPC.

The analysis of the properties of the cotton CPC as a subject of the processing and drying.

The main reason of the lack of the production of the powder cellulose is in the lack of the study of separate technological processes during its production. One of which is the process of drying and grinding, because irrelevant of the method of production of the powder cellulose, its particles are prone to sticking and forming the clogs that must be grinded. Herewith, the process of drying should be revised in the context with other technological treatments of the material and their mutual influence must be considered as well. This will not only intensify and improve the drying process, but recommend the advisable technology of other processing treatments, thus increasing their technical and economical indices as well as the quality of the final product.

Based on the properties of the powder cellulose as the subject of drying, the method of its drying can be proposed observing the sufficiently high degree of activity of the hydrodynamic condition in the drying unit and the time of processing of the material in it.

Different drying equipment is used for drying the disperse polymeric material, that differ in the range of the characteristics constituting the base of their classification [2].

Big quantity of the being dried products together with different methods of heat supply and hydrodynamic conditions, contribute to the variety of the design of the driers thus create the problem of selection of the right drier, type of the drying equipment and the mode of the drying process [3]. The general requirements to the selection of the rational method of drying take to the more profitable technical and economical indices of the operation of the unit, provide reliability of tis operation in producing the high quality product, provide convenient control and maintenance, implementation of the requirements of the safety rules, sanitary and hygiene conditions of labor, exclude the emissions to the atmosphere.

The selection of the rational method of drying shall be supported with the complex of the material characteristics as the subject of drying and the technological peculiarities of production. The main conditions are the technological and economic requirements, as well as the requirements to the quality of the being dried product.

For drying of the disperse materials in the range of the industries, mechanical and low capacity drying units are widely used. Their critical deficiencies are conditioned by the incompliance of the design and operational characteristics used for drying of a particular material.

At the present time, the progressing method of drying in the boiling layer becomes widely used, that allows to sharply intensify the process. However, in many cases, the use of this method largely wears the products, generates static electric charges, as well as create the canals and dead zones. That is why the drying is recommended to be conducted in the flash driers, that are free from the most of the deficiencies that the driers with the boiling layers have. The wide introduction of the mentioned driers was prevented by the lack of the correct methods of calculation and rational schemes of installation that allow effective drying of the loose materials. Hence, the proposed methods of calculation provided satisfactory results only in separate cases.

For the first time the prof. Mukhitdinov D.N. mentioned the specifics of the flash driers [4]. The interest to the flash driers in the industry considerably increased after the publication of the works of P.G. Romakova and N.B. Rashkovskoy [8] that demonstrated big possibilities of the mentioned driers.

There are two types of flash driers: without the grate and with it.

The unit without the grate, in the narrow part of it, the gas speed is accepted equal to the stable speed of the pneumatic transportation of big particles, in other words:

$U_г = (1,5 -:-2,0)U_B$

U_B – speed of the particles.

The cone angle is usually accepted within the range of 30-70°. Based on the data of Elperin I.T. and Ephremtseva V.S., the optimal value of the cone angle shall be calculated based on the formula:

$\Theta_0 = 180^0 - 2(\varphi + \gamma);$

φ – the angle of the natural slow of material; γ – the angle equal to 30-40°.

The study of the balanced condition of the wet material with the environment facilitates the valuable interest for the evaluation of the moving force of the drying process. Hence the importance is in the analysis of the isotherms of adsorption - desorption. Based on the isotherms of adsorption-desorption define whether the final wetness of the material complies with the even one and hence recommend the conditions of storage of the material, as well as evaluate the porous structure of the material and define the forms of cohesion of the moist with the material.

In order to obtain the isotherms of adsorption and desorption of the materials, we used the vacuum sorption scales. The character of the obtained at 25°C isotherms, it is necessary to underline the presence of the node of the sorption hysteresis in the relevant humidity of the air, that justifies the soft structure of the material which is peculiar to the capillary-porous colloid bodies.

Our aim, besides of the analysis of the cohesion forms of the moist in the being studies material, was to determine the energy of the cohesion of water. The energy of cooperation of this material is set within the interval below 3.5x10J/kg, which characterizes relevantly weak forms of cohesion of the physic-chemical or physics-mechanical types.

To evaluate the porous structure of the material based on the sorption measurements on the Kelvin-Thomson equity, the differential and integral functions of distribution of the volume of the pours along the radiuses were calculated. On the curves the vital diameter of the pours of 6.0Nm was defined, in other words the diameter of the thinner pours out of which it is required to remove the moist from the material to obtain the required residual wetness. On the values of the vital diameter of the pours taking into account the adhesion-cohesion properties of the material and based on the forms of cohesion of the moist with the material, the belonging of the material to this or that classification of the disperse materials as the subjects of the drying is defined.

The knowledge of the thermal characteristics is required during processing of the results of the experimental researches of the heat exchange processes during the drying in criterial form, to analyze the mechanism of heat transfer in the wet material, as well as to analyze the forms and types of cohesion of the moist with the material.

In order to determine the TPC (thermal and physical coefficient) of the cotton powder cellulose, we selected the method that is based on the solutions of the nonstationary problems of the heat transfer with the inner source of the heat of constant force.

As seen in the results of the experiments, the heat transfer factor of the being studied material, increases with the increase of the moist content with the tendency to fade at the presence of the moist of the level close to the complete saturation, when the sedimentation of the particles is observed and the difficulties of their stabilizing in the required volume of the liquid occur.

The heat transfer factor of the wet materials increases at the beginning, reaches the maximum at the particular wetness, and then decreases. The explanation of this interdependencies is the following: the dry material has lower thermal and temperature conductance due to the weak thermal contact between its particles. The presence of water improves the thermal contact and considerably eases the transfer of the heat from one particle to the other through thin water film.

In the technological process of production of the cotton powder cellulose, the drying takes an important place. The selection of the drier for this material depends on its properties.

The microscopic research of the structural and mechanical properties of the cotton powder cellulose revealed that its particles represent the spiral fiber, that easily form the aggregates (clogs) that during the drying must be broken into small pieces.

By the thermographic analysis of the hypothermic properties of the cotton powder cellulose it is defined that within the range of the moist content 2.3-:-0.06 kg/kg the capillary moist in the pours is removed (83% of the total content of the moist in the material), osmotic moist (10%) and the moist of poly-molecular adsorption (2%).

The adsorption method of study of the material structure allows to receive the curves of distribution of the volumes of the pours on the radiuses, out of which, based on the Kelvin-Thomson equation, the vital diameter of the pours was calculated of 6.0Nm, in other words the smallest diameter of the pours, the removal of moist from which allows reaching the required residual wetness of the material.

Based on the complex analysis of all the properties of the cotton powder cellulose, 3 subgroups were referred to the materials of the II group based on the classification of Mukhitdinov D.N., for which the combined flash drier is recommended.

During the production of the cotton powder cellulose in the industry, the special requirements are set forth to its dispersiveness, hence the process of drying was combined with the grinding of the material. The intensification of the process was reached as a result of renewing and activation of the surface of vaporization during grinding of the cotton powder cellulose. However the time of presence of the particles of the material in the unit is not sufficient to reach the required final wetness of the powder cellulose. That is why for the process of drying and grinding of the cotton powder cellulose in the flash drier, the winded stream of the gas mixtures and neutral additives. The use of the vortex in the supply orifice of the heating service allows to increase the retention capacity and decrease the hydraulic resistance of the layer, which prevents the formation of the canals and dead zones in the chamber.

The peculiar property of such device is that the drying agent at high speed can be injected into the drying chamber, where the vortexed stream is developed. A convection heat exchanging is intensified in it and at intensification of which the special role is assigned to the centrifugal forces that provide not only active mixing, but better grinding. Herewith, the degree of winding of the heat transfer medium plays a vital role for the highly wet materials which stick to the walls of the device. The role of the neutral agents is in "grinding" the particles of the material and in prevention of formation of the clogs, sticking of the particles to the wall of the device, thus providing the movement of the layer. The drying device of this type operates in autonomous mode, in other words, the yet wet and under grinded material is not taken out of the device until it reaches the earlier set parameters of the process.

In this way, the hydrodynamic mode of the flash drier of cone form with the winded stream of the gas mixture and with the neutral additive for the drying and grinding, fully complies with the required property of the material as the subject of the drying.

During the drying of the CPC, the hazardous substances are the dust of the powder cellulose. The powder cellulose is a combustible and explosive substance. The ignition temperature is 210^{0}C. The lower explosion limit is 50g/m3 in the air. The dust of CPC can damage the health. The base of the hazard: content of (SiO_2), (CaO), (Fe_2O_3) in the dust. The maximum allowable concentration of the dust in the air is 6mg/m3 (CH 246-63). The process name of the product - electrode cellulose. Empirical formula - $(C_6H_{10}O_5)_n$.

Let's calculate the temperature and pressure at the explosion of the cotton powder cellulose. The initial temperature of the mixture is +20 ^{o}C.The initial pressure of the mixture is 760mm.mrc.clm.Let's write the thermochemical equation of the combustion reaction of the powder cellulose in the air:

$$C_6H_7(OH)_3O_2 + 6\,3,76N_2 =$$

$$6CO_2 + 5H_2O + 6\,3,76N_2;$$

The value Q_r, that equates to 4200: 6.173 = 680.38 kcal/mole, calculated from [5]. The calculation of the explosion temperature is done based on the assumption that the explosion is an adiabatic process that occurs at constant volume. The heat produced during explosion is used only for heating the products of the explosion, based on this assumption, the temperature of the explosion can be calculated by the formula:

$$Q_v = C_v \cdot T_v. \qquad (1)$$

Where: (C_v) – average heating capacity of the products of explosion in the range from (0) to T ^{0}K (υ = *const*); (T_v) – being searched explosion temperature, (^{0}K); Q_v – explosion heat, (kcal/mole).

Taking into account, that (T_v) is not known, and (C_v) depends on the temperature, the explosion temperature is calculated based on comparison of the values of the inner energy of the products of the explosion (combustion) with each other from one side and the heat of the explosion (combustion) of the initial substances from the other side. The heat, emitted at the moment of explosion, is contained in the products of the explosion and equates to the increase of the inner energy. Based on this condition, it can be written down:

$$Q_v = Q_г + Q_{в.э.} = \sum m_i\, \Delta E_{m.взр.}; \qquad (2)$$

Where: $Q_г$ – heat of combustion of one mole of the initial substance, taken to the temperature (T_i), *[kcal/mole]*; $Q_{в.э.}$ – value of the inner energy of the initial mix at the temperature (T_i), [kcal/mole]; m_i – number of moles of i component in the products of explosion (combustion); $\Delta E_{m.взр.}$ - value of the inner energy of the products of explosion at the temperature $[T_{взр.}{}^0K]$.

The inner energy $(Q_{в.э.})$ of the mixture is calculated by the formula:

$$Q_{в.э.} = n_1C_{v1}T_1 + n_2C_{v2}T_1 + \cdots + n_iC_{vi}T_1; \quad (3)$$

Where: n_i – number of moles of each component of the initial mixture; C_{vi} - average heat capacity of each component of the mixture at the temperature $(T\ ^0K)$, $(T_1 = 293\ ^0K)$; $C_{v.(цел.)} = 44,152$ [cal/mole.degr]; $C_{v(O2)} = 5,006$ [cal/mole.degr]; $C_{v(v2)} = 4,268$ [cal/mole.degr]; and:

$$Q_{в.э.}=12936,536+8800,548+32838,877=$$

$$54575,961[cal.]=54,576\ [kcal.]$$

the heat of the explosion of the mixture is calculated by the formula (2)

$$Q_v = Q_г + Q_{в.э.} = 680,38 + 54,576 = 734,956\ [kcal.]$$

The inner energy of the products of the explosion is calculated for the preliminarily set value of the temperature of the explosion. The explosion is the adiabatic process, that is why:

$$\Delta E = 6\int_0^T (C'_{v1})dT + 5\int_0^T (C'_{v2})dT + 6\cdot 3,76\int_0^T (C'_{v3})dT; \quad (4)$$

Where: C'_{v1}, C'_{v2}, C'_{v3} - number of products of the explosion (carbon dioxide, water, nitrogen), [mole]

The preliminary heat of the explosion ($Q_{взр.}$) is defined on the approximate value of the inner energy of the gases (products of the explosion), obtained by division of (Q_v) to the sum of the products of the explosion;

$$Q_{взр.}=Q_v/C'_{vi} =734956/25,46=28867\ [cal\ /mole.]=$$

$$28,867\ [kcal/mole]$$

This value is within the range of $(2000\text{-}3400)^0K$ [1]. Accepting the $T_1=2800\ ^0K$. Average values C'_{v1}, C'_{v2}, C'_{v3} are calculated based on the pressure of the inner energy [1] to the value of the accepted temperature. For example, for the carbon dioxide gas:

$$C'_{v1}= Q_{взр} / T_1 = 28867 /2800 = 10,31\ [cal\ /mole.degr];$$

$$C'_{v2} = 8,57\ [cal\ /mole.degr];$$

$$C'_{v3} = 6,04\ [cal\ /mole.degr];$$

Hence the average values for water and nitrogen are calculated. By inserting the values C'_{v1}, C'_{v2}, C'_{v3} in the formula [4] we receive:

$$E_1=6\cdot 10,85\cdot 2800 + 5\cdot 8,57\cdot 2800 + 6\cdot 3,76\cdot 6,04\cdot 2800 = 683788\ [cal].$$

Hence $(T=3000\ ^0K)$ and calculating $(E_2=740941\ cal.)$

By the linear interpolation the explosion temperature is found:

$$T_{взр.}= 2800 + [(734956 - 683788)/(740941 - 683788)]\cdot 200 = 2979\ ^0K.$$

The pressure generated at the explosion is calculated. The calculation is by the formula:

$$P_{взр.}= (P_0 T_{взр} / T_0);\ (m / n); \quad (5)$$

$$P_{взр.}= 11,54\ [atm].$$

Conclusion

The properties of the cotton powder cellulose were researched as the subject of the technological processing and drying. The selection of the type of the drier was made. The calculation method of the device was developed applicable to the CPC, as well as a range of the similar clogging materials. Based on the conducted researches, the pilot device for the drying and disaggregation of CPFC was designed and manufactured. Taking into account the pressure of the explosion, calculated above, the structure of the devices has anti-explosion valve installed in it and in the case of the explosion it discharges the developed pressure.

References

1) B.Kh. Yunusov* M. B. Aripjanova, N.A. Yunusova. Use of artificial neural network in the process of drying of the cotton powder cellulose. Problems of energy and resource preservation No.1-2, Tashkent, 2015, p.111.

2) B.Kh. Yunusov, D.N. Mukhitdinov, A.A. Mukolyannts, Sh.B. Yunusov. "Sorption-structural characteristics of the cotton powder cellulose (CPC)", International science conf. "INNOVATION-2011", assembly of science articles, Tash.-2011, 25-27 of October, p.108-109.

3)I.N. Pavlov, V.A. Kunichan. Grinding of the micro-crystal cellulose during drying. Chemicstry of the plant raw material. 1999, No.2, p.159-162.

4) B.Kh. Yunusov, D.N. Mukhitdinov. Drying unit with the winded stream of gas mixtures with the neutral additives for drying and grinding of highly wet, clogging materials. TshDTU, "Catalogue of innovation ideas, developments and projects of the professors, teachers and young scientists", Tashkent 2011, p.68.

5) V.S. Saushev, P.P. Sheglov. "Examples and tasks on special chemistry". Moscow-1971, g.55.

AUTHOR INDEX

AUTHOR INDEX